2nd International Conference on Environmental and Energy Engineering (IC3E 2018)

IOP Conference Series: Earth and Environmental Science
Volume 146

Xiamen, China
1 – 3 March 2018

ISBN: 978-1-5108-6262-3
ISSN: 1755-1307

Printed from e-media with permission by:

Curran Associates, Inc.
57 Morehouse Lane
Red Hook, NY 12571

CURRAN ASSOCIATES INC. proceedings.com

Some format issues inherent in the e-media version may also appear in this print version.

Printed by Curran Associates, Inc. (2018)

For permission requests, please contact the Institute of Physics
at the address below.

Institute of Physics
Dirac House, Temple Back
Bristol BS1 6BE UK

Phone: 44 1 17 929 7481
Fax: 44 1 17 920 0979

techtracking@iop.org

Additional copies of this publication are available from:

Curran Associates, Inc.
57 Morehouse Lane
Red Hook, NY 12571 USA
Phone: 845-758-0400
Fax: 845-758-2633
Email: curran@proceedings.com
Web: www.proceedings.com

Table of contents

Volume 146

2018 2nd International Conference on Environmental and Energy Engineering (IC3E 2018)

1–3 March 2018, Xiamen, China

Accepted papers received: 11 April 2018
Published online: 4 May 2018

Preface

Papers

Energy and Enviroment

Advanced Energy Technology

IOP Conf. Series: Earth and Environmental Science **146** (2018) 011001 doi:10.1088/1755-1315/146/1/011001

PREFACE

IC3E 2018 was held in held in Xiamen University of Technology, China during March.1-3, 2018. IC3E 2018 was organized by Xiamen University of Technology, International Society for Environmental Information Sciences (ISEIS) and Asia Pacific Institute of Science and Engineering (APISE). The conference provides a useful and wide platform both for display the latest research and for exchange of research results and thoughts in enviromental and energy engineering. The participants of the conference were from almost every part of the world, with background of either academia or industry, even well-known enterprise. The success and prosperity of the conference is reflected high level of the papers received.

The proceedings are a compilation of the accepted papers and represent an interesting outcome of the conference. This book covers 3 chapters: Energy and Enviroment, Energy Research and Application and Advanced Energy Technology.

We would like to acknowledge all of those who supported IC3E 2018. Each individual and institutional help were very important for the success of this conference. Especially we would like to thank the organizing committee for their valuable advices in the organization and helpful peer review of the papers.

We sincerely hope that IC3E 2018 will be a forum for excellent discussions that will put forward new ideas and promote collaborative researches. We are sure that the proceedings will serve as an important research source of references and the knowledge, which will lead to not only scientific and engineering progress but also other new products and processes.

Gordon Huang, University of Regina, Canada

IOP Conf. Series: Earth and Environmental Science **146** (2018) 011002 doi:10.1088/1755-1315/146/1/011002

COMMITTEES

Conference Chair

Gordon Huang, University of Regina, Canada

Zhifeng Yang, Beijing Normal University, China

Local Chair

Yongping Li, Beijing Normal University, China

Haiyan Fu, Xiamen University of Technology, China

Local Organizational Committee

Tian Chai, Xiamen University of Technology, China

Xiaoqing Chen, Xiamen University of Technology, China

Huajing Cheng, Xiamen University of Technology, China

Huiyan Cheng, Xiamen University of Technology, China

Qian Cheng, Xiamen University of Technology, China

Zineng Dai, Xiamen University of Technology, China

Panfeng Gao, Xiamen University of Technology, China

Yanhong Jiang, Xiamen University of Technology, China

Bangbang Jiu, Xiamen University of Technology, China

Wenchao Liao, Xiamen University of Technology, China

Guangsheng Liu, Xiamen University of Technology, China

Jing Liu, Xiamen University of Technology, China

Jiangxue Long, Xiamen University of Technology, China

IOP Conf. Series: Earth and Environmental Science **146** (2018) 011002 doi:10.1088/1755-1315/146/1/011002

Yicheng Wu, Xiamen University of Technology, China

Aili Yang, Xiamen University of Technology, China

Mengxiang Zeng, Xiamen University of Technology, China

Qiwei Zhang, Xiamen University of Technology, China

Mulan Zhu, Xiamen University of Technology, China

International Technical Committee

Rajasekhar Balasubramanian, National University of Singapore, Singapore

Sayed M. Bateni, University of Hawaii at Manoa, USA

Masoud D. Champiri, University of Houston, USA

Ali Cheknane, Amar Telidji University of Laghouat, Algeria

Changping Chen, Xiamen Ocean Vocational College, China

Huichao Chen, Southeast University, China

Zhi Chen, Concordia University, Canada

Xiaowen Ding, North China Electric Power University, China

Binlin Dou, University of Shanghai for Science and Technology, China

Qingyun Duan, Beijing Normal University, China

Liang Fang, Jiangping Bio-Substrate, China

Hossein Ganjidoust, Tarbiat Modares University, Iran

Cheung Sai Hung, Joseph, Nanyang Technological University, Singapore

Katsuyuki Kawai, Kindai University, Japan

Michael K. H. Leung, City University of Hong Kong, Hong Kong

Liu Liangying, Jinan University, China

IOP Conf. Series: Earth and Environmental Science 146 (2018) 011002 doi:10.1088/1755-1315/146/1/011002

Peer review statement

All papers published in this volume of *IOP Conference Series: Earth and Environmental Science* have been peer reviewed through processes administered by the proceedings Editors. Reviews were conducted by expert referees to the professional and scientific standards expected of a proceedings journal published by IOP Publishing.

Comparison and Application of MPDI and MSMMI for Drought Monitoring in Desert Mining Area

Ying Liu[1], Yao Li[1], Yang Lu[1] and Hui Yue[1]

[1]College of Geomatics, Xi'an University of Science and Technology, Xi'an 710054, Shaanxi, China

E-mail: 13720559861@163.com

Abstract. Soil moisture is connected one of the most important environmental factors and participation as an important parameter in hydrological, meteorological and agricultural production process. Soil moisture monitoring by remote sensing has become the focus of research at home and abroad. This paper takes Shendong mining area as an example and choose 2015-10-5 SPOT and TM images for remote sensing data source, comparative analysis of MPDI and MSMMI which used for soil moisture monitoring of desertification and their effects. Both MPDI and MSMMI soil moisture monitoring methods were verified by 0-5 cm, 10 cm and 20 cm deep soil moisture (SM) data measured by 2015-10-5.

1. Introduction

Soil moisture is one of the most important environmental variables and participation as an important parameter in land surface hydrology, meteorological and agricultural production process. Soil moisture change on land surface energy budget, runoff and vegetation productivity have a profound impact [1]. With the rapid development of today's remote sensing technology, soil moisture monitoring by remote sensing has become the focus of research at home and abroad. There are many methods of soil moisture remote sensing monitoring, and the spectral feature space method based on NIR-Red has been further developed and applied because of its simple and easy application [2-3]. Spectral feature space based on NIR-Red was established as soil moisture monitoring index mainly has two kinds: one kind is based on the arbitrary feature space point to a straight line through the origin points of vertical distance size to characterization research district soil humidity status (the line perpendicular to the vertical distance from the baseline soil), such as Perpendicular Drought Index (PDI) and Modified Perpendicular Drought Index (MPDI); the other is directly use of any point to the origin of distance in the feature space size, such as shortwave infrared moisture index, Soil Moisture Monitoring Index (SMMI), Modify Soil Moisture Monitoring Index (MSMMI) to indicate the status of soil moisture in study area [4]. Both of these methods are based on spectral characteristics of NIR-Red space, calculation methods are different, both of these methods can be used for soil moisture monitoring and how it works and need further validation.

2. Study area and data

2.1 Study area

IOP Conf. Series: Earth and Environmental Science **146** (2018) 012001 doi:10.1088/1755-1315/146/1/012001

Shendong mining area (Figure 1) is located in southeast of the Ordos Plateau, northern edge of the Loess Plateau in northern Shaanxi and the South-East edge of Mu Us Desert (110°18′30″E, 39°11′30″ N). This mining area is one of the major coal production bases in China, belonging to the arid and semi-arid desert mine area. Wind erosion area accounted for more than 70% of the total area, the original vegetation was monotonous, and the vegetation coverage was only 3-11%, the average annual rainfall is only 362 mm. Water resources are extremely poor, so it is listed as the key monitoring area of soil and water loss in China.

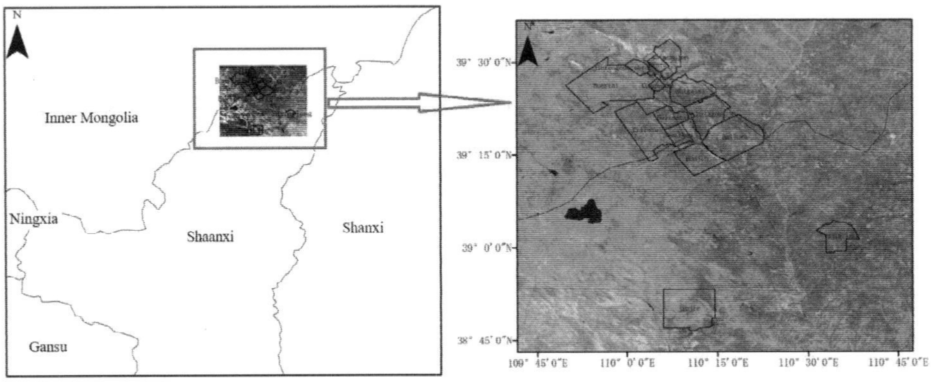

Figure 1. Location of study area

2.2 Data

Satellite data used in this study was SPOT-6 (Systeme Probatoire d'Observation de la Terre) image with 6m resolution registered on October 5, 2015 over Shendong mining area and its surrounding region in China. Landsat 8 OLI of October 5, 2015 was also used for data validation. After the geometric correction, digital numbers (DNs) values were converted into spectral radiance and FLAASH in the ENVI software was carried out to eliminate the atmospheric perturbation and obtain the reflectance at ground level. Data on relative soil moisture to compare with MPDIS and MSMMI were obtained by field sample on October 5, 2015. We measured 5 cm, 10 cm and 20 cm deep soil moisture (SM) data of 31sample points. Soil samples from the Shendong site were taken back to the laboratory, and a traditional weighing method was used to obtain relative soil moisture.

3. Method

3.1 Modified Perpendicular Drought Index (MPDI)

MPDI is a modified perpendicular drought index (PDI), which was introduced by Ghulam et al (2007). On the basis of the original perpendicular drought index, PDI is only a good monitoring effect of soil drought degree in bare soil area, but the effect of vegetation coverage is not obvious [5-6]. However, the vegetation coverage is the ratio of the vertical projection area to the total area of the vegetation, which is an important factor to describe the canopy reflectance. When the vegetation coverage is less than 100%, the soil has a great influence on the vegetation reflectance spectrum. Therefore, the modified perpendicular drought index (MPDI) with vegetation coverage can eliminate the limitation of PDI to the greatest extent.

$$\mathrm{MPDI} = \left(\mathrm{R_{red}} + \mathrm{M} * \mathrm{R_{nir}} - \mathrm{F_v} * (\mathrm{R_{v,\ red}} + \mathrm{M} * \mathrm{R_{v,\ nir}})\right) / \left((1 - \mathrm{F_v}) * \sqrt{(\mathrm{M}^2 + 1)}\right) \quad (1)$$

Where R_{nir} and R_{red} are the atmospherically corrected surface reflectance of Red and Near Infra-Red (NIR) bands of data, M represents the slope of the soil line BC in the NIR–Red space. The greater the vegetation coverage (Fv), the smaller the value of MPDI, the results show that the more humid the surface; the smaller the vegetation coverage, the greater the value of MPDI, the results indicate that the surface is more arid.

3.2 Modified Soil Moisture Monitoring Index (MSMMI)

The physical meaning of MSMMI is that in the characteristic space of near infrared red band reflectance, the MSMMI monitoring index line and the soil line form a right angle triangle. The moisture content of soil in the direction perpendicular to the soil line is similar to that of MPDI, and with the increase of the number of vegetation, the reflectance of the near infrared band will increase [7]. The greater the vegetation coverage, the smaller the value of MSMMI, the more moist the soil, and vice versa.

$$\mathrm{MSMMI} = 1/\left(\sqrt{2} * (1 - \mathrm{F_v})\right) * \left(\sqrt{(\mathrm{R_{nir}} - \mathrm{F_v} * \mathrm{R_{v,\ nir}})^2 + (\mathrm{R_{red}} - \mathrm{F_v} * \mathrm{R_{v,\ red}})^2}\right) \quad (2)$$

Where R_{nir} and R_{red} are the surface reflectance of the NIR and red bands, respectively. Fv represents vegetation coverage. $R_{v,\ nir}$ and $\mathrm{R_{v,\ red}}$ refers to the vegetation in the red and near infrared reflectance, approximate value of 0.05 and 0.5.

4. Results

The validity of MPDI and MSMMI soil moisture monitoring methods was verified by 0-5cm, 10cm and 20cm deep soil moisture (SM) data measured by 2015-10-5. The soil moisture (SM) as the ordinate, MPDI and MSMMI as the abscissa respectively, the construction of SM-MPDI and SM-MSMMI scatter plot (Figure 2) and calculate the coefficient of determination.

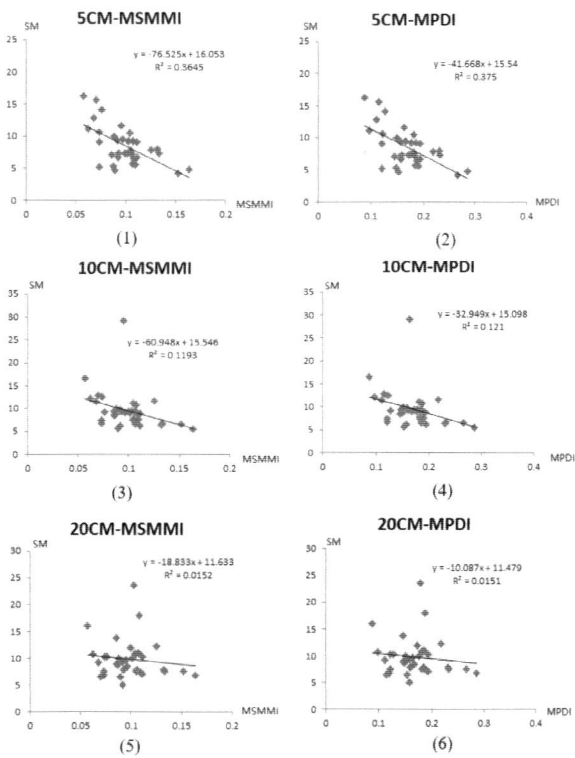

Figure 2. Relationship between soil moisture (5 CM) and MPDI/MSMMI of Landsat 8 OLI in 2015-10-5

From the decisive monitoring results of different depth of soil drought remote sensing drought monitoring, the simulation results of soil depth soil layer for mining area of the 0~5 cm is relatively good, and the simulation results on the 10 cm and 20 cm depth of soil drought monitoring are relatively poor. This phenomenon indicates that the closer to the surface layer, the better the effect of monitoring soil drought of both indexes.

The validity of MPDI and MSMMI soil moisture monitoring methods was verified by 0-5cm, 10cm and 20cm deep soil moisture data measured by 2015-10-5. The soil moisture (SM) as the ordinate,

MPDI and MSMMI as the abscissa respectively, the construction of SM-MPDI and SM-MSMMI scatter plot (Figure 3) and calculate the coefficient of determination.

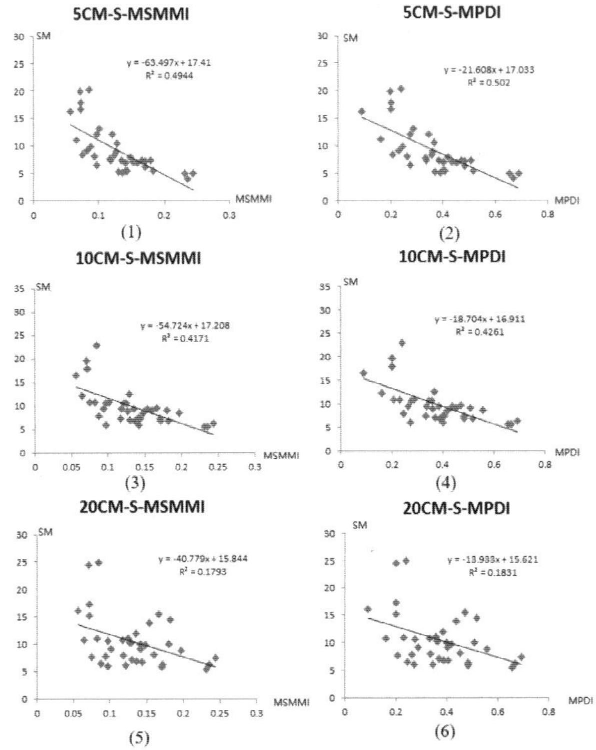

Figure 3. Relationship between soil moisture (20 CM) and MPDI/MSMMI of SPOT 6 in 2015-10-5

From the decisive monitoring results of different depth of soil, the simulation results of soil depth soil layer for mining area of the 0~5cm and 10cm are relatively good, and the simulation results on the drought monitoring 20cm soil depth is relatively poor. This phenomenon indicates that the closer to the surface layer, the better the effect of monitoring soil drought.

The coefficients of determination between MSMMI, MPDI and 0-5-cm-depth of Landsat 8 relative soil moisture are the highest, with R^2=0.364 and 0.375, respectively. The coefficients of determination between MSMMI, MPDI and 10-cm-depth of Landsat 8 relative soil moisture is higher than the 20-cm-depth relative soil moisture. The results showed that SMMI and PDI all can reflect the surface soil moisture and they are more suitable for assessing 0-5-cm-depth soil moisture conditions.

The coefficients of determination between MSMMI, MPDI and 0-5-cm-depth of SPOT 6 relative soil moisture are the highest, with R^2=0.494 and 0.502, respectively. The coefficients of determination between MSMMI, MPDI and 10-cm-depth of Landsat 8 relative soil moisture is higher than the 20-cm-depth relative soil moisture, with R^2=0. 417 and 0.426. The results showed that SMMI can reflect the surface soil moisture and they are more suitable for assessing 0-5-cm-depth and 10-cm-depth soil moisture conditions. However, MPDI is slightly better than MSMMI for monitoring 0-5-cm-depth soil moisture conditions.

5. Conclusion

This paper taking Shendong mining area as an example, using SPOT and Landsat 8 OLI image data validating two different drought monitoring index of mining area, comparison and analysis of two different monitoring models for the same region of the monitoring results of the main conclusions are as follows: (1) From the different depth of soil drought monitoring by remote sensing (R^2) decisive results, the simulation results of 0~5cm drought monitoring soil depth is relatively good, and the simulation results on the drought monitoring 10cm and 20 cm soil depth is relatively poor. This phenomenon indicates that the closer to the surface soil layer, the better the effect of monitoring soil drought index.

(2) Compared with MSMMI and PDI, including the improvement of soil type and land cover, the spatial resolution is improved, and the soil type and surface coverage of MPDI are higher. Generally speaking, MPDI is based on the vertical distance of the soil line from any point in the feature space to the coordinate origin to represent the soil moisture, MSMMI uses the distance from any point in the feature space to the origin to represent the soil moisture condition. We should make full use of them to achieve the best results in the monitoring service of the drought in the mining area. When comparing the monitoring degree of each kind of monitoring index to the drought, we can increase the quantity of the mining area, and improve the comparison result from the different soil moisture.

Acknowledgements
This work was jointly supported by The National Natural Science Foundation of China (Grant No. 41401496) and China Postdoctoral Science Foundation Funded Project (Grant No. 2016M592815).

References

[1] Shahabfar, A., Ghulam, A., Eitzinger, J 2012 *Int. J. Appl. Earth. OBS.* **18** 119-127.
[2] Shahabfar, A 2014 *IEEE J. S. T. Appl. EORS.* **7** 983-993.
[3] Ghulam A, Q Q M, Teyip T. 2007 *ISPRS. J. Photogramm.* **49** 150-164.
[4] Liu Y, Wu L X, Yue H 2015 *Can. J. Remote Sens.* **41**159-169.
[5] Ghulam A ,Q Q M, Zhan Z M . 2007*Environ Geol.* **48**1045-1052.
[6] Ghulam A, Christopher C.2014 *Remote Sens Society.* **48** 983-993.
[7] Han Y, Wang Y, Zhao Y. 2010 *IEEE T Geosci Remote*, 2010, **48** 2509-2515.

Promising Potentials of Agrivoltaic Systems for the Development of Malaysia Green Economy

N F Othman[1], A S Mat Su[1] and M E Ya'acob[2]

[1] Department of Agriculture Technology, Faculty of Agriculture, University Putra Malaysia, 43400 Serdang, Malaysia
[2] Department Process and Food Engineering, Faculty of Engineering, University Putra Malaysia, 43400 Serdang, Malaysia

Corresponding author's: fadzlin013@gmail.com

Abstract. Combination of solar photovoltaic (PV) systems with crops cultivation creates the need for renewable synergy known as Agrivoltaic (AV). The AV system is in line with the Kyoto Protocol which promotes the reduction of carbon emission and usage of fossil-fuel. The integration of these two resources would optimize the yield, improving AV system efficiency and solving the issue of land scarcity. Currently PV landscapes are transformed to a new transdisciplinary design of land used and extended to ecological performances and beneficial impacts to the surroundings. Malaysia has been actively promoting the adaptation of PV technology as an alternative energy mix for the country since 2010. This study provides some insights of techno-economic aspects with respect to the increasing number and size of PV installations. Detail calculation based on Monte Carlo Simulation for growth rate are also discussed. The large scale solar PV farms are a good indicator of inculcating the AV concept for pathway in developing Malaysia green economy.

1. Introduction

Over the last decade, Malaysia as an emerging developing country continues to pursue and actively promote Renewable Energy by launching several policies, programs, incentives and funding for the sustainability of future generation. Other efforts such as in the 7th Malaysia Plan, the Malaysian government has allocated large amounts for rural electricity programs to help provide solar energy installation in urban and remote communities such as powering individual homes, longhouses (culture house), rural schools and clinics [1].

Mekhilef et al., 2012 [2], in his recent study on solar energy in Malaysia also highlighted the strategies and actions taken by the Malaysian government and Non-Government Organization (NGOs) to implement solar PV systems and applications in the future as announced in the 9th Malaysia Plan. The Malaysian government has introduced the Feed-In Tariff programmed, Malaysia Building Integrated Photovoltaic (MBIPV) project, Rural Electrification Programme (REP) and financial support for the development of solar photovoltaic in Malaysia [3]. This effort, mostly are based on the fact that solar energy has great potential to be used efficiently in energy services like power generation, heating and transport. It is naturally available, no noise or pollution in its operation and provide clean electricity in a longer service lifetime. In a tropical country like Malaysia, average solar radiation is about 1643kWh/ m^2/ year, with rates above 10 sun hours per day [4].

A PV system or solar power system is a power system that consists of several components i.e solar panel, solar inverter, mounting, cabling and other electrical components commonly known as Balance of Systems (BOS). This PV system converts light energy from the sun into electricity. The PV module are structured from a single cell structure to an array formation from different materials such as crystalline, multi-crystalline, polycrystalline, microcrystalline and amorphous type as shown in Figure 1.

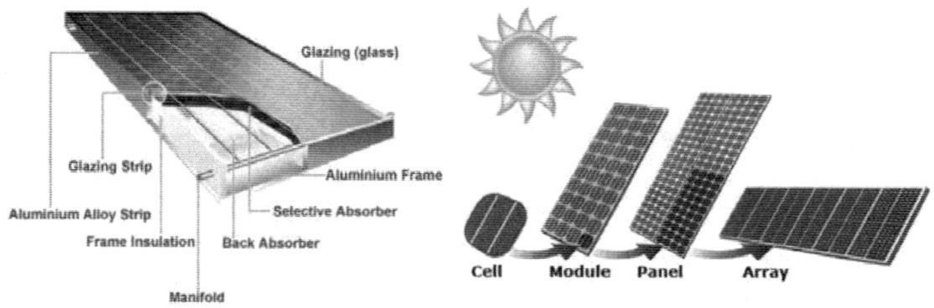

Figure 1: Illustration of PV module construction from a single cell. Source: http://sunmetrix.com

2. Agrivoltaic Technology Adaptation for Malaysia Green Economy

Photovoltaic landscape has been transformed not only to cater for clean energy generation but also a new transdisciplinary design of land used and extended to ecological performances and beneficial impacts [5]. The increasing of number and size of installation especially large scale solar PV farm may impacts in terms of the land scarcity and ecosystem services. One kiloWatt peak (1kWp) solar PV arrays covers about 6 to 9 square meter of land would simply create the suitable environment by making use of this vacant space under the solar panels. Most of the solar PV installation today are ground-mounted arrays and implies the fix structure PV panel. The adaptation of solar PV in large scale area need to be more future oriented by improving their design and the cost effective installation [5]. The idea of the AV or solar farming can be defined as an alternative conversion of agriculture (farming) into photovoltaics (solar panel) in a single land use system to optimize the yield [6]. The AV system considers appropriate plant characteristics based on crop height, sustainability, water requirements and shading tolerance. This new way of growing crops beneath solar panel is an experiment in integrating renewable energy with agriculture and is a part of bringing a new dimension to community agriculture [7]. Some of the published results relating to AV projects are listed in Table 1. Figure 2 shows the concept of AV system applied in some countries using the concept of utilizing unused land area with different kind of crop.

Table 1: Successful integration of PV Farms with agriculture cultivation.

Author/ Year	**Country**	**Highlights/ Findings**
Scognamiglio, [5]	Italy	The AV system allowing better ecological performance.
Dupraz et al., [6]	France	The AV system is able to provide more efficient use of light and space for the dual production of energy and food.
Dinesh & Pearce, [7]	USA	AV system can improve the technical potential for both PV production and agricultural production by propose a coupled simulation model.
Malu, Sharma, & Pearce, [8]	India	The AV system may increase the crop production as compared to conventional farming.

Figure 2: Typical Agrivoltaic projects in France and Japan (left to right).

Green economy is an alternative economy that can improve well-being and social equity, reduce environmental damage, climate change, help ecological scarcity issues and even incorporate natural resources into economic policy [9]. The green economy aims to enhance a sustainable and efficient economy for a better quality of life. Figure 3 shows the green growth effect by implementing the green economy. In Tanzania, the concept of green economy has been redesigned and reinterpreted green implementation to suit their basic aspects and governance in agri-business [10].

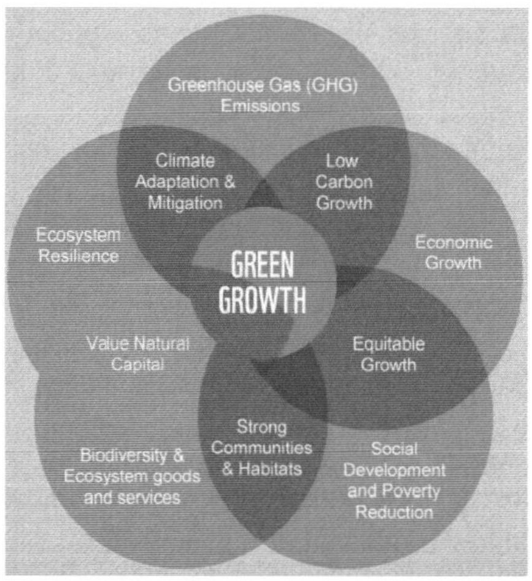

Figure 3: The concept of green economy towards sustainable development and efficient economy

The agricultural sector has good potential in the future, but there are some problems that arise such as problems of workers, capital and land. Most large companies rely on foreign workers while independent smallholders have limited capital resources. In line with the support of the Malaysian government's efforts in agriculture and renewable energy, the Agrivoltaic (AV) system is recommended for the Malaysian agribusiness industry as one of the methods for pre-industrial agricultural production systems such as viticulture, arboriculture, gardening market and plant crops[11].

Locally in Malaysia, one of the Agrivoltaic pilot project was set up in Universiti Putra Malaysia integrating Java Tea as the selected high-value herbal crops. Figure 4 shows the launching ceremony by the Ministry of Science Technology and Innovation (MOSTI) as to creating catalyst in promoting Green Energy to the society.

Figure 4: Launching ceremony by the Minister on December 2015

The coverage area under PV array is 1 acre with 7 strings of 1 kWp PV array with consideration of 20% space buffers from the structure of PV and cabling works. This concept of integration shows some good growth progress to the crops which can sustain under PV shaded condition with increasing value of profit return. [12].

3. Monte Carlo Simulation for Agrivoltaic Growth Rate

Monte Carlo (MC) simulation is a statistical tool to accurately estimate the probabilities of uncertain events as illustrated in Figure 5, where the simulation model situations that present uncertainty and then play them out on a computer thousands of times within the range of 0 to 1 [13]. This estimation uses random numbers to represent uncertainty of Agro-PV annual income. Based on cost observation, the solar PV production is the highest operational cost component where to maintain good growth, the cost outlays must be lower than the benefits obtained. In the final accounting stage, the impact of applying herbal plant under the solar PV array shows positive cash returns. This finding supports the hypothesis of utilizing the unused space under the solar PV array for Agro-PV integration.

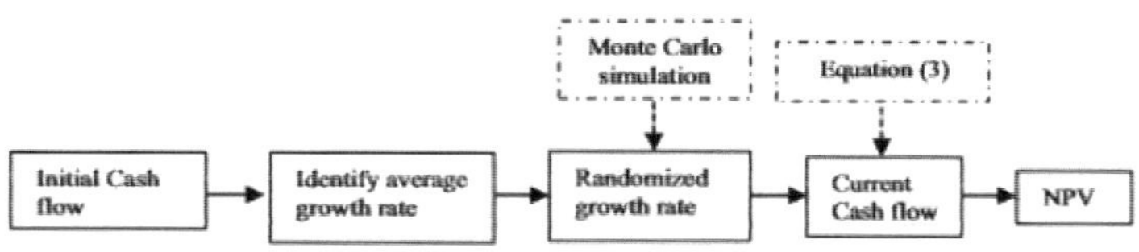

Figure 5: NPV Monte Carlo simulation using randomized growth rate.

The MC-Simulation process for Agro-PV data are projected in Table 2 and Figure 6 where the growth rate varies between -10% (economic downtime) up to 10 % (economic boost). Negative NPV of RM (1,794) shows the level of positive NPV projection can only withstand economic downtime not more than -10% of growth rate fluctuation. The value of IRR maintains at the range of 13 to 15 % with a good payback period of 6 years (at 3.36% discount rate constant).

Table 2: NPV-sensitivity analysis on varying growth rate for 10 years duration of Agrivoltaic operation

	Agro-PV (Monte Carlo Simulation)				
Growth rate	-10%	-5%	0%	5%	10%

NPV (RM million)	RM(1,794)	RM 2,678	RM 7,456	RM 15,162	RM 24,123
IRR (%)	13% - 15%				
Payback period	5-6 years				

The NPV mean value for Agrivoltaic (after 1000 random simulation) are calculated at RM14.3 million with minimum value of RM12.6 million and maximum value of RM16.2 million. The simulation projects a good payback period of 6 years with IRR values at ranging 13% to 15% as compared to 13.5% as previously calculated. The effect of cash flow after initial operation cost does not significantly affected by the fluctuating growth rate (varying range of 3.54% - 5.3% based on a good economic condition) which is justified by the linear cumulative cash flow projection.

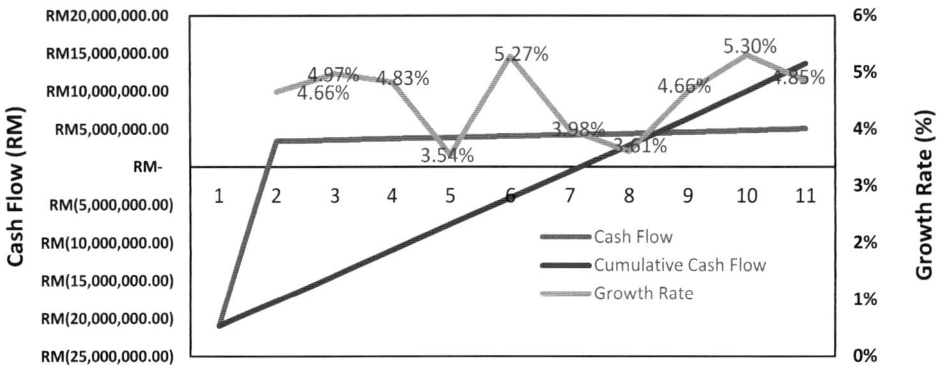

Figure 6: Monte Carlo Simulation based on random growth rate.

Based on these two simulated conditions, the MC simulation shows that when the growth rate decrease to -10% and below, the investment is showing negative net present value (-ve NPV). The output highlights that, higher probability of success from this project happen in 10 years of the project lifetime. Thus, this work highly suggest that Solar PV farms with integrated agriculture elements is the most suitable arrangements for a clean development projects.

4. Conclusion

The green economy approach for Malaysia development are presented with a typical Monte Carlo simulation for Agrivoltaic projects. The transdisciplinary design of land used and extended to ecological performances provides additional monetary benefits to the existing Solar Farm operators thus, will strengthen the platform for a mutual collaboration towards Malaysia Green Economy.

Acknowledgement

The authors delegate our thanks to the Research Management Centre (RMC), Universiti Putra Malaysia for the approval of research funding under the IPB Putra Grants Scheme (Vote no: 9515303)

Reference

[1] M. Z. A. Ab Kadir, Y. Rafeeu, and N. M. Adam, "Prospective scenarios for the full solar energy development in Malaysia," *Renew. Sustain. Energy Rev.*, vol. 14, no. 9, pp. 3023–3031, Dec. 2010.

[2] S. Mekhilef, A. Safari, W. E. S. Mustaffa, R. Saidur, R. Omar, and M. A. A. Younis, "Solar energy in Malaysia: Current state and prospects," *Renew. Sustain. Energy Rev.*, vol. 16, no. 1, pp. 386–396, 2012.
[3] F. Muhammad-Sukki, A. B. Munir, R. Ramirez-Iniguez, S. H. Abu-Bakar, S. H. Mohd Yasin, S. G. McMeekin, and B. G. Stewart, "Solar photovoltaic in Malaysia: The way forward," *Renewable and Sustainable Energy Reviews*, vol. 16, no. 7. pp. 5232–5244, 2012.
[4] S. C. Chua, T. H. Oh, and W. W. Goh, "Feed-in tariff outlook in Malaysia," *Renew. Sustain. Energy Rev.*, vol. 15, no. 1, pp. 705–712, Jan. 2011.
[5] A. Scognamiglio, " Photovoltaic landscapes ' : Design and assessment . A critical review for a new transdisciplinary design vision," *Renew. Sustain. Energy Rev.*, vol. 55, pp. 629–661, 2016.
[6] C. Dupraz, H. Marrou, G. Talbot, L. Dufour, A. Nogier, and Y. Ferard, "Combining solar photovoltaic panels and food crops for optimising land use : Towards new agrivoltaic schemes," *Renew. Energy*, vol. 36, no. 10, pp. 2725–2732, 2011.
[7] H. Dinesh and J. M. Pearce, "The potential of agrivoltaic systems," *Renew. Sustain. Energy Rev.*, vol. 54, pp. 299–308, 2016.
[8] P. R. Malu, U. S. Sharma, and J. M. Pearce, "Agrivoltaic potential on grape farms in India," *Sustain. Energy Technol. Assessments*, vol. 23, no. August, pp. 104–110, 2017.
[9] E. Loiseau, L. Saikku, R. Antikainen, N. Droste, B. Hansjürgens, K. Pitkänen, P. Leskinen, P. Kuikman, and M. Thomsen, "Green economy and related concepts: An overview," *J. Clean. Prod.*, vol. 139, pp. 361–371, 2016.
[10] J. T. Buseth, "The green economy in Tanzania: From global discourses to institutionalization," *Geoforum*, vol. 86, no. August, pp. 42–52, 2017.
[11] Q. Mastoi, A. Rahman, and A. Dahlan, "Agro-tech in Malaysia and Literature Review Knowledge Sharing from Agro-Technology Using Nations," vol. 5, no. 12, pp. 606–611, 2014.
[12] N. F. Othman, M. E. Ya'acob, A. S. Abdul-Rahim, M. Shahwahid Othman, M. A. M. Radzi, H. Hizam, Y. D. Wang, A. M. Ya'acob, and H. Z. E. Jaafar, "Embracing new agriculture commodity through integration of Java Tea as high Value Herbal crops in solar PV farms," *J. Clean. Prod.*, vol. 91, pp. 71–77, Mar. 2015.
[13] S. S. Shaffie and S. H. Jaaman, "Monte Carlo on Net Present Value for Capital Investment in Malaysia," *Procedia - Soc. Behav. Sci.*, vol. 219, pp. 688–693, May 2016.

Water conservation within planetary boundaries: residents' perception of recycled water use

PF Tunji-Olayeni[1], LM Amusan[1], RA Ojelabi [2] and OG Abina[1]

[1]Covenant University, P.M.B 1023, Ota, Nigeria
E-mail: pat.tunji-olayeni@covenantuniversity.edu.ng

Abstract: Fresh water use is one of the nine planetary boundaries that must not be transgressed if the earth must continue to exist. Climate change has necessitated the need for conservation of natural resources, including fresh water. Infrastructure provision for recycling water is one thing: accepting to use the recycled water is another. This paper assessed residents' perception on recycled water use. It adopted a qualitative research design. Findings reveal that residents are willing to use recycled water, however, motivations of residents to use recycled water is different from the motivations suggested by the scientific community which focuses on climate change issues such as drought and conservation. The study recommends that water authorities should possess the required competencies for treating recycled water so that health risks associated with the use of recycled water can be minimized.

1. Introduction

Fresh water use is one of nine planetary boundaries proposed by [1,2]. Climate change has altered weather patterns leading to reduced amount of rainfall and diminishing underground water. Moreover, there is great pressure on the existing water resources due to population growth and unsustainable water consumption patterns. This situation calls for conservation of available water resources.

Recycling water has been suggested as one of the means of conserving water [3,4]. Recycled water is community waste water that is treated to a re-usable quality [3]. It involves the treatment of community waste water for the replacement and conservation of available fresh water [5]. In some parts of the world, like the US, UK, Belgium, Australia, South Africa and Namibia recycled water has been used to augment available water [3,6,7]. Even so, only a minute portion is used to supplement available fresh water. In the US for example, only one-tenth of 1% of community waste water is recycled [8].

The full potential of water recycling is however hampered by stiff oppositions from end users. The literature is replete with evidences of public resistance to use of recycle water mainly because of the perceived health risks associated with recycled water [9, 3,6] and the thought that recycled water originates from human waste i.e. the 'yuck'/ 'disgust' factor [9,10].

Empirical evidences [11, 12, 10] show that the quality of recycled water is better than water from conventional sources. Moreover, water recycling systems have been operating for decades without any reported cases of epidemics traceable to the use of recycled water [9]. Yet, public use of recycled water has remained relatively low; not necessarily because end users do not believe that recycled water can be of higher quality but because they perceive that decision makers may compromise the quality of recycled water at some point, or may not take them seriously or fail to consult with them [13, 14, 15].

On the Lagos Island in Nigeria, much land is reclaimed to meet the demand of teeming population and this has a toll on available water resources. [16] revealed that the main source of residents' dissatisfaction with public housing was poor access to services and infrastructure like water. However, public opinion about re-use of treated waste water to augment fresh water is unclear. Since the public is a key stakeholder as the intended user of recycled water, their attitude towards recycled water use is important in order to avoid futile investment in water recycling infrastructure and to ultimately promote water conservation. Hence, the aim of this paper was to assess residents' perception on recycled water for domestic use.

2. The Need for Recycling Water

Global population is increasing at an alarming rate with about 7.6 billion people on the earth today and an estimated 9.8 billion by the year 2050 [17]. Humans depend on water for sustenance and for almost every endeavour. Water is used for domestic purposes, agriculture and manufacturing.

However, the earth's growing population is placing great pressure on the existing global water resources of which just 1/3 is fresh. Climate change has further worsened the water challenge by altering weather patterns resulting in reduced rainfall and drought in some parts of the world. According to [18] the prevalence of strong climactic seasonal variations worsens water challenges in tropical regions. Moreover, underground water is used up more than it is being replaced.

Besides, icecaps (where much of the world's fresh water is trapped) is melting at an alarming rate due to global warming. Furthermore, unsustainable waste disposal constitutes an environmental threat because large quantities of waste end up in streams, lakes other rivers thereby, contaminating available water. With the development of advanced water technologies, waste water can be treated to high quality.

The growing demand for fresh water fuelled by population increase, climate change , unsustainable disposal of waste water and developments in water technology has made researchers and world agencies such as the United Nations to seek ways of conserving existing fresh water system of which recycling is one.

2.1.Public Perception on the Use of Recycled Water

Public support and agreement is vital for implementing effective waste water reuse projects. Although some recycled water reuse programmes have been successful in the UK, US, Australia, Belgium, South Africa and Namibia [17], many of such schemes witnessed stiff resistance from the public.

Some proposed water recycling projects were aborted prematurely while others were abandoned after completion because of fierce resistance from the general public [18]. [3] noted that public resistance is a major challenge in recycled water programmes. Some factors responsible for public resistance of recycled water use are as follows: the disgust factor, perceived health risks, use and source of recycled water, trust, demographic factors, environmental stewardship, scarcity of fresh water.

3.Methodology

Quantitative research paradigm with the use of questionnaire survey was the primary data collection strategy used. Some researchers [6, 5] used a similar approach. A 5 point Likert scale questionnaire type was used to elicit information randomly from 250 residents in Ajah, Lagos- Nigeria. Descriptive statistics of frequencies and mean item score were used for data analysis.

3.1.Data Presentation, Analysis and Results

IOP Conf. Series: Earth and Environmental Science **146** (2018) 012003 doi:10.1088/1755-1315/146/1/012003

3.1.1.Respondents' Characteristics

Table 1. Respondents Characteristics

Characteristics	Frequency
Gender	
Male	153
Female	97
Total	250
Age	
<25	39
25-35	67
36-45	94
46-55	27
>55	23
Total	250
Educational Level	
Secondary school education	26
Graduates of higher institutions	138
Advanced degree holders	86
Total	250
Religion	
Christianity	147
Islam	103
Total	250

Table 1 shows the demography of respondents that participated in the survey. Male respondents were 153 while females were 97. The respondents fell within the age bracket <25, 25-35, 36-45, 46-55, >55 with 39, 67, 94, 27 and 23 respondents respectively. For educational level, 26, 138 and 86 respondents had received secondary school, undergraduate and post graduate education respectively. One hundred and forty seven of the respondents were Christians while the remaining 103 were Muslims. In summary, there were more males than females in the sample. The greatest age bracket studied was ages 36-45. Majority of the respondents had obtained at least an undergraduate degree and most of them were Christians.

3.1.2Respondents' Willingness to use recycled water (in percentage)

Table 2. Willingness to use recycled water (in percentage)

Recycled water reuse option	SA	A	NS	DA	SD	MIS*
Toilet flushing	69.6	28.3	3			0.94
Gardening	57.2	38.8	0.4	1.2	2.4	0.89
Water fountain	49.2	33.6	2.0	10.8	4.4	0.82
Car washing	41.2	36.4	2.8	12.4	7.2	0.78
Clothes washing	38.3	29.6	2.8	18.4	10.4	0.74
Bathing	8.8	10.4	10.4	29.6	40.8	0.43
Dish washing	2.8	6.4	1.2	41.2	48.4	0.35
Cooking	2.0	4.0	1.6	41.2	51.2	0.33
Drinking	0.80	3.2	0.8	30.0	65.2	0.29

*MIS - mean item score

Table 2 shows the agreement level of the respondents in recycled water use. Recycled water use for toilet was ranked highest with an MIS of 0.94. This is followed by gardening (MIS = 0.89) and water fountain (MIS = 0.82). The three least uses of recycled water by the residents surveyed are dish washing (MIS=0.35), cooking (MIS = 0.33) and drinking (MIS = 0.29). Even though residents are willing to use treated waste water, their willingness is limited to the source of the treated waste water. Most residents support recycled water reuse to the extent that it is used indirectly without having direct contact with it as in toilet flushing, gardening, water fountain. [19] also found a sharp decline in recycled water use once it is becomes too personal.

3.1.3.Strategies for improving the use of recycled water

Table 3 Strategies for improving the use of recycled water

Strategy	MIS*
Public awareness	0.93
Evidence of successful recycled water project	0.85
Public participation in recycled water use discourse	0.70
Recycled water project simulation	0.53

*MIS - mean item score

From table 3, it can be seen that the most effective strategy for improving residents' use of recycled water is public awareness (MIS=0.93). Other strategies are evidence of successful water recycling project (MIS=0.85) public participation (MIS=0.70) and project participation (MIS=0.85). The most effective strategy for improving residents' use of recycled water is public awareness. The second most effective strategy for improving residents' use of recycled water is evidence of successful water recycling project. [20] also discovered that evidence of successful water recycling projects can be a useful strategy for improving residents' use of recycled water. Public participation in recycled water use discourse was found to be the third most significant strategy for improving residents' use of recycled water. [21,22] also found that public engagement and participation in recycled water reuse discourse can improve residents' use of recycled water

3.1.4Factors influencing use of recycled water

Table 4Factors influencing use of recycled water

Factors	MIS*
Source	0.96
Use	0.90
Competence	0.86
Health Risk	0.81
Cost	0.62
Drought	0.60
Reports from the scientific community	0.45
Conservation	0.43

*MIS - mean item score

Table 4 indicates the factors that will influence residents' use of recycled water. The most significant factor influencing residents' use of treated waste water is the source of the recycled water (MIS=0.96). This is followed by the usage of the recycled water and competence of water authorities (MIS=0.90 and 0.86) respectively. The 3 least significant factors influences residents' use of recycled water are drought (MIS = 0.60), reports from the scientific community (MIS = 0.45) and the need for conservation (MIS = 0.43). The most significant factor that will influence residents' use of recycled water is the source of the recycled water. The 'dirtier' the source of the recycled water, the more averse residents will be towards using it. People have always seen recycled water as disgusting because of the belief that it has a faecal origin. Hence, the source of the recycled water is the most significant factor that will influence residents' use of recycled water. The second most important factor that will likely influence residents use of recycled water is the use of the recycled water. Residents are willing to use recycled water provided it does not come in contact with them. Many residents prefer to use recycled water for toilet flushing, gardening and water fountain. The third most significant factor is competence of water authority. Residents believe that a competent water authority will be able to recycle water efficiently and this will reduce the risk of health challenges arising from the use of recycled water. The least factors affecting residents' use of recycled water are drought, reports from scientific community and need for conservation. Whilst these factors are the major drivers of water recycling by global agencies like the United Nations, local residents are influenced by other issues that boarder around safety of recycled water such as source and use of water and competencies of water authority.

4. Conclusions

Residents are willing to use recycled water provided it is limited to non-portable uses. Residents' willingness to use recycled water is not influenced by global issues like drought, reports from scientific community and need for conservation. It is recommended that public awareness at local levels be carried out to increase residents' awareness of global issues like population growth and climate change and how these issues exert pressure on existing water systems. Hopefully, better understanding of the global pressures driving water recycling will improve residents' agreement to use recycled water.

Appreciation

The writers are grateful to Covenant University for sponsorship

References

[1] Rockstrom, J. *et al.,* (2009). Planetary boundaries: exploring the safe operating space for humanity. *Ecology and Society*, 14 (2).
[2] Steffen, W. et al., (2015). Planetary boundaries: guiding human development on a changing planet. *Science*, 347 (6223).
[3] Ormerod, K.J. (2016) Illuminating elimination: public perception and the production of potable water reuse *WIREs Water* 2016, 3:537–547. doi: 10.1002/wat2.1149
[4] United States Environmental Protection Agency.(2012) *Guidelines for Water Reuse*. Washington, DC and Cincinnati, OH: U.S. Environmental Protection Agency, Office of Research and Development; U.S. Agency for International Development
[5] Dolnicar, S. & Schafer, A. (2009). Desalinated versus recycled water — public perceptions and profiles of the accepters. Journal of Environmental Management, 90 (2), 888–900.
[6] Chen, W, Bai, Y, Zhang,W, Lyu, S and Jiao, W .(2015). Perceptions of Different Stakeholders on Reclaimed Water Reuse: The Case of Beijing, China. *Sustainability* **2015**, *7*, 9696-9710; doi:10.3390/su7079696
[7] Crovello, S., Davidson, J and Keller, A. (2010). Perception and communication of water reclamation for the sustainable future of Windhoek. *BSc Thesis*. Worcester Polytechnic Institute, City of Windhock, Namibia

IOP Conf. Series: Earth and Environmental Science **146** (2018) 012003 doi:10.1088/1755-1315/146/1/012003

[8] Drewes J.E and Khan S.J.(2012). Water reuse for drinking water augmentation. In: Edzwald JK, ed, American Water Works Association. *Water Quality & Treatment a Handbook on Drinking Water*. New York: McGraw- Hill.
[9] Rock, C., Solop, F.I and Gerrity, D. (2012). Survey of statewide public perceptions regarding water reuse in Arizona. Journal of water supply: Resaerch and Technology-AQUA, 61(8), 506-517
[10] Gerrity, D and Snyder, S. (2011). Review of ozone for water reuse applications: Toxicity, regulations, and trace organic contaminant oxidation. *Ozone Sci. Eng.* 33, 253–266.
[11] National Research Center.(2012). *Water Reuse: Potential for Expanding the Nations Water Supply through Reuse of Municipal Wastewater*. Washington DC: National Academies Press.
[12] Gerrity, D., Trenholm, R and Snyder, S. (2011) Temporal variability of pharmaceuticals and illicit drugs in wastewater and the effects of a major sporting event. Water Res. 45, 5399–5411.
[13] Ormerod K.J and Scott C.A. (2013). Drinking wastewater: public trust in potable reuse. *Sci Technol Hum Val*, 38:351–373.
[14] Russell S, Lux C, Hampton G. (2009). Beyond "Information": integrating consultation and education for water recycling initiatives. *Soc Nat Resour* , 22:56–65.
[15] Marks J. (2006). Taking the public seriously: the case of potable and non-potable reuse. *Desalination* , 187:134–147.
[16] Ibem, E.O., Opoko, P.A and Aduwo, B.E. (2015). Satisfaction with neighbourhood environments in public housing: evidence from Ogun State, Nigeria. *Social Indicators Research*, 130(2), 733-757
[17] UN DESA, (2017). Accessed from www.un.org/desa/publications on 16/11/17
[18] Badejo, A.A., Ndambuk, J.M., Kupolati, W.K., Adekunle, A.A., Taiwo, S.A and Omole, D.O. (2015). Appraisal of access to safe drinking water in south west Nigeria. *African Journal of science, technology and development*, 7(6), 441-445
[19] Sydney Water (1999). *Community Views on Re-cycled Water*. Syndey: Australia.
[20] Kemp B, Randle M, Hurlimann A and Dolnicar S. Community acceptance of recycled water: can we inoculate the public against scare campaigns? *J Public Aff* 2012, 12. 337–346.
[21] H urlimann A and Dolnicar S. (2010). When public opposition defeats alternative water projects - the case of Toowoomba. Australia. *Water Res* , 44:287–297.
[22] Nancarrow B.E, Leviston Z, Po M, Porer N.B, Tucker D.I. (2008). What drives communities' decisions and behaviours in the reuse of wastewater. *Water Sci Technol* 2008, 57:485–491. 64.

IOP Conf. Series: Earth and Environmental Science **146** (2018) 012004 doi:10.1088/1755-1315/146/1/012004

Sustainability strategies in the construction industry: implications on Green Growth in Nigeria

PF Tunji-Olayeni[1], TO Mosaku[1], OO Oyeyipo[2] and AO Afolabi[1]

[1]Covenant University, P.M.B 1023, Ota, Nigeria
[2]Bells University of Technology, P.M.B 1015, Ota, Nigeria

pat.tunji-olayeni@covenantuniversity.edu.ng

Abstract. The construction industry is a major driver of economic growth. However, activities from the construction industry contribute significantly to environmental pollution and unsustainable consumption of depleting natural resources. Green growth on the other hand, is an innovative growth which motivates businesses to adopt environmental friendly activities. This paper assessed sustainability strategies in the construction industry. A quantitative research design was adopted with the use of questionnaires distributed to construction organizations in Lagos, Nigeria. Findings reveal that many of the construction organizations surveyed do not have sustainability strategies because of low awareness about sustainability issues in construction. However, of the three pillars of sustainability (environmental, economic and social sustainability), environmental sustainability was found to yield greatest benefit. For the construction industry to contribute to Nigeria's green growth agenda there should be increased awareness of sustainability issues and institutional policies to drive sustainability.

1. Introduction

Construction activities facilitate both social and economic development in all countries. The social and economic progress of a nation is determined by the quality of infrastructure provided by the country [1]. Construction businesses provide the manpower, competencies, materials, plant and equipment required for infrastructure development: a key goal of the sustainability agenda. Sustainability ensures that the present generation is able to meet its need without depriving future generation of meeting their needs [2].

However, in a bid to meet developmental needs, the construction industry consumes a large quantity of natural resources [3] particularly, fresh water, timber, sand and limestone. Buildings for instance, use about 50% of fresh water [4]. Moreover, the over dependence of fossil fuels [5]as alternative sources of power in buildings produce large quantities of atmospheric carbon, sulpur, nitrogen and other volatile organic compounds (VOCs) which pollute the environment. Embodied energy in building materials also releases toxic gases which are detrimental to the environment. Moreover, the construction industry generates large amount of waste deposited on landfills, which results in environmental degradation and CO2 emission.

Construction professionals need to adopt environmentally friendlily techniques to combat environmental issues from construction activities. [6]also reiterated the need for reliable and clean energy sources for Nigeria's sustainable development. Green growth on the other hand, is an agenda of governments and international agencies, which fosters development, and ensures that natural

IOP Conf. Series: Earth and Environmental Science **146** (2018) 012004 doi:10.1088/1755-1315/146/1/012004

resources are used sustainably, and continue to provide the resources and environmental services on which our well-being relies [7].

To achieve green growth in Nigeria, sectors such as construction which significantly pollute the environment and use considerable amount of natural resources, need to develop strategies for combating environmental challenges arising from the industry. The present study assessed the sustainability strategies adopted by firms in the construction industry. The study objectives are:

i. To assess the core elements in organizations mission statement
ii. To identify the factors militating against the development and implementation of sustainability strategies
iii. To identify the benefits of incorporating sustainability into construction activities

2. Literature Review

2.1.Sustainability strategies of construction organisations

For sustainability to be entrenched into the fabrics of construction organizations there has to be a sustainability agenda driven by construction businesses. Good sustainability practices within organisations stems from the integration of sustainability concerns into core business strategies [8]. Organisational strategy flows from top management down to operational staff. [9] noted that sustainable practices can only be made possible by executives at the top strategic level of management. Once strategy has been delivered then, operational staff can align with the strategic consciousness of the organizations they work with. The anticipated strong commitment to sustainable practices will require that sustainability be incorporated in the organizational strategy [9]. For sustainable construction to thrive, managers and leaders of construction businesses have to develop and implement sustainability plans and provide requisite resources to handle changes arising from such implementation [10].

2.2Factors militating against the development and implementation of strategies in construction organisations

One of the major hindrances to the development and implementation of sustainability strategies in the construction sector is poor awareness level. [11] noted that many stakeholders in the construction industry do not have sufficient information on sustainable. [8] also asserted that mangers who are responsible for charting the sustainability cause have little or no information about sustainability. In Nigeria, the knowledge of sustainable construction is below average and the construction sector still oblivious of sustainable construction [12]. It is difficult to develop and implement a concept that is not well understood.

Fear of the cost of implementing sustainability construction also inhibits the development of sustainability strategies in construction businesses. This includes perception of higher investment costs, long pay-back period, client fears about viability, poor knowledge of life cycle cost and difficult access to financial resources [13]. Perceived high cost of undertaking sustainable construction is a barrier to its implementation. Higher costs of implementing Sustainable construction maybe perceived as more expensive than conventional construction because of increase in consultancy fee[13]. Moreover, the long term gains of sustainability are difficult to express in terms of financial gains [14]

Poor support from the government is also another factor militating against the development and implementation of sustainability strategies. [13] noted that intensive support from the government is required to implement sustainable construction. [14] also asserted that sustainable construction will thrive if the government puts in legislations in place that will drive corporate sustainability in construction organizations. Successful implementation of sustainable construction requires government commitment. The government plays a major function in the construction industry. Governments have the responsibility of providing adequate support for sustainable construction to thrive.

Low demand of sustainable construction by client has also been identified as a factor militating against the development and implementation of sustainability strategies. [13] reported that low demand of sustainable construction by clients possess a great challenge in the implementation of sustainable construction .

Poor expertise for sustainable construction also inhibits the implementation of sustainable strategies. [14] noted that unavailability of sustainable building materials, non-existence of sustainability measurement tools andabsence of exemplar demonstration projects are issues bothering around expertise for sustainable construction and these have negative implications for the implementationof sustainable construction. [9] further noted resistance to cultural change is another hindrance to implementing sustainable construction.

3.Methodology

The quantitative research approach was used in this research. Target population for this research comprised of owners and managers of construction businesses in Lagos, Nigeria. Sixty copies of the questionnaire were distributed to owners and managers of construction businesses in Lagos, Nigeria. The questionnaire had four sections. Section one covered questions on the core elements in a organizations' mission statement, section two was about the factors militating against the development and implementation of sustainability strategies in construction organizations, Section three sought to elicit information on the benefits of incorporating sustainability in construction, while the fourth section covered demographic information of the respondents.

In sections one, two and three likert scale type questions where 1 signified strongly disagree, 2 signified disagree, 3 signified neutral, 4 signified agree and 5 signified strongly agree was used to obtain information from respondents.

The descriptive statistics (tables and mean score) were used to analyze the data obtained for this study.

3.1.Core element in company's mission

The core element in the mission statement of majority of the construction businesses surveyed is quality in project and product delivery (Table 2). Profit maximization (Table 2) was also found as the second core element in the mission statement of the construction businesses surveyed. Client satisfaction (Table 2) was the third core element in the mission statement of the construction businesses surveyed. Recycling waste was the least element in the mission statement of the construction businesses surveyed.

Table.1. Core element in company's mission

Company's core mission	Mean
To ensure quality in project/product delivery	4.65
To maximize profit	4.48
To ensure client satisfaction	4.42
To ensure occupational health and safety for all workers	4.17
To ensure the use of sustainable building products/materials	4.15
To reduce waste in operations	4.15
To reduce carbon foot print in all operations	4.13
To ensure cost reduction in all operations	4.13
To ensure better work life balance for all workers	3.96
To ensure that all workers develop capacity	3.94
To ensure equality for all workers	3.83
To conserve energy in operations	3.62
To conserve water in operations	3.52
To reach out to the immediate community as part of corporate Social responsibility	3.48
To recycle waste	3. 25

3.2. Factors militating against the development and implementation of sustainability strategies
The most significant factor militating against the development and implementation of sustainable construction is low awarenessof sustainable construction (Table 3). Poor government support for sustainable construction and lack of leadership support for sustainable construction were the second and third most significant factors respectively militating against the development and implementation of sustainable construction (Table 3).

Table 2. Factors militating against the development and implementation of sustainability strategies

Factor	Mean
Low level of awareness of sustainable construction	4.42
Poor government support for sustainable construction	4.35
Lack of leadership support for sustainable construction	4.21
Lack of relevant laws and regulation to drive Sustainable construction	4.12
Lack/poor expertise for sustainable construction	3.85
Fear of the cost of adopting sustainable construction	3.83
Resistance to change	3.38
Low demand for sustainable construction from clients	3.17

3.3. Benefits of incorporating sustainability in construction
Table 4 indicates the benefits of sustainable construction. The greatest benefit of sustainable construction is the belief that sustainable construction makes a company environmentally sustainable (Table 4). The second greatest benefit of sustainable is it perceived economic benefits while the third benefit of sustainable construction is that sustainable construction improves a company's reputation (Table 4).

Table 3. Benefits of Sustainable Construction

Benefits of Sustainable Construction	Mean
Sustainable construction improves organizations' environmental sustainability	4.56
Sustainable construction has economic value	4.48
Sustainable construction improves company's reputation	4.31
Sustainable construction is a business strategy	4.17
Sustainable construction creates a humane environment	4. 20
Sustainable construction creates a dignifying environment	3.79
Sustainable construction creates an equitable working environment	3.73

4. Discussion
Quality performance was found to be the most common element in the mission statement of the organizations studied. This is followed by profit maximization and then client satisfaction. The mission statement of most of the construction organizations surveyed is more inclined towards project performance indicators.

Sustainability elements in mission statements of the organizations studied ranked low. This may be because of poor awareness of the concept of sustainable construction and poor government and leadership support for sustainable construction.

The greatest deterrent to developing and implementing sustainable construction is poor awareness. It is difficult to drive a concept that is not well understood by clients and business operators. This probably explains why sustainable construction is not a core element in the mission statement of most construction organizations surveyed.

Another factor militating against the development and implementation of sustainable construction is poor support from the government in driving the sustainability agenda. This is evident in the absence of relevant laws and regulations to drive sustainability especially in the construction sector. Inadequate support for sustainable construction by leaders of construction organizations also militates against the development and implementation of sustainable construction. This could be because the concept is not well understood.

The major perceived benefit of sustainable construction is that sustainable construction improves organizations' environmental sustainability. Respondents view sustainability more in terms of environmentally sustainability. This may be because of greater emphasis on environmental sustainability than economic or social sustainability.

Since majority of the firms surveyed do not have strategies for tackling sustainability challenges from construction activities, Nigeria's green growth will be hampered. Moreover, since the sustainability ethos is yet to be enshrined in construction organizations, it may be difficult to achieve the much needed green growth in Nigeria.

5. Conclusion

The mission statement of most of the construction organizations surveyed is more inclined towards project performance indicators. Sustainability elements in mission statements of the organizations studied ranked low. Sustainability awareness and sustainability polices are recommended in order to increase interest and implementation of sustainability in the construction industry. Development and implementation of sustainability strategies will foster achievement of the green growth agenda in Nigeria.

Appreciation

The authors are grateful to Covenant University for sponsorship

References

[1] Ogunlana, S.O, Li, H, and Sukhera, F.A. (2003) System Dynamics Approach to Exploring Performance Enhancement in a Construction Organization.*Journal of Construction Engineering and Management* 129 (5): 528–536.

[2] World Commission on Environment and Development, 1987. Our Common Future, London.

[3] OmoleD.O and Ndambuki, J.M. (2014). Sustainable living in Africa: case of water sanitation, air pollution and energy. *Sustainability* 6(8), 5187-5205

[4] Dixon, W. (2010). The impacts of construction and the built environment, Briefing Notes, Willmott-Dixon Group

[5] Oyedepo, S.O. (2012). On energyfor sustainable development in Nigeria.*Renewable and Sustainable Energy Review*, 16, 2583-2598

[6] Oyedepo, S.O. (2014). Towards achieving energy for sustainable development in Nigeria. *Renewable and Sustainable Energy Review*, 34, 255-272

[7] OECD (2011). *Inclusive green growth: for the future we want* , Paris

[8] Elmualim, A., Valle, R and Kwawu, W. (2012). Discerning policy and drivers for sustainable. facilities management practice *International Journal of Sustainable Built Environment* 16–25

[9] Ikediashi, D.I., Ogunlana,S.O., Oladokun, M.G and Adewuyi, T. (2012). Assessing the level of commitment and barriers to sustainable facilities management practice: A case of Nigeria.*International Journal of Sustainable Built Environment*. 1(2), 167–176

[10] Osaily, N.Z. (2010). The key Barriers to Implementing Sustainable Construction in West Bank – Palestine, Robert Kennedy College / Zurich University of Wales / UK, March – 2010.

[11] Williams, K., &Dair, C. (2007). What is stopping sustainable building in England Barriers experienced by stakeholders in delivering sustainable developments. *Sustainable Development, 15*(3), 135-147 http://dx.doi.org/10.1002/sd.308
[12] Abolore, A.A. Comparative study of environmental sustainability in building construction in Nigeria and Malaysia.*Journal of Emerging Trends in Economics and Management Science (JETEMS)*. 3(6), 951-961
[13] Alsand, S., Gale, A and Edwards, R. (2011). Challenges of sustainable construction in Kuwait: Investigating level of awareness of Kuwait Stakeholders. *World Academy of Science, Engineering and Technology* , 59, 2197-2204
[14] Dzokoto S. D. &Dadzie J. (2013). Barriers to sustainable construction in the Ghanaianconstruction industry: consultants perspectives *In:* Laryea, S. and Agyepong, S. (Eds) *Procs 5th West Africa Built Environment Research (WABER) Conference,* 12-14 August 2013, Accra, Ghana, 223-234.

Plant litter effects inversely on air and soil temperatures in *Leymus chinensis* grassland, Northeast China

Y T Song[1], L L Li[1], Q Li[2], Wuyunna[1] and D W Zhou[2]

[1]College of Environment and Bioresources, Dalian Minzu University, Dalian 116600, Liaoning, China
[2]Northeast Institute of Geography and Agroecology, Chinese Academy of Sciences, Changchun 130102, Jilin, China.
yantaosong@dlnu.edu.cn

zhoudaowei@neigae.ac.cn

Abstract. Plant litter is an important component of grassland ecosystem, and has significantly reduced by human activities. These changes in natural systems cause variations in air and soil temperatures. This study described an experiment that different amounts of plant litter effect on air and soil temperatures in Songnen grassland, Northeast China. Plant litter reduction resulted in decreasing air temperature at 10 cm above surface and increasing soil temperature at 20 cm below soil surface in the growing season. The mean, maximum, and amplitude of air temperatures at 10 cm above soil surface of plots with plant litter were significantly higher than plots without litter. The mean, extreme, and amplitude of soil temperatures at 20 cm depth increased significantly as plant litter decreased in the growing season. Temperature differences between air and soil temperatures decreased significantly as plant litter decreased in the growing season. The results show that the soil and air temperatures respond differently to plant litter, and the changing temperature difference between air and soil temperatures has the potential to produce complex effects on grassland ecosystem.

1.Introduction

Grasslands are one of the most important constituents of the earth's vegetation and they are characterized by diverse types with a very broad range of environmental conditions in all climate types [1-2]. Grasslands in China cover nearly 4 million km^2 (more than 40 %) of the total land area, and play an important role in livestock farming and environmental conservation [3]. However, grassland degradation is advancing over wide areas through human activities, such as overgrazing, excessive cutting and unregulated collection of fuel and medical plants [3]. One of the characteristics of grassland degradation is aboveground biomass decreasing which results in plant litter becoming less and less [4-5].

Plant litter is by the definition as "the dead plant material of small size laying loose on the ground" [4], which is an import component in grassland ecosystem. Besides significantly effecting on net primary production [5-6] and nutrient cycles [7], plant litter also modifies the local physical environment by providing cover, changing soil surface microclimate [5], especially soil temperature which is one of the most important factors for the grassland ecosystem.

Soil temperature has significant impacts on the rates of biochemical reactions and has a strong influence on the growth of plants and soil biota [8-11]. Extreme soil temperature also is particularly important related to plant growth, for example, seed germination and early seedling growth are more highly correlated with daily maximum temperature of the soil than with air temperature [12]. The latest research showed that maximum and minimum warming asymmetrically effected on vegetation growth and carbon sequestration [13]. Therefore, a small changes of soil temperature would have great influence on ecosystems in all aspects [14].

Plant litter intercepts solar radiation and insulates the soil from air temperature, thus, modifies soil temperature. Many studies of the litter effecting on soil temperature focus on forest ecosystem [9, 15-17] or tillage systems [18-20], only few studies show that litter removal increased the soil temperature in grassland by 5-8 °C at the beginning of the growing season [21-22]. Furthermore, there is general consensus that air temperature warming is asymmetric because of minimum temperature having increased at twice the rate of maximum temperatures [23-24]. However, it is unclear that how the maximum and minimum soil temperature changes by plant litter.

The objective of this study was to examine the influence of plant litter to soil temperatures of grassland in the growing season. First, we measured the daily average air temperature and soil temperatures of our study site to find the differences of them in the growing season, and second, we compared the maximum, minimum temperature and diurnal daily temperature range between different litter amounts.

2.Material and methods

2.1.Site description

The experiments were conducted at the Ecological Research Station for Grassland Farming (ERSGF) in Jilin Province, Northeast China (44°33' N, 123°31' E, elevation 145 m). The study area has a semi-arid, continental climate. Mean annual temperature is 4.9 °C. Annual precipitation is 300-500 mm [25]. The grassland was dominated by *Leymus chinensis*. The main soil type is meadow chernozem at the study site.

2.2.Experimental design

Experimental plots were set up following a randomized block design in 2007. Plant litter was simulated by adding hay on clipped grassland. The amounts were 0, 200, 600, 1000 g/m^2 with three replicates. Each plot was 1.5 m by 1.5 m, with 1 m intervals between plots.

In 2008, the temperatures at 10 cm above the soil surface and 20 cm soil depth were measured from 17th April to 29th August. EM50 sensor (Decagon, USA) was used to record temperatures automatically with 1 hour intervals.

2.3.Statistical analysis

The variations in daily mean (Mean), minimum (Min), maximum (Max), and amplitude (Max-Min) of air and soil temperatures during the experiment were evaluated in every fifteen days by repeated measure ANOVA with plant litter amount as the main between-subject factors and time as the within-subject factor. Univariate analysis in general linear model was used to detect the effect of the different amount litter treatment on averaged temperatures during the experiment, LSD was used to compare the multiple treatment mean values, all statistical analyses were carried out using SPSS 13.0 (SPSS Inc., Chicago, USA) at 0.05 level.

3.Results

3.1.Mean, extreme and amplitude of daily air and soil temperatures

Mean daily air and soil temperatures had a similar changing pattern from 17 April to 29 August. However, mean daily air temperatures were more fluctuating compared with soil temperatures, especially at the beginning of the measuring time (Figure 1a and 2a). Mean daily air temperatures were 2-5 °C higher than soil temperatures during the observation time. Temperature difference between treatments decreased at the last 15 days for both mean daily air and soil temperatures (Figure 1a and 2a). There was no significant difference between the treatments of daily air temperatures at the beginning 30 days (Figure 1a). Mean daily soil temperatures were significantly different from 17 April to early August and decreased from 0 to 1000 g/m^2 (Figure 2a). Plant litter significantly affected on daily minimum, maximum and amplitude of air and soil temperatures (Figure 1 and 2).

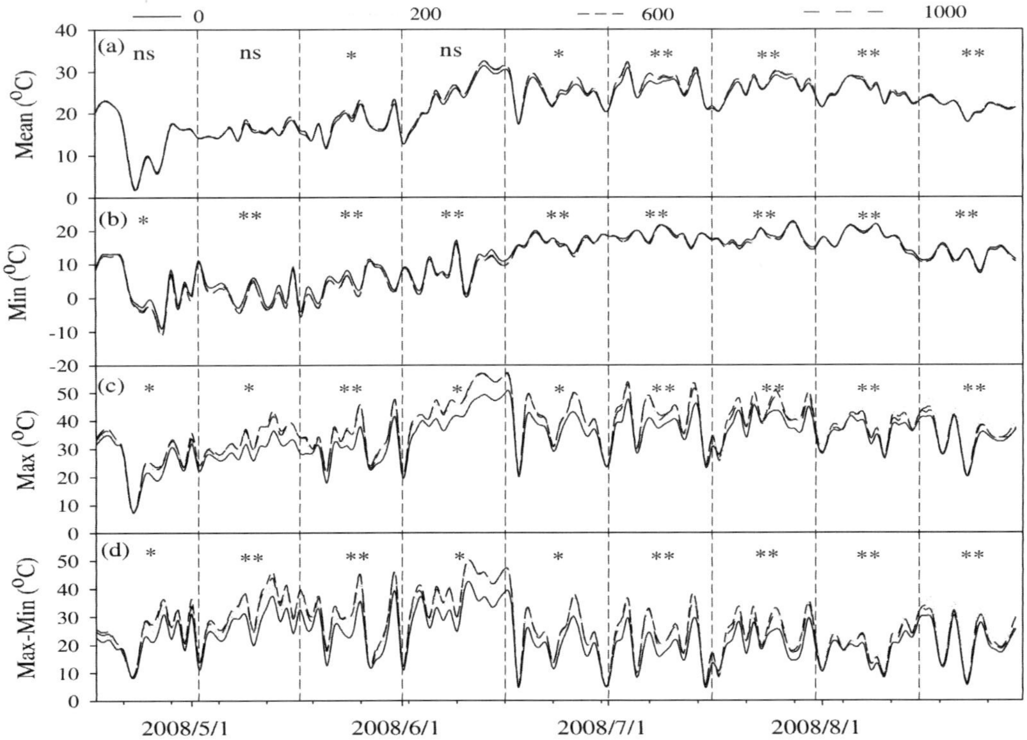

Figure 1. Changes in daily (a) Mean, (b) Min, (c) Max, and (d) Max-Min temperatures (°C) from 17 April to 29 August 2008 at 10 cm above soil surface. ** $P < 0.01$, * $P < 0.05$, ns non-significant.

3.2.Mean, extreme and amplitude of averaged daily air and soil temperatures

The averaged daily mean, maximum, and amplitude of air temperatures of the plots with plant litter were significantly higher than the plots without plant litter. However, there were no significant differences between treatments with 200, 600, and 1000 g/m^2 plant litter. The averaged daily mean air temperature of plots with plant litter was 0.5°C higher than plots without plant litter (Figure 3a). However, the averaged daily minimum air temperature of plots with 1000 and 600 g/m^2 plant litter were significantly lower than the treatments with 200 and 0 g/m^2 plant litter (Figure 3a).

The averaged daily mean, minimum, maximum, and amplitude of soil temperatures at 20 cm depth increased significantly as plant litter decreased (Figure 3b). Both averaged daily mean and minimum soil temperatures were significant difference between the treatments. No significant differences were detected with regard to averaged daily maximum and amplitude of soil temperatures between the 1000 and 600 g/m^2 (Figure 3b).

The averaged daily mean, maximum and amplitude of air temperatures at 10 cm above soil surface were higher than that of soil temperatures at 20 cm depth (Figure 3c). Temperature difference between averaged daily air and soil temperatures decreased significantly as plant litter decreased except for the treatments with 1000 and 600 g/m^2 (Figure 3c). The temperature differences of averaged daily temperatures were 4.3, 4.0, 3.2, and 2.0 °C for 1000, 600, 200 and 0 g/m^2 treatments, separately. The averaged daily minimum air temperature was about 6.5 °C lower than that of soil temperature. However, there was no significant temperature difference between the treatments (Figure 3c). Temperature differences between the mean daily maximum air and soil temperatures with plant litter were significantly higher than plots without plant litter (Figure 3c), and there was also the same pattern of the temperature differences between averaged daily amplitude of air and soil temperatures between the treatments (Figure 3c).

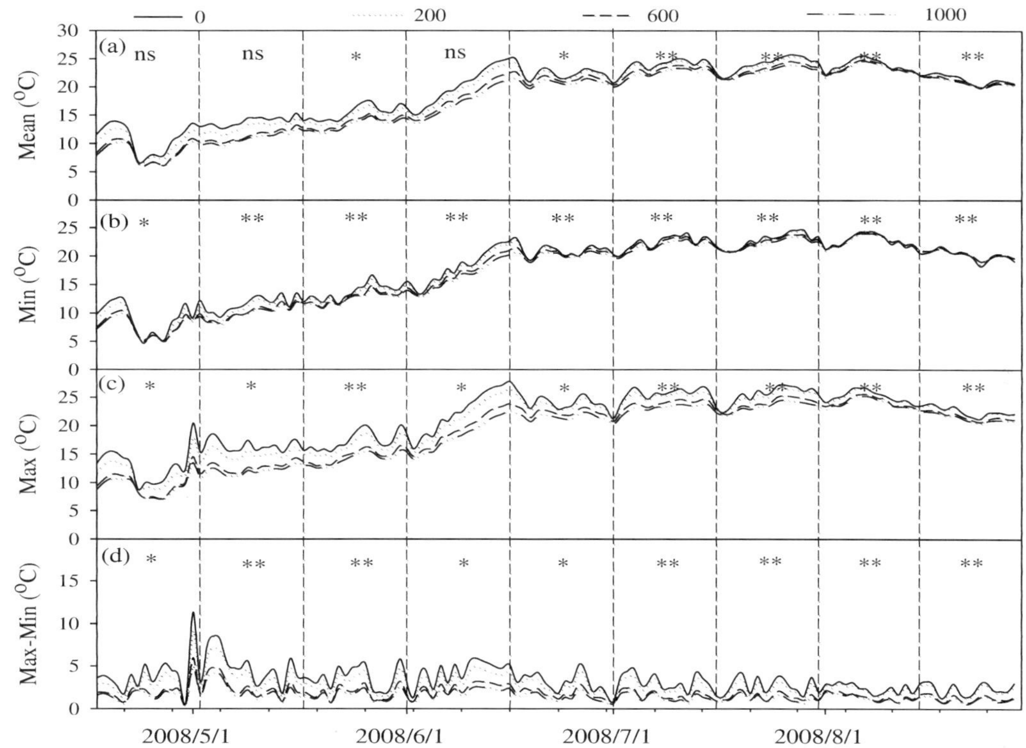

Figure 2. Changes in daily (a) Mean, (b) Min, (c) Max, and (d) Max-Min temperatures (°C) from 17 April to 29 August 2008 at 20 cm below soil surface. ** $P < 0.01$, * $P < 0.05$, ns non-significant.

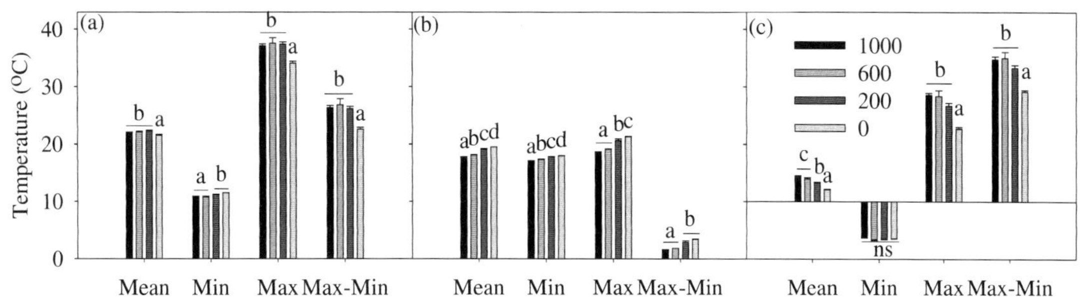

Figure 3. The average of daily Mean, Min, Max, and Max-Min temperatures (°C),from 17 April to 29 August 2008 in different treatments at (a) 10 cm above soil surface, (b) 20 cm below soil surface, (c) temperature difference between air and soil temperatures. Bars (mean ± SE) with different letters are significantly different ($P < 0.05$), "ns" non-significant ($P > 0.05$).

4.Discussion

Plant litter reduction by human activities has caused change in solar radiation energy fluxes between air and soil temperatures [4]. Our experiment showed that the plots with plant litter has significant higher air temperature and lower soil temperature than the plot without plant litter in the growing season (Figure 1a and 2a). It agreed with results of other studies involving soil temperature responding to grassland litter. Weaver and Rowland [21] found that soil temperature in a grassland with litter was 8 °C lower than that in a bare one. Hulbert [22] showed that both burning and litter removal by hand increased soil temperature in a grassland by 1-5 °C during the growing season. Our results found that plant litter removal increased the averaged daily mean, minimum, maximum, and amplitude of soil temperatures by 0.5-2 °C in the growing season (Figure 3).

It is well known that most researches are focused on comparing the soil temperatures between different vegetation types [9, 12] or just comparing covered plots with bare land [21-22]. However, within grasslands a degradation gradient may influence soil temperature significantly. Our results showed that different amounts of plant litter had significant impacts on soil temperature (Figure 3b). Linear regression of the histogram of averaged daily soil temperature showed that soil temperature increased 0.16 °C as plant litter decreased 100 g ($P = 0.004$, Figure 3b). Thus, litter reduction increased soil temperatures of grassland in the growing season.

Overgrazing and/or high production has degraded grasslands in spite of the suggestion of grassland studies that advised about necessarily of their conservation [3]. In the past decades, e.g. Songnen grassland has experienced remarkable land use/cover changes as a result of human activities. The grassland had the most dramatic decrease from 1986 to 2000, when his area decreases in area of 6,127 km2. Nowadays, only 25.6% of the original area of 1986 remained [25]. The decrease in grassland cover caused by unreasonable development of stockbreeding has increased evaporation, soil temperature, and soil organic decomposition [26]. In this area, the dominant plant species of grasslands were wiped out, meanwhile the average vegetation coverage, height, density, and productivity decreased to a great extent [27]. The reduction of grassland contributes soil warming which could produce complex potential effects on grassland ecosystems.

The global climate has changed rapidly, the warming pattern shows that the minimum temperature increased two order of magnitude of the maximum temperature and the amplitude of temperature decreased, i.e. an asymmetric diurnal temperature increase [23-24]. However, plant litter reduction modified the air and soil temperatures near-surface in our study by altering the maximum temperature (Fig 3a, b). Interestingly, the amplitude of soil temperature became larger as plant litter decreased in the growing season (Figure 3c). Plant litter created a barrier to heat exchange between air and soil [4]. The amplitude of soil temperature increase was mainly due to the increase of the maximum soil temperature as plant litter reduction (Figure 3c). The changing pattern of soil temperature is likely to affect grassland ecosystem, such as seed germination especially in early growing season [4, 12].

Both air and soil temperatures are important abiotic factors of grassland ecosystem. Vegetation types also influence air temperature [12]. Air temperatures were significantly different between plots with and without plant litter, but there was no difference between treatments of 200, 600, 1000 g/m^2 plant litter (Figure 3a). This may be associated with the same vegetation type and similar community structure. Interestingly, temperature differences between air and soil temperatures were significantly different among the treatments except for the minimum temperature (Figure 3c). In other words, grassland degradation will decrease temperature difference between air and soil temperatures in growing season. The variation of temperature patterns may alter the microenvironment and affect the structure and dynamics of plant communities. For example, growth rates of shoots and roots of plants will change [12].

Grassland degradation leads to vegetation variations in height and density [28]. We simulated grassland degradation through decreasing the amount of plant litter on the plots. Unfortunately, our treatments may not reflect the gradient of grass height and density variations. We measured soil temperature of the unmoved grassland during the growing season from 17th April to 29th August 2008;

the mean soil temperature at 20 cm depth was 18.9 ± 0.18 °C which was between 200 and 600 g/m^2 treatment (Figure 3b). The standing biomass of the plant community was 196.5 g/m^2, and plant height was about 30 cm. Thus the height of the standing plant litter also influences mean soil temperatures [12].

5.Conclusions

Plant litter reduction resulted in decreasing mean, maximum, and amplitude of air temperature at 10 cm above surface and increasing mean, minimum, maximum, and amplitude of soil temperatures at 20 cm below soil surface in the growing season. The amplitude of air temperature decrease was due to the maximum air temperature decrease and the minimum air temperature increase as plant litter reduction. The amplitude of soil temperature increase was due to the increasing rate of the maximum soil temperature higher than the minimum soil temperature as plant litter reduction. The temperature difference between air and soil temperature become less and less as plant litter reduction. Future work is needed to study the change of temperature patterns effect on grassland ecosystem.

Acknowledgements

This study was funded by the National Natural Science Foundation of China (31500366), the National Key Research and Development Program of China (2016YFC0500707), the Prior Sci-Tech Programs of Overseas Chinese Talents Funds (2015), and the Fundamental Research Funds for the Central Universities. Thanks to Kun Tian for help in carrying out the experiment. Thanks also to Guangdi Li for helping to check the language.

References

[1] Tuba Z, Kaligaric M 2008 *Community Ecol.* **9** 3-12
[2] Luo C, Xu G, Chao Z, Wang S, Lin X, Hu Y, Zhang Z, Duan J, Chang X, Su A 2010 *Glob. Change. Biol.* **16** 1606-1617
[3] Kang L, Han X G, Zhang Z B, Sun O J 2007 *Phil. Trans. Roy. Soc. B-Biol. Sci.* **362** 997-1008
[4] Facelli J M, Pickett S T A 1991 *Bot. Rev.* **57** 1-32
[5] Wang J, Zhao M, Willms W D, Han G, Wang Z, Bai Y 2011 *J. Veg. Sci.* **22** 367-76
[6] Xiong S, Nilsson C 1999 *J. Ecol.* **87** 984-94
[7] Liu P, Huang J, Sun O J, Han X 2010 *Oecologia* **162** 771-80
[8] Jimenez C, Tejedor M, Rodriguez M 2007 *Eur. J. Soil. Sci.* **58** 445-49
[9] Scull P 2007 *Phys. Geogr.* **28** 360-73
[10] Knapp A K, Seastedt T R 1986 *Bioscience* **36** 662-68
[11] Jacobs A F G, Heusinkveld B G, Holtslag A A M 2011 *Agr. Forest. Meteorol.* **151** 774-80
[12] Green F H W, Harding R J, Oliver H R 1984 *J. Climatol.* **4** 229-40
[13] Peng S, Piao S, Ciais P, Myneni R B, Chen A, Chevallier F, Dolman A J, Janssens I A, Penuelas J, Zhang G, Vicca S, Wan S, Wang S, Zeng H 2013 *Nature* **501** 88-92
[14] Richter D 2007 *Soil Sci.* **172** 957-67
[15] MacKinney A L 1929 *Ecology* **10** 312-21
[16] Bonan G B 2001 *J. Climatol.* **14** 2430-42
[17] Haskell D E, Flaspohler D J, Webster C R, Meyer M W 2012 *Restor. Ecol.* **20** 113-21
[18] Gauer E, Shaykewich C, Stobbe E 1982 *Can. J. Soil Sci.* **62** 311-25
[19] Grant R, Izaurralde R, Chanasyk D 1995 *Agr. Forest Meteorol.* **73** 89-113
[20] Flerchinger G, Sauer T, Aiken R 2003 *Geoderma* **116** 217-33
[21] Weaver J, Rowland N 1952 *Bot. Gaz.* **114** 1-19
[22] Hulbert L 1969 *Ecology* **50** 874-77
[23] Easterling D R, Horton B, Jones P D, Peterson T C, Karl T R, Parker D E, Salinger M J, Razuvayev V, Plummer N, Jamason P, Folland C K 1997 *Science* **277** 364-67
[24] Karl T R, Jones P D, Knight R W, Kukla G, Plummer N, Razuvayev V, Gallo K P, Lindseay J, Charlson R J, Peterson T C 1993 *Bull. Amer. Meteorol. Soc.* **74** 1007-23

[25] Liu D W, Wang Z M, Song K S, Zhang B, Hu L J, Huang N, Zhang S M, Luo L, Zhang C H, Jiang G J 2009 *Chinese Geogr. Sci.* **19** 299-305
[26] Foley J A, DeFries R, Asner G P, Barford C, Bonan G, Carpenter S R, Chapin F S, Coe M T, Daily G C, Gibbs H K, Helkowski J H, Holloway T, Howard E A, Kucharik C J, Monfreda C, Patz J A, Prentice I C, Ramankutty N, Snyder P K 2005 *Science* **309** 570-74
[27] Li X Y, Wang Z M, Song K S, Zhang B, Liu D W, Guo Z X 2007 *Environ. Monit. Assess.* **131** 421-37
[28] Song Y T, Zhou D W, Zhang H X, Li G D, Jin Y H, Li Q 2013 *Chinese .Sci. Bull.* **58** 907-12

IOP Conf. Series: Earth and Environmental Science **146** (2018) 012006 doi:10.1088/1755-1315/146/1/012006

Shading Effect and Heat Reflection Performance of Green Façade in Hot Humid Climate Area: Measurements of a Residential Project in Guangzhou, China

H Lin[1], Y Xiao[1] and F Musso[2]

[1] Room 816, State Key Laboratory of Subtropical Building Science (Liwu Building), South China University of Technology, Tianhe District, 51000 Guangzhou, China
[2] Lehrstuhl für Baukonstruktion und Baustoffkunde, Technische Universität München, Arcisstr. 21, 80333 München, Germany

yqxiao@scut.edu.cn

Abstract. In response to the problem of high-density urbanism and urban heat island (UHI), vertical greenery system (VGS) applying in building design and construction is regarded as one of the solutions. Green Façade (GF), with the climber plants growing on the building façade, is one branch technology of VGS. This research is focused on the shading effect and the heat reflection performance of the GF in Hot humid Climate Area. A high-rise residential project with the GFs on east and south façade is chosen to measure and analyze. Results reveal that the air Ave. temperature (Temp) shading by the GFs in sunny days is reduced by 0.6℃(E) and 0.7℃(S), and the Max. reduction is up to 3.2℃(E) and 3.8℃(S). The surface Ave. Temp of GF is reduced 4.7℃ comparing to the bare wall, and the Ave. Temp of the whole façade is decreased about 3-4℃ with GFs. This research has proved the effects of GFs and also show the GFs' potentials for energy reduction and urban heat environment improvement.

1. Introduction

High-density urbanism is one of the trends in the fast-developing countries and areas. The spread of the cities reduces a large amount of the green land, which is important for the air cleaning[1], rainwater collecting[2] and UHI reducing[3]. As a response to this trend, VGS applying in building design and construction is more and more regarded as one of the technologies of the UHI reducing[4][5]. Recent researchers have also pointed out that the VGSs have good effects on building thermal environment improvement[6], energy saving[7], urban acoustic reducing[8] and urban biodiversity[9].

VGS could be divided into two typologies as living wall system (LWS) and green facade (GF) [10] [Figure 1]. The LWS is a complex technology which includes the construction system on the existing wall, the waterproof layer, the module plant and soil containers or plant substrate, the irrigation and drainage system and the water control machine. Various small shrubs could be used in the LWS in order to shape different patterns. The GF is much simpler technology supporting the climber plants to grow on the building façade, which normally includes the support structure, the plant and soil container, the irrigation system and the drainage system[11]. Compare to the LWS, the GF is less paid attention to because of the weaker thermal insulation effect and lower influence to the environmental

Temp [12]. In fact, the GF has a much longer history in application, especially growing on the walls or fences in a natural situation[11]. As the performance of shading devices (SDs), more and more buildings with GF are built up and concentrated by researchers lately[13]. With the lower cost of construction and maintenance, the GF still has the potential in the application area[14].

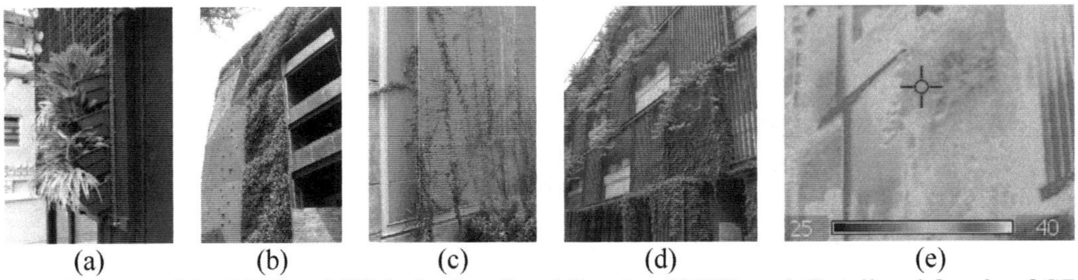

Figure 1. Cases of the LWS and GF (a, b. Detail and façade of LWS; c, d. Detail and façade of GF; e. Thermal image showing that the Temp of the SDs is higher than the GF)

This research is focused on the shading effect and the heat reflection performance of the GF in Hot humid climate area through measurements in situ. Measurements of the LWS and the GF have been tested by several researchers, including tests with experimental walls, contrasting rooms and building facades [15-17]. However, the results are various because of the influences of several factors such as differences of the façade construction, the building height and orientation, plant typologies, the growth situation of plants and so on. Thus, in order to evaluate the effects of GF in application for a typical climate area, the measurements in situ are still necessary. This research takes the measurements in Guangzhou, China. Guangzhou lies in southern China (23°08′N, 113°16′E), whose Ave. max Temp is 33.3 ℃ and Ave. relative humidity(RH) is 82% in summer [18] [Figure 2]. It also belongs to the humid subtropical climate according to the Köppen climate classification [19].

In Guangzhou, SDs are widely used in order to reduce the solar radiation on the building façade [Figure 1, d-e]. Various of SDs, which are installed on the edge of the openings, balconies, and windows on the buildings, offer efficient effects of the surface Temp reduction. However, with the physical characteristics of the SDs materials such as aluminium, the heat reflection of which still could affect the Temp of the surrounding environment. Contrast with this, the GF could change the hydrothermal performance of the surrounding area as the results of the evapotranspiration of plants, reducing the heat reflection to the building facade and the environment [20]. Thus, the GF is much friendlier to the building environment and could be even considered as one of methods to optimize the function and performance of the SDs. In this research, the effects of the GF could be tested and analysed through the measurements in situ.

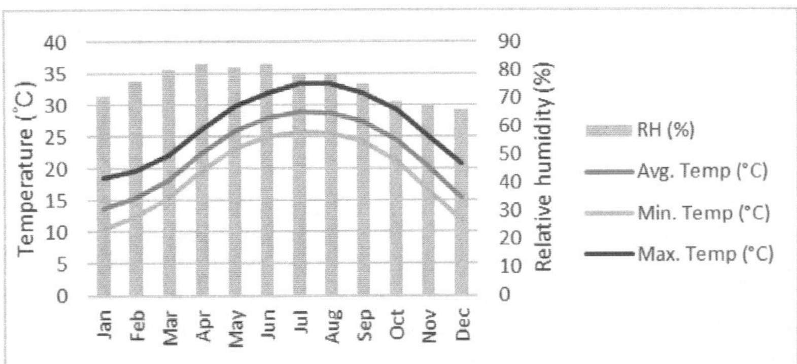

Figure 2. Climate data of Guangzhou, China (1986-2015) [18]

2.Research methodology

2.1. Object of the measurement

This research selected a high-rise residential project (Ave. height of 54m) in Guangzhou, China. With the architectural concept of vertical gardens, the project is designed with serials of public balconies, planting with shrubs and trees. Besides, climber plants are designed to cover the bare walls of the building façade. In the investigation before the measurement, the climber plants on the east and south façade facing to the external environment grew better than which were shaded by the buildings. Thus, three external GFs were chosen to measure, two of them face to the east (A&B) and one faces to the south (C) [Figure 3].

The construction of the GFs is the additional component of the building. Vertical and horizontal aluminium pipes (Φ=20mm, horizontal distance=200mm, distance to the bare wall=800mm) are set to support and guide the plants to grow up. Soil containers are designed to 800mm width and 800mm depth, setting at every 2 floors. Irrigation systems are set to auto control, providing water twice per day. Water store tanks are set on the roof of the buildings, connecting to the water supply system of the buildings. Climber plants are combined with Qiusqualis indicia and Lonicera japonica. The project is finished in 2013, thus the plants on the GFs have grown for 4 years till 2017.

(a) (b) (c) (d)

Figure 3. Project photos (a. Site-plan with the test GF A, B and C; b. Model of the project; c. Photo of the GF B; d. Construction detail of the GF)

2.2. Tools and methods of the measurement

In order to evaluate the shading effect of GFs and the influences on the environment, the measurement is divided into two parts. One is about the air Temp and RH behind the shading area of the GF, one is about the surface Temp change of the building façade.

For the measurement A, two HOBO loggers [Table 1] were set unmovable behind the GF with the distance of 400mm to the GF, one on the east façade and the other one on the south facade. The measure points were chosen at the places fully shading by leaves and whose foliage layers thicknesses were over 200mm. Besides, one logger was set on the roof of the building as a contrast point of the external environment. During the measurement, the HOBO loggers were set in solar radiation shields in order to avoid the interference of the direct solar radiation and rains. The measurement lasted for 6 days (2/08/2017-08/08/2017) with the test frequency of 5min.

For the measurement B, the thermal imagers [Table 1] were invited to record the thermal images of three building facades (GF A and GF B on the east and GF C on the south). The measure points were set in front of the building facades with the distance of 50m. The measurement was lasted for 10h in the daytime (08/08/2017) and the test frequency is 30min. Before the measurement, the environmental Temp and the material emissivity (μ value) were set as default values (Temp=20℃, μ=0.95). After the recording, the thermal images were input to the software FLIR QuickReport to reset the background parameters (the environmental Temp and RH), which have been recorded on the same day by the HOBO logger. Because the thermal images are composed of pixels of the Temp data, thus they could be counted and analysed in the statics software in further (Figure 4). Besides, the thermal images could be also analysed directly using the data points of the climber plants and the bare wall (Figure 11, a2, b2, c2).

Table 1. Range and Accuracy of the instruments

Instrument	Accuracy	Test frequency
HOBO data logger (U23 Pro v2)	Temp: ±0.21 °C (from 0 to 50 °C) RH: ±2.5% (from 10% to 90% RH)	5 min
FLIR B50 thermal imager	Temp: ±2% or 2 °C	30 min

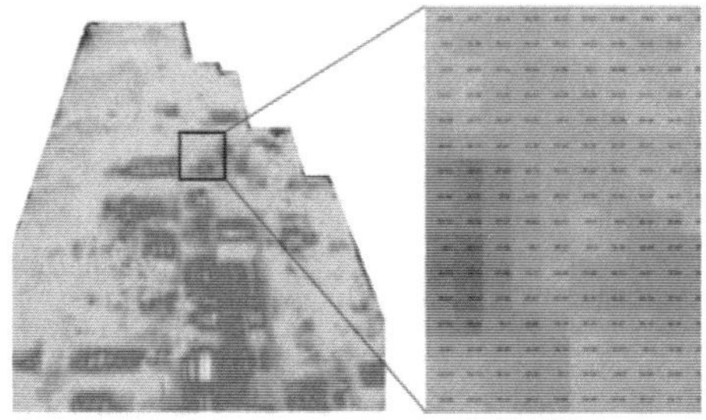

(a) (b)

Figure 4. Diagram of pixels analysis of the thermal images (a. Thermal image; b. Thermal image data input to the Excel file for value statics)

3.Results and discussion

3.1. Results of measurement A- Shading effect analysis

Firstly, the Temp and RH data recorded by the HOBO loggers were compared to which recorded by an official weather station setting at about 5km away from the test project.

Figure **5** and

Figure **6** show that two data sources had the same variation trend in 6 days of the measurement. Temp of 08/02/2017 to 08/04/2017 is lower because of the continuous rain. The average of the difference of Temp is 0.9℃, with the average error percentage of 3.3%. And the average of the difference of RH is 4.3%, with the average error percentage of 5.5%. The comparison verifies the credibility of the measurement.

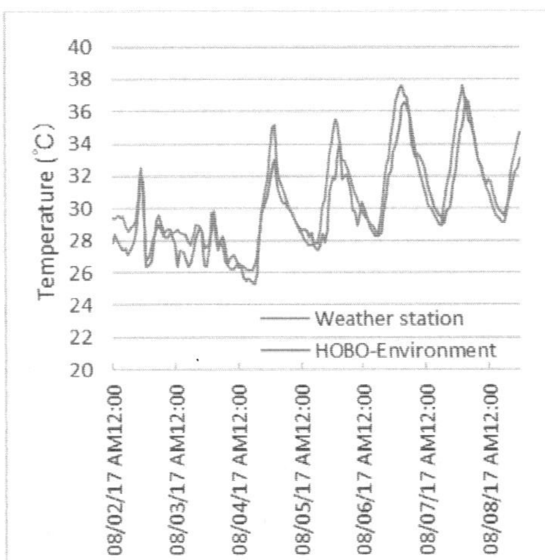

Figure 5. Environmental air Temp data comparison (08/02/2017 AM12:00-08/08/2017 PM04:00)

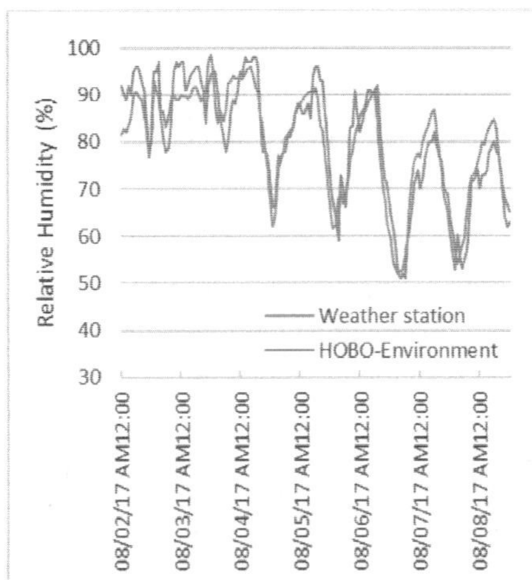

Figure 6. Environmental air RH data comparison (08/02/2017 AM12:00-08/08/2017 PM04:00)

Secondly, the Temp and RH data of three HOBO loggers are compared.

Figure **7** and

Figure **9** show the data variation of the Temp and the RH of the measure days.

Figure **8** and

Figure **10** show the Temp and the RH differences between the outdoor environment and the shading area behind the GFs. The results could be divided into two groups to discuss because of the obvious differences between the rainy days (8/2/2017 to 8/3/2017) and the sunny days (8/4/2017 to 8/8/2017). The extreme values and wave ranges are lower in the rainy days than the sunny days.

From 8/2/2017 to 8/3/2017, the Ave. Temp is 28.3℃(E-east) and 28.0℃(S-south), comparing to the outdoor environment of 28.4℃. And the Ave. RH is 77.3%(E), 88.8%(S) and 88.6%(Environment). Thus, the Ave. difference values between the outdoor environment and the shading area are not so obvious. However,

Figure **8** shows that even in the rainy days, the Temp of the shading area of the east and the south GF are still lower than the outdoor environment about 0-2℃ in most time of the day. From 8/4/2017 to 8/8/2017, the Ave. Temp is 31.0℃(E), 30.9℃(S) and 31.6℃(Environment). And the Ave. RH is 73.5%(E), 76.7%(S) and 75.4%(Environment). Compare to the rainy days, the differences of Ave. Temp in sunny days are clearer with 0.6℃(E) and 0.7℃(S). Table 2 shows that the difference values between the shading area and the environment are about 0-2.5℃ in most time of the day, and the maximum is up to 3.2℃(E) and 3.8℃(S) at noon. However, the differences of the RH are still similar in both the rainy and sunny days, mostly in the range of 0-5%.

Furthermore, analysis of the data of daytime (06:00-18:00) and night time (18:00-06:00) is given in Table 3, revealing that the differences of the Temp are higher in the daytime than in the night time. The Ave. Temp differences are up to 1℃(E) and 1.2℃(S) in the daytime, but only 0.0℃(E) and 0.1℃ (S) in the night time. However, the Ave. RH differences are less obvious, which are only 0.8%(E) and -2.8%(S) in the daytime, while 3.3%(E) and 0.9%(S) in the night time.

Results reveal that as the GF is not a closed system, the air Temp and RH of the shading area and the outdoor environment have the same variant trend. The GFs presented a better shading effect in the daytime in the sunny days. There is nearly no difference of the Temp in the night time. Besides, the difference of RH between the shading area and the outdoor environment is not so obvious as the Temp, whenever in the rainy days or the sunny days. With the contrast of the maximum of the Temp, the shading effect of the GF is presented clearly.

Table 2. Ave. and Max. value of Temp and RH

Date typology	East	South	Environment
Ave. Temp (℃)	30.2	30.0	30.7
Ave. Temp of 8/2/2017-8/3/2017 (℃)	28.3	28.0	28.4
Ave. Temp of 8/4/2017-8/8/2017 (℃)	31.0	30.9	31.6
Max. Temp difference compare to environment (℃)	3.2	3.8	-
Ave. RH (%)	77.3	80.3	79.4
Ave. RH of 8/2/2017-8/3/2017 (%)	86.2	88.8	88.6
Ave. RH of 8/4/2017-8/8/2017 (%)	73.5	76.7	75.4
Max. RH difference compare to environment (%)	6.7	9.9	-

Table 3. Ave. value of Temp and RH (divide into daytime and nighttime)

Date typology	East	South	Environment

Ave. Temp of 06:00-18:00 (℃)	30.9	30.7	31.9
Ave. Temp of 18:00-06:00 (℃)	29.5	29.4	29.5
Ave. RH of 06:00-18:00 (%)	73.8	77.4	74.6
Ave. RH of 18:00-06:00 (%)	80.3	82.7	83.6

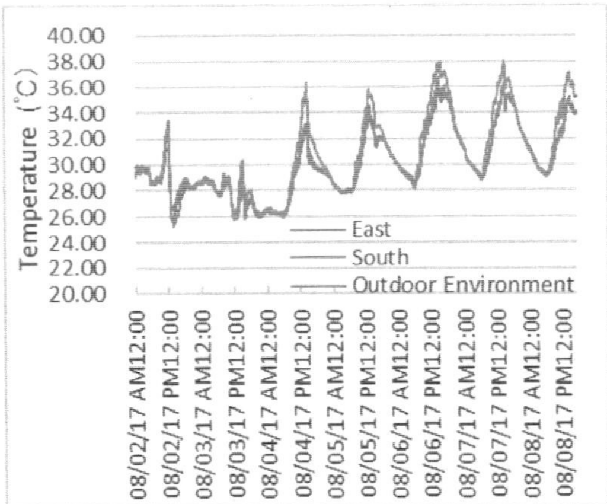

Figure 7. Air Temp data of three HOBO loggers (08/02/2017 AM12:00-08/08/2017 PM04:00)

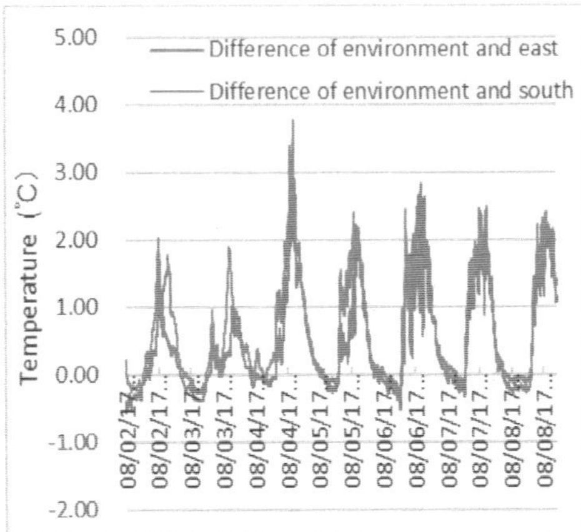

Figure 8. Differences of the air Temp data of three HOBO loggers (08/02/2017 AM12:00-08/08/2017 PM04:00)

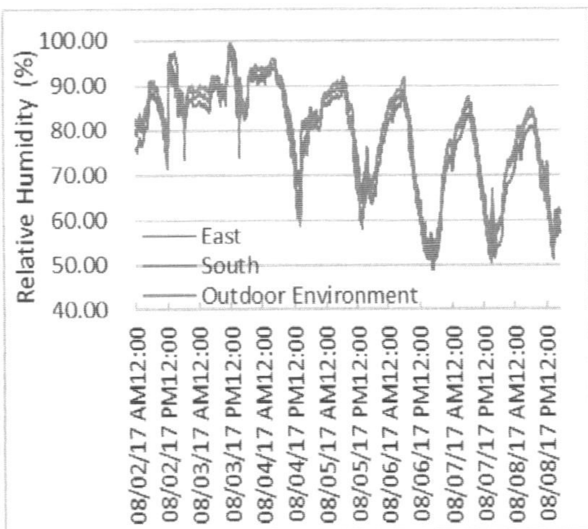

Figure 9. Air RH data of three HOBO loggers (08/02/2017 AM12:00-08/08/2017 PM04:00)

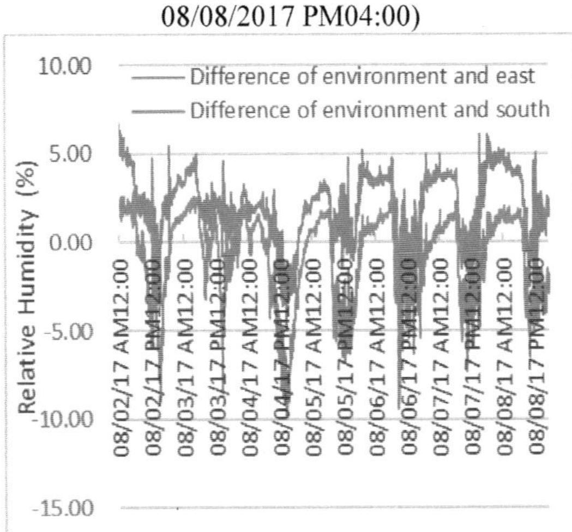

Figure 10. Difference of air RH data of three HOBO loggers (08/02/2017 AM12:00-08/08/2017 PM04:00)

3.2Results of measurement B- Heat reflection performance analysis

Measurement B is focused on the surface Temp of the GF because the heat reflection of the building surface could influent on the surrounding environment. Two analyses are invited to evaluate the surface Temp, one is the statist of Temp pixels on the thermal image, one is the continuous measurement on the same points of the GF and the bare wall (Figure 11, a1-c1). Lastly, two results could be compared with each other.

On the first analysis, the thermal images should be dealt with to remove the data of the background sky. Firstly, the Temp values (<30℃), which released the Temp of the background but not the building façade, were excluded in the thermal images. Secondly, the values (≥30℃) were calculated with the interval of 1℃.

Figure **12** and

Figure **13** show the percentage variation of different Temp of the whole façade surface, revealing that the largest percentage of the Temp distribution is around 33℃-36℃.

On the other side, the foliage coverage ratio (FCR) of the GFs is calculated through the photo pixel colour filtrating in the software of Adobe Photoshop (Figure 11, d). As a result, the FCR of the GF is 37.5% (GF A), 54.4% (GF B) and 95.4%(GF C, without the bare wall behind). Combine with the distribution ratio of the Temp and FCR, the surface Temp of the GF could be read again in

Figure **12** and

Figure **13** with the value of 34℃-36℃ (GF A) and 33℃-36℃ (GF B).

On the second analysis, the Temp data of three points of the GF and the bare wall of each façade are recorded in the thermal image (Figure 11, a2, b2, c2).

Figure **14** and

Figure **15** show the point Temp variation and reveal that the surface Temp of the GFs is lower than which of the bare walls of 3-7℃. Table 4 presents the results of the Ave. Temp of the GF, the bare wall, and the whole façade. Results show that the surface Temp of the GF is lower than the bare wall, the Ave. difference is 4.6℃ (GF A) and 4.8℃ (GF B). Because the GF C is all covered with climber plants, thus there is no comparison to the bare wall. The Ave. Temp of the façade on GF C is the lowest in three test facades. In general, thanks to the GFs, the surface Temp of the whole façade is decreased about 3-4℃.

Both of two analyses reflect the Temp distribution on the building façade. Via the Ave. Temp comparison, they could be proved to each other at the end. The decreases of the façade Temp were recorded obviously. On the discussion of the micro-climate of the urban environment, the decrease of the surface Temp of the building could also reduce the heat reflection to the surrounding environment and provide a much friendlier public space.

(a1) GF A, east (b1) GF B, east (c1) GF C, south

(a2) thermal image of GF A (b2) thermal image of GF B (c2) thermal image of GF C

(d) FCR analysis, take GF B as an example (FCR=54.4%)

Figure 11. Photos, thermal images and FCR analysis of the GFs

Table 4. FCR and Ave. surface Temp of the GFs

GFs	GF FCR (%)	GF Ave. Temp (℃)	Bare wall Ave. Temp (℃)	Whole façade Ave. Temp (℃)
GF A	37.5	34.1	38.7	35.8
GF B	54.4	33.7	38.5	34.7
GF C	95.4	33.5	-	33.5

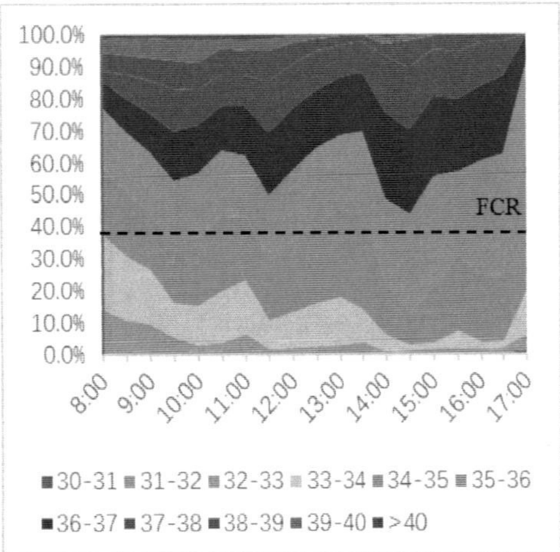

Figure 12. Percentage of pixels of the facade Temp (East façade on GF A, 8/8/2017 8:00-17:00)

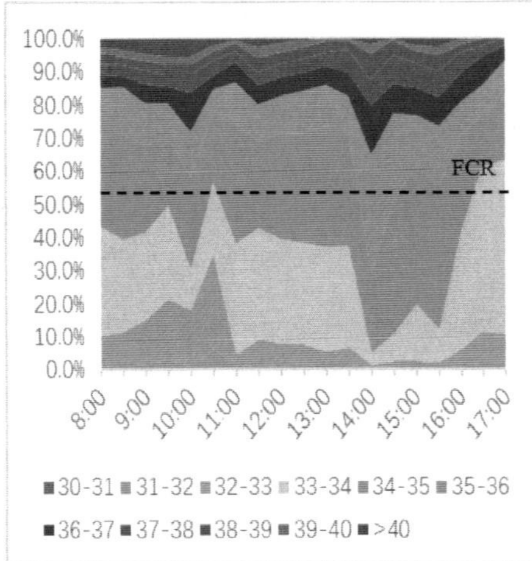

Figure 13. Percentage of pixels of the facade Temp (East façade on GF B, 8/8/2017 8:00-17:00)

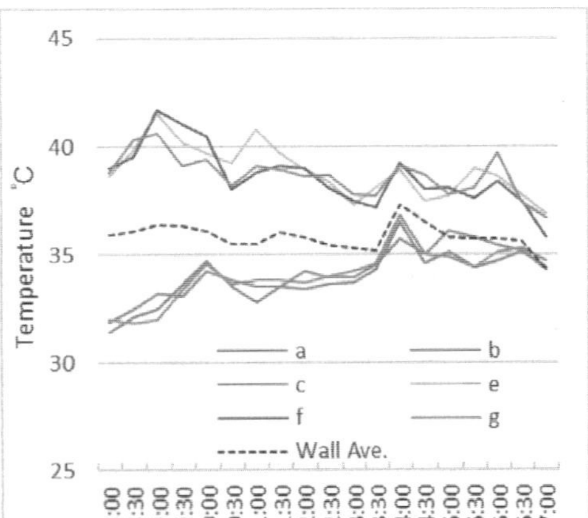

Figure 14. Surface Temp of GF A (a-c: GF; d-e: the bare wall; Wall Ave.: the whole wall Ave. Temp; 8/8/2017 8:00-17:00)

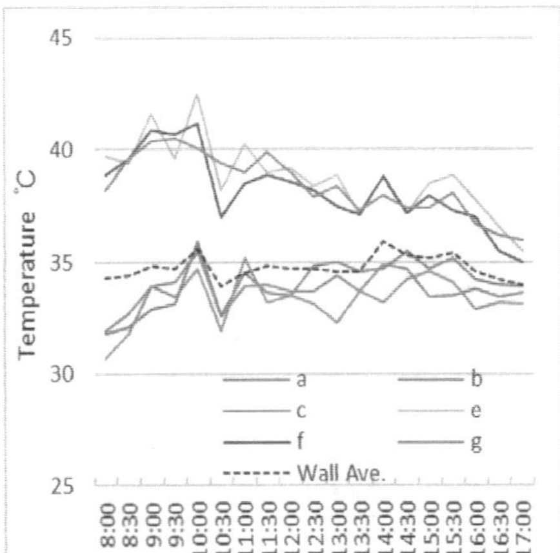

Figure 15. Surface Temp of GF B (a-c: GF; d-e: the bare wall; Wall Ave.: the whole wall Ave. Temp; 8/8/2017 8:00-17:00)

4.Conclusion

Basing on the measurement in situ of the GFs, the shading effect and the decrease of the heat reflection are proved. The air Ave. Temp shading by the GFs in sunny days is reduced by 0.6℃(E) and 0.7℃(S), and the Max. reduction is up to 3.2℃(E) and 3.8℃(S) at noon. On the other side, the surface Ave. Temp of the GFs is reduced 4.7℃ comparing to the bare wall, and the whole façade is decreased about 3-4℃ with the GFs, whose FCRs are 37.5-54.4%. Via the complete record of the project information, construction details, plant typologies, FCR, and the hydrothermal data, this research could be a valuable report for the research area of VGS and for GFs application in the hot humid climate area. Further studies will be focused on the GFs' influence on building energy consumption and the urban canyon. As the GF is a dynamic system that the plants are always growing, a continuous tracking of the variation is still necessary in future.

Acknowledgements

This research is supported by National Natural Science Foundation of China (No. B5141490) and Science and Technology Planning Project of Guangzhou, China (No. B3160760).

Reference

[1] Feng H and K Hewage 2014 Lifecycle assessment of living walls: air purification and energy performance *Journal of Cleaner Production*. **69** 91-99
[2] Prodanovic V, et al. 2017 Green walls for greywater reuse: Understanding the role of media on pollutant removal *Ecological Engineering*. **102** 625-35
[3] Mitterboeck M and A Korjenic 2017 Analysis for improving the passive cooling of building's surroundings through the creation of green spaces in the urban built-up area *Energy and Buildings*. **148** 166-81
[4] Tsilini V, et al. 2015 Urban gardens as a solution to energy poverty and urban heat island *Sustainable Cities and Society*. **14** 323-33
[5] Aflaki A, et al. 2017 Urban heat island mitigation strategies: A state-of-the-art review on Kuala Lumpur, Singapore and Hong Kong *Cities*. **62** 131-45
[6] Afshari A 2017 A new model of urban cooling demand and heat island—application to vertical greenery systems (VGS) *Energy and Buildings*. **157** 207-17
[7] Poddar S, D Y Park and S Chang 2017 Simulation Based Analysis on the Energy Conservation Effect of Green Wall Installation for Different Building Types in a Campus *Energy Procedia*. **111** 226-34
[8] Pérez G, et al. 2016 Acoustic insulation capacity of Vertical Greenery Systems for buildings *Applied Acoustics*. **110** 218-26
[9] Collins R, M Schaafsma and M D Hudson 2017 The value of green walls to urban biodiversity *Land Use Policy*. **64** 114-23
[10] Perini K, et al. 2011 Greening the building envelope, facade greening and living wall systems *Open Journal of Ecology*. **01** 1-8
[11] Manso M and J Castro-Gomes 2015 Green wall systems: A review of their characteristics *Renewable and Sustainable Energy Reviews*. **41** 863-71
[12] Wong N H, et al. 2010 Thermal evaluation of vertical greenery systems for building walls *Building and Environment*. **45** 663-72
[13] Coma J, et al. 2017 Vertical greenery systems for energy savings in buildings: A comparative study between green walls and green facades *Building and Environment*. **111** 228-37
[14] Tedesco S, R Giordano and E Montacchini 2016 How to Measure the Green Façade Sustainability? A Proposal of a Technical Standard *Energy Procedia*. **96** 560-67
[15] de Jesus M P, et al. 2017 Green façades and in situ measurements of outdoor building thermal behaviour *Building and Environment*. **119** 11-19
[16] He Y, et al. 2017 A detailed investigation of thermal behavior of green envelope under urban canopy scale in summer: A case study in Shanghai area *Energy and Buildings*. **148** 142-54
[17] Pérez G, et al. 2017 Green facade for energy savings in buildings: The influence of leaf area

index and facade orientation on the shadow effect *Applied Energy*. **187** 424-37
[18] http://www.cma.gov.cn/en2014/
[19] Köppen W 1884 (Translated by Volken E and Brönnimann S 2011) Die Wärmezonen der Erde, nach der Dauer der heissen, gemässigten und kalten Zeit und nach der Wirkung der Wärme auf die organische Welt betrachtet *Meteorologische Zeitschrift*. **20** 351–60
[20] Sheweka S M and N M Mohamed 2012 Green Facades as a New Sustainable Approach Towards Climate Change *Energy Procedia*. **18** 507-20

IOP Conf. Series: Earth and Environmental Science **146** (2018) 012007 doi:10.1088/1755-1315/146/1/012007

A new real-time flood forecasting model based on robust kalman filter method

Chao Zhao, Guangsheng Liu and Mulan Zhu

School of Environmental Science and Engineering, Xiamen University of Technology, Xiamen 361024, PR China

zhaocfang@163.com

Abstract: The telemetry inevitably brings some outliers in observation data. The real-time flood forecasting based on conventional Kalman filter is sensitive to the outliers, that cannot guarantee the reliable forecasting results. In this paper a new real-time flood forecasting model based on a robust Kalman filter is described. The proposed robust filter assigns the less weight to the outliers by inserting the weight factors, and the unity weight to the normal data. That makes the robust Kalman filter method is insensitive the outliers, meanwhile keeps the properties of the conventional Kalman filer. The feasibility of the robust approach is demonstrated with unbiased and biased data.

1. Introduction

It is a trend that the real-time water regime data are automatically got by use of the telemetry. It can transmit massive and timely data, meanwhile it also inevitably brings some outliers because of sensor noise, instrument degradation and/or process disturbances, in addition to random errors normally distributed with zero mean and a small variance [1-2]. The Outliers have an unknown distribution with a much bigger variance, and appear to be inconsistent with the remainder of the data set and relative large in magnitude [3-4]. The real-time flood forecasting models usually are interested in speedy and convenient telemetric data. However in order to maintain the real-time flood forecasting accuracy and reliability, the models must be robust to these outliers and tolerant to random errors during updating process.

Kalman filters(KF) are the most popular technique in hydrological models[5-8]. It can take models and data uncertainties into account in the estimation process and provides an estimation of the model prediction uncertainty. If errors involved in model and observation are Gaussian, KF is the best (i.e. minimum variance) linear unbiased estimator[9]. However, for hydrological systems, especially real-time flood forecasting system, the distribution of observation errors is often biased and inconsistent with the Gaussian assumption, that can make KF get biased estimations [10].

This paper will combine robust theory with KF, and propose a new robust KF(RKF) for real-time flood forecasting system. RKF assigns less weight to a small portion of outliers, so that they will not greatly affect the predicting results. Therefore, the robustness of the RKF estimation could be guaranteed effectively.

We first give the formulation of RKF with focus on its robust objective function, in order to show the essential distinction between RKF and conventional KF. Then, we compare the flood predicting results based on RKF and KF by use of synthetic data.

2. Methodology

2.1 Kalman filter (KF)

For the real-time flood forecasting model, the predicting errors of by means of hydrological models can be written by an autoregressive function:

$$e_t = \phi_t e_{t-1} + \varepsilon_t \tag{1}$$

Where e_t is the predicting errors got by flood forecasting models at the t th time point. $e=(e_1\ e_2\ \dots\ e_t)$. ϕ_t is the state vector. ε_t is error

The state transition function can be showed by the following function:

$$\phi_t = I\phi_{t-1} + \omega_t \tag{2}$$

The real-time flood forecasting model is to find a state estimator $\hat{\phi}_t$ based on e_t, then get forecasting results of $(t+1)$ th time point. ω_t is error.

The performance criterion of KF is to minimize the cost function which is the expected value of a weighted two-norm of estimation error:

$$\left(\hat{\phi}_t e_{t-1} - e_t\right)\left(Z_{t-1}\right)^{-1}\left(\hat{\phi}_t e_{t-1} - e_t\right) + \left(\hat{\phi}_t - \hat{\phi}_{t-1}\right)\left(P_{t-1}\right)^{-1}\left(\hat{\phi}_t - \hat{\phi}_{t-1}\right) = \min \tag{3}$$

Where Z_{t-1} is the covariance of the time series e at $(t-1)$ th time point, and is set as the weight to ε_t. P_{t-1} is the covariance of the state vector at $(t-1)$ th time point, and is set as the weight for ω_t. When ε_t and ω_t are Gaussian, KF possesses the smallest possible standard deviation.

2.2 Robust Kalman filter (RKF)

When there are the outliers in ε_t got by telemetry, the Gaussian assumption is violated. KF is affected by the outliers, the results are biased. The robust theory has been brought to modify the conventional KF in order to reduce the influence of outliers [11-12]. By use of weight factor $\overline{p}_t$, change the objective function of KF to the following the new objective function:

$$\left(\hat{\phi}_t e_{t-1} - e_t\right)\overline{p}_t\left(\hat{\phi}_t e_{t-1} - e_t\right) + (\hat{\phi}_t - \hat{\phi}_{t-1})\left(P_{t-1}\right)^{-1}(\hat{\phi}_t - \hat{\phi}_{t-1}) = \min \tag{4}$$

Where $\overline{p}_t$ is the weight factor, which suppresses the outlier.

The desired value $\hat{\phi}_t$ can be obtained by solving the equation

$$e_{t-1}\overline{p}_t\left(\hat{\phi}_t e_{t-1} - e_t\right) + \left(P_{t-1}\right)^{-1}(\hat{\phi}_t - \hat{\phi}_{t-1}) = 0 \tag{5}$$

then

$$\hat{\phi}_t = \hat{\phi}_{t-1} + \frac{e_{t-1}\overline{p}_t P_{t-1}(e_t - \hat{\phi}_{t-1}e_{t-1})}{1 + e_{t-1}\overline{p}_t P_{t-1} e_{t-1}} \tag{6}$$

$$\hat{\phi}_t = \hat{\phi}_{t-1} + K_t(e_t - \hat{\phi}_{t-1}e_{t-1}) \tag{7}$$

Where $K_t = \dfrac{e_{t-1}\overline{p}_t P_{t-1}}{1 + e_{t-1}\overline{p}_t P_{t-1} e_{t-1}}$

The update step is given by

$$P_t = (I - K_t e_{t-1})P_{t-1} \tag{8}$$

The robust ability of RKF is focus on the $\overline{p}_t$, therefore, it is import to choose desired function for $\overline{p}_t$. In the paper, we refer the previous results [1], and select IGG I estimator

$$\bar{p}_t = \begin{cases} 1.0 & M \le 1.5 \\ 1.5/M & 1.5 < M \le 2.5 \\ 0 & 2.5 < M \end{cases} \tag{9}$$

Where

$$M = \frac{(e_t - \hat{\phi}_t e_{t-1})}{\sigma_v}$$

$$\sigma_v = \frac{\sqrt{\sum_{i=1}^{t}\left(e_t - \hat{\phi}_t e_{t-1}\right)\bar{p}_i\left(e_t - \hat{\phi}_t e_{t-1}\right)}}{t-m-1}$$

Where m the number of outliers

Using function (9) is to assign less weight to a small portion of outliers so that they can not greatly affect on the final results, while giving unity weight to the bulk of small moderate residuals. When the errors are Gaussian($M < 1.5$), the RKF contains the properties of KF. When outliers inevitably exist, the less weight is assigned, the estimation is robust.

3. Experiments

The proposed RKF and KF were tested based on discharge data with and without outliers. The observations without outliers from the Qilijie basin of Fujian province of China, from 2007 to 2008, were selected.

3.1. The results of two filters with unbiased errors

When the errors are Gaussian, the comparison of RKF and KF was shown in Table 1.

Table 1. Comparison of Dc using two filers with outlier free data

Flood	$Dc(KF)$	$Dc(RKF)$	$Er(KF)$	$Er(RKF)$
080808	0.966	0.962	-0.002	-0.001
080609	0.958	0.942	0.006	0.041
080401	0.976	0.978	0.004	0.003
080315	0.965	0.959	0.001	0.002
070901	0.976	0.976	-0.002	-0.003
070718	0.969	0.978	0.006	0
070505	0.889	0.905	-0.002	0.024
mean	0.957	0.957		

Dc is the deterministic coefficient, $Dc = 1 - \frac{\sum (Q - Q_c)^2}{\sum (Q - \bar{Q})^2}$, where Q is observing discharge;

Q_c is forecasting discharge; and $\bar{Q}$ is the average of Q. $Er = \frac{\sum(Q-Q_c)}{\sum Q}$

The index Dc displays the fit degree between the forecasting discharge hydrograph and observing discharge hydrograph. It was obviously that the forecasting hydrograph of the two filters were closed without outliers. Furthermore, the Dcs both were greater than 0.8, which demonstrated the performance of two filters were satisfied without outliers.

3.2. The performances of two filters with outliers

The following abnormal errors were added to the above observing discharge in order to format the non-Gaussian distribution:

$$k_i = \begin{cases} (r-0.5)\bar{Q}TT & i = \text{int}(i/L)L \\ 0 & i \neq \text{int}(i/L)L \end{cases} \tag{10}$$

where, r is random numbers; $\bar{Q}$ is the average of observed discharge; TT is a constant that controls the maximum of k, and L is the frequency of outliers. Adjusting TT and L, outliers of different numbers and sizes can be generated. In the paper, TT and L were generated randomly.

The comparison of the two filters were displayed in table 2. The Dc(KF) was obviously smaller than that of table 1. That showed the KF was sensitive to outliers and was affected by outliers. However, RKF assigned smaller weights even zero weight to the outliers, that makes it inensitive to outliers and gave greater deterministic coefficient. So when outliers inevitably existed, the robust KF was a better method to get robust forecasting results.

Table2. Comparison of Dc using two filers with abnormal data

Flood	$Dc(KF)$	$Dc(RKF)$	$Er(KF)$	$Er(RKF)$
080808	0.876	0.967	-0.031	-0.002
080609	0.940	0.964	0.007	0.021
080401	0.908	0.976	0.017	0.006
080315	0.822	0.961	0.015	0.002
070901	0.886	0.976	-0.060	-0.003
070718	0.890	0.927	0.017	0.006
070505	0.721	0.839	0.003	0.009
mean	0.863	0.944		

4. conclusion

In modern real-time flood forecasting, outliers inevitably exist in the speedy water regime data. The proposed robust KF dose not only maintain the property with unbiased data, but also are robust to outlier data. RKF is especially suitable for the real-time flood forecasting

Acknowledgements

The authors wish to acknowledge the financial assistance provided by National Nature Science foundation of China (50909084) and Nature Science Foundation of Fujian Province (2017J01721,

2015J05114) and Xiamen Science and Technology Planning Project (3502z20150051) and the Education Department of Fujian Province (JAT170410).

References

[1] Zhao C, Hong H S, Bao W M, Zhang L P 2008 Robust recursive estimation of auto-regressiveupdating model parameters for real-time flood forecasting. Journal of hydrology 349 376
[2] Liu H C, Shah S, Jiang W 2004 On-line outlier detection and data cleaning. Computer & chemical engineering 28 1635-
[3] Barnett V, Lewis T 1995 Outlier in statistical data（New York: John Wiley and Sons）
[4] Rousseeuw M, Huber M 2011 Robust statistics for outlier detection. Wiley Interdis-ciplinaryReviews: Data mining and knowledge discovery 1 73
[5] Kalman R 1960 A new approach to linear filtering and prediction problems. Journal of basic engineering. 82 35
[6] Tahir H 1985 Kalman filter estimation model in flood forecasting. Advances in water resouseces.15
[7] Eppstein M J, Dougherty D 1996 Simultaneous estimation of transmissivity values and zonation Water resources research 32 3321
[8] Moradkhani H, Sorooshian S, Gupta H V, Houser P R 2005 Dual state-parameter estimation of hydrological models using ensemble Kalman filter. Advances in water resources. 28 135
[9] Sorensensen J V T, Madsen H, Madsen H 2004 Data assimilation in hydrodynamic modelling: on the treatment of nonlinearity and bias. Stochastic Environmental Research and Risk Assessment. 8 228
[10] Kavetski D, Franks S W, Kuczera G. 2003. Confronting input uncertainty in environmental modeling, in calibration of watershed models. Water science and applications series, 6: 49–68.
[11] Huber P J. 1981. Robust Statistics. (New York : John Wiley)
[12] Rey W J. 1977. Robust Statistical Methods. (New York: Springer)

The Practice of Water Rights Transaction in China

Hongzhi Wu[1],Dong Yang[1], Rui Qiao[2], Feng Shi[1]

[1].Institute for Development Strategy of science and Technology,Qilu University of Technology(Shandong Academy of Sciences), Jinan, China
[2].Department of Statistics and Actuarial Science, University of Waterloo, 200 University Avenue West, Waterloo, Canada N2L3G1

Sihan_shelly@163.com

Abstract. The allocation and transaction of water rights have been widely concerned in China. It is essential to carry out the theory innovation and practice of water rights transaction. Based on the corresponding practices of water rights trading system at home and abroad, the paper put forward some discussions on the ownership of water, the water allocation system, the water transaction system and Legal norms and institutional system.

1. Introduction

Water is one of the important resources in industry and agriculture production process. But along with continuing population growth and rapid economy development, China faces prominent challenges: water shortage,water pollution,over-exploitation of underground water and low efficiency of water usage and the imbalance between water supply and demand in China has been continuously sharpened and the conflict of water rights has been a serious social problem undoubtedly.In recent years, with scarcity of water resources increasingly serious, the allocation and the transaction of water rights and water management have been widely concerned in China. However, the theory and practice of water right is few, especial in legislation on water rights system. The article mainly summarized the advanced water rights management systems and current water rights management systems actuality at home and abroad and gave some discussions and suggestions on water rights transaction system.

2. The Practice of Foreign Water Rights

Currently, world water right trading market mainly concentrated in places such as America, Australia and Asia, among which the United States, Australia, Chile, Mexico and south Asian countries is the most representative. They execute the separation of water resource management, that is to say, power shift down and attach great importance to environmental protection and the protection of the third party for the deal.

In USA, the water right system is the earliest established and its theory is relatively complete. Water right system in the United States is generally defined by the state laws and regulations. Based on perfect legal system, water rights systems differ in the states. In the regions of the east, the southeast and west, most of them adopt the riparian rights. Riparian rights system refers to the rational use of land connected with the shore of water bodies but do not affect other riparian landowners who use the right to the reasonable use of water. However, the preemption system is established and developed in the arid and semi-arid western states of the United States to solve the water problems in these water-scarce areas. Within the specified boundary, water resources belong to the public or the state government. Under the ownership of water resources, the allocation of water rights refers to the allocation of water resources. Although water rights trading started earlier in America, there are only several western states which established the water rights trading system. The water right system in the western states is perfect, but water market has not been set up yet. The reason is that in the water rights trading, administrative transaction costs for the buyer and the transaction cost for seller are too high to bear so as to affect the benefits and the activities of water right[1-4].

Similar to the United States, Australia's water management focus on the state as the unified management of water resource. Water rights trading can be divided into four forms in the inner state, interstate temporary water right trading and the inner state, interstate permanent water right transaction. The Murray darling basin of the south Australia (Murray-DathngBasin) is the representative water right trading [2-4].

Developing countries such as Mexico and Chile step in advance in this respect. After nearly a century of development, water resources in Mexico gradually formed a relatively complete legal system. In 1975, the first national water resources plan was implemented in Mexico. In 1992, the national water law put forward the outline of a set of integrated water resources management which included the goal of sustainable management of water resources and strengthened that water commission is the only the water management authority. It also encouraged that more private enterprise participated in water resources management and development. In addition, it introduced the market mechanism, promoted the effective management of water resource, maintained the national water commission as the management of the water right transfer and awarded the national water resources commission to licensing adjustment, water restrictions, updating licensing, and the parameters determination for the water quality protection and so on[3-4].

2.1. The Ownership and Use of Water Rights

Countries adopt different water rights regulations on the basis of their respective characteristics of water resources. Even within the same country, the ownership of water is different. For example, the owner of water belongs to property right in USA. In the eastern United States, such as Arkansas, Delaware, Florida, Georgia, and so on, the ownership and use of water adopt the Riparian rights (Riparian Doctrine) due to relatively abundant water, whereas they adopt the prior appropriation doctrine because of drought and water shortage in the western United States [5-8]. In Japan, the 1995 law regulated the river belongs to the national industry and implemented the appropriate protection, utilization and management of the rivers and strive to achieve the objectives. Rivers may not belong to private ownership. So to speak, the licensed owners of water rights can't again transfer water rights to others and the only government can award the water rights to new users[4]. Compared with the early water law, the 1981 national water law regulated that the water rights completely was separated from the ownership of land. This is the first time in the history of Chile. And the law regulation can free to trade and guaranty the water right just like any other property [9].

2.2. The Categories of Water Rights

The categories of water rights are different in many countries. For example in Australia, water rights are commonly divided into three kinds; the first kind is bulk entitlements as is granted with the administrative institutions of irrigation and water supply, electric power company; the second kind is license, namely to award individuals to the water right of rivers, underground water and the management engineering institution within a certain time limit; The third kind is water rights, which is related to the land, namely to use for life, irrigation and livestock. In USA, water rights are divided into riparian right, exclusive priority right, mixed water right and public water right. Riparian right indicates that it applies to rich water resources in the east and awards the owners of the land associated with adjacent to the river to a limited share of water right. The exclusive priority used for water shortage in the western region include three principles: one is that who occupies first , who has the priority; two is that the use of water can not damage the interests of others; Three is that it is valid for the utilized water right; it is invalid for the no having utilized water right. Mixed water right is namely used with riparian rights and priority comprehensively. In general, riparian rights prior to priority. Public water right, used for shipping, fishing, commercial purposes, such as swimming, water entertainment, leisure, scientific research, and to meet the requirements of ecological and environmental protection of water resources, has become an important factor of water resource utilization in this respect of the interests of the public. In Chile they are divided into consumptive and not consumptive water right, permanent and temporary water right [6-9].

2.3. The Distribution of Water Rights

Since 1980s, the state government began to implement the water rights auction in Austrilia. Water rights may transfer temporarily or permanently; or transfer within the state or across the state; or transfer all or part. At present, they no longer approval to issue new water rights. Water rights are obtained only through the water rights trading.

Also Chile adopts permanent and temporary water rights registration system in the national water right management department [9]. In America, the distribution of water rights refers to the application of priority exclusive right which is a priority to man-made waterway such as reservoirs, canals and pumping stations. Namely it is 'time preference, right priority'. In general, there are the following procedures to access to the exclusive right: (1) a written application; (2) administrative institutions or judicial authority acceptance; (3) the acceptance of application, permits issued by administrative organizations. If the objections have been received, a hearing will been convened. Within the stipulated time, the priority right proved as uselessness will be confiscated[4,10].

3. The Practice of Water Rights Transaction in China

Due to the vast territory and big differences in natural conditions, historical development, economic, social and cultural factors, China is not feasible to establish a unified national water rights market. Limited to a certain basin or region, water right market is of watershed or regional characteristics. That is to say, different river basins or regions have own water right trading characteristics respectively.

For example, Ning(Ningxia Hui Nationality Autonomous Region)Meng(Inner Mongolia Autonomous Region) water rights succeeded to be transferred through high strength water saving investment and without increasing water indexes in the Yellow River watershed[8]. Zhangye city(Gansu province) is a pilot site of water saving society construction in China. After the reform of water rights system, water tickets have become the carrier of water exchange and water use[10]. Zhangye provincial river basin water rights trading practice achieved remarkable social and economic benefit [10]. Through the optimal allocation of water resource, it effectively alleviated the contradiction between the upstream and downstream water. Also it optimized the allocation of water resource and fostered the Zhangye water rights market. It was to break through the traditional concept and explore a new way of solving water disputes using economic means and market mechanism.

The practice of water right market in Jiangxi Province shall be helpful to carry out theoretical studies in water rich areas in southern part of China. There are two categories in water right market. one is a delegate with reservoir or irrigation region. It is the market transaction behaviour as agricultural water is used for industry or urban public water supply; another is inter-basin water rights trading as a representative of Pingxiang mountain rock reservoir. It is the only real deal in Jiangxi province[11].

Changji state, as one of the shortage of water in Sinkiang, actively carried out the pilot project of agricultural water right transaction in early 2014. There are mainly two types of transactions: one is internal between agricultural water trading, that is, between farmers; another is cross-industry water rights trading, namely water right trading transfers from agriculture to industry and public water supply[12].

In Shaanxi province, the basic work of water right transaction has already been carried out to promote environmental protection and water saving as the usage of market mechanisms. Although in three regions split into as Shanbei (northern Shaanxi), Guanzhong (central Shaanxi) and Shannan (southern Shaanxi) from north to south, Shaanxi have the same demand of water right transaction, the way of the trade and the design of the mechanism are different. Also the purpose of the transaction is different. So in Shanbei, the inter-basin water rights conversion is used to solve the problem of water resources shortage; in Guanzhong, cross-industry water trading system is applied to improve the water efficiency of irrigation areas; in Shannan, cross-regional water rights trading system helps to establish the ecological compensation mechanism[13].

4. Discussion

With the various kinds of exploration at home and aboard, it has become necessary to apply market means to manage the utilization efficiency of water rights, water rights trading and water market development. The discussion mainly involves the following points:

4.1. The use of Water right

Clear water rights are the fundamental requirement of optimizing the water resource allocation under the market system. The innovation of water right system in China should focus on the separation of ownership and use of water resources. Now in China and many countries in the world, the water resource belongs to the public, but for the right to use it does not be regulated. In order to establish water rights trading market to accumulate more experience, we should seriously carry out the theory innovations in water resources property right system and actively encourage various kinds of water rights trading.

In the process of water rights system construction, we should clearly not only develop water rights for approval to register, but also focus on developing agricultural water approval. Also we should set up different types of water right transaction trading premise condition and rule, thus play a role of market mechanism.

4.2. Water Resource Allocation System

The practices of water allocation above showed that water resource allocation can be divided into three types: administrative (government) configuration, user participation and negotiation allocation respectively. Administrative (government) configuration indicates that water resources are allocated through the control of water volume or water price between the industries and the regions as is a top-down approach. On the contrary, in the process of user participation or negotiation allocation, the users or related interest groups control the water distribution. The management of the farmer's participation in the irrigation system is one of the most typical examples of this approach.

The water market uses the market mechanism to allocate water resource under the market condition. According to relevant experience and China's current market level, the construction of water right allocation system in China should innovate and adjust measures to local conditions step by step.

The domestic pilot regions showed the allocation form or model of water rights has presented a diversified development trend. To establish the perfect water market, new forms of water rights allocation are actively explored and adjust measures to local conditions.

4.3. Water Right Transaction System

The practices of water rights transaction above show that water resource transaction present the characteristic of diversification and flexibility. There are permanent and temporary transactions; intra regional and cross regional transactions; intra industry and cross industry transactions. From the current practice of water rights trading, the water rights trading system has the following important advantages: (a) to be a more mature and feasible water right system in which market and government play important roles; (b) to save water resources, improve the efficiency of water resource and optimize the allocation of water resources; (c) to increase the participation in the management and distribution of water resources; (d) to be a better way of water resources distribution using market mechanism; (e) to effectively control the over-exploitation of groundwater.

In China, water rights trading system has not yet regulated in the concrete execution links. We should draw lessons from foreign advanced experience, such as Chile and Australia, and so on, and adjust measures to local conditions to satisfy national and local water conditions. In the construction process of water rights transaction system, government management and two-way market regulation should be strengthened so as to sustainable utilization of water resource.

4.4. Legal Norms and Institutional System

Along with the improvement of the market, some countries such as the United States, Australia, Chile, Mexico and so on, have formulated the corresponding water law according to their water situation one by one. In China, with respect to management model, transfer implementation rules, and intermediary services of water rights, and so on, they must be considered now. To solve the distribution of water resource, corresponding legal procedures of water rights allocation, reasonable water price setting mechanism, water rights trading market supervision and management system, water right transaction registration system and should be established and developed.

Acknowledgments

The authors gratefully acknowledge the financial supports from the National Natural Science Foundation of China (No.41701627) and *the Taishan Scholar Program of of Shandong Province, China.*

References

[1] Wenhua Yu.: Water right system legislation revelation abroad. J. The rule of law and society, (3):510-511(2007)
[2] R. Quentin G., James H.: Water markets in the Murray-Darling Basin. J. Agricultural water management.145,61-71(2013)
[3] Jacinta Palerm-Viqueira,Colegio de Postgraduados.: Water rights and water institutions in Mexico.Worldbank,http://siteresources.worldbank.org/EXTWAT/Resources/4602122-1213366294492/5106220-1213804320899/10.2_Water_Use_Rights_Mexico.pdf
[4] Jiaoyan Song.: The study on water rights trade system: In: Master thesis in northwest agriculture and forestry university of science and technology (2010).
[5] Go Palkarishnna C.: The doetrine of prior appropriation and its impact on water development:a critical survey, J. American Journal of economics & society, 32(1973)
[6] Huffaker, R., et al.: The role of prior appropriation in allocating water resources into the 21st century. J. Water Resource Development,16 (2000)
[7] Colby,B.: Transactions costs and efficiency in western water allocation. J. American Journal of Agriculture Economics,72 (1990)
[8] Stephen B., Annara H.: Water trade and extenralities of water use in Australia. In: Common wealth of Australia(2002)
[9] Jie C., Changxin X.: The influence of Chile's water law to China's water rights trading system. J. Yellow river, 27,:47-48,54(2005)
[10] Jie C.: Theory, method and application of transaction of water rights option, In: Doctoral Dissertations in Hehai University(2006)
[11] Zhou Ying, Li Hongren.: Studies of institutional system for water right trading in Jiangxi Province, In: China Water Resources,30-32(2017).
[12] Qinhui Jiang.: Xinjiang agricultural water rights trading legal thinking: the case of Changji state, In: Journal of Kashgar University,29-31(2017).
[13] Qian Zhang,: Water rights trading mode analysis in Shaanxi, In: Environmental protection and circular economy, 17-19(2016).

IOP Conf. Series: Earth and Environmental Science **146** (2018) 012009 doi:10.1088/1755-1315/146/1/012009

Analysis on Temporal Change and Grey Relation of Transportation Carbon Emissions in Jilin Province

Biao GAO[1, 2],Xiao-dongSUN[2*]

[1]College of Earth Sciences, Jilin University, No. 2199, building street, Changchun, Jilin 130061, China

[2]College of Tourism and Geography Science, Baicheng Normal University, No. 57, Zhongxing West Road, Baicheng, Jilin137000, China

Corresponding author: niatgb@126.com

Abstract. Based on basic datum of transportation energy consumption,this paper measured transportation carbon emissionsfrom 1999 to 2015 in Jilin Province by using measurement model of carbon emissions, and analysed its temporal change.On this basis, grey relational analysis model was used to investigate evolution relationshipof transportation carbon emissionsand relevant factors in Jilin Province. The results indicated that temporal change of transportation carbon emissions was divided into three phases: smooth progression and slight elevationphase, rapid-growth phase, slow-growth phase, the quantities of transportationcarbon emissions increased from 993750t to 3592514t. Diesel, raw coal, electricity power and gasoline were the main factors that affecting total carbon emissions because they had a larger proportion of carbon emissions, in the above factors, temporal change tendency of diesel carbon emissions was basically the same as the total carbon emissions. There was a close relationship betweentransportation carbon emissionsand all relevant factors in Jilin Province, their order of grey relational valuewas GDP > urbanization rate > population number > unit GDP energy consumption > transportation investment > private cars quantity. On the basis of prediction model, according to the situation of existing economic development, transportation carbon emissions will show a low growth tendency in next five years in Jilin Province.

1. Introduction

In recent years, greenhouse effect caused by excessive energy consumption of the human has become a focus of academic circles. Therefore, the issue of carbon emissions in different areas of production of human life has become a hot research topic.Some studies show thatChinese transportation industry has become the fastest growing sector of China's carbon emissions[1]. Transportation carbon emissions have accounted for about 22% of total carbon in China, and restrictedseriouslyrapid development of low-carboneconomy andconstruction ofecological civilization in our economic and social[2]. Jilin Province is an important part of the northeast old industrial base, its construction level of ecological civilization is low, andconstraints of resources and environment is becoming more and more obvious. Therefore, in Jilin Province, there is important significance that researchingtransportation carbon emission in the background of development of low carbon economy andaccelerating construction of ecological civilization. Based on the above, this research measurestransportation carbon emissionsfrom 1999 to 2015, and analyses its temporal change.On this basis, grey relational analysis model is used to investigateevolution relationshipof transportation carbon emissionsand relevant factors. Prediction model of transportation carbon emissions is constructed, and is used to predicttransportation carbon emissions from 2016 to 2020. This study can provide reference for other areas to develop related study of carbon emissions and provide scientific basis for carbon emission reduction policies made bygovernment or relevant organizations.

2. Temporal change of transportation carbon emissions

2.1. Calculationmodel

Transportation carbon emissions were due to greenhouse gas emissions caused byenergy consumption in transportation process[3]. In this study, basic datum of energy consumption of transportation industry in Jilin Province were collected in the statistical yearbook of Jilin Province (2000-2016), energy types were raw coal, crude oil, gasoline, kerosene, diesel, natural gas, and electricity power, the specific datumare shown in figure 1.According to the guidelines on greenhouse gas inventories of IPCC, calculation model of transportation carbon emission was selected based on previous studies [4, 5], the formula is shown in formula (1).

$$C = \sum N_i \times F_i \quad (1)$$

Type of: $C$was transportation carbon emissions, t. $N_i$was energy consumption of i sort,tec. $F_i$was carbon emission coefficient of i sort energy, t/t, raw coal: 0.7559, crude oil: 0.5857, gasoline: 0.5538, kerosene: 0.5714, diesel: 0.5921, natural gas: 0.4483, electricity power: 2.2132.

2.2. Calculationresults and temporal changes

Formula (1) was used to calculate transportation carbon emission in Jilin Province from 1999 to 2015 based on the basic datum of energy consumption in figure 1. The specific results are shown in figure 2.

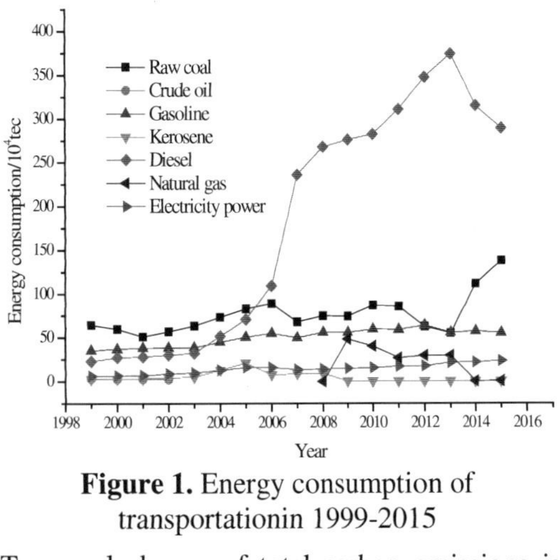

Figure 1. Energy consumption of transportationin 1999-2015

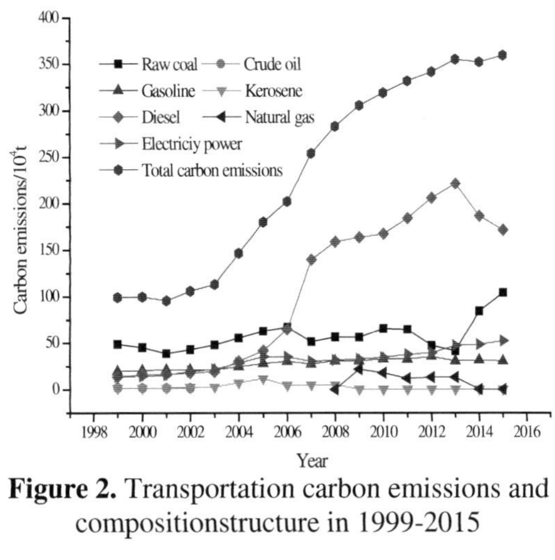

Figure 2. Transportation carbon emissions and compositionstructure in 1999-2015

Temporal change of total carbon emissions is shown in figure 2. Transportation carbon emissions present a growth state in 1999-2015.Total carbon emissions increased from 993750t to 3592514t, total growth rate was 262.51% and annual growth rate was 16.34%.The overall evolution process could be divided into three phases which includedsmooth progression and slight elevation phase, rapid-growth phase and slow-growth phase. In 1999-2003, temporal change of total carbon emissionschanged smoothly, and showed a steady increase in status. Carbon emissionsincreased from 993750t to 1133526t, during this period, net increment was 139776t, annual growth rate was 3.52%, and the minimum value was 957286t that appeared in 2001. In 2003-2009, temporal change of total carbon emissionsshowedthe status of rapid-growth, withincreased from 1133526t to 3054279t, net increment was 1920753t, and annual growth rate was 28.24%. After 2009, growth rate of total carbon emission became smaller, showed a low-growth rate that increased from 3054279t in 2009 to 3592514t in 2015, net increment was 538235t, and annual growth rate was 2.94%, during this period, the range of change was small in 2013-2015, which appeared first drop and then rise.

Composition structure of transportation carbon emissions was analysed according toconsumed energy categories. In figure 1 and figure 2, the application time of crude oil and natural gas was short, and their carbon emissions were small. In addition,the amount of kerosene had continued to decline and its carbon emissions were maintained at around 2000t because of energy structure optimization of transportation industry in Jilin Province, accounted for a small proportion of total carbon emissions. Therefore, the temporal changes of the above three types of energy were not analysed in detail.

In figure 2, diesel, raw coal, electricity power and gasoline were the main factor that affectingtotal carbon emissions. Among them, in 1999-2013, carbon emissions of raw coal showed a fluctuation state, but the overall amplitude was not large, increased rapidly after 2013, and reached 1044200t in 2015; in 1999-2006, carbon emissions of electricity power showed an increasing tendency, which increased from 138768t to 351899t, then dropped to 303208t in 2007. After 2008, carbon emissions of electricity power showed continuous growth, and reached 522758t in 2015;carbon emissions of gasoline fluctuated in study time, and the maximum value was 354543t in 2012, and thengradually decreased to 3306251t in 2015;dynamic change of diesel carbon emissionscorresponded to total carbon emissions that explained the main cause of carbon emissions growth of transportation was a lot of diesel energy consumption in Jilin Province. In 2007-2013, due to diesel consumption growth rate decreased, total carbon emissions showed slow growth state. After 2013, total carbon emissions fluctuated because diesel consumption reduced greatly and coal consumption increased.

3. Grey relational analysis of transportation carbon emissions

3.1. Model of grey relational analysis

Grey relational analysis wasa method of measuring development tendency of factors according to similarity or dissimilarity between dynamic processes of factors[6].The analytic model is shown in formula (2).

$$R_{0i} = \frac{I}{N}\sum_{k=I}^{N} L_{0i}(k) = \frac{I}{N}\sum_{k=1}^{N}\frac{\Delta_{\min} + \rho\Delta_{\max}}{\Delta_{0i}(k) + \rho\Delta_{\max}} \tag{2}$$

Type of: R_{0i} was grey relational value. N was data number of relevant factor sequence. $L_{0i}(k)$was relational coefficient of$k$ moment in relevant factor sequence. $\Delta_{0i}(k)$was absolute difference of k moment in relevant factor sequence.$\Delta_{\max}$, $\Delta_{\min}$were maximum and minimum of absolute difference at every moment in relevant factor sequence.$\rho$was resolution ratio, ρ=0.5 in this study.

3.2. Results of grey relational analysis

According to relativity analysis, this study selected population number, GDP, unit GDP energy consumption, transportation investment, urbanization rate, and private cars quantityasrelevant factors of transportation carbon emissions in Jilin Province, the detailed data are shown in table 1. Formula (2) was used to calculate grey relation between transportation carbon emissions and relevant factors. The specific results are shown in table 2.

Table 1. Relevant factor of transportation carbon emissions in Jilin Province from 1999 to 2015

Year	Population number / ten thousand people	GDP / hundred million Yuan	Unit GDP energy consumption (t standard coal/ 10000 Yuan)	Transportation investment / hundred million Yuan	Urbanization rate / %	Private cars quantity / 10^4sets
1999	2616.1	1682.07	2.1956	72.8169	48.41	13.5194

2000	2627.3	1979.22	1.7824	82.3315	49.66	13.9785
2001	2637.1	2122.86	1.7489	57.3836	49.80	16.1010
2002	2649.4	2363.13	1.7811	59.8639	50.88	18.0601
2003	2658.6	2646.86	1.6883	62.5857	51.77	24.4576
2004	2661.9	2981.90	1.6026	169.4971	52.30	29.1269
2005	2669.4	3406.70	1.5436	169.4971	52.52	45.3037
2006	2679.5	3967.38	1.4799	175.6815	52.97	49.1724
2007	2696.1	4679.65	1.3817	213.3675	53.16	77.9748
2008	2710.5	5414.26	1.3114	260.0979	53.21	88.4680
2009	2719.5	6126.52	1.2329	396.9592	53.32	109.8360
2010	2723.8	7035.20	1.1617	621.7350	53.36	134.4251
2011	2726.5	8154.36	1.0898	483.2225	53.36	159.2421
2012	2701.5	8987.02	1.0046	528.2772	53.40	185.3028
2013	2678.5	9543.64	0.8955	584.8165	54.20	217.8388
2014	2671.3	9899.23	0.8570	812.5400	54.81	249.4115
2015	2662.1	9917.09	0.8095	955.5200	55.31	279.6332

Table 2. Grey relational values between transportation carbon emissions and relevant factors[a]

Relevant factors	Population number	GDP	Unit GDP energyconsumption	Transportation investment	Urbanization rate	Private cars quantity
Raw coal	0.9753	0.9047	0.9454	0.7803	0.9773	0.6875
Gasoline	0.9442	0.9246	0.9345	0.7665	0.9465	0.6780
Diesel	0.6741	0.6529	0.6647	0.7781	0.6756	0.8555
Electricity power	0.9038	0.8542	0.8934	0.7888	0.9061	0.7088
Total carbon emissions	0.8757	0.9239	0.8454	0.8273	0.8803	0.7149

[a]Note:application time of crude oil and natural gas was short, and use quantity of kerosene was Less, so there was no analysis on the above three kinds of energy.

Grey relational values between transportation carbon emissionsand relevant factorsare shown in table2. There was a close relationship betweentransportation carbon emission and all relevant factors in Jilin Province, their order of grey relational valuewas GDP > urbanization rate > population number > unit GDP energy consumption > transportation investment > private cars quantity. Among them, grey relational value between transportation carbon emissionsand GDP was largest, and its value was 0.9239, which explained economic development level was the primary factor that influencingtotal carbon emissions of transportation, the raising of economic development level would inevitably lead to an increase of carbon emissions. The second factor was urbanization rate, and its value was 0.8803,with the pace ofurbanizationhad beenaccelerating in recent years, the level of social production and service industry had been developing rapidly, such as railway, highway, waterway, and aviation industry, thereby increased carbon emissions caused by energy consumption. The third factor was population number, and its value was 0.8757, which explained the expansion of population scale would lead to direct and indirect energy consumption of transportation industry, and then had an important impact on transportation carbon emissions. The fourthfactor was unit GDP energy consumption, and its value was 0.8273, which explained the upgrading of energy utilization technology could play an important role in reducing transportation carbon emissions. Thefifthfactor was transportation investment, and its value was 0.8273, which explained capital investment would promote the increase of transportation industry that had an important effect on transportation carbon emissions. Thesixthfactor was private cars quantity, and its value was 0.7149. In recent years, carbon emissions of private cars had been widespread concerned. Private carsquantityincreased substantially, which driven by the improvement of people's living standards, but also produced a large number of greenhouse gas emissions and strengthened greenhouse effect. With the implementation of private cars control policy and the increasing use of small carbon car, electrical cars and other clean energy cars, the carbon emission effects of private cars will be further controlled.

4. Prediction of transportation carbon emissionsbased on economic growth

Based on the above analysis, SPSS software was used, this studybuiltprediction model, whichtotal GDP of 1999-2015 in Jilin Province (comparable price, using 1999 as a baseline) were independent variables (x), and transportation carbon emissions were dependent variable (y),because GDP was the first relational factor of transportation carbon emissions, and in response torelevant national policies. The decision coefficient and test parameters were used to optimize the model, the prediction model is shown in formula (3).

$$y = 168.3026 \ln x - 1184.1570 \quad R^2 = 0.9762 \quad F = 615.7454 \quad P(Sig.) < 0.0001 \tag{3}$$

In order to respondnational macroeconomic development policy and ensure GDP and per capita income of urban and rural residents will be doubled from 2010 to 2020, the bottom line of economic growth should be above 6.5% in 2016 -2020. Therefore, in this study, GDP growth rate was calculated by 6.5% innext five years, and then formula (3) was usedto predict transportation carbon emissions of Jilin Province in 2016-2020. The specific predicted results are shown in table 3 and figure 3.

Table 3. Predicted value of GDP and transportation carbon emissions in 2016-2020

Year	Predicted value of GDP / hundred million Yuan	Predicted valueof transportation carbon emissions/10^4 t
2016	10561.70	375.1469
2017	11248.21	385.7457
2018	11979.35	396.3445
2019	12758.01	406.9434
2020	13587.28	417.5422

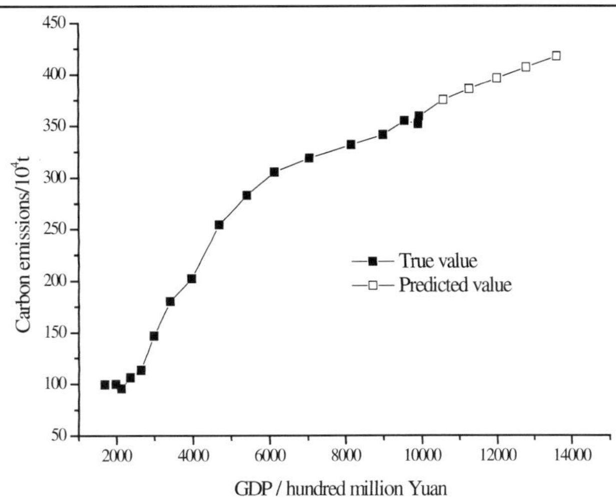

Figure 3. Transportationcarbon emissions in 1999-2020

According topredicted results, in 2015-2020, GDP will increase from 9917.09 hundred million Yuan to 13587.28 hundred million Yuan, andtotal carbon emissions of transportationwill increase from 3592514t to 4175422tin Jilin Province.During this period,net increment of carbon emissions will be 582908 t,total growth rate will be 16.23% andannual growth rate will be 3.25%.

5. Conclusions

This study measured transportation carbon emissions ofJilin Province andanalysed its temporal change in 1999-2015. It applied grey relational analysis model to investigateevolution relationship between transportation carbon emissionsand relevant factors. Then, prediction model of transportation carbon emissions was constructed, and was used to predict transportation carbon emissions from 2016 to 2020.

According to the results, in 1999-2015, total carbon emissions increased from 993750t to 3592514t, total growth rate was 262.51% and annual growth rate was 16.34%. The overall evolution process could be divided into three phases which includedsmooth progression and slight elevation phase, rapid-growth phase and slow-growth phase. Carbon emissions caused by a lot of diesel consumption was the first factorofcontinued growth of transportation carbon emissions, which mainly because agricultural transportation, logistics, transportation, rail transport, urban passenger transport and othershad large demand on diesel fuel.

In transportation industry, increasing the usage of electricity and natural gas could cut the usage of diesel, gasoline, and reduce carbon emissions.

Government had introduced a series of laws and regulations on energy saving and emission reduction, as well asnew transportation regulations, which effectively limited consumption of diesel and gasoline, and slowed downrapid growth of carbon emissions.

There was a close relationship betweentransportation carbon emission and all relevant factors in Jilin Province, their order of grey relational grade was GDP > urbanization rate > population number > unit GDP energy consumption > transportation investment > private cars quantity. Among them, grey relational value between transportation carbon emissionsand GDP was largest.

According toprediction model and current situation of economic development innext five years, transportation carbon emissions in Jilin Province will continue to show slowing growth tendency.In order to ensureeconomic growth rate and realizeearly negative growth of transportation carbon emissions, Jilin Province must take some effective measures.

References

[1] Bastani P, Heywood J B, Hope C 2012 The effect of uncertainty on US transport-related GHG emissions and fuel consumption out to 2050*Transportation Research Part A: Policy and Practice.***46** L517

[2] ZHANG Jian-zhen, WANG Xiao-chen, TAI Qi-long, et al 2017 Forecastingof Energy Demands and Carbon Emissionof Transportation in Hainan Province*Natural ScienceJournal of Hainan University.***35** L164

[3] GAO Biao, XU Qing-tao, LI Yu-bo, et al 2013 Calculation and Driving Factors Analysis of Carbon Emissions from Traffic and Transportation Energy Consumption in Jilin Province*Economic Geography.***33** L25

[4] YI Peng, DUAN Pei-li, CHEN Cai 2016Pattern of transportation carbon emission and relationship with transportationeconomic growth in China*Journal of Arid Land Resources and Environment.* **30** L7

[5] XIE Shou-hong, CAI Hai-ya, XIA Gang-xiang 2016 Calculation of the carbon emissions of Chinese transportation industry and thedriving factors*Journal of Arid Land Resources and Environment.***30** L13

[6] GAO Biao, XU Qing-tao 2015 Analysis on Evolution and Gray Correlation between Ecological Footprint and Economic Growth in Jilin Province*Journal of Baicheng Normal University.* **2** L59

IOP Conf. Series: Earth and Environmental Science **146** (2018) 012010 doi:10.1088/1755-1315/146/1/012010

Considerations on Rural Energy and New Countryside Construction in Beijing Deputy Center Area

WANG ZHI[1,*] and GAO XIANG[2]

[1]Institute of Agricultural Information and Economics, Beijing Academy of Agricultural and Forestry Sciences, Beijing 100097. [2]Beijing's Tongzhou District Lucheng Towns Government

angelwang1009@163.com

Abstract."Beijing Deputy Center Construction" is an important part of one of the three major national development strategies in China. The Deputy Center is located in Tongzhou District in Beijing. At present, the construction of the Beijing deputy center is in its infancy and there are many rural areas that need to make strategic design for rural energy and new countryside construction in deputy center. The government can make great efforts to make the key links in the new countryside construction by establishing a digital management system, promoting new energy technologies and changing farmers' traditional concept of using energy. If we can make a new type of urban agriculture in deputy center through active selection, and adjust the structure of rural energy use. In the future, Beijing's deputy center will form a seamless connection between "international village" and "cosmopolitan city."

1. Introduction

"Beijing Deputy Center Construction" is an important part of one of the three major national development strategies in China[1-3]. The Deputy Center is located in Tongzhou District in Beijing. At present, the construction of the Beijing deputy center is in its infancy and there are many rural areas that need to make strategic design for rural energy and new countryside construction in deputy center. Developing and building a rural energy source in Beijing's sub-center area is a powerful means of promoting regional integration and accelerating the coordinated development of Beijing, Tianjin and Hebei. At the same time, it is also an effective measure to improve air quality and smog in the capital and its surrounding areas.

2. Beautiful countryside construction is an important basis for the construction of ecological civilization and beauty of China

Chinese capital of Beijing, an area of 16,400 square kilometers in rural areas accounted for 1.5 million square kilometers, if rural development is not good, ecological Beijing will be no barrier, there will be no space environment[3,4], and sustainable development will be on the lack of foundation[5].

3. Major policies and measures to promote the construction of beautiful countryside

We recognize the importance and urgency of the beautiful countryside construction, but also, perceive that the complexity and difficulty to build a beautiful countryside. In recent years, Beijing launched a series of livelihood projects, including farmhouse seismic energy-saving projects, risk village and insurance households from mountain relocation project, Clean Air Initiative in rural areas reducing coal, Agricultural non-point source pollution control projects, Rural Ecological Environment

Construction Engineering, and so on, rural environment continues to improve, rural infrastructure and public service facilities gradually improved, the rural energy structure was further optimized, farmer housing conditions and income levels are rising[5,6]. New Rural Construction of Beijing has been built a beautiful countryside and laid a good foundation. At the same time we should also see that although the new rural construction of Beijing has made remarkable achievements, yet, with the urgent needs of the majority of farmers, and the requirements of building world-class city, there is still a big gap, but also need to be improved in the following areas[7,8].

3.1. Rural energy structure adjustment and optimization

In recent years, the Beijing government has been to actively promote the use of new energy sources, new technologies in rural areas, and started to implement the " Reducing Coal and Replaced Coal, Clean air" Initiative. Table 1 reflects the research team's research on rural energy use in deputy center.

Table 1. Beijing Deputy Center Rural Living Energy Data.

Energy form	**Statistics sample total (4235households)**		**Sample per household (tce)**	**The city's total(Ten thousand tce)**	**proportion (%)**
	The actual amount (t)	**Conversion standard coal (tce)**			
Commodity energy	—	11573.73	2.73	588.13	87.01
Bulk coal	9250.8	6607.85	1.56	335.78	49.68
Briquettes	726.5	259.51	0.06	13.19	1.95
Briquette	1311.87	468.6	0.11	23.81	3.52
Liquefied gas	361.78	620.2	0.15	31.52	4.66
Electricity	111.071 (ttkWh)	3617.57	0.85	183.83	27.20
Biomass energy	—	1727.31	0.41	87.78	12.99
Straw	1245.03	684.76	0.16	34.80	5.15
Tree branches	1895.56	1042.55	0.25	52.98	7.84
Total energy consumption of life	—	13301.03	3.14	675.89	100
Total coal consumption	—	7335.95	1.73	372.30	55.15

The data in the above table shows that the energy consumption in rural areas in Beijing'sdeputy center is mainly commodity energy, reaching 87.01%. Coal is the main source of domestic energy in rural areas, accounting for 55.15% of the total energy consumption in the country.

"Heating and cooking," these two links is an important part of rural energy, they are also the main part of the extensive use of coal as well as air pollution. In the rural energy structure adjustment, we must really control the proportion of coal in heating and cooking in order to effectively improve the quality of the atmospheric environment.

Through the survey of rural areas in Beijing found that the residents of the current application of the main heating technologies are electric heating, heat pump heating, gas heating, clean coal heating and solar heating in five ways[5]. According to the survey data, combined with the application of the Beijing area and product technology maturity, the existing mainstream rural areas in Beijing can be used for rural use of clean energy technologies in Table 2.

Table 2. Commonly used in rural areas heating technology.

Serial number	**Heating mode**			
	category	**Heat source (priority)**	**Auxiliary heat source (alternate)**	**Terminal**

1	Heat pump	Low temperature air source heat pump	——	Floor heating	Low temperature hot water terminal
2		Ground source heat pump	——	Floor heating	
3	Solar composite heating category	Solar hot water heating	Clean coal boilers	Floor heating	
4		Solar hot water heating	Gas boiler	Floor heating	
5		Solar hot water heating	Low temperature air source heat pump	Floor heating	
6		Solar hot water heating	Ground source heat pump	Floor heating	
7		Solar hot water heating	Electric heating	Floor heating	
8	Gas category	Gas boiler	——	Radiator / floor heating	High temperature / low temperature hot water terminal
9	Coal-fired class	Clean coal heating stove	——	Radiator / floor heating	
10	Electric heating	Electric hot water boiler	——	Radiator / floor heating	
11		Heat storage type electric boiler	——	Radiator / floor heating	
12		Carbon crystal plate class	——	——	——
13		Energy storage electric heating	——	——	

3.2. Rural Residential Seismic energy-saving projects

Beijing government started farmhouse seismic energy-saving projects since 2007, it has supported 380,000 rural residents in the implementation of energy-saving residential earthquake. Through the implementation of seismic energy-saving projects, farmers housing winter indoor temperature has increased by 4 to 6 degrees, saving more than 30% of coal-fired[9,10]. In summer, obviously feeling the cool interior, indoor temperature is relatively low outside temperatures 5 to 6 degrees[11]. According to statistics, only this thermal insulation, has an annual saving of coal-fired nearly 55 million tons, a significant reduction in carbon dioxide emissions, effectively improve air quality. According to surveys, there are still 530,000 rural residents in Beijing have seismic energy-saving requirements. Therefore, to further increase the Farmhouse seismic energy-saving efforts, using of 5 years, at a rate of about 100,000 a year, it will complete the city's Farmhouse seismic energy-saving task.

3.3. Sewage treatment works in rural areas

In recent years, new rural construction in Beijing, more than 500 villages have been built sewage treatment facilities in 1010, but due to complex process, facilities scattered layout, unfocused, factory network is not complete, the high cost of operation and maintenance, management operation is not in place and other reasons, rural sewage treatment facilities, 30% did not run or intermittent operation. In Beijing rural area, annual output of sewage is more than 100 million tons, but, treatment rate is only about 50%. large amounts of sewage is discharged directly[12]. With the rapid development of urbanization and rural economic and social, supply and demand contradictions in rural city sewage treatment facilities have become increasingly prominent. Therefore, the reality requires us to rural sewage treatment as an important part of the construction of the beautiful countryside. For facilities that have been built to run on to assess the situation, according to local conditions to transform and strengthen the operation and maintenance, and ensure that the newly built sewage treatment facilities operating normally, the basic objectives of rural sewage discharge standards may successfully achieve in 2020.

4. Beijing 's rural energy demand forecast

4.1. Analysis of energy - using factors

4.1.1 Farmers' own needs. With the development of rural economy, farmers' concept of energy use is gradually changing. The higher income level of farmers, the more emphasis on the convenience and comfort with energy, and then health, the last is the economy. Therefore, in cooking, the farmers are more willing to use the cleanliness of liquefied petroleum gas or biogas, rather than coal. In the heating, according to affordability, farmers will focus on electricity, solar and other clean energy heating.

4.1.2 The Impact of Urbanization Policy. With the acceleration of urbanization in the rural areas and the implementation of ecological resettlement, the rural energy structure has also been greatly affected.

(1) Heating will be converted into centralized heating. The Beijing municipal government proposed that by 2020 the city will make about two-thirds of the farmers into urban residents, the living conditions of farmers will be dispersed from the living, into a centralized way to live, therefore, the corresponding heating will focus on the form Heating mode change, which will help save energy consumption of heating[13-15].

(2) Farmers' choice of energy will become more commoditized and cleaner. Due to the concentration of living forms and the difficulty in obtaining biomass resources such as straw, farmers will tend to favor commercialized energy sources for energy selection[13-15]. With the popularization and cleaning of public facilities such as heat and gas, The proportion of the use of energy will be a substantial increase.

(3) Housing construction will become more energy-saving. With the process of urbanization, the new farmers will focus on energy-saving housing and comfort[16], thus directly reduce the heating load and reduce the use of heating energy.

(4) Energy use will become more low-carbon and comfortable[17]· Urbanization directly promote the increase of farmers' income, and with the construction of new houses, solar energy and geothermal energy and other low-carbon new energy and renewable energy in the proportion of domestic energy consumption will gradually increase.

4.2. Potential evaluation of coal adjustment in rural energy structure

In the heating, through the existing energy-saving transformation of rural households, the annual heating coal can be saved about 62.76 million tons of standard coal (in accordance with the energy-saving transformation to save 40% of the heating coal to calculate). On this basis, by 2020, 952,000 farmers will live in the cities and towns of residence, so that the original decentralized heating will transform into a central heating. This way can reduce heating coal-burning 94,100 tons of standard coal per year.

In addition, for the remaining decentralized heating farmers, the Government can actively promote solar heating, electric heating and other new green energy-saving technologies, which can reduce the heating coal-fired 25.10 million tons of standard coal.

From the adjustment potential, based on full use of existing resources and actively promote green energy-saving technology, respect for heating and cooking, the total coal pressure reduction will be more than 1.1451 million tons of standard coal per year.

5. Some thoughts on rural energy policy and new rural construction in Beijing

In the capital, to promote therural energy construction process, the author has deeply thinking for rural energy policies and beautiful countryside construction as follows:

First, the establishment of Beijing rural information service platform and the beautiful countryside of information resource library. It will cover many aspects of rural infrastructure, rural economy, industry characteristics, population health, culture, education, grass-roots organizations and other resources for effective integration of information and data, and establish a cover Beijing, connecting Beijing, Tianjin city, county, township and village "digital village" information network platform.

The second is the establishment of "Beijing Digital Village" platform, its function should have a strong practical and advanced. It can form in a variety of vivid visual image, data, icons and text, etc., it can provide information on all aspects of rural work for the government at all levels in Beijing, so as to provide a reliable basis for the scientific and democratic decision-making of new rural construction. Beautiful countryside construction in every measure can be demonstrated and information management, the establishment of relevant information platform. Moreover, with the continuous improvement platform, which will provide reliable first-hand information for all agricultural sectors and investment units, it lay a solid foundation for the economic development of Beijing Rural Development and Investment, the same time, the Present Situation can be shown by the platform the beautiful countryside as a window, it can be to promote the rural areas this emerging e-commerce business model, so as to promote the exchange of agricultural products and accelerate rural economic development has laid a good foundation.

The third, pay attention to the establishment of several information platform, focusing on people's livelihood and ecological protection. The main measures "optimal adjustment of rural energy structure, agricultural nonpoint source pollution control, rural residential seismic energy-saving projects, sewage treatment works in rural areas" and other beautiful countryside construction are to be monitored and managed through the information platform to maximize for the construction of ecological civilization provide effective protection, as well as the national new rural construction to enhance digital play a leading and exemplary role.

Acknowledgement

This research was financially supported by the Beijing Municipal Rural Work Committee Technology Services Special Fund, meanwhile, by Beijing Academy of Agriculture and Forestry Youth Research Fund.

References

[1] LIU Jun-ping. The Pathof Green Developmentin Beijing Country [J]. Chinese Journal of Agricultural Resources and Regional Planning. 2016(11):81-86.
[2] https://baike.so.com/doc/24305453-25755444.html.
[3] http://www.bjtzh.gov.cn/n95/n534/n725/
[4] http://baike.baidu.com/view/2282762.htm.
[5]WANG Zhi. Thoughts on Beijing's Long-Term Rural Infrastructure Management and Perspective of the Government to Effectively Perform Their Duties.Mechanical Engineering and Control Systems.2016,PP:490-493
[6] CHEN Zhang, YAN Ping, SHAO Wei. Beijing, Tianjin Economic Contacts[M].DATA, 2011, (3).
[7] Weng Bo-qi, Ma Hong-min, Su Han-fang, etc. Developing Paths and Promoting Strategies for Characteristic Modern Agriculture [J]. Journal of Agriculture. 2015,5(12):127-132.
[8] Wu Qian, Li Qian.Green Development of Agriculture Based on Endogenous Development: Taking Beijing, Tianjin, Hebei Province for Example [J]. Chinese Agricultural Science Bulletin,2017,33(2):155-159.

[9] Al-Homoud M S. Performance characteristics and practical applications of common building thermal insulation materials. Building and Environment, 2005, 40: 353–366.
[10] Du Liming, Zhu Ming. Big Agriculture New Urbanization Strategy:Dayong Town, Harbin Example[J]. China Academic Journal.2016,1(32):40-44.
[11] Erlandsson M. Energy and environmental consequences of an additional wall insulation of a dwelling. Building andEnvironment,1997,32(2):129-136.
[12] WANG Zhi, ZHOU Lian-di, LIU Shi-rong, et al. The Effect on Abating Pollution from Farmland Runoff by Using Biogas Manures[J]. JOURNAL OF ANHUI AGRICULTURAL SCIENCES, 2009, 37(10).
[13] Ahmad.A, Saini JS, Varma HK. Effect of geometrical and thermophysical characteristics of bed materials on the enhancement of thermal performance of packed bed solar air heaters. Energy Conversion and Management, 1995, 36(12): 1185-1195.
[14] Hegazy AA. Performance of flat plate solar air heaters with optimum channel geometry for constant/variable flow operation. Energy Conversion and Management,2000, 41(4):401-417.
[15] Varol Y, Oztop HF. A comparative numerical study on natural convection in inclined wavy and flat-plate solar collectors. Building and Environment, 2008, 43(9): 1535-1544.
[16] Varol Y, Oztop HF. Buoyancy induced heat transfer and fluid flow inside a tilted wavy solar collector. Building and Environment, 2007, 42(5): 2062-2071.
[17] Wang Gehua, Chen Yanbini, Shi Deming. The effect assessment of integrated rural energy development and the application of Grey System Method. Transactions of the CSAE.1992(9):60-66.

Evapotranspiration change under Short-term Experimental Warming in a Swamp Meadow Ecosystem of the Qinghai-Tibet plateau

G S Liu[1] and C Zhao[1]

[1] College of Environmental Science and Engineering, Xiamen University of Technology, Xiamen 361024, PR China

liugs@xmut.edu.cn

Abstract: The impact of climate change on the cold region ecosystem is deepening, especially in the Qinghai-Tibet plateau(QTP), with its impacts on the water cycle. Evapotranspiration (ET), as an important part of the hydrological cycle in terrestrial ecosystems. Thus, we analyzed the influence of experimental warming on ET through observed data of 2008 under two open top chambers (OTC-1 and OTC-2) plots and ambient air (Control) plot. The study showed that the potential and actual ET of OTCs were higher than that of Control, due to increase in vapor pressure deficit, warmer air temperature, increase in ground surface soil moisture and temperature compared to the outside. ET is largest component of water loss from alpine ecosystem, accounted for 74.6% of the annual total precipitation. Elevated CO_2 concentration had no significantly affected at ET rate in our study. The results will be helpful for future climate change study in the QTP.

1. Introduction

The impact of climate change on the cold region ecosystem is deepening, especially in the Qinghai-Tibet plateau(QTP). [1,2]In addition, climate warming significantly influences the stability of the alpine swamp meadow ecosystem.[3]Evapotranspiration (ET), reflects the dry and wet condition of the eco system, as an important part of the hydrological cycle in terrestrial ecosystems, have a huge impact on the local climate and even the global water cycle, and directly determines how much water dissipated in the ecosystem. [4] Therefore, quantifying the ET change under climate change is important for upstanding the impact of global climate change on water budgets in the cold regions. [5]

Fei et al. conducted a field experiment to investigate infrared radiation experiment warming on ET in an alpine meadow ecosystem in the QTP,[6] results showed that annual ET enhanced with the increase of air temperature and plant transpiration accounted for most of the ET change due to soil surface drying. Sorokin et al. conducted chambers experiment with elevating CO_2concentration in a semi-arid grassland,[7] results showed that there had no significant differences in ET during the growing season. Zhang et al. suggested that ET are projected to increase under climate warming but vary across different ecosystems.[8]

However, rare studies focus on the ET change under experimental warming in the swamp meadow ecosystem, especially in QTP, due to poor test environment and shortage of observation. The purpose of this study was to clarify the ET change and its impact factors under experimental warming in a swamp meadow ecosystem in the QTP.

2. Materials and Methods

2.1. Site description

The study site located in the Fenghuoshan Watershed(92°50'-93°3' E and 34°40'-34°48' N), a tributary of the Yangtze River, with typical swamp meadow ecosystem, was chosen for study. [9]The alpine swamp meadows accounted for a large proportion of the main vegetation type in the QTP. [10]As shown in Table 1, the main soil type is swamp meadow soil, and the parent material is mostly quaternary sediment).[9]

Table 1. Soil property in different manipulations plots of the swamp meadow ecosystem

Soil profile /cm	Bulk density /mg.cm^{-3}	Granularity >0.05 mm /%	Granularity <0.002 mm /%	Organic matter /%	Available nitrogen /g.kg^{-1}
0-10	0.83(±0.11)	47.77	11.55	8.75(±2.2)	112(±7.6)
10-20	0.94(±0.21)	50.94	11.58	4.32(±1.3)	97(±4.4)
20-30	1.19(±0.12)	47.96	12.94	2.46(±2.1)	23(±5.3)

2.2. Methodology

The open-top chamber (OTC),used by the International Tundra Experiment (ITEX) for experimental warming study in the high altitude and latitude region. [11] There have no significant differences ($p>0.4$) of vegetation coverage (90%) and soil property between the three experimental plots (Table1).

The materials of OTC were PMMA fiber. The structure of OTC presented the cube structure(1.5 m × 1.5 m× 1.7 m height). OTC-1 plot, with a partial closed top chamber, have a 0.6 m diameter hole on the top, which was used by Havstrom *et al.* in Abisko and Svalbard. [12] OTC-2 plot, with a complete open top, represents a traditional OTC. Control plot, keeping the original stage, was taken as an outdoor control group.

In the three plots, temperature sensors and moisture probes were placed to observe soil temperature and moisture at 5 cm and 20 cm depth, respectively. In the OTCs, the thermal and hygrometer sensors was installed to measure air temperature and relative humidity. The probes and sensors were connected to Campbell CR1000 dataloger (Campbell Scientific Ltd., USA).The meteorological data was obtained from a standard meteorological station, situated near the experiment plots. The observed data was collected in one-year period of 2008.

The lysimeter method was used to observe actual ET of alpine meadow under 92% vegetation coverage near the experiment plots. The CO_2concentration of the atmosphere under OTCs and Control were also observed using CO_2 sensor (Veisala GMT222, Finland) in the growth season from May 1st, 2008 to September 29th, 2008.

The Penman-Monteith equation is used to calculate the ET rates under different manipulations. [13]

3. Results and discussion

3.1. Microclimate

3.1.1. Precipitation and topsoil moisture. The total precipitation of 2008 was 490.1mm and the precipitation of rainy-warm season from May to September accounted for 87.4% of the total precipitation. During early spring and winter, only several rainfall events occurred with rainfall less than 3mm (Figure 1). Soil moisture at 5 and 20cm depth was related to precipitation. The response of soil moisture to precipitation and the response time of OTCs and the control were almost identical to each other. The fluctuation cycle of OTCs was slightly larger than Control and the rainfall replenishment was slightly higher than the Control at 5cm depth. They may account for higher surface moisture in OTCs.

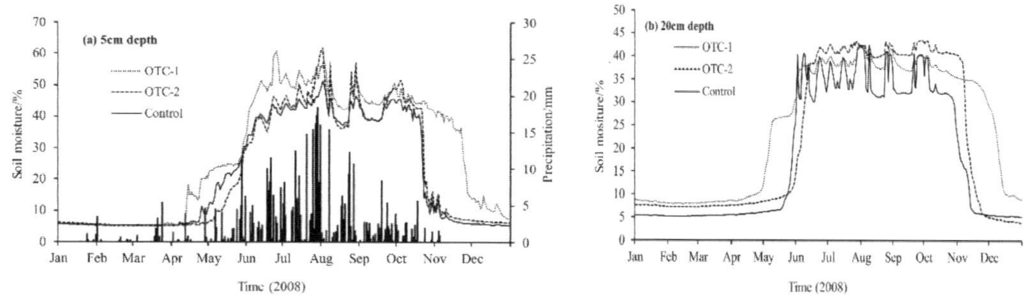

Figure 1. Annual variations of precipitation, soil moisture at 5 and 20cm depth under Control and OTCs warm treatments

3.1.2. Air temperature and soil temperature in 5 cm depth. Annual mean air temperatures of OTC-1 and OTC-2 were 6.70℃ and3.48℃warmer than the Control, respectively (Figure 2a). Kimballand Marion reported that the air would be warmer than outside about 6℃ under the full sun, roughly equal to the OTC-1. [14,15] due to generally transpiring rate, [16] as this is the case with OTC-2.

As shown in Figure 3b, the annual mean soil temperature of Control, OTC-2, and OTC-1 were -1.71℃, -1.64℃ and -0.05℃ in 5cm depth, respectively. The annual mean soil temperature of OTCs was warmer than Control in all periods.

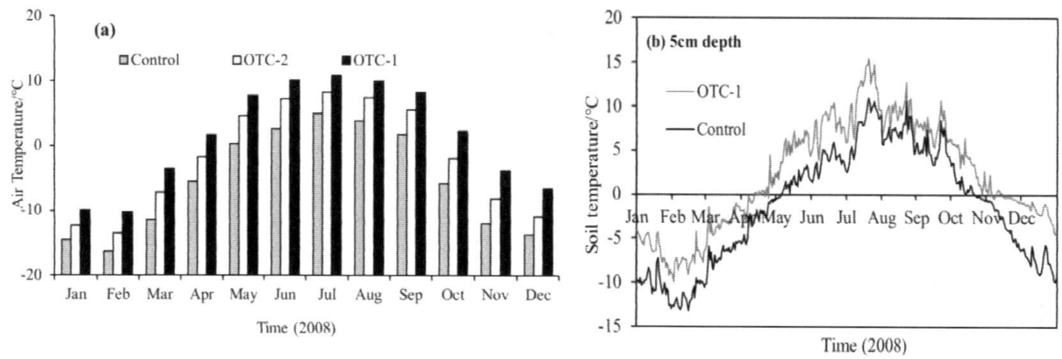

Figure 2. Monthlyvariations of air temperature (a) and annual variation soil temperature in 5 cm depth (b)under different manipulations.

3.1.3. Relative humidity and air vapor pressure. Annual mean relative humidity of OTCs was less than Control (Figure 3a).Marion also suggested that there had a universality lower relative humidity inside chambers. [15] This is due to the increasing in air vapor pressure caused by air warming.

Annual mean actual air vapor pressure of OTC-1 and OTC-2 were 41% and 16%larger than that of Control (Figure 3b), respectively. Weinstock found that the air vapor pressure in OTC is about 14% higher than the outside, [17] which is the case with OTC-2. We suppose that the higher air vapor pressure in OTC-1 can be attributed to its higher air temperature and comparably low water vapor passing through the chamber in a relatively enclosed space.

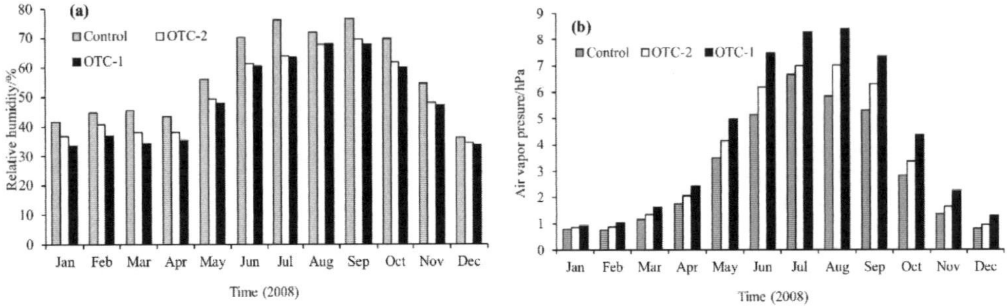

Figure 3. Distribution of Relative humidity (a), air vapor pressure (b) under different manipulations.

3.2 Potential evapotranspiration

More measurements on the fluxes of net radiation, soil heat flux, wind speed, and other environmental factors in OTCs and Control would be helpful to estimate evapotranspiration more accurately. However, that would not be practical, since they were costlier and would disturb the environment in OTCs. Therefore, the meteorological factors measured in a standard meteorological station near the experiment plots were adopted for replacing some environmental meteorological elements of the OTCs.

The calculated potential evapotranspiration (ET_p) using PM-equation under OTCs and Control is shown in Figure 4. The annual potential evapotranspiration of Control, OTC-2 and OTC-1 were 688.0, 785.6 and 811.6mm, respectively. The maximum potential ET of OTC-1, OTC-2 and Control were 74.1, 87.3 and 92.9 mm, respectively, all appearing in June. Meanwhile, the minimum ET of OTC-1, OTC-2 and Control were 34.4, 39.5 and 43.1 mm, respectively, all appearing in February.

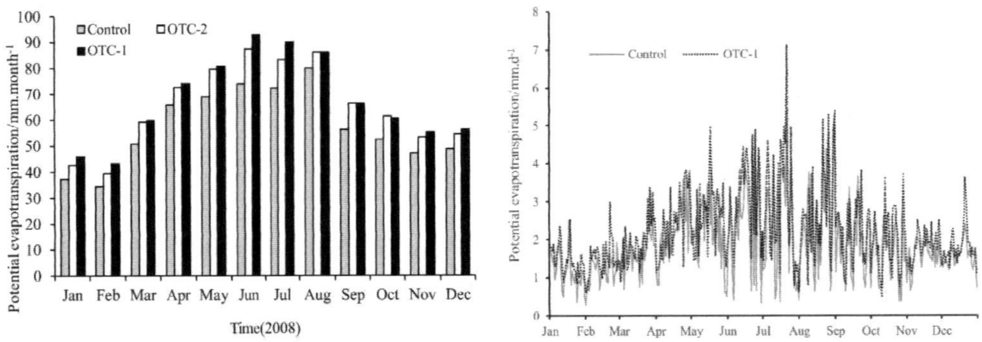

Figure 4. Annual variation of potential evapotranspiration under different manipulations.

3.3 The relationship between actual ET, water surface evaporation and potential ET

The lysimeter method is used to observe actual ET of alpine meadow under 92% vegetation cover plot near the experiment plot. The total actual ET (ET_a) under 92% coverage of 2008 was 365.6 mm (Figure 6a), approximated to be the actual ET of Control.

The Penman hypothesis is a common method to reduce potential ET to actual ET in response to the water availability, given by a function of soil water stress and vegetation conditions.[13] The formula is $ET_a=f_c\times ET_p$, where f_c is soil water supply capacity and vegetation coefficient. Therefore, the relationship between the actual ET and potential ET is compared in the Control plot (Figure 6b). The annual actual ET is about 0.54 times of the potential ET and the f_c is 0.54. Although there is no observed data under OTCs manipulations, based on the $f_c$of the Control, the actual ET of OTCs could be preliminary estimated. The actual ET of OTC-2 and OTC-1 were approximately 424.4 mm and 438.2 mm.

Water surface evaporation (E_{ws}) is an important index to measure the evaporation capacity of land surface and is an indispensable factor in the study of regional water balance and water circulation. In addition, water surface evaporation, is an indicator of potential evaporation. The relationship of potential ET and water surface evaporation was shown in Figure 6c, and the regression equation is proposed.

$$ET_p=0.44E_{ws}+0.39 \quad (1)$$

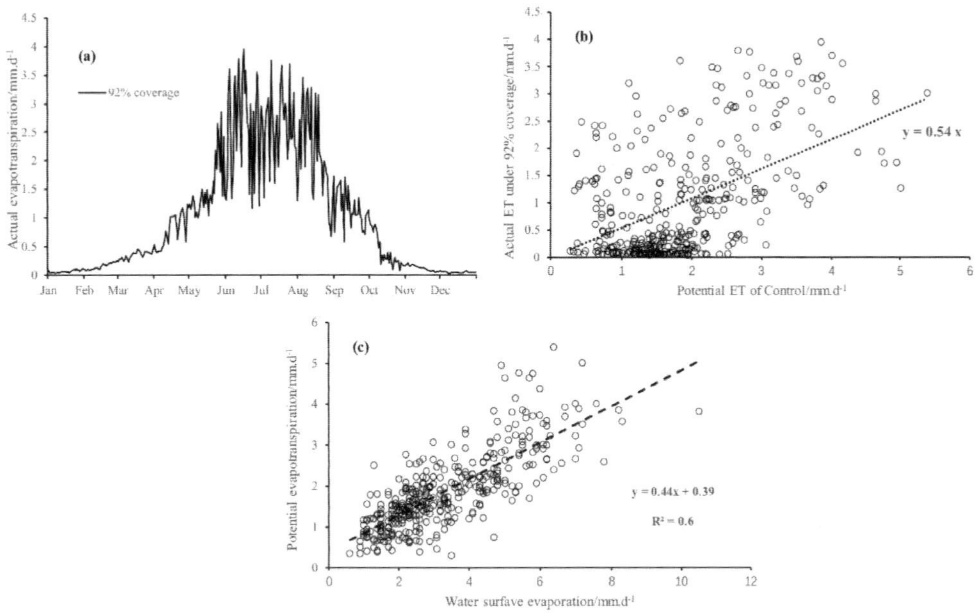

Figure 5. Annual variation of actual evapotranspiration under 92% coverage (a) which is similar with Control Plot, and the relationship between actual ET under 92% coverage and potential ET of Control plot (b), and the relationship between potential ET of Control plot and water surface (pan) evaporation.

4. Discussion

4.1 The ET change response to experimental warming

The ET change under experiment warming depends on air temperature, relative humidity, stomatal conductance, soil moisture, and plant phenology. [6]The observed actual ET (365.6 mm under 93% coverage, similar with Control) was higher than the study in an alpine meadow ecosystem of QTP (311 mm), [6] where annual precipitation (318 mm) is much lower. Our study confirmed that ET is maximum component of water loss from swamp meadow ecosystem, [18] accounted for 74.6% of the annual total precipitation of our study (490.1 mm). ET was assumed to increase under climatic warming in mountainous regions, which is consistent with the results of this study. As shown in Figure 7, the increasing in ET could be explained by increase in vapor pressure deficit(VPD), due to which is positively correlated with ET. [6] The 14.2% and 17.9% increase in actual ET of OTC-2 and OTC-1 with 3.48℃ and 6.7℃ experimental warming were lower than relative ET increase (29.8%) with 3.19℃ experimental warming in an alpine meadow ecosystem. As the air temperature increase, it is expected that near surface air should be drier, which would result in increasing in evaporation rate from ground surface.

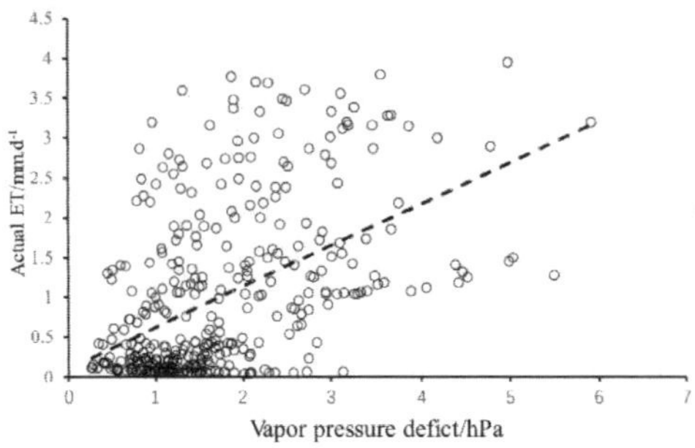

Figure 6. The relationship between actual ET under 92% coverage and VPD

4.2 The effect of CO_2 concentrations on ET

As shown in Figure 8, the CO_2 concentrations inside the OCTs are found to be slightly larger than outside, [15] because air movement inside OTCs was less than outside, resulting in CO_2 accumulation in OTCs. High CO_2 concentration and high temperature made ET increase in high（55% ~ 65% field water capacity）soil moisture regimes, which similar with the swamp meadow system(Figure 2).Costa and Foley reported that ET decreased by 2.8% in a doubling CO_2 concentrations experiment. [21] The OTC conditions favor growth of vegetation, which would prompt the plant transpiration. In general, elevated CO_2 concentration and temperature had not significantly affected at ET rate in our study.

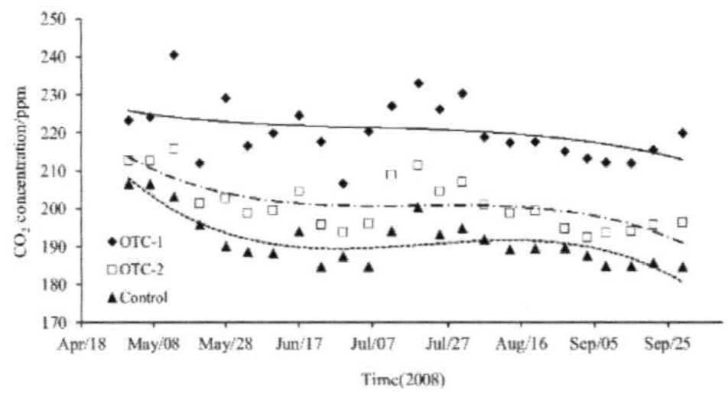

Figure 7. Distribution of CO_2 concentrations under different manipulations.

5. Conclusions

This study describes the evapotranspiration change and its impact factors under experimental warming in a swamp meadow ecosystem in the QTP. The results showed that annual mean air temperatures of OTCs were warmer than the Control， and a universality lower relative humidity inside chambers. The annual mean potential ET of Control, OTC-2 and OTC-1 were 688.0 , 785.6 and 811.6 mm, respectively. The total actual ET wereapproximately365.6 mm, 424.4 mm and 438.2 mm under Control, OTC-2 and OTC-1 manipulations.ET is largest component of water loss from alpine ecosystem, accounted for 74.6% of the annual total precipitation.

The increasing in ET could be explained by increase in vapor pressure deficit(VPD), due to which is positively correlated with ET. Elevated CO_2 concentration and temperature had not significantly affected at ET rate in our study. The increasing in ground surface soil moisture and temperature might reinforce the positive effect on ET.

Acknowledgements

This study was funded by The Natural Science foundation of Fujian Province (No.2015J05114,2017J01721) and Xiamen science project (No.3502z20150051).

References

[1] Liu XD, Chen BD2000 Climatic warming in the Tibetan plateau during recent years*Int. J. Climatol.***20**1729-42

[2] Wang G, Li Y, Wang Y, et al 2007 Typical alpine wetland system changes on the Qinghai-Tibet plateau in recent 40 Years *Acta. Geogr. Sin.* **62** 481-91

[3]Walker D, Jia G, Epstein H, et al 2003 Vegetation-soil-thaw-depth relationships along a low arctic bioclimate gradient, Alaska: synthesis of information from the ATLAS studies Permafrost. *Periglac.* **14** 103-23

[4] Hu H, Wang G, Liu G, et al 2009 Influences of alpine ecosystem degradation on soil temperature in the freezing-thawing process on Qinghai–Tibet Plateau*Environ. Geol.***57** 1391-97

[5] Jung M, Reichstein M, Ciais P, et al 2010 Recent decline in the global land evapotranspiration trend due to limited moisture supply *Nature***467** 951

[6] Fei P, You Q, Xian X, et al 2015Evapotranspiration and its source components change under experimental warming in alpine meadow ecosystem on the Qinghai-Tibet plateau. *Ecol. Eng.***84** 653-59

[7] Sorokin Y, Pendall E, Brennan A, et al 2012 Responses of evapotranspiration to experimental warming and elevated CO_2 in a semi-arid grassland*ESA Convention*

[8] Zhang Y, Liu C, Tang Y, et al 2007 Trends in pan evaporation and reference and actual evapotranspiration across the Tibetan plateau *J. Geophys. Res.-Atmos.* **112(D12)** 113-20

[9] Wang G, Cheng G, and Shen Y 2010*Research on ecological environmental changes in Yangtze and Yellow Rivers source regions and their integrated protection*(Lanzhou: Lanzhou University Press) p 33

[10] Zhou Y, Guo D, Qiu G, et al 2000*Geocryology in China* (*Beijing: Science Press*) p 65

[11] Klein J, Harte J, Zhao X 2005 Dynamic and microclimate responses to warming and grazing manipulations*Global. Change. Biol.***11** 1440-51

[12] Havstrom M, Callaghan T, Jonasson S 1993 Differential growth responses of Cassiope tetragona, an arctic dwarf-shrub, to environmental perturbations among three contrasting high and subarctic sites*Oikos***66** 389-402

[13] Penman H 1948 Natural evaporation from open water, hare soil and grass. *Proceedings of the Royal Society of London. Series A, Mathematical and physical sciences*, **193** 120

[14] Kimball B, Pinter P, Gerard W, et al 1997Comparisons of responses of vegetation to elevated carbon dioxide in free-air and open-top chamber facilities*asaspecialpubli***61**113-30

[15] Marion G, Henry G, Freckman D, et al1997Open-top design for manipulating field temperature in high latitude*Global. Change. Biol.***3(S1)**20-32.

[16] Rustad L, Huntington T, Boone R 2000 Controls on soil respiration: implications for climate change *Environ. Earthl Sci.***48** 1-6

[17] Weinstock, L, Kender WJ, Musselman RC 1982 Microclimate within open top chambers and its relation to grapevine physiology *J.Am.Soc.Hortic.Sci*, **107** 923-29.

[18] Lu Z, Wu Q, Yu S, et al 2006Heat and water difference of active layers beneath different surface conditions near Beiluhe in Qinghai-Xizang Plateau. *J. Glaciol.Geocry.***28** 642-47

IOP Conf. Series: Earth and Environmental Science **146** (2018) 012012 doi:10.1088/1755-1315/146/1/012012

A study of lifestyle changes in Xavier Institute of Development Action & Studies, India to mitigate global warming

Daitri Tiwary

GL Bajaj Institute of Management & Research

Greater Noida, Knowledge Park III, India

daitri.tiwary@gmail.com

Abstract. The Sustainable Development Goals (SDGs) outlined by the United Nations as tools to *transform the world by 2030,* outline Goal 13 to *"Take urgent action to combat climate change and its impacts by regulating emissions and promoting developments in renewable energy"*. But the seventeen goals are all interrelated and it is our cumulative efforts that can make them achievable in the next 13 years. As a civilization we have thrived upon decades of denudation and degradation, gradually, but immensely, yet quite indifferently. This indifference might be debated upon as the recent times has seen ever-growing concerns to address the crisis at hand, but this is no more than burning the night oil for a saving grace. If we look back, then it will not be difficult to analyze that the exploitation of the resources and the processes of degradation accelerated from the era of Industrial Revolution, dating back to 1750, but it took us more than a century to feel the repercussions of our actions. The first time ever the terms "global warming" and "climate change" was addressed as an internal issue of concern was in the Rio Summit, held in Rio-de-Janeiro in Brazil in 1992. Since then, the world has witnessed a series of environmental protocols, treaties and summits, each one raising a doubt on the validity and effectiveness of the previous. Meanwhile, the ever-impending population surge has compounded the environmental stresses. To combat the changes we are introducing in our ecosystem, the research focuses on implementation of minor and doable actions in the mundane yet regular activities of our daily life, which have a direct impact on individual carbon footprint. The research establishes the success of these measures in Xavier Institute of Development Action and Studies (XIDAS) by comparing the scenarios before and after implementation of a "Green Lifestyle", which will help us achieve sustainability beyond environment protocols and beyond international treaties.

1. Introduction: Need of Life-Style Changes as a Mitigation Measure

1.1. Highlighting Concerns over Climate Change

Human beings, by nature tend to take beneficial things for granted and it is by the virtue of this tendency that the natural environment in which we have thrived upon for decades has been denuded and degraded, gradually, but immensely, yet quite indifferently. This indifference might be debated upon as the recent times has seen ever-growing concerns to address the crisis at hand, but this is no more than burning the night oil for a saving grace. If we look back, then it will not be difficult to analyze that the exploitation of the resources and the processes of degradation accelerated from the era of Industrial Revolution, dating back to 1750, but it took us more than a century to feel the repercussions of our actions. The first time ever the terms "global warming" and "climate change" was addressed as an internal issue of concern was in the Rio Summit, held in Rio-de-Janeiro in Brazil in

1992. The melting polar ice-caps were accidentally recorded by remote sensing satellites and the temperature changes all over the world were moving towards an all time high. The unprecedented incidents gave global climate change the due concern it deserved and it was the most highlighted issue of United Nations Conference on Environment and Development (UNCED) in Rio. [1] A landmark achievement towards bringing the majority of nations, 171 countries to be precise, to accept the issue of climate change as an imminent threat to the bare existence of human race has been the ratification of The Paris Agreement (CMA 1) on October 5, 2016 in Marrakech, Morocco.[2] Beyond environment protocols and beyond international treaties, there is a dire need to focus upon implementation of minor and doable actions in the mundane yet regular activities of our daily life, which have a direct impact on individual carbon footprint. The research establishes the success of the measures by comparing the scenarios before and after implementation of the measures to a *"Green Lifestyle"*.

1.2. The Rio Summit and Its constitution

The Rio Summit by the United Nations had 172 governments participating. The statistics are not as staggering as that of the recent Copenhagen Summit (COP15, 2009 UNFCC) or the Paris Agreement (Paris Climate Agreement, UNFCC, with effect from 4 November 2016), but it was still impressive, considering that fact that the issue was not as highlighted as it is today.

The details of the conference are as follows: [1] [2]

It was organized by Conference United Nations Conference on Environment and Development (UNCED) in Rio de Janeiro, 3-14 June 1992. The Earth Summit Host nation was Brazil. Number of Governments participating was 172, 108 at level of heads of State or Government. Conference Secretary-General was Maurice F. Strong, the Canada Organizers UNCED secretariat. Principal themes were “environment” and “sustainable development”. It was under the issue of environmental crisis that climate change was first ever addressed as a global issue. NGO presence was somewhere around 2,400 representatives of non-governmental organizations (NGOs), with 1700 people participating parallel.

The outcomes of the Rio “Earth” Summit can be summarized as the following:

1. Resulting document Agenda 21
2. The Rio Declaration on Environment and Development
3. The Statement of Forest Principles
4. The United Nations Framework Convention on Climate Change
5. The United Nations Convention on Biological Diversity Follow-up mechanisms

But the follow-up mechanism of the conference, which include the Commission on Sustainable Development, Inter-agency Committee on Sustainable Development and High-level Advisory Board on Sustainable Development proved to be both inefficient to force eco-friendly standards and irrelevant with the changing economic conditions.

1.3. The Protocols That Followed

“*Rules are meant to be broken*”. It is unanimously accepted by us when most of the times we find the rules too stringent to obey. The situation is identical for the developed nations. The only difference is that, the rules are not broken but bended around. After the much talked-about Rio Summit, a number of protocols followed. The most significant of them are as follows: [3] [4]

1. Kyoto Protocol on Carbon Emission Norms (1997, Kyoto, Japan):

It brought into existence the concept of “Carbon Trading” and constituted “Carbon Funds” which promoted the Annex-I countries to purchase carbon credits from the third world nations. Though its inception took place as early as 11th December 1997, yet it came into existence in 16th February 2005 as it required minimum 55 countries with minimum 55% of the total Carbon Dioxide (CO_2) emission.

In November 2009, it has 187 states who have signed it, aiming to reduce emission of 4 GHGs and other PFCs and HFCs.

2. Montreal Protocol on Substances that Deplete Ozone Layer (1989, Montreal):

The biggest ever ozone-hole was about the size of the surface area of North America and the population was subjected to the harmful UVB wavelengths (270–315 nm). The alarming rate of this depletion gave rise to the Montreal protocol which came up with projects to replenish the ozone layer. It came up with the following norms for CFCs:

A. From 1991 to 1992 the levels of consumption and production of the controlled substances in Group I of Annex A do not exceed 150 percent of its calculated levels of production and consumption of those substances in 1986.

B. From 1994 the calculated level of consumption and production of the controlled substances in Group I of Annex A does not exceed, annually, twenty-five percent of its calculated level of consumption and production in 1986.

C. From 1996 the calculated level of consumption and production of the controlled substances in Group I of Annex A does not exceed zero.

3. Vienna Convention on Civil Liability for Nuclear Damage (1963, Vienna, Austria):

The Vienna Convention came into being when nuclear energy was being considered as a probable energy alternative. It was highlighted specially after the nuclear leak in Chernobyl, Germany.

2. Expanding our Horizons beyond Protocols

2.1. Increased Environmental Stressors

The word "Stressor" is defined as *an activity, event or other stimulus that causes stress.* Outlined by the Encyclopaedia of Biodiversity, second edition, *Environmental Stress* is:

An action, agent or condition that impairs the structure or function of a biological system. Environmental stress can be either natural or anthropogenic (i.e. resulting from human actions) in origin.[5]

Amongst the basis of *characterization* of Environment Stress[6] the utmost challenging one has been the "*novelty*" of the stress; it signifies the damages are greater since the preparedness of the ecological system to combat a "novel" environmental stress is comparative much lower than widely known phenomena. With the world population estimated at 7.3 Billion in the year 2015[7], the impending surge is illustrated in figure.1.

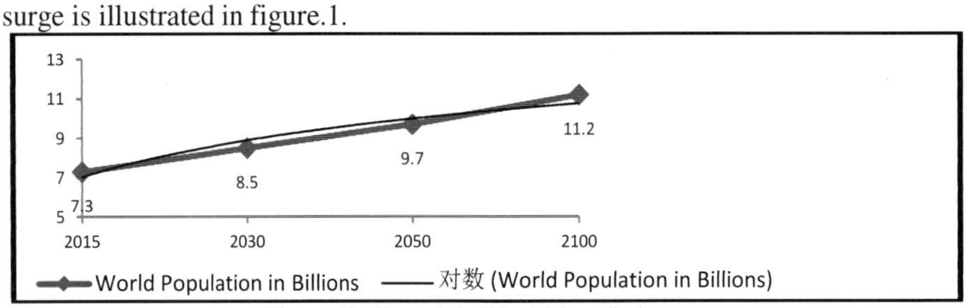

Figure 1. Growth in World Population

The surge illustrated in figure.1 in the world population will be compounding the "novelty" factor of Environmental Stresses since the natural balance mechanism of our ecosystem will be unable to cope with multiple changes *introduced* in the system; these may include generation of non-biodegradable wastes, green-house gasses, release of chemical compounds, exploring radioactive compounds for generating nuclear energy and various other sources of disturbing the equilibrium beyond scope of restoration. In such a hypothetical yet probable scenario in the not-so-distant future, protocols and preventive measures, which have existed beyond early 1960s, may result in being fractionally effective even with 100% implementation.

2.2. Objectives & Hypothesis of the Study

Instrumental in framing sustainable livelihood promotion and eco-friendly initiatives in the state of *Madhya Pradesh, India,* Xavier Institute of Development Action and Studies have been in a pioneering role to frame action plans which promote sustainable development. In an attempt to check the reality at ground zero, the research has taken the aforementioned institute as the universe of the study with the following objectives:

I. To identify areas of habitual and unintentional wastage of resources in XIDAS

II. To estimate and quantify the extent of wastage for impactful visualization

IOP Conf. Series: Earth and Environmental Science **146** (2018) 012012 doi:10.1088/1755-1315/146/1/012012

III. To recommend habitual changes designed to be implemented at individual level to develop a Green Lifestyle to mitigate Global Warming

The root cause of the extensive impact of global warming has been attributed to the human desire to keep raising the standard of living above subsistence level[5]. The Hypothesis hence formulated is:

H_0: There is exists no area to be identified, in which individual efforts can bring impactful changes to mitigate climate change

H_1: Designing a quantitatively impactful *Green Lifestyle* will result in measurable reduction of Carbon Footprint of an individual, result in mitigation of Global Warming.

3. Trends to Transform[≠]

3.1. Research Methodology

The research has been limited to Xavier Institute of Development Action & Studies[9] as the universe of the study, with a total population of 500. Sample size is 300, 60% of the population. Simple random sampling has been applied; data collection has been made using focused group discussions and questionnaire-based survey method.

3.2. Food Habits

We can incorporate an eco-friendly consumption pattern of edibles in our life-style which promotes sustainability in the following ways:

3.2.1. Eat to live and don't live to eat: The total population of the city of Jabalpur is 932484[10] and it classifies under C-Grade cities of India. Xavier Institute of Development Action and Studies is located in the *Tilhari Region* of the city with an on-campus population of 300 people.

Recommendation.1: Estimating our Diet: while the institute has a detailed meal-plan for its mess, it is recommended that a consumption plan is formulated as per the calorie required in an individual's diet, i.e. 2100 calorie/ day considering an urban population.

3.2.2. Distinguish between needs and wants: The consumption pattern of the canteen portrays despite 3-regular meals provided to the respondents per day, the 24X7 operational canteen provides quick-service refreshments and packaged food which can be curtailed if not absolutely abolished.

Recommendation.2: Student diet charts can be created, without the institute limiting itself to the mess food. By reflecting on the wasteful ways, we can reduce the rate of overconsumption. An institute promoting rural development and sustainability shouldn't be at the front-foot serving packed chips and cold-drinks. Thus there is a need to review the policies and the nutrition of the food being served in the canteen.

Figure.2 depicts the expenditure pattern and figure.3 depicts the consumption patterns of the respondents during November 2017 for the sample size of 300 respondents:

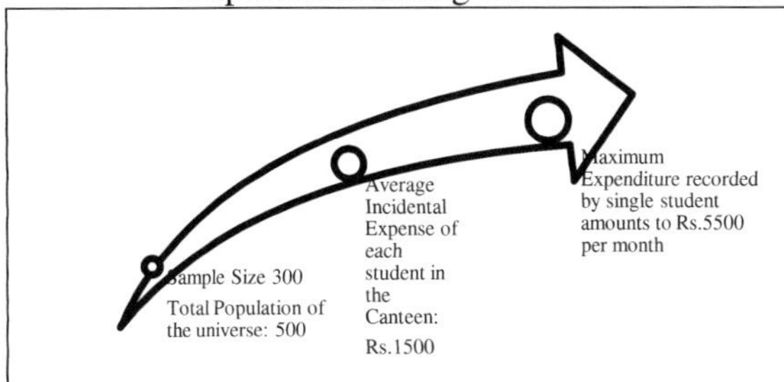

Figure 2. Average Consumption Pattern observed for a respondent in a month

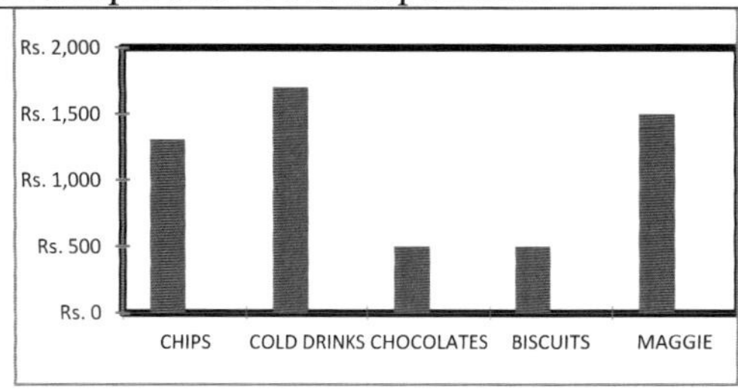

Figure 3. Monthly per Student Consumption Pattern of Specific Item

[≠]Trends mentioned are designed for individual efforts and not depicted for institutional policies.

3.2.3. Prevent wastage: When was the last time we didn't, though unknowingly, leave a few morsel of food on our plate? Mostly the quantity is so less, it doesn't pay or even matter to waste. But it seems immaterial when it is on an individual level. Simple mathematics and approximation can help us decide the magnitude of this problem:

Case-I

1. Total population of India = 1,156,897,766[10]
2. Collectively, if every Indian wasted a single grain of rice just for a single day of the year, it will be around = 1,156,897,766 grains.
3. The number of grains of rice per gram = 50[11]
4. Hence, 1,156,897,766 grains = 23137955.32 grams = 23138 Kilograms
5. The amount of *Annapurna* [12] Rice distributed to Public Distribution Scheme Card Holder, a Government of India initiative for Below Poverty Line population in India is 10 Kilogram per month.

Thus, a day's wastage of a single grain of rice can boost the PDS distribution of 2313 people for a month.

Case-II

1. Considering the present population of XIDAS, Jabalpur:
2. Total number of female respondents having dinner in the Ladies' mess every day = 38 (sample size of a single batch allowed entry)
3. The capacity of the waste bucket where food is dumped = 20 Litres
4. On an average, almost every day the bucket is half full, the daily approx. wastage amounting to 10 litres i.e. weighing 20 Kilogram of food on an average per day when weighed for 30 days consecutively.

Recommendation.3: Apart from sharing and sensitizing the respondents about the gravity of the situation as illustrated above, the institute can devise segregation of the wasted food from metal/plastic/non-biodegradable items and utilize it in portable compost bins. This can be of instrumental in maintaining the in-house plantation of the campus.

3.3. Water Crisis

It is a common site to see reservoirs and tanks overflowing almost every day and we never pay much heed to it. This wastage of fresh water can amount to a few gallons per household per year.

Case III:

The water wasted (per student in XIDAS, after lunch, once a day) while washing dishes, when stored in a container, amounted up to 200 ml, even if the tap is kept on in a medium speed.

This amounts up to a whopping 200 ml * 7 days = 1.4 litre per week per student.

Reccomendation.4: Instead of using normal taps, if fail-safe mechanism taps are used, only the amount of water required will be used by the respondents.

3.4. Implementing Energy Saving Initiatives

The institute has been a flag-bearer of renewable energy in the Jabalpur City, being the first to install solar water heaters in the year 2005. Another laudable initiative has been the *"Don't Stand-By, Just Turn It Off"* slogan, applied to the computer lab and the data center of the institute. Nevertheless, in a "*Green Lifestyle,* there remains a wide scope for further improvement.

Recommendation.5: Presently running with a 100% power back-up in the power-deficit state of Madhya Pradesh, India, it is recommended that the on-campus population is alerted through the on campus alarm/hooter once there is a switch-over from grid power to diesel generator sets. Individuals can thereafter selectively use electronic appliances which are required on priority; this will not only reduce the consumption of diesel but alert each user about the necessity to save power.

4. Conclusion

4.1. Calculating Our Contribution:

Carbon Footprint, by definition, is the amount of CO_2 or any other GHG emission caused by an organization, event or product. But as life-style changes focus on initiatives taken on an individual level, the technique of carbon foot-printing can be used to evaluate how much an individual is contributing towards the global phenomena. The carbon footprint of an individual can be broken down into primary and secondary footprint. Primary footprint of an individual is the direct emission of CO_2 by burning fossil fuels for consumption of energy. Secondary footprint is the indirect measure of CO_2 and other GHGs produced during the entire life-cycle of a product which is being used by an individual.

The recommendation made in the study can be mapped with the corresponding decrease in carbon footprint using the Carbon Footprint Calculator [13] as per the illustrations of Table.1.

Table 1. Estimation of Reduction in Annual CO_2 Emission with reference to Recommendations

Sr. No	Recommendation	Reduction in Annual CO_2 Emission
1	Recommendation.1: Estimating our Diet with Calorie Count	540 metric tons for avoiding food wastage / offsetting wasted food in to compost
2	Recommendation.2: Avoiding Overconsumption	
3	Recommendation.3: Converting Food-wastage to compost	
4	Recommendation.4: Installing Fail-Safe Taps	2 metric tons for avoiding 40 leaking taps
5	Recommendation.5: Minimum use of Power Back-ups	163 metric tons for avoiding 6 hours of Diesel-fired power back-up per day

4.2. Impactful Green Lifestyle

The quantification of CO_2 reductions as an effect of the recommendations, illustrated above nullify H_0 that there is exists no area to be identified, in which individual efforts can bring impactful changes to mitigate climate change.

Supported by a reduction of 3.25 metric tons of CO_2 per respondent per year, the working hypothesis H_1, stating that designing a quantitatively impactful Green Lifestyle will result in measurable reduction of Carbon Footprint of an individual and result in mitigation of Global Warming, is hence accepted. The cumulative effect of the recommendations made in the study show an estimated reduction of 705 metric tons of CO_2 emission for 300 respondents over a period of one year. The figure can be hypothecated as carbon sequestered by 18,271 tree seedlings grown for 10 years. [17]

Acknowledgments

I am thankful to **Dr. Uma Saha, Dean, Xavier Institute of Development Action and Studies** for her valuable mentorship, guidance and encouragement. Most importantly, I am thankful to all the respondents in XIDAS for their involvement and support during conducting the study.

References

[1] Mintzer I, Leonard A, 1994, *Negotiating Climate Change: The Inside Story of the Rio Convention,* (Cambridge: Cambridge University Press) p 17-24
[2] Durant R, Fiorino D, O' Leary R, 2017, *Environmental Governance Reconsidered: Challenges, Choices and Opportunities,* Second Edition, (Cambridge: The MIT Press) preface xix
[3] Zürn M, 1998, *The Rise of International Environmental Politics: A Review of Current Research World Politics*, Vol. 50, No. 4, **pp. 617-649**

[4] Goldberg D, 1991, *Technological Cooperation and the Montreal Protocol Multilateral Fund: A Brief Description*, (California: CIEL Publication) p 9

[5] Cairns J, 2013, *Encyclopedia of Biodiversity,* Second Edition, Vol. 7, **pp. 39-44**

[6] Harwell M., Kelly J, 1989, *Environmental Management: Indicators of Ecosystem Recovery*, Vol.14, p 527-545

[7] Department of Economic and Social Affairs of the United Nations Secretariat, 2015, *World Population Prospects: The 2015 Revision, Key Findings and Advance Tables*, (New York: United Nations), **pp. 1-3**

[8] World Energy Council, 2016, *World Energy Resources: Coal 2016*, (London: World Energy Council) **pp. 3**

[9] Girija, Rai C, 1998, *SME Sector Policy Perspective: Proceedings of The International Conference on Promotion, Policy, and Sustainability Of Micro, Small, and Medium Enterprises*, (Jabalpur: National Institute of Small Industry Extension Training), **pp. 61-64**

[10] Banthia J, 2009, *Census of India, 2001*, Vol. 1, (India: Controller of Publications), **pp. 151**

[11] Fageria N, Baligar V, Jones C, 1997, *Growth And Mineral Nutrients of Field Crops*, Third Edition, (U.S: CRC Press), p 303

[12] The Department of Food & Civil Supplies, Govt. of Delhi, 2009, *Economic Survey of Delhi 2008-2009*, (Delhi: Govt. Of Delhi), **pp. 224-227**

[13] United States Environmental Protection Agency, December 2017, retrieved from https://www.epa.gov/energy/greenhouse-gas-equivalencies-calculator

Construction Waste Prevention as a Sustainable tool in Building Mega Cities: A Theoretical Framework

AO Afolabi[1], PF Tunji-Olayeni[1], RA Ojelabi[1] and OI Omuh[1]
[1]Covenant University, P.M.B 1023, Ota, Nigeria

adedeji.afolabi@covenantuniversity.edu.ng

Abstract. In building of mega cities, the construction industry is contributing to alarming volumes of construction materials waste being generated causing an imbalance in the world's ecosystem. The study developed a framework for construction waste prevention strategies as a sustainable tool in building mega cities. Using content analysis of various literatures reviewed, the study developed a framework for preventing waste generation in construction projects in mega cities. In the building of present and future mega cities, there is need to submit mandatory construction waste prevention strategies as part of the contract documentation to the client, client representatives and the town planning authorities in order to build liveable and sustainable cities.

1.Introduction

Construction material resources go through several transformation from source before they are utilized in civil and building structures. These transformation exposes the drawbacks of construction materials in that not all these entities are utilized for the reasons for which they are requested. In this we realise that construction material waste is generated in construction delivery process, although, this study posits that construction waste may exist before the construction operation begins. General consensus has defined construction waste as the consequence of the processes of new developments, refurbishments and maintenance works in meeting the needs of mega cities[1].

As cities around the world are expanding, the magnitude and intricaciesin construction material waste is mounting [2]. The world as at 2017 has over 47 mega cities and intends to graduate many other cities to this status soon due to high urban migration. The state of these mega cities require sustainable solutions to manage the rising social and environmental challenges arising from the mega city status. Mega cities due to their uncontrollable population are plagued with high slum dwellers, high crime, homelessness, urban sprawl, traffic congestion, gentrification, air pollution, high energy and material consumption leading to uncontrollable waste generation. Therefore, there is need to consider the hindrances in creating liveable mega cities such as construction waste generated while meeting the needs of housing and urban development projects. In building more mega cities of the future, there is need to tackle waste prevention rather than dealing with waste generation. Just like the common adage "Prevention is better than cure".

The unbecoming image of the construction industry as the major consumer of earth's resources thereby generating 40% of the green-house gas (GHG) and producing unsatisfactory levels of waste is a major reason for discourse. For example, in Nigeria; a developing country generated over 6 million tonnes of construction waste in five years, in England; a developed country, the industry generates the highest waste compared to other sectors from the over 400 million tonnes consumed in the construction industry.This is evident from other international studies on the alarming rate of waste generation from the sector [3, 4], which is mostly linked to the loss of control in the design to construction processes. There is a necessity of treating materials wastage of the construction industry as part of the environmental problem, which, if not prevented or managed in a dedicated way, could lead to eco-imbalance in the world. Therefore, by understanding the sources of construction waste in the construction project delivery and the prevention techniques, the study would develop a framework for construction waste prevention as a sustainable tool in building mega cities.

2.Methodology

This paper brings together an in-depth review of ground breaking works on construction material waste by [1] and [2]. For this study, a desktop review was carried out which made use of content analysis to depict the result of the findings.

The study developed a framework for preventing construction material waste in the construction delivery processes of structures in mega cities. A lot of studies have been on waste generation and dealing with the waste generated in form of decreasing waste from source, reclaiming of construction waste, reprocessing, incineration and use of landfills. However, this study posits that in building mega cities, the best sustainable solution is hinged on prevention. As such, the study focused on prevention strategies that can be used as a sustainable tool. In other to understand the prevention strategies, the study identified the stages in the construction processes where waste could be generated as a result can be prevented using several strategies highlighted in this study.

3.Discussion

The study observed that waste prevention can be carried out from the root sources of waste generation based on six (6) major processes in building structures for mega cities. The stages were classified as design stage, procurement processes, construction material handling, construction operation, residual and other construction related causes of material waste. Figure 1 showed a network framework for construction waste reduction as a sustainable tool in building mega cities. Figure 1 showed that the construction processes from design to completion have certain inadequacies that could generate material waste but these can be prevented using the waste prevention strategies depicted in the figure.

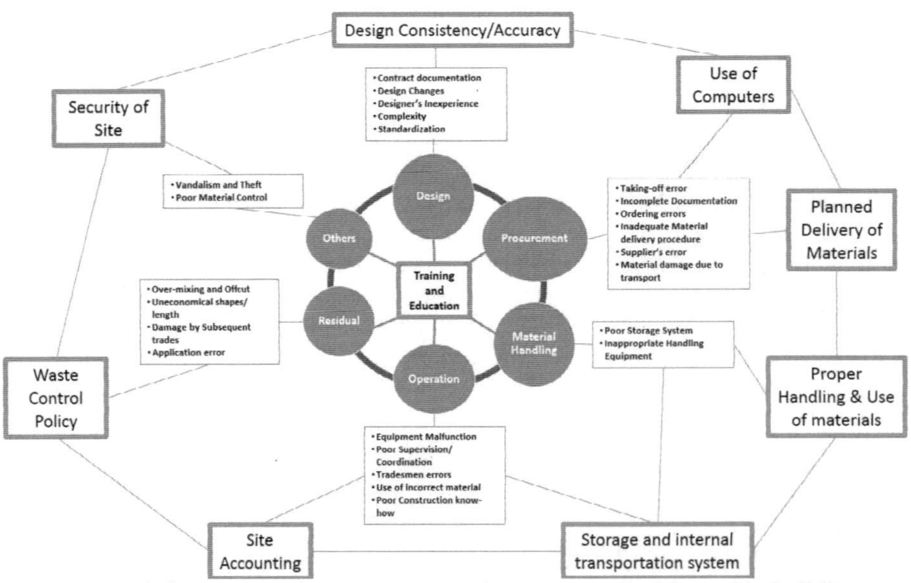

Figure 1: Framework for construction waste prevention as a sustainable tool in building mega cities.
Source: Author's Design

The research framework is explained further under the following headings;

3.1 Causes of materials wastage on construction sites.
[1]opined that the entire construction project delivery process is plagued with the production of construction wastes which can be linked to several factors creating them. Some of these factors are standalone while some of them are interwoven resulting in the generation of waste in the delivery process. However, for this study the causes that would be adopted will be the one proposed by [2]. The causes of material wastage whereby prevention techniques can be applied will include the following: (1) Design Stage (2) Material Procurement Process (3) Construction Material Handling (4) Construction Operation (5) Residual Related processes and (6) Other Construction activities.

3.1.1 Design Stage Design is the process of planning the form of a new object. It also refers to drawings of the proposed new object. The construction waste at this stage can be attributed to design errors due to designer's inexperience or carelessness, of which, site operatives or contractors may not be able to correct during the construction process. The carelessness at the design stage by the designer (either architectural, structural or services) can lead to demolition, rework or use of incorrect materials within the structure thereby resulting in huge pile of construction waste on the construction site [1]. Furthermore, errors in the details or specification can lead to huge economic loss to the client and the contractor. The following are means in which waste occurs from design: Failure to recognize undesirable consequences of a design decision; the inexperience of the specialist and lack of knowledge on the construction delivery process; difficulty by site operatives in understanding and translating complex details in working drawings; non-use of standardized products and dimensional variety of products; and time-constrained designs. The negligence and carelessness of the specialists at this stage in building structures for mega cities can also lead to building collapse which is a monumental waste,leading to loss of lives and economic value as witnessed in developing countries. Materials from collapsed buildings due to defective designs are a harm to the environment due to the resultant effect on landfills.

3.1.2 Material Procurement Process Procurement of materials is about obtaining the materials and monitoring of the project successfully. This stage is interwoven between the errors of the specialists, the estimators and the constructors. [1]highlighted the means material procurement errors result in waste as when estimators prepare incorrect bill of quantities due to inexperience or errors in the detailing of structures by the specialists. The critical nature of material procurement errors can be observed in that, over-ordering or under-ordering of construction materials have dire consequence on the project performance as well as the effect of generating construction waste. At this stage, experience and accuracy is key for construction professionals. It is worth to note that during transportation of construction materials that is ordered for use on the construction site, some of these materials may become damaged due to reckless driving, sub-standard quality of the materials or poor offloading techniques.

3.1.3 Construction Material Handling Materials wrongly handled or mishandled, even well stacked can easily be damaged. When construction workers are oblivious of the value a component of construction materials removes from the economic value of a structure it would lead to increase in construction waste. Apart from storage, when materials are stock piled in large quantities around the site, construction operatives may mishandle such materials. It is also important to use the right handling equipment for the appropriate material. The right equipment must be chosen as well as the operator must know how to handle such equipment in relation to the construction materials so as to avoid damage during movement. [1] stated that the level of construction waste increases when there are no storage spaces for materials on the construction site leading to more construction material handling errors. Due to the confined space, construction materials stored anywhere are subjected to adverse weather conditions such as rain washing away fine aggregates (sand), rusting of steel reinforcement and damage of other wood elements. Other expensive materials poorly stored can be imperilled to sabotage and pilfering by construction operatives. In addition, some of the construction materials could be improperly used by tradesmen.

3.1.4 Construction Operation Operational is used to tag the stage of a project by which material is directly applied and used to make up the components of the building. This segment is the main part of construction where proper controlling is needed. Where there is no control, the construction waste is bound to be high. [1]referred to this stage as the Construction or Renovation waste. Therefore, the contractor must apply a large portion of control over the available construction materials. At this construction stage, the materials are going through conversion and therefore there must be utmost care in ensuring human error as a result of using the construction materials is brought to the barest minimum. Project Managers, site supervisors and general foremen on the building or civil engineering projects must be up-to-date on their project management skills in supervising the activities of tradesmen and coordinating the material flow.

3.1.5 Residual The residual related processes are an off-shoot of the construction operation. These are more specific to the job location. At this stage there is need for close monitoring of materials that are brought out of the store and the purpose for which are to be used. Material substitution should be strictly avoided as this may lead to construction waste. There must be adequate protection for works already completed so as to avoid damage by subsequent tradesmen on the same spot. It is common sight on construction site to witness over-mixing of wet materials as in the case of concrete or mortar which are indiscriminately disposed.

3.1.6 Others Others in this context refers to other factors which may cause material wastage on a construction site. This may include: poor project information, site stacking and storage, inadequate material planning and control techniques and poor waste management policies by the constructor. There are other subtle construction activities that construction professionals are not aware that can result in waste generation on construction sites and during design.

3.2 Waste Preventive Strategies

Waste prevention is about sound judgement in practical matters, desire to generate results and a cultural change in design to construction of structures in mega cities rather than new technologies or a new field of endeavour. Often waste prevention options have no price tag to implement but give results straightaway when implemented with little or no exertion on the construction delivery process. Waste preventive measure would include the following:

3.2.1 Planned Delivery of Materials When material is delivered to the construction site, proper material handling is essential to reduce loss or decreased caused by exposure of material to external weather e.g. bags of cement to be stored should be covered with either polythene bag or plastic to ensure the protection of the material agent agents of moisture and they can also be placed on pallets to prevent moisture from the ground from penetrating the bags of cement. [5]suggested that there should be an inclusive use of innovative tools or technologies in planning construction activities and materials as well as storing the stock levels as work progresses on site. The project manager must ensure that the materials to be used on site are well calculated before ordering to avoid shortage or over-supply. By engaging computing technologies, construction managers can estimate construction materials to least waste generation on construction projects[5]. Some construction managers having arrived at an accurate material quantities may use 'just in time' deliveries to reduce the need for storage space and material losses due to theft and vandalism. In case of storage, construction materials should be stored away from areas of heavy traffic, passers' by and workers; allow only single handling for fragile materials and stick to manufacturer's instruction.

3.2.2 Site Activities Proper control of material used on site requires controlled supervision at every stage of operation. Many large company employ material quantity control supervisors who visit the site regularly to check on materials in store and the use of materials during construction or the client will engage a clerk of work to supervise the work (especially under a direct labour contracts) and ensure that all materials are judiciously utilized. [6, 7] noted that the use of unskilled labour in handling construction materials increases the risk of materials wastage to a critical level due to their lack of construction technical know-how andless value of the components. Besides waste of materials on site, there are lots of damages done due to lack of proper supervision. This is obvious in the average construction site is not the responsibility of anyone, as it is impossible to keep with the overall situation on site, particularly a large contract. Construction operatives should be encouraged to work together as a team in building structures in mega cities. Site operatives should discipline operatives for untidy stockpiles and mishandling of materials. The supply and distribution of materials should be discussed regularly at site meetings.

3.2.3 Training and Education The orientation of all stakeholders in the construction industry should be directed towards effective material utilization to minimize construction cost. This is a pivotal role in waste prevention from design to completion. The designer, contractors, construction managers and site operatives need to be educated on the best practices in building structures that are sustainable. Continuous professional development should be at the core of professional bodies and trade association in preventing waste on construction sites. Construction stakeholders should be educated on issues such as climate change, rising sea levels and the environmental consequences of their actions. The roles of supervision by qualified professionals on construction sites for structures in mega cities should not be neglected. The qualified construction professionals should possess up-to-date technical know-how on preventive measures to tackle waste generation on construction projects.

3.2.4 Design Consistency/ Accuracy Most construction waste arise from the faulty design erroneously prepared by an inexperienced or careless designer or specialist. In dealing with design inconsistencies, the design stage should be a collaborative stage where all construction stakeholders including the client make tangible contributions before the construction begins. The final design should be trouble shoot through a buildability analysis in order to detect deficiencies which will help prevent subsequent construction waste. A common waste prevention technique at the design stage is by optimizing the building dimensions to correspond to set dimensions available in the market i.e. standardization, making sure that the detail design is prepared before construction work commences to avoid a lot of variation orders been issued.

3.2.5 Construction Waste Management Policy Every construction project in existing and upcoming mega cities should have a construction waste management policy. The primary target of this waste prevention strategy is to at first ensure that contractors and sub-contractors are aware about construction waste, how to prevent them and how to deal with the resultant waste during the construction process. When the prior strategies discussed above are not judiciously carried out it may defeat the essential goal of the construction waste management policy. Although, when construction waste is eventually generated on the construction site, the aim of the policy is to ensure that three-quarter of the volume of waste is redirect away from landfills or incineration by reuse or recycling.

4. Conclusion

The study examined construction waste prevention strategies as a sustainable tool in building mega cities. It was observed that construction material waste can protrude from design stage, material procurement process, construction material handling, construction operation, residual related processes and other construction activities.In the design process, the study noted that preventive measures of design consistency and accuracy can be put in place against errors and inadequacies in contract documentation, design changes, designer's inexperience, design complexity and non-use of standardized products and dimensions. The study noted that in the procurement process the use of computers and planned delivery of materials can help curtail taking off errors by estimators, incomplete documentation, ordering errors, inadequate material delivery procedure, supplier's error and material damage due to transportation. During the material handling process, issues of poor storage system and inappropriate handling equipment can be prevented by using proper handling and correct use of materials and provision of adequate storage and internal transportation of materials on the site. In the construction operation, waste that could be generated from equipment malfunction, poor supervision and poor coordination, tradesmen errors, use of incorrect materials and poor construction know-how can be prevented through site accounting and increased training and awareness of the construction operatives. In the event of an application error, over-mixing of wet materials, offcut of materials and damage of materials, an effective waste management policy should be in place to prevent further materials wastage. The study showed that material wastage can arise from vandalism and theft of construction materials and this can be prevented through provision of adequate site security.

The study recommended that in the building of mega cities, there is need to put in place construction materials waste preventive strategies. There should be a conscious application of the waste preventive strategies. The framework developed showed that all the strategies revolve around training and education, therefore, construction stakeholders which include clients and professionals need to be adequate brought to speed on preventing waste generation rather than a "fire brigade approach" in dealing with construction waste. Seminars and workshops should be organized for construction professionals as a continuous professional development strategies. In the building of present and future mega cities, there is need to submit mandatory construction waste prevention strategies as part of the contract documentation to the client, client representatives and the town planning authorities.

Appreciation
The researchers appreciate the financial contribution of the Centre for Research, Innovation and Discovery (CUCRID), Covenant University, Ota, Nigeria in the publication of this article.

References

[1] Akinkurolere O O and Franklin S O 2005 Investigation into Waste Management on Construction Sites in South Western Nigeria *American J. of Applied* Sciences 2 5 pp 980-984

[2] Urio A F and Brent A C 2006 Solid waste management strategy in Botswana: The reduction of construction waste *J. of the South African Institution of Civil Engineering*48 2 pp 18–22

[3] Olofinnade O M Ndambuki J M Ede A N and Booth C 2017 Application of Waste Glass Powder as a Partial Cement Substitute towards more Sustainable Concrete Production *International J. of Engineering Research in Africa* 31 pp 77 - 93

[4] Akinwumi I and Booth C 2015 Experimental insights of using waste marble fines to modify the geotechnical properties of a Lateritic soil *J. of Environmental Engineering and Landscape Management* 23 2 pp 121-128

[5] Afolabi A Fagbenle O I and Mosaku T O 2017 IT Management of Building Materials' Planning and Control processes Using Web-Based Technologies Á Rocha et al eds*Recent Advances in Information Systems and Technologies Advances in Intelligent Systems and Computing* 570 p 12 – 19

[6] Afolabi A Emeghe IOyeyipo O and Ojelabi R 2016 Professionals' Preference for Migrant Craftsmen in Lagos State*Mediterranean J. of Soc. Sciences* 7 1 pp 501 – 508

[7] Ogunde A O Olaolu O Afolabi AOwolabi J and Ojelabi R2017 Challenges confronting construction project management system for sustainable construction in developing countries: Professionals perspectives (a case study of Nigeria)*Journal of Building Performance* 8 1 pp 1 – 11

IOP Conf. Series: Earth and Environmental Science **146** (2018) 012014 doi:10.1088/1755-1315/146/1/012014

A Multi-objective Optimal Model for Groundwater Remediation under Health Risk Assessment in a Petroleum contaminated site

A.L. Yang [1]*, Z.N. Dai [1], Q. Yang [2], E.A. Mcbean [3] and X.J. Lin[1]

[1]College of Environmental Science and Engineering, Xiamen University of Technology, Xiamen 361024, China.
[2]Huzhou Port and Shipping Authority, Nanhua road No.106, Huzhou 313000, China.
[3]School of Engineering, University of Guelph, Guelph, Ontario, Canada

* Correspondence: yangaililoy@xmut.edu.cn; Tel:+86- 0592-6291620

Abstract. This study proposes a multi-objective model (i.e. MOHRA) for groundwater remediation in a petroleum-contaminated site under health risk assessment. The optimization system is designed based on the pump and treat technology considering both cost and human health risk, which can provide reliable groundwater remediation strategies.

1. Introduction
Groundwater contaminated by petroleum products poses significant threats to human health and natural environment. Pump-and-treat (i.e. PAT) is one of the most widely used technologies in the petroleum polluted groundwater remediation process [1]. However, remediation of such contaminants involves large expenditure and time [2]. Therefore, simulation/optimization models have been used for determining optimal remediation strategies [3-7]. The simulation models are used to simulate contaminants whereabouts under various conditions, while, the optimization models are used to enhance design of management actions [8].

Challenge may emerge because any remediation decision can hardly ensure the constraints to be satisfied in health risk and the minimization of remediation costs. Multiple objectives optimization method may be applied to address this problem[9]. Thus, under this consideration, a multi-objective optimal model under health risk assessment (i.e. MOHRA) is developed in this study for optimal remediation design in western Canada. In the approach, the contaminant transport processes under remediation operations will be simulated and subsequently is replaced with a regression model. A multi-objective programming model considering health risk will be formulated to seek the optimal remedial design strategies.

2. Methods and materials

2.1. Optimization Objective
The optimization objective of a PAT system includes the system fixed cost, operation costs,

contaminant environmental standards and human health risk. In the PAT system, the costs of one-time drilling and installation are significant less when comparing with operation costs. Hence, the optimization function can be described as follow:

$$\text{Min} \quad f_1 = \sum_{i=1}^{I} q_i^{In} + \sum_{j=1}^{J} q_j^{Ex}$$

$$f_2 = \left(\sum_{k=1}^{K} ELCK_k \right) \Big/ K \tag{1}$$

where f_1 is total pumping rate at all injection and extraction wells, m^3/hr; I , J are numbers of injection, extraction wells respectively; q_i^{In} is pumping rate for the ith injection well, m^3/hr; q_j^{Ex} is pumping rate for the jth extraction well, m^3/hr.

2.2. Constraints of the PAT Optimization Model

In addition to human health risk, PAT system may be impacted by engineering techniques and hydrogeological factors. Hence, the constraints for the PAT optimization model can be described as follow:

$$c_k(q_i^{In}, q_j^{Ex}) \le c_{\max} \quad \text{k=1,2, \ldots K} \tag{2}$$

$$0 \le q_i^{In} \le q_{\max}^{In} \quad \text{i=1,2, \ldots I} \tag{3}$$

$$0 \le q_j^{Ex} \le q_{\max}^{Ex} \quad \text{j=1,2, \ldots J} \tag{4}$$

$$\sum_{i=1}^{I} q_i^{In} = \sum_{j=1}^{J} q_j^{Ex} \quad \text{i=1,2, \ldots I; j=1,2, \ldots J} \tag{5}$$

$$ELCR_k \le ELCR_{\max} \quad \text{k=1,2,\ldots K} \tag{6}$$

$$ELCR_k = SF \times CDI_k \tag{7}$$

$$CDI_k = CW_k / 1000 \times IR \times EF \times ED / (AT \times BW) \tag{8}$$

$$c_k = a_{0,k} + \sum_{i=1}^{n} a_{i,k} q_i + \sum_{i=1}^{n} \sum_{j=1}^{n} a_{ij,k} q_i q_j \left(i \ne j\right) + \sum_{i=1}^{n} a_{ii,k} q_i^2 + e_k \tag{9}$$

where $q_{\max}^{Ex}$ for the ith injection well and $q_{\max}^{Ex}$ for the jth extraction well,are maximum pumping rates, m^3/hr; k is the number of monitoring wells; c_k is contaminant concentration of the kth monitoring well which has been remediated, μg/L, and by a inference, a simulation, and the optimization method proposed by He et al. (2008)[10], the alternative of c_k would be obtained; c_{max} is the maximum receivable contaminant concentration, μg/L; $ELCR_k$ is carcinogenic human health risk at k^{th} monitoring well, $ELCR_{\max}$ is the maximum acceptable carcinogenic human health risk value; *SF* is slope factor which is related to carcinogenic health risk, $(mg/kg\cdot day)^{-1}$; CDI_k is daily intake of given contaminant at exposure location k, mg/kg·day; CW_k is concentration of exposed contaminant at exposure location k, μg/L; *IR* is ingestion rate, L/d; *EF* is exposure frequency, d/year; *ED* is exposure duration, year; AT is average time , d; BW is body weight, Kg; $a_{0,k}$ is an intercept term of surrogate k; $\sum_{i=1}^{n} a_{i,k} q_i$ are linear terms of surrogate k; $\sum_{i=1}^{n}\sum_{j=1}^{n} a_{ij,k} q_i q_j \left(i \neq j\right)$ are interaction terms of surrogate k; $\sum_{i=1}^{n} a_{ii,k} q_i^2$ are quadratic terms of surrogate k; e_k is the error of surrogate k; n is the number of explanatory variables.

2.3. Model solution

In order to solve the multi-objective problem, the liner weighting method may be adopted. Each single objective's weight can be obtained by the following steps: 1) normalize the objective functions; 2) find the pseudo nadir f_i^+ and anchor point f_i^-, for each single objective; 3) calculate each objective's weight by using the following formulas[11]:

$$\alpha_1 = \frac{f_2^+ - f_2^-}{f_2^+ - f_2^- + f_1^+ - f_1^-} \tag{10}$$

$$\alpha_2 = \frac{f_1^+ - f_1^-}{f_2^+ - f_2^- + f_1^+ - f_1^-} \tag{11}$$

And then, the multi-objective model can be transformed into the single objective, which can be solved through MATLAB.

$$\text{Min} \quad U(q) = \frac{(f_1^+ - f_1^-) f_2(q) + (f_2^+ - f_2^-) f_1(q)}{f_2^+ - f_2^- + f_1^+ - f_1^-} \tag{12}$$

3. Case Study

In western Canada, the MOHRA model is applied to a petroleum-contaminated site. The aquifer has been remediated many years with duel phase vacuum extraction method[10]. Since contaminated plume still found in the monitoring well, a PAT system, is applied, with 2 injection and 4 extraction wells, in or around the contaminant plume. The simulation domain, with an area of 270×225 m^2 and a depth of 10 m, is shown in Figure 1, which is composed of 9720 finite difference grids (in the directions of x, y, z, each grid has dimensions of 5, 5, and 2.5 m, respectively).

IOP Conf. Series: Earth and Environmental Science **146** (2018) 012014 doi:10.1088/1755-1315/146/1/012014

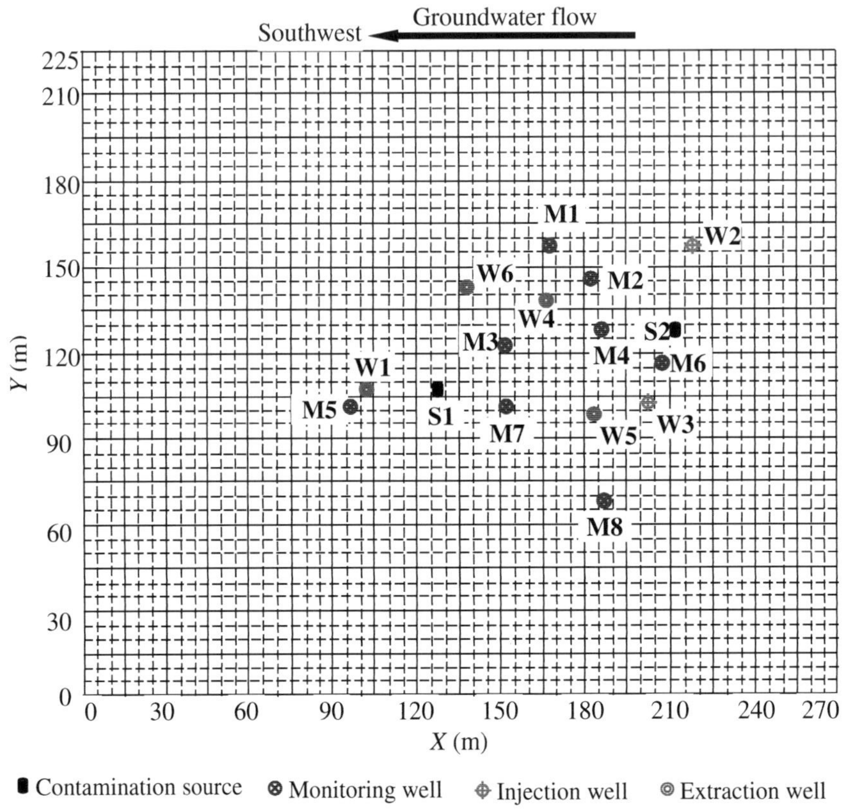

Figure 1. The simulation domain (He et al. 2008)

In this study, 5 years remediation duration is selected and benzene contaminant is considered as target contaminant in the optimization model. The acceptable carcinogenic health risk of benzene is 0.00002, *AT* is 10750 days and *BW* is 70Kg.

4. Results analysis

When the environmental standard is 150 μg/L and the benzene carcinogenic health risk level is 0.00002, at six wells, the optimization pumping rates are 100, 89.4, 84.4, 0.0, 73.2, and 0.0 m^3/hr, respectively. It shows that at wells 4 and 6, the pumping rates are zero, which means that 4 wells are enough, and also leads to less cost through this optimization model.

Figure 2 shows the predicted benzene carcinogenic risk level under 5-year remediation durations. The risk level is higher in the center area compared with the boundary area, which may be impacted by the contamination sources. The peak health risk is 0.000019 (approximate to 0.00002). It means that, the contradiction between remediation cost and remediation effect is effectively solved by the multi-objective model.

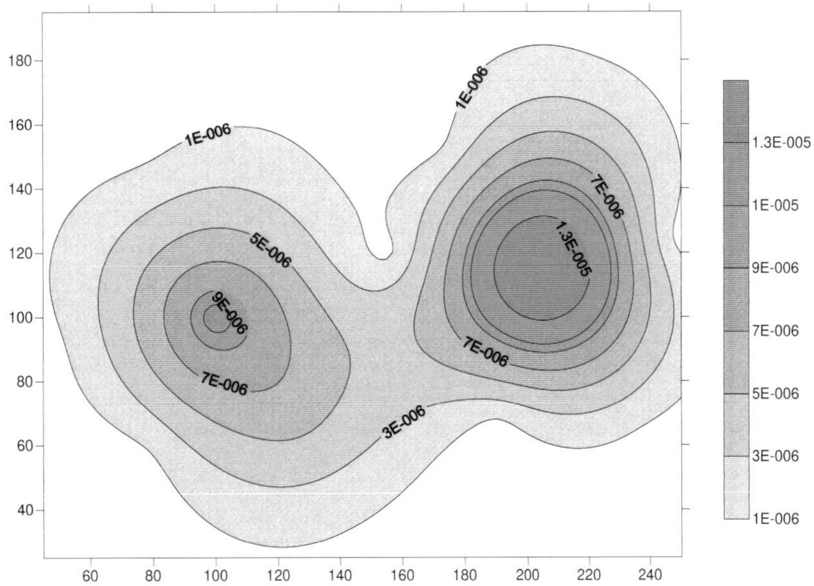

Figure 2. Predicted risk levels for 5-year remediation duration when environmental standard is 150μg/L and acceptable health risk level is 0.00002

5. Conclusions

In this study, a multi-objective model (i.e. MOHRA) was developed under health risk assessment for groundwater remediation in a petroleum-contaminated site. There are two advantages in comparison with other models: (1) the relationship between contaminant concentrations and pumping rates were obtained by a series of surrogate models; (2) both health risks and remediation cost were considered in the model. MOHRA can help systematically analyze interaction among contaminant transport, remediation technology, environment and health factors, and provide useful optimization information for decision makers.

Fund:
Type A from Fujian Educational Committee with Grant JA15371; Xiamen Science and Technology Project 3502Z20179035

References

[1] Kim, J.W. Optimal pumping time for a pump-and-treat determined from radial convergent tracer tests. Geosciences Journal, 2014, 18: 69-80.
[2] Singh T.S., Chakrabarty D. Chance-constrained multi-objective programming for optimal multi-layer aquifer remediation design. Engineering Optimization, 2011,43 (4): 417-432.
[3] Zheng C., Wang P.P. A field demonstration of the simulation–optimization approach for remediation system design. Ground water, 2002, 40 (3): 258–66.
[4] Kalwij I.M., Peralta R.C. Simulation/Optimization Modeling for Robust Pumping Strategy Design. Ground Water, 2006, 44 (4): 574-582.
[5] Baú D.A., Mayer A.S. Stochastic management of pump-and-treat strategies using surrogate functions. Advances in Water Resources, 2006, 29 (12): 1901–17.
[6] He L., Huang G.H., Lu H.W. A A coupled simulation-optimization approach for groundwater remediation design under uncertainty: An application to a petroleum-contaminated site. Environmental Pollution, 2009,157: 2485-2492.

[7] He L., Huang G.H., Lu H.W. A stochastic optimization model under modeling uncertainty and parameter certainty for groundwater remediation design-Part 1. Model development. Journal of Hazardous Materials, 2010, 176: 521-526.
[8] Qin, X.S., Xu, Y. Analyzing Urban Water Supply through an Acceptability-Index-Based Interval Approach, Advances in Water Resources, 2011, 34(7): 873-886.
[9] Thangjam S.S, Dibakar C. Multi-objective optimization for optimal groundwater remediation design and management systems, Arabian Journal of Geosciences, 2010, 14 (1): 87-97.
[10] He L., Huang G.H., Zeng G.M. An integrated simulation, inference, and optimization method for identifying groundwater remediation strategies at petroleum-contaminated aquifers in western Canada. Water Research 42: 2629-2639
[11] Yang Q., Shen J., Fan X., Shen Y.S. et al. A multi-objective optimal model for a petroleum-contaminated aquifer groundwater remediation in western Canada. Advanced Materials Research, 2012, 518-523: 2054-2057.

Measurement and Evaluation of Residual Disinfection by Products in Tap Water from Xiamen

Lin Xiajing, Yang Aili*, Chen Fuxiang

Xiamen University of Technology, College of Environmental Science and Engineering, Xiamen, China 361024

* Correspondence: yangaililoy@xmut.edu.cn

Abstract. This paper has tested typical anions including F^-, Cl^-, NO_3^-, BrO_3^-, ClO^-, SO_4^{2-}in tap water from Xiamen in order to understand the residual status of disinfection byproducts. Ion chromatography is used to analyze anionsconcentrations from residential and industrial zones in six Districts of Xiamen. The results demonstrate that: (1)BrO_3^- is not detected in all samples, (2) NO_3^- and SO_4^{2-} exceed 4.046mg/L and 2.223mg/L, the peakconcentrations are 14.342mg/L and 39.345mg/L, respectively, (3) NO_3^- and SO_4^{2-} are quite high in Haicang, Siming and Huli districts, which exceed 14.031mg/L and 36.708mg/L, respectively, (4) Smallerresidual disinfection byproductsare found in the tap water near the industrial areas of Jimei, Tongan and Xiangan Districts compared to other districs, all criteria meet the national standard, which means water quality is in a good condition in these three areas.

1. Introduction

According to the WHO survey: 50% of the world's children's death, and 80% of the disease is more or less related to the quality of tap water[1]. Therefore, it is self-evident to control the effluent quality of tap water. Typically, most of the anions in tap water are closely related to physical health, including fluoride ion, chloride ion, bromate, nitrate, sulfate and hypochlorite ion[2]. These ions are also the routine hygienic indexes of drinking water [3]. The main water source in Xiamen is JiuLong River water[4], and the water quality of primary water supply, secondary water supply, and tap water near the industrial areas in six districts are mainly analyzed in this paper.

2. Materials and Methods

2.1. Instruments and Reagents

2.1.1. Instrument Ion chromatograph (metrohm 883), analytical balance, pipette, ultra pure water machine, suction machine.

2.1.2.Main Reagent Sodium chloride (analytical purity), sodium nitrate (analytical purity), sodium bromate (analytical purity), sodium fluoride (analytical purity), sodium sulfate (analytical purity), sodium hypochlorite (analytical purity), sulfuric acid, sodium carbonate, sodium bicarbonate, 0.22 M filter membrane.

2.2.Methodology

2.2.1.Ion chromatography Chromatographic conditions[5-7] : Swiss Wantong type 883 chromatograph, chromatographic column is Metrosep ASUPP 5150 anion chromatographic column, eluent 3.2mmol/LNa_2CO_3+1mmol/L $NaHCO_3$, 1mL/min flow rate, 27 degrees column temperature, 20 L injection volume.

2.2.2.Sampling point distribution principle According to the characteristics of six districts in Xiamen, the principle of random sampling was adopted to ensure that sampling points should have a certain representation and could reflect the water quality of the region. The distribution of sampling point can be shown in Figure 1.

Figure 1. Water Sampling

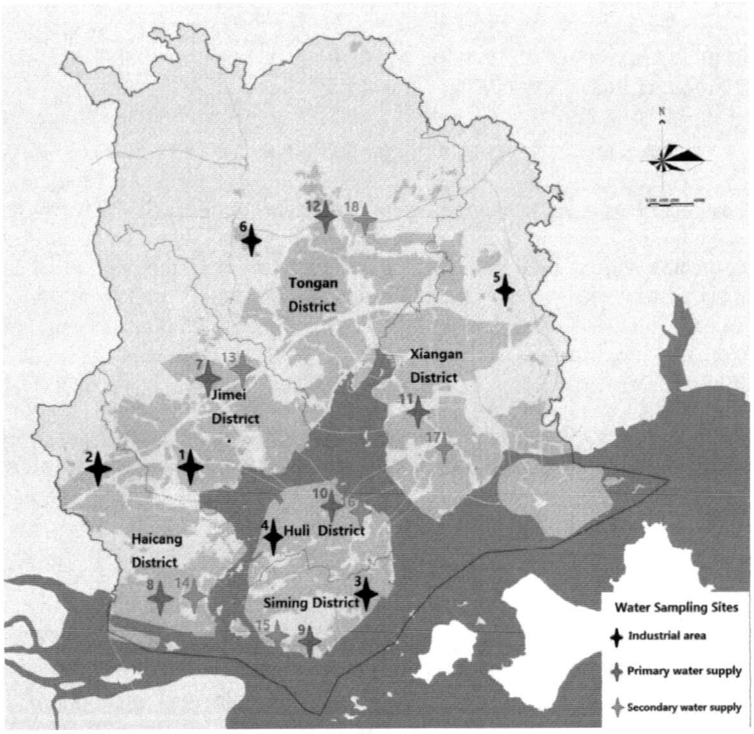

Primary water supply sample, secondary water supply sample, and tap water sample near the industrial areas were collected in six districts. Primary water supply sample was collected in the residential areas(50% of the whole). In Siming and Huli Districts, the software parks or various office buildings were considered as the water sample points of the Industrial areas, while the water samples from the center of the entire industrial areas are the figure of the rest of the industrial areas from the other four districts. The secondary water supply was defined as centralized water supply mode, which must be stored again before entering the household, and then with pressure and disinfection or advanced treatment, transferring water into the users' residence through pipes or containers[8].

2.2.3.Establishment of standard curve Standard solution preparation: accurately weighting NaCl, $NaNO_3$, $NaBrO_3$, NaF, Na_2SO_4 and NaClO each 1g in 6 dry beakers, placed in dryer drying 2h. Took out the beaker and cooled it. Dissolved in ultrapure water, then moved into 1000mL capacity bottle, diluted to the capacity bottle mark, shook well. The concentration of the reserve solution of the anion was 1g/L, and the solution was the reserve one. The standard solution of six anions was diluted from the stock solution.

Linear equation: Linear regression was carried out for six different anionic standard solutions, After the ion chromatography conditions were selected. Taking the concentration as abscissa, the peak area as ordinate, drew the linear regression equation of different anions.

3. Results and Analysis

3.1.Result and deviation of standard curve

When the eluent concentration was 3.2mmol/LNa_2CO_3+1mmol/L$NaHCO_3$, the flow rate was 1mL/min, the column temperature of the Swiss Wantong type 883 ion chromatography was generally set to27 ℃ , and the injection volume was generally set to 20 L. The linear equation and correlation coefficient are shown in Table 1.

Table 1. Six anionic linear regression equation

anion	Range/ (mg/L)	linear equation	correlation coefficient/ R
F^-	0.01-0.5	y=4.7741x-0.0403	0.9995
BrO_3^-	0.5-1.5	y=0.5364x-0.253	0.9998
ClO^-	0.1-2	y=3.7678x-0.2414	0.9997
Cl^-	0.1-5	y=0.1164x+0.0062	0.9999
NO_3^-	0.25-10	y=0.0709x-0.0039	0.9999
SO_4^{2-}	0.2-16	y=0.0629x+0.0968	0.9997

Table 2. The detection limit of six anions

anion	detection limit / (mg/L)
F^-	0.001
BrO_3^-	0.002
ClO^-	0.004
Cl^-	0.001
NO_3^-	0.002
SO_4^{2-}	0.001

Different concentrations of anions were analyzed in the same chromatographic conditions. It was not difficult to observe the test results through the standard curve of Table 1 and various anions: F^- at 0.01-0.5mg/L, BrO_3 at 0.5-1.5mg/L, ClO^- at 0.1-2mg/L, Cl^- at 0.1-5mg/L, NO_3^- at 0.25-10mg/L, SO_4^{2-} had a good linear relationship in the range of 0.2-16mg/L. The correlation coefficients of each ion were 0.9995-0.9999.

Detection limit: the concentration of the standard solution was close to the blank value. Injected the solution continuously for 10 times , the detection limit, was the standard deviation of 3 times response value. The results are shown in Table 2.

The experimental results showed that the detection limit of the anion was nearly 10 times lower than that of the national drinking water standard (GB/T5750.5-2006) [9].

Accuracy: the 1.00mg/Lstandard mixture solution was used to calculate the relative standard deviation of anions. The final results are shown in Table 3.

Table 3. Accuracy

anion	Measured value (mg/L)								relative standard deviation / (%)
F^-	1.032	1.025	1.013	1.025	1.003	1.007	1.032	1.026	1.10
BrO_3^-	0.994	1.011	0.998	1.012	1.015	1.027	1.012	1.015	1.02
ClO^-	0.998	0.987	1.005	1.028	1.032	1.011	1.013	1.002	1.49
Cl^-	1.026	0.987	0.998	1.003	1.015	1.036	1.019	1.021	1.58
NO_3^-	1.024	1.008	1.036	1.003	1.027	1.002	1.024	1.015	1.21

SO_4^{2-}	1.011	1.012	1.003	1.022	1.026	1.013	0.998	0.994	1.10

Through the table above, it was found that the method which was used in this paper met the requirements of drinking water according to GB/T 5750.5-2006.

3.2.Overview of residual disinfestion byproducts in tap water

After the pretreatment of collected eighteen samples, making use ofsuction filtrationwith 0.22m membrane to filter the water. The filtrationsamples were analyzed by ion chromatography and the anions concentrations were determined by the peak height, peak area and peak time of the six anions. The analysis results are shown in Table 4.

Table 4. Water sample

region	Ion concentration / （mg/L）					
	F^-	BrO_3^-	ClO^-	Cl^-	NO_3^-	SO_4^{2-}
Water supply samples of Jimei Industrial Zone	0.014	Below detection limit	0.346	0.068	5.936	5.826
Secondary water supply samples in Jimei district	0.014	Below detection limit	0.313	0.068	5.569	4.270
Primary water supply samples in Jimei district	0.014	Below detection limit	0.437	0.094	5.795	5.089
Water supply samples of Haicang Industrial Zone	0.029	Below detection limit	0.536	0.077	16.175	36.708
Secondary water supply samples in Haicang district	0.030	Below detection limit	0.551	0.068	16.104	38.362
Primary water supply samples in Haicang district	0.029	Below detection limit	0.565	0.068	16.302	39.345
Water supply samples of Siming Industrial Zone	0.030	Below detection limit	0.546	0.068	14.342	38.198
Secondary water supply samples in Siming district	0.023	Below detection limit	0.572	0.094	14.186	38.624
Primary water supply samples in Siming	0.030	Below detection limit	0.568	0.068	14.031	38.264

district						
Water supply samples of Huli Industrial Zone	0.026	Below detection limit	0.587	0.077	14.313	38.100
Secondary water supply samples in Huli district	0.029	Below detection limit	0.564	0.068	14.553	37.396
Primary water supply samples in Huli district	0.027	Below detection limit	0.596	0.008	14.962	37.822
Water supply samples of Xiangan Industrial Zone	0.012	Below detection limit	0.257	0.059	4.766	3.009
Secondary water supply samples in Xiangan district	0.031	Below detection limit	0.551	0.068	13.933	37.298
Primary water supply samples in Xiangan district	0.026	Below detection limit	0.569	0.068	14.045	39.181
Water supply samples of Tongan Industrial Zone	0.012	Below detection limit	0.277	0.025	4.342	2.698
Secondary water supply samples in Tongan district	0.013	Below detection limit	0.441	0.068	5.767	6.055
Primary water supply samples in Tongan district	0.013	Below detection limit	0.255	0.042	4.046	2.223

The concentration of F^- and BrO_3^- was extremely low, so the data of ClO^-, Cl^-, NO_3^- and SO_4^{2-} were extracted from Table 4. The chart example results are as follows.

3.3.Sampling from industrial zones

Figure2. Concentrations of ClO^- and Cl^- in industrial area

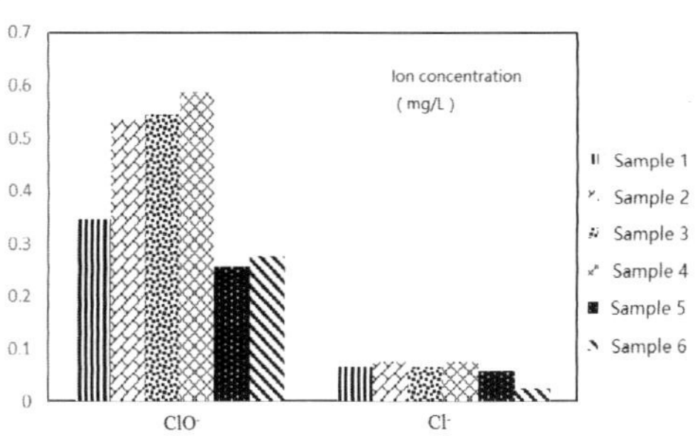

Figure3. Concentrations of NO^{3-} and SO_4^{2-} in industrial area

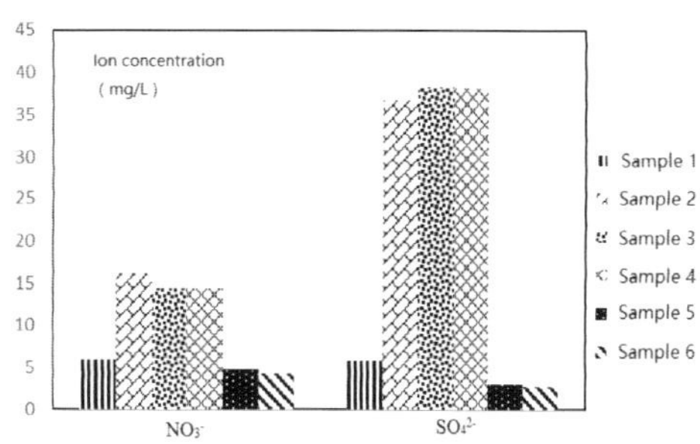

Figure 2 and figure 3 show that the water pollution in industrial areas of Siming, Haicang and Huli Districts was relatively serious. SO_4^{2-} in Siming District was comparatively high (38.198mg/L) compared with other districts. Besides, the highest NO_3^- was found (16.175mg/L) in Haicang District, which had exceeded 50% of the national standard, while the Chlorine was evenly distributed in different industrial areas.

3.4.Sampling of residential areas

Figure 4. Concentrations of ClO^- and Cl^- in Secondary water supply

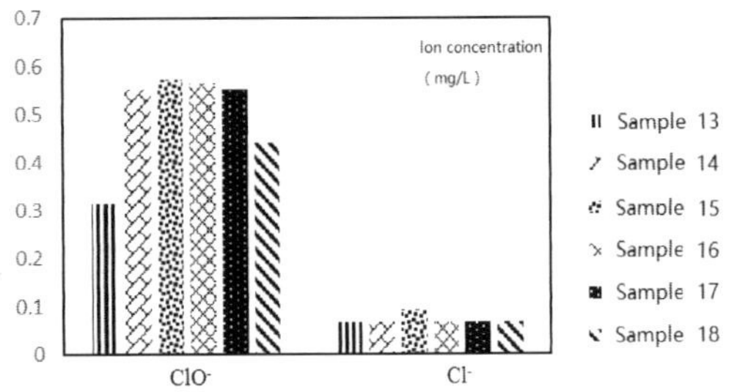

Figure 5. Concentrations of NO^{3-} and SO_4^{2-} in Secondary water supply

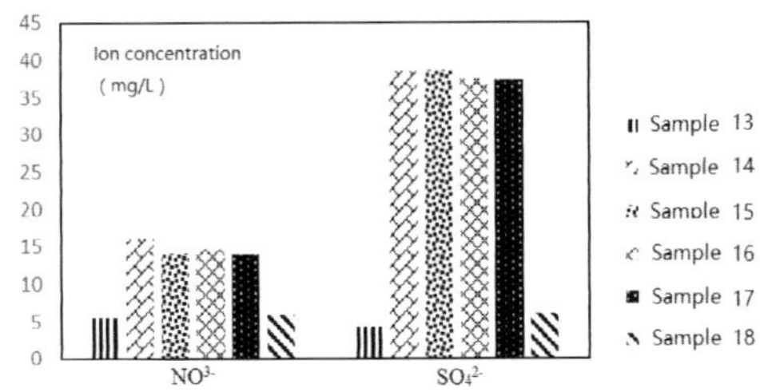

Through figure 4 and 5, the lowest concentration of SO_4^{2-} and NO_3^-were found in the secondary water supply samples of Tongan and Jimei districts, which had met the national water quality standard. $ClO^-$and Cl^- in the secondary water supply samples was in average level.

In particular, little difference of parameters can be found in water samples between primary water supply and secondary water supply.

3.5.Distribution characteristics of anions in different districts

1) In Jimei District, the disinfection byproducts were much less, all the indexes met standard, water quality was in good condition;

2) In Haicang District, the concentration of SO_4^{2-} and NO_3^-in primary water supply was relatively high, up to 39.345mg/L and 16.302mg/L, respectively, while NO_3^-exceeded the standard (10mg/L) ;

3) In Siming District, the concentration of four kinds of anions was in average level, but SO_4^{2-} was comparatively high, three samples were all above 38.194mg/L;

4) In Huli District, Cl^- in primary water supply was much lower (0.08mg/L), which was suitable for residents as drinking tap water. But the number of NO_3^- was still higher than 10mg/L, overtaking the standard;

5) In Xiangan District, Industrial Zone, where 4 kinds of ions had low amount of SO_4^{2-} (3.009mg/L) and NO_3^-(4.766mg/L), which illustrated that the water quality was much better than the figure for primary or secondary water samples;

6) In Tongan District, SO_4^{2-}, NO_3^- and Cl^- detected in secondary water supply samples were significantly higher than that of primary water supply and industrial zone.

3.6.Reasons for the difference in distribution

Research shows that the concentration of ClO^- and Cl^- is more reasonable, this is may due to the fact that the tap water is disinfected by the amount of Cl^-which is controlled by the waterworks, and then through the network transmission, the secondary compression process, residual chlorine is consumed in the interaction of chemical conversion, but not less than 0.05mg/L, which can reach the standard of drinking water.

SO_4^{2-} detected in Haicang, Siming and Huli district is significantly high, which may be caused by : these three districts was developed relatively early, the pipeline buried in the construction areas had been damaged to a certain extent. It could cause the waste water and some special materials infiltrate into the pipeline. At the same time, SO_4^{2-} as industrial waste, by the form of atmospheric deposition, is discharged into the water body, which lead to a significant number of SO_4^{2-} in these three districts.

The three districts of $NO_3^-$is much higher which may be caused by: The main tap water sources of Haicang, Siming and Huli District originate from 40km outside of JiuLong River water, mainly for surface water supply. And there is a little NO_3^- in this type of water, which is difficult to remove in purification.

4. Conclusion

Under the evaluation of the residual status of disinfection by-products including F^-, Cl^-, NO_3^-, BrO_3^-, ClO^-, SO_4^{2-} in tap water of main districts, in Xiamen city, by using ion chromatography, the results of the figure illustrate that:the disinfection byproducts in three kinds of water samples in all districts of Xiamen show the trend : SO_4^{2-}> NO_3^-> ClO^-> Cl^-> F^-> BrO_3^-.

1) the concentrations of BrO_3^- in all sampling sites of six districts are less than the detection limit, which means BrO_3^-in Xiamen city meet the standard. 2) The residual chlorine in tap water is free residual chlorine (ClO^-), the concentration of which should be controlled at 0.05~0.8mg/L, and the test results are in line with the standard. 3) Although SO_4^{2-} is quite high, especially in Haicang, Siming and Huli districts, where all the figure exceed 36.708 mg/L, and one of which reach its peak at 39.345 mg/L in Haicang District, all SO_4^{2-} detected in these areas are within the standard value of 250mg/L. 4) F^-, BrO_3^-, ClO^-, Cl^-, SO_4^{2-}, these five kinds of ions are within the standard range, while NO_3^- overtakes the national standard concentration (10 mg/L) in Haicang, Siming and Huli districts, which exceeds 14.031mg/L, and the peakconcentration is 14.342mg/L. 5) Smallerresidualare found in the tap water near the industrial areas of Jimei, Tongan and Xiangan Districts compared to other districts, all criteria meet national standard, which means water quality is ingood condition in these three areas.

Therefore, from the result of test, the relevant departments should attach importance to it, who shall arrange personnel to renovate and manage pipe network regularly, so as to reduce the possibility of tap water pollution. Besides, dispatching appropriate water sources in different areas to ensure the water safety should be taken seriously.

Acknowledgments:
Type A from Fujian Educational Committee with Grant JA15371; Xiamen Science and Technology Pr oject 3502Z20179035

Reference

[1]WHO, 2012 World Health report[R]. WHO press, 2012
[2]Liu Z., Cui Y., Hu Z.J., etal. Determination of Trace Bromate in water by ultraviolet spectrophotometry[J]. Chinese inorganic analytical chemistry, 2012,2 (2): 14-16.
[3]Ministry of Health, PRC, China National Standardization Administration Committee,

GB/T5749-2006 sanitary standard for drinking water, Beijing, China Standard Press, 2007.
[4]Survey of surface water in JiuLongjiang River Basin, Xiamen municipal water resources bulletin[N]. Xiamen Water Conservancy Bureau, Xiamen , China, 2015,5 (1): 8-9
[5] Zeng M.J., Yan T.H.Z., Determination of halogen oxide anion in water by ion chromatography [J]. Chinese inorganic analytical chemistry, 2012,2 (z): 19-21
[6]Qi C.H., Zhang Y., Wang H.Y., Simultaneous determination of bromate, chlorite and chlorate in drinking water by ion chromatography[J]. Chinese Journal of hygiene inspection, 2012,22 (10): 2284-2287
[7]Sun W.M., Sun J.F., Wang Q.S., Determination of bromateand chlorite in drinking water by ion chromatography[J]. Occupational and health, 2013,29 (7): 824-826
[8]Ministry of Health, PRC, China National Standardization Administration Committee, GB/T5749-2006 sanitary standard for drinking water, Beijing, China Standard Press, 2007.
[9]GB/T5750.5-2006Determination of fluoride, chloride, nitrate and sulfate in drinking water and its source water by ion chromatography, Beijing, China Standard Press, 2006

IOP Conf. Series: Earth and Environmental Science **146** (2018) 012016 doi:10.1088/1755-1315/146/1/012016

Appraising the Barriers to Public-Private Partnership as a Tool for Sustainable Development of Infrastructures in a Developing Economy

RAOjelabi[1], OI Fagbenle[1], AO Afolabi[1], PF Tunji-Olayeni[1] and LM Amusan[1]
[1]Covenant University, P.M.B 1023, Ota, Nigeria

rapheal.ojelabi@covenantuniversity.edu.ng

Abstract. The quest for sustainable development has been a subject discourse across the globe. Sustainable development has been deemed to be the pathway to all that is good and desirable in the society. Undoubtedly, social and economic infrastructure desirability in any economy is second-to-none and its sustainability needs to be prioritised. The worsened state of infrastructures and wide gap in its supply in developing economy have warranted this study. The study identified the sustainable tool (PPP) through which infrastructural supply can be enhanced and sustained and it revealed through literature review the major barriers to the tool performance in delivering infrastructures in developing country. Among the major challenges identified from literatures include inadequate consultation of stakeholders for greater acceptance of PPP, conflict of interest among PPP stakeholders, negative behaviour of the people towards PPP, lack of confidence and mistrust in PPP by stakeholders, poor enabling environment for PPP, weak or poor regulatory frameworks, law and regulation changes and weak and poor enabling policies. The barriers identified are due to more to the public than the private and the people. Therefore, the study recommends that beyond the need for stronger collaboration between the public and private sector, government should integrate the people in planning phase of the sustainable tool adoption for public infrastructures delivery. Also, government should build confidence and trust in the parties to PPP by creating enabling environment that can guarantee investors security.

1. Introduction

It is no more news that the world is driving towards sustainability as the sustainability ideal is gaining prominence in the global economy.[1] asserted that there have been an increase in awareness and concern for sustainedsocial and economic growth across the globe. [2] also attested that the introduction of sustainability concept have witnessed a geometric increase in the awareness on sustainability development agenda across the globe. The world quest for sustainability is evidenced in the United Nations sustainable development goals which aim at making the world a better place. To make the world a better place, the idea of sustainability must gain prominence in all areas of life. The concept of sustainability is encompassing as it cut across every spheres of life. Its application include and not limited to the following sectors; medicine, construction, manufacturing, education and governance.The adaptation of sustainability concept in all walk of life have birthed different sustainable terms which include sustainable consumption, sustainable production, sustainable policy and sustainable development among others.

This paper will dwell on sustainable development as it relates to the construction industry in infrastructural provision. The term sustainability and sustainable development are two inseparable terms in which the latter is used to achieve the former. [3] affirmed that by describing sustainable development as that kind of development that must be attained in a bid to achieve sustainability. The sustainable development idea like sustainability concept is not limited to any sector of the economy as it is applicable in any good and desirable part of the society. [4] attested that sustainable development is a pathway to all that is good and desirable in the environment. One of the good and desirable things in any society or environment is infrastructural development. The United Nations affirmed this by prioritizing infrastructures among the seventeen (17) sustainable development goals key to achieving sustainability of our environment. In a bid to achieve infrastructural provisions and other sustainable development goals, United Nations identified partnership as a vital tool crucial in achieving sustainable infrastructural development. Likewise, [5]admitted that Public-Private Partnership is an effective sustainable tool capable of achieving sustainable goals in the global world.

Public-Private Partnership is considered as one of the sustainable route through which infrastructural development challenge can be resolved across the globe. The partnership identified has PPP has to do with public institutions and private bodies' collaboration in a bid to provide social and economic infrastructures. The demand for supply of social and economic infrastructures in developing economy is second to none as the gap in infrastructural strength in the region is very porous. [5] attested that countries from developing economy of the world falls at the bottom of the chart of countries in infrastructural strength. The necessity for adoption of the PPP tool lies in its capacity to close the existing gap in infrastructures supply among countries in the developing region. However, despite the contribution of the sustainable tool in developed world, its impact has not been significant in developing nations due to some perceived challenges confronting its adoption and implementation. The objective of this study is to identify the challenges to the adoption and implementation of the sustainable tool (PPP).

2. Public-Private Partnership as a Sustainable Tool

PPP is the trending household name in social and economic infrastructural provisions across the globe. The gap in infrastructural supply in different parts of the world has called for a sustainable medium through which the gap can be bridged. [6]asserted that Public-Private Partnership (PPP) is a sure and effective sustainable tool that can enhance and sustain infrastructural provisions. [7] revealed that government across the globe have found PPP as a sustainable tool through which their infrastructural deficit can be addressed. Public-Private Partnership according to [8] is described as a tool that allow for collaboration between the private sector and the public agencies (central, state and local governments) with the aim of delivering and sustaining public works and services with private sector resources and technical-know-how. It is evidenced that the call for sustainable development of infrastructures is beyond the reach of the governments alone, as the need for synergy with the private sector towards forming formidable sustainable development agents through the platform of PPP is very pivotal.

Prior to the introduction of Public Private Partnership tool, the government is saddled with the responsibility of infrastructural provision. The capacity of the government for continuous supply of infrastructures was threating in the face of global financial deficit which is key to infrastructural provision. The adoption of the Public-Private Partnership tool has helped boosts the supply of infrastructures during the cash crumb period. [9] revealed that the Public-Private Partnership tool has been instrumental in the successful delivery of 600 construction projects which ranges from hospitals, schools, prisons, bridges, roads and the likes. Likewise, [10] also reported that 1,121 PPP infrastructure projects were delivered between 1985 and 2004 across Europe, Asia and the Far East. It is evidenced from the forgoing that the Public-Private Partnership is a sustainable tool for infrastructural provisions when other tools failed.

2.1 Establishing the theory for adapting PPP as a Sustainable Tool for Infrastructural Provision

The quest for partnership in the 21st century is second-to-none due to the wide gap of needs which exceed the capacity of the individual organization, body or firm. Social partnership theory evolves to address the form of partnership that is birthed due to social and economic needs. Social partnership according to [11] is described as a form of partnership that solved social related problems through the synergy of organizations from more than one sector of the economy. [12] also defined social partnership as a networks of parties including industry, government and local community toward solving a social problem in the local environment. [13] emphasised that social partnership is a form of partnership initiated to cause societal improvements. It is evidenced from theese descriptions that social partnership takes its root from the social economy which focuses on meeting the social needs of the people [14]. The basis for interaction among different social groups is evidenced in the social partnership model in Figure 1. The social partnership model proposed by [15]stressed the need for partnership among different social groups which include the governments, private bodies and others in a bid to create a social project. The model distinguished the government position from the private counterpart. The government occupy the base of the pyramid due to their significance in any social relationship while the private body occupy the tip of the pyramid and the mid is the centre of the partnership deal. The paradigm shift from individualism to social partnership experienced globally is viewed by government and non-government agencies as a means capable of enhancing and achieving sustainable development ofinfrastructures. It is evidenced from the forgoing that Public-Private Partnership takes its root from social partnership theory.

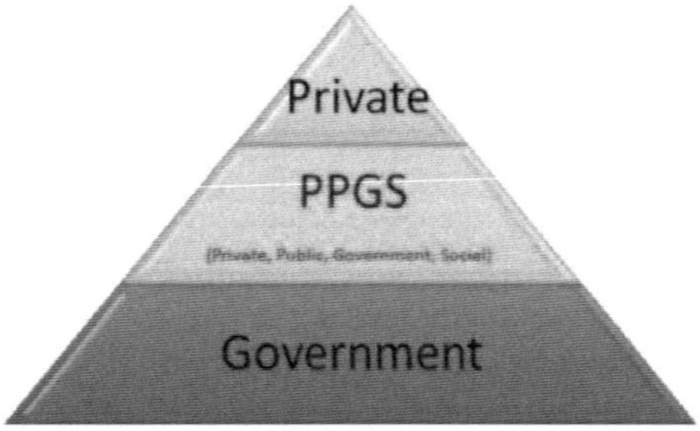

Figure 1. Social Partnership Model
Source: [15]

2.2 Public-Private Partnership in Developing Economy

The world 'developing economy' is a pointer to a need gap awaiting development. [16] affirmed that developing economy is one that is searching for means of enhancing public infrastructures and services due to measurable underdevelopment within her territory. [5] also identify developing economy as such that is weak in infrastructure strength. [17] estimated that the cost of maintaining existing infrastructures and building new ones in developing countries in capital is around 7 percent of their GDP, however, the spending capacity of the developing economies for public infrastructures is about 3 percent which shows a clear gap in infrastructural supply. It is evidenced that the infrastructural state of the developing economy is such that is beyond the capacity of the government, therefore, there is a need for partnership in this environment.

The introduction of Public-Private Partnership is a blessing to the developing country due to its capacity to meet the environment needs. [16] revealed that PPP tool has wide range of application which include construction of physical infrastructures, provision of health and social services among others which are in short supply and high demand in developing countries. [18] however revealed that the private investment garnered from collaboration with the public sector in Africa which is one of the region with high numbers of developing countries is 2 percent despite the massive demand for more private investment in this region. [6] asserted that the low numbers of private investment in the region is due to some challenges or perceived risk discouraging investors in collaborating with the government under the platform of PPP.

2.3 Challenges of PPP Adoption and Implementation in Developing Economy

The need for adoption of PPP tool for infrastructural supply in developing country is of high essence. However, its adoption and successful implementation cannot be compared to the developed region of the world. [19] report revealed the wide gap in PPP adoption for infrastructural provisions between the developed and developing economies. [20] however questioned the wide gap in the adoption of the sustainable tool (PPP) between the developed and developing economies. The slow adoption of PPP in the developing country can be attributed to some inherent challenges due to the peculiar nature of the environment.

[21] asserted that one the most significant challenges confronting PPP adoption in developing economy especially in Africa is inadequate consultation of stakeholders for greater acceptance of the tool. The adaptation and successful implementation of any tool for public benefit without doubt require deep consultation among the concerned stakeholders and the public at large. The need for this is for the acceptability and support of the public at large. Negligence in consulting stakeholders and different interest group will no doubt spell doom for the adoption of the sustainable tool (PPP) because its implementation is beyond personal or private control.

[22, 23] likewise identified conflict of interest among stakeholders as a barrier to PPP growth in developing economy. Disagreement among the stakeholders have contributed to the issues with PPP tool adoption in the developing country especially in Africa. [8] further recognized the negative behaviour of the people towards PPP as a challenge militating against the sustainable tool adoption. The position of the people under PPP arrangement cannot be silenced as their role is highly significant. The people are also stakeholder in any PPP arrangement as such, their opinion should be sorted before implementation of PPP for infrastructural provisions. Isolation of the people in developing economy in the planning process of the sustainable tool adoption has caused the people to develop a thick skin towards the implementation of the tool due to their negative perception that have not been addressed.

In the same vein, [24] also identified public disapproval or public resistance as one of the significant barriers to the use of PPP for sustainable development of infrastructures. This can be traced to the lack of trust between the government and the public in the developing economy. In the same vein, [25] revealed that lack of confidence and mistrust in PPP by stakeholders has impacted the adoption of the sustainable tool negatively. The lack of confidence and mistrust are however due to weak and poor enabling policies [26], weak or poor regulatory frameworks [27], law and regulation changes [28] and poor enabling environment [29]. It is evidenced from the foregoing that these challenges are interrelated and are under the government jurisdiction and the private sector. However, the challenges are more in the government territory. The government as a public custodian are charged with the responsibility of adoption of sustainable public tool (PPP) for sustainable development of infrastructures. [30, 31] also attested that the government and private parties are not all together equal in PPP adoption, as such, the successful adoption and implementation of PPP tool is more of her responsibility. However, [32] revealed that political interference in the process of PPP adoption from the government has contributed to the dwindling state of the sustainable tool in the developing economy. The aggregate of the challenges identified have all contributed to the slow adoption of the sustainable tool (PPP) in developing economy despite its urgent need in the region as a result of worsening or widening infrastructural gap.

3. Conclusion

This paper focused on challenges to the adoption of sustainable tool like PPP in developing economy. It identified the major challenges causing the slow adoption of PPP option for infrastructural supply or sustainable development in the developing region like Africa. Among the major challenges identified confronting the sustainable tool adoption and implementation include inadequate consultation of stakeholders for greater acceptance of PPP, conflict of interest among PPP stakeholders, negative behaviour of the people towards PPP, lack of confidence and mistrust in PPP by stakeholders, poor enabling environment for PPP, weak or poor regulatory frameworks, law and regulation changes and weak and poor enabling policies. Majority of the identified challenges are under government jurisdiction, therefore, the study recommended that the government should specifically address these challenges in the developing regions. The government should ensure that all the stakeholders including the people are well engaged prior to the adoption and implementation of the sustainable tool. Likewise, government need build confidence and trust in the PPP parties by creating enabling environment that can guarantee the security of the parties.

Appreciation

The authors are grateful to Covenant University for sponsorship

References

[1]Oyedepo, S.O. (2014). Towards achieving energy for sustainable development in Nigeria. *Renewable and Sustainable Energy Review, 34, 255-272.*

[2]Abidin, Z.A. (2010). Investigating the awareness and application of sustainable construction and concept by Malaysian developers. *Habitat International, 34 (421-426).*

[3]Du Plessis, C. (2002). Agenda 21st for sustainable construction in developing countries. Capture press, Pretoria, South Africa.

[4]Holden, E., Linnerud, K and Banister, D. (2014). Sustainable Development: Our Common F Future Revisited. *Global Environmental Change, 26, 130-139.*

[5]World Economic Forum, (2016). The global competitiveness report 2016-2017. Available at www.weforum.org/gcr.

[6]Hamilton, G., and Holcomb V. (2013). Public Private Partnership for sustainable development. Retrieved 3 February, 2016, from http//www.commonwealthministers.com/articles/public-private partnerships for sustainable development/

[7]Ibem, E.O. (2010). Assessment of the role of government agencies in Public-Private Partnerships in housing delivery. *Journal of Construction in Developing Countries*, 15(2), 23-48.

[8]Gunnigan, L and Rajput, R. (2010). Comparison of Indian PPP Construction Industry and European PPP Construction Industry: Process, Threshold and Implementation. *Proceedings of CIB World Congress,* Salford, May 10th-13th.

[9]Treasury HM. (2013). Public-Private Partnership signed project list. Available at http://www..hm-treasury.gov.uk/ppp_stats.htm

[10]Sambrani, V.N. (2013). PPP for Asia and African perspective towards infrastructure development: A case study of Greenfield Bangalore international airport, India. *Social and Behavioral Sciences, 157, 285-295.*

[11]Waddock. S. (1989). Understanding social partnerships: An evolutionary model of partnership organizations. *Administration and Society,* 2(1), 78-100.

[12]Seddon, T and Billet, S. (2004). Social partnership in vocational education: building community capacity. NCVER, Adelaide.

[13]Flyvberg, B., Bruzelius, N., & Rothengatter,W. (2003). Megaprojects and risk: An anatomy of ambition. Cambridge, UK: Cambridge University Press.

[14]Evans, M., & Syrett, S. (2007). Generating Social Capital? The Social Economy and Local Economic Development. European Urban and Regional Studies, 14(1), 55–74.

[15]Parthasarathy, V.S. (2016). Making India future ready for technology innovation. Available at: http://computer.expressbpd.com/features/making-india-future-ready-for-technology-innovation/19836/

[16]Pessoa, A. (2006). Public-Private Sector Partnerships in Developing Countries: Prospects and Drawbacks. Being a presented at the 11th International Conference on Public-Private Partnerships: Enhancing Sustainable Development in Romania, 25th -28th May.

[17]Kateja, A. (2012). Building Infrastructure: Private Participation in Emerging Economies. *Procedia-Social and Behavioural Sciences, 37, 368-378.*

[18]Sanni A.O., and Hashim, M. (2014). Building infrastructure through public private partnership in Sub-Saharan Africa. *Procedia-Social Behavioral Sciences, 143, 133-138.*

[19]Organization for Economic Commission Development (2012). Public Governance of Public-Private Partnerships. Available at: www.oecd.org/governance/budgeting/PPP-Recommendation.pdf.

[20]Babatunde, S.O., Perera, S., Zhou, L., and Udeaja, C. (2015). Barriers to Public Private Partnership projects in developing countries: A case of Nigeria. *Engineering, Construction and Architectural Management,* 22(6), 669-691.

[21]Chen,C. (2007),"Institutional barrierstoprivate participation in infrastructure: the case ofelectronictoll collectioninTaiwan",Proceedingsof23rdAnnualARCOMConference, Belfast,3-5September.

[22]Akintoye,A.,Hardcastle,C., Beck,M., Chinyio,E.andAsenova,D.(2003). Achievingbest valueinprivatefinanceinitiativeprojectprocurement. *ConstructionManagementand Economics, 21(5),461-470.*

[23]Afolabi, A., Emeghe, I., Oyeyipo, O. and Ojelabi, R. (2016). Professionals' preference for migrant craftsmen in Lagos State. *Mediterranean Journal of Social Sciences,* 7 (1), 501-508.

[24]Gibson,H.andDavies,B.(2008).Theimpactofpublicprivatepartnershipsoneducation: acase studyofSewellGroupPlc.andVictoriaDockprimaryschool.*InternationalJournalof EducationalManagement,22(1), 74-89.*

[25]Mahalingam, A. (2010).PPP experiencesinIndian cities: barriers, enablers and the wayforward.*JournalofConstructionEngineeringand Management, 136(4),419-429.*

[26]UnitedNationsEconomicandSocialCommissionforAsiaandthePacific(UNESCAP). (2007). Public-PrivatePartnershipsforinfrastructuredevelopment:anintroductiontoissuesfrom different perspectives.TheHigh-levelExpert Group MeetingJointlyOrganized byUNESCAPandtheMinistryofPlanningandBudget,Seoul,2-4October.

[27]AsianBusiness(1996). Special reportonAsia'sinfrastructureboom. *AsianBusiness,60-69.*

[28]Yang,J.B.,Yang,C.C and Kao,C.K.(2010). Evaluatingscheduledelaycausesforprivate participatingpublicconstructionworksunderthebuild-operate-transfer model.*InternationalJournalof ProjectManagement,*28, 569-579.

[29]Akampurira, E.,Root,D.and Shakantu, W. (2006). Stakeholderperceptionsinthefactors constrainingthedevelopmentandimplementationofpublicprivatepartnerships in the Ugandanelectricitysector. *JournalofEnergyinSouthernAfrica,*20(2),2-9.

[30]Namblard, A. (2000). In favour of pragmatic approach towards Public-Private Partnership, In Perrot, J.Y and Chatelus G. Financing of major infrastructure and public service projects: Public Private Partnership, presses points.

[31]Oyeyipo, O., Odusami, K. T., Ojelabi, R. and Afolabi, A. (2016). Factors Affecting Contractors' Bidding Decisions for Construction Projects in Nigeria. *Journal of Construction in Developing Countries,* 21 (2), 21-35.

[32]Gidado, K. (2010). PFI implementation and evaluation model for developing economies: example of Nigeria, Proceedings of the 2010 International Conference on Engineering, Project and Production Management. National Pingtung University of Science and Technology, Pingtung, Taiwan.

Virtual reality for showcasing sustainable engineering design

Walter Chen, Dong-Huang Li, Yeou-Fong Li

National Taipei University of Technology, Taipei 10608, Taiwan

E-mail: waltchen@ntut.edu.tw

Abstract. One of the goals of national parks is to promote sustainable development and environmental education. This paper presents a study to further the goal by demonstrating the value of virtual reality and augmented reality in the construction of a FRP (Fiber Reinforced Polymer or Plastics) pedestrian bridge in Taiwan's national park. The purposes are three-fold. First, the bridge serves to educate visitors about the infrastructure of the park. Second, it attracts the attention of the public and raises interest in conservation and sustainability. Third, it allows to showcase sustainable engineering design with the latest virtual reality technology. The example described in this paper is based on the same principles of e-learning and mobile learning. The experience is valuable in encouraging and promoting tourism opportunities and sustainable resource use in national parks.

1. Introduction

National Parks of Taiwan have been in existence since 1984. Sharing the same vision as the conservationist John Muir, Taiwan established nine national parks throughout Taiwan and its affiliated islands, accounting for 8.63% of the total land area [1]. As indicated by the Construction and Planning Agency of the Ministry of the Interior (Taiwan), the goals of the national parks are [2]:

(i) to protect the natural environment,
(ii) to preserve species and genes,
(iii) to provide public recreation while developing the local economy, and
(iv) to promote academic research and environmental education.

To reach these objectives, construction activities in the national parks usually require an environmental impact assessment or special permits to protect environmentally sensitive areas. Man-made structures in the parks are required to perform multiple functions. Aside from their structural purposes, the structures need to be visually non-intrusive to the surrounding environment, and provide educational values. Construction materials should be specially selected and be compatible and harmonious with the surrounding areas. For example, the District of Columbia Water and Sewer Authority (DC Water) used fiber reinforced polymer (FRP) composite material to retrofit an eroding sewer pipe on US NPS land [3]. DC Water's decision to select FRP not only solved the environmental issues, but also cut cost and construction time significantly.

2. Materials and Methods

Taijiang National Park is located on the southwest shoreline of Taiwan (in greater Tainan city). It consists of mainly coastal areas and wetlands filled with abundant wildlife resources and cultural significance. To no one's surprise, four wetlands in the Taijiang National Park were decreed as protected areas of national and international importance to protect the wetland's biodiversity: Zengwen Estuary Wetland, Sicao Wetland, Qigu Salt Pan Wetland, and Yanshui Estuary Wetland (using the standard English translations used by Taijiang National Park). Among them, the Zengwen Estuary Wetland and Qigu Salt Pan Wetland are famous as winter habitats of the globally endangered black-faced spoonbills (*Platalea minor*). In recent years, there has been a marked increase in interest in wildlife observation (such as bird watching) in national parks, and ecological tourism is on the rise. One famous location frequented by visitors is the Qigu lagoon in the Qigu Salt Pan wetland. The Qigu lagoon is an inland sea surrounded by sandbanks. It is the largest in Taiwan and is made famous for its oyster stands and the fixed fishing nets in the lagoon. Visitors can take a guided raft tour in the lagoon and visit the nearby saltponds. As pointed out by Zorrilla-Pujana and Rossi [4], it is important for national parks to use environmental education to move from the mere creation of protected areas to the effective management of the protected areas, at the same time satisfying the needs of both people and nature. To preserve the wetland ecosystems and promote environmental education, the Taijiang National Park Management Office established the Taijiang Wetland School in 2012, greatly enhancing their education outreach program [5].

2.1. Description of the Bridge

The Taijiang National Park has a total area of 39,310 hectares, among which 34,405 are shallow seas [6]. The FRP pedestrian bridge is located near the Liukong Service Station and Visitor Center, south of the Qigu lagoon. It is ideal for serving as an example to test environmental compatibility and public education. First, the material is unique. Unlike traditional construction materials, FRP is resistant to corrosion brought about by local high humidity and high chloride in the air. Second, the guidelines of national parks prohibit unwarranted construction, and structures in the parks should be long-lasting to avoid excessive repairs and renovation. In that regard, the long-lasting nature of FRP and its easy installation (usually by manual labor) suits the national parks very well. Therefore, the use of FRP to replace traditional construction materials such as reinforced concrete in national parks and environmentally sensitive areas should be encouraged. Aside from the FRP material, the pedestrian bridge itself also serves multiple functions.

2.2. Construction of the Bridge

The FRP pedestrian bridge is located south of the Qigu lagoon dike. It was analyzed by a Finite Element software for allowable deflections and installed at the site in 2012 [7], but so far the bridge does not appear on Google Maps, which is used by the Taijiang National Park to show the general public how to arrive at the park. As can be seen from Google Maps, quite a few channels and waterways pass through the area. The pedestrian bridge is an essential crossing for people visiting the Liukong Visitor Center and the lagoon. It is conceivable that endemic amphibians and other small animals could use the bridge as a corridor to move between habitats during the night. How wildlife uses the bridge as a corridor necessitates a new study. One additional benefit from the introduction of the FRP bridge is that people learn more about FRP material and how it can be used to retrofit damaged buildings, a need exemplified by frequent earthquake occurrences in Taiwan. Since the pedestrian bridge was built with FRP, it avoided usage of heavy machinery and reduced the ecological footprint.

3. Results and Discussion

As has been noted in previous sections, the role of national parks has shifted from establishment to management and education. To satisfy the shifting needs of the Taiwanese public, national parks should encourage and promote tourism opportunities and sustainable resource use. Putting part of the parks online is a good first step. The process of placing the FRP bridge in virtual reality (VR) comprises several stages. First and foremost, a computer model of the bridge (and the surrounding area) needs to be built. Then the model needs to be imported into an immersive virtual environment. For visual display purposes, the model needed is not the same as the one previously used in the Finite Element analysis. We chose SketchUp for this task and created a replica of the FRP bridge in virtual space. The model bridge allows views of multiple perspectives. The purposes are twofold. First, the bridge serves to educate visitors about the infrastructure of the park. Second, it attracts the attention of the public and raises interest in conservation. In some circumstances there is even a third function, which is to showcase sustainable engineering design. Deformation diagrams obtained by the Finite Element analysis can be overlaid onto the surface of the model bridge to show the stress-strain relationship. Interested persons can view the model to examine the structural behavior from different perspectives, assuming the bridge is subjected to the specified loading condition.

3.1. Immersive Virtual Reality Display

Notice that the FRP bridge has a minimal design and non-intrusive profile. To assess the compatibility of the bridge with the environment, we examine the bridge model in virtual reality with its surrounding area (also with the Liukong Service Station and Visitor Center). The bridge model has excellent details, including posts, guardrails, balusters, edge beams, bridge decks, a wheelchair ramp, and an information board. To further enhance the appearance of the model bridge, which was essential as a tourist attraction and public education ([8], [9], [10], [11]), we added textures closely matching the real FRP to the model, and pasted high resolution photographs onto the information board at the entrance of the bridge. Viewers can read the text and see the figures on the information board without any difficulty. This was an important component of a successful 3D immersive exhibition. Finally, the bridge model was imported to Sketchfab for future 3D printing (if needed).

The FRP bridge was built to minimize its impact on the environment and to emphasize its adherence to the landscape ecological principle. For everyone to see the results and to deliver them in a timely manner, we put the bridge model on display in a 3D immersive virtual environment. This action highlights our emphasis on the material selection and bridge design. Compared to mere design drawings or photographs, the novelty in this contribution is the use of VR in the process of viewing, interpretation, and discussion. For this particular project, the FRP bridge has already been built. The VR model adds to the value of a real and functional bridge. For other projects, in which the structures have not been built, VR models could be used beforehand to assess whether the structures are environmentally compatible or not, and whether some of the placement and configurations are optimal or not. This increases the value returned by creating a virtual model.

To view VR models, computer screens are not the best choice. In the past few years, big tech companies such as Facebook, HTC and Samsung have all released VR helmets or goggles that can be worn to experience VR. The only problem is that these devices are too expensive. A low cost alternative solution for VR viewing is the Google Cardboard, the design for which was released by Google, and it has evolved into many derivative designs on the market. Users can acquire the material they need with little money and assemble a VR goggle. When the FRP bridge is viewed with a Google Cardboard, details of the bridge are preserved. Viewers can turn their heads to see different angles of the bridge or turn around to see their environment (figure 1). They may also pretend to visit the park and stand in front of the information board. The text and figures on the information board can be viewed clearly in virtual reality. It is as if they were at the national park, which is a true immersive experience that has high educational value and attracts tourist attention.

3.2. Augmented Reality and Mobile Display

Virtual reality is not the only thing that one can experience with the bridge model. Without wearing a VR goggle, users can still see the bridge and its surroundings on any surface using a mobile device such as a phone or a tablet. This is accomplished by augmented reality (AR). Using a specially designed AR tag for the bridge, the bridge will come to life and show up on the screen of the mobile device when the device is pointed at the AR tag (figure 2). The viewer can rotate the bridge to see and examine different perspectives. They can also see more or less details of the bridge by pinch zooming in or out of the model. This AR viewing works on various surfaces such as the pages of a book, fliers, posters, and stickers. It is fun for education and a great supplement to courses such as those at the Wetland School.

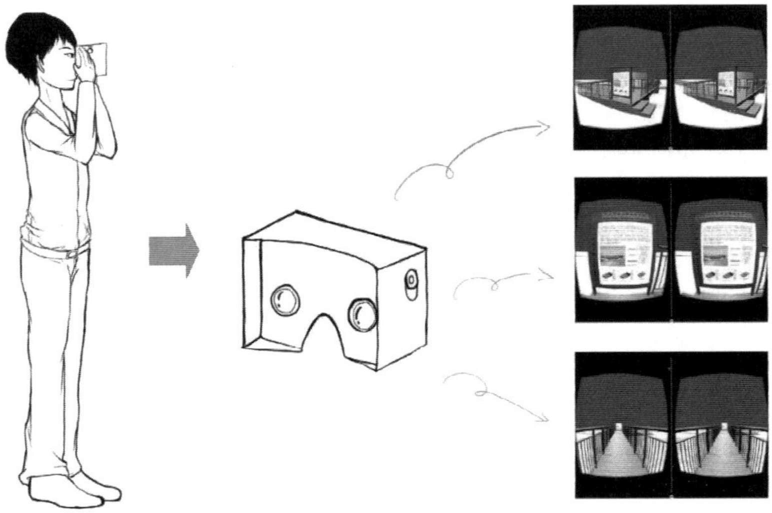

Figure 1. Conceptual illustration of viewing the virtual FRP bridge using a Google Cardboard.

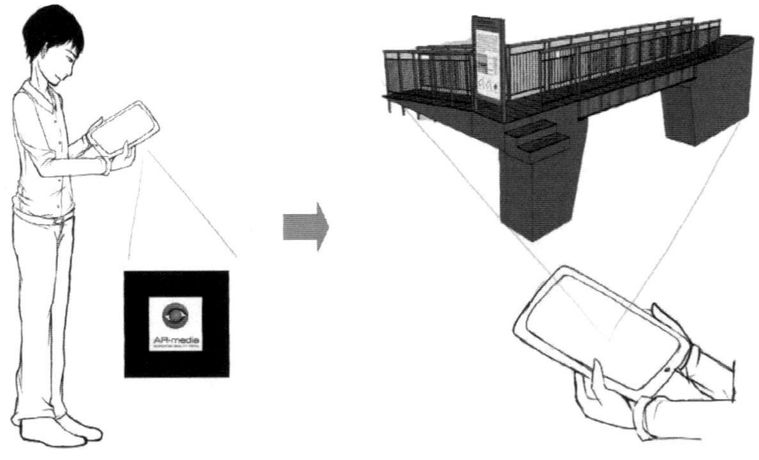

Figure 2. Conceptual illustration of augmented reality display using a mobile device.

4. Conclusions

In this study, a 3D model of the FRP bridge and its surroundings was built. Realistic details were preserved, including structure members of the bridge and an information board. Using VR goggles, visitors can tour the bridge in virtual reality. The experience is lifelike and the viewer can even read the text on the information board. In addition to 360 VR immersion, the FRP bridge model can also be displayed on any surface using AR. The viewer can rotate the bridge to see and examine different perspectives, or see more or less details of the bridge by pinch zooming in or out of the model. This is a great method of disseminating structural and material knowledge to a wider audience. It shows that substantial educational value can be returned from creating a bridge model alone. Simply stated, placing a bridge in VR is every bit as important and useful as placing it in a national park. It should be considered by national park authorities.

Acknowledgments
The authors are grateful to Mr. Ming-Chieh Chen, who created the original bridge model for Taijiang National Park. A very special acknowledgement goes to Mr. Jing- Yuan Li for his service during the study. Finally, kudos are due to Miss Chi Chang for her assistance in preparing the excellent illustrations used in this paper.

References

[1] National Parks of Taiwan: Introduction of national parks 2009 (in Chinese) http://np.cpami.gov.tw/chinese/index.php?option=com_content&view=article&id=1&Itemid=128&gp=1
[2] National Parks of Taiwan: About the parks 2014 (in Chinese) http://np.cpami.gov.tw/english/files/15-1000-1548,c90-1.php
[3] Bian S, Pridmore A, Loera R and Marshall J 2011 Pipeline Rehabilitation amidst Environmentally Sensitive Location *Proc. Pipelines 2011 Conference* (Seattle WA) p 523–532
[4] Zorrilla-Pujana J and Rossi S 2016 Environmental education indicators system for protected areas management *Ecological Indicators* **67** p 146–155
[5] Taijiang National Park: Taijiang wetland school is now officially in operation 2012 (in Chinese) http://www.tjnp.gov.tw/Eng/NewListsDetailE002000.aspx?Cond= eaedd6ff-6998-4b34-8c2a-9d656f29079a
[6] Wang J-A, Hsu C-K, Jhan Y-K, Chen W and Li Y-F 2013 Scanning of FRP passenger bridge in Taijiang national park *Proc. 34th Asian Conference on Remote Sensing* (ACRS 2013) (Bali Indonesia)
[7] Li Y-F, Badjie S, Chen W and Chiu Y-T 2014 Case study of first all-GFRP pedestrian bridge in Taiwan *Case Studies in Construction Materials* **1** p 83–95
[8] Neubert B, Bellian J, Xu X, McGowan K, Schindel G and Alexander E Jr 2008 3-D photo real modeling of Devil's sinkhole in Rocksprings, Texas *Proc. 11th Multidisciplinary Conference on Sinkholes and the Engineering and Environmental Impacts of Karst* (Tallahassee Florida)
[9] Tseng Y-T, Chou H-T, Jhan Y-K, Wang J-A and Chen W 2013 Laser scanning of the award-winning NTUT green gate *Proc. 34th Asian Conference on Remote Sensing* (ACRS 2013) (Bali Indonesia)
[10] Marsico A, Sabato L, Spalluto L, Infante M and Nitti A 2016 Integrated geological and digital surveys to know, preserve and promote a geoheritage: the sea cliff of Vignanotica Bay (Gargano Promontory, southern Italy) *Geoheritage* p 1–10
[11] Yang K-J, Chen Y-J, Xu K-W and Chen W 2015 Application of Google cardboard in fire hazard simulation *Proc. 36th Asian Conference on Remote Sensing* (ACRS 2015) (Quezon City Metro Manila Philippines)

IOP Conf. Series: Earth and Environmental Science **146** (2018) 012018 doi:10.1088/1755-1315/146/1/012018

Physiological Responses of Three Kinds of Street Trees to Acute Stress of Ozone

J X Long*, J F Liu and H Y Cheng

School of environmental science and engineering, Xiamen University of Technology, 600 Ligong Road, Xiamen, 361024, China

2011121601@xmut.edu.cn

Abstract: The stress impacts of ozone pollution on the common street trees in tropical cities were investigated in this study. Three kind of common trees(*C.camphora*, *M.indica* and *F.altissima*) in Xiamen were artificial ozone fumigation stressed for 8h, with the ozone concentration of 0.43 mg/m^3 and 0.86 mg/m^3 respectively, and their physiological index responses were also analyzed. The results showed that the cell membrane permeability and malondialdehyde (MDA) content of leaves increased but chlorophyll content decreased, the contents of proline and soluble sugar content to regulate osmotic potential, while the activity of superoxide dismutase (SOD) was inhibited when 3 kinds of street trees were subjected to acute ozone stress. In addition, the change amplitude of 6 indexes increased with the increment of ozone stress concentration, which can be used as reference indexes for plant ozone resistance. Based on the performance of plants after stressed, the resistance intensity of 3 species of street trees to ozone stress was: *M.indica*>*C. camphora*> *F.altissima*. The results could provide the reference for screening of ozone resistance trees in the tropics.

1. Introduction

Surface ozone (O_3) concentrations have more than doubled since the Industrial Revolution[1]. Among air pollutants, ozone is classified as a secondary pollutant which has a strong oxidation and has gradually become a public enemy of air pollution. According to the report of the Ministry of Environmental Protection, the major air pollutants are ozone in large and medium-sized cities in China, followed by PM2.5[2]. The harm of ozone to human health is closely related to its concentration and exposure time in the environment[3]. The plant photosynthesis would be inhibited and hence lead to reduce crop yield under the stress of ozone[4-9]. However, few studies were focused on urban landscaping plants of ozone pollution[10-12].

Xiamen city, based on its characteristics of the geographical location and climate in the southeastern of China, is selected as the studing zone in this work, *Cinnamomum camphora*, *Mangifera indica* and *Ficus altissima* representative for street trees in tropical city are selected and theirs physiological indexes response to ozone stress under different concentration were investigated[13]. The resistance of 3 kinds of street trees to ozone pollution in surface layer ground is evaluated by the monitored physiological indexes, which provide an important reference to select street trees species for the greening construction of urban road.

2. Materials and methods

2.1 Experimental materials

Three kinds of street trees: The potted plants of *Mangifera indica*, *Cinnamomum camphora* and *Ficus altissima* were bought from the local gardening market, about 1 year old and 0.5 m height respectively.

2.2 Experimental methods

3 kinds of plants were first placed in the semi-shading state to adaptive cultivate for 1 weeks, and then 3 plants with good growth and uniform height were selected as experimental subjects from each kinds plant. The laboratory hood was utilized to model the open top stress device for acute ozone stress.

Specific operations: The potted plants were placed inside the laboratory hood (volume: 120 cm × 120 cm × 200 cm), and the ventilator and ozone generator (ZA-D5G) were turn on for the formation of partially enclosed convection space in the laboratory hood. The humidity, temperature and light intensity in the laboratory hood are 70%-80%, room temperature and 24000 lx, respectively. The ozone stress with a low concentration of 0.43 mg/m^3, a high concentration of 0.86 mg/m^3 and non-ozone stress were carried out, and the ozone concentrations inside laboratory were online monitored by an ozone detector (Z-1200, USA ESC). Error control is within ±0.05 mg/m^3. The actue stress was treated for 8 h from 8:00 to 16:00. The 3-4 round mature leaves of the plants were selected in the next day after stressed, and theirs physiological and biochemical indexes, including cell membrane permeability, malondialdehyde (MDA), proline, soluble sugar, cholrophyll and superoxide dismutase (SOD) were determined. The specific methods of determination can refer in the book of Experiment Guidance on Plant Physiology edited in chief by Zhiliang Zhang, et al. All experiments were repeated three times. The experiment datas were analysed using the software of SPSS 16.0 and Excell 2010, and theirs average values were determined as the final results. Meanwhile, these 3 kinds of plants were calculated again for 2 weeks and theirs appearance changes were observed during the calculation period.

3. Results and discussion

3.1 Effects of ozone stress on the cell membrane permeability of the plant

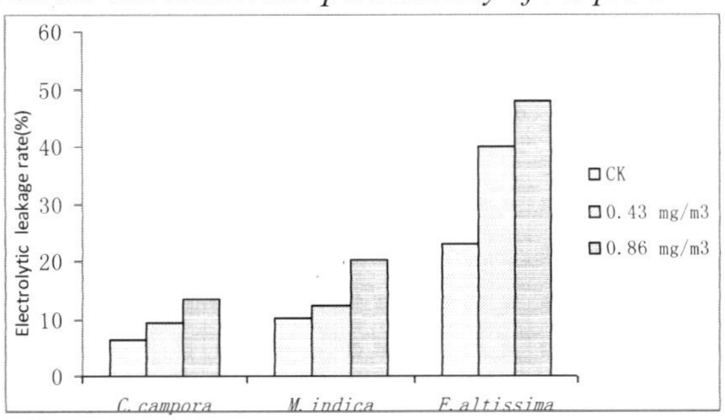

Figure 1. The electrolytic leakage rate in three tree species

When plant were placed in a stress environment, its cell membrane permeability usually becomes large which leads to the partial electrolyte exudation from cells and increases the conductivity of external liquid. Therefore, the more of the conductivity of external liquid indicates the more serious damage of cell membrane as well as the lower resistance of plant but higher resistance on the contrary side. As shown in Fig.1, the cell membrane permeability of 3 kinds of plants became large after actue ozone stressed and increased with the increment of ozone concentrations. Compared to conventional state, no significant difference in the relative extravasation rate of cell membranes for *M.indica* and *C.camphora* ,whlie more significant difference for *F.altissima* were observed under low ozone concentration actue stress, which indicates that the low ozone concentration has a low damage on *M.indica* and *C.camphora*. However, significant differences in the relative extravasation rate of cell membranes for 3 kinds of plants were observed under high ozone concentration, which are 2.16, 1.84 and 2.08 times higher than that of the control one, respectively. This result indicates that the cell membranes are damaged in varying degrees under high concentration of ozone acute stress. In summary, 3 tree species are resistant to low concentrations of ozone, and in high concentration, *M.indica* and *C.camphora* resistance are stronger relatively.

3.2 Effects of ozone stress on the MDA content of the plant

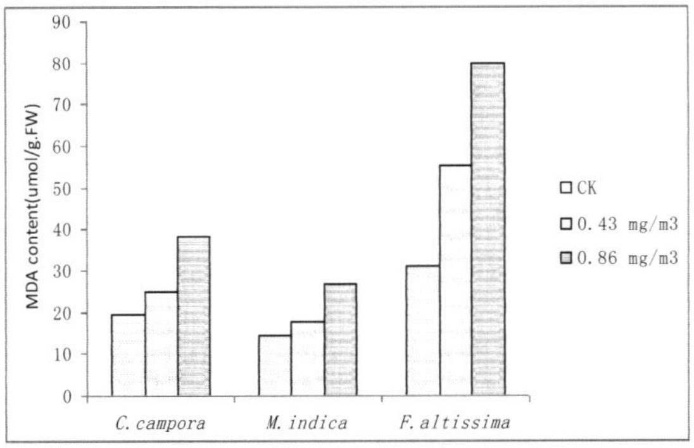

Figure 2. The malondialdehyde (MDA) content in three tree species

The content of MDA, an important indicator of the degree of leaf senescence, indicates the peroxidation degree of membrane lipid and the response intensity of plant to adverse conditions. As shown in Fig.2, the MDA content of the leaves for 3 species of plants increased compared to the control group under ozone stress. Among them, *F.altissima* has an obvious increment magnitude with 1.8 and 2.6 times to control under low and high ozone concentration stress, respectively. Under the same conditions, MDA in both *M.indica* and *C.camphora* have a low increment magnitude in the content respective with 1.29/1.96 and 1.22/1.84 times to the control group. In summary, *F.altissima* is more sensitive to ozone stress, with free radical accumulation and damage appeared in cell under low ozone concentration stress obviously.

3.3 Effects of ozone stress on the proline content of the plant

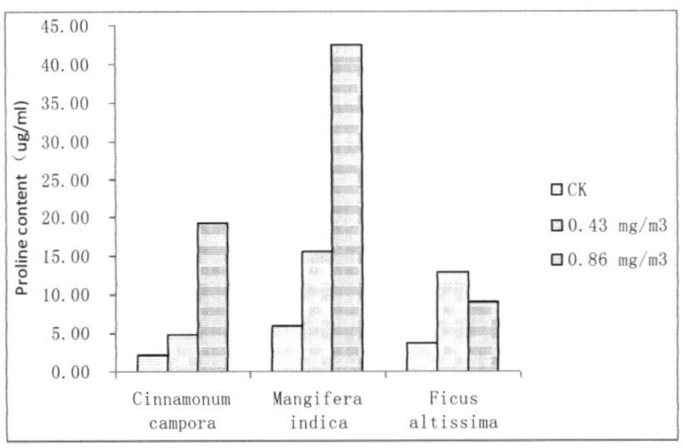

Figure 3. The proline content in three tree species

Proline is one of the important components of plant protein and usually exists with a free state in the plant. When the plant is under adversity stress, the body tends to accumulate proline and increasing significantly, which is beneficial for plant to resist to external stress. Therefore, the plant that its proline contents can increase after stressed has a relative strong resistance. As shown in Fig.3, the proline contents in the cells of 3 kinds of street trees exhibited an increased in various degree, which indicates all the plants has a certain resistance to ozone stress. Under low ozone concentration stress, no significant differences were observed on the relative accumulation rates of proline between 3 kinds of plants and theirs value were 2.26, 2.64 and 3.44, respectively. However, the relative accumulation rates of proline for *M.indica* and *C.camphora* can respectively reach to 9.1 and 9.2 under high ozone concentration stress, which indicate that they have a good emergency response to ozone stress. On the contrary, the proline concent of *F.altissima* is obvious lower than former two which indicates an increased damage on *F.altissima* under high ozone concentration stress.

3.4 Effects of ozone stress on the soluble sugar content of the plant

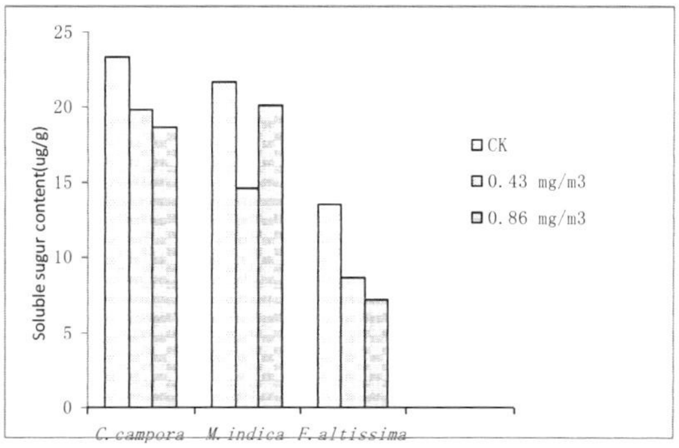

Figure 4. The soluble sugar content in three tree species

The content of soluble sugar is closely related to the resistance of plants and the solube sugar as an osmotic regulator and an anti-dehydrating agent plays a role in the plant growth. As shown in Fig.4, a common reduction tendency existed in the accumulation of soluble sugar contents in the body of 3 kinds of plants. Among them, the change magnitudes for *C.camphora* were close between low and high ozone concentration stress, and respectively reduced by 15% and 22% compared to the control one. The no obvious difference indicates the *C.camphora* had a strong resistance to ozone stress. The change tendency of *F.altissima* was similar to that of *C.camphora*, however, the change magnitudes of *F.altissima* was significant when compared to that of the control one and can reach to 35% and 46%, respectively. The no obvious difference between the low and high concentration experiment indicated the limitation of emergency response capability of *F.altissima* under low ozone concentration as well as the relative weak resistance of F.altissima. The soluble sugar content of *M.indica* exhibited an obvious decline with a rate of -33% under low ozone concentration stress while recoveried to the level of that of the control one with a rate of -8% under high ozone concentration stress, which may be attributed to the formation of adaptability to the external stress and the start-up of resistance mechanism of *M.indica* and hence induced the content of soluble sugar to recovery to the level before stressed.

3.5 Effects of ozone stress on the chlorophyll content of the plant

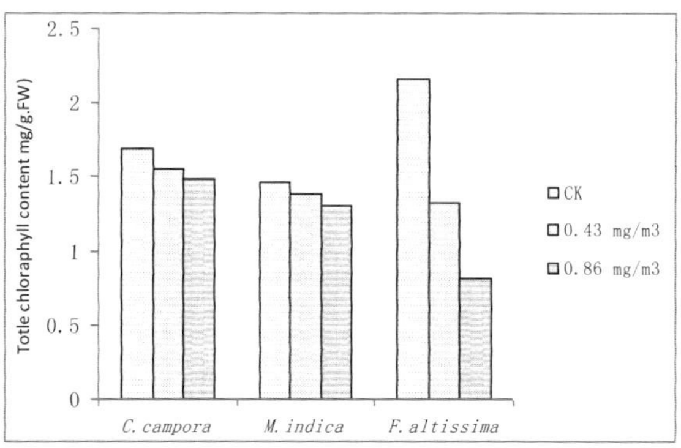

Figure 5. The total chlorophyll content in three tree species

The structures of chloroplast organelles will be subjected to damage after stressed, and the more decline of chlorophyll content implies the more damage from stress, the more sensitive and the more weak resistance of the plant and vice versa. As shown in Fig. 5, the chlorophyll content of 3 kinds of plants generally exhibited a decline tendency. Compared to control, the decline magnitudes for *C.camphora* and *M.indica* were minuscule under low ozone concentration stress, while relative high decline magnitudes with a respective rate of 12% and 11% for *C.camphora* and *M.indica* were observed under high ozone concentration stress which indicated that the chloroplast was damaged slightly under high ozone concentration. Furthermore, significant difference were observed between the ozone-stressed and the control group on *F.altissima*, and the decline magnitudes were respectively up to 39% and 62% under high and low ozone concentration stress. This result indicated the damage of chloroplast for F.altissima was more serious than that of the former two.

3.6 Effects of ozone stress on the SOD activity of the plant

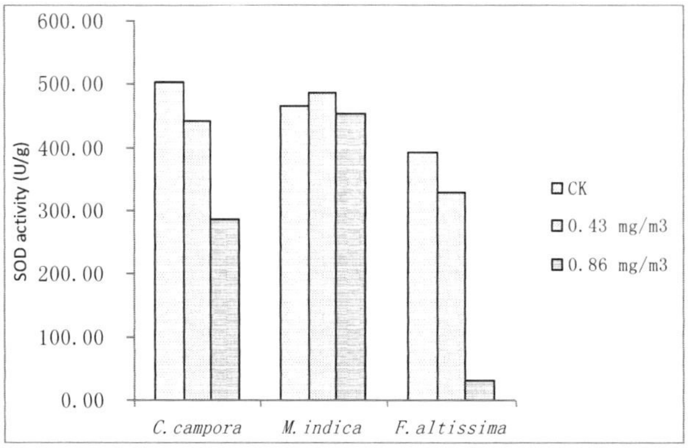

Figure 6. SOD activity in three tree species

SOD plays an important role in the antioxidant system and is a key enzyme for eliminating the reactive oxygen species produced by the stress of external ozone. The activity of SOD can reflect the adversity resistance of the plant. As shown in Figure 6, besides the slight increase of SOD activity for *M.indica* under low ozone concentration stress, all of SOD activity exhibited the decline tendency under ozone stressed. These results indicated that the activity of SOD was inhibited and the inhibition intensity became strong with the increasing of ozone concentration. The decline magnitudes of SOD activities for *C.camphora* and *F.altissima* were obvious when stressed under high ozone concentration and the change rates for them were respectively reach to 43% and 92% compared to the control group.

3.7 Apparent damage symptoms of plant after ozone stressed for 2 weeks

According to the literature, ozone can directly diffuse into the cell through the stomata of plant leaves and destroy the cell, and hence the apparent changes in leaves are the direct responses of plants to the physiological effects of ozone stress[14-15]. After 2 weeks of observation, the damage symptoms and recovery state as follows.

C.amphora and *M.ndica*: No obvious changes on leaves were observed when stressed with 0.43 mg/m^3 ozone concentration. After stressed with 0.86 mg/m^3 for 1 week, small gray spots appeared around the main vein of *C.camphora* leaves with an area ratio of 15%, and small brown spots with an area ratio of 10% appeared at the tip and edge of the leaves of *M.indica* leaves. Both leaves of *C.camphora* and *M.indica* did not shed during 2 weeks.

F.altissima: After stressed with 0.43 mg/m^3 ozone concentration for 1 week, a number of black punctate spots with an area ratio of 20% appeared in the leaves. Furthermore, the area of black spots was further enlarged to 45% after stressed with 0.86mg/m^3, and the tip of leaves curled with dehydration. Subsequently, the leaves fell off but the apical bud maintained the vitality. After that, new buds could grow again in the axils of deciduous leaves and plants recovered to the vitality.

4. Conclusions

Based on the analysis of the response of 3 kinds of trees to ozone stress, the following conclusions could be obtained. When trees were stressed with ozone, the indexes representative for the damaged degree of cell appeared different changes. Increased in the relative membrane permeability and the content of MDA, declined in the content of chlorophyll, inhibited to SOD activity, and increased proline and soluble sugar content to regulate osmotic potential. The index change regulation of each was consistent with the observation feedback to damage after being stressed, which could be used as reference index for evaluating the resistance intensity to ozone of garden plants.

Comprehensive analysis showed that 3 kinds of street trees have relative strong resistance to ozone and the intensity in the following order: *M.indica>C.camphora>F.altissima*. *M.indica* and *F.altissima* are typical tropical evergreen plants and *C.camphora* is a subtropical broad-spectrum plant, which exhibited excellent anti-typhoon abilities during the period of typhoon "Meranti" in 2016 and have good ecological effects. Therefore, it is suggested to give priority to use these 3 kinds of plants for the planning of urban street trees in Xiamen city.

Acknowledgments
This study was financially supported by Key Laboratory of Environmental Biotechnology, Fujian Province University ; Fujian provincial education department science and technology project (No. JB13156).

References

[1] Monks, P.S., Archibald, A.T., Colette, A. et al. Tropospheric ozone and its precursors from the urban to the global scale from air quality toshort-lived climate forcer[J]. Atmos. Chem. Phys. 2015, 15, 8889–8973.
[2] Xu Hui,Zhang Han,Xing Zhen-yu. Pollution characteristics and ozone formation potential of ambient VOCs in winter and spring in Xiamen[J]. Environmental Science, 2015,01：11-17.
[3] Kong Qin-xin , Liu Guang-ren ,Li Gui-chen.Surface ozone contentration variation and possible influenees on human health[J].Climati and Envionmnatl Reseach, 1999,4(1)：61-66.
[4] Lie Gan-wen,Ye Long-hua,Xue Li. Effects of ozone stress on major plant physiological functions[J]. Acta Ecologica Sinica,2014,34(2) : 294-306.
[5] Henryk Debski,Wiesław Wiczkowski, Dorota Szawzra-Nowak, et al. The effect of tropospheric ozone on flavonoids and pigments content in common buckwheat cotyledons[J]. Ecological Chemistry and Engineering S. 2017,24(3):457-465.
[6] Elizabeth A, Ainsworth. Understanding and improving global crop response to ozone pollution[J].The Plant Journal ,2017, 90, 886–897.
[7] Courtney P, Leisner E, Ainsworth A. Quantifying the effects of ozone on plant reproductive growth and development[J].Global Change Biology, 2012, 18: 606 –616.
[8] Fitzagerald B, Russell M, Margaret M, et al. The Ozone component of global change: potential effects on agricultural and horticultural plant yield, product quality and interactions with invasive species[J].Journal of Integrative Plant Biology, Volume 2009,51(4):337–351.
[9] Kivi M, Sirka S, Per E K,et al. Cell structural changes in the seedlings of Norway spruce exposed to long-term ozone and drought[J].Annals of Botany,2003,92(6):779-793.
[10] Niu J F, Zhang W W,Li L,et al. Effects of elevated ozone on foliar chlorophyll content and antioxidant capacity in leaves of *Cinnamomum camphora* under enhanced nitrogen loads[J].Acta Ecologica Sinica,2012,32(16) : 5062-5070.
[11] Lu Guang-chao, Huang Yu-yuan, Chen Hong-yue, et al. Effects of ozone on membrane lipid peroxidation and protective enzyme activities of *Thevetia peruviana* and *Mangifera* leaves[J].Ecology and Environmental Sciences, 2012,07：1235-1240.
[12] Zhang Hong-xing,Sun Xui,Yao Yu-hui, et al. Ground-level ozone distribution pattern in summer of Beijing and its foliar injury[J].Effect upon Plants Acta Ecologica Sinica, 2014,34(16)：4756-4765.
[13] Liu Qi-ming, Fang Yue-min, Huang Zhi-yong ,et al. Biological indicators for air quality assessment effects of ozone[J]. Ecology and Environmental Sciences, 2011, 20(4): 612-615.
[14] Laisk A, Kull O, Moldau H. Ozone concentration in leaf intercellular air spaces is close to zero[J]. Plant Physiol,1989,90:1163-1167.
[15] Calatayud V, Garcia-Breijo FJ, Cerwero J, et al. Physiological, anatomical and biomass partitioning responses to ozone in the Mediterranean endemic plant *Lamottea dianae*[J]. Ecotox. Environ.Safe.2011,74:1131-1138.

IOP Conf. Series: Earth and Environmental Science **146** (2018) 012019 doi:10.1088/1755-1315/146/1/012019

Adsorption Properties of Microwave Modified Diatomite on Heavy Metals in Landfill Leachate

H Y Cheng*, A L Yang, Y T Zhang and X M Kong

School of environmental science and engineering, Xiamen University of Technology, 600 Ligong Road, Xiamen, 361024, China

2012111601@xmut.edu.cn

Abstract: Adsorption properties of diatomite modified by microwave on lead, chromium and cadmium in landfill leachate under different conditions were studied with diatomite as adsorbent, and the best microwave modification conditions were determined in this study. Meanwhile, the adsorption and removal effects of diatomite on Pb, Cr and Cd in landfill leachate under different adsorption conditions were investigated and which including adsorbent dosage, adsorption time, temperature, pH and the ion strength. The results showed that the best adsorption removal efficient of heavy metal ion in landfill leachate can be obtained under the adsorption conditions of low ionic strength, alkaline leachate, with adsorption temperature of 25℃, the adsorption time of 40 min, and the modified diatomite dosing quantity of 5~10g/L.

1. Introduction

Landfill leachate is generated at landfill sites as a result of precipitation, infiltration, compaction and waste degradation. The concentration of organic and inorganic components is higher in landfill leachate, and it contains more than ten kinds of heavy metal ions[1,2]. Thus, it should be treated before discharge in the environment or to a sewage system for treatment[3,4].

The diatomite is a natural nano-materials with a light weight, large specific surface area, super strong adsorption, noise resistance, wear resistance, thermal properties, corrosion resistance and other outstanding performance[5]. A large number of silicon hydroxyl groups (Si-OH) are distributed on the surface and the inner surface of the micropores. Generally, the hydrogen ions H^+ of silicon hydroxyl in aqueous solution will be dissociated. When combined with the adsorbed water, the particles will have a certain surface negative electricity to adsorb positive charges[6]. Therefore, the application of diatomite is widespread increasingly in the treatment of heavy metal pollution.

After sulfated modification or alkalized modification of diatomite, the maximum adsorption capacity of primary diatomite for Pb^{2+} and Cd^{2+} increased compared with that without modification[7-9]. While modified the diatomite with sixteen alkyl trimethylamine bromide solution, manganese dioxide and calcium carbonate, the effects of diatomite dosage, adsorption time and initial pH value on the adsorption properties of modified diatomite were studied[10-12].

Although diatomite and modified diatomite have been used to adsorb heavy metal pollutants in waste water, little research has been done on the adsorption of heavy metals in landfill leachate by microwave modification of diatomite. In this study, the adsorption effect of modified diatomite on heavy metals in landfill leachate under different conditions was investigated by determining the optimum conditions for microwave modification of diatomite. The application of this repair technology has a certain guiding significance for the application of modified diatomite in the treatment of landfill leachate.

2. Materials and Methods

2.1 Reagents and instruments
Hydrochloric acid, sodium hydroxide, sodium chloride, analysis pure. Experimental diatomite is purchased from Linjiang Yiruishi diatomite Co. Ltd. The landfill leachate is taken from East solid waste disposal center in Xiamen city. The deionized water is made by the laboratory of the Department of Environment Engineering.

Galanz microwave oven (P70D20TJ-D3), electric thermostat blast dryer (DHG-9146A), constant temperature oscillator (TH2-320), low speed large capacity centrifuge (TDL-40C), ICP (OPT/MQ 700DV), electronic balance(TE 124S), pH meter(PB-10).

2.2 Preparation of microwave modified diatomite
Diatomite was immersed in hydrochloric acid and deionized water according to a certain ratio of solid to liquid about (1:5~20), and was modified by microwave under microwave irradiation. The power was set to 490 W and the time was in the range of (0~10min). Precipitated the mixture, poured out the supernatant and being filtered, the microwave modified diatomite was prepared. The filtered solid was dried in the oven with the temperature range from 65℃to 85℃, and the dried solid was activated for 0.5-2 h at constant temperature of 100℃. Put the soil sample into mortar and stirred evenly, then sieved with 100 mesh, the microwave modified diatomite is obtained.

2.3 Adsorption experiment
The microwave modified diatomite was added into landfill leachate from which heavy metal will be removed. Then put it in a incubator, set a certain reaction temperature (25~45℃), rotate speed (200rpm) and time (20~120min) to fully contact reaction. Made Centrifugal separation and the supernatant is a landfill leachate after removing heavy metals.

2.4 Analysis method
The standard curves of Pb, Cr and Cd are obtained by using OPT/MQ 700DV inductively coupled plasma emission spectrometer at 220.353 nm, 267.716nm and 228.802nm respectively. Landfill leachate and landfill leachate treated by modified diatomite were also measured at the same wavelength. The concentrations of Pb, Cr and Cd were determined from the standard curve by Optima 7000 DV ICP emission spectrometer.

3. Results and Discussion

3.1 Effect of microwave irradiation time on the modified diatomite
The microwave power was controlled at 490 W and the microwave irradiation time was set in 0min, 5min, 8min and 10min respectively. A certain amount of microwave modified diatomite was prepared by mixing diatomite, 1 mol/L hydrochloric acid and deionized water. 0.1g modified diatomite was added to 10mL landfill leachate of, placed in the incubator at 25℃, and the reaction time was 40min at about 200rpm. After that, centrifuged the supernatant and the concentration of heavy metals was measured by ICP.

IOP Conf. Series: Earth and Environmental Science **146** (2018) 012019 doi:10.1088/1755-1315/146/1/012019

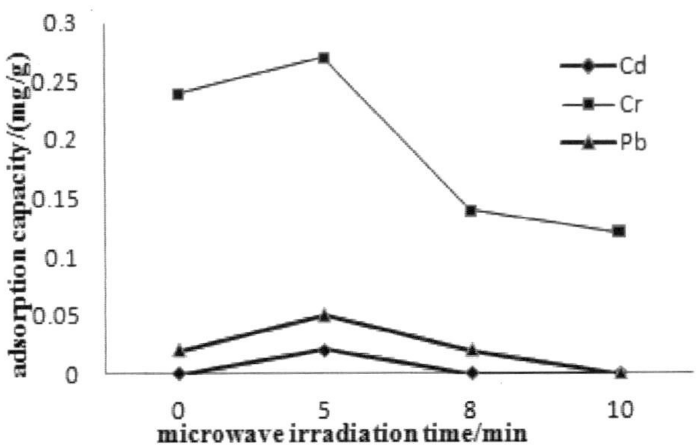

Figure 1. The changes of adsorption of modified diatomite under different microwave irradiation time.

As shown in figure 1, in the range of 0~5min, increase the time of microwave irradiation can improve the adsorption properties of modified diatomite. When the irradiation time is 5min, the adsorption property of modified diatomite to heavy metals is the best. While the irradiation time is longer than 5min, the adsorption capacity decreased with increasing of irradiation time and even less than the diatomite without microwave irradiation. This is mainly because the part of impurities and moisture in diatomite can be removed with appropriate microwave irradiation, promote the formation of the pore and improve the adsorption performance of diatomite. Further prolonging the time of microwave irradiation will increase the temperature of diatomite, destroy the structure and even cause the pulverization of diatomite, thus reducing the adsorption ability of diatomite to heavy metal ions.

3.2 Effect of solid - liquid ratio on the modified diatomite

The mixture of diatomite, hydrochloric acid and deionized water in the mass ratio of 1/5, 1/7, 1/10, 1/20 are prepared. The microwave modification conditions was set for microwave power 490 W and irradiation time 5 min. 0.1 g modification diatomite was added into the 10mL landfill leachate. The mixture was placed in a incubator at 25℃ and rotated for 40 min at 200 rpm. The supernatant was separated by centrifugation and the concentration of heavy metals was measured by ICP. The result is shown in figure 2.

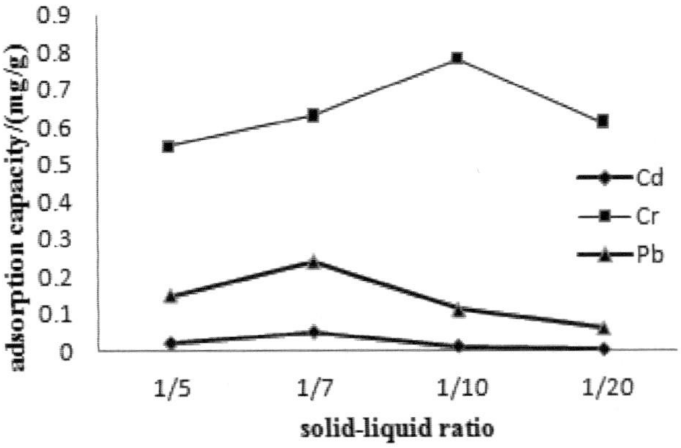

Figure 2. The changes of adsorption of modified diatomite under different solid-liquid ratios.

It can be seen from figure 2 that the best solid-liquid ratio of diatomite in 3 heavy metal ions is also different. For Pb and Cd, the best solid-liquid ratios of modified diatomite are both 1/7, as for Cr, it is 1/10. The hydrochloric acid will increase the surface roughness of diatomite particles, form more micropores and grooves. The greater the specific surface area, the stronger the adsorption capacity of diatomite.

3.3 Effect of adding amount on adsorption of heavy metals

0.05 g, 0.07 g, 0.1 g, 0.15 g and 0.2g modified diatomites were added to 10 mL landfill leachate respectively. The mixture was placed in a incubator at 25℃ and rotated for 40 min at 200 rpm. The supernatant was separated by centrifugation and the concentration of heavy metals was measured by ICP. The result was shown in figure 3.

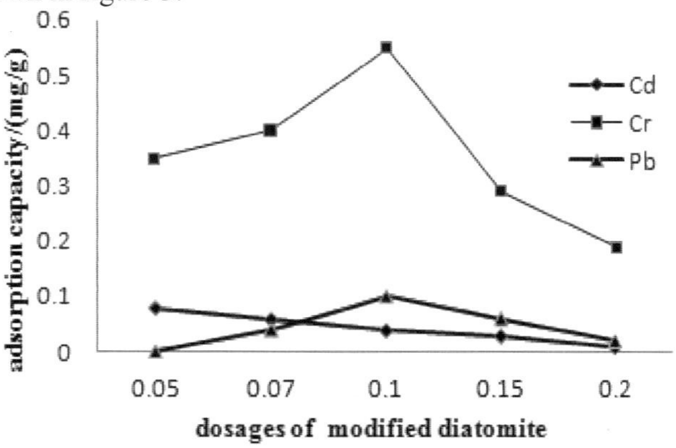

Figure 3. The changes of adsorption of modified diatomite under different dosages.

As shown in figure 3 we can see, the adsorption capacity of Cd would decreased with the modified diatomite dosage increased. The best dosage of modified diatomite was 0.05 g, i.e. 5 g/L. As for Pb and Cr, the adsorption performance was the best with the modified diatomite dosage of 0.1 g, i.e., the modified diatomite dosage was 10 g/L.

3.4 Effect of adsorption time on adsorption of heavy metals

0.1 g modified diatomite was added to the landfill leachate and put in 25 ℃ incubator. A series of reaction time, i.e., 20 min, 40 min, 60 min, 90 min, 120 min, were set with the 200 rpm respectively, The supernatant was separated by centrifugation and the concentration of heavy metals was measured by ICP. The result was shown in figure 4.

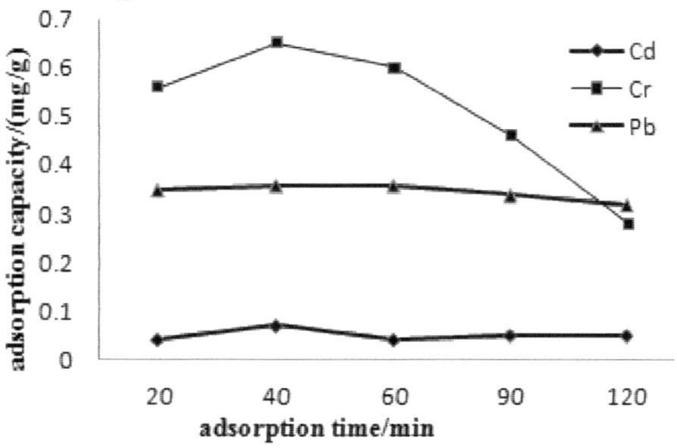

Figure 4. The changes of adsorption of modified diatomite under different adsorption time.

It was shown from figure 4, when the adsorption time was less than 40 min, the modified diatomite adsorption curve of heavy metals increased with the extension of adsorption time. When the adsorption time reached 40min, the modified diatomite adsorption properties for heavy metal ions was best. While the adsorption time was prolonged, the adsorption properties of modified diatomite for heavy metal ions decreased. Therefore, the best adsorption time of the modified diatomite to heavy metal ions was 40min.

3.5 Effect of adsorption temperature on adsorption of heavy metals

0.1 g modified diatomite was added to the landfill leachate and put the mixture in an incubator. A series of reaction temperature, i.e., 25℃, 35℃, 45℃, were set and rotated for 40 min at 200 rpm. The supernatant was separated by centrifugation and the concentration of heavy metals was measured by ICP. The result was shown in figure 5.

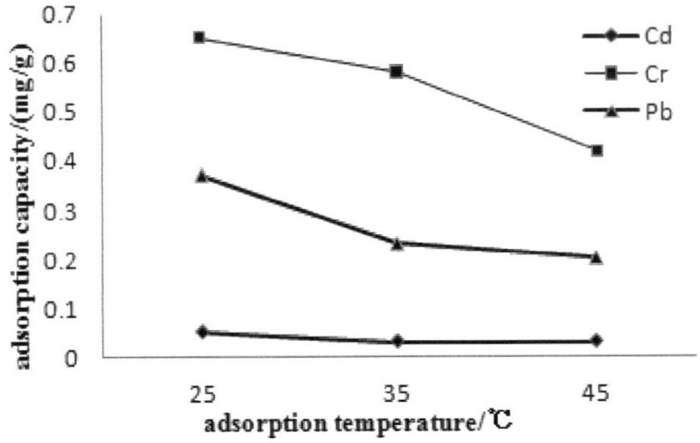

Figure 5. The changes of adsorption of modified diatomite under different adsorption temperature.

The adsorption temperature is one of the important factors affecting the adsorption properties of Modified Diatomite on heavy metal ions. The adsorption of heavy metal ions by diatomite will increase with the increase of temperature, but overheated is not conducive to the adsorption. According to figure 5, the best adsorption temperature of modified diatomite on heavy metal ions is 25℃.

3.6 Effect of pH on adsorption of heavy metals

The pH of landfill leachate was regulated by adding some hydrochloric acid and sodium hydroxide. Added 0.1g modified diatomite to the 10ml landfill leachate with different pH respectively, put the mixture in an incubator and rotated for 40 min at 200rpm. The supernatant was separated by centrifugation and the concentration of heavy metals was measured by ICP. The result was shown in figure 6.

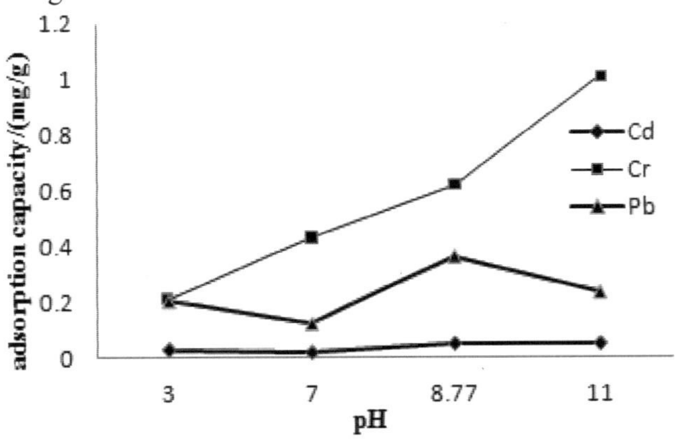

Figure 6. The changes of adsorption of modified diatomite under different pH.

As shown in figure 6, for Cr, the landfill leachate was acidic when the pH was low and H^+ will produce competitive adsorption, which was not conducive to the adsorption. With the increase of pH, the competitiveness of H^+ was reduced, and the removal efficiency and adsorption performance of modified diatomite on Cr were obviously improved. The main reason lies in that heavy metal ions are removed mainly by binding to the OH^- in the solution under alkaline conditions. As for Pb and Cd, the removal efficiency of the modified diatomite can reach the highest because the pH value of the landfill leachate was weak alkaline (pH 8.77) in the natural environment.

3.7 Effect of ionic strength on adsorption of heavy metals

A different amount of sodium chloride solid was added to the leachate to change its ionic strength, 0.01 mol/L, 0.1 mol/L, 1 mol/L, respectively. Added 0.1g modified diatomite to the 10ml landfill leachate with different pH respectively, put the mixture in an incubator and rotated for 40 min at 200rpm. The supernatant was separated by centrifugation and the concentration of heavy metals was measured by ICP. The result was shown in figure 7.

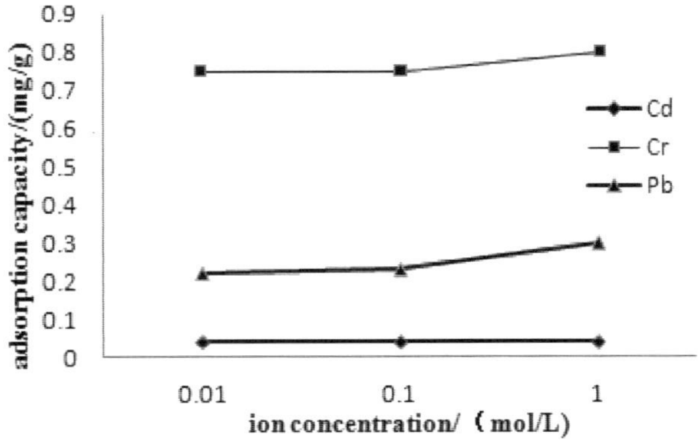

Figure 7. The changes of adsorption of modified diatomite under different ion concentrations.

As shown in figure 7, compared with landfill leachate, adding sodium chloride solid to change the ionic strength of landfill leachate improved the adsorption performance of modified diatomite to Cr, while it don't have much influence on Pb and Cd . So a certain amount of sodium chloride solid can be added to improve the removal efficiency of modified diatomite on Cr.

4. Conclusion

The adsorption properties of microwave modified diatomite on heavy metal ions such as lead, chromium and cadmium in landfill leachate under different conditions were studied, and the best microwave modification conditions were determined.

The removal efficiency of heavy metal ions such as lead, chromium and cadmium in landfill leachate under different adsorption conditions was also investigated, including the dosage of adsorbent, adsorption time and temperature, pH and ionic strength of landfill leachate. Some conclusions can be drawn as follows:

(1) The best microwave modification conditions for diatomite are diatomite and 1mol/L hydrochloric acid and deionized water, among which the solid-liquid ratio is 1:(7~10), the microwave power is 490W, and the microwave irradiation time is 5min.

(2) The best adsorption conditions for removing heavy metal ions such as lead, chromium and cadmium in landfill leachate by modified diatomite were low ionic strength and alkaline landfill leachate. The adsorption temperature was 25℃, the adsorption time was 40min, and the dosage of modified diatomite was 5~10g/L.

Acknowledgments:

This research was supported by funding (type A) from Fujian Educational Committee with Grant JA15371; funding from XMUT with Grant 0900200611; Key Laboratory of Environmental Biotechnology, Fujian Province University.

References

[1] Renou S, Givaudan JG, Poulain S, Dirassouyan F, Moulin P. Landfill leachate treatment: Review and opportunity. J Hazard Mater 2008, 150(3):468–493.

[2] Aktas O, Cecen F. Addition of activated carbon to batch activated sludge reactors in the treatment of landfill leachate and domestic wastewater. J Chem Technol Biotechnol 2001, 76:793–802.

[3] Zazouli MA, Yousefi Z. Removal of heavy metals from solid wastes leachate by Coagulation-Flocculation process. J Appl Sci 2008, 8:2142–2147.

[4] Zazouli MA, Maleki A, Izanloo H. Assessment of raw leachate characteristics and its

pretreatment by Lime. Asian Journal of Chemistry 2010, 22:6155–6163.
[5] G. Gustavo, C. Edgar, C. Saúl, H. Jonas, M. Johanne, Synthesis of zeolite Y from diatomaceous earth as silica source, Microporous Mesoporous. Mater. 2016, 219:29–37.
[6] Wu JL, Yang YS, Lin JH. Advanced tertiary treatment of municipal wastewater using raw and modified diatomite. Journal of Hazardous Materials B. 2005,127(1-3):196-203 .
[7] Su YW, Yang ZJ, Zhou YZ. A review of the adsorption of heavy metal ions by diatomite and its application to the remediation of environmental heavy metal pollution. Non-metallic mineral industry in China. 2007, 28(1): 94-101.
[8] Liu P, Zhao QR, Yuan LB. Adsorption of Pb (II) by Modified Diatomite. Yunnan Chemical Technology. 2003,30(5):11-13.
[9] Patricia M, Carolina M, Enrique CU. Cd^{2+} adsorption on alkaline-pretreated diatomaceous earth: equilibrium and thermodynamic studies. Environmental Chemistry Letters. 2011, 9(1):55-63.
[10] Luo DC, Liu JF. Study on the adsorption properties of modified datomite on Pb^{2+}, Cu^{2+} and Zn^{2+} in Wastewater. China Mining. 2005, 14(7): 69-71.
[11] Cai HL, Wang ZJ. Study on the preparation of calcium carbonate modified diatomite and its adsorption properties. Guangzhou Chemical Industry. 2011, 39(17): 43-46.
[12] Aldegs Y, Khraisheh MA, Tutunji MF. Sorption of lead ions on diatomite and manganese oxides modified diatomite. Water Research. 2001, 35(15):3724-3728.

A Type-2 Fuzzy Based Interval-Stochastic Risk Management Programming Model for Identifying Sustainable Water Resources Allocation Policies

J Liu[1*] and Y P Li[1]

[1]Department of Environmental Engineering, Key Laboratory of Environmental Biotechnology (XMUT), Fujian Engineering and Research Center of Rural Sewage Treatment and Water Safety, Xiamen University of Technology, Xiamen 361024, China

[*]Corresponding Author, Department of Environmental Engineering, Xiamen University of Technology, Xiamen 361024, China; E-mail: zyljing@126.com

Abstract. In this study, a type-2 fuzzy based interval-stochastic risk management programming (TISRM) method is proposed for optimal irrigation water resources allocation associated with multiple uncertainties expressed as interval numbers, probability distributions, and type-2 fuzzy sets. TISRM can also reflect authorities' attitudes towards system risk using a financial risk management measure by controlling the variability of the recourse cost. The developed method is applied to irrigation water resources allocation for the Zhangweinan River Basin in China. Solutions of irrigation water allocation under different probability distributions and confidence levels are generated. Results reveal that system benefit and satisfaction degree would decrease with the raised λ levels; the CVaR values would change with the increment of λ and β levels as well as the varied cases; the irrigation targets of all crops in all subareas are reach their lower bounds; the irrigation deficits would be different with risk levels and cases. The results can help the local authority to adjust the current food security policy.

1. Introduction

The optimal allocation of irrigation water resources is of great importance to guarantee food security taking into consideration the burgeoning global population which is expected to reach 9.7 billion in 2050 [1]. A number of mathematical techniques have been proposed to analyze economic, social and environmental impacts of alternative allocation actions, and thus aid the authorities in determining sustainable water resources allocation policies. Besides, the irrigation water resources allocation systems are often complicated due to the random available water and cultivated land conditions, varied water requirements, and changed market situations and their interrelationships [2]. Effective irrigation water resources allocation integrating uncertainties into a general optimization framework is essential.

Previously, a range of optimization methods were developed for irrigation water resources allocation in terms of interval-parameter programming (IPP), fuzzy mathematical programming (FMP), and stochastic mathematical programming (SP) [3-6]. For instance, Srivastava and Singh [7] proposed a fuzzy multiobjective-based goal programming (GP) model for optimizing crop cultivation patterns, where fuzzy solutions provide the necessary aspiration and tolerance levels for goal programming. Chen et al., [8] developed an interval multistage water allocation model to obtain the maximum food production in reservoir irrigation systems, where rainfall was divided into five levels with probability distributions in each growth stage and other parameters were presented as intervals. Robert et al. [9] employed a stochastic dynamic programming model to figure out decisions on crop selection, irrigation investment and water allocation rates with different probabilities of borewell failure over a long-term planning horizon. Generally, the above models have advantages in their effectiveness for coping with single and/or dual uncertainties in water resources allocation system. Nevertheless, in many irrigation water resources allocation problems, several types of uncertainties often exist in a complex system. The conventional models have difficulty in dealing with multiple uncertainties. Specifically, SMP only take into account the expected loss/penalty due to the violated policies and ignores the risk caused by randomness [10]. Financial risk management (FRM) is an effective tool to cope with the above shortcoming. Yamout et al. [11] coupled FRM and two-stage stochastic programming for water resources allocation. Soltani et al. [12] employed a FRM named conditional value at risk (CVaR) for simultaneous agricultural water and return flow (waste load) allocation in rivers. Hu et al. [13] developed a multi-objective model to help water managers mitigate problems due to water scarcity, where CVaR was integrated into the model constraints to control the economic efficiency loss risk corresponding to variations in water availability.

Therefore, this study aims to develop a type-2 fuzzy based interval-stochastic risk management programming (TISRM) method for supporting irrigation water resources allocation. Through integrating IPP, TSP, FRM and T2FS into a general framework, TISRM cannot only deal with uncertainties expressed as interval numbers, probability distributions, and type-2 fuzzy sets, but also effectively quantify risks due to randomness. Then, the TISRM method will be applied to a real-world case for irrigation water resources allocation (surface water and groundwater) in the Zhangweinan River Basin, North China. A number of cases associated with different attitudes of authorities towards the objective function value and constraints as well as varied available water levels will be analyzed to obtain optimal water allocation patterns. The relationship among cropland planning and irrigation water utilization will be disclosed to obtain a maximized system benefit and a minimized system-failure risk.

2. Case study

2.1 Overview of the study area

The Zhangweinan River Basin (with longitude of 112~118°E and latitude of 35~39°N) is a tributary of the Haihe River (in North China), which covers parts of Shanxi, Henan, Hebei, and Shandong Provinces as well as Tianjing Municipality; total area of the basin is approximately 37,700 km^2. The basin is one of the most productive and intensively cultivated agricultural regions in North China; main cultivated crops include wheat, maize, and cotton. In the study area, more than 75% of the cropland needs to be irrigated (consuming 70-80% of the total water resources allocation) because the insufficient annual precipitation cannot fulfill the crop water requirement. Conjunctive use of surface water and groundwater is practiced in the study area to irrigate more croplands, increase agricultural production, and maintain agricultural system balance. Yuecheng Reservoir is the main reservoir for flow regulation for surface water supply in this basin [2]. The estimated amount of renewable groundwater resources in the study area is 3.635 billion m^3/year.

2.2 Problem statement

Decreasing precipitation and increasing reservoirs and diversion channels in the upper reaches of the river lead to much less inflow of Yuecheng Reservoir now than that in decades ago.

In addition, the local authority initiates strategic changes to transfer most agricultural water to other sectors (e.g., municipal and industrial sectors). As a result, crop production relies heavily on groundwater irrigation. Consequent excessive groundwater pumping has led to continuous drawdown (averagely 1.5 m per year) in the water table during the last few decades. The implications of over-pumping include higher pumping cost, wasteful energy use, salt water intrusion, increased soil salinity and waterlogging, soil compaction, and land subsidence. These issues have made the regional agricultural water system degrade (which cannot guarantee food security and can bring economic loss) and brought about severe damage to the local ecological and environmental sustainability. Therefore, advancing optimization models for finding the optimal strategy for the crops cultivation as well as the amount of surface water allocation and groundwater extraction with the consideration of environmental sustainability and economic development is essential.

3. Methodology

A type-2 fuzzy based interval-stochastic risk management programming (TISRM) model can be formulated by incorporating TSP, IPP, T2FS, and FRM into a general framework [14-17]:

Max $\xi^{\pm}$

Subject to:

$$(1-\lambda)\sum_{j=1}^{n_1} c_j^{\pm} x_j^{\pm} - \sum_{j=1}^{n_2}\sum_{s=1}^{S} p_s e_j^{\pm} y_{js}^{\pm} + \lambda \cdot (\eta^{\pm} - \frac{1}{1-\beta}\sum_{s=1}^{S} p_s V_s^{\pm}) \geq \underline{f}^{\pm} + \Delta\tilde{f}^{\pm}\xi^{\pm} \quad (1.1)$$

$$\sum_{j=1}^{n_1} a_{rj}^{\pm} x_j^{\pm} \leq \bar{b}_r^{\pm} - \Delta\tilde{b}_r^{\pm}\xi^{\pm} \quad (1.2)$$

$$\sum_{j=1}^{n_1} a_{tj}^{\pm} x_j^{\pm} + \sum_{j=1}^{n_2} a_{tj}'^{\pm} y_{js}^{\pm} \leq \bar{\omega}_{ts}^{\pm} - \Delta\tilde{\omega}_{ts}^{\pm}\xi^{\pm}, \ t = 1, 2, ..., m_2; \ s = 1, 2, ..., S \quad (1.3)$$

$$V_s^{\pm} \geq \eta^{\pm} - [\sum_{j=1}^{n_1} c_j^{\pm} x_j^{\pm} - \sum_{j=1}^{n_2} e_j^{\pm} y_{js}^{\pm}], \ \forall s = 1, 2, ..., S \quad (1.4)$$

$$x_j^{\pm} \geq 0, \ j = 1, 2, ..., n_1 \quad (1.5)$$

$$y_{js}^{\pm} \geq 0, \ j = 1, 2, ..., n_2; \ s = 1, 2, ..., S \quad (1.6)$$

$$V_s^{\pm} \geq 0, \ \forall s = 1, 2, ..., S \quad (1.7)$$

$$0 \leq \xi^{\pm} \leq 1 \quad (1.8)$$

Where $\xi^{\pm}$ is a control variable corresponding to the membership grade of the fuzzy decision.

A $\xi^{\pm}$ level close to 1 corresponding to a solution with a high degree of satisfaction for the fuzzy goal/constrains; conversely, a $\xi^{\pm}$ level approaching 0 relates to a solution that has a low degree of satisfaction for the fuzzy goal/constraints. $x_j^{\pm}$ is the first-stage anticipated decisions made before the random variables are observed; $\omega_s^{\pm}$ is the random variable which takes interval discrete value with a probability level p_s; Denote p_s as the probability of realization of scenario *s*, with $p_s > 0$ and $\sum_{s=1}^{S} p_s = 1$; $y_{js}^{\pm}$ is the second-stage decision variables (i.e. recourse variables) that depend on the realization of the first-stage random vector. β is the confidence level; typically β often range from 0.9 to 0.99. In order to maximize the system benefit and $V_s^{\pm}$ is the auxiliary decision variables, which is used to compute the conditional value-at-risk under level *s*.

4. Modeling formulation

Optimization of cropland planning and water resources allocation which considers the balance of socio-economic development, food security, resources conservation and environment protection can effectively support local sustainable development. In this study, the authorities are responsible for allocating surface water and groundwater to three main crops (i.e. wheat, maize, and cotton). The study area is divided into fifteen irrigation districts. A one-year planning horizon is subdivided into four periods on the basis of the water shortage response, including winter of last year, spring, summer and autumn of current year. The growth stages of wheat and maize cover periods one to three, while the growth stage of cotton spans across periods three and four. Therefore, the TISRM model for the agricultural system management can be formulated as follows:

$$\text{Max } \xi^{\pm} \tag{2.1}$$

(1) Constraint for objective function:

$$\begin{aligned}&(1-\lambda)\times\sum_{i=1}^{I}\sum_{j=1}^{J}\sum_{t=1}^{T}TS_{ijt}^{\pm}\times SB_{ijt}^{\pm}-\sum_{i=1}^{I}\sum_{j=1}^{J}\sum_{t=1}^{T}\sum_{s=1}^{S}p_s\times CS_{ijt}^{\pm}\times SA_{ijts}^{\pm}\\&+(1-\lambda)\times\sum_{i=1}^{I}\sum_{j=1}^{J}\sum_{t=1}^{T}TG_{ijt}^{\pm}\times SG_{ijt}^{\pm}-\sum_{i=1}^{I}\sum_{j=1}^{J}\sum_{t=1}^{T}\sum_{s=1}^{S}p_s\times CG_{ijt}^{\pm}\times GA_{ijts}^{\pm}\\&+\lambda\times(\eta^{\pm}-\frac{1}{1-\beta}\sum_{s=1}^{S}p_sV_s^{\pm})\geq \underline{f}^{\pm}+\Delta\tilde{f}^{\pm}\xi^{\pm}\end{aligned}\tag{2.2}$$

(2) Surface water supply constraints:

$$\sum_{i=1}^{I}\sum_{j=1}^{J}WA_{ijt}^{\pm}\times(TS_{ijt}^{\pm}-SA_{ijts}^{\pm})-\varepsilon_{(t-1)s}^{\pm}\leq \bar{R}_{ts}^{\pm}-\Delta\tilde{\hat{R}}_{ts}^{\pm}\times\xi^{\pm},\ \forall t,\ s \tag{2.3}$$

$$\varepsilon_{(t-1)s}^{\pm}+\sum_{i=1}^{I}\sum_{j=1}^{J}WA_{ij(t-1)}^{\pm}\times(TS_{ij(t-1)}^{\pm}-SA_{ij(t-1)s}^{\pm})-\varepsilon_{(t-2)s}^{\pm}=\bar{R}_{(t-1)s}^{\pm}-\Delta\tilde{\hat{R}}_{(t-1)s}^{\pm}\times\xi^{\pm},\ \forall t,\ s \tag{2.4}$$

(3) Groundwater supply constraint:

$$\sum_{i=1}^{I}\sum_{j=1}^{J}(TG_{ijt}^{\pm}-GA_{ijts}^{\pm})\times WA_{ijt}^{\pm}\leq\bar{Q}_{ts}^{\pm}-\Delta\tilde{\hat{Q}}_{ts}^{\pm}\times\xi^{\pm},\ \forall t,\ s \tag{2.5}$$

(4) Irrigation requirement constraint:

$$\sum_{j=1}^{J}WA_{ijt}^{\pm}\times[(TS_{ijt}^{\pm}-SA_{ijts}^{\pm})+(TG_{ijt}^{\pm}-GA_{ijts}^{\pm})]\leq TWD_{it\ \max}^{\pm},\ \forall i,\ t \tag{2.6}$$

$$\sum_{j=1}^{J}\delta_j^{\pm}\times WA_{ijt}^{\pm}\times\left[\alpha_j^{\pm}\times(TS_{ijt}^{\pm}-SA_{ijts}^{\pm})+(TG_{ijt}^{\pm}-GA_{ijts}^{\pm})\right]\geq TWB_{it\ \min}^{\pm},\ \forall i,\ t \tag{2.7}$$

$$WA_{ijt}^{\pm}=ETc_{ijt}^{\pm}-RI_{ijt}^{\pm}+\theta_{ij(t+1)}^{\pm}D_{ij(t+1)}^{\pm}-\theta_{ijt}^{\pm}D_{ijt}^{\pm}-\theta_{ijt\max}^{\pm}(D_{ij(t+1)}^{\pm}-D_{ijt}^{\pm})+DP_{ijt}^{\pm},\ \forall i,\ j,\ t \tag{2.8}$$

$$\theta_{ijt\ \min}^{\pm}\leq\theta_{ijt}^{\pm}\leq\theta_{ijt\ \max}^{\pm},\ \forall i,\ j,\ t \tag{2.9}$$

$$D_{ijt}^{\pm}\leq D_{ij(t+1)}^{\pm},\ \forall i,\ j,\ t \tag{2.10}$$

(5) Aquifer hydrologic balance Constraints:

$$\begin{aligned}&\sum_{i=1}^{I}\sum_{t=1}^{T}rcl_j^{\pm}\times WA_{ijt}^{\pm}\times[(TS_{ijt}^{\pm}-SA_{ijts}^{\pm})+(TG_{ijt}^{\pm}-GA_{ijts}^{\pm})]\\&+\sum_{i=1}^{I}\sum_{t=1}^{T}rrf_j^{\pm}\times RF_{ijt}^{\pm}\times(TS_{ijt}^{\pm}+TG_{ijt}^{\pm})/100+\sum_{i=1}^{I}\sum_{t=1}^{T}ral_j^{\pm}\times(TS_{ijt}^{\pm}-SA_{ijts}^{\pm})\times WA_{ijt}^{\pm}\\&-\sum_{i=1}^{I}\sum_{t=1}^{T}(TG_{ijt}^{\pm}-GA_{ijts}^{\pm})\times WA_{ijt}^{\pm}\geq PMA_j^{\pm},\ \forall j\end{aligned}\tag{2.11}$$

$$PMA_j^{\pm}=\sum_{i=1}^{I}\sum_{t=1}^{T}\Delta h^{\pm}\times(TS_{ijt}^{\pm}+TG_{ijt}^{\pm})\times SY_j^{\pm},\ \forall j \tag{2.12}$$

(6) Land area constraints:

$$0\leq SA_{ijts}^{\pm}\leq TS_{ijt}^{\ \pm}\leq TS_{ijt\ \max}^{\pm},\ \forall i,\ j,\ t,\ s \tag{2.13}$$

$$0\leq GA_{ijts}^{\pm}\leq TG_{ijt}^{\ \pm}\leq TG_{ijt\ \max}^{\pm},\ \forall i,\ j,\ t,\ s \tag{2.14}$$

$$TS_{ijt}^{\pm}=TS_{ij(t-1)}^{\pm}\ \ \forall i,\ j,\ t \tag{2.15}$$

$$TG_{ijt}^{\pm}=TG_{ij(t-1)}^{\pm}\ \ \forall i,\ j,\ t \tag{2.16}$$

(7) Risk management constraint:

$$\begin{aligned}&V_s^{\pm}\geq\eta^{\pm}-\sum_{i=1}^{I}\sum_{j=1}^{J}\sum_{t=1}^{T}(TS_{ijt}^{\pm}\times SB_{ijt}^{\pm}+TG_{ijt}^{\pm}\times SG_{ijt}^{\pm})\\&+\sum_{i=1}^{I}\sum_{j=1}^{J}\sum_{t=1}^{T}(CS_{ijt}^{\pm}\times SA_{ijts}^{\pm}+CG_{ijt}^{\pm}\times GA_{ijts}^{\pm}),\ \forall s=1,\ 2,\ \ldots,\ S\end{aligned}\tag{2.17}$$

(8) Non-negativity constraints:

$$SA^{\pm}_{ijts},\ WA^{\pm}_{ijt},\ GA^{\pm}_{ijts} \geq 0,\ \ \forall i,\ j,\ t,\ s \tag{2.18}$$

The detailed nomenclatures for the variables and parameters are presented in the "**Appendix**". The planning horizon is divided into periods according to growth stages of the three crops. Interval numbers are adopted to handle other uncertain parameters such as unit benefit of crop, irrigation target, and economic penalty.

5. Results and discussion

5.1 Satisfaction degree and system benefit

In this study, six risk weight levels (λ = 0, 0.2, 0.4, 0.6, 0.8, and 1) and three confidence levels (β = 0.9, 0.95, and 0.99) as well as six cases (which represent different attitudes of authorities towards the objective value and constraints) are taken into account to identify optimal agricultural water allocation policies within multiple subareas for irrigating multiple crops over a multi-period planning horizon. Figures 1 and 2 show the satisfaction degrees ($\xi^{\pm}_{\text{opt}}$) and system benefits ($f^{\pm}_{\text{opt}}$) under different risk levels (i.e. λ and β) and cases, respectively. It is indicated that no feasible solutions can be obtained under cases 3, 5, and 6. Case 3 represents that authorities are conservative to achieve high system benefit but, at the same time, they take optimistic attitudes towards constraints (e.g., available water resources), revealing a conflicting relationship between system benefit and constraints. Under cases 5 and 6, authorities are neutral in terms of system benefit, while they respectively hold optimistic and pessimistic attitudes towards constraints. It is revealed that the neutral attitudes to objective value are not appropriate to the allocation problem in this study. The solutions under cases 1, 2 and 4 are mostly presented as intervals, reflecting potential system condition variations caused by uncertain modeling inputs. It is indicated that $\xi^{\pm}_{\text{opt}}$ and $f^{\pm}_{\text{opt}}$ would varied with the changed cases. For instance, when λ = 0.2 and β = 0.9, the value of $\xi^{\pm}_{\text{opt}}$ would be [0.578, 0.661] (with $f^{\pm}_{\text{opt}}$ = [430.2, 1309.2] × 10^6 RMB¥), [0.536, 0.791] (with $f^{\pm}_{\text{opt}}$ = [532.1, 1284.5] × 10^6 RMB¥), and [0.541, 712] (with $f^{\pm}_{\text{opt}}$ = [488.1, 850.7] × 10^6 RMB¥) under cases 1, 2, and 4, respectively. Generally, f^{-}_{opt} would be ranked as: $(f^{-}_{\text{opt}})_{\text{case 1}} < (f^{-}_{\text{opt}})_{\text{case 4}} < (f^{-}_{\text{opt}})_{\text{case 2}}$; f^{+}_{opt} would be ranked as: $(f^{+}_{\text{opt}})_{\text{case 1}} > (f^{+}_{\text{opt}})_{\text{case 2}} > (f^{+}_{\text{opt}})_{\text{case 4}}$. It is disclosed that the values of $f^{\pm}_{\text{opt}}$ would be highly dependent on the authorities subjective attitudes. Under case 1, authorities would be aggressive to obtain high system benefit and take positive attitudes towards related constraint, leading to wider interval value of $f^{\pm}_{\text{opt}}$. It is also displayed that system benefit and satisfaction degree would decrease with the raised λ levels. For example, under case 1 and β = 0.99, the values of $\xi^{\pm}_{\text{opt}}$ and $f^{\pm}_{\text{opt}}$ would be [0.662, 0.921] (with $f^{\pm}_{\text{opt}}$ = [560.5, 1426.8] × 10^6 RMB¥), [0.449, 0.478] (with $f^{\pm}_{\text{opt}}$ = [324.5, 1169.2] × 10^6 RMB¥) , [0.148, 0.352] (with $f^{\pm}_{\text{opt}}$ = [244.5, 992.24] × 10^6 RMB¥), and [0.012, 0.289] (with $f^{\pm}_{\text{opt}}$ = [173.9, 516.38] × 10^6 RMB¥) under λ = 0, 0.4, 0.8, and 1, respectively. This is because that higher λ levels correspond to higher values of risk weight to minimize the financial risk, reflecting the risk aversion attitudes of authorities. When λ = 0, possess a risk-neutral attitude and would not consider the variability of the uncertain recourse costs. Besides, system benefits and satisfaction degrees would also reduce with the raised β levels. This is due to the fact that lower β levels reflect higher system reliability, leading to lower satisfaction degree and system benefits.

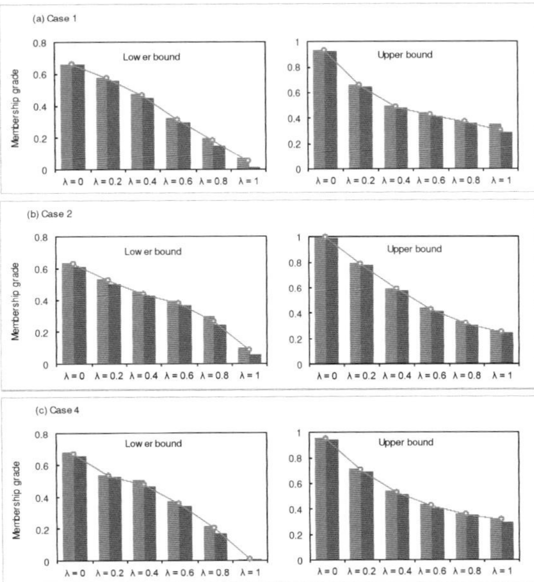

Figure 1. Satisfaction degree

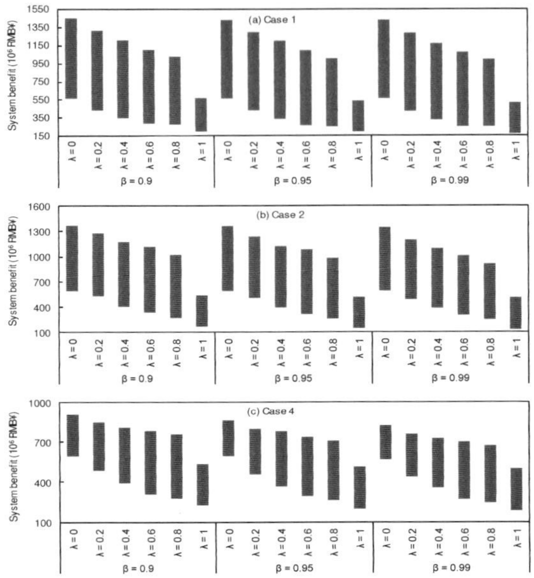

Figure 2. System benefit

5.2 *Optimal irrigation target*

Figure 3 depicts the optimal targeted crop area irrigated by surface water and groundwater under different risk levels and cases. It is observed that optimal irrigation target would decrease with the raised λ and β levels. When λ = 0 and β = 0.9, the irrigation targets of all crops in all subareas are set as their upper bounds, revealing authorities' positive attitudes towards water available to crops. The optimal targeted crop area irrigated by surface water and groundwater would be 186109.4 ha and 924647.6 ha. When λ = 1 and β = 0.99, the irrigation targets of all crops in all subareas are reach their lower bounds, presenting authorities' negative attitudes towards water available to crops. The optimal targeted crop area irrigated by surface water and groundwater would be 172919.5 ha and 569222.5 ha. As the λ and β levels rise, the optimal target of wheat would firstly shrink due to its low benefit and penalty, revealing its less competiveness than maize and cotton. Moreover, the optimal irrigation area would be ranked as: case 1 > case 4 > case 2, which is mainly attributed to the different attitudes of authorities towards constraints (e.g., available water resources, canal capacity and total water shortage). Generally, optimal irrigation targets could be identified under varying system conditions. Variation in the values of crop area targets reflects different policies for managing water resources and planning agricultural activities under uncertainty.

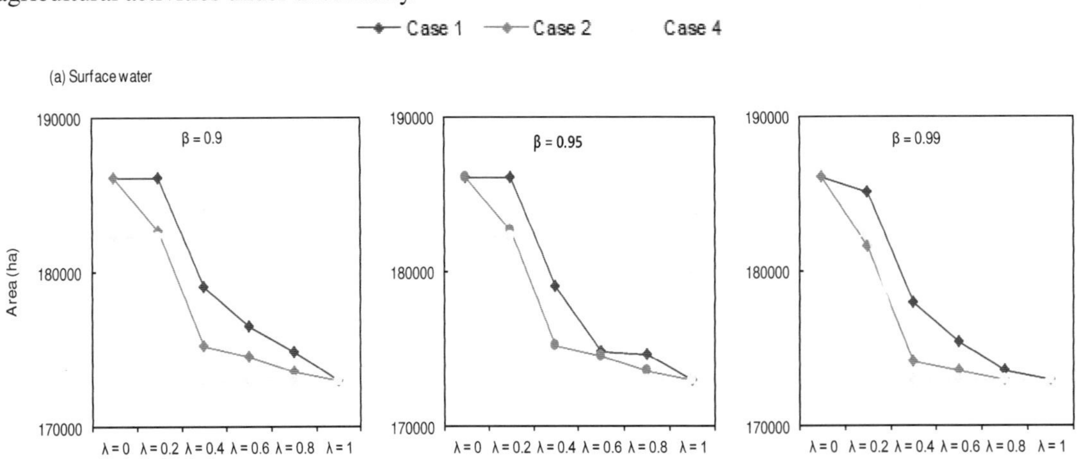

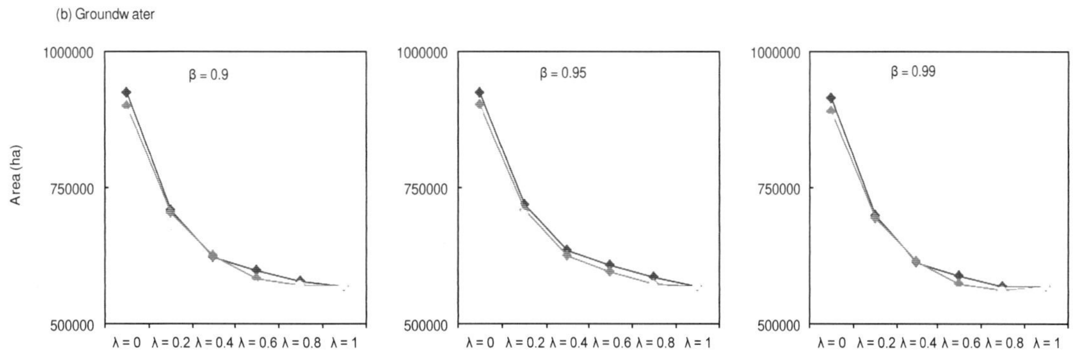

Figure 3. Optimal target of crops irrigated by surface water and groundwater

6. Conclusions

In this study, a type-2 fuzzy based interval-stochastic risk management programming (TISRM) method is proposed for optimal irrigation water resources allocation associated with multiple uncertainties expressed as interval numbers, probability distributions, and type-2 fuzzy sets. TISRM can also reflect authorities' attitudes towards system risk using a financial risk management measure by controlling the variability of the recourse cost. The developed method is applied to irrigation water resources allocation for the Zhangweinan River Basin in China. Solutions of irrigation water allocation under different probability distributions and confidence levels are generated. Results reveal that system benefit and satisfaction degree would decrease with the raised λ levels; the CVaR values would change with the increment of λ and β levels as well as the varied cases; the irrigation targets of all crops in all subareas are reach their lower bounds; the irrigation deficits would be different with risk levels and cases. The results can help the local authority to adjust the current food security policy.

Acknowledgement

This research was supported by the National Key Research Development Program of China (2016YFC0502803 and 2016YFA0601502) and the 111 Project (B14008).

References

[1] Singh A 2016 Optimal allocation of resources for increasing farm revenue under hydrological uncertainty Water Resources Management 30(7) 2569-2580
[2] Liu J, Li Y P, Huang G H, Zhuang X W, Fu H Y 2017 Assessment of uncertainty effects on crop planning and irrigation water supply using a Monte Carlo simulation based dual-interval stochastic programming method Journal of Cleaner Production 149 945-967.
[3] Halliburton S S, Abbara S, Chen M Y 2011 An inexact rough-interval fuzzy linear programming method for generating conjunctive water-allocation strategies to agricultural irrigation systems Applied Mathematical Modelling 35(9) 4330-4340.
[4] Nikoo M R, Kerachian R, Poorsepahy-Samian H 2012 An interval parameter model for cooperative inter-basin water resources allocation considering the water quality issues. Water Resources Management 26(11) 3329-3343.
[5] Peña-Haro S, Pulido-Velazquez M, Llopis-Albert C 2011 Stochastic hydro-economic modeling for optimal management of agricultural groundwater nitrate pollution under hydraulic conductivity uncertainty. Environmental Modelling & Software 26(8) 999-1008.
[6] Koichi U, Macarius Y, Abul H M, Badiul A, Gordana K B 2013 Stochastic control of a micro-dam irrigation scheme for dry season farming Stochastic Environmental Research and Risk Assessment 27 (1) 77-89.
[7] Srivastava P, Singh R M 2017 Agricultural land allocation for crop planning in a canal command area using fuzzy multiobjective goal programming. Journal of Irrigation & Drainage

Engineering 143(6) 04017007.
[8] Chen S, Shao D, Gu W 2017 An interval multistage water allocation model for crop different growth stages under inputs uncertainty Agricultural Water Management 186 86-97.
[9] Robert M, Bergez J E, Thomas A 2018 A stochastic dynamic programming approach to analyze adaptation to climate change-application to groundwater irrigation in India. European Journal of Operational Research 265(3) 1033-1045.
[10] Zhang C, Guo P 2017An inexact CVaR two-stage mixed-integer linear programming approach for agricultural water management under uncertainty considering ecological water requirement Ecological Indicators *Available online*.
[11] Yamout G M, Hatfield K, Romeijn H E 2007 Comparison of new conditional value-at-risk-based management models for optimal allocation of uncertain water supplies Water Resources Research 430(7) 116-130.
[12] Soltani M, Kerachian R, Nikoo M R 2016 A conditional value at risk-based model for planning agricultural water and return flow allocation in river systems Water Resources Management 30(1) 427-443.
[13] Hu Z, Wei C, Yao L 2016 A multi-objective optimization model with conditional value-at-risk constraints for water allocation equality. Journal of Hydrology 542 330-342.
[14] Ahmed S, Elsholkami M, Elkamel A, Du J, Ydstie E B, Douglas P L 2014 Financial risk management for new technology integration in energy planning under uncertainty Applied Energy 128 75-81.
[15] Li Y P, Huang G H 2009 Fuzzy-stochastic-based violation analysis method for planning water resources management systems with uncertain information Information Sciences 179 4261-4276.
[16] Liu J, Li Y P, Huang G H, Zeng X T, Nie S 2016 An integrated optimization method for river water quality management and risk analysis in a rural system Environmental Science and Pollution Research 23 477-497.
[17] Birbil S I, Kaynar B, Frenk J B G, Noyan N 2008 Risk measures and their applications in asset management In: Gregoriou, G.N. (Ed.) The VaRImplementation Handbook The McGraw-Hill Companies New York 311-337.

Appendix: Nomenclatures

i	crop, i = 1, 2, ..., I (I = 3 in this study)
j	subarea, j = 1, 2, …, J (J = 15 in this study)
t	planning time period, t = 1, 2, ..., T (T = 4 in this study);
s	level of water availability, s = 1, 2, …, 5 with s = 1 representing low level (L), s = 2 representing low-medium level (LM), s = 3 representing medium level (M), s = 4 representing medium-high level (MH), s = 5 representing high level (H).
$f^{\pm}$	expected system benefit, which presented as fuzzy goal with type-2 membership function (RMB¥);
$SB_{ijt}^{\pm}$	unit benefit of crop i in subarea j per unit of planting area irrigated by surface water (RMB¥/ha);
$TS_{ijt}^{\pm}$	allocation target for surface water that is promised to crop i in subarea j during period t (ha);
$CS_{ijt}^{\pm}$	economic penalty to crop i in subarea j per unit of area not irrigated by surface water (RMB¥/ha);
$SA_{ijts}^{\pm}$	cropland that not irrigated by surface water when available water level is s (ha);
$SG_{ijt}^{\pm}$	unit benefit of crop i in subarea j per unit of planting area irrigated by groundwater (RMB¥/ha);
$CG_{ijt}^{\pm}$	economic penalty to crop i in subarea j per unit of area not irrigated by groundwater (RMB¥/ha);
$GA_{ijts}^{\pm}$	cropland that not irrigated by groundwater when available water level is s (ha);

IOP Conf. Series: Earth and Environmental Science **146** (2018) 012020 doi:10.1088/1755-1315/146/1/012020

$\eta^{\pm}$	targeted benefit (RMB¥)
β	confidence level
p_s	probability of occurrence
V_s	auxiliary decision variables
$\xi^{\pm}$	control variable corresponding to the membership grade of the fuzzy decision
$WA_{ijt}^{\pm}$	irrigation quota for crop i in subarea j in period t (m^3/ha);
$\varepsilon_{(t-1)s}^{\pm}$	surplus flow when water is delivered in period t-1 under scenario s (m^3) , and assuming no spilling for reservoir;
$\varepsilon_{(t-2)s}^{\pm}$	surplus flow when water is delivered in period t-2 under scenario s (m^3) , and assuming no spilling for reservoir;
$\tilde{\hat{R}}_{ts}^{\pm}$	available water for irrigation in Yuecheng Reservoir under inflow level s during period t (m^3)
$\tilde{\hat{Q}}_{ts}^{\pm}$	available groundwater for irrigation under inflow level s during period t (m^3)
$TWD_{it\ max}^{\pm}$	maximum water demand of crop i during period t (m^3);
$\delta_j^{\pm}$	conveyance efficiency of canal system (fraction);
$\alpha_j^{\pm}$	field water application efficiency (fraction);
$TWB_{it\ min}^{\pm}$	minimum water demand of crop i during period t (m^3);
$ETc_{ijt}^{\pm}$	actual amount of evaporation of crop i in subarea j in period t (m);
$RI_{ijt}^{\pm}$	amount of effective precipitation of crop i in subarea j in period t (m);
$\theta_{ijt}^{\pm}$	soil moisture content of crop i in subarea j in period t (m/m);
$\theta_{ij(t-1)}^{\pm}$	soil moisture content of crop i in subarea j in period t-1(m/m);
$\theta_{ijt\ max}^{\pm}$	maximum soil moisture content of crop i in subarea j (m/m);
$\theta_{ijt\ min}^{\pm}$	minimum soil moisture content of crop i in subarea j (m/m);
$D_{ijt}^{\pm}$	average root depth of crops in period t (m);
$D_{ij(t-1)}^{\pm}$	average root depth of crops in period t-1 (m);
$DP_{ijt}^{\pm}$	deep percolation for crop i in subarea j during the period t (m);
$rcl_j^{\pm}$	recharge factor for irrigation application losses in subarea j (fraction);
$ral_j^{\pm}$	recharge factor for conveyance losses of canal water in subarea j (fraction);
$rrf_j^{\pm}$	recharge factor for rainfall in subarea j (fraction);
$PMA_j^{\pm}$	permissible annual groundwater mining allowance of the aquifer (m^3);
$\Delta h^{\pm}$	annual average groundwater table fluctuations (m);
$SY_j^{\pm}$	specific yield of aquifer (fraction);
$CQ_{ij}^{\pm}$	capacity of the canal for delivering surface water to crop i during period t (m^3)
$QS_{ij}^{\pm}$	capacity of the canal for delivering groundwater to crop i during period t (m^3);
$MS_t^{\pm}$	maximum total water shortage for all crops during period t (m^3);
$MG_i^{\pm}$	maximum water shortage for crop i over the planning horizon (m^3);
$\sigma_1^{\pm}$, $\sigma_2^{\pm}$	Coefficient to guarantee water supply to food crops;
$TS_{ijt\ max}^{\pm}$	maximum allowable planting area of crop i in subarea j irrigated by surface water (ha);
$TG_{ijt\ max}^{\pm}$	maximum allowable planting area of crop i in subarea j irrigated by groundwater (ha);
γ_t	Gini coefficient
$RMD_{ijt}^{\pm}$	labor force requirement of crop i in subarea j during period t (unit/ha);
$LA_{jt}^{\pm}$	labor force availability in subarea j during period t (unit)

Computational study of small-scale laminar coflow diffusion flames: influences of fuel dilution on the negative buoyant flame

Nasreldin M Mahmoud [1, 2,*] and Balla M Ahmed[1, 2]

[1]School of Automotive Engineering, Wuhan University of Technology, Wuhan 430070, China
[2]Mechanical Engineering Department, Faculty of Engineering, Sinnar University, Sinnar, Sudan

*Correspondence: Nasrmohmaed53@whut.edu.cn;

Abstract. The effects of adding H_2O and CO_2 on the hydrodynamic structure of laminar diffusion hydrocarbon flame was investigated at atmospheric pressure, focusing on the recirculation zone in the vicinity of nozzle exit with buoyancy force applied to the fuel. H_2O and CO_2 diluent were used to dilute the propane fuel focusing on how these diluents affect the vortices inside the recirculation zones. To clarify the influence of diluents, H_2O and CO_2 content (in mole basis %) in the fuel mixture was gradually increased from 0 to 20%. The numerical computations were performed using one-step gas-phase chemistry, thermal, and transport properties. The computational results of non-diluent propane fuel found to be in good agreement with available experimental data. The results showed that the fuel dilution notably influences the flow-field. In particular, the dilution of fuel with H_2O and CO_2 weakens the vortices formation. With increasing the diluents ratio, the scale of vortices structure becomes smaller. The results also showed that H_2O -diluent was more effective in reducing the vortex scale compared with CO_2-diluent.

1. Introduction

Diffusion flame exists in most practical combustion devices, and an accurate understanding of their structure is essential in the development of various combustion systems. The need to study a very simple geometric configuration is well fulfilled in the configuration of co-flow diffusion flame, where interactions between fluid dynamics, heat transfer and chemical reactions can be controlled and studied easily [1]. Therefore, co-flow diffusion flame is considered as the most frequently used. In addition, it provides a basic understanding of turbulent diffusion flames [2] and, easier to validate. Therefore, it is commonly used to improve the combustion models and characteristics such as gas-phase reaction [3], radiation [4], hydrodynamic structure [5], instability [6], polycyclic aromatic hydrocarbons (PAHs) [7,8] and soot [9,10].

It is a known fact that The formation of soot particles enhanced by flame flickering motions [11,12], Flame flickering motions are closely associated with the buoyancy-induced vortices near nozzle exit region [13,14], whereas these motions occur by accelerating the velocity of flow-field due to the buoyancy force [15,16]. Furthermore, buoyancy force distorted the parabolic velocity profile at an exit of fuel nozzle [13]. The relation of soot with the recirculation zone has also been addressed by many studies, Mueller et al. [17] conducted an experimental and computational study in turbulent nonpremixed bluff body flame, and they showed that the dynamics of the recirculation generate long residence time which supports surface growth, resulting in further soot growth.

Large-scale vortex structure was observed in both turbulent and laminar diffusion flame, according to the experimental study [18] which identified two of vortices structures: the first one observed on the oxidizer side close to the flame bulge, and the second inner vortex is due to shear layer instability. Furthermore, the toroidal vortex which identified at the end of a potential core, this vortex identified in heavy fuel when the coflow air velocity was lighter than that of the fuel jet [13,19,20]. The experimental study of propane jets with lifted flames reported that the flow-field in the vicinity of the nozzle could significantly influence when notable difference in density existed between air and fuel [21]. Recently in the experimental study of small laminar coflow diffusion flame [22] with different types of hydrocarbon fuels methane, ethylene, propane, and n-butane, This study shows that a new recirculation zones were observed near the exit of fuel nozzle exit with fuels notably heavier than coflow-air (propane and n-butane).

Dilute the fuel stream with the combustion products technology have been developed to lower flame temperatures and reduce both soot and NO_x formation [23,24]. Since H_2O and CO_2 are a major component of combustion products, they are most frequently used as a fuel diluent. It is fundamental and practical importance to understand how the addition of H2O and CO2 to fuel affects the near-nozzle flow fields, in particular, on the formation of vortices near the fuel nozzle exit.

Given the above considerations, the purpose of the current study seeks to assess the influences of adding H2O and CO2 on the near-nozzle flow field in small-scale laminar coflow propane diffusion flames at atmospheric pressure, focusing on the recirculation zone in the vicinity of nozzle exit.

2. Numerical model and computational details

The computational study was performed for small-scale laminar coflow flames, ANSYS Fluent software package was used [25]. Two-dimensional non-uniform mesh was used covering a rectangular domain with a length of 110 mm and width of 30 mm. Mesh with 11800 computational cells with finest spacing in the regions directly above the fuel nozzle exit while coarser mesh points are used in the air coflow region to avoid expensive of computation time. Since the accuracy of simulation results may be sensitive to inlet boundary conditions [22,26], therefore the inlet boundary condition of fuel specified at Z_i= -70 mm far upstream of the nozzle exit by imposing a fully developed parabolic velocity profile, while a flat velocity profile was assumed for the coflow stream at Z_i= -10 mm. A schematic of computational domain and details of connected boundary conditions implemented in this study are shown in Figure 1.

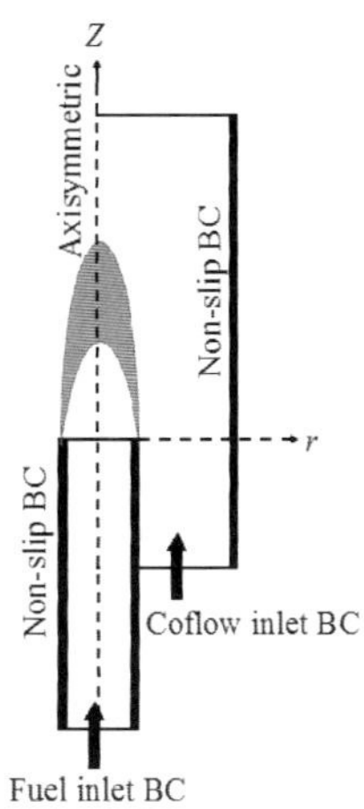

Figure 1. Computational domain and connecting boundary conditions.

The governing equations for mass, momentum, energy, and species are solved in PISO pressure-velocity coupling fashion [27]. A second-order upwind scheme was adopted for spatial discretization.

One-step overall reaction is employed for propane [28]. The thermal and transport properties were taken from CHEMKIN III and TRANSPORT Package [29,30], respectively. The reference flame condition was specified as the same as that conducted by Xiong et al. [22] experimentally, such that the inlet temperature of the fuel and air was set as ambient temperature [298 K] with pure propane fuel. In fuel dilution set of simulations, the mole fraction of H_2O - and CO_2-diluents were set as 0.1 and 0.2. The mean velocity of propane fuel was set as 1.40 cm/s based on experimental condition [22], while the velocity of the air stream was set as 6.2 cm/s.

3. Results and Discussions:

3.1 Validation of simulations:

Before discussing the effects of fuel diluents on the flow field, it is necessary to validate the simulation results with available experimental data. In this regard, the simulated radial velocity profile was compared with available experimental data in [22]. As presented in Fig.2 the predicted radial profile of axial velocity at height Z=1 mm above the nozzle exit for pure propane show excellent agreement with experimental data of Xiong et al. [22].

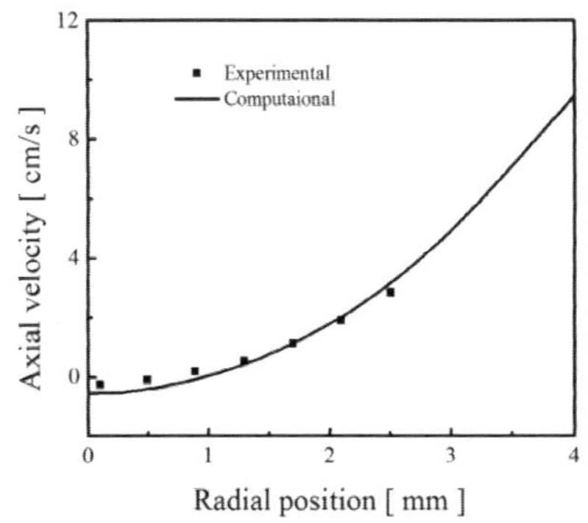

Figure 2. Comparisons of the radial profiles of axial velocity above nozzle exit at 1 mm; (symbols) experimental data of [22]; (solid line) current computational data.

3.2 Velocity Profiles

The predicted axial velocity profiles at centerline for propane with and without diluents are displayed in Figure 3. The predicted recirculation zone with negative axial velocity component appears in the near-nozzle region for around $0.45 < Z < 1.52$ mm with pure propane fuel, the negative component of the axial velocity decrease with added H_2O -or CO_2- diluent to the fuel stream. Dilution by either 10% H_2O or 10% CO_2 is effective to reduce the recirculation tendency.

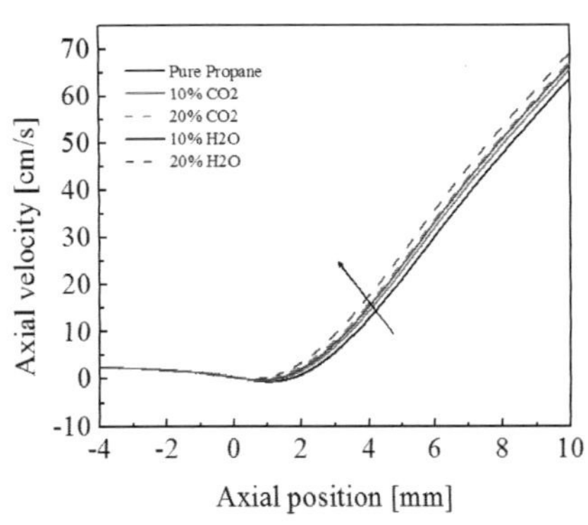

Figure 3. Show axial velocity profiles along the centerline.

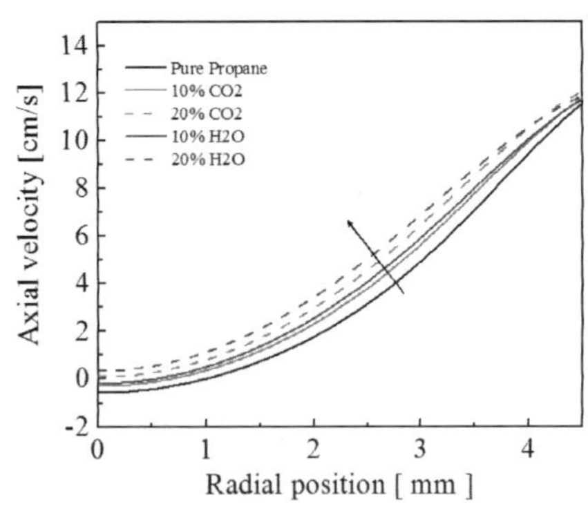

Figure 4. Show radial profiles of axial velocity at height Z=1mm.

As seen in Fig. 4 the axial velocity of the H_2O -diluted flame is slightly higher than that of the CO_2-diluted one. In addition, the region of negative axial velocity component diminish around 0.38 < Z 1.09 mm. with 20% of H_2O the negative component of axial velocity shift to the positive component for H_2O -diluted flame, while in 20% CO_2 it is limited to a small region just around 0.066 < Z < 0.076 mm. The radial profiles of the axial velocity at height Z = 1 mm above the nozzle exit are shown in Figure 4, For pure propane, the axial velocity is below zero at the nozzle centerline, this is due to the propane fuel being heavier than air i.e., negative buoyancy. As the H_2O - and CO_2-diluent increased by 10%, the axial velocity at centreline increased about 51%. Again, the radial profile of axial velocity of the H_2O -diluted flame is higher than that of the CO_2-diluted one. For instance, the axial velocity of 20% H_2O - and 20% CO_2-diluted flame at centerline is 0.36 and 0.10 cm/s respectively, these results indicate that H_2O is more effective in increasing the axial velocity than CO_2-diluent. This may attribute to the faster diffusion properties of H_2O compared with CO_2.

3.3Effect of Fuel Dilution on Vortices

It is interesting to compare the results in the form of streamline between pure, H_2O - and CO_2 diluted flames as in Figure 6. With pure propane, two large-scale vortices structure can be identified near the exit of fuel nozzle as seen in Figure 5 (a), and that was found to be in satisfactory agreement with experiment [22].

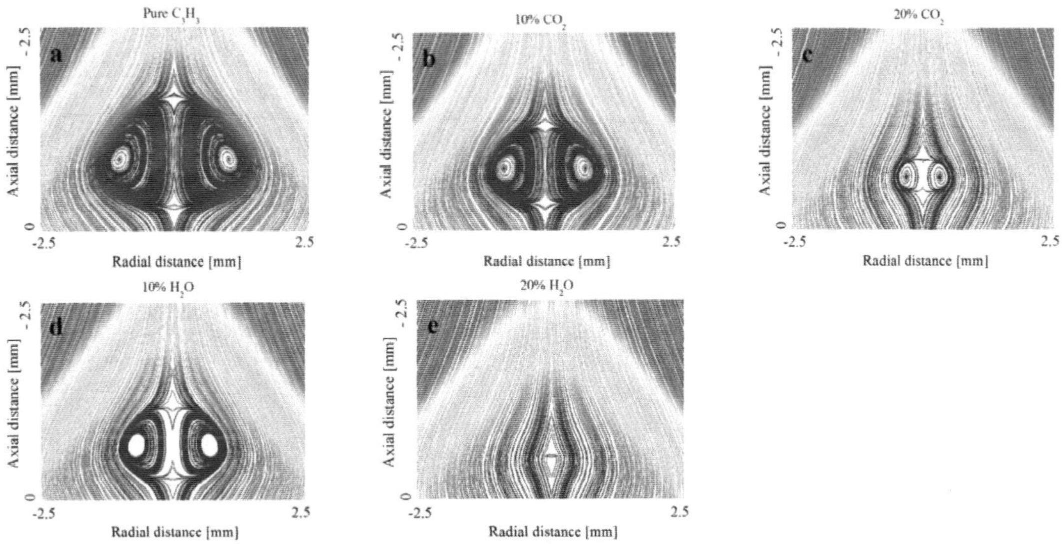

Figure 5. Flow streamline visualization near nozzle exit for : (a) pure C_3H_8; (b) 10% CO_2 ; (c) 20% CO_2; (e) 10% H_2O ; (f) 20% H_2O.

It is speculated that these vortices are generated by the flow disturbance at the nozzle exit and amplified due to the appreciable difference in density between propane fuel and air [31]. With adding 10% H_2O and 10% CO_2-diluent to the fuel stream, the size of the vortices becomes smaller as seen in Figure 5(b) and (d) respectively, adding 10% of H_2O presents an even more significant reduction of the vortices size compared with 10% CO_2. As the H_2O increased to 20%, the vortices structure are dissipated as seen in Figure 5(e). While with 20% CO_2 the vortices structure still existed with small size. It is clear that the simulation result demonstrates the H_2O –diluent is more effective than CO_2-diluent. The changes in the density of the unburned fuel mixture (fuel and diluents) are the main responsible for the reduction of the vortices size near the nozzle. The differences in the results between the two diluent flames due to the differences in density between H_2O and CO_2-diluents. For example, the decrease in the fuel mixture density at nozzle exit for 20% CO_2 and 20% H_2O diluted flame is 5.4% and 16.9% respectively. However, the CO_2 diluent effect has a smaller influence than H_2O when considering the higher density of CO_2 than H_2O.

4. Concluding remarks

We investigate computationally the effects of fuel dilution on the near-nozzle flow-fields focusing particularly on the vortices inside the recirculation zone in small-scale laminar buoyant coflow diffusion flames with propane fuel. The vortices scale inside recirculation zone decrease with increasing H_2O - or CO_2- diluent, the H_2O -diluent is more effective in reducing the vortex scale compared with CO_2-diluent. The diminishing of vortices scale is due to a significant decrease of the unburned fuel mixture density. Future work will focus on the dependence of sooting behavior on growth/decay of vortices inside the recirculation zone

References:

[1] Howard D Ross 2001 *Microgravity combustion: Fire in free fall.* Academic press.
[2] Bilger R 1977 Reaction rates in diffusion flames, *Combust. Flame.* **30**, pp 277-284.
[3] Mitchell D Smooke, Reginald E Mitchell and David E Keyes 1986 Numerical solution of two-dimensional axisymmetric laminar diffusion flames. *Combust. Sci. Technol.* **67**, pp 85-122.
[4] Viswanath R Katta, William M Roquemore, Arvind Menon, Seong-Young Lee, Robert J Santoro and Thomas A Litzinger 2009 Impact of soot on flame flicker. *Proc. Combust. Inst.* **32**, pp 1343-1350.
[5] Trees D, Brown T, Seshadri K, Mitchell D, Balakrishnan G, Pitz R, Giovangigli V and Nandula S, 1995 The structure of nonpremixed hydrogen-air flames. *Combust. Sci. Technol.* **104**, pp 427-439.
[6] Won SH, Kim J, Shin MK, Chung SH, Fujita O, Mori T, Choi JH and Ito K 2002 Normal and microgravity experiment of oscillating lifted flames in coflow. *Proc. Combust. Inst.* **29**, pp 37-44.
[7] Hanfeng Jin, Alessio F, Yizun W, Xiaoyuan Z, Meirong, Yuyang L, Fei Q, Alberto C and Tiziano F 2015 Kinetic modeling study of benzene and PAH formation in laminar methane flames. *Combust. Flame.* **162**, pp 1692-1711.
[8] Hanfeng J, Wenhao Y, Yizun W, Yuyang L, Fei Q, Alberto C, Alessio F, and Tiziano F 2015 Experimental and kinetic modeling study of laminar coflow diffusion methane flames doped with 2-butanol. *Proc. Combust. Inst.* **35**, pp 863-871.
[9] Victor Ch, Murray J Thomson, Seth B Dworkin, Nadezhda A Slavinskaya and Uwe R 2014 Soot formation with C_1 and C_2 fuels using an improved chemical mechanism for PAH growth. *Combust. Flame.* **161**, pp 592-601.
[10] Ali Khosousi, Fengshan L, Seth B Dworkin, Nick A Eaves, Murray J Thomson, Xu H, Yujie D, Yongli G, Fushui L and Shijin Sh 2015 Experimental and numerical study of soot formation in laminar coflow diffusion flames of gasoline/ethanol blends. *Combust. Flame.* **162**, pp 3925-3933.
[11] Kermit C Smyth, Joel E Harrington, Erik L Johnsson and William M Pitts 1993 Greatly enhanced soot scattering in flickering CH_4/air diffusion flames. *Combust. Flame.* **95**, pp 229-239.
[12] Shaddix, C.; Harrington, J.; Smyth, K. 1994 Enhanced soot formation in flickering CH_4/air diffusion flames. *Preprints of Papers-American Chemical Society Division Fuel Chemistry.* **39**, pp 152-152.
[13] Davis R, Moore E, Roquemore W, Chen L-D, Vilimpoc V and Goss L 1991 Preliminary results of a numerical-experimental study of the dynamic structure of a buoyant jet diffusion flame. *Combust. Flame.* **83**, pp 263IN3267-3266IN4270.
[14] Xiong Y, Min S Cha and Suk H Chung 2015 Numerical and experimental study on negative buoyance induced vortices in n-butane jet flames, *ASME/JSME/KSME Joint Fluids Engineering Conference, American Society of Mechanical Engineers.*

[15] Buckmaster J and Peters N 1988 The infinite candle and its stability a paradigm for flickering diffusion flames, *Symp. (Int.) Combust.* pp 1829-1836.
[16] Yuan TM Durox D and Villermaux E 1994 An analogue study for flame flickering. *Exp. Fluids.* **17**, pp 337-349.
[17] Michael E Mueller, Qing N Chan, Nader H Qamar, Bassam B Dally, Heinz P, Zeyad T Alwahabi and Graham J Nathan 2013 Experimental and computational study of soot evolution in a turbulent nonpremixed bluff body ethylene flame. *Combust. Flame.* **160**, pp 1298-1309.
[18] Chen L-D, Seaba J, Roquemore W, Goss L 1989 Buoyant diffusion flames, *Symp. (Int.) Combust.* pp 677-684.
[19] Chen L-D and Roquemore W 1986 Visualization of jet flames. *Combust. Flame.* **66**, pp 81-86.
[20] Chen L-D, Vilimpoc V, Goss L, Davis R, Moore E and Roquemore W 1992 Time evolution of a buoyant jet diffusion flame, *Symp. (Int.) Combust.* pp 303-310.
[21] Sang H Won, Chung S, Cha M and Lee B 2000 Lifted flame stabilization in developing and developed regions of coflow jets for highly diluted propane. *Proc. Combust. Inst.* **28**, pp 2093-2099.
[22] Xiong Y, Min S Cha and Suk H Chung 2015 Fuel density effect on near nozzle flow field in small laminar coflow diffusion flames. *Proc. Combust. Inst.* **35**, pp 873-880.
[23] Wang Q, Legros G, Bonnety J and Morin C 2016 Experimental characterization of the different nitrogen dilution effects on soot formation in ethylene diffusion flames. *Proc. Combust. Inst.* **36**, pp 3227-3235.
[24] Mingyan G, Huaqiang Ch, and Fengshan L 2016 Effects of simultaneous hydrogen enrichment and carbon dioxide dilution of fuel on soot formation in an axisymmetric coflow laminar ethylene/air diffusion flame. *Combust. Flame.* **166**, pp 216-228.
[25] ANSYS FLUENT 17.0 user's guide http://www.ansys.com.
[26] Hongsheng G, Fengshan L, Gregory S and Ömer L Gülder 2002 The flame preheating effect on numerical modelling of soot formation in a two-dimensional laminar ethylene-air diffusion flame. *Combust. Theory Model.* **6**, pp 173-187.
[27] Raad I Issa 1986 Solution of the implicitly discretised fluid flow equations by operator-splitting. *J Comput Phys.* **62**, pp 40-65.
[28] Charles K Westbrook and Frederick L Dryer 1981 Simplified reaction mechanisms for the oxidation of hydrocarbon fuels in flames. *Combust. Sci. Technol.* **27**, pp 31-43.
[29] Robert J Kee, Fran M Rupley, Ellen M and James A Miller 1996 Chemkin-III: A Fortran chemical kinetics package for the analysis of gas phase chemical and plasma kinetics," sandia national laboratories report.
[30] Robert J Kee, Warnatz J and James A Miller, 1983. A FORTRAN computer-code package for the evaluation of gas-phase viscosities, conductivities, and diffusion coefficients. [Chemkin]; Sandia National Labs., Livermore, CA (USA).
[31] Lee J, Sang H, Seong H and Suk Ch 2003 Lifted flames in laminar jets of propane in coflow air. *Combust. Flame.* **135**, pp 449-462.

IOP Conf. Series: Earth and Environmental Science **146** (2018) 012022 doi:10.1088/1755-1315/146/1/012022

Numerical Study of Smoke Emissions of an Air-Assisted Elevated Flare

Qi Guan[1,*], Chao Guo[1] and Shoukun Hou[1]

[1]Shanghai Marine Diesel Engine Research Institute, Shanghai, 201108, China

*mech_guanqi@163.com

Abstract. In the petroleum industry, flares are used to burn off unwanted waste gas generated during upstream processes. For elevated flares where a huge amount of waste gases is gathered and burned at a single flare header, visible smoke emission is a common problem. It is known that smoke emission is a complex thermochemical process and is affected by many factors. To suppress smoke production, air or steam is often injected into the header to supply additional oxygen and enhance mixing. In this paper, computational fluid dynamics (CFD) is used to study the smoking behavior of an air-assisted elevated flare of a domestic petroleum company. The Moss-Brookes model is used to simulate the soot generation. The impacts of fuel composition, fuel flow rate, and crosswind are investigated by varying the fuel type, waste gas mass flow rate and crosswind speed. Our results suggest that soot concentration is strongly affected by the fuel composition and flow rate. The crosswind speed also greatly affects the flame shape and size and the overall soot generation.

1. Introduction

In petroleum companies, waste hydrocarbon gases are centralized in the flare system and burned before releasing to the atmosphere. According to the estimation by the World Bank, there were nearly 150 billion cubic meters of waste gases burned through flaring annually [1]. Successful disposal of the waste gas requires complete combustion and maintains emissions of NOx, SOx, poisonous gases and/or smoke at permissible levels. For elevated flares, combustion normally takes place above 100 meters or higher. If not carefully controlled, toxic emissions can spread downstream to nearby cities and cause serious health problems to local residents.

Smoke emission usually happens when the waste gas has a large flow rate or contains unsaturated hydrocarbon molecule, e.g. toluene or butadiene, which leads to soot precursors during combustion. Visible smoke emissions cause air pollution which leads to respiratory problems and crowd panic and therefore is strictly prohibited by environmental law enforcement. In order to control smoke emission, one often injects assisted air or steam into the combustion zone to provide additional oxygen and improve the fuel-air mixing.

Numerous work has been devoted to the study of soot formation mechanisms and emission measurements. However, most of these works have been performed in the laboratory environment where combustion is well controlled. Due to the unpredictability of the ambient and operating conditions as well as the open-air combustion mode, a quantitative field measurement of the smoke emission is nearly impossible.

In recent years, with the development of computational fluid dynamics (CFD) and combustion mechanisms, numerical studies of flare operation have become a popular approach [2, 3], from which smoke emissions can be quantitatively studied by applying appropriate models.

2. Methodologies

2.1. Conservation Equations

In order to determine the distributions of temperature and molecular species in a fluid dynamics problem, one needs to solve the Navier-Stokes equations which includes the conservations of mass, momentum and energy. With the interactions of the fuel injection and the crosswind, the fluid dynamics is turbulent. In this study, the standard k-ε model is chosen to solve the turbulence fluctuation. Neglecting the contribution of buoyancy and compressibility of fluid, the transportations of k and ε are expressed as:

$$\frac{\partial(\rho k)}{\partial t}+\frac{\partial}{\partial x_i}(\rho u_i k)=\frac{\partial}{\partial x_i}\left[\left(\mu+\frac{\mu_t}{\sigma_k}\right)\frac{\partial k}{\partial x_i}\right]+G_k-\rho\varepsilon \tag{1}$$

$$\frac{\partial(\rho\varepsilon)}{\partial t}+\frac{\partial}{\partial x_i}(\rho u_i\varepsilon)=\frac{\partial}{\partial x_i}\left[\left(\mu+\frac{\mu_t}{\sigma_\varepsilon}\right)\frac{\partial\varepsilon}{\partial x_i}\right]+C_{1\varepsilon}\frac{\varepsilon}{k}G_k+C_{2\varepsilon}\rho\frac{\varepsilon^2}{k} \tag{2}$$

Where G_k represents the turbulence generation due to the mean velocity gradients. σ_k and σ_ε are turbulent Prandtl numbers for k and ε. $C_{1\varepsilon}$ and $C_{2\varepsilon}$ are constants.

2.2. Moss-Brookes Soot Model

Smoke generated in a diffusion flame is also called soot. Its formation follows four important stages: (a) nucleation; (b) molecular growth; (c) aggregation and (d) oxidation.

Several soot models are available in ANSYS Fluent, which includes the one-step Khan and Greeves model, two-step Tesner model, Moss-Brookes model and Moss-Brookes-Hall model. The kinetic mechanism of the Moss-Brookes model is based on data from lab experiments, so it is chosen to solve the soot generation and destruction.

The Moss-Brookes model solves transport equations for normalized radical nuclei concentration b_{nuc}^* and soot mass fraction Y_{soot}:

$$\frac{\partial(\rho Y_{soot})}{\partial t}+\frac{\partial}{\partial x_i}(\rho u Y_{soot})=\frac{\partial}{\partial x_i}\left(\frac{u_t}{\sigma_{soot}}\frac{\partial Y_{soot}}{\partial x_i}\right)+\frac{dM}{dt} \tag{3}$$

$$\frac{\partial(\rho b_{nuc}^*)}{\partial t}+\frac{\partial}{\partial x_i}(\rho u_i b_{nuc}^*)=\frac{\partial}{\partial x_i}\left(\frac{u_t}{\sigma_{nuc}}\frac{\partial b_{nuc}^*}{\partial x_i}\right)+\frac{1}{N_{norm}}\frac{dN}{dt} \tag{4}$$

Where Y_{soot} is soot mass fraction, M is soot mass concentration (kg/m^3), b_{nuc}^* is the normalized radical nuclei concentration (particles × 10^{-15}/kg) and N is the soot particle number density (particles/m^3). $N_{norm} = 10^{15}$ particles [4].

3. Numerical Modelling

3.1. Geometry and Mesh

A typical air assisted flare header is shown in Figure 1, where flare gas and assisted air are separately conveyed to the flare header through inner and outer cylinders. The outer diameter of the flare header is about 3.3 m and the diameter for the fuel tunnel is about 1.7 m. A flare with this size runs a maximum disposal flow rate of 250 kg/s. The assisted air is injected into the header through frequency conversion blowers and the maximum air supply is approximately 118 kg/s. On top of the header, vent tunnels of flare gas are evenly distributed among the air tunnels to enhance the fuel and air mixing. In our study, the flare structure, combustion domain and mesh are all created in the pre-processing software Gambit 2.6. Figure 2(a) shows the computational domain and the flare position while Figure

2(b) shows a simplified header geometry where the waste gas inlet is marked in blue and assisted air inlet in red.

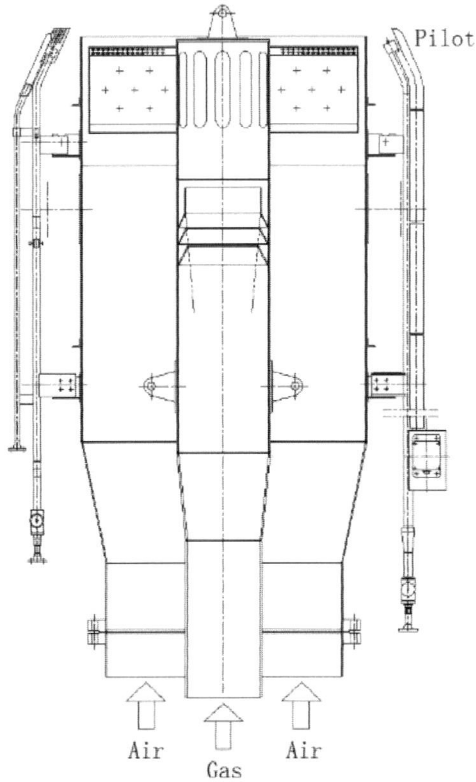

Figure 1. Air-assisted flare header

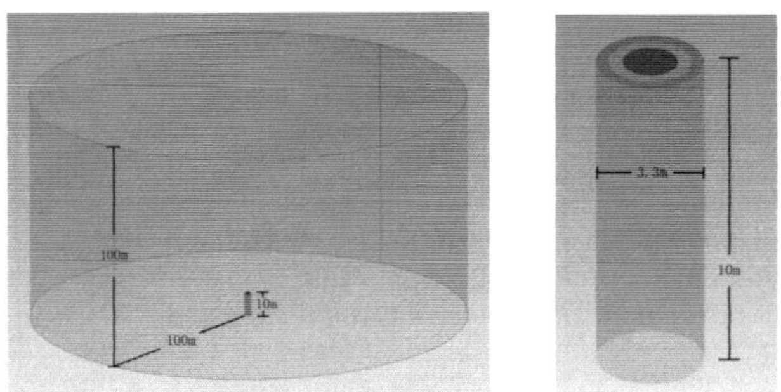

(a) Computational domain (b) Flare geometry

Figure 2. Geometry model used in this study

Meshing is critical factor in CFD simulation. A good quality mesh can not only lead to more accurate results but also save iterations on re-model and re-mesh. In our work, the cell number for the entire computational domain is about 2.21 million. Denser mesh is built near the flare header where mixing and combustion take place, while coarser mesh is built near the domain boundary. Size functions are applied to adapt the cell size variation between dense mesh and coarse mesh. Multiple iterations of mesh improvements are performed along the simulation to reach converged results. The series of this study are performed on a DELL Precision T7600 workstation with a processor of 16 Intel® Xeon® CPU at 3.1 GHz and a RAM of 128 GB.

3.2. Numerical Settings and Boundary Conditions

All simulations are performed in the steady state pressure-based solver. The SIMPLE algorithm is chosen for velocity-pressure coupling, Standard k-ε model is used for turbulence and the P1 model is used for radiation. The combustion chemistry is solved using a non-premixed combustion species model (PDF model) [4]. The flamelets are generated using USC 2.0 mechanism and thermodynamic data [5]. The soot precursors are C_2H_2, C_2H_4, and C_6H_6, while C_2H_2 and C_6H_6 were used as the surface growth species. Fenimore-Jones was chosen as the soot oxidation model. Gravity is considered by applying a gravitational acceleration of -9.81 m/s^2. All transport equations are applied to a 2nd order upwind scheme except for the turbulent terms to a 1st order upwind scheme.

The inlets of waste gas and assisted air are applied with mass inlet boundary conditions. The surrounding and the upper faces of the domain have pressure outlet boundary conditions. The temperatures of the air and the waste gas are 276 K and 473 K, respectively.

4. Results and Discussion

In this section, results for all series of simulations are presented. We compare the contour plots of temperature, flame shape and soot distribution to analyse the impacts of the fuel composition, the flow rate of waste gas and the crosswind speed.

4.1. Impact of fuel composition

Table 1. Waste gas fuel composition (by mole fraction)

Composition		Fuel 1	Fuel 2	Fuel 3
Methane	CH_4	5%	50%	80%
Ethylene	C_2H_4	10%	20%	10%
Propylene	C_3H_6	85%	30%	10%

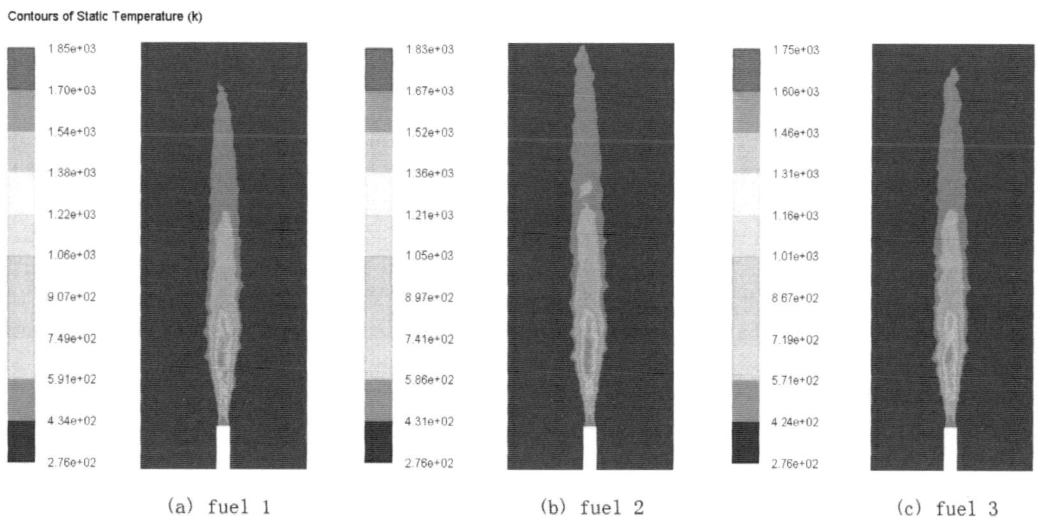

Figure 3. Contours of temperature for different fuel types

IOP Conf. Series: Earth and Environmental Science **146** (2018) 012022 doi:10.1088/1755-1315/146/1/012022

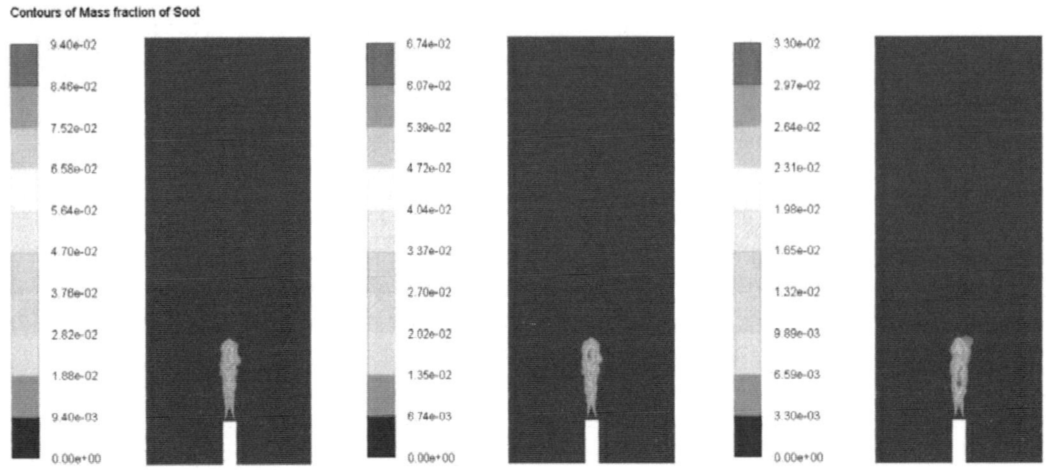

(a) fuel 1 (b) fuel 2 (c) fuel 3

Figure 4. Contours of soot mass fraction for different fuel types

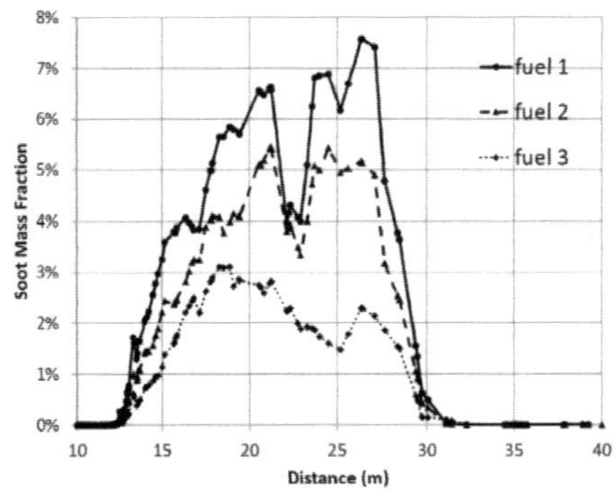

Figure 5. Centerline profile of soot mass fraction for different fuel types

Figure 3 shows the temperature contours on the z=0 plane of the domain for the three cases. Since all three cases have the same mass flow rates of fuel and assisted air and the heating values of fuels are similar, comparable temperature distributions are obtained. One can also notice that the maximum temperature when using fuel 1 is about 100 K higher than when using fuel 2, which in turn is slightly higher than fuel 3. This is probably because fuels 2 and 3 have lighter molecular weights, and the ejecting speeds are higher, thus more air is entrained into the combustion zone to cool down the temperature.

On the other hand, smoke emissions of the three flares behave quite differently. Figure 4 shows the contours of soot mass fraction on the z=0 plane, where the range of soot yield of the three cases are similar to each other. Figure 5 is a plot of the soot mass fraction centreline distributions for the three fuel cases. As shown, the incipient and terminating smoking points of the three cases lie closely on the axis of distance, but the soot mass fraction of fuel 1 is about twice of fuel 3. In real observation [6-7], since most of the visible light is emitted from the hot soot particles, the flame of fuel 1 will look much brighter than the fuel 2 and fuel 3 cases, even though their temperature values are close.

4.2. Impact of fuel flow rate

Ideally, it would be nice to have the correlation between smoke visibility and air supply so that smoke can be accurately and timely captured and suppressed. This is hard to achieve because the waste gas flow rate often varies significantly during operation. In this work, we study the impact of fuel flow rate by comparing the temperature and soot distributions under the fuel flow rates of 5 kg/s, 15 kg/s and 25 kg/s, i.e. 2%, 6%, and 10% of the full designed disposal capacity. Fuel 1 in Table 1 is used as the waste gas and the assisted air flow rate is maintained at 39.4 kg/s.

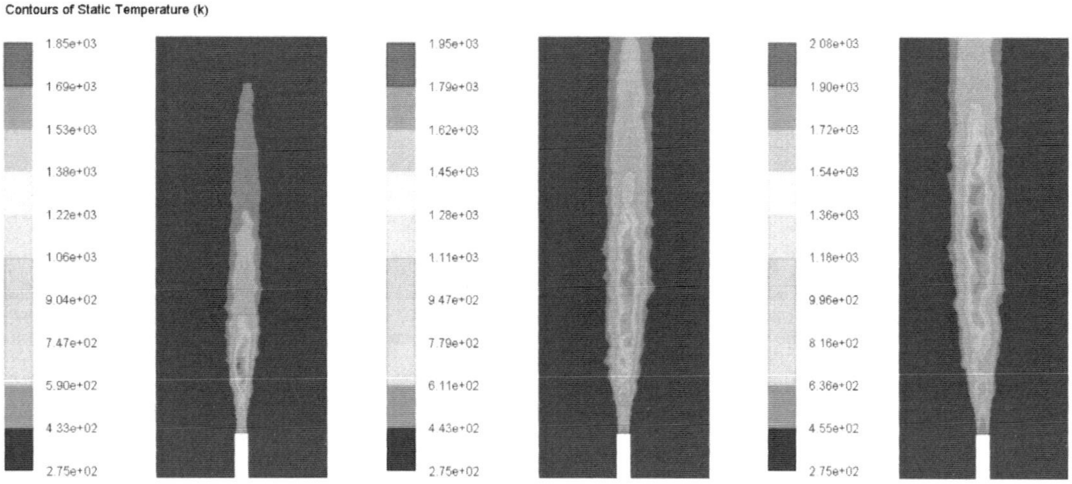

(a) fuel = 5kg (b) fuel = 15kg (c) fuel = 25kg

Figure 6. Contours of temperature under different fuel flow rates

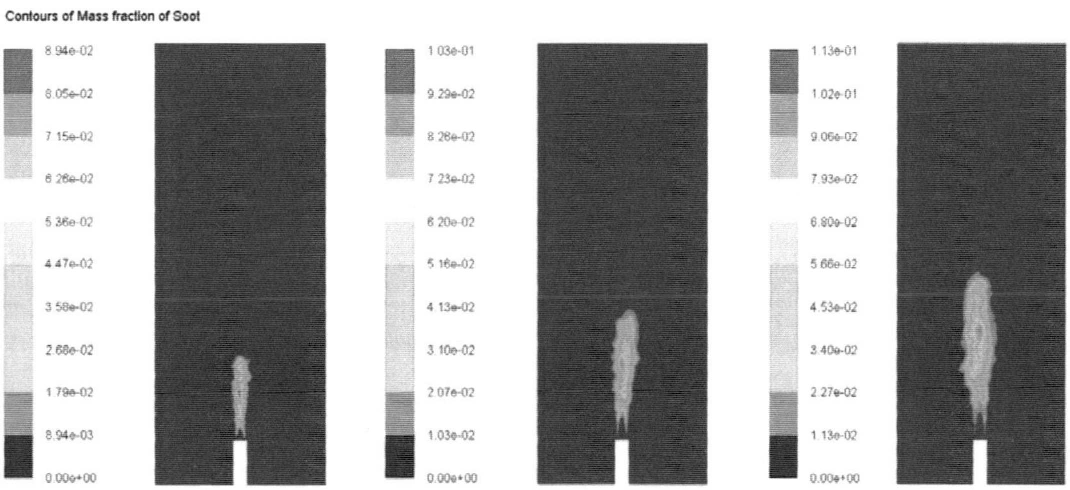

(a) fuel = 5kg (b) fuel = 15kg (c) fuel = 25kg

Figure 7. Contours of soot mass fraction under different fuel flow rates

As is shown in Figure 6, the flame length increases with the flow rate of the waste gas in addition to the maximum temperature. As shown in Figure 7, the soot mass fraction follows a similar pattern of the temperature distribution. When the flow rate of the fuel increases, the soot yield area becomes significantly longer and wider. This agrees with laboratory observations: more fuel needs more air to combust and therefore the flame becomes longer and wider. One may also notice that the maximum soot mass fraction only slightly increases with the fuel flow rate, from about 8% to 10%. This means the brightness of the flame is only slightly affected by the fuel flow rate.

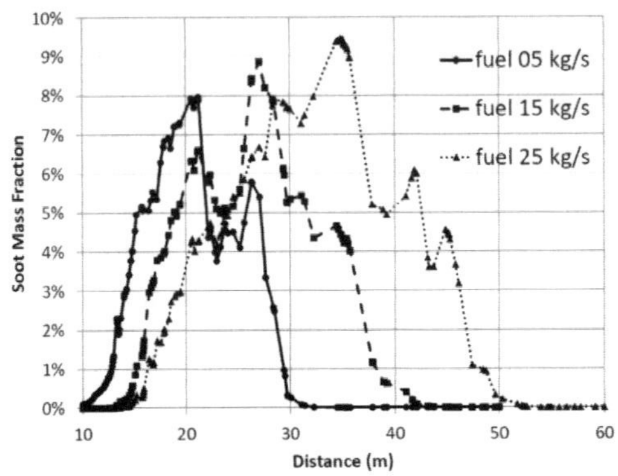

Figure 8. Centerline profile of soot mass fraction under different fuel flow rates

Figure 8 plots the soot mass fraction along the flame centerline. As shown, when flare runs with higher fuel flow rates, the range of soot emission becomes wider. The noticeable soot emission range increases from 20 m for the 5 kg/s case to about 35 m for the 25 kg/s. Therefore, the overall soot production also increases with the fuel flow rate. To suppress the smoke emission during operation, one often needs to increase the assisted air supply or enhance the fuel-air mixing such as partially premixed combustion.

4.3. Impact of crosswind

During operation, elevated flare combustion is also always affected by the ambient wind. Air entrainment and mixing can be very different depending on the wind speed. To classify the flare flame, one often uses the momentum flux ratio R [8]:

$$R = \frac{\rho_j V_j^2}{\rho_\infty V_\infty^2} \tag{5}$$

Where ρ_j and ρ_∞ represent the density of the waste gas and the air, while V_j and V_∞ represent the velocity. When the wind momentum is smaller than the fuel momentum, the flame tends to be vertical, as previously shown in section 4.1 and 4.2 where wind speed is zero. Under quiescent or mild wind condition, the combustion is said in the "lifted jet flame" region. The surrounding air is entrained continuously into the combustion zone. However, when the wind momentum is greater than the fuel, the flame tends to shift into the wake. In this case, the combustion is in the "wake-stabilized flames" region where the fluid dynamics is dominated by the wind momentum.

In this study, we investigate the effect of crosswind by simulating the flare flame under five crosswind speeds, i.e. 1 m/s, 3 m/s, 5 m/s, 7 m/s and 10 m/s, while the result of quiescent condition is used as the base case. The flow rates of the fuel and the assisted air are maintained at 25 kg/s and 39.4 kg/s, respectively, and fuel 1 in Table 1 is used as the waste gas.

Figure 9 shows the temperature contours under different crosswind speeds. As the crosswind gets stronger, the flame shape tends to shift with the wind and the flame length becomes significantly shorter. When the wind speed is smaller than 5 m/s, flare jet momentum still dominates, as most of the products flow out through the upper boundary. The combustion can be classified in the "lifted jet flame" region. When the wind speed reaches 7 m/s, crosswind becomes a more dominant force, as most of the products flow to the right. At a wind speed of 10 m/s, the flame becomes even shorter and close to the ground. In this case it is considered in the "wake-stabilized flame" region.

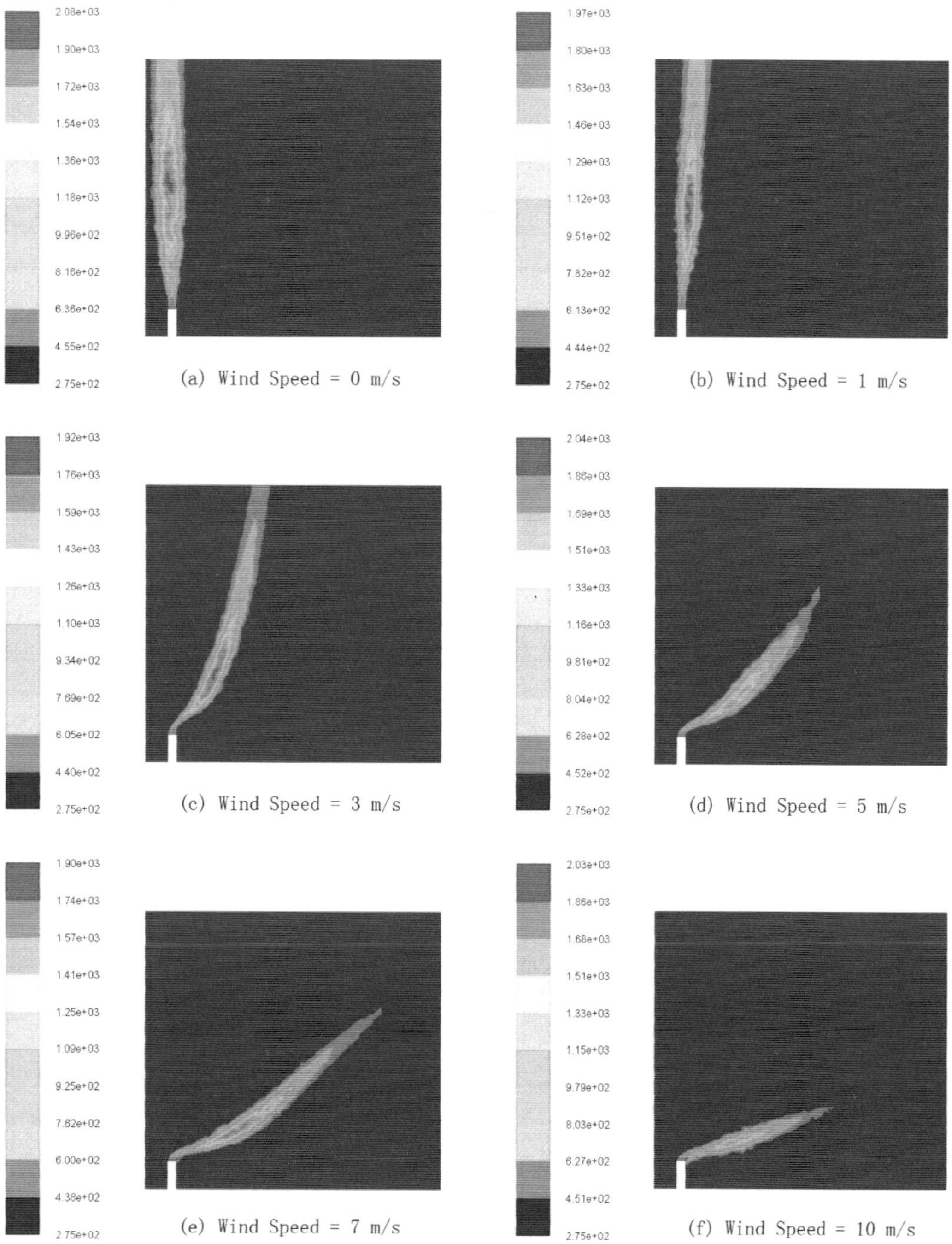

Figure 9. Contours of temperature under different crosswind speeds

IOP Conf. Series: Earth and Environmental Science **146** (2018) 012022 doi:10.1088/1755-1315/146/1/012022

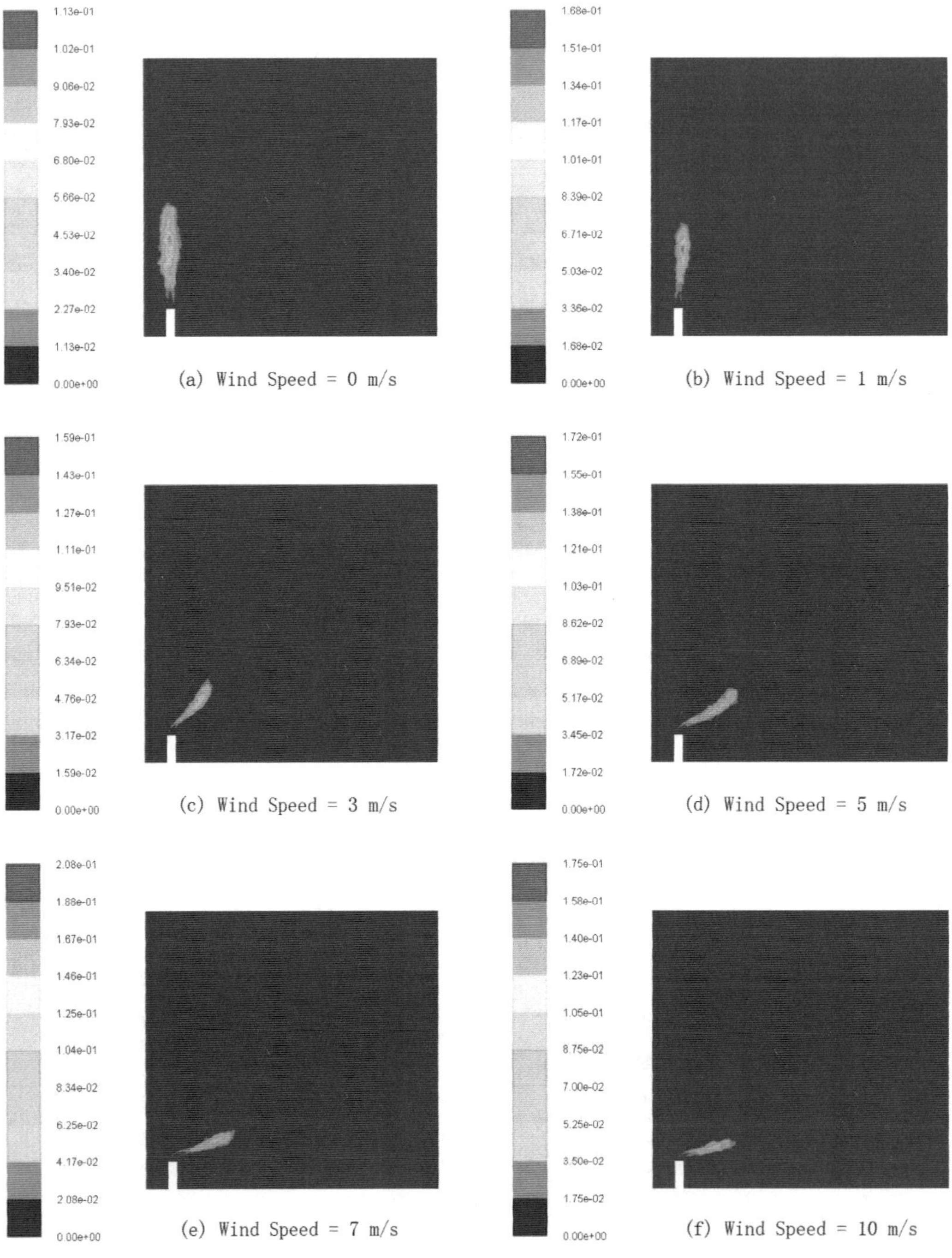

Figure 10. Contours of soot mass fraction under different crosswind speeds

Figure 10 shows the contours of the soot mass fraction. As indicated, soot distribution generally follows the flow direction of the flame. And although the maximum soot mass concentration for each case is about the same level, one cannot infer the same conclusion for the total smoke emission. By summing the dot product of the flow density ρ, soot mass fraction x_{soot}, normal velocity v and area A over the outlet boundary, one can get the overall soot production rate [3],

$$\dot{m}_{soot} = \sum_{boundary} \rho x_{soot} v A \tag{6}$$

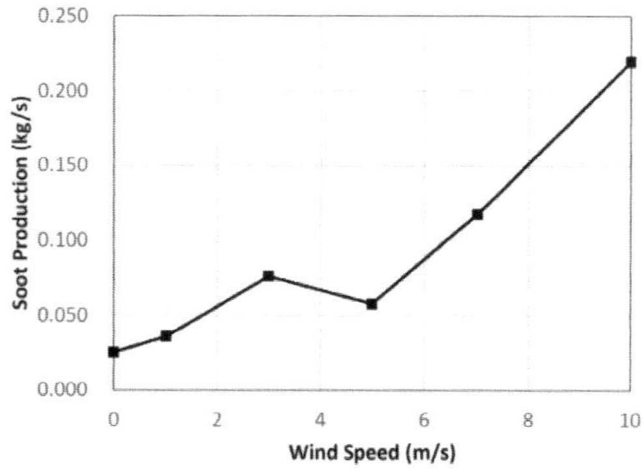

Figure 11. Overall soot production rate under different crosswind speeds

As is shown Figure 11, the overall soot production rate generally increases with the wind. Soot production in the 10 m/s case is about 10 times that of the quiescent case. This is somewhat contradictory to the perceived impression that stronger ambient wind leads to less visible smoke. One explanation might be that smoke diffuses more quickly under high wind conditions but the fuel in the wake region has a low momentum which leads to less air entrainment and incomplete combustion. Our modeling study suggests that stronger ambient wind can cause more total smoke production and therefore one needs to increase the fuel jet momentum under high crosswind conditions.

Acknowledgments
This work was supported by Shanghai Pujiang Program (No.16PJ1432000).

5. Conclusions

The combustion and smoke emission of an air-assisted flare under various conditions are numerically simulated. The impacts of fuel composition, fuel flow rate and crosswind speed on smoke emission is analyzed using CFD software ANSYS Fluent. Based on the simulated results, it can be concluded that soot mass concentration is strongly affected by the fraction of unsaturated hydrocarbons in the fuel. This study also indicates that higher fuel flow rates leads to higher flame temperatures and wider soot production range. Strong crosswind can greatly vary the flame shape and length which can significantly increase the overall soot production.

References

[1] Global Gas Flaring Reduction Partnership (GGFR), World Bank, October 2011 Brochure p. 3
[2] Castiñeira D and Edgar T F 2006 CFD for simulation of steam-assisted and air-assisted flare combustion system. *Energy & Fuels* **20**(3) 1044-56
[3] Castiñeira D and Edgar T F 2008 CFD for simulation of crosswind on the efficiency of high momentum jet turbulent combustion flames. *J. Environ. Eng.* **134**(7) 561-71
[4] Fluent Inc 2011 FLUENT 13.0 User's Guide
[5] Information on http://ignis.usc.edu/USC_Mech_II.htm
[6] Wagner H G 1981 Soot formation – an overview. *Particulate Carbon.* **8**(4) 1-29
[7] U.S. EPA. 1995 AP 42 – compilation of air pollutant emission factors, Vol. 1, 5th ed. – section 13.5 industrial flares; AP-42; United States Environmental Protection Agency (U.S. EPA): Research Triangle Park, NC
[8] Huang R F and Wang S M 1999 Characteristic flow modes of wake-stabilized jet flames in a transverse air stream. *Combust. Flame* **117** 59-77

IOP Conf. Series: Earth and Environmental Science **146** (2018) 012023 doi:10.1088/1755-1315/146/1/012023

Application of Fenton Process in Industrial Wastewater Treatment Plant

Fang Liu[1] and Jun Li[1]

[1]Beijing University of Technology, Beijing, China

*Email: liufang@capitalwater.cn

Abstract. In an industry zone wastewater treatment plant where bio-pharmaceutical enterprises are concentrated, the inlet water quality is far beyond the standard with poor biochemical purification ability. The Fenton treatment is mainly used to deal with refractory biodegradable COD_{cr}, while flocculation by dosing chemicals is used to remove TP. In order to optimize the dosage of Fenton process, it was validated by laboratory test and productivity test respectively, and the cost of Fenton dosage was calculated. In this paper, the experimental process and results are described in detail, which provides the basis for the dosing of industrial wastewater treatment plants using Fenton process.

1. Introduction

In an industry zone where bio-pharmaceutical enterprises are concentrated, the scale of wastewater treatment plant is 20,000 tons per day, while the inlet and outlet water standards are designed as shown in Table 1.

Table 1. Design of water standard

Index	COD_{cr}(mg/L)	BOD(mg/L)	SS (mg/L)	NH_3(mg/L)	TN (mg/L)	TP (mg/L)	PH
influent water quality	≤500	≤250	≤220	≤40	≤50	≤4.0	6-9
effluent standard	≤50	≤10	≤10	≤5(8)	≤15	≤0.5	6-9

The implementation of emission standard is first grade A standard with “urban wastewater treatment plant pollutant emission standards" (GB18918-2002). The wastewater treatment plant adopts the process of "pretreatment + modified A^2/O process + advanced oxidation tank (Fenton process) + advanced treatment system (flocculation sedimentation tank + fiber dial filter + contact disinfection tank)" [1] ~ [4] and the process details are shown in Figure 1.

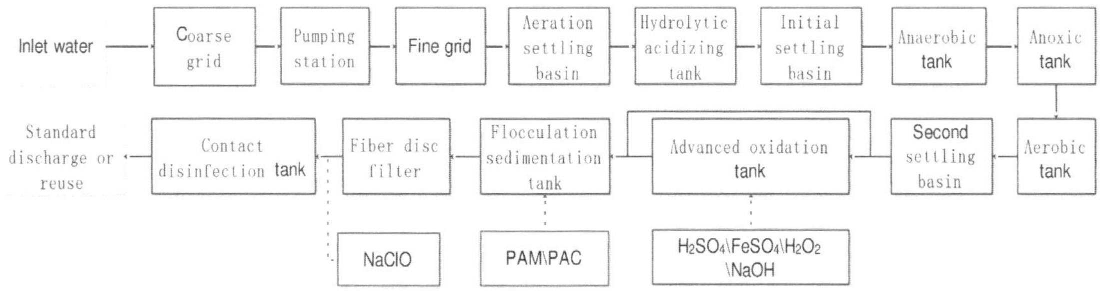

Figure 1. Process flow sheet.

The main dosing chemicals for advanced oxidation tank as shown in Figure 2 are sulfuric acid(H_2SO_4), ferrous sulfate($FeSO_4$), hydrogen peroxide(H_2O_2), caustic soda(NaOH) and so on. The Fenton reaction is one of the advanced oxidation processes, which generates hydroxy radicals (·OH) by the reaction of H_2O_2 and ferrous ions (Fe^{2+}) to realize oxidative degradation of pollutants in wastewater because (·OH) can interact with most organic matter and degrade it. The Fenton process is one of the more applied technologies in the treatment of refractory organic wastewater in recent years [5] ~ [8].

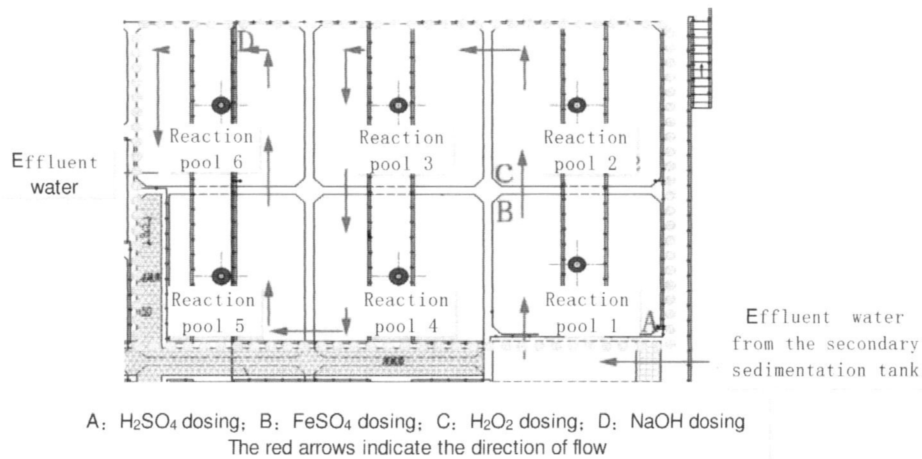

Figure 2. Detail diagrammatic sketch of advanced oxidation tank.

During the operation of the plant so far, many problems are found, of which the inlet water quality is the most serious problem. The influent COD_{cr} has reached the highest value of 2126mg / L, and carbon nitrogen ratio is about 0.2 with poor biodegradability [9], [10]. In order to make the outlet water quality to meet the effluent standards, firstly the carbon source is added to optimize the effect of biological processes, and then the dosage of Fenton process is optimized. In this paper, an overall Fenton process dosage optimization experiment is introduced, which provides the basis for the design and operation of industrial wastewater treatment plant.

2. Laboratory test of chemical optimization in Fenton process

The experiment is divided into two stages of laboratory test and productive test. The laboratory test is used to find out more suitable initial pH, end pH, the dosage of ferrous and hydrogen peroxide tons of water, etc. The influent water is from the effluent water of secondary sedimentation tank which use concentrated sulfuric acid (mass fraction 98%), ferrous sulfate (mass fraction 14%), hydrogen peroxide (mass fraction 27.5%) and caustic soda (mass fraction 32%).

2.1. Experiment 1

IOP Conf. Series: Earth and Environmental Science **146** (2018) 012023 doi:10.1088/1755-1315/146/1/012023

Take 41L water sample once, divide it into four beakers of 10L, leave 1L water as original water sample, add 5ml, 6ml, 7ml and 8ml concentrated sulfuric acid respectively to each 10L beaker to adjust PH value. Add 10ml of ferrous sulfate and 2ml of hydrogen peroxide to start the Fenton oxidation reaction, and mix the chemicals by stirring during the reaction. After one hour of reaction time, add 8ml, 9ml, 10ml and 11ml caustic soda respectively to adjust the PH value. Take the supernatant liquid to measure the value of COD and evaluate Fenton oxidation effect. The starting PH value and ending PH value could be determined with experiment 1. The experimental results are shown in Table 2. According to the experimental results, water sample 1 # is best. The best starting PH value is 4, while the ending PH value is 6.8.

Table 2. The starting PH value and ending PH value.

Reagent	**5ml H_2SO_4 8ml NaOH**	**6ml H_2SO_4 9ml NaOH**	**7ml H_2SO_4 10ml NaOH**	**8ml H_2SO_4 11ml NaOH**	**Original water**
No.	**Sample 1#**	**Sample 2#**	**Sample 3#**	**Sample 4#**	**Sample 5#**
starting PH value	4	3.8	3.5	3	
ending PH value	6.8	7	7.2	7.5	
COD value **mg/L**	39.9	45	42.5	41.6	83.5
Removal rate	52.2%	46.1%	49.1%	50.2%	
TP value **mg/L**	0.03	0.06	0.05	0.05	2.34
Removal rate	98.7%	97.4%	97.8%	97.8%	

2.2. Experiment 2

According to the results of experiment 1, take 31L water sample once, divide it into three beakers of 10L , leave 1L water sample as original water sample, add 5ml concentrated sulfuric acid to each 10L beaker to adjust PH value. After adding 10ml ferrous sulfate to each beaker, then add 1ml, 1.2ml, 1.5ml hydrogen peroxide respectively to start Fenton oxidation reaction and mix the chemicals by stirring during the reaction. After one hour of reaction time, add 8ml caustic soda respectively to adjust the PH value. Take the supernatant liquid to measure the value of COD_{cr} and evaluate Fenton oxidation effect. The dosage of hydrogen peroxide could be determined with experiment 2. The experimental results shown in Table 3. According to the experimental results, water sample 1 # reaches the effluent water standard, and the dosage of hydrogen peroxide is 1ml.

Table 3. Effect of hydrogen peroxide dosage on removal efficiency.

Reagent	**1ml H_2O_2**	**1.2ml H_2O_2**	**1.5ml H_2O_2**	**Original water**
No.	**Sample 1#**	**Sample 2#**	**Sample 3#**	**Sample 4#**
COD_{cr} value **mg/L**	28.5	29.8	31.1	64.4
Removal rate	55.7%	53.7%	51.7%	
TP value **mg/L**	0.06	0.06	0.02	2.40
Removal rate	97.5%	97.5%	97.5%	

2.3. Experiment 3

Take 41L water samples once, divide it into four beakers of 10L, leave 1L water sample as original water sample, then add 5ml concentrated sulfuric acid to each 10L beaker to adjust PH value. After adding 1ml, 1.5ml, 2ml, 2.5 ml ferrous sulfate, then add 1ml hydrogen peroxide respectively to start Fenton oxidation and mix the chemicals by stirring during the reaction. After one hour of reaction time, add 8ml caustic soda to adjust PH value. Take the supernatant liquid to measure the value of COD_{cr} and evaluate Fenton oxidation effect. The dosage of ferrous sulfate could be determined with experiment 3. The experimental results are shown in Table 4. According to the experimental results, water sample 1 # reaches the effluent water standard, and the dosage of ferrous sulfate is 1ml, but TP concentration of effluent water is a little higher, the stability of effluent water need to be further verified during actual production test.

Table 4. Effect of ferrous sulfate dosage on removal efficiency.

Reagent	**1ml $FeSO_4$**	**1.5ml $FeSO_4$**	**2ml $FeSO_4$**	**2.5ml $FeSO_4$**	**Original water**
No.	**Sample 1#**	**Sample 2#**	**Sample 3#**	**Sample 4#**	**Sample 5#**
COD_{cr} value **mg/L**	40.1	36.6	34.5	33.2	62.3
Removal rate	35.6%	41.3%	44.6%	46.7%	
TP value **mg/L**	0.31	0.28	0.24	0.21	4.82
Removal rate	93.6%	94.2%	95.0%	95.6%	

3. Productive test of chemical optimization in Fenton process

Based on the conclusion of laboratory test, productive test is validated in the actual production system. According to the field situation, the order and dwell time of adding chemical are shown as follow, the first step is adding concentrated sulfuric acid, the second step is adding ferrous sulfate after 0.5 hour, the third step is adding hydrogen peroxide after another 0.5 hour, the fourth step is adding liquid caustic after last 0.5 hour. The productive test started from 9 am on May 28, 2017 to 10 am the next day, which took 25 hours. The influent volume was about $314m^3/h$, and the dosage of concentrated sulfuric acid was 150L/h, while the dosage of ferrous sulfate was 56L/h. The dosage of hydrogen peroxide was 42L/h, and the dosage of liquid caustic soda was 180L/h. By sampling and testing COD_{cr} and TP of effluent water from second sedimentation tank and efficient sedimentation tank every 4 hours , the quality of influent and effluent water is shown in Figure 3.

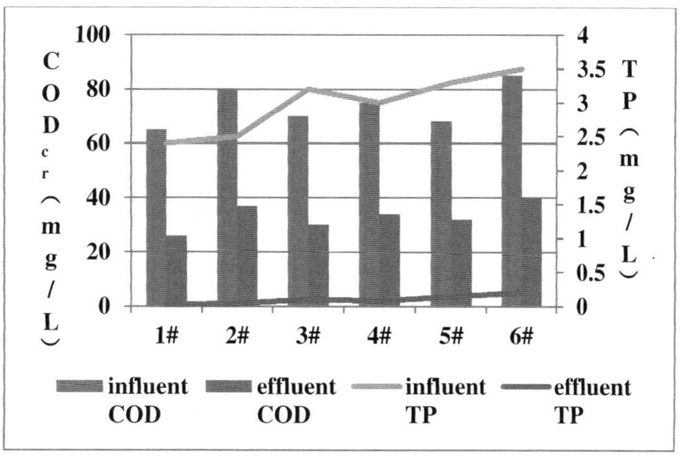

Figure 3. Productive test results.

It is verified that Fenton technology could be stable operation to meet requirements of effluent water standard according to the above dosage.

4. Cost accounting

The cost of chemical was calculated according to the dosage of productive test. As shown in Table 5, the Fenton chemical cost was 0.99 yuan/m³.

Table 5. Cost accounting results.

No.	Chemical	Actual dosage L/hour	Actual dosage ton/day	Unit cost yuan/ton	Cost yuan/day	Wastewater unit cost yuan/m³
1	H_2SO_4	150	3.60	450	1620	0.21
2	$FeSO_4$(solid)	56	1.35	240	324	0.04
3	H_2O_2	45	1.08	900	972	0.13
4	NaOH	180	4.32	1050	4536	0.60
				Total:	7453	0.99

5. Conclusion

1) In laboratory test, the condition of Fenton reaction is the best when the starting PH value is 4-5 and the ending PH value is 6-7. The dosage of sulfuric acid and caustic soda could be adjusted according to the above conditions.
2) The COD_{cr} of influent water is 60-80mg/L and the TP is 2-5mg/L, while ferrous sulfate dosage is 0.1-0.25ml/L and the dosage of hydrogen peroxide is 0.1-0.15ml/L. The COD_{cr} and TP values of treated water samples could reach the first grade A standard. With the increase dosage of ferrous sulfate and the decrease dosage of hydrogen peroxide, the treatment effect is further improved.
3) In productive test, the amount of influent water is about 314m³/h and the COD_{cr} of influent water is 65-85mg/L, while the TP is 2.4-3.5mg/L. The dosage of sulfuric acid is 150L/h and the dosage of ferrous sulfate is 56L/h, while the dosage of hydrogen peroxide is 42L/h and the dosage of liquid caustic soda is 180L/h. The COD_{cr} of effluent water is 26-40mg/L and the TP is 0.02-0.2mg/L which meet requirements of the effluent water standard. According to the above dosage of chemicals, the cost was accounted for 0.99 yuan/ton.
4) Laboratory test and productive tests verify the effluent stability of the Fenton process and the dosing amount of each chemicals of the Fenton process, which provides a basis for dosage of the industrial wastewater treatment plant adopting the Fenton process.

Acknowledgments

Thanks to the support of national water pollution control and governance science and technology major special projects: Study on pollutant depth reduction technology of sewage treatment plant in middle and small town of Yinma river. (2014ZX07201-011-004-3)

References

[1] Antonin VS, Santos MC, Garcia-Segura S, Brillas E. Electrochemical incineration of the antibiotic ciprofloxacin in sulphate medium and synthetic urinematrix. Water Res 2015;83:31-41
[2] Zupanc M, Kosjek T, Petkovsek M, Dular M, Kompare B, sirok B,et al. Removal of pharmaceuticals from wastewater by biological processes, hydrodynamic cavitation and UV treatment. Ultrason Sonochem 2013;20:1104-12
[3] Romero V, Acevedo P, Marco J, Esplugas S. Enhancement of Fenton and photo-Fenton processes at initial circumneutral PH for the degradation of the β ; blocker metoprolol. Water Res 2016; 88:449-57
[4] Esplugas S, Bila DM, Krause LGT, Dezotti M. Ozonation and advanced oxidation technologies to remove endocrine disrupting chemicals(EDCs) and pharmaceuticals and personal care products(PPCPs) in water effluents .J Hazard Mater 2007;149:631-42

[5] Zhao Changshuang,Zhang Jiankun. Research Progress and Application on Fenton Oxidation Technology in Wastewater Treatment [J].ENVIRONMENTAL SCIENCE AND MANAGEMENT, 2014, 39 (5):83-87

[6] Wang Chunxia,Xiao Shuhu,Zhao Xu,et. Study on the treatment of landfill leachate by photo-electro-Fenton oxidation [J].chinese Jounal of Environmental Engineering,2009,3(1):11-15

[7] Neyens E, Baeyens J. A review of classic Fenton's peroxidation as an advanced oxidationtechnique[J].Journal of Hazardous materials,2003,98(1):33-50

[8] Cheng Lihua,Huang Junli,Wang Li,et. Property and Application in Wastewater Treatment of Fenton,2001,84(3):24-25

[9] Liu Rui,Fan Juhong,Mu Chunfang, et. Study on Biodegradability and Biotoxicity of Pharmaceutical Wastewater[J]. Asian Journal of Ecotoxicology, 2011, 6(5):485-490

[10] Brillas E, Martinez-Huitle CA. Decontamination of wastewaters containing synthetic organic dyes by electrochemical methods. An updated review.Appl Catal B: Environ 2015:166-167:603-43

IOP Conf. Series: Earth and Environmental Science **146** (2018) 012024 doi:10.1088/1755-1315/146/1/012024

Microwave-Enhanced Photolysis of Rifampicin Resistant Bacteria

Wenchao Liao*, Su Xu, Qingsong Li, Xiaoqing Chen and Xiaohui Zhu

1. School of Environmental Science and Engineering, Xiamen University of Technology, Xiamen, Fujian 361024, China

*Corresponding author e-mail address: liaowenchao@xmut.edu.cn

Abstract. Antibiotic resistant bacteria (ARB) and antibiotic resistant gene (ARG), which pose great threaten to human health and ecological environment safety, have been widely recognized as new type of environmental pollutants. The microwave-enhanced photolysis process (MW/UV) was applied for the rifampicin resistant bacteria (RRB) disinfection in this paper. The effects of microwave power, irradiation time, initial concentration and pH of RRB were investigated, and the reactivation of the disinfected RRB in lake water, sea water and tap water was also studied. The RRB (40 mL, 3.8×10^7 CFU/mL, pH = 7) disinfection rate could reach 100% after 60 s irradiation of 500 W microwave and 18.54 mW/cm^2 UV light. However, the RRB after MW/UV disinfection all experienced reactivation when discharged into the actual water matrices. The reactivation rate followed the sequence: Lake water ＞Sea water＞ Tap water. Results revealed that the MW/UV process was an effective RRB disinfection method, but further research on the change of rifampicin resistant genes in the MW/UV process was urgently needed.

1. Introduction

The production and consumption of antibiotics are huge in China. When large amounts of antibiotics enter into the environment, it will cause long-term selective pressure on microbes in the environment and lead to the generation of antibiotic resistant gene (ARG) and the increase of bacterial antibiotic resistance. Antibiotic resistant bacteria (ARB) and ARG, which were widely recognized as new environmental pollutants [1], have been frequently detected in various water matrices such as surface water, groundwater, drinking water, sewage [2,3] and pose great threat to human health and the safety of ecological environment [4]. However, ARB and ARG are unbiodegradable, replicable and have strong persistence, which make them more difficult to control than traditional chemical pollutants.

Disinfection was the main way to kill microorganisms in wastewater treatment plant. The most widely used disinfection methods were chlorine disinfection, ozone disinfection and ultraviolet disinfection. Among them, UV disinfection was highly recommended because it had high disinfection efficiency, broad-spectrum antimicrobial property, and would not produce toxic or harmful disinfection by-products. UV disinfection was also considered to be one of the most effective ways for the treatment of ARB and ARG [5]. When the DNA of the ARG matched with the UV emission spectrum, the UV light could effectively photolytic degrade the DNA fragments containing resistant gene [6]. However, when the UV dosage was insufficient, the DNA structure might be restored and the bacterial cells might be reactivated with the help of the bacterial reactivation enzyme [7].

Microwave is an electromagnetic wave with wavelength from 1mm to 1m. The thermal and non-thermal effect of microwave irradiation could cause the abnormal permeability of bacterial cell membrane and the degeneration of protein, resulting in bacterial metabolic abnormalities and irreversible damage of bacteria. The combination of microwave and UV disinfection could effectively make up the shortage of ultraviolet disinfection and improve the disinfection efficiency. Zhang et al. [8] studied the disinfection of the clarifier effluent in a wastewater treatment plant by microwave induced electrodeless UV irradiation. It was found that the coupling of microwave and ultraviolet light could take the advantages of both sides, and lead to rapid sterilization and irreversible damage to the bacteria.

Rifampicin is a broad-spectrum antibiotic, which has strong antibacterial property against Mycobacterium *tuberculosis* and is also effective for positive or negative gram bacteria and viruses. It was widely used in medical and pharmaceutical fields, and the rifampicin resistant bacteria (RRB) were also frequently detected in pharmaceutical factories and hospital wastewater. The microwave-enhanced photolysis process (MW/UV) was used for the disinfection of RRB in this paper. The effects of different operating parameters on the RRB disinfection were investigated and the operation conditions were optimized. The reactivation of the sterilized RRB in different actual water matrices was studied to explore the feasibility of the MW/UV disinfection of RRB.

2. Materials and methods

2.1. Materials
The RRB was provided by Institute of Urban Environment, Chinese Academy of Science. Rifampicin was purchased from Aladdin Reagent Co., Ltd. (Shanghai, China), while dimethylsulfoxide (DMSO), glycerinum and anhydrous ethanol were received from Sinopharm Chemical Reagent Co., Ltd. (Shanghai, China). The biochemical reagents tryptone and agar powder were also purchased from Sinopharm Chemical Reagent Co., Ltd. (Shanghai, China), while yeast extract powder was provided by Beijing double spin microbiological medium products factory (Beijing, China). All the chemicals were of analytical grade and used without further purification.

2.2. Experimental methods

2.2.1 Preparation of mediums. The LB solid medium was prepared by putting 10 g tryptone, 5 g yeast extract, 10 g NaCl and 15 g agar into 800 mL distilled water, using 1 mol/L NaOH to adjust the pH to 7.0±0.1, adding distilled water until the volume reaching 1 L, then packing the medium and sterilized under high pressure (121 ℃, 20 min).

The antibiotic medium was prepared by adding sterilized rifampicin solution (3.2 mg/mL DMSO) into the cooled down LB solid medium (50-60 ℃), and the final rifampicin concentration in the antibiotic medium was 4 μg/mL.

2.2.2 MW/UV disinfection procedure. The static MW/UV disinfection experiments were carried out in an APEX microwave system produced by Preekem Scientific Instruments Co., Ltd. (Shanghai, China), and the UV light was a microwave discharged electrodeless lamp (MDEL) prepared by Shanghai Jiguang Special Illumination Instruments Factory (Shanghai, China). The MDEL had maximum wavelength at 254 nm, and the UV light intensity linearly increased with the microwave power. The relation was shown in equation (1).

$$y = 0.0435x - 1.146,\ R^2 = 0.96401 \quad (1)$$

Where y was the UV light intensity, mW/cm^2, x was the microwave power.

In the MW/UV disinfection procedure, 40 mL RRB solution was added into a 250 mL Erlenmeyer flask. The MDEL was inserted into the RRB solution, and then the whole flask was put in the center of the microwave oven. After certain time of irradiation, the RRB solution was withdrawn for analysis.

2.2.3 RRB reactivation experiment. Sea water (sampled from the sea near Xiamen Jiageng Park, China), Lake water (sampled from the Ligong Lake in Xiamen University of Technology, China), and Tap water (sampled from the lab of Xiamen University of Technology, China) were involved in the RRB reactivation experiments to investigate the RRB reactivation in different water matrices. The compositions of these three water samples were illustrated in Table 1.

Table 1. The composition of the sampled water

Water sample	COD_{Cr} (mg/L)	DO (mg/L)	Conductivity (mS/cm)	Temperature (℃)
Lake water	397.72	10.63	0.13	22
Sea water	696.00	11.38	58.60	21
Tap water	47.00	10.80	0.07	23

In the RRB reactivation experiments, different dosages of the MW/UV disinfected RRB solution were added into the three water samples, and separately cultivated under the same temperature with the sampled water matrices for 48 h. The Sea water and Lake water samples were cultivated in light, but the Tap water sample was cultivated in dark to simulating the dark water pipe environment.

2.3. Analytical method

2.3.1. Determination of the bacteria population. According to GB/T4789.2-2008, 100 μL of the MW/UV disinfected RRB solution was withdrawn using a sterilized pipette and gradually diluted to a proper concentration through ten-times dilution method. Then, two of the diluted 100 μL disinfected RRB solution was separately injected into the LB medium and rifampicin medium. After uniformly spreading, the mediums were cultivated in constant-temperature incubator at 28℃ for 48h. The number of the bacterial colonies on the LB medium was counted as the total bacteria amount (CFU/mL), and the number of the bacterial colonies on the rifampicin medium was counted as the total RRB amount (CFU/mL). Each experiment was conducted in triplicate, and only the result between 30-300 CFU/mL was regarded as effective.

2.3.2. The calculation of the relative RRB abundance. The relative RRB abundance was calculated by the total RRB amount divided by the total bacteria amount.

3. Results and discussion

3.1. Effect of microwave power on the disinfection of RRB

Microwave was the only energy source in the MW/UV process. Increasing the microwave power would linearly increase the UV light intensity (see in equation (1)). Therefore, the effect of microwave power on the disinfection of RRB was initially studied, and the results were presented in figure 1.

The disinfection rate of RRB was significantly enhanced when the microwave power increased. The main reason might be the improved UV light intensity caused by the increase of microwave power. The higher UV light intensity would provide more energy for the deconstruction of RRB [9]. On the other hand, the enhanced microwave power would accelerate the molecular movement in the RRB solution, which would induce the temperature rises and finally improve the disinfection of RRB. In all, the microwave and UV light worked together to achieve a higher reduction of RRB. Considering the energy consumption, 500 W was suggested to be a suitable microwave power for the MW/UV disinfection of RRB.

3.2. Effect of irradiation time on the disinfection of RRB

The irradiation time was then extended to investigate the effect of continuous irradiation on the disinfection of RRB. As shown in figure 2, extending the irradiation time improved the reduction of RRB. When the irradiation was prolonged to 120 s, the disinfection rate of RRB could reach 100%. The investigation revealed that the reduction of RRB had close relation with the accumulation of microwave and UV light irradiation. The more irradiation RRB received, the worse injury the cell and DNA got. Therefore, suitable irradiation time is critical to the reduction of RRB in the MW/UV process.

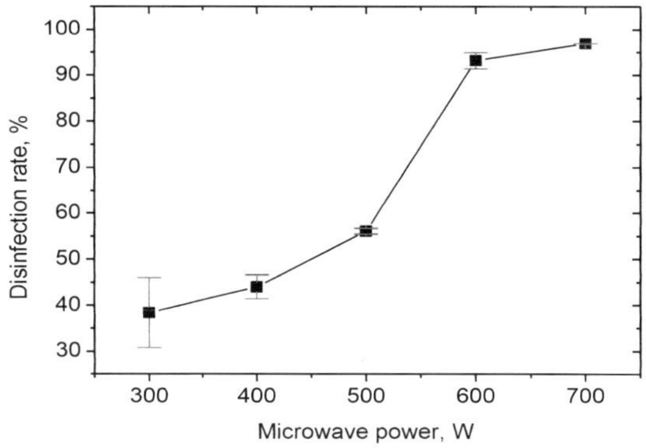

Figure 1. The influence of microwave power on the disinfection of RRB
(Experimental conditions: RRB volume, 40 mL; initial RRB concentration, 3.3×10^9 CFU/mL; reaction time, 10 s)

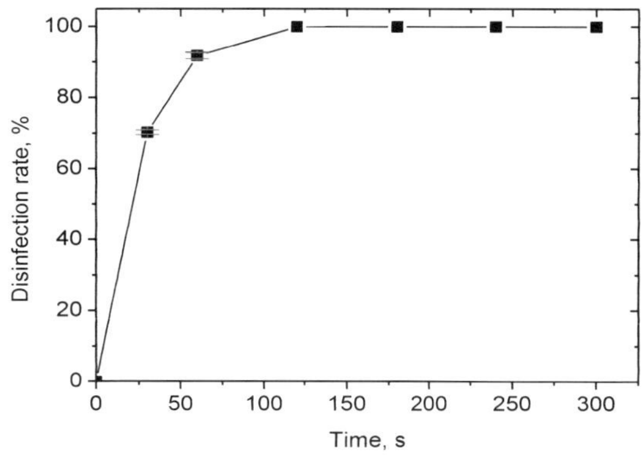

Figure 2. The influence of irradiation time on the disinfection of RRB
(Experimental conditions: RRB volume, 40 mL; initial RRB concentration, 4×10^9 CFU/mL; microwave power, 500 W; UV light intensity, 18.54 mW/cm^2)

3.3. Effect of initial concentration on the disinfection of RRB

The initial concentration was another factor which might influence the removal of RRB. The raw RRB solution with the concentration of 3.77×10^9 CFU/mL was diluted 10, 100, 1000 and 10000 times to study to effect of initial concentration on the removal of RRB. Results in figure 3 illustrated that the higher the initial concentration of RRB was, the lower the disinfection rate was gained. It was because the short wavelength UV lights were easy to be absorbed and attenuated when they were penetrating the RRB solution. The higher initial concentration meant more RRB cells existed, which would absorb more UV lights and block the penetration of UV lights [10]. Thus, more RRB couldn't receive enough UV irradiation and couldn't be effectively reduced. Besides, microwave irradiation was also prevented to a certain extent by the higher concentration RRB, which could be another reason for the drop of the disinfection rate under higher initial RRB concentration. In practical application, proper dilution would help the disinfection of RRB.

3.4. Effect of pH on the disinfection of RRB

Microbes need a suitable pH environment for growth, so the pH of the raw RRB solution was adjusted to 3, 5, 7, 9, 11 to investigate the effect of pH on the disinfection of RRB in the MW/UV process. Under the experimental conditions, the lowest disinfection rate occurred at pH = 5, and the removals of RRB under more acidic and alkali environment were much more effective (shown in figure 4). That was because the most suitable pH for the growth of the *Escherichia. coli* was around 5.5. When the pH of the water matrice was around 5.5, the RRB cultivated from *Escherichia. coli* had the strongest resistance to the unfavorable external conditions and was most likely to survive. On the contrary, when the water matrice went more acidic or alkali, the acid-base balance of RRB was destroyed and the RRB was more easily to be damaged in the MW/UV process. Therefore, adjusting the pH properly would improve the disinfection of RRB in practical application.

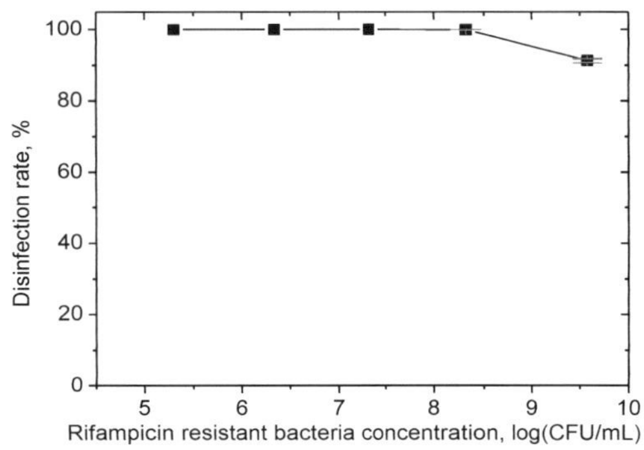

Figure 3. The influence of initial concentration on the disinfection of RRB
(Experimental conditions: RRB volume, 40 mL; microwave power, 500 W; UV light intensity, 18.54 mW/cm^2; irradiation time, 60 s)

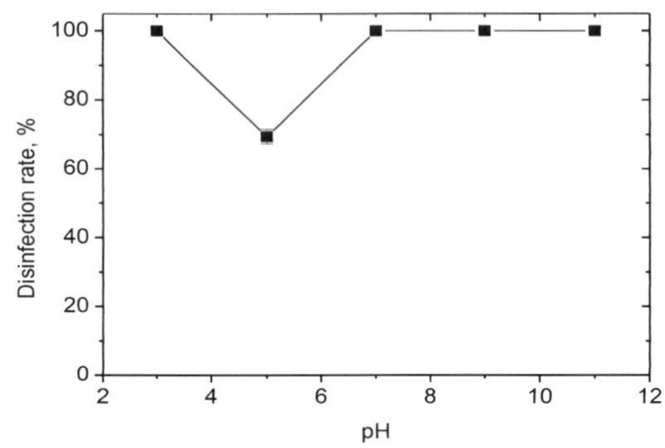

Figure 4. The influence of pH on the disinfection of RRB
(Experimental conditions: RRB volume, 40 mL; initial RRB concentration, 3.8×10^7 CFU/mL microwave power, 500 W; UV light intensity, 18.54 mW/cm^2; irradiation time, 60 s)

3.5. The reactivation of the RRB in different water matrices
Although RRB could be disinfected under short time irradiation of microwave and UV light, it was still not clear whether the antibiotic resistance in *Escherichia. coli* was completely removed. It was not sure whether the RRB would be resurrected under appropriate natural environment. Based on this consideration, simulated resurrection experiments were conducted by cultivating the disinfected RRB solution in different collected water samples. For the Lake water and Sea water case, the samples were cultivated under light and constant temperature for 48 h. For the Tap water case, the sample was cultivated in dark to simulate the dark environment in water pipe. The specific cultivation conditions and the composition of different water samples were presented in section 2.2.3.

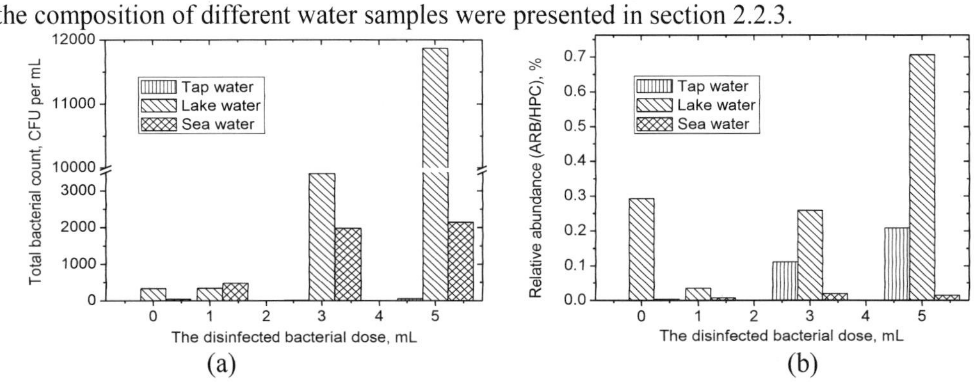

Figure 5. Effect of the disinfected RRB dose on the resurrection of RRB in different water samples. (a) Total resurrected RRB counts; (b) Relative resurrected RRB abundance

Figure 5 demonstrated the RRB resurrection in different water samples when accepted different doses of the MW/UV disinfected RRB solution. After 48 h cultivation, RRB resurrection occurred in all the three water samples, and both the total RRB counts and the relative abundance increased with the increase of the disinfected RRB dose. Those meant the rifampicin resistant gene was not thoroughly degraded after the short time MW/UV irradiation. The RRB might be damaged, and the rifampicin resistant gene might be released from the cell. When the outside environment was suitable, the rifampicin resistant genes would be expressed again.

Comparatively, the resurrected RRB in Lake water was more serious than other two water samples. It might be ascribed to the higher RRB counts and relative abundance in the raw Lake water. On the other side, the presence of large amounts salt in Sea water might inhibit the growth of RRB. As for the Tap water condition, it almost had no RRB in the raw Tap water, and the Tap water sample was cultivated in dark after injected with the disinfected RRB. The water samples cultivated in light, such as the Lake water and Sea water cases, might experience photo-reactivation, while the Tap water sample cultivated in dark might occur dark-reactivation. In photo-reactivation, a photo-reactivating enzyme formed in the bacteria, which restored the replicative ability of the cell and finally increased the RRB amounts. Meanwhile, the repair in the dark-reactivation belonged to the excision repair, which might take a long time to recover, so the RRB reactivation in Tap water was not significant.

Considering the resurrection of the disinfected RRB, further research on the change of rifampicin resistant gene in the MW/UV process must be carried out to evaluate the ecological risk.

4. Conclusions

RRB could be effectively reduced by the coupled irradiation of microwave and UV light. When the microwave power was 500 W, UV light intensity was 18.54 mW/cm^2, RRB volume was 40 mL, initial RRB concentration was 3.8×10^7 CFU/mL, and pH was 7, the RRB disinfection rate could reach 100% after 60 s reaction. Enhancing the microwave power, diluting the RRB solution, and adjusting the RRB solution to acidic and alkali condition would all improve the disinfection of RRB. But unfortunately, the RRB after MW/UV disinfection all experienced reactivation when discharged into Lake water, Sea water and Tap water samples. The reactivation rate followed the sequence: Lake water > Sea water > Tap water. The MW/UV process was an effective RRB disinfection method, but further research on the change of rifampicin resistant genes in the MW/UV process was urgently needed.

5. Acknowledgements

The authors are grateful for the financial support from the National Natural Science Foundation of China (No. 51408518, 51378446).

References

[1] Pruden A, Pei R, Storteboom H and Carlson KH 2006 Antibiotic resistance genes as emerging contaminants: studies in northern Colorado *Environ. Sci. Technol.* **40**(23) 7445-50

[2] Qiao M, Ying GG, Singer AC and Zhu YG 2018 Review of antibiotic resistance in China and its environment *Environ. Int.* **110** 160-72

[3] Gao H, Zhang LX, Lu ZH, He CM, Li QW and Na GS 2018 Complex migration of antibioticresistance in natural aquatic environments *Environ. Pollut.* **232** 1-9

[4] Page GL, Gunnarsson L, Snape J and Tyler CR 2017 Integrating human and environmentalhealth in antibiotic risk assessment: a critical analysis of protection goals, species sensitivity and antimicrobial resistance *Environ. Int.* **109** 155-69

[5] Zhang CM, Xu LM, Wang XC, Zhuang K and Liu QQ 2017 Effects of ultraviolet disinfectionon antibiotic-resistant Escherichia coli from wastewater: inactivation, antibiotic resistance profiles and antibiotic resistance genes *J. Appl. Microbiol.* **123** 295-306

[6] Gao P, Munir M and Xagoraraki I 2012 Correlation of tetracycline and sulfonamide antibiotics with corresponding resistance genes and resistant bacteria in a conventional municipal wastewater treatment plant *Sci. Total Environ.* **421-422** (3) 173-83

[7] Jungfer C, Schwartz T, Obst U 2007 UV-induced dark repair mechanisms in bacteria associatedwith drinking water *Water Res.* **41**(1) 188-96

[8] Zhang Y, Zhang GS and Wang P 2016 Disinfection for the municipal secondary effluents under microwave induced electrodeless ultraviolet irradiation *China Environ. Sci.* **36** (5) 1463-68

[9] Sommer R, Pribil W, Appelt S, Gehringer P, Eschweiler H, Leth H, Cabaj A and Haider T 2001 Inactivation of bacteriophages in water by means of non-ionizing (UV-254.7nm) and ionizing (gamma) radiation: a comparative approach *Water Res.* **35**(13): 3109-16

[10] Loge FJ, Emerick RW, Thompson DE, Nelson DC and Darby JL 1999 Factors influencing ultraviolet disinfection performance part I: Light penetration to wastewater particles. *Water Environ. Res.* **71**(3) 377-81

IOP Conf. Series: Earth and Environmental Science **146** (2018) 012025 doi:10.1088/1755-1315/146/1/012025

A Stochastic-Fuzzy Optimization Model for Adjustment of Urban Industrial Layout and Environmental Regulation in Beijing City under Uncertainties

X.T. Zeng[1]*, G.L. Gao[1], J. Liu[2] and A.L. Yang[2]
[1].Capital University of Economics and Business, Beijing 10070, China;
[2] Xiamen science and technology, Ximen, Fujian 361024, China

zxt1231@sina.com

Abstract. In this study, a stochastic-fuzzy optimization model (SFM) is developed for adjusting current urban industrial layout and environmental regulation in Beijing under uncertainties. The SFM can not only handle uncertainties expressed as probability and possibility distributions, but also reflecting the infeasibility risks between expected targets and random second-stage recourse penalties. The results of adjustment of production, industrial layout pattern, pollutant mitigation and system benefit under various Laplace criterion scenarios are analyzed. Tradeoff between industrial development and pollution mitigation can support policymakers generating a sustainable mode to alleviate air pollution issue.

1. Introduction

In some rapid-developed city such as Beijing City, disorder and drastic exploitation of industry and production would result in pollutant emission exceeding the limits of environmental load, leading air crisis. It has brought about huge pressures for human being to confront unprecedented risk of live and development [1]. On the pressures of hazard air pollution issues, the pollutant emission-permit trading can be introduced to provide incentives to adopt pollution abatement, which can drive industrial companies to improve the efficiency of pollutant treatment technology. Therefore, an optimized strategy associated with the adjustment of industrial layout and environmental regulation under emission-permit trading mechanism (OSET) is desired to coordinate relationship between economic development and environmental protection [2].

However, an OSET system is complicated with a variety of uncertainties [3-5]. Previously, many research works have been developed for handling uncertainties in the decision process of an OSET issue, which can reduce the difficulties and risk levels of decision-making [6-7]. Among them, two-stage stochastic programming (TSP) can solve decision-making problems associated with objective randomness, which can rectify initial decision with probabilistic event. However, in a practical OSET issue, the economic and environmental strategies with scenario assumption can be also expressed as probabilistic distributions, leading more stochastic situations. Thus, a Laplace criterion is introduced to handle the probability of each scenario occurrence under the supposition of no data available [8-10]. Meanwhile, numbers of fuzzy occurrences can be handled by a fuzzy credibility constraint programming (FCP) [11]. Previously, few works have focused on multiple uncertainties in hybrid formats in an OSET planning issue.

Therefore, a stochastic-fuzzy optimization model (SFM) is developed for adjusting current urban industrial layout and environmental regulation in Beijing under uncertainties. The SFM can not only handle uncertainties expressed as probability and possibility distributions, but also reflecting the infeasibility risks between expected targets and random second-stage recourse penalties. Results of emission-permit transactions, production reductions, pollution mitigation schemes, adjusted industrial structures and system benefits under various Laplace criterion scenarios are analyzed. These findings can support policymakers to identify optimized industry-environment policies for coordinating relationship between economic development and environmental protection

2. Application

Beijing city as a capital of China, which has undergone rapid urbanization/industrialization, corresponding to accelerated population expansion and high-speed economic growth. The population of city has exceeded 21.14 × 106 people; the growth rate of population has reached 8.93‰ per year [12]. Until 2013, the GDP of Beijing city has reach 1215.3 billion Yuan, which maintains high growth rate about 9% in recent years. With the industrial expansion, large-scale pollutant emission has exceeded the self-purification capacity of atmosphere. Therefore, a comprehensive strategy associated with a combination of a series of strategies is desired, which concludes adjustment of industrial structure, command of production scale and reduction of pollutant emission.

The enterprises would reduce the scale of production to confront the environmental penalty. However, a single political plan cannot conserve the environment at the extreme due to "governance failure". Thus, a market approach can be introduced to correct these weaknesses, which can reallocate emission-permit from lower- to higher-value to increase economic efficiency but also provide incentives to adopt pollution abatement measures [13-14]. However, in an OSET system, numbers of factors such as on-site survey and monitoring, analysis of main affected factors, determination of pollution-source emission standards, partition of functional zones and design of their respective environmental capacities, as well as generation of control measures. Thus, a stochastic fuzzy model is introduced to handle the uncertainties and complexities. Based on the SFM method, an OSET is formulated as follows:

$$\max f_{Laplace} = \{\frac{1}{m_1} * \left[\begin{pmatrix} LPS_{11} & LPS_{12} & \dots & LPS_{1n} \\ LPS_{21} & LPS_{22} & \dots & LPS_{2n} \\ \dots & \dots & \dots & \dots \\ LPS_{m_1 1} & LPS_{m_1 2} & \dots & \end{pmatrix} + \frac{1}{m_2} * \begin{pmatrix} LCS_{(m_1+1)1} & LCS_{m_1 2} & \dots & LCS_{(m_1+1)n} \\ LCS_{(m_1+2)1} & LCS_{(m_1+2)2} & \dots & LCS_{(m_1+2)n} \\ \dots & \dots & \dots & \dots \\ LCS_{m_2 1} & LCS_{m_2 2} & \dots & \end{pmatrix} \right] * \tag{1}$$

$$[(1)-(2)-(3)-(4)+(5)-(6)]$$

(1) Income from various industrial production activities based on expected economic development:

$$\sum_{t=1}^{3}\sum_{j=1}^{4}(\sum_{m=1}^{5} BFH_{tmj} \cdot WH_{tmj} \cdot hr_{tmj} + \sum_{n=1}^{3} BFM_{tnj} \cdot WM_{tnj} \cdot mr_{tnj} + \sum_{i=1}^{2} BFL_{itj} \cdot WL_{tij} \cdot lr_{tij}) \tag{2}$$

(2) Penalty for excessive emission from various industrial production activities:

$$\sum_{h=1}^{3}\sum_{t=1}^{3}\sum_{j=1}^{4}[\sum_{m=1}^{5} PH_{tmj} \cdot (WH_{tmj} - RH_{htmj}) \cdot hr_{tmj} + \sum_{n=1}^{3} PM_{tnj} \cdot (WM_{tnj} - RM_{htnj}) \cdot mr_{tnj} + \sum_{i=1}^{2} PL_{tij} \cdot (WL_{tij} - RL_{htij}) \cdot lr_{tij}] \tag{3}$$

(3) Loss for reduced production activities based on adjustment of industrial structure:

$$\sum_{h=1}^{3}\sum_{t=1}^{3}\sum_{j=1}^{4}(\sum_{m=1}^{5} LLH_{tmj} \cdot RH_{htmj} \cdot hr_{tmj} + \sum_{n=1}^{3} LLM_{tnj} \cdot RM_{htnj} \cdot mr_{tnj} + \sum_{i=1}^{2} LLL_{tij} \cdot RL_{htij} \cdot lr_{tij}) \tag{4}$$

(4) System benefit from emission-permit transaction:

$$\sum_{h=1}^{3}\sum_{t=1}^{3}\sum_{j=1}^{4} (\sum_{n=1}^{5} p_{tjh} \cdot TCM_{htnj} \cdot BM_{tnj} hr_{tnj} + \sum_{m=1}^{2} p_{tjh} \cdot TCH_{htmj} \cdot BH_{tmj} \cdot mr_{tnj} + \sum_{i=1}^{3} p_{tjh} \cdot TCL_{htij} \cdot BL_{tij} \cdot lr_{tij}) \tag{5}$$

(5) Cost for emission-permit transaction:

$$\sum_{h=1}^{3}\sum_{t=1}^{3}\sum_{j=1}^{4} (\sum_{n=1}^{5} p_{tjh} \cdot TCM_{htnj} \cdot BM_{tnj} hr_{tnj} + \sum_{m=1}^{2} p_{tjh} \cdot TCH_{htmj} \cdot BH_{tmj} \cdot mr_{tnj} + \sum_{i=1}^{3} p_{tjh} \cdot TCL_{htij} \cdot BL_{tij} \cdot lr_{tij}) \tag{6}$$

IOP Conf. Series: Earth and Environmental Science **146** (2018) 012025 doi:10.1088/1755-1315/146/1/012025

There are numbers of constraints as follows:

(1) Constraint of available emission-permit for transaction:

$$\sum_{h=1}^{H}\sum_{t=1}^{3}\sum_{j=1}^{4}[\sum_{m=1}^{2}TCH_{htmj}\cdot hr_{tmj}+\sum_{n=1}^{5}TCM_{htnj}\cdot mr_{tnj}+\sum_{i=1}^{3}TCL_{htij}\cdot lr_{tij}]\leq[(\sum_{h=1}^{H}\sum_{t=1}^{3}\sum_{j=1}^{1}PA_{htj}^{SO_2} +\sum_{h=1}^{H}\sum_{t=1}^{3}\sum_{j=2}^{2}PA_{htj}^{NO_2}+\sum_{h=1}^{H}\sum_{t=1}^{3}\sum_{j=3}^{3}PA_{htj}^{PM10}+\sum_{h=1}^{H}\sum_{t=1}^{3}\sum_{j=4}^{4}PA_{htj}^{PM2.5})] \quad (7)$$

(2) Constraint of ambient air quality requirement:

$$Cr\{[\exp(-\frac{\delta+\mu}{2\delta})]/\{\pi\cdot(\tilde{\sigma}_y\tilde{\sigma}_z r[\frac{\delta H^2}{s^2(\delta+\mu)}]\cdot[\sum_{h=1}^{H}\sum_{t=1}^{3}\sum_{j=1}^{4}(\sum_{m=1}^{2}(WH_{tmj}-RH_{htmj}-TCH_{htmj})\cdot hr_{tmj}+ \sum_{n=1}^{5}(WM_{tnj}-RM_{htnj}-TCM_{htnj})\cdot mr_{tnj}+\sum_{i=1}^{3}(WL_{tij}-RL_{htij}-TCL_{htij})\cdot lr_{tij}] \leq\{[\sum_{h=1}^{H}\sum_{t=1}^{3}\sum_{j=1}^{1}\tilde{Q}A_{htj}^{SO_2}+\sum_{h=1}^{H}\sum_{t=1}^{3}\sum_{j=2}^{2}\tilde{Q}A_{htj}^{NO_2}+\sum_{h=1}^{H}\sum_{t=1}^{3}\sum_{j=3}^{3}\tilde{Q}EA_{htj}^{PM_{10}}+\sum_{h=1}^{H}\sum_{t=1}^{3}\sum_{j=4}^{4}\tilde{Q}EA_{htj}^{PM_{2.5}}]\}\geq\alpha,\quad\forall i,j,h \quad (8)$$

(3) Constraint of industrial production scale:

$$CME_{tmj\min}\leq\sum_{m=1}^{2}WH_{tmj}\cdot hr_{tmj}\leq CME_{tmj\max} \quad (9)$$

$$CDE_{tnj\min}\leq\sum_{n=1}^{5}WM_{tnj}\cdot mr_{tnj}\leq CDE_{tnj\max} \quad (10)$$

$$ICR_{tij\min}\leq\sum_{i=1}^{3}WL_{tij}\cdot lr_{tij}\leq ICR_{tij\max} \quad (11)$$

(4) Constraint of non-negativity:

$$WH_{tmj},\ WM_{tnj},\ WL_{tij}\geq 0,\qquad hr_{tmj},\ mr_{tnj},\ lr_{tij}\geq 0, \quad (12)$$

Model (7) presents available emission-permit for transaction, where the demand of air pollutant emission can not excess the maximum emission-permit allowance. Model (8) shows that the requirement of ambient air quality in study region, which is built on Gaussian dispersion model (GDM). Modes (9) to (11) are industrial development scales, which can reflect the minimum / maximum development scale of production in study region. Model (12) is non-negativity restrictions. The parameters for model are calculated based on government reports, statistical yearbooks, and related research works [12]. There are 10 cases associated with Laplace criterion to be assumed in this study as follows: (a) case 1 is basic plan, which equal to current plan; (b) cases 2, 3, 4 and 5 are conservative plans, which adopt to environmental regulation; (c) cases 5, 6, 7 and 8 are progressive plans, which focus on economic development; (d) case 10 is plan under Laplace criterion.

3 Results analysis

Figure 1 presents optimized reduced production among various industrial sectors under cases 1, 4, 8 and 10 when ε and α levels are 0.95 and 0.99. The results show that the higher wind velocity would be suitable for pollutants diffusion, thus the reduced production scale would be lower; vice-versa. The relative higher reduction amount would occur in petroleum refining (denoted as "PR") and power generation (denoted as "PG") due to their higher pollutant emissions.

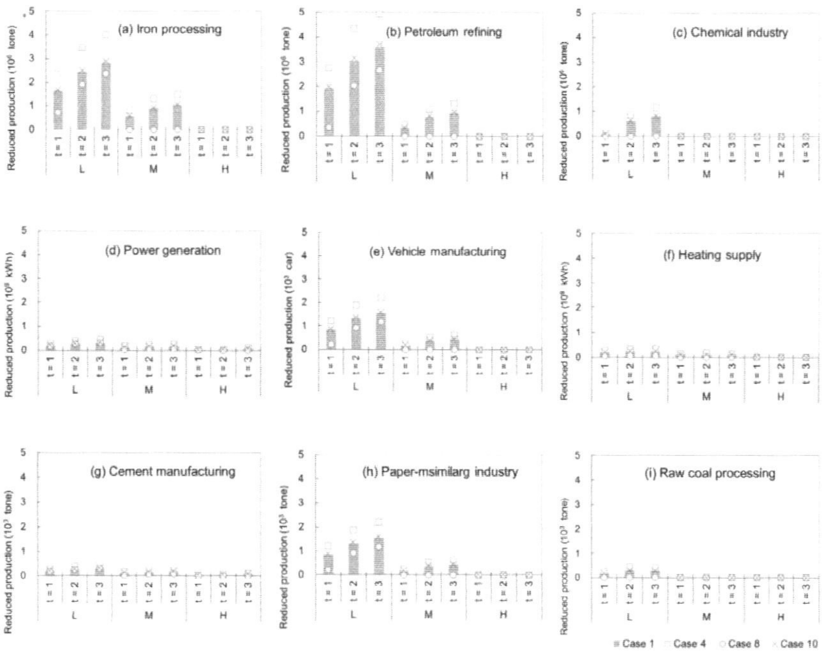

Figure 1. Optimized reduced production among various industrial sectors under cases 1, 4, 8 and 10 when ε and α levels are 0.95 and 0.99

Figures 2 presents total excess pollution emissions under various cases when α level is 0.99. In this study, excess pollutant emission would result in ultra concentration of pollution in the air. The results present that under case 1 in period 1, when meteorological condition is bad for diffusion, the ultra concentration ratio of SO_2 and NO_2 would be more than 6 times, and PM_{10} and $PM_{2.5}$ would reach 3 times at highest. It implies that current pollution emission pattern and industrial production mode in Beijing city would not suitable for future higher environmental regulation.

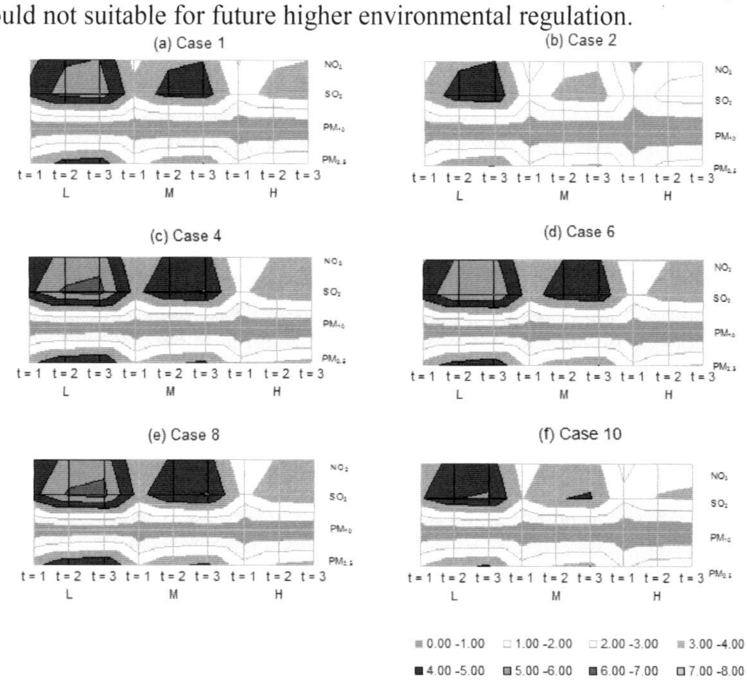

Figure 2. Total excess pollution emissions under various cases when α level is 0.99

Figure 3 presents the system benefits under cases 1, 5, 7, 9, 10 and optimal state when α level is varied (ε = 0.95). The results indicate that a lower α level can result in a lower benefit; vice versa. The results present that case 1 (basic plans) would generate higher system benefit than conservative and progressive cases. For example, when α level is 0.9, system benefit would be RMB ¥ 3.87 × 1012 under case 1, while system benefits are RMB ¥ 3.45 × 109 and \$ 1.05 × 109 under case 5 and case 7. The highest system benefit would be obtained under Laplace scenario (case 10), which can generate a reliable and optimal results with consideration of all risks of scenarios.

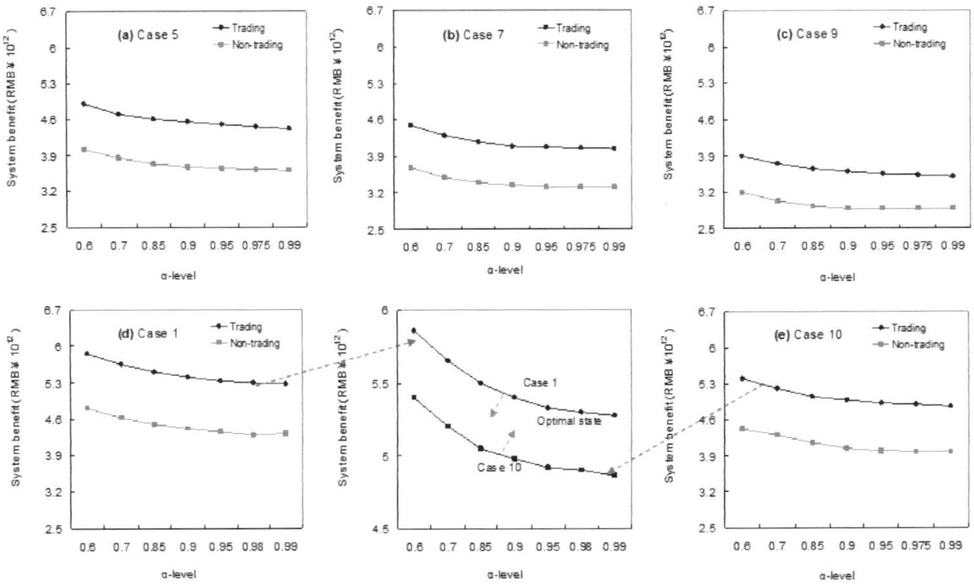

Figure 3 System benefits under cases 1, 5, 7, 9, 10 and optimal state when α level is varied (ε = 0.95)

4 Conclusions

In this study, a stochastic-fuzzy optimization model (SFM) embedded into an optimized strategy associated with the adjustment of industrial layout and environmental regulation under emission-permit trading mechanism (OSET) is proposed for confronting air crisis in Beijing City under uncertainties. The developed SFM model will be applied to a real case study of OSET issue in Beijing city of China. The results disclose that emission-permit trading would be an effective manner to adjust industrial layout and air quality management, which should be encouraged to coordination of economy and environment in the future. Meanwhile, the improper industrial layout is a main reason for air crisis in study region, which requires adjustment of local industrial structure and improvement of treatment technologies could lessen pollutant emissions to reduce environmental penalties, leading higher system benefits.

Acknowledgements
This research was supported by the National Social Science Foundation of China (Grant Nos. 14BFX107), Beijing Natural Science Foundation (L160011), National Key Research and Development Program (2016YFC0207800) and National Natural Science Fund (51608422).

Reference

[1] Jiang C, Wang H, Zhao T, Li T, Che H. 2015. Modeling study of PM2.5 pollutant transport across cities in China's Jing-Jin-Ji region during a severe haze episode December 2013. Atmospheric Chemistry and Physics 15: 5803-5814. DOI: 10.5194/acpd-15-3745-2015

[2] Clemens B, Bamford CE. and Douglas TJ. 2008. Choosing strategic responses to address emerging environmental regulations: size, perceived influence and uncertainty. Business Strategy and the Environment 17: 493-511. DOI: 10.1002/bse.601

[3] ApSimon HM, Warren RF, Kayin S. 2002. Addressing uncertainty in environmental modelling: a case study of integrated assessment of strategies to combat long-range transboundary air pollution. Atmospheric Environment 36: 5417-5426. DOI: 10.1016/S1352-2310(02)00659-3

[4] Haar LN, Haar L. 2006. Policy-making under uncertainty: Commentary upon the European Union Emissions Trading Scheme. Energy Policy, 34: 2615-2629. DOI: 10.1016/j.enpol.2005.07.003

[5] Wagner M. 2009. National culture, regulation and country interaction effects on the association of environmental management systems with environmentally beneficial innovation. Business Strategy and the Environment 18: 122-136. DOI: 10.1002/bse.641

[6] Chen WT, Li YP, Huang GH, Chen X, Li YF. 2010. A two-stage inexact-stochastic programming model for planning carbon dioxide emission trading under uncertainty. Applied Energy 87:1033-1047. DOI: 10.1016/j.apenergy.2009.09.016

[7] Nahorski Z, Horabik J. 2007. Compliance and emission trading rules for asymmetric emission uncertainty estimates. Climatic Change 103:303-325. OI: 10.1007/s10584-010-9916-4

[8] Laplace PS 1951. A Philosophical Essay on Probabilities. Dover, New York.

[9] Peterson, GD, Graeme SC, Stephen RC. 2003. Scenario Planning: a Tool for Conservation in an Uncertain World. Conservation Biology 17: 358-366. DOI: 10.1046/j.1523-1739.2003.01491

[10] Zeng XT, Huang G. 2016. A simulation-based resource-environment management model for regional sustainability in compound wetland ecosystem under uncertainties. Ecological Modelling 334: 60-77. DOI: 10.1016/j.ecolmodel.2016.04.021

[11] Inuiguchi M. 2012. Robust optimization by fuzzy linear programming in Managing Safety of Heterogeneous Systems. Lecture Notes in Economics & Mathematical System 658: 219-239. DOI: 10.1007/978-3-642-22884-1_11

[12] Beijing statistical yearbook (BSY) 2000-2013. Beijing Municipal Bureau of Statistics, Beijing, 2001-2014.

[13] Li YP, Huang GH. 2008. Interval-parameter two-stage stochastic nonlinear programming for water resources management under uncertainty. Water Resources Management 22: 681-698. DOI: 10.1007/s11269-007-9186-8

[14] Huang GH, Loucks DP. 2000. An inexact two-stage stochastic programming model for water resources management under uncertainty. Civil Engineering & Environmental System 17: 95-118. DOI: 10.1080/02630250008970277

IOP Conf. Series: Earth and Environmental Science **146** (2018) 012026 doi:10.1088/1755-1315/146/1/012026

Analysis of wind environmental characteristics around a square building

Q H Chen[1], H T Hu[1], C Z Qian[1] and C P Chen[1,2 *]

[1] West Coast Wind Engineering Research Center, Xiamen University of Technology, Xiamen, Fujian, 361024, China
[2]Xiamen Ocean Vocational College, Xiamen, Fujian, 361100, China

*Corresponding Author: C P Chen, +865927769301, cpchen@126.com

Abstract. The simulation of the wind environment around a square building is conducted in this paper. Two obvious flow features and their law are presented. When the airflow comes to the building, it separates into four branches, namely up-and-down flows and lateral flows. Firstly, the up-and-down flows are demarcated at 3/4 of the height with varied incoming flow speed. The upward flow reaches maximum speed at the roof and runs over the building, forming vortex in the leeside. The downward flow also achieves the maximum value near the bottom. Secondly, the lateral flows separate symmetrically about the axis and the lateral speed increases from zero along the width direction till the stream reaches the edges of the building and then decreases gradually. The study of the wind flow around a single building plays a basic role in the wind environment research of multi-buildings.

1 Introduction

Along with the population aggregation, the contradiction between the rapid expansion of urban and limited land resources intensified, the urban construction becomes denser, greatly increasing the flow complexity near the ground and making adverse effect to the wind environment outside the buildings. These adverse impacts become worse especially in the weather with strong winds or typhoon.

Therefore, the wind environmental characteristics of buildings already become essential component of the wind engineering, and prediction of the wind environment around buildings has been carried out at the practical design stage[1-9]. To be mentioned, the working group for CFD prediction of the wind environment around one building was organized by the Architectural Institute of Japan(AIJ)[10]. Based on the studies above, the complexity features of the flow around buildings are clarified gradually. But the research focusing on the wind environment of a single building still needs to be further deepened, which is the basic step for wind environment research of architectural complex and has not been specified clearly yet, namely the detail laws of the airflow, such as the vortex separation, reflux and reattach status. Therefore, the flow characteristics surrounding a square building is studied further in this paper with Computational Fluid Dynamics (CFD) approach.

2 Validation of numerical method

Attempts to provide statements about the accuracy of CFD method adopted, the first step of this paper is to validate the accuracy of simulation approach and the experimental results by Meng and Hibi[11] are used to validate the results of the CFD simulation.

2.1 Numerical setup

The shape of experimental model is with 2:1:1(height: width: depth) and the height 160mm (see figure 1a). The height is represented as h and the width/depth is represented as b. In the experimental case, the incoming flow velocity u varies with height and u_r (reference velocity at $z=h$) is 4.4m/s with Reynolds number 2.4×10^4. The origin is set in the center of the building and three components of velocity are described as u_x, u_y and u_z, as shown in figure 1b.

The software Fluent is adopted for the simulation and the setting details are as follows:

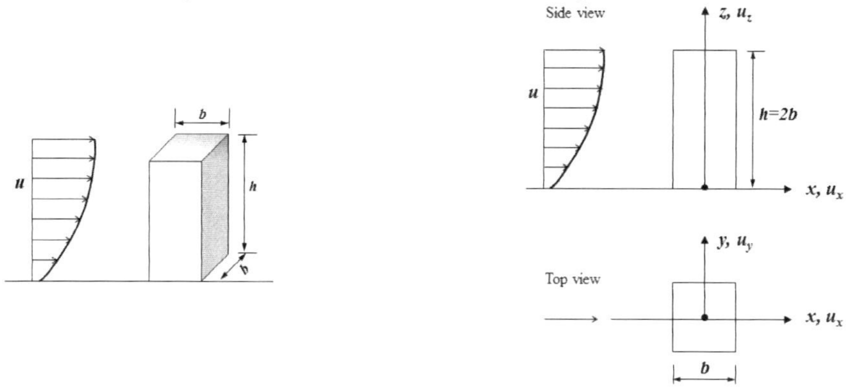

(a) Configuration of test model and notations (b) Origin set diagram

Figure 1. Model building sketch[11]

- Size of the computational domain: $21b(x)\times13.75b(y)\times11.25b(z)$
- Boundary conditions:

Inflow boundary condition	Based on the values tested in the wind tunnel[11]
Outflow boundary condition	Pressure outflow
Surface of the building model	Wall
Surface of the wind tunnel floor	Wall with roughness length z_0 ($z_0=1.8\times10^{-4}$m)
Surface of wind tunnel side walls and ceiling	Symmetry

- Turbulence model: Standard k-ε

The hexahedral structured grid is employed for the discrete of the whole calculation domain and in order to improve calculation accuracy, the mesh refinement is considered near the building. Aiming to determine the appropriate meshing plan, three grid schemes are conducted (see table 1).

Table 1. Grid schemes

	Minimum grid size	Grid number
Coarse mesh	0.005	1.04×10^6
Medium mesh	0.002	2.50×10^6
Fine mesh	0.001	5.25×10^6

2.2 Comparison of simulating and experimental results

The distributions of average wind velocity component u_x on the vertical cross-section at the center of the building ($y=0$) is shown in figure 2. The values are normalized and only the velocity profiles of u_x at the positions of $x/b=-0.075$ and $x/b=0$ are presented in the figure due to space limitations. Generally, the calculated values agree well with the experimental values except the position near the roof. The explanation could be that when approaching the roof surface, the flow slows down quickly and the calculation might not able to reflect this phenomenon. Meanwhile, the distributions of average wind velocity component u_x on the horizontal plane near the ground surface ($z=1.25b$) is shown in figure 3. The calculated data agree relatively well with the experimental values, except the value near the building boundary. Based on these diagrams, the simulation approach is proved with good accuracy.

Regarding to the grid independence, three different mesh schemes all presents good accuracy and the simulated results show some differences only in the region quite near the building, which also means that the conducted grid arrangements are sufficient and can be employed for further study. Moreover, this paper focuses on the detailed flow characteristics around the building, thus the finer mesh scheme is adopted in the research.

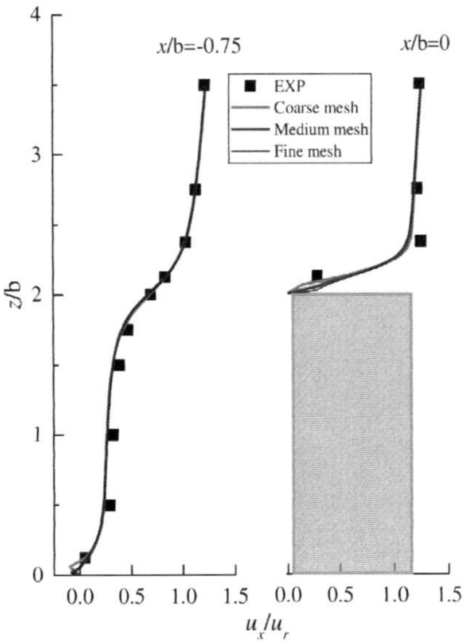

Figure 2. Comparison of u_x distributions along z-direction in vertical section y=0

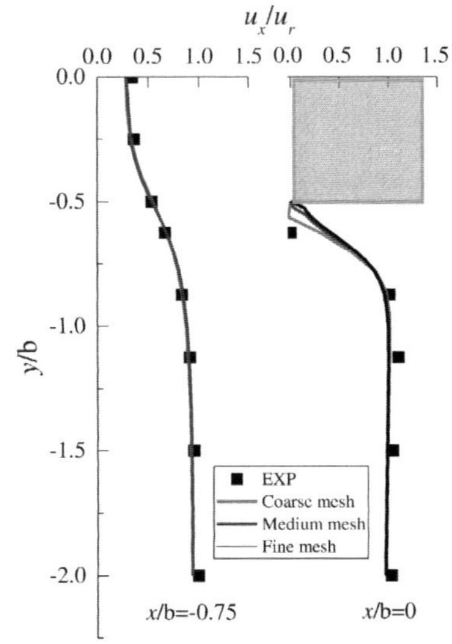

Figure 3. Comparison of u_x distributions along y-direction in horizontal section z=1.25b

3 Results of the flow field simulation around a single building

The simulation setup above is then adopted. The incoming wind speed at reference height is varied and set up as u_r=0.1m/s, 4.4 m/s, 10 m/s and 20m/s respectively(see table 2).

Table 2. The incoming flow conditions

u_r (m/s)	*Re* (-)
0.1	0.054×10^4
4.4	2.40×10^4
10	5.40×10^4
20	10.8×10^4

3.1 Up-and-down flow

3.1.1 Streamlines of up-and-down flow around the building. From the streamline chart of y=0 plane (figure 4), the block effect of the building on the flow making obvious change on the flow field both the windward side and leeward side of the building.

It can be observed that the phenomenon of flow spilt occurs just before the building. The majority part of the airflow moves down to the bottom area and mixes with the incoming flow, forming reflux. In this region, the pedestrian are affected not only by the incoming flow but also reflux and the dive flow. Wind environment uncomfortableness should be considered.

On the other side, upward flow climbs to the roof height and then separation happens. When the incoming flow wind velocity is small, such as u_r=0.1m/s (figure 4a), two obvious irregular vortices formed above and behind the building, and the phenomenon of backflow also appear around the bottom of leeside. With the increase of wind speed, the flow above the roof reattaches again and one large flow vortex is formed behind the building (figure 4b~d). The vortex impact range extents with the increase of wind speed. Additionally, the flow field behind the building is with relative low wind speed, in such case the pollution is not easy to diffuse.

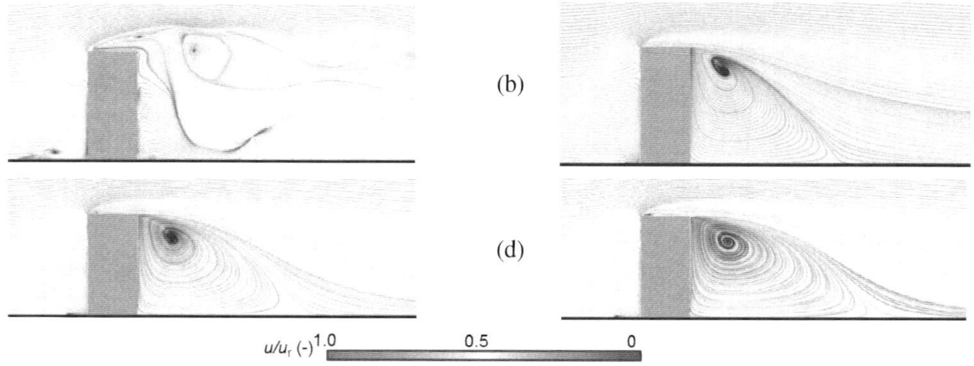

Figure 4. Streamlines of relative mean velocity on the y=0 plane with incoming flow variations (a)u_r=0.1m/s, (b) u_r =4.4 m/s, (c) u_r =10 m/s, (d) u_r =20 m/s

3.1.2 u_z profile before the building. The phenomenon of up-and-down flow can be found from the u_z profile more obviously shown in figure 5. The values of velocity component u_z along the height are divided into evident positive and negative partitions and the demarcation points with different incoming flow almost coincident at z/b=1.5, namely the height position $3h/4$.

The flow reinforcements are found in the regions near the ground and roof. To be more explicit, in the region z/b>1.5(or z/h>0.75), the flow climbs up with increasing velocity to u_z/u_r=1.0(at the roof z/b=2.0) and then goes upward with gradually less values. On the other hand, in the region z/b<1.5(or z/h<0.75), the air dives and with the most significant enhancement at z/b=0.3. Based on the u_z profile, three key height positions should be highlighted, respectively z/b=0.3, 1.5 and 2.0.

3.1.3 u_z profile after the building. In figure 6, the values of velocity component u_z behind the building are shown. Besides, a significant speed loss area is observed. When the coming flow reference velocity is 0.1, which is low, the flow pattern behind the building is disorder (see figure 4a) and the velocity speed fluctuates round zero. As the incoming wind speed increases, the u_z values are positive within the scope of building height, namely the air flows from down to up, presenting the vortex behind the building is rotating clockwise (see figure 4b~d). The u_z value increases along the height and reaches a maximum at z/b=1.0, then it weakens gradually till z/b=2.0. Since the separation vortex appears on the roof, a slight fluctuation also turns up between 2.0 <z/b<2.5.

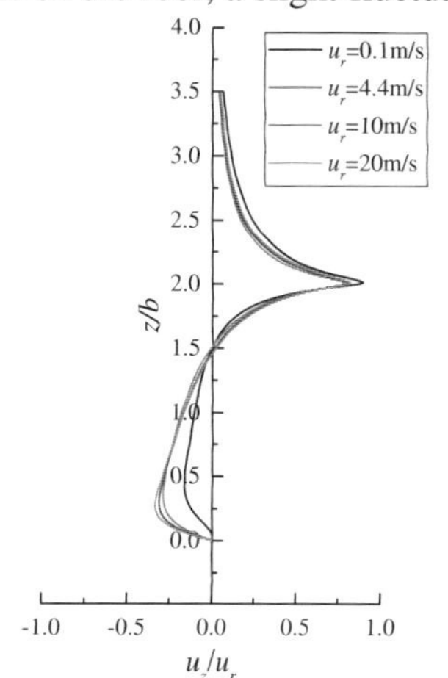

Figure 5. u_z/u_r profile right before the building with incoming flow variations

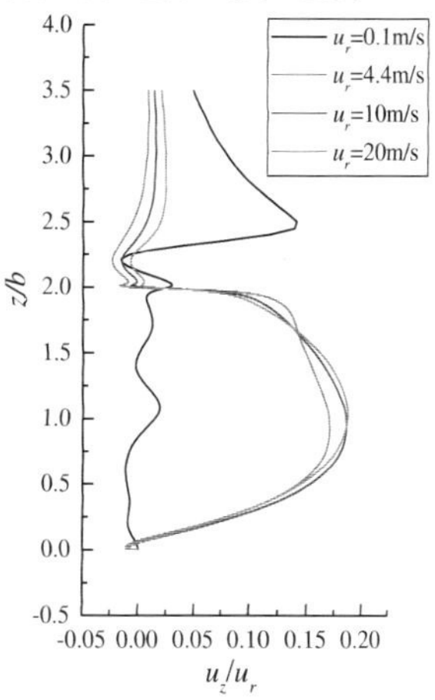

Figure 6. u_z/u_r profile right after the building with incoming flow variations

IOP Conf. Series: Earth and Environmental Science **146** (2018) 012026 doi:10.1088/1755-1315/146/1/012026

3.2 Lateral flow

3.2.1 Streamlines of lateral flow around the building. Based on the horizontal view of flow field in figure 7, when the airflow approaches the construction, the stream flows to the both sides and reaches maximum speed at the edges of building (at $y/b=\pm0.5$), then gets over with expansion, cladding the shedding vortices. The wake vortices generate and move downstream with the evolution of time, but only one flame is shown here.

When the flow is laminar (such as u_r=0.1m/s), three groups vortices generate on the leeside. When the flow rate increases and turns to turbulent, two symmetric vortices can be observed and the vortex extension expands along with the increase of flow speed.

3.2.2 u_y profile before the building. In order to investigate the lateral flow quantitatively, the profiles of u_y component right before the building at different horizontal positions on the condition of same incoming flow, such as z/b=0.5, 1, 1.5 and 2 are depicted in figure 8 and the profiles of u_y component at z/b=1 on the condition of different incoming flow are shown in figure 9.

From figure 8, the symmetrical velocity distribution can be found with the condition u_r=4.4m/s. As the airflow gets close to the building, it divides into left and right halves and the u_y values begin to increase from the symmetry axis. The maximum values appear at almost the same y positon ($y/b=\pm$ 0.5). The u_y/u_r values creep up to 0.8 to the great extent at z/b=0.5, 1 and 1.5, but 0.5 at the height of roof(z/b=2). When the flow moves laterally, the u_y data decrease to zero gradually.

From figure 9, although the incoming flow is varied with different speeds, the normalized values are almost coincident at the height z/b=1. That is, the flow field forwarding to the lateral sides of building is proportional to the incoming flow.

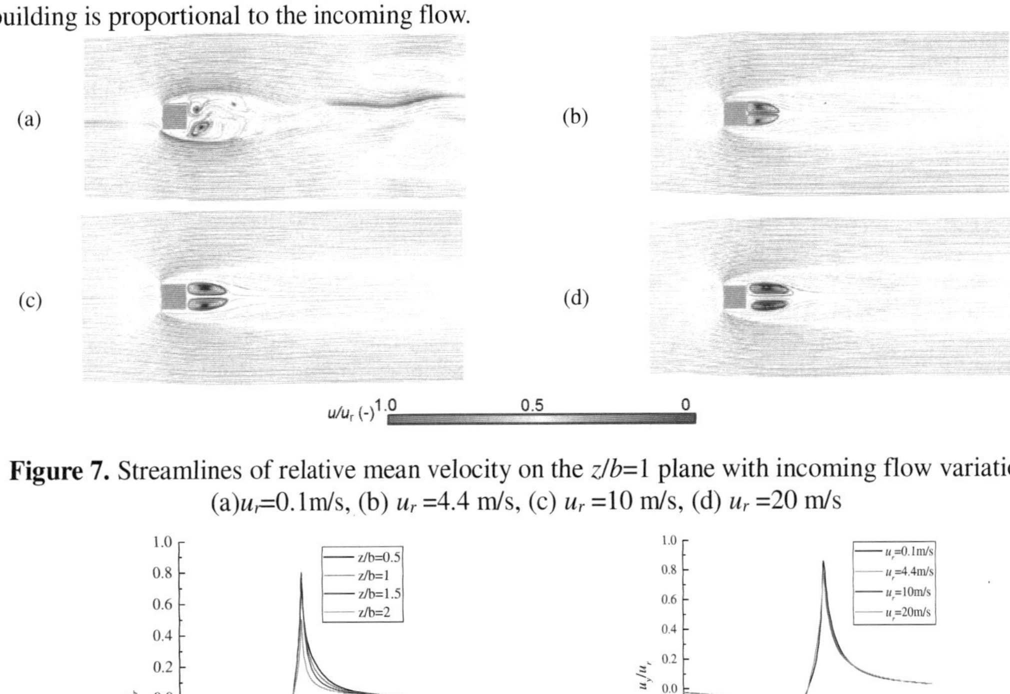

Figure 7. Streamlines of relative mean velocity on the z/b=1 plane with incoming flow variations (a)u_r=0.1m/s, (b) u_r =4.4 m/s, (c) u_r =10 m/s, (d) u_r =20 m/s

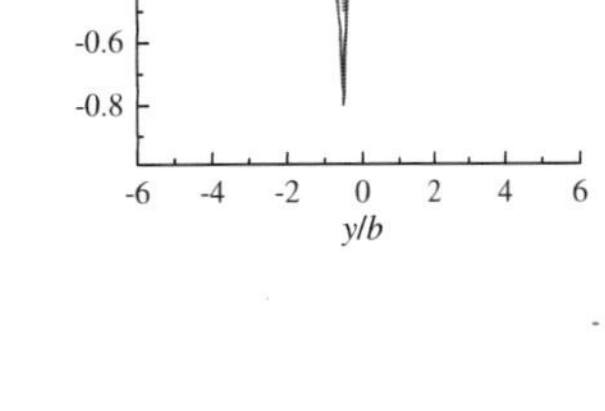

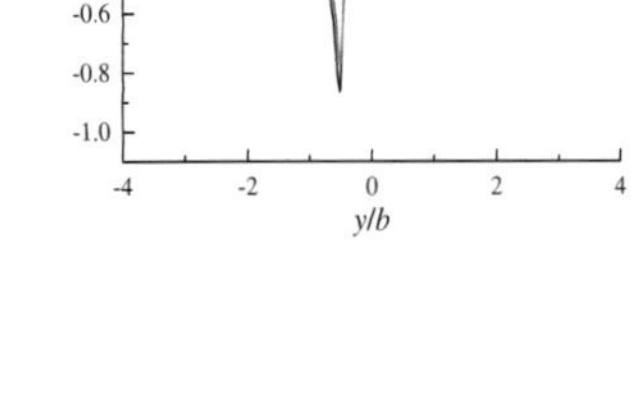

Figure 8. Distributions of u_y along y-direction at different heights with u_r=4.4m/s

Figure 9. Distributions of u_y along y-direction at z/b=1 with varied incoming flow

4 Conclusion

The three-dimensional flow field around a single building is analyzed with Fluent in this paper and the simulating approach is validated with experimental data. The flow characteristics are revealed and when the incoming flow approaches the building, it splits into up-and-down flow and lateral flow.

1) Right before the building, the up-and-down airflow generates at the demarcation point z/b=1.5. The majority part of the airflow moves down directly and mixes with the incoming flow near the bottom, obtaining its most significant velocity enhancement at z/b=0.3. The left climbs up to the roof and the u_z component starts to increase and reaches the maximum at the height of roof. When the upward flow moves over the building, the vortex appears in the leeward side.

2) Lateral flow due to the building block also causes the rise of wind speed. The stream flows to the both sides symmetrically and reaches maximum u_y speed at the edges of building ($y/b=\pm 0.5$). The u_y/u_r values creep up to 0.8 to the great extent at z/b=0.5, 1 and 1.5, but 0.5 at the height of roof (z/b=2). Then lateral flow gets over with expansion, cladding the shedding vortices.

Acknowledgement

The authors would like to acknowledge with great gratitude for the supports of the National Science Foundation of China(Grant No: 11272270, 51778551), the Science and Technology Plan Project of Xiamen(Grant No: 3502Z20161016), the Young Teachers Education Science Research Project of Fujian(Grant No: JA15384, JAT170410), the High-level Talent Project of Xiamen University of Technology(Grant No: YKJ14025R, YKJ15042R) .

References

[1] Wu Y, Zhang X, Li H, Wang C, Nie G, Cheng Z 2011 Numerical simulation of the pedestrian level wind environment around high-rise buildings J. Zhengzhou Univ.(Nat. Sci. Ed.) 04 110-115

[2] Wu X, Li Q, Li Y 2014 Test on surface wind pressure distributions and wind load characteristics for complex shape high-rise building *Journal of Architecture and Civil Engineering* **01** 76-82

[3] Zhao Y, Su H, Kang Y, Liu Z 2016 CFD Simulations of the wind environment of an industrial project in Wenzhou *Building Energy Efficiency* **02** 125-128

[4] Shi X, Zhu Y, Duan J, Shao R, Wang J. 2015 Assessment of pedestrian wind environment in urban planning design Landscape and Urban Planning 140 17-28

[5] Li L, Zhang L, Liu G, Liu S, Zheng S 2015 Wind environment simulation of green building. *Journal of Civil Architecture & Environment Engineering* **S2** 215-218

[6] Neofytou P, Venetsanos A G, Vlachogiannis D, Bartzis J G, Scaperdas A 2006 CFD simulations of the wind environment around an airport terminal building *Environmental Modelling & Software* **21** 520-524

[7] Mochida A, Tominaga Y, Murakami S, Yoshie R, Ishihara T, Ooka R 2002 Comparison of various k-e models and DSM applied to flow around a high-rise building *Wind and Structures* **5** 227-244

[8] Hussein A S, El-Shishiny H. 2009 Influences of wind flow over heritage sites: A case study of the wind environment over the Giza Plateau in Egypt *Environmental Modelling & Software* **24** 389-410

[9] Blocken B, Carmeliet J. 2008 Pedestrian wind conditions at outdoor platforms in a high-rise apartment building: generic sub-configuration validation, wind comfort assessment and uncertainty issues *Wind and Structures* **11** 51-70

[10] Yoshie R, Mochida A, Tominaga Y, Kataoka H, Harimoto K, Nozu T, Shirasawa T. 2007 Cooperative project for CFD prediction of pedestrian wind environment in the Architectural Institute of Japan Journal *of Wind Engineering and Industrial Aerodynamics* **95** 1551-78

[11] Meng Y, Hibi K. 1998 Turbulent measurements of the flow field around a high-rise building *Journal of Wind Engineering* **76** 55-64

Facade measurement of building along the roadway based on TLS and GIS of project supervision

He Yuanrong[1], Dai Jiaquan[1], Cui Shenghui[2] and Peng Degui[1]

[1]College of Computer and Information Engineering, Xiamen University of Technology, Fujian Xiamen 361024, China

[2]Key Laboratory of Urban Environment and Health, Institute of Urban Environment, Chinese Academy of Sciences, Xiamen 361021,China

E-MAIL： heyuanrong@126.com.

Abstract. In view of the traditional surveying and mapping, such as total station measuring tape measure method is applied to building facade measurement of low operation, data precision, incomplete information, etc. as well as large quantities of urban and rural landscape building facade renovation project fund accounting, planning management and project controls. Research based on point cloud data map facade map site plan, based on the UAV remote sensing image, using c#+AE integration facade renovation management in the whole process of data and business processes to develop GIS system, for the urban road construction project planning, management, provides a quick and precise integration technology platform, "city double repair" in the current urban renovation related engineering information management has practical value.

1. Introduction

Facade reconstruction measurement of building along the road is to measure of old buildings, provide the original information for design data[1-2], and use traditional tape measure with hand painting, total station, and other surveying and mapping methods in the face of the problems such as the old buildings' complex structure, difficult collection, incomplete information, etc. Unmanned Aerial Vehicle (UAV) remote sensing technology for spatial data acquisition is a new method developed after spaceflight, aviation remote sensing technology, which can achieve real-time transmission of image, with the remote sensing image to fast acquire the high-resolution and high-definition ground target objects. At the same time, it has low cost and good mobility, which effectively supplements and enhances the shortage of the satellite remote sensing and manned aerial remote sensing[3]. Terrestrial Laser Scanning (TLS) is measuring the time needed for laser transmission or phase difference between two points for spacing, and using the principle of for the conversion of triangular geometry for scanners space position and scanning angle to get the cloud data of the surface point of the object, and it is a non-contact measurement method with wide[4-5]. There are important values in architectural measurement modeling, deformation monitoring, fine mapping of ancient buildings, digital cultural heritage and digital urban planning, etc [6-8].

In recent years, the application of UAV technology to architecture has become an important research topic by domestic and foreign scholars. Metni and Hame successfully used the UAV system to monitor and repair bridges and construction applications[9]; Rathinam et al. used the visual technology of UAV to detect the linear structures of pipeline, road and bridge, canal, power grid, etc.[10] Du Caihong took the application of UAV to Ningxia region of the Yellow River as example, using the UAV photogrammetry data processing system (GodWork) to verify the current UAV image accuracy can meet the needs of the current image processing [11]. Stamos and Aleln from Switzerland used 3D laser scanning techniques to reconstruct the St. Pierre's cathedral[12]. P.K.Allen and I.Stamos et al. successfully developed a relatively comprehensive 3D laser measurement system, which uses the 3D data information obtained to produce the results needed for the construction project [13]. Zhao Xu et al. used 3D laser scanning data to extract and construct the 3D information model of large cultural relics in Yungang grottoes[14]. Li Bijun used 3D laser scanner to acquire the space geometric data of buildings, and filter the information on the parts of the buildings from the point cloud data for analysis, assisting in smooth progress of construction[15]. These research work has received good results, but so far, there is no the report on database system management based on aerial image and point cloud data that is applied in the building reconstruction project.

In March 2017, Xiamen was in the second batch of ecological restoration city repair(hereinafter referred to as "city double repair") pilot among 19 cities. To realize quick measurement of building facade and serve the whole cycle management of the facade, based on the actual work of the building facade renovation on the both sides of the 30 km part of Tongji Road and Binhai West Avenue in Tong'an District, Xiamen, this paper took the UAV remote sensing images combined with 3D laser point cloud data to map site plan, facade, etc. and eventually build facade measurement and GIS platform of supervision with the complete process materials including integrated UAV remote sensing maps, facade measurement results, planning design and reform effect pictures, etc, with the reference for the city "double repair" building facade renovation project.

2. Site-plan drawing based on aerial-photo data

2.1. Advantages of site-plan drawing based on aerial-photo data

In the process of drawing the site plan of Tongji Road in Tong'an District, Xiamen, using low air UAV aerial photography can obtain the remote sensing image data quickly, efficiently and extensively, and the drawing of the site plan is carried out on the image graph. At the same time, the measurement and drawing are made for the buildings on the both sides of Tongji North Road in Tong'an District, Xiamen. The purpose of measuring the site plan, on the one hand, the corresponding edge of each building can take matching data management for 2D graph. On the other hand, through the correspondence between Facade and site plan of the road buildings, more intuitive understanding of the whole situation of the road can be realized, providing the data source with good integrity for the Facade reconstruction of the buildings in the future.

2.2. Site-plan drawing based on aerial-photo data

Based on the orthogonal projection of UAV aerial data's spatial resolution at centimeter level, it can fully display the distribution and surrounding environment of the buildings with facade renovation, and reproduce the position of each facade clearly at the same time, providing positioning information basis for the implementation and result integration of facade measurement. In this paper, the facade measurement for the buildings on the both sides of Tongji Road in Tong'an District, Xiamen was taken as work object. Since the renovation objects are the Facade of the buildings on either side of the main road, it is planned that the route acquired by UAV remote sensing image is 200m along the road as the aerial-photo area, as shown in figure 1 that is about aerial-photo area and the aerial-photo route plan.

Figure 1. Aerial-Photo area

3. Acquisition of facade results based on point cloud data

3.1. Fieldwork measurement of the facade of 3D laser buildings

The 3D laser scanner can obtain the complete data of the facade of the buildings and ensure the accuracy and quality of the data, which cannot be obtained by traditional measurement methods. For each side of a building, the objects of each side include doors and windows, shop recruit, billboards, air conditioning, water pipes, power lines and building texture elements, etc. As shown in figure 2, point cloud data obtained through 3D laser scanner can fully display all kinds of information on the building facade.

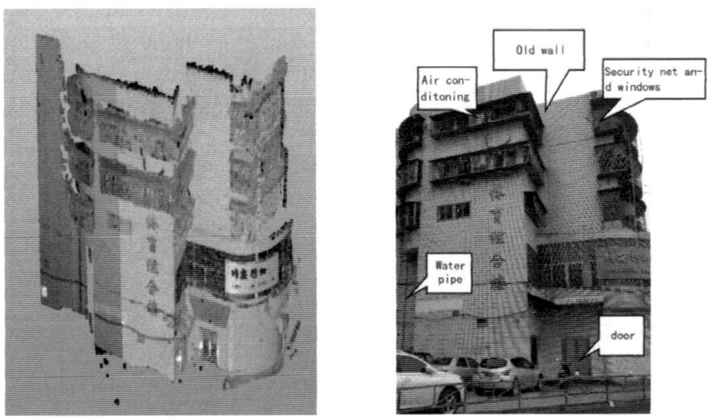

（a）Point cloud data of buildings （b）Physical photo of buildings

Figure 2. Objects of the building facade reforming

3.2 Interior work of the facade of 3D laser buildings

3.2.1 Preprocessing of point cloud data. The preprocessing of point cloud data collected by 3D laser scanner is mainly used in FARO scene software. Preprocessing process mainly includes the import of original point cloud data, noise point removal, registration and connection, filtering simplification and the export of original Facade's point cloud data, etc., of which the purpose is to make the point cloud data of different scanning sites accurately match and unified to a specific coordinate system, so that various kinds of information extraction, 3D physical modeling and quantitative analysis can be made for the research object.

(1) Denoising simplification. The main procedure for the removal of noise points is to project the point cloud data onto the orthogonal plane, and based on elevation information of the point cloud, to

adopt the way of human interaction to eliminate the scattered points floating around the building and in the air, as shown in figure 3 for the effect before and after denoising simplification of point cloud data of the building facade.

(a) Original point cloud data (b) Point cloud after removing noise points

Figure 3. Removal of the noise points of laser point cloud

(2) Registration and connection. In this paper, the method of artificial registration with target is adopted, and the registration method is fast and accurate. After the registration of building facade on the point cloud data, both have three-dimensional measurable coordinate information of the space point, which can be more accurate, more intuitive for all kinds of spatial information of building facade including windows and doors, billboards and the position of air conditioning and so on.

3.2.2 Two-dimensional drawing of building Facade. In the drawing of building facade, first, the coordinates of the point cloud data should be defined to make the point cloud Facade as orthographic projection image, and then the two-dimensional information extraction is carried out, to ensure the surface integrity and accuracy of structural line. After that, the image data of the corresponding buildings collected from the actual scene can be used to produce the two-dimensional drawing of each surface of the building by Tangent software.

RGB color information of each pixel of the live-action images integrates with laser point cloud data, realizing the display of true color of the target's point cloud data. As shown in figure 4 (a), it is the orthographic point cloud of the true color of the building Facade in some commercial area on Tongji North Road. Based on point cloud data of the building facade, the two-dimensional line drawing of the building Facade is carried out, and this method not only is characterized by high precision, high efficiency. The 1:100original facade diagram of a building based on point cloud data is as shown in Figure 4 (b).

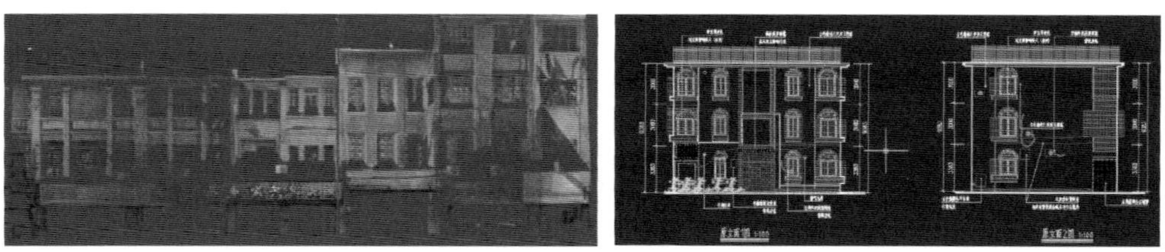

(a) The orthographic point cloud of the true color of the building (b) Original facade of building

Figure 4. Acquisition of the two-dimensional line drawing of the building facade

3.2.3 Appearance design of building facade. The main idea of 3D modeling based on 2D line drawing is to use the structure line to generate the plane, and import the CAD file of 2D line drawing into 3DMax software to complete the real 3D modeling of the building. Based on point cloud data, it can be concluded that the 2D line drawing of building components have the associated relationship in location. he complete building structure is obtained by rotating capture, creating 3D physical model one by one. In accordance with the requirements of project planning, the field photos of the building Facade are completed through photoshop, making the building facade more clear and beautiful. Then, by using the processed photos, the three-dimensional physical model is made for stickers, finally obtaining the real 3D model design of the complete building construction. As shown in figure 5, it is the design of the facade of the building with historical features.

（a）Effect on the historical building facade Reform

（b）Effect on the commercial building facade reform

Figure 5. Effect on the building facade reform

Through the site plan of the buildings on both sides of the road, the overall distribution of the buildings on both sides of the road can be found. By looking at the graphic plane of the house, the serial number of all facades of the building can be viewed. Then, in the corresponding effect drawing of the building Facade, all Facade pictures of the building can be quickly and accurately found. Through the "integral-detailed" "general plane – detailed Facade", the results of the reconstruction and measurement of the road buildings' Facade are completely and accurately filed. Through the corresponding filing of the building facade pictures and the road site plan, it makes measurement results more complete, intuitive, and vivid, and also provides the most accurate and fastest query data source and important technical support for subsequent building facade reform.

4. Design and construction of GIS platform of facade construction supervision

4.1 Design of GIS platform of facade construction supervision

4.1.1System function design. The main purpose to develop the building facade data management system is to conveniently manage a huge amount of facade data results, and make name, point cloud data, CAD plane and CAD 2D elevation lines of each building correspond one by one to constitute a complete data management pattern, so as to be convenient for management and archiving, provide data for subsequent work.

Based on the system, in terms of the functions of browse query and data management to achieve the goal of building Facade management information system position in initial stage and present stage, the main function is divided into three types, including remote sensing image browsing, data query, graphics library management, and so on.

（1）Remote sensing image browsing: load remote sensing image of reconstructed Facade building complex; browse, zoom in, zoom out, wander, change base map, and select layer for selected images.

（2）Data query: By clicking on the point coordinate of the reconstructed Facade building, the system can select the corresponding layer elements to carry out related object data query, and cooperate with the related fields to make related queries.

（3）Graphics library management: The image data is stored in the database; through the system operation the external graphics image is imported, and the image is deleted and replaced; the progress of the project is reviewed.

4.1.2 System structure design. Platform selection: to meet the project requirements, the system design uses Windows operating system, PC equipment, ArcGIS Engine Runtime as the current computer hardware and platform of software operation.

Application environment selection: the system design uses Microsoft Access database as the data management center of the system and make the ArcGIS secondary development tool and Access database connected to achieve the system design function demand.

Application module design: the system design is divided into image browsing module, data query module, image library module. Through the functional requirements for the rationalized distribution, it makes the system functional modular with integrity.

After consideration, for the design of the system structure, it selects the Microsoft Access database, ArcGIS engine secondary development tool, Visual Studio platform and ArcGIS 10.2 operating environment to meet the system structure design requirements. The functional structure of the GIS platform of building Facade supervision is as shown in figure 6.

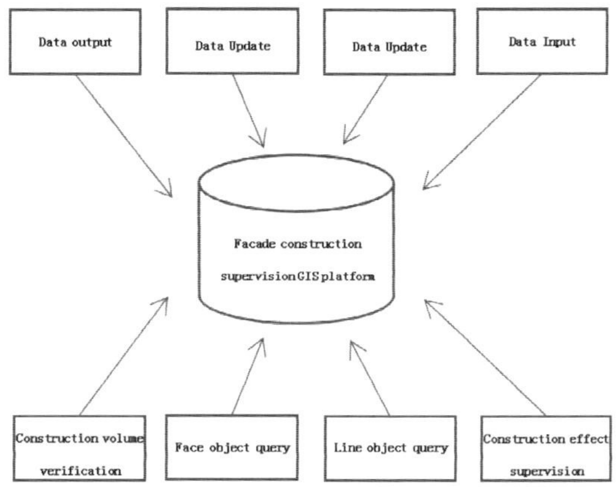

Figure 6. The functional structure of the GIS platform

4.2 Construction of GIS platform of facade construction supervision

4.2.1 Source of system database. The system mainly uses the geodatabase model, and the map objects in the graph are designed by using the main interfaces such as Imap and IGraphicsContainer, etc. The main database of system design includes basic database, spatial database, resource database, planning database, etc.

(1) Basic database: Basic data includes vector data and raster data.. The specific data directory is shown in table 1.

Table 1. Basic data

Type of data	Thematic types	Thematic elements
Vector data	Building distribution	Towns and villages
	traffic	Tongji Road
	Building facade point cloud	Doors, windows, pipes, billboards
	Building facade two-dimensional line drawing	Doors, windows, air conditioners, walls
	Building type	Hut, building in the room, broken room
Raster data	Digital remote sensing image	Aerial video along the route
	Building facade image data	Picture

（2）Spatial database: Spatial database mainly includes UAV orthophotoquad of housing complex, subject image after cutting, plane design, etc. Based on the spatial database construction, it provides the foundation of spatial data for the system, and is associated with the function development of the system. The function of the database system is realized based on spatial database as benchmark.

（3）Resource database: Resource database mainly manages various data on the building including the name, the geographical position and the house number and so on of the building, as well as the building surface information including scanning data, CAD elevation data, construction audit and so on, to realize the visual display of the facade information. At the same time, the information of this part is displayed in the data interface of the system.

（4）Planning database: Planning database will manage and store building data before reform, planning pictures, pictures after reform, etc. to realize the system of audit work for building facade renovation. This part in the system interface is displayed and operated in query function module.

4.2.2 Realization of system database. The external facade database management system is divided into three parts: image interface module, data management module and data interface module.

（1）Image interface module: The main interface of the system shows the remote sensing image of the building with the reformed external facade, and users can browse, enlarge, narrow, wander, change base map and select layer for the selected maps.

In image interface module of the system, users can choose to view the building needing reform the facade, and by clicking on the location of building, the system will relate the coordinate data of the building to spatial database information, resources database information, planning database information of the building. Figure 7 is the operation diagram of the system image interface module.

Figure 7. Operation image of system's image interface module

Upon operation in image interface, by clicking on the building, the system generates round taps in the related position in images. By calculating the coordinate range of the buildings in the coordinates, the system can retrieve the database information of selected buildings, and jump into the GIS database.

In the interface, it is to click on the wall to edit layer elements, load ArcGIS function, achieve the popup of corresponding metope information windows with the display of realistic picture. On the left side of data interface of the system, it is to input the name of the building to be enquired and the metope direction of the building, the Facade picture before and after reform can pop up.

（2）Data management module: This module is mainly responsible for the input and export of data and images in data interface module. To guarantee the normal work of facade renovation, after clarifying the building with facade renovation work, the basic information is acquired and the information is recorded in the database. When new data is added, data interface module takes the synchronous updating of the form information. Click the upload button to import the data into the database. This function is shown in figure 8 (a).

(3) Data interface module: The module shows the information of the building with Facade reform in remote sensing images. According to the building name and the wall direction of the building, users will set the parameters, and then click the query system to jump to the query function of facade project. The operation of data interface module is shown in figure 8 (b).

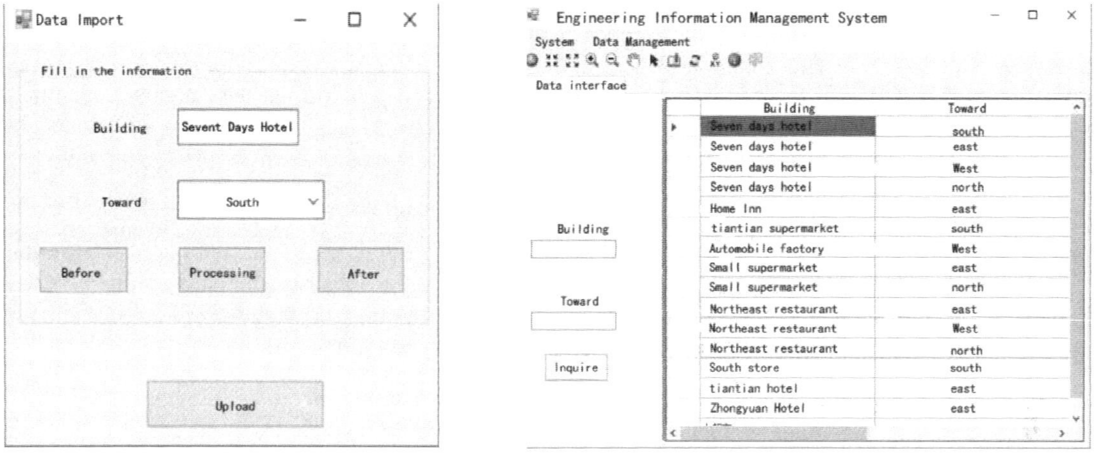

(a)Data import (b)Operation of data interface module

Figure 8. Data import and query interface

The query function of facade project is mainly used to display the facade pictures of the building selected in data interface before reform, in planning, and after reform. The function also provides the building point cloud data and backup data of CAD facade data for the system, and the audit of the Facade reforming work by the audit personnel is also completed in this part. Figure 9 shows the pictures of the building before reform, in planning, and after reform.

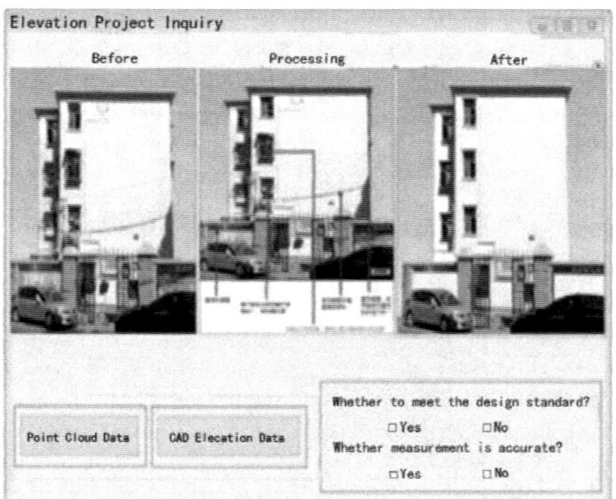

Figure 9. Pictures of the building before reform, in planning, and after reform

The point cloud data and two-dimensional facade data can be also downloaded. By clicking the two buttons above, the required data can be downloaded respectively. The system also provides a verification of the building Facade reform. On the one hand, it is whether the supervision and reconstruction achieve the design standards. On the other hand, it is whether the supervision construction work calculation is accurate. After the operation on the facade of each building, by clicking update, the data can be saved. Realizing the objectification management of the three facades is the characteristic of this system.

5. Conclusion

With the buildings on Tongji Road in Tong'an District, Xiamen as research object, this study carried out the research on the management of the measurement data database system of the facade reconstruction project of the road building, with main design of the road building's site plan drawing and the acquisition of 2D data for building Facade, mapping Facade's site plan based on point cloud data and UAV remote sensing image, using C#+AE to integrate the whole-process data in the facade renovation and business process to develop GIS system, so as to provide fast and precise integrated technique platform for the urban road construction project planning and management, with practical value for current "city double repair" take engineering information management of urban renovation.

6. Acknowledgments

Fund program: key techniques of geography national condition monitoring and its system integration in mine subject (Fujian natural fund projects 2016J01199); Typhoon disaster analysis mapping based on Unmanned Aerial Vehicle Remote Sensing and emergency response integration system (science and technology projects of Fujian Bureau of Surveying and Mapping 2018JX02)

References

[1] Wang Feng, Chen Huangran and Cheng Xiaojun.Digitalizing Building Based on Ground Laser Scanner[J]. Bulletin of Surveying and Mapping,2011(6):39-43.
[2] LIN Zengnan.Introduction to building facade measurement method-Based on the building elevation measurement of the house on the side of entrance road of traveling scenic area[J].GUANGDONG SCIENCE & TECHNOLOGY,2013,22(12):161-161.
[3] HE Yuanrong. High resolution remote sensing monitoring of mining area environment and information resources development and utilization[M].Beijing:Science Press Co.Ltd, 2014.
[4] Zhang Yi, Yan li and Cui Chenfeng. Application of terrestrial 3D laserscanning to highway modeling[J]. Science of Surveying and Mapping, 2008,33(5):100-102.
[5] He Yuanrong, Zheng Yuanmao. Program Development and Implementation of Roughness Measurement Based on Wind Tunnel Point Cloud[J].Highway Engineering,2016,41(6):73-77.
[6] Tang Kun, Hya Xianghong, Wei Cheng and Xia Zhen.The Experimental Study of Buildings Deformation Monitoring Method Based on 3D Laser Scanning[J].2013,38(2):54-56.
[7] Zhou Junzhao, Zheng Shumin and Hu Song. Application of Terrestrial 3D Laser Scanning to Surveying and Mapping of Cultural relics Protection in Grottoes Carved Stone[J]. Bulletin of Surveying and Mapping,2008(12):68-69.
[8] He Yuanrong, Zheng Yuanmao and Pan Huoping. Real Three-dimensional Modeling and Application of Complex Construction based on the Point Cloud Data[J]. Remote Sensing Technology and Application, 2016, 31(6): 1091-1099.
[9] Metni N and Hamel T. A UAV for bridge inspection: Visual servoing control law with orientation limits[J]. Automation in Construction, 2007, 17(1):3-10.
[10] Rathinam S, Zu W K and Sengupta R. Vision-Based Monitoring of Locally Linear Structures Using an Unmanned Aerial Vehicle[J]. Journal of Infrastructure Systems, 2008, 14(1):52-63.
[11] Du Caihong. Application of low altitude UAV photogrammetry technology in the Yellow River Ningxia river project[J].Shanxi Architecture, 2014, 40(32): 217-218.
[12] Allen P K, Troccoli A and Smith B. New methods for digital modeling of historic sites[J]. IEEE Computer Graphics and Applications, 2003, 23(6): 32-41.
[13] Stamos I and Allen P K.Geometry and texture recovery of scenes of large scale[J].Computer Vision and Image Understanding,2002,88(2):94~118.

[14] Zhao Xu, Zhou Keqin,YAN Li and DENG Fei.3D Reconstruction Method for Large Scale RelicLandscape from Laser Point Cloud[J]. Geomatics and Information Science of Wuhan University,2008,33(7)：684-687.
[15] LI Bijun, FANG Zhixiang and RENJuan. Extraction of Building.s Feature from Laser Scanning Data[J]. Geomatics and Information Science of Wuhan University,2003,28(1):65-70.

IOP Conf. Series: Earth and Environmental Science **146** (2018) 012028 doi:10.1088/1755-1315/146/1/012028

Occurrence of Antibiotics in Water in Xiaodian Sewage Irrigation Area, Northern China

Jiale Li [1,2], Yihui Dong [2,*], Fawang Hu[2] and Jiangming Wang[2]

[1] Jiangxi Key Laboratory for Mass Spectrometry and Instrumentation (East China University of Technology), Nanchang, 330013
[2] School of Water Resources and Environmental Engineering, East China University of Technology, Nanchang, 330013

*Corresponding author's e-mail: yihuisnoopy@163.com

Abstract. The Xiaodian sewage irrigation area in northern China has a long history of sewage irrigation. The surface water and groundwater samples were collected and analyzed in this region. All the 25 kinds of antibiotics have been detected in the surface water in this study area. The order of average detection frequency is Macrolides (MLs, 98.2%) > Fluoroquinolones (FQs, 91.3%) > Tetracyclines (TCs, 87.5%) > Sulfonamides (SMs, 75.0%). The sample with the highest concentration of antibiotics (730.19 ng/L) is located at Taiyu Drainage Canal while that with the lowest concentration (245.93 ng/L) at Xiaohe River. The type and concentration of antibiotics in SW-3 and SW-4 were almost the same, indicating the measures to prevent canal banks from seepage were able to retard their transport into the soil and groundwater system via lateral recharge. Along the Beizhang Drainage Canal, the sewage from the swine farm, the cow farm, the irrigation field and domestic usage were all discharged into the canal, so the concentration kept increasing. Along the down gradient direction of Taiyu Drainage Canal, the concentration decreased. Transport of the contaminants via lateral recharge due to lack of seepage prevention measures, therefore, poses a potential threat to the soil-groundwater system. 15 of 25 kinds of antibiotics were detected in the groundwater. The detection frequency was lower than other study areas in China. The concentrations ranged from 2.75 ng/L to 114.38 ng/L with the average concentration of 27.60 ng/L.

1. Introduction

Because of lack of water resources, sewage irrigation has had a long history in Taiyuan city[1] and this sewage irrigation area has became one of the major sewage irrigation areas in the north of China. By 1995, the city's total irrigation area was 834,000 mu while effective irrigation area was 812,000 mu[2], the annual irrigation requirement was 178.2 million m^3. Under the situation of water shortage, it is difficult to ensure that all of them irrigated by clean water. At present, More than 200 million m^3 of industrial sewage and domestic sewage from the urban area is mostly discharged into Fenhe River. With the increase of sewage irrigation frequency and quantity of the sewage irrigation, environment quality of the surface water and groundwater were affected in some regions. The environmental conditions in Taiyuan City sewage irrigation area were studied by some researchers. However, the report for organic pollution of sewage irrigation area was much less. Zhang et al. studied the influence of long-term sewage irrigation to the distribution of organochlorine pesticides in water and soil[3]. Li et al. reported the distribution and sources of polycyclic aromatic hydrocarbons in the topsoil of the sewage irrigation area[1]. Liao researched the adsorption behavior of nonylphenol in sewage irrigation area[4]. However, there has been no systematic study on the environmental geochemical behavior of antibiotics in sewage irrigation area[5]. The objectives of this study are systematically study about the environmental geochemistry behavior of antibiotics in the surface water and groundwater, analyse the distribution, source and migration of these organic contaminates in the environment.

2. Materials and methods

2.1. Site description

Because of lack of water resources, sewage irrigation has had a history of nearly 40 years in Taiyuan city, and this sewage irrigation area has became one of the major sewage irrigation areas in the north of China. By 1995, the city's total irrigation area was 834,000 mu while effective irrigation area was 812,000 mu, the annual irrigation requirement was 178.2 million m^3. Under the situation of water shortage, it is difficult to ensure that all of them irrigated by clean water. At the same time, low cost is another major reason promoting the development of sewage irrigation. At present, More than 200 million m^3 of industrial sewage and domestic sewage from the urban area is mostly discharged into Fenhe River. According to statistics, in the "Eighth Five Year Planning" period (1991-1995), Taiyuan sewage irrigation area was 394,500 mu, accounting for 47.3% of the total irrigation area; the annual irrigation water capacity was 95.85 million m^3, accounting for 53.7% of total agricultural irrigation water.

2.2. Sampling

A summary of 14 surface water samples and 9 groundwater samples were collected in the study area. Fig. 1 shows locations of surface water samples and groundwater samples. Surface waters included Fenhe River water, Xiaohe River water, water from East Main Canal and water from drainage canals. Groundwaters were all collected from wells with depth between 15 m and 35 m after 10 minutes' pumping to get fresh samples, and then stored in 4 L amber glass bottles at 4°C and treated as soon as possible.

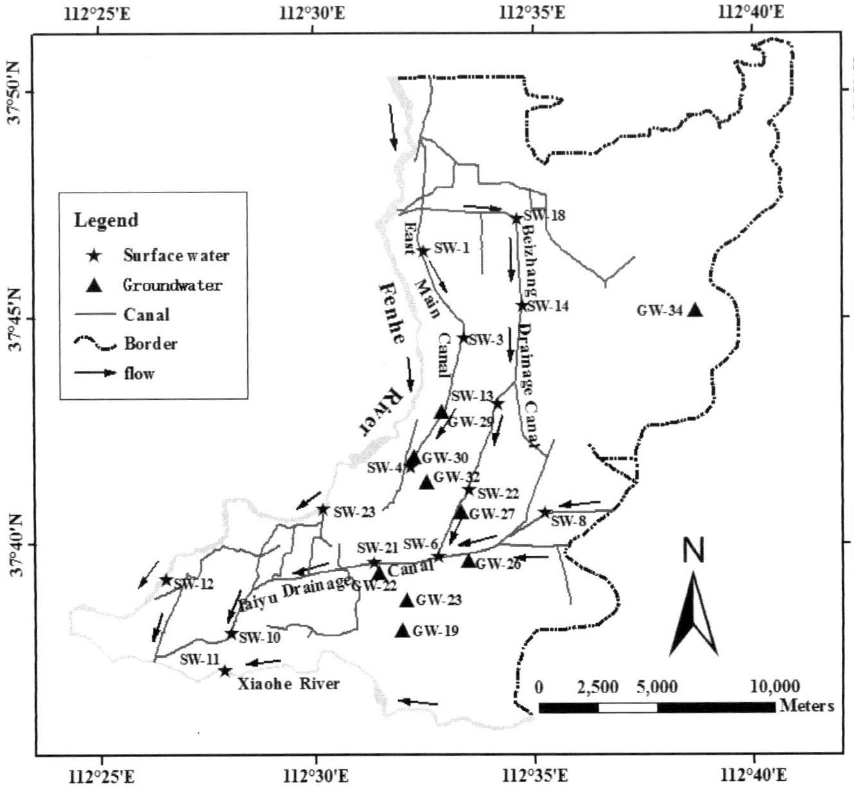

Figure. 1 Location of water samples collected from Xiaodian sewage irrigation area

2.3. Analysis

The target antibiotics can be devided to 4 categories: (1) sulfonamides (SMs): Sulfapyridine (SPD), Sulfathiazole (STZ), Sulfamethoxazole (SMX), Sulfamerazine (SMR), Sulfamethazine (SMZ), Sulfamethoxypyridazine (STP), Sulfameter (SFM), Sulfdiazine (SDZ); (2) fluoroquinolones (FQs): Ofloxacin (OFL), Lomefloxacin (LOM), Ciprofloxacin (CIP), Enrofloxacin (ENR), Norfloxacin (NOR), Sparfoxacin (SPA), Gatifloxacin (GAT), Fleroxacin (FLE), Enoxacin (ENO); (3) macrolides (MLs): Roxithromycin (RTM), Clarithromycin (CTM), Azithromycin (AZM), erythromycin (ERY); (4) tetracyclines (TCs): Tetracycline (TC), Oxytetracycline hydrochloride (OTC), Chlorotetracycline hydrochloride (CTC), Doxycycline hyclate (DOC).

The samples were extracted using the methods of preview studies[6]-[7].

The antibiotics were analyzed by high-performance liquid chromatography - electrospray ionization tandem mass spectrometry (HPLC-ESI-MS/MS), using the MRM mode. Simetone was used as the internal standard compound. The concentrations were calculated based on the internal calibration curves.

3. Results and discussion

3.1. Antibiotics in surface water

All the 25 kinds of antibiotics have been detected in the surface water. Among them, RTM, CTM, ERY, CTC, OFL, GAT, FLE, OFL, ENR, NOR, SMZ, SMX, SDZ and SPD were 100% detected. AZM, CIP and ENO were 92.9% detected. OTC and DOC were 85.7% detected. TC and SPA were 78.5% detected. STP and STZ were 64.3% detected. SFM were 50.0% detected. Only 21.4% of surface water contains SMR. Order of average detection frequency is as follows: MLs (98.2%) > FQS (91.3%) > TCs (87.5%) > SMs (75.0%). Luo reported the order of the first-order attenuation rate coefficients (K values) for these 4 classes antibiotics is TCs (K ranged from 0.39 to 1.22 h^{-1}) > MLs (0.09 to 0.48 h^{-1}) > FQs (0.08 to 0.45 h^{-1}) > SMs (0.01 to 0.44 h^{-1})[8]. Comparing with the detected ratio and attenuation rate coefficients, it could be indicated that there are lots of TCs enter into the surface water.

The highest concentration of antibiotics (730.19 ng/L) occurs at SW-6, this is because SW-6 is located at the intersection of Beizhang Drainage Canal and Taiyu Drainage Canal, the antibiotics of both the canals integrate to this site. The site SW-11 has the lowest concentration (245.93 ng/L). FQs is the main contaminant in the East Main Canal. TCs, FQs and MLs is the main contaminant in the beginning, middle and end of the Beizhang Drainage Canal, respectively. MLs is the main contaminant in the Taiyu Drainage Canal except for the beginning site SW-8, which indicates the concentration of antibiotics in Taiyu Drainage Canal has been strongly affected by the sewage of Beizhang Drainage Canal after they intersected. FQs and MLs is respectively the main contaminant in Xiaohe River and Fenhe River. Summarily, the concentration of antibiotics in surface water shows FQs and MLs are frequently used at the Xiaodian irrigation area, which corresponds to the order of the detected ratio.

Fig. 2 shows the type and concentration of antibiotics in SW-3 and SW-4 were almost the same, indicating that measures to prevent canal banks from seepage were able to retard their transport into the soil and groundwater system via lateral recharge, this is probably the reason why there is a slight increased along the direction of East Main Canal. The site SW-18 has a high concentration of antibiotics, this is because of this site close to the sewage treatment plant. The irrigation water flow the irrigation field from the East Main Canal to Beizhang Drainage Canal, the water quantity has been increased in the Beizhang Drainage Canal, so the concentration was decreased from SW-18 to SW-14. Along the Beizhang Drainage Canal, the sewage which came from the swine farm, the cow farm, the irrigation field and domestic usage were all discharged to the canal, so the concentration kept increasing. Compare with E zone and B zone, we could deduce that the compound NOR is more likely to be adsorbed in the soil or groundwater in the irrigation process. After the intersection of Beizhang Drainage Canal and Taiyu Drainage Canal, the concentration reached the peak at site SW-6. Along the down gradient of Taiyu Drainage Canal, the concentration decreased, transport of contaminants via lateral recharge due to lack of the seepage prevention measures therefore pose a potential threat to the soil-groundwater system.

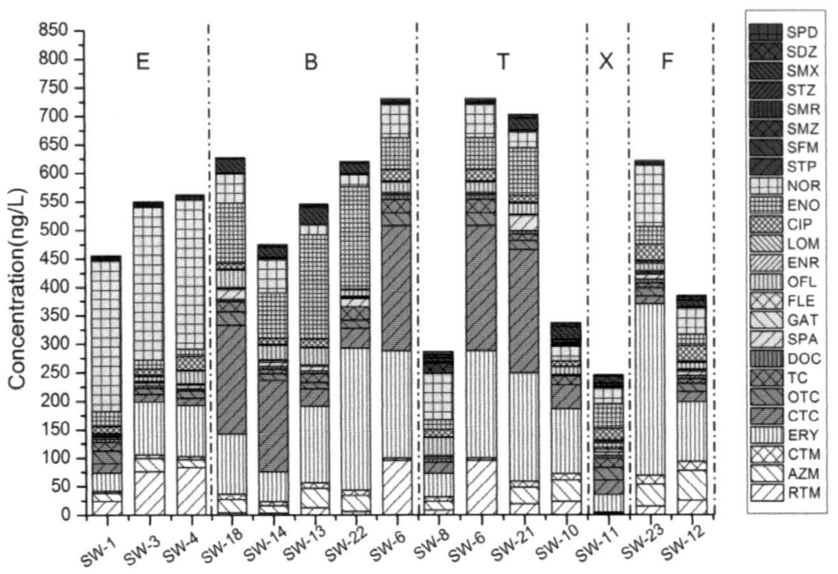

Figure. 2 The distribution of antibiotics along with the rivers and canals. E: East Main Canal; B: Beizhang Drainage Canal; T: Taiyu Drainage Canal; X: Xiaohe River; F: Fenhe River. The direction of water flow in the rivers and canals are from left to right. Yellow: MLs; Magenta: TCs; Cyan: FQs; Blue: SMs.

3.2. Antibiotics in the groundwater

As Fig. 3 shows, 15 of 25 kinds of antibiotics were detected in the groundwater. The detection frequency was lower than other study areas in China[7]. The concentrations ranged from 2.75 ng/L to 114.38 ng/L with the average concentration of 27.60 ng/L, which is much lower than the surface water. However, the concentration is higher than some other study areas[9]. MLs are detected at 22.2% of the groundwater samples with the range from below the limit of quantification (BLQ) to 44.51 ng/L. DOC and CTC have the highest detection frequency, 88.9% and 44.4%, respectively, and with the range from BLQ to 22.06 ng/L and BLQ to 15.06 ng/L, respectively. TCs is also the main contaminant which is 100% detected with the concentration of 1.90 ng/L to 35.06 ng/L. Only 3 of 9 FQs has been detected with the concentration between BLQ to 36.30 ng/L. 7 of 8 SMs were found in the groundwater, their highest concentration was not greater than 1.60 ng/L. MLs were only detected at sites GW-19 and GW-27.

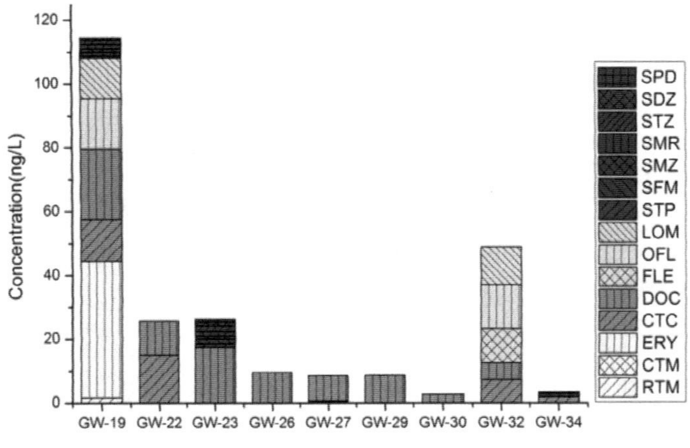

Figure. 3 The distribution of antibiotics in the groundwater

4. Conclusions

The antibiotics are ubiquitous in the soil and water environment. All the 25 kinds of antibiotics have been detected in the surface water in this study area. The order of average detection frequency is MLs > FQs > TCs > SMs. The concentration of antibiotics in drainage canals is higher than rivers. 15 of 25 kinds of antibiotics were detected in the groundwater. The detection frequency was lower than other study areas in China. The concentrations ranged from 2.75 ng/L to 114.38 ng/L with the average concentration of 27.60 ng/L. MLs and FQs were the main antibiotics detected in the shallow vadose zone. 22 and 20 kinds of 25 antibiotics were detected respectively in the sediment core of sewage irrigation area and groundwater irrigation area. The order of concentration of 4 categories of antibiotics in both cores was: FQs > TCs > SMs > MLs.

The antibiotics in the environment mainly came from the sewage which from the sewage treatment plants and livestock breeding farms which along the drainage canals. Transport of the contaminants via lateral recharge due to lack of seepage prevention measures, therefore, poses a potential threat to the soil-groundwater system. The measures to prevent canal banks from seepage were able to retard antibiotics transport into the soil and groundwater system via by lateral recharge.

5. Acknowledgments

This study was financial support by Open Fund of Jiangxi Key Laboratory for Mass Spectrometry and Instrumentation, East China University of Technology (No. JXMS2017016), Water Resources and Water Environment Research Center Fund of Water Resources Ministry in China (No. ZXKT201709), Science and Technology Project Fund of Jiangxi Provincial Education Department (GJJ170443), International Atomic Energy Agency Coordinated Research Project (IAEA No. 21122) and Program for Changjiang Scholars and Innovative Research Team in University (PCSIRT)(No. IRT_17R20).

6. References

[1] Li J L, Wang Y X, Zhang C X, et al. The source apportionment of polycyclic aromatic hydrocarbons (PAHs) in the topsoil in Xiaodian sewage irrigation area, North of China. Ecotoxicology, 2014, 23 (10): 1943-1950.

[2] Yang X F. Thought about the construction and management of Fenhe River irrigation area. Soil and Water Conservation Science and Technology in Shanxi, 2012, 3: 28-30 (in Chinese).

[3] Zhang C X, Liao X P, Li J L, et al. Influence of long-term sewage irrigation on the distribution of organochlorine pesticides in soil-groundwater systems. Chemosphere, 2013, 92 (4): 337-343.

[4] Liao X P, Zhang C X, Yao L L, et al. Sorption behavior of nonylphenol (NP) on sewage-irrigated soil: kinetic and thermodynamic studies. Science of the Total Environment, 2014, 473-474: 530-536.

[5] Li J L, Dong Y H and Sun Z X. Fate of antibiotics in the soil-groundwater system in sewage irrigation area. Proceedings of 44th Annual Congress of the International Association of Hydrogeologists, 2017.

[6] Azzouz A and Ballesteros E. Combined microwave-assisted extraction and continuous solid-phase extraction prior to gas chromatography-mass spectrometry determination of pharmaceuticals, personal care products and hormones in soils, sediments and sludge. Science of The Total Environment, 2012, 419: 208-215.

[7] Tong L, Huang S B, Wang Y X, et al. Occurrence of antibiotics in the aquatic environment of Jianghan Plain, central China. Science of the Total Environment, 2014, 497-498: 180-187.

[8] Luo Y, Xu L, Rysz M, et al. Occurrence and transport of tetracycline, sulfonamide, quinolone, and macrolide antibiotics in the Haihe River basin, China. Environ Sci Technol, 2011, 45 (5): 1827-1833.

[9] Hannappel S, Balzer F, Groeneweg J, et al. Incidence of veterinary drugs in near-surface groundwater below sites with high livestock density in Germany. Hydrologie Und Wasserbewirtschaftung, 2014, 58 (4): 208-220.

On the Water Environment System in Water Sensitive Area —Building of Sponge Airport Stormwater System in Beijing New Airport

Kai Guo[1], *Lubing Wang[1,a]

[1] Beijing's New Airport Construction Headquarters (BNAH),102602,China

[a]ustb_wlb@126.com

Abstract. This paper intends to study the water environmental system within water sensitive area of Beijing New Airport, further to research how it realizes the objectives of effectively controlling urban drainage and flood, scientifically utilizing and managing stormwater resource and efficiently reducing the total runoff volume and pollution, then building the stormwater system of "sponge airport". Through analyzing the problems in the water sensitive area and some limiting factors of airport building, this paper will emphasize the system of stormwater drainage and storage, the system of controlling runoff volume and pollution, the system of recycling stormwater and the system of digitalized stormwater management. The design philosophy and construction method of "sponge airport" as far as functional division of airport and mutual connection of various systems are concerned, and the construction goal of "sponge airport" is put forward. Based on these researches, this paper will state how the construction objectives of new airport are realized and how the different systems provide guarantee to its goal of being an advanced international aviation hinge.

Key words: Beijing New Airport; Sponge city; Digital model; Stormwater management system

1. Overview of Beijing New Airport

1.1. Site and positioning

Beijing New Airport is situated at the junction among Daxing District, Beijing and Langfang City, Hebei Province. The airport is about 50km away from the center of Beijing, which is south to Yongding River, north to Tiantang River, and across both Beijing and Hebei administrative regions (Figure 1) [1]. It is oriented to a large international airport hub, which will provide a critical guarantee for the sustainable development of regional economy in Beijing area, secure sustainable high-speed and safe development of civil aviation industry, international aviation hub construction and integrate Beijing-Tianjin-Hebei construction after completed. It is scheduled to construct and run by phases. In the current phase (in 2020), the planned land area is about 27.5km^2 and the annual passenger throughput is about 45 million. In the recent phase (2025) and the long-term phase (in 2040), the planned land area will be about 28.8km^2 and 45.8km^2, the annual passenger throughput will be about 72 million and 100 million.

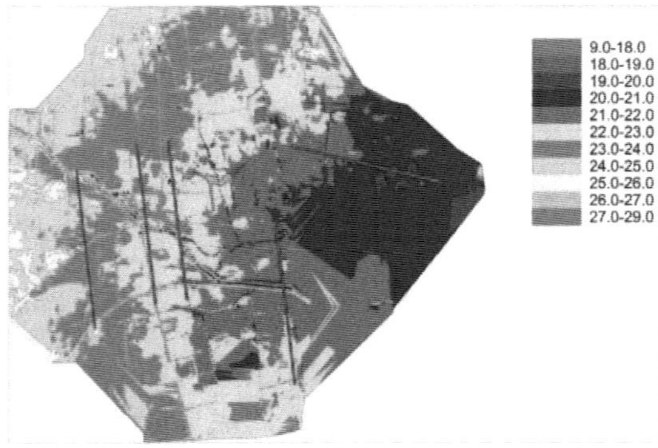

Figure 1. Sketch map of the airport **Figure 2**. Sketch map of terrain

1.2. Terrain feature

The existing l terrain within the planned land in the airport is dipping from west to east and from north to south with about 3m of height difference from east to west and about 1m of height difference from north to south. Within the site, the original gradient is about 0.3 ‰, the original elevation is about 20.0~25.0m, and the top elevation of sand dunes and sand ridge is 27.0 ~ 29.0m (Figure 2) [2].

The location, weather, terrain, flood control and other multiple factors are taken into accounts for decision of the airport site. Based on the existing terrain conditions of the new airport, a variety of new airport terrain design solutions were proposed, such as the earth self-balancing solution, flood control solution and gravity drainage solution. In addition to the earth self-balancing solution, the other solutions require a large volume of earthwork, however there is no suitabl3 source of earthwork near the geographical location of the site. If a large volume of earthwork is excavated, the vegetation and the ecological environment in the earthwork source area will be destroyed, a large amount of energy will be consumed, the pollution will be imposed on the ecological environment along the way, a huge amount of investment will be yielded and the project is thus not feasible. So, the earth self-balancing solution is selected for the terrain solution.

In the earth-building self-balancing solution, the relevant factors are taken into accounts. In addition to the areas other than such areas as runways, airport platforms, taxiways, terminal, etc., which are highly demanded, the site terrain tendency is the same kept as the original one within the site as far as possible. To meet the use requirement and relevant criteria, the cut and fill of the land shall be reduced. As the new airport is south to the Yongding River, and north to the Tiantang River, the designed terrains shall be lower than the outer flood level so that the new airport area becomes a water-sensitive area in addition to the areas of terminal and runway, etc.

2. Analysis on Water Environment in the airport

2.1. Water sensitivity analysis

2.1.1. Analysis on outer flood. The airport is north to the new Tiantang River, and south to the Yongding River. Because the construction land occupies a part of the new Tiantang River, it is an important basis for the safe operation of new airport to coordinate the relationship between water level and terrain among the Yongding River, new Tiantang River and the new airport. In order to protect the new airport from external floods, the new embankment against the 1-in-100-years floods will be built on the north side of Yongding River and the south side of the new Tiantang River. By then, the new airport will be lower than the external terrain elevation so that it will form an independent on-site drainage system, and the airport will be less subject to the external floods.

2.1.2. Analysis on floods. As a large number of hardened areas, such as runway and terminals, will be built, the total runoff coefficient will be greatly increased in the whole site, and the total amount of stormwater runoff volume and the peak runoff volume will increase and the peak runoff time will be shortened, and other sensitive factors will yield after the development of the new airport. As the floods occurred before have caused great loss,[3] the new airport will form an independent internal drainage system, the stormwater cannot be discharged into the downstream by gravity. The factors, such as accidents in pumping station, non-timely pumping, incapable pumping, etc., will cause the stagnant water inside the new airport and form the waterlogging disasters in the rainstorms after the embankments of Yongding River and the New Tiantang River are built.

2.1.3. Analysis on runoff pollution. The villages, houses, ditches and roads are mainly distributed on the existing land type in the new airport without large industrial and mining enterprises and tall buildings. The existing overall runoff coefficient is low, and there is no excessive pollution source. The stormwater runoff can be received and purified through the soil and tunnels and have a good water environment self-purification capacity. A large number of complexes, roads and other non-point source pollution will appear after the completion of the new airport. Relevant researches [4] have shown that the roads and building roofs are the main source of non-point source pollution in cities and towns, where SS and COD values are higher. For this purpose, the airport will inevitably bring the runoff pollution to an extent.

2.2. Analysis on limiting factors

2.2.1. Limiting factors for drainage peak runoff volume. The diverted New Tiantang River will be used as the drainage outlet in the new airport according to the overall plan of the airport, the provincial and municipal agreements, the riverbed and water level elevation in the New Tiantai River, transformation solution of Tiantang River, the flood evaluation report of the airport and external water systems. According to the flood control in the Yongding River basin, the permitted external drainage flow rate is merely 30 m^3/s in the new airport when the drainage flow rate of the New Tiantang River and the Yongding River meets the design standards. The new airport only uses the land area about $27.5km^2$ in the current phase and will fail to meet the drainage and flood control requirement by merely relying on the external drainage flow rate of 30 m^3/s in the traditional way in storms.

2.2.2. Requirement for utilization of non-traditional water. Beijing is a city with serious water shortage, where the sewage that is treated in-depth and made as a renewable resource is an important mean to relieve the shortage of water resources, and it is an inevitable requirement for the construction of new airport (green airport). For this purpose, in the construction of the airport, all sewage and partial stormwater stored in the primary and secondary water retention facilities are made as a reclaimed water source for extensive treatment. The treated water meets the quality standards of the renewable water and is used to sprinkle green land, wash vehicles, clean roads, make up the landscape water and circulating and cooling water for air-conditioners. The surplus of reclaimed water will be discharged into the inner landscape lake within the site.

These factors shall be taken into accounts, such as the water system evaporation, leakage, rainfall, water quality maintenance and other factors for the make-up water in the landscaping river and lake within the airport. The main factors include water evaporation, leakage loss, exchanged and circulated water amount for maintaining the water quality. In the new airport, the surface area of water system is about 110 hectares, the annual average daily evaporation rate is about $3000m^3/d$ in the current phase; as the summer temperature is high, if the water quality requirement is to be met, the water must be exchanged once every 10 days with a large make-up water volume. Therefore, in addition to recycled water, the stormwater will be the main source of water for the water system in the new airport, and the necessary stormwater collecting facilities must be applied.

2.2.3. Requirement of water environment policies. As the water problems, such as water shortages, water pollution, and flooding, are becoming increasingly prominent,[5] Chinese government has always attached importance to the protection of the water environment. Relevant documents, norms and standards have been issued by the state and various regions in view of the major water environment problems brought about by the intensified urbanization in recent years. The water environment construction has been strictly enforced during the regional development. In the water environment construction of the airport, we should plan and design the supporting stormwater infrastructures based on the runoff control requirement as described under General Office of the State Council's Guidance on the Recommended Sponge City Construction *(State Office [2015] No.75),* Technical Guidance for Sponge City Construction-Low-Impact Development of Stormwater System Construction *(Hereinafter referred to as the "Guide")issued by the Ministry of Housing and Urban Construction, Beijing municipal specification DB11/685-2013:* Code for Stormwater Control and Utilization Engineering Design.

2.3. Analysis on countermeasures

Conclusively, the main problem facing the construction of water environment system in the airport is to meet the regional flood control requirements, to avoid the waterlogging due to the unsmooth internal independent drainage, to achieve the effective control of the total stormwater runoff volume, and to maintain the water quality in the new airport. Stormwater control and the facilities should be used to regulate the peak runoff, the total storage and retention volume of runoff water, etc. Thanks to the limited external drainage capacity, drainage and waterlogging demand large water supply and make-up water storage capacity.

The core to construct the sponge city is to realize the overall target of pollution control, disaster prevention, stormwater resources and urban ecological restoration, etc. With the mechanism building, planning control, design implementation, construction operation and management, multi-professional coordination and control, protection and utilization of urban green space and water system, it's priority to use green infrastructure scientifically combining gray stormwater infrastructure, and a flexible stormwater infrastructure is built jointly to achieve the goal of "infiltration, retention, storage, purification, use, drainage" for the stormwater runoff water. It's necessary to cope with the extreme storm and climate change, restore the urban benign hydrological cycles, protect or restore the urban ecosystems [6]. The coincidence of existing conditions, challenges and concept of building the sponge city just necessitates the building the sponge airport.

3. Overall frame of sponge airport

3.1. Overall objective

The determined overall control objectives are as followed: total runoff volume, runoff pollution, drainage and waterlogging, stormwater resource management and water environment protection according to the objectives of drainage, flood prevention and sponge airport construction, as well as characteristics of water system in the airport (Figure 3).

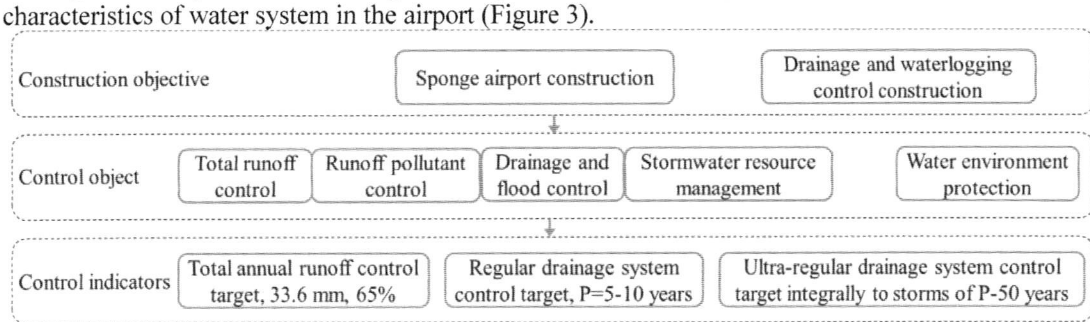

Figure 3. Major control objectives of the sponge airport

IOP Conf. Series: Earth and Environmental Science **146** (2018) 012029 doi:10.1088/1755-1315/146/1/012029

3 major control indicators are determined to fulfill the above control objectives for the water environment in the airport.

- According to the requirement under the "Guide", as the control rate of total annual runoff volume is 80%~85% in Bejing, the control rate of total annual runoff volume is 85% in the airport, namely, the designed rainfall is 33.6mm; these will meet the requirements for the storage volume equipped with less than 30 m^3 stormwater storage facilities per 1000m^2 hardening area under Beijing municipal specification DB11/685-2013: *Code for Stormwater Control and Utilization Engineering Design*.
- In the stormwater planning in the master plan of the airport, the designed recurring period of stormwater for the regional stormwater pipe in general area is P=5 years, the independent stormwater system is used in the terminal area and the stormwater recurring period of the stormwater pipe is P=10 years.
- In addition, with the integrated function of water environment system control measures, the airport should have a comprehensive capability of response to 1-in-50-years extreme rainfall.

3.2. Total runoff volume and pollution control solution

The "sponge airport" will be constructed involving a diversity of land types, such as construction area, roads, rivers and lakes. Regional drainage system planning, water system planning, vertical planning, etc. should be combined, and linking relationship between the areas and system should also be considered. To implement the overall goal of "sponge airport", the sponge airport stormwater system consists of three parts: source, midway and end controls (Figure 4). Among them, the source control mainly means that the total runoff is controlled and the runoff drained downwards is reduced by using the biological retention facilities and stormwater detention tank and other facilities in the area around the hardened area within the airport building area, roads, etc; the midway control measures mainly means that the runoff peak is adjusted and the runoff is purified by configuring the retention tank, grassing tunnel, etc in the small drainage system; the end control measures mainly mean that the requirements for total runoff, runoff peak and runoff quality control are met by using the landscape water, ecological embankment and other measures based on the source and midway control measures. With the organic combination of 3 parts, the airport city construction is promoted, the safety at the airport flood season is ensured, the quality of the new airport water environment is increased and the regional heat island effect is improved.

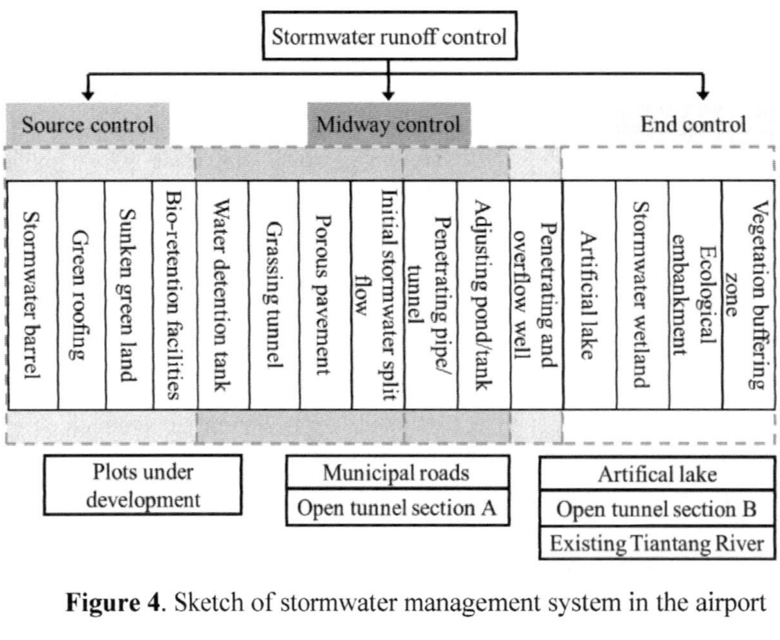

Figure 4. Sketch of stormwater management system in the airport

IOP Conf. Series: Earth and Environmental Science **146** (2018) 012029 doi:10.1088/1755-1315/146/1/012029

The airport is divided into five zones: flight, terminal, work, freight and maintenance according to the function of land use (Figure 5). According to its land use characteristics in the different areas, the different patterns and measures for stormwater runoff control are used to determine different stormwater runoff control solutions. The work, freight and maintenance zones are hereinafter referred to as the work zone to discuss as they have the similar conditions.

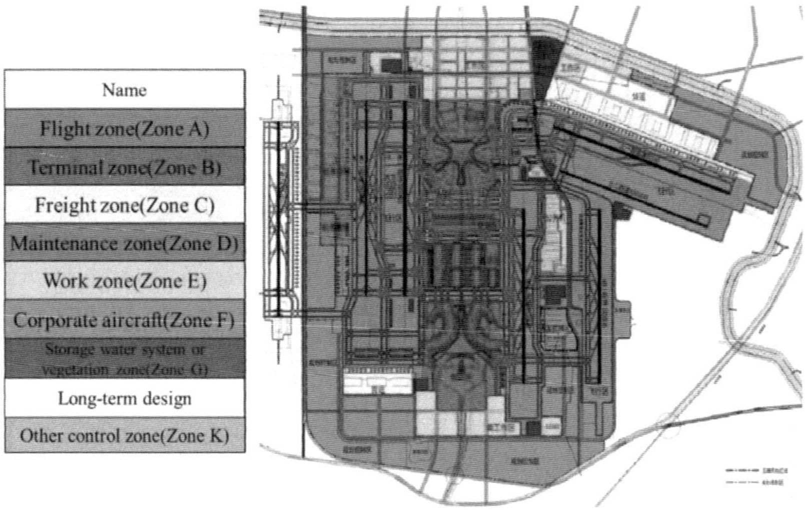

Figure 5. Sketch of the forward land division in the airport

The work zone is one of the core areas in the new airport. Its site conditions include the construction area, roads, green space, water, etc. Among them, the site construction area is the main land use type of the work zone. The area should be given full play to the storage capacity within the stormwater runoff area, and control the total runoff and clean runoff water quality by combining the land conditions and using the sunken green space, stormwater garden and other facilities. During the stormwater transfer, the runoff water quality is purified through the environment-friendly stormwater outlet, initial stormwater pool, etc. and the large public green space is used to regulate the rainfall runoff. Finally, the total runoff volume and peak runoff volume are stored with the end water detention tank. In the freight and other local zones, the rainfall coefficient is higher, and there is no adequate space to build the large-scale or distributed water storage facilities, the large-scale ecological stormwater detention tank must be built in the drainage downstream to centrally infiltrate and purify the runoff water through the infiltration area at the pool bottom, control the total runoff volume and adjust the peak runoff with the start and shutdown of pumping station in addition to the usage of the inner green land, the usage of the infiltration pavement, local stormwater detention tanks, etc. so as to reduce the rainfall runoff coefficient and receive the stormwater runoff.

The flight zone is based on runway and green space for the site conditions. Around the runway and the roads, there is a large number of green space and a good control condition of source stormwater runoff. A large number of source green spaces directly regulate the total amount of stormwater runoff volume, purify the runoff water quality, indirectly reduce the rainfall runoff coefficient and peak runoff volume in the flight zone by the sunken form. Meanwhile, the stormwater runoff in the flight zone is mainly drained through the open channel and regulated through the detention tank inside the flight zone. During the transfer of stormwater runoff, some stormwater runoff seeps into underground and makes up the groundwater and the runoff water can be purified. The stormwater runoff finally drains into the lower reaches of rivers after regulated through the large-scale detention tank and some stormwater runoff is received by seeping into the underground, evaporating and making up the landscape water and so forth.

The terminal zone is mainly based on the terminal building in the north side. In addition to the independent stormwater drainage system with the rainfall recurrence period of P=10 years, according to the stormwater runoff control principle in the work zone, the source control requirements are fully implemented in other areas, the water converges into the downstream detention tank at the end before draining into the landscape lake. The terminal zone is mainly based on the north terminal building. The terminal zone is mainly based on the air side area in the south side. The site condition is similar to the flight zone. The principle of stormwater runoff control is the same as that of the flight zone.

3.3. Storage and drainage system solution

The airport has a large catchment area, the smooth terrain, and merely $30m^3/s$ allowable external drainage flow rate is. By considering the impact of stormwater pipe & ditchdesign on the investment, pipeline depth, pipeline layout, drainage safety, etc, the stormwater drainage pipe & ditch and detention facilities shall be planned as a whole [7]. The stormwater system in the airport will use a 2-level drainage system (Figure 6). The runoff from each land block will be diverted into the municipal road stormwater drainage channel or the open drainage tunnel in the flight zone before the runoff water drains into the primary detention facilities in each drainage area. The runoff water drains into the open tunnel and landscape lakes for secondary detention in the field after pumped in the primary pumping station and finally drains into the planned New Tiantang River through the secondary pumping station at the river end. 2-level water storage volume reaches 2 million m^3 above.

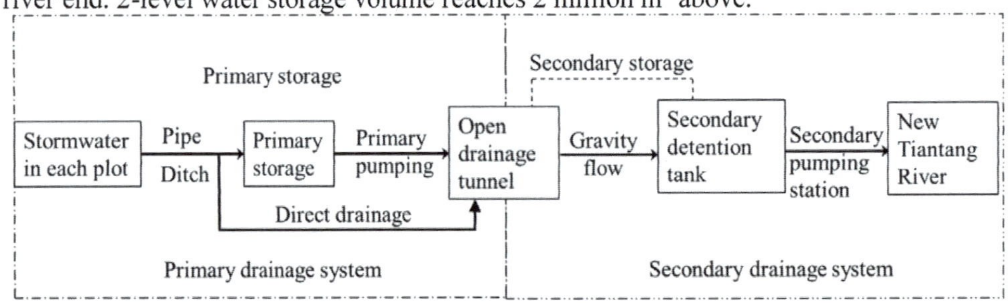

Figure 6. Sketch of simplified 2-level drainage system in the airport

The regional terrain in the airport is lower than the external flood level, and the drainage cannot be achieved through the gravity flow, the rational layout of pumping stations and storage facilities is essential in the airport to build the water storage system. According to the light, medium and heavy rainfall levels, the pumping station implements 3 different operating plans. In the case of light and medium rainfall levels, the pumping station has a relatively high pumping water level and the water storage facilities meet the total runoff control requirements. In the case the forecast heavy rainfall or heavy storms, the water detention tank and the landscape lake will lower the water level, reduce the pumping water level in advanced time, and keep the adequate time for regulating the peak stormwater flow rate.

3.4. Build the digital stormwater management system

3.4.1. Stormwater management system framework. Research has shown that not only the basic infrastructure construction such as monitoring system, drainage facility and demonstration area should be strengthened, but the water management system should be performed.[8] The digital stormwater management system to be built in the new airport is the key technology that the "sponge airport" is distinguished from the other stormwater system in general areas. The digital stormwater management system in the new airport uses digital model to simulate stormwater system, achieves comprehensive control and utilization of stormwater with the intelligent management goal, the model prediction and remote control mean, and has the ability to respond to emergencies rapidly.

The digital stormwater management system framework in the airport is shown in Figure 7, and divided into five layers. The 1st and 2nd layers are mainly used to provide information through the field equipment. In the open drainage channel, stormwater pipe & ditch outlets, etc. the measurement devices for liquid level, flow rate, etc. are set to monitor/control the running state of pumping station,

IOP Conf. Series: Earth and Environmental Science **146** (2018) 012029 doi:10.1088/1755-1315/146/1/012029

water level before and after grates, as well as the water quality; the 3rd layer is to create the digital model to analyze and predict the drainage system operation state in differential precipitations in the new airport; the 4th and 5th layers are for the managerial staff to schedule and control the stormwater system in the new airport through the information feedback in the former 3 layers, integral real-time data and monitoring videos, etc.

With the construction of digital stormwater management system, the airport will integrate the monitoring, operation and management, flood control and scheduling organically, base on the basic data sharing, analysis and exchange platform of database management layer and automated control layer, realize the real-time flood control data collection, storage and optimization processing, visually display the production state, analyze and guide the operation, production and schedule, and timely generate the accurate statistical analysis report. With modeling and analysis, the rainfall scope and extent are derived from the weather report rainfall data or real-time data so as to provide the valuable info for decision-making of the upper management. With the automatic control means, the remote control of pumping station with the control scope is realized so as to overcome the defects of traditional manual tour inspections, such as large workload, low efficiency and slow response.

	Main contents	
ODSS Intellectual decision	Risk simulation and evaluation /decision-making expert system	Command and coordinating layer
COLLABORATION command and scheduling	Real time data integration/ 3D trend display/ video monitor integration/Rainstorm analysis system/flood prediction system/flood control material management/flood command &scheduling/ video conference system/geological info system/mobile flood control command/flood emergency plan/flood info release/flood season traffic evacuation	
SIMULATION system simulation	Regional drainage model/stormwater prediction	Analysis and prediction layer
SCADA real-time monitoring	Network communication/ data collection/real time monitoring	Sensing and executing layer
DCS/PLC equipment & meters	Onsite equipment/monitoring meters/electric applauses	

Figure 7. Sketch of conceptualization for digital stormwater management system in the airport

3.4.2. Effects of stormwater management system. The construction of the digital model is one of the core construction contents in the digital stormwater management system for the airport. The modeling and simulation results, if the stormwater pipe& ditch system in the new airport that meets the design criteria is verified, the flood risk and effect of extreme rainfall is evaluated on the new airport flood, the capacity of the stormwater system solutions is optimized in response to the flood risk, the countermeasure against flood risk and emergency plans against flood are proposed and the scientific and rational stormwater system guideline is determined based on the rational formula computing.

The airport is modeled based on the data, such as stormwater ditches, pumping station, landscape lakes, detention tank, planned terrain elevation and other data all together. Beijing 24-hour rainfall pattern and Chicago rainfall pattern are respectively used for modeling. The modeling results need to meet the pipe's non-pressure flow state for 1-in-5-years rainfall and no-stagnant water state on the surface in the flood control design standard for 1-in-50-years rainfall. It is found from the modeling that the stormwater pipe & ditches meet the drainage requirements for 1-in-5-years rainfall basically except for pressure flow in a partial of pipes & ditches; the manhole overflows the water in local areas for 1-in-50-years rainfall. We propose the following correction plans after the reasons are analyzed:

- Adjust the riverbed water level, reduce the flood pipe & ditch flooded outflow level, and avoid the raising of water level in stormwater pipe & ditches;
- Increase the stormwater pipe & ditch design standards, discharge the water of high water-level at high speed and avoid the extraneous water inflow;

- Increase the number of stormwater outlets at the indented road section and improve the capacity of pumping station;
- Optimize the road elevation design.

The drainage system of Beijing New Airport is remodeled to see whether it meets the above criteria after the stormwater system solution is adjusted. After the completion of the new airport, with a period of monitoring the condition of runoff generation and convergence, the model is calibrated and verified, the running solution of drainage system in the airport is proposed against different rainfalls so as to ensure the safe operation of drainage system in the new airport.

4. Conclusion

The "sponge airport" will be constructed by combining the traditional stormwater pipe& ditch system planning, drainage and flood control planning and water system planning, etc. The source control system, small and large drainage systems will be coordinated scientifically and reasonably. Through analyzing problems in the water sensitive area and some limiting factors in face of airport building, as well as the different land characteristics, the construction of "sponge airport" of New Beijing Airport is discussed respectively from the systems of controlling the total runoff volume, the system of pollution control, the system of stormwater drainage and storage, the system of recycling stormwater and the system of digitalized stormwater management. With mutual supplementing forms between the source and end, surface and underground, green and gray spaces, the 3-character policy of "infiltration, retention, storage, purification, use, drainage" is fully implemented. With the construction of optimal digital stormwater management solution, we can improve the capacity of operation and management, improve the ability of decision and analysis, ensure the safety and stability of production, and achieve the goals of total runoff volume control, runoff peak control, runoff water quality control, stormwater utilization, etc. for the construction of sponge city.

Acknowledgement

We are grateful to Beijing New Airport Construction Headquarters, China Civil Aviation Airport Construction Group Corporation and Beijing Municipal Engineering Design and Research Institute Co., Ltd. for their endeavors.

References

[1] The master plan of Beijing New Airport (2015) [M]. 2015 Beijing: China Airport Construction Group Corporation of CAAA.
[2] The feasibility study report of Beijing New Airport [M]. 2014 Beijing: China Airport Construction Group Corporation of CAAA.
[3] Zheng Z, Qi S, Xu Y. Questionable frequent occurrence of urban flood hazards in modern cities of China [J]. Natural Hazards, 2013, 65(1):1009-1010.
[4] Che Wu, Ou Lan, Wang Huizhen et al. The quality and major influential factors of runoff in Beijing urban area[J].Techniques and Equipment for Environmental Pollution Control,2002, 3（1）: 33-37.
[5] Liu H, Jia Y, Niu C. "Sponge city" concept helps solve China's urban water problems[J]. Environmental Earth Sciences, 2017, 76(14):473.
[6] Che Wu, Zhao Yang, Li Jun Qi et al. Explanation of Sponge City Development Technical Guide: Basic Concept and Comprehensive Goals. [J].Journal of China Water& Wastewater, 2015, 31(8):1-5.
[7] Che Wu, Wu Yan Jie, Yang Zheng et al. Explanation of Sponge City Development Technical Guide: Rational Building of Urban Stormwater Detention and Retention System[J].Journal of China Water& Wastewater,2015,31(8):13-17.
[8] Xia J, Zhang Y Y, Xiong L H, et al. Opportunities and challenges of the Sponge City construction related to urban water issues in China[J]. Science China Earth Sciences, 2017, 60(04):1-7.

The Research on Economic Development Based on Entropy Method and Grey Relational Analysis – from the Perspective of Residents' Consumption Level

Y Y Li[1], B Y Tao[1] and D Y Si[2]

[1]Department of Industrial engineering, Sichuan University, Chengdu, 610065, China
[2]Department of Logistics engineering, Sichuan University, Chengdu, 610065, China

Email: 2016225025033@stu.scu.edu.cn

Abstract. This article is based on the consumption function theory, with the development of domestic economic, consumption is the main driver of the economic and the leading factor on the growth of GDP. Obviously, the residents' consumption level can represent the trend of consumption of residents. Finding the incentive and analysing the reasons can contribute to sustainable economy. Five influencing factors are selected including the per capita fiscal expenditure, the consumer price index, commodity retail price index, the urban and rural residents' per capita deposits, residents' per capita income. Then, the entropy method is adopted to evaluate each index and assign weight. Combining with the grey relational analysis, we prefer to use the entropy weight of each factor rather than use the average coefficient for more comprehensive and reasonable results. And the comprehensive correlation degree is calculated. Finally, we select the statistical yearbook of Chengdu from 2007 to 2015, and the empirical analysis is carried out on the factors which influencing the residents' consumption level. The results show that the residents' per capita income not only takes the most important role in these factors, but also develops a gap gradually. The fiscal expenditure promotes the improvement of the residents' consumption level and the economy has been boosted. However the consumption of rural residents is prevented by the high deposits and it has led to an economic depression in some areas. We combine the two methods and analyse from the perspective of macro and micro angle, the influence of consumption on sustainable economic growth has been analysed based on these five factors, and the comprehensive analysis and reasonable suggestions are given in order to promote sustained economic development.

1. Introduction

Since the reform and opening-up, it's a great growth on the gross domestic product (GDP) and the national fiscal revenue. The public services and the per capita income has been improved significantly. Even so, the rapid development of China's economy also has some disadvantages. For example, the domestic economy depends on external investment and exports excessively, and the domestic consumer market has not been realized. What's more, the urban and rural residents have burden heavier pressure than before such as housing, education and health care. The topics of narrowing the income gap between city and countryside and improving the residents' consumption level have become popular in recent decades.

Considering the researches on residents' consumption, the Keynesian theory of consumption is a starting point abroad and he elaborates the relationship between consumption and income. It means that the consumers' consumption is determined by income [1-2]. Based on this, the overseas economists carry on it until now. Duesenberry puts forward relative income hypothesis of consumption function. Friedman propose the persistence of income hypothesis [3-5]. A variety of consumption function forms have been built, and the development of economy has been promoted by them. Otherwise, they also provide reference for the research of our country.

Based on the foreign researches, researchers consider the domestic consumption market in China such as population structure and regional difference etc. They constantly enrich the study of residents' consumption level. Here are researches that are mainly divided into the following several aspects:

1) In macro perspective of view, to explore the influencing factors of urban and rural residents' consumption level, Zhang focus on the influence of the social security expenditure and analyses 15 typical provinces. He holds the point that the level of social security status is also crucial in addition to the traditional concept of people's income [6]. Zhao combines static panel data model with dynamic panel data model, three influencing factors are provided include the income, the consumption and the price index. The reasonable suggestions are given for the regulation of macroeconomic too [7].

2) Discussing the different parts of the consumption level from the perspective of regional differences. Dang analyses the annual consumption of urban households in China, and explained the impact of regional differences on the consumption level [8]. Wu and Chen do surveys by using the data of 31 provinces based on the theory of consumption economics theory and economic convergence. The analysis focus on the country residents' consumption, and some rational proposals are put forward in order to narrow the gap of regional income and coordinate the geographic policies [9].

3) Some literatures merely refer to the urban or rural residents. Deng and Zhou have carried on the statistical analysis about the urban residents' consumption structure. Based on a data analysis model, the changing consumption structure has been analysed and the influencing factors of consumption level are presented [10-11]. Li and Yang discuss the rural residents' consumption structure. They adopt the method of principal component analysis and give suggestions to improve the rural consumption structure and the level of consumption [12-13].

From the perspective of literatures at home and abroad, many articles only consider the one-sided influence factors. They lack of comprehensive analysis on the urban and rural residents' consumption level. Otherwise, the single angle of view still has some limitations. In view of this, combining with macro and micro factors, we selects five influencing factors and makes a comprehensive analysis and evaluation. Through the analysis of comprehensive correlation degree, a new research perspective is provided and some referential suggestions are provided to improve the residents' consumption level.

2. The summarize of residents' consumption level

2.1 The analysis of influencing factors

With the rapid development of economy in recent years, the residents' income has been driven by the domestic economy. From the influencing factors of residents' consumption level, the per capita fiscal expenditure affects the construction of public service and the residents' consumption psychologically. The consumer price index reflects the changing purchase of family in the prices of consumer goods and services generally. In the same way, commodity retail price index reflects the degree of changes on retail prices of commodities in certain period of time. The consumer price index and commodity retail price index are based on the influence of consumer goods and services, and provide the consumer spending judgment and choice advice. The urban and rural residents' per capita deposits is planning for the future by saving process. It has to do with the current level of economic development, income status and the change rates of interest. The behaviours of residents' consumption are influenced by the fluctuation of deposits directly. Obviously, changing unilateral factor can cause the change of other factors. They all lead to the fluctuation of residents' consumption level.

2.2 Grey relational analysis

Grey relational analysis is a quantitative analysis of the development trend in the research system. By comparing the characteristics of the statistical sequence curve, the relational grade is determined between each elements [14]. It takes the grey system with some uncertain information as the research object, and finds out the main influencing factors. The grey relational analysis can grasp the essential characteristics of things. The detailed calculation steps are as follows [15]:

1) Reference sequence and comparative sequence

In order to make a reasonable comparison and evaluation, the influencing factors of residents' consumption level are selected as the reference sequence as follow:

$$X0 = \{X01, X02, \cdots\cdots, X0\mathrm{n}\} \tag{1}$$

The comparison sequence has m comparison objects, each sequence has n measurement data:

$$Xm = \{Xm(1), Xm2, \cdots\cdots, Xm(\mathrm{n})\} \tag{2}$$

2) Data normalization

Because of the differences in the physical meaning and the unit of the evaluation index, it is necessary to do the dimensionless and normalization.

$$Y0 = \{X0(1)1nk = 1nX0(k), \cdots\cdots, X0(n)1nk = 1nX0(k)\} \tag{3}$$

3) The correlation coefficient

After transforming the data, in the k moment, the correlation coefficient of the residents' consumption level is calculated below:

$$\varepsilon 0i = \Delta min + \rho\Delta max\Delta 0ik + \rho\Delta max \tag{4}$$

In this formula, ρ is the resolution coefficient. The function is to improve the difference in the correlation coefficients. It is usually taking 0.5. Δ_{0i} (k) is the absolute difference of the two sequence at k time. Δ_{min} and Δ_{max} are the minimum absolute difference and the maximum absolute difference.

4) Comprehensive relational grade

$$r0i = i = 1m\varepsilon 0i(k)\omega i \tag{5}$$

In this formula, ω_i is named weight. Entropy weight method is used to determine the weight, instead of the average coefficient. The better relational grade can be calculated.

2.3 *Entropy weight method*

The entropy method is an objective weighting method. During the process of using the indicators, the information entropy and entropy weight can be calculated according to the data. By modifying the weight of each index, the objective weight of index is calculated. Otherwise, it can provide more reasonable reference for subsequent evaluation. The detailed steps are calculated as follows [16]:

Here are k evaluative indexes $(X1, X2, \cdots\cdots, Xk)$. Each evaluation indexes contains m data.

1) Data standardization

$$Yij = Xij - \min(Xj)\max Xj - \min(Xj) \tag{6}$$

2) The contribution degree of index

$$Pij = Yiji = 1mYij \tag{7}$$

3) The information entropy

$$Ej = -1\ln mi = 1mPij\ln Pij \tag{8}$$

$$\lim Pij = 0Pij\ln Pij = 0 \tag{9}$$

4) Entropy weight

$$Wij = 1 - Ejm - i = 1mEj \tag{10}$$

The grey relational analysis aims to calculate the relational grade ultimately. By optimizing the grey relational analysis, this article uses entropy weight instead of average coefficient in order to obtain the comprehensive relational grade. Sorting the size of the relational grade, Results can be analyzed analysed and some reasonable suggestions can be given.

3. Case analysis

The study of residents' consumption level includes the selection of influencing factors and the establishment of evaluation index. According to the above analysis, this article selects five influencing factors including the per capita fiscal expenditure (Y_1), the consumer price index (Y_2), commodity retail price index (Y_3), the urban and rural residents' per capita deposits (Y_4) and residents' per capita income (Y_5). According to the residents' consumption structure, five indexes are selected as the indicators of grey relational analysis including food (X_1), clothing (X_2), housing (X_3), transportation and communication (X_4), health care (X_5). The comprehensive data analysis is given below [17-21].

3.1 Data analysis

According to the statistical yearbook of Chengdu from 2007 to 2015, the influencing factors and evaluation indexes of the residents' consumption level are counted. Through the screen of yearbook data, the original analysis data are obtained as Table 1 and Table 2.

Table 1. Relevant data of urban influencing factors and evaluation indexes

	2007	2008	2009	2010	2011	2012	2013	2014	2015
Y_1	6811.0	8334	9225	15377	15510	15186	18207	19452	18247
Y_2	105	104	100	103	105	103	103	101	101
Y_3	104	104	99	102	104	101	101	98	99
Y_4	22263	29186	37391	44317	51723	61260	69042	74848	81368
Y_5	14849	16943	18659	20835	23932	27194	29968	32665	33476
X_1	4576	4806	5283	5739	6584	6745	7147	7555	7508
X_2	111	124	131	148	163	186	205	219	194

	2	6	0	9	7	7	7	3	2
X_3	1006	1054	1043	1117	1619	1734	1853	1997	4409
X_4	1720	1940	2296	2652	2972	3182	3360	3582	2554
X_5	690	848	944	822	925	1029	1120	1216	1157

Table 2. Relevant data of rural influencing factors and evaluation indexes

	2007	**2008**	**2009**	**2010**	**2011**	**2012**	**2013**	**2014**	**2015**
Y_1	6811	8334	9225	15377	15510	15186	18207	19452	18247
Y_2	105	104	100	103	105	103	103	101	101
Y_3	104	104	99	102	104	101	101	98	99
Y_4	22263	29186	37391	44317	51723	61260	69042	74848	81368
Y_5	5642	6481	7129	8205	9895	11501	12985	14478	17690
X_1	1851	1890	2045	2359	2954	3313	3582	3859	4881
X_2	336	425	441	516	682	790	869	950	1220
X_3	432	539	611	626	724	798	860	950	199

									6
X_4	548	662	697	823	1111	1338	1472	1600	1538
X_5	204	247	306	325	436	500	594	688	788

Note: the data of table 1 and table 2 are selected from the statistical yearbook of Chengdu from 2007 to 2015.

1) The original data is used by entropy weight method and the weight of each evaluation index is determined. The corresponding information entropy and entropy weight are obtained.

Table 3. Information entropy and entropy weight of urban and rural evaluation indexes

	city					**countryside**				
	X_1	**X_2**	**X_3**	**X_4**	**X_5**	**X_1**	**X_2**	**X_3**	**X_4**	**X_5**
information entropy	0.870	0.871	0.671	0.888	0.903	0.799	0.847	0.788	0.853	0.846
entropy weight	0.163	0.161	0.413	0.140	0.122	0.232	0.177	0.244	0.170	0.178

2) Looking at the Table 3, the high proportion of information entropy shows that entropy weight can be used instead of the average coefficient. Combining with the method of grey relational analysis and entropy method, the comprehensive relational grade can be calculated between the influencing factors and the evaluation indexes.

Table 4. Comprehensive relational grade of urban residents' consumption level

	X_1	**X_2**	**X_3**	**X_4**	**X_5**	**comprehensive relational grade**	**rank**
Y_1	0.824	0.843	0.749	0.842	0.781	0.793	2
Y_2	0.836	0.802	0.735	0.820	0.847	0.788	3
Y_3	0.831	0.797	0.731	0.817	0.842	0.784	4
Y_4	0.741	0.810	0.708	0.768	0.740	0.742	5
Y_5	0.870	0.938	0.763	0.882	0.860	0.837	1

Table 5. Comprehensive relational grade of rural residents' consumption level

	X_1	**X_2**	**X_3**	**X_4**	**X_5**	**comprehensive relational grade**	**rank**
Y_1	0.824	0.856	0.729	0.864	0.834	0.814	3
Y_2	0.732	0.695	0.734	0.697	0.678	0.711	4
Y_3	0.73	0.69	0.73	0.69	0.67	0.710	5

	1	5	2	6	9		
Y_4	0.847	0.865	0.709	0.868	0.879	0.826	2
Y_5	0.905	0.910	0.744	0.818	0.872	0.846	1

3.2 Result

From the perspective of table 4 and table 5, the rank of urban and rural comprehensive relational grade can be calculated above. We can draw the following conclusions:

1) Most comprehensive relational grades are the residents' per capita income compared with the urban and rural residents' consumption level. The income of residents still play a leading role currently. The rising incomes not only increases the consumer confidence, but also taps the potential consumption and stimulates consumer desire. The steady rise of residents' consumption level is contributed.

2) Compare with the income of urban and rural residents, the income has been increased yearly, while the gap is gradually expanding in less than ten years. The overall situation is not optimistic. Some obstacles still affect the rural residents' income level such as low technological level, laggard production mode. In contrast, urban residents could enjoy services and related goods expediently because of favourable location and higher income.

3) Considering the urban and rural residents' per capita deposits, the comprehensive relational grade of the rural areas is much higher than the urban areas. The deposits in rural areas have a lot to do with the food, clothing, transportation, communication and health care in the consumption structure. Compared to the consumption of urban residents, on the one hand, rural residents tend to make choices in order to solve the problem of life. On the other hand, the medical service and the facilities of transportation have not been prepared well. Residents are too worried to reserve deposits because of uncertainty. That's why the precautionary deposit is so high in rural residents.

4) In terms of the per capita fiscal expenditure, the rural areas with higher comprehensive relational grade are reflected in transportation and health care. It also reflects that the government construction plays an important role in countryside. The government will invest more fiscal expenditure on rural public services including the highway system and the rural cooperative medical. Excellent infrastructure promotes the development of rural areas and improve the residents' consumption level.

5) Comparing to rural residents, the consumption level of urban residents is much related to the consumer price index and commodity retail price index. On the one hand, the majority of rural residents have their own agricultural products, and they can be self-sufficient. However the city residents rely on consuming. On the other hand, the city's infrastructure construction is relatively perfect. It promotes urban residents to consume and enjoys better medical services.

4. Suggestion

With the development of China, investment, consumption and export have been the "three carriages" which stimulate the economy. The economy of China relies on investment and exports so much in the early period of economic development. However, the problem solving domestic consumption has not been solved until now even though it's vital. The residents' consumption could influence the economic growth radically by adjusting consumption structure. It also provides a powerful impetus to the sustainable development of the economy. The following suggestions are put forward:

1) Government should narrow the income gap between urban and rural areas and improve the income level of rural residents. The imbalance economic development has restricted the pace of rural development and the output of rural manpower. Measures should be taken to ensure stable income and drive consumption expenditure. The government appropriately adjust the structure of distribution and increase the proportion of workers' initial distribution. It also could encourage external investment to set up factories which can solve the problem of surplus manpower. Adjusting the structure of the agriculture, eliminating the limitation of the single agricultural products, strengthening professional agricultural production can improve the competitiveness of agricultural products too.

2) The positive behaviour sof consumption should be guided normatively. The current market consumption is still faced with difficult to meet the supply of consumer demand. The government should organically consider the three effects of production, circulation and exchange. At the same time, the fair market environment is important particularly. It can not only guide consumers to rationally adjust the consumption structure, establish a scientific and healthy consumption, but also protect the rights of residents in the channel of consumption.

3) The fiscal expenditure should be inputted more than before and the government need to improve the social security system too. In addition to income, the expectations for the future still restrict consumption, and good expectations stimulate the residents' spending ahead of time. Otherwise, the inclination of fiscal expenditure on social security system aims to play the critical role. Social security system makes sure the basic living security. It eliminates uncertain worries in the future. It also helps the residents to form reasonable consumer expectations, and release the consumption potential.

5. Conclusion

The article focus on the evaluation of residents' consumption level and selects the statistical yearbook of Chengdu from 2007 to 2015. By using the methods, the curve of the sample data can be portrayed and the economic development trend can be predicted better. The results show that:

1) The influencing factors are selected integrally. The research of economic development has been enriched by analysing the residents' consumption level and consumption structure.

2) The article combines the entropy method and gray relational analysis, from the perspective of macro and micro angle, the comprehensive consideration of the influence factors and the consumption expenditure structure may provide comprehensive relational grade objectively. At the same time, it also provides a new perspective to stimulus economic growth.

3) On the evaluation of the consumption lever, the government should make reasonable control on consumptive goods and services. It provides complete public facilities and services and narrows the gap between city and countryside. The consumption ability of residents is released further.

Considering the macro and micro factors, the article refer to many literature. However the selection of factors is not comprehensive, and there are some limitations. Therefore, the more reasonable evaluation methods and influencing factors should be the focus of research in the future.

References

[1] D'Orlando F, Sanfilippo E 2010 *J.* Behavioral foundations for the Keynesian consumption function. *Journal of Economic Psychology*. **31(6)**:1035-1046.
[2] Murota R, Ono Y 2010 *J.* A Reinterpretation of the Keynesian Consumption Function and Multiplier Effect. *Social Science Electronic Publishing*.
[3] Sanders S 2012 *J.* A Model of the Relative Income Hypothesis. *Journal of Economic Education*. **41(41)**:292-305.
[4] Park S 2016 *J.* How to Consume and Invest for Retirement: Revisiting Friedman's Permanent Income Hypothesis. *Social Science Electronic Publishing*.
[5] Apostolou T, Jr V K 2011 *C.* Permanent Income and Adaptive Expectations: Estimation of a EU Consumption Function. *International Atlantic Economic Conference*. **33(23)**:2782–2797.
[6] Zhang Q 2013 *J.* Comparative Study of Expenditure Impact on the Level of Consumption in Urban and Rural Social Security --The empirical analysis of the 15 provinces Panel Data. *Social*

security studies. **2** 71-82.

[7] Zhao H C, Jeffrey F 2013 *J.* The influence of income, price and interest rate on the consumption level of urban residents in China -- Based on the static and dynamic panel data model analysis. *Mathematical statistics and management*. **32(2)** 211-220.

[8] Dang Y J 2013 *J.* The influence of regional differences on the consumer level of urban residents in China. *China securities futures*. **3** 136-136.

[9] Wu Y M, Chen Z J 2009 *J.* The spatial correlation and regional convergence of people's consumption lever. *World economic analysis*. **5** 76-89.

[10] Deng X, Ge T T 2014 *J.* Panel Data analysis of the consumption structure of urban residents in China. *Economic system reform*. **1** 193-197.

[11] Zhou H 2015 *J.* Influence factors and development trends of consumption structure of urban residents in China. *China market*. **21** 13-14.

[12] Li C Y 2014 *J.* Analysis on the principal component of consumption structure of rural residents in China. *Commodity and quality: consumption research*. **5** 6-6.

[13] Yang Q 2015 *J.* The analysis of consumption structure of rural residents in Sichuan Province Based on principal component analysis. *Jiangsu business theory*. **23** 105-106.

[14] Liang B S 2009 *M. Applied mathematics*. Beijing: Meteorological Press.

[15] Yi D S, Guo P 1992 *M. Grey system theory and method of solution, abstract, procedure and application*. Beijing: Petroleum Industry Press.

[16] Feng Y Q, Li X M, Li X W 2014 *J.* Comprehensive evaluation of railway safety based on entropy weight method and grey relational analysis. *Journal of safety and environment*. **14(2)** 73-79.

[17] Wang J H, Li M, Meng F 2012 *J.* The influencing factors of China's residents' consumption level. *Open herald*. **2** 44-47.

[18] Miao Z 2013 *J.* The study on the influencing factors of per capita consumption level of Sichuan residents. *Wuhan Finance*. **8** 55-57.

[19] Wang L C, Zhang N 2016 *J.* The consumption structure of urban and rural residents in Henan Province, the comparative analysis. *Chinese Agricultural Science Bulletin*. **32(1)** 193-199.

[20] Zhang Z L, Song Q 2016 *J.* Trends and Countermeasures of residents' consumption structure change under the new normal. *Macroeconomic management*. **6** 68-71.

[21] Ma H 2016 *J.* The consumption structure of rural residents China trend analysis. *Gansu theory*. **2** 135-139.

IOP Conf. Series: Earth and Environmental Science **146** (2018) 012031 doi:10.1088/1755-1315/146/1/012031

Study of Literatures for the Impact on PM2.5 Purification by Architectural Design under Haze Conditions

Cao Jian[1,2], Hu Ningning, Yang Lei[1,2], Yan Wei and Chu Minghao

[1] Zhongyuan University of Technology, Zhengzhou 451921, China
[2] Key Laboratory of Heating and Air Conditioning, The Education Department of Henan Province, Zhengzhou 451921, China

Email: 6092@zut.edu.cn; 994859160@qq.com; xsc_yl@163.com ; yanwei1003@126.com; Xl6278907@sina.cn

Abstract. This paper focuses on literatures reviewing with key words of architectural and PM2.5 based on the China Academic Journal Network Publishing Database (CKNI) and Elsevier SD database. Statistical analysis on the searched 184 related literatures show the following results. (1) 20.6% of the all targeted literatures showed that PM2.5 concentration was one of the important impact parameters of indoor & outside air quality. (2) 16.3% of the literatures illustrated that the architecture designing contributed significantly to indoor and outdoor fine particulate pollution by controlling particle generation and diffusion. It can be concluded that the three-dimensional architectural design, the total layout of the building designing strategy and the architectural space design based on green building-oriented architectural designing concept will be the future focused research and the frontier question.

1. Introduction

Industrial development has brought about economic development and environmental pollution in China in the past years. The high concentration of fine particles in the air has been a serious threat to people's health, especially with haze weather. From 2001 to 2006, the National Aeronautics and Space Administration (NASA) had plotted the global pollution particulate matter (PM2.5) average concentration satellite map as shown in figure 1. China's north and east were the most polluted areas in the mainland of Asia. Haze destroys the ecological environment and seriously affects human life. Haze can induce respiratory diseases and cardiovascular diseases and increase the prevalence of cancer and so on. People are apt to haze weather caused traffic accidents.

Pope et al. conducted a 16 year study of nearly 500 thousand adults in 50 states in the United States. Research findings showed that 10μ g/m^3 increasing of the average concentration of PM2.5 might contribute to 8%, 6% and 4% increasing of human lung cancer mortality, heart lung disease mortality and annual total mortality, respectively [1]. People spent more than 90% of the time in building. How to keep indoor air clear is important, especially in the environment condition of haze.

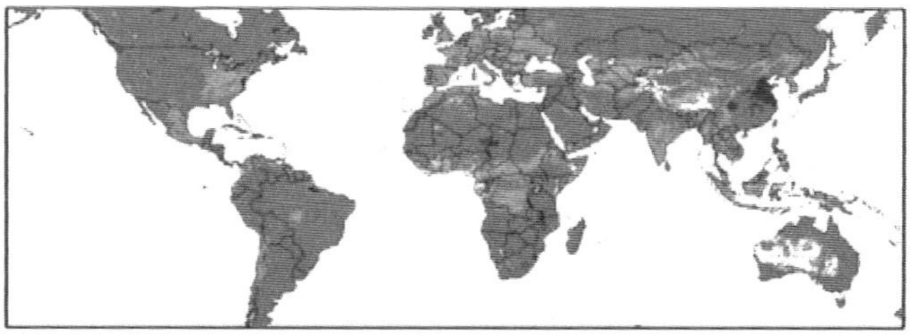

Figure.1. satellite remote sensing test chart

2. Research background of haze concentration control by Architectural design

2.1. Literature research status

The authors of this paper took the "building" and "PM2.5" as key words in CNKI and Elsevier SD database. The author compiled the data, and the author drew a broken line diagram. The figure 2 displays a literature review of PM2.5 control in the field of architecture both at home and abroad. There are few literatures in china before 2011. The number of domestic related literatures is not the state of growth. After 2011, the relevant domestic literature had a significant upward trend. This shows that people know and understand PM2.5 in recent years. Relevant scholars pay attention to PM2.5. As can be seen from the diagram, the international research on the control of PM2.5 had been in a steady growth trend. Our country has a wide development space for the control measures of fine particulate pollution in the ambient air.

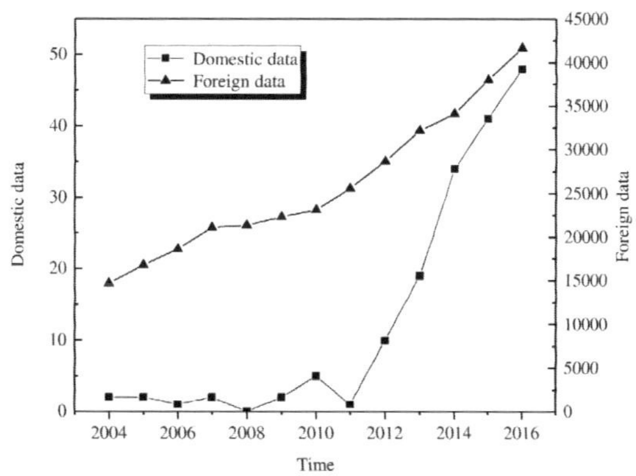

Figure.2. The literature research data of fine particles on the buildings at home and abroad

2.2. Direction of literature research

The author of this paper reviewed the domestic and foreign literatures in recent years. Building control of indoor PM2.5 concentration will be the main direction of future researches. On the views of the current literature analysis. In the construction field, the control strategies of the concentration of fine particles in buildings can be roughly divided into two parts. One is the combination of indoor ventilation system and air purification equipment of active control measures. Another is to control the indoor PM2.5 concentration passive control measures through green building design methods. The former is one of the main focuses of the current literature research, and the latter is in a gradual development trend.

3. Domestic and foreign standards for PM 2.5

Haze has serious threat to environment, human life and health. Relevant laws and regulations have been formulated for haze control all over the world. PM2.5 concentration limits for second classes of countries are illustrated in figure 3.

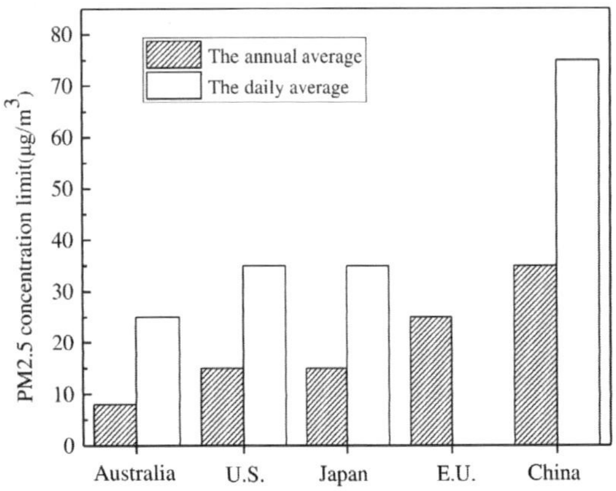

Figure 3. PM2.5 concentration limit requirements of China and some other countries or areas.

It can be seen from figure 3, there was still a big gap between China's PM2.5 concentration limits and international standards. Hence, China has a long way to improve the domestic air environmental quality to acceptable level.

4. Literatures on the influence of architectural design on indoor and outdoor PM2.5 concentration distribution

4.1. *Overall layout of building*

The overall layout of the building and the green around of building can affect the PM2.5 concentration around building. Wu et al. (2013) found that the height of the building layout and the way of enclosure affect the concentration of PM2.5 in the air. Studies had shown that the avenue in the multi-storied courtyard layout could effectively reduce the concentration of PM2.5 in the air[2]. Figure 4 displays a multi-storied courtyard residential district layout map.

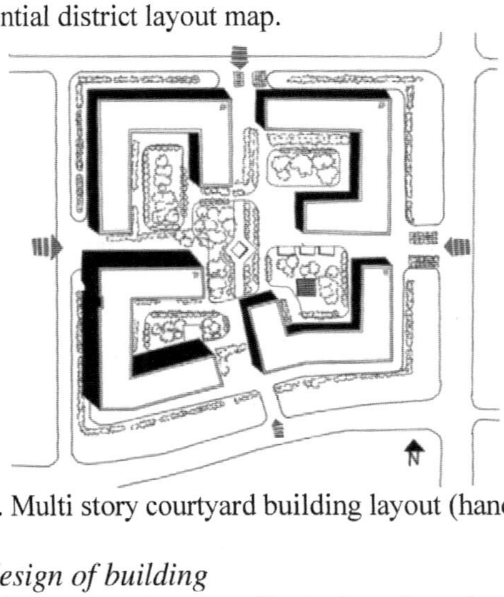

Figure.4. Multi story courtyard building layout (hand-painted)

4.2. *Natural ventilation design of building*

The three-dimensional greening construction can effectively reduce the concentration of fine particles around the building. The three-dimensional greening of the building includes green vegetation around the building, building wall greening, roof garden and so on. Three dimensional greening can beautify

IOP Conf. Series: Earth and Environmental Science **146** (2018) 012031 doi:10.1088/1755-1315/146/1/012031

the building environment, in which green plants have obvious ecological effects. Three dimensional greening is one of the important ways to control air pollution passively, As figure 5 shown that one of the three-dimensional greening methods.

According to Zhang (2010), each cubic meter of lawn can absorb 146kg of carbon dioxide, 0.0031kg of sulfur dioxide, and release 105.85kg of oxygen per year [3]. Therefore, increasing the green area is of great significance to improve air quality and reduce air pollution.

Figure.5. three-dimensional greening (hand-painted)

4.3. *Natural ventilation design of building*

4.3.1 The thermal pressure ventilation. Brown (2008) believed that combining the sun shaded courtyard with the open courtyard could lead the air from one courtyard to another. He also thought that the principle of hot pressing could drive indoor ventilation. For example, Mareuil designed Alice Spring House. The designer had two residential courtyards into a central courtyard. The sun heated the air in the courtyard which contributed to the hot air rising in the courtyard. Moreover, Cold air entered the room from the entrance of the sunshade. The air entered the room complemented the rising hot air in the courtyard. This resulted in hot pressing ventilation, the hot pressure ventilation increases indoor natural ventilation efficiency [4], as shown in figure 6 (a, b).

When the concentration of fine particles outside the room is low, natural ventilation can take away a lot of harmful fine particles in the room and bring fresh air into the room.

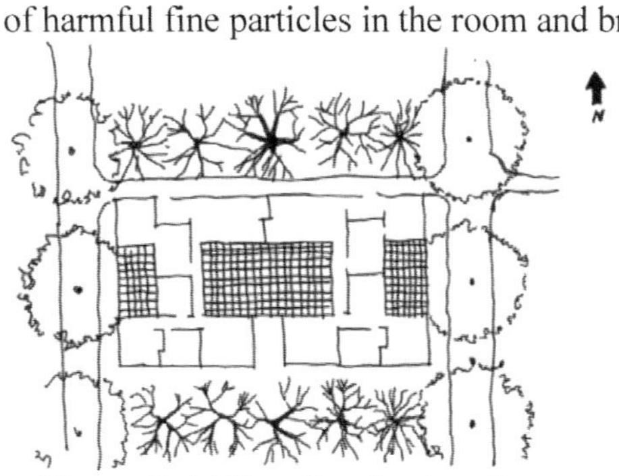

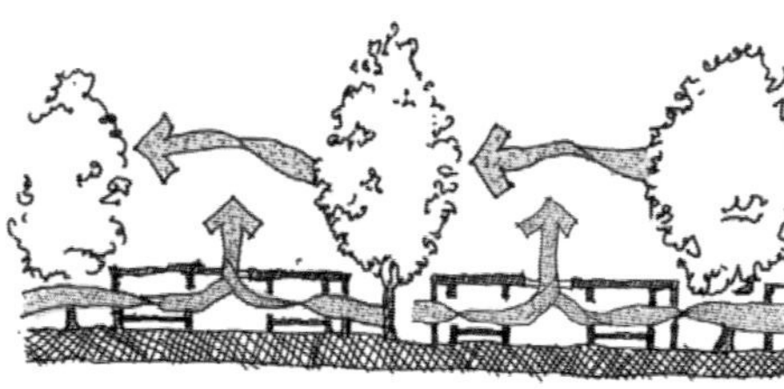

Figure.6. (a)Plan (hand-painted) **Figure.6.** (b)Profile analysis chart (hand-painted)

Additionally, air ventilation by hot pressure can provide fresh air for buildings and effectively reduce energy consumption of buildings. The authors analyze the documents and monographs, author sketch below several common building hot air ventilation schematic diagram figure 7 (a-e). Figure 7 (a) is commonly used in single storey low rise buildings. It's a single building space using hot air ventilation. Figure 7 (b) is a compound architectural space. The air flow in many low spaces converged into large spaces. The airflow in the large space leads to outdoor with the principle of hot pressing. Such large spaces can be arranged in conjunction with halls and corridors. Figure 7 (c) is a common form of atrium space design. Green vegetation can be planted in the atrium. Planting green vegetation can effectively reduce the concentration of indoor fine particles. Figure 7 (d) is one of the main forms of public buildings. Atrium can effectively organize indoor air flow and effectively organize indoor air flow with only one disadvantage, which is that the large space in the atrium increases the difficulty of building fire protection. Moreover, Atrium has a high demand for building fire protection.

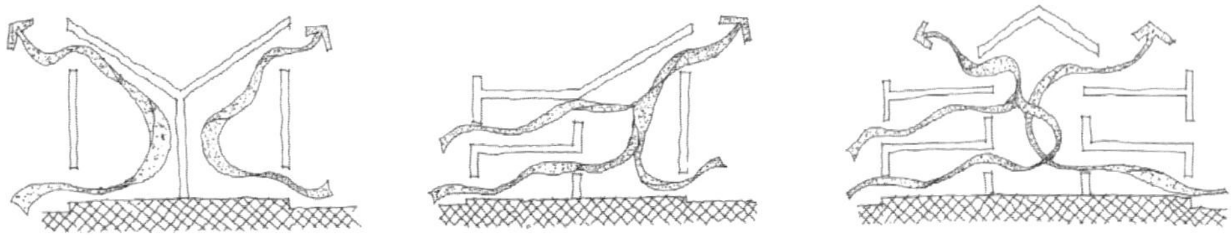

Figure.7.(a)Space with two sides high (b)High space around (c)Space in the middle-atrium

Multiple hot-pressing ventilation is a way of ventilation by chimney effect. This method can greatly improve the effect of indoor natural ventilation. Multiple hot air ventilation can remove the indoor PM2.5 in time. Multiple hot pressed ventilation adds additional building structures and is usually used in buildings with special requirements for ventilation.

The architectural space of figure 7 (e) is often arranged in conjunction with building staircases. It is also a common form of architectural design. Indoor and outdoor temperature pressure to the indoor hot air rises. The outdoor cold air enters the room from the lower level and good circulation ventilation is formed which plays an important role in reducing the concentration of indoor particulate matter.

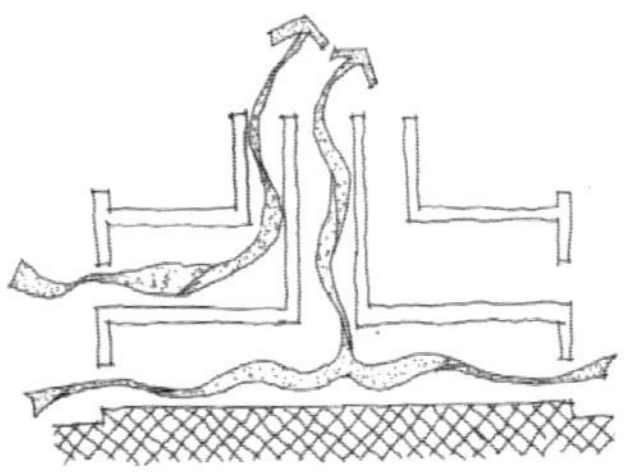

Figure.7.(d) Multiple hot pressing ventilation (hand-painted)

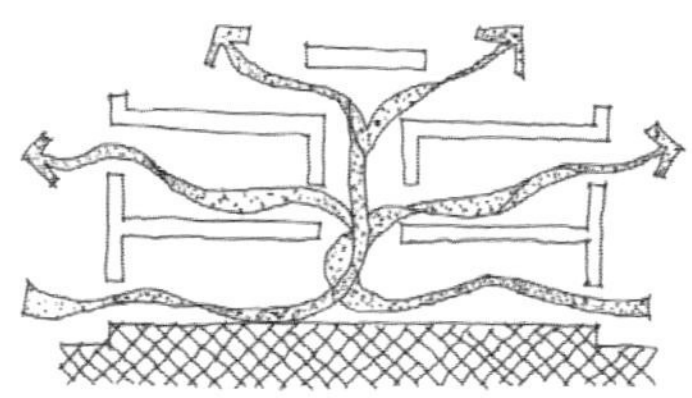

Figure.7.(e)Staircase for hot pressing ventilation

4.3.2 Wind pressure ventilation. In spring and autumn, air pressure ventilation is a good ventilation method without energy consuming. This ventilation method is more common in architectural design. For example, when the outdoor haze pollution is small in spring and autumn season, Indoor fine particulate matter is mainly due to people's activities in the indoor. Open the window at this time and air pressure ventilation can take away a lot of indoor contaminated particles. Wind pressure ventilation ventilates outdoor clean air into the room and reduces indoor PM2.5 concentration.

Wind pressure ventilation mode is easy to implement in architectural design. At the same time, the window can also deal with the artistic effect of the building facade to achieve beautiful visual enjoyment.

The authors analyze and study the effect of reducing the concentration of fine particles in the room by using the ventilation space. Figure 8 (a) is a common indoor ventilation design which is often said as the draught. Windows are arranged on the opposite sides of the same space. This kind of space should be generally combined with the wind direction of the regional leading wind. Smaller windage under the influence of dominant wind causes the local wind speed larger and impacts on indoor activities because the space is straight.

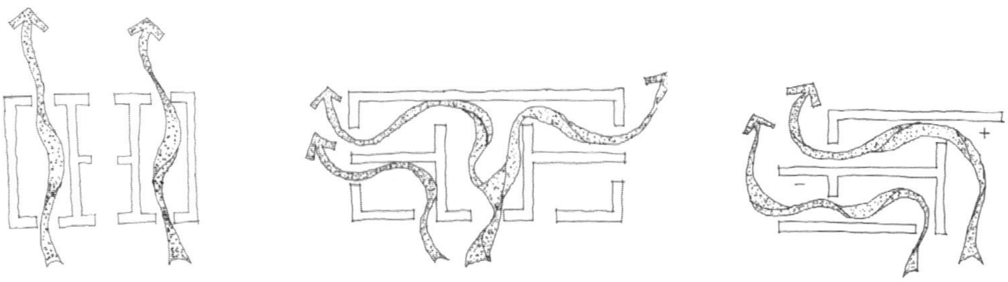

Figure.8.(a)Single transparent (hand-painted) (b)Interlaced space (hand-painted) (c)Wind-wall space space (hand-painted)

Figure 8 (b) is a common residential space in summer locating at the northern hemisphere. Open the window on the windward side of the wall (south side). South side of the window is to accommodate the dominant wind in summer. Outdoor fresh air is introduced to improve indoor air quality. Indoor temperature around south window is effectively reduced in summer. Figure 8 (c) formed a lively architectural space by windfall. Combined with architectural form to form a space staggered architectural style. Achieve flexible and diverse architectural art is a common form of architecture in modern architecture. Positive pressure appears on the windward side of building and negative pressure locates on the leeward side of the building. The building ventilation effect is good and is beneficial to reduce the concentration of indoor particulate matter.

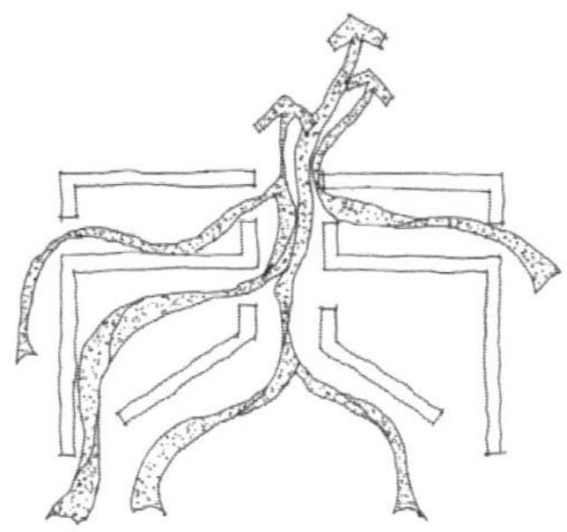

Figure.8.(d)Space using Venturi effect (hand-painted)

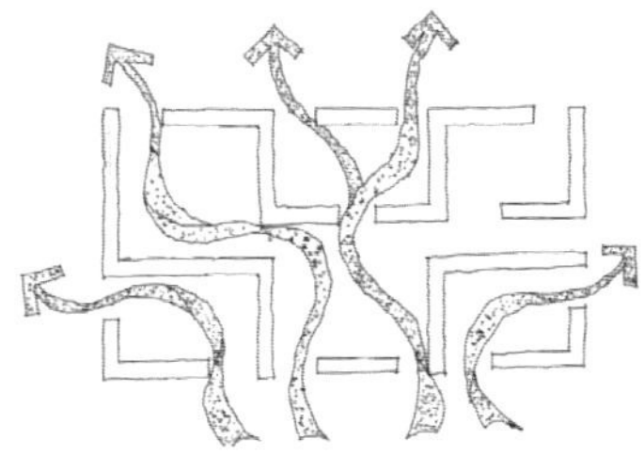

Figure.8.(e)The space of circular ventilation with each other(hand-painted)

Venturi effect is that, a low pressure is produced on the lee side of the barrier when the wind blows through the barrier duing to the adsorption of low pressure zone. Venturi effect is a typical ventilation mode of wind pressure. Ventilation effect of venturi is very good but the space is at the end of the air flow. The flow rate is faster and have adverse effects on the daily life of indoor people. Figure 8 (e) is a space connected with each other which is more suitable for residential buildings. The interconnected building space can avoid the discomfort caused by the local gas flow. This uniform and steady airflow can take away the PM2.5 gathered in each room as to achieving a healthy and comfortable living environment.

4.4. Ventilation of building machinery

The research shows that the most serious period of haze pollution in China is from January to March. At this point, a large number of fine particles in the atmosphere is difficult to disperse[5]. Moreover, outdoor PM2.5 concentration is greater than indoor. Natural ventilation is clearly not appropriate. It is necessary to combine mechanical ventilation with air purifier to reduce the concentration of fine particles in the room. That's now called active control. The main way is to use fresh air equipment purified and introduce outdoor air into the room. When outdoor haze is serious the tightness of building is also an important factor affecting the concentration of indoor fine particles. Li (2015) proposed that building tightness is bad for indoor personnel when exposed to outdoor PM2.5 high concentrations. Cold air infiltration will increase heating energy consumption [1]. Window should be closed to avoid severe haze penetrating into the room in winter. Indoor air purification equipment and fresh air system can be used to purify indoor air.

4.5. Building water system
Water is one of the most important parts of the building environment. Water can change the microclimate around the building because the surface of the water can take away a lot of heat in the process of evaporation and reduce temperature and increase humidity.

Water can produce negative ions which has certain adsorption effect and adsorb fine particles in the air to achieve the purpose of air purification. According to the experts from the United Nations air environmental protection field research confirmed that small particles with negative ions in the air can capture PM2.5 in the air and agglomerate it.

4.6. Building water system
At present, the study for the adsorption particles of building materials research showed appropriate environmental protection building wall material could also slow down the increase of PM2.5 concentration in air to some extent. Such as new fly ash block and cork material were relatively environmentally friendly building materials and had very good absorptive capacity for carbon dioxide in the air. Cork material was renewable and degradable. Excellent performances were shown in heat insulation, sound insulation, vibration resistance, fire prevention and so on. Wang (2015) believed that cork made of wood can absorb dust. It can also resist bacterial and fungal adverse effects [6]. Using diatomite as wall material in the laboratory can not only decorate indoor environment and fine particles of negative ions generated on the interior of the diatomite with certain absorption to purify indoor air well.

The dust pollution caused by the construction is also an important reason for the increase of outdoor PM2.5 concentration. Ni (2014) think that the reduction of fine particles produced by construction can be improved mainly from the following aspects[7]:

Firstly, reasonable architectural function should be designed to avoid construction secondary pollution and reducing the potential waste of building materials.

Secondly, using environmentally friendly building materials to avoid dust generated during construction to aggravate the pollution of PM2.5 in the air.

Thirdly, promoting green construction with reasonable planning for the construction site. Appropriate water should be sprayed before and after the construction of the construction area to ensure the construction site humidity and reduce dust diffusion,

5. Conclusions
Comprehensive analysis above can be summed as the following. Through the layout design of the building and the construction of three-dimensional greening, it can reduce the concentration of fine particles around the building. The ventilation design of building space can control the concentration of PM2.5 indoors. For example, the atrium with hot pressing ventilation. Wind pressure ventilation is conducive to the rational arrangement and use of building space. Wind pressure can effectively increase indoor ventilation effect and reduce the aggregation of fine particles in the room. These methods are passive control of indoor and outdoor particulate matter concentration. In addition, the concentration of indoor particulate matter directly and effectively can be controlled through the construction machinery ventilation and indoor purification equipment combination of active control measures. However, active control measures increase building energy consumption relative to passive measures. In addition, the building water system. the construction of three-dimensional green similar and the negative ions produced can not only adsorb PM2.5 but also beautify the environment. The negative ion on people's physical and mental health has an important contribution. This is also an important part of architectural design. Overall, Passive control measures with green building design can effectively control the concentration of indoor air particulate matter. At the same time, the use of

natural ventilation principle greatly reduces the energy consumption of buildings and indirectly controlling the production of PM2.5 in the atmosphere.

Acknowledgement
This study is supported by the Research Funds of Key Laboratory of Heating and Air Conditioning, The Education Department of Henan Province (2017HAC204).

References

[1] Li Zhemin, Zhou Tiantian, Lin Han, Lu Yao. Discussion on architectural design and indoor PM2.5 control[J]. *Housing industry*. 2015 (7): 48-51(in Chinses).
[2] Wu Zhengwang, Ma Xin, Yang Xin. Comparison of PM2.5 and investigation of air pollution in the layout of the building of several haze weather conditions[J]. *Architectural practice*. 2013 (10): 46-48(in Chinses).
[3] Zhang Shenlin. Economic analysis of roof greening construction project in Beijing[D]. Beijing: Beijing Forestry University, 2010(in Chinses).
[4] Brown, G. Z, Dekay, M. Sekiguchi, T. Kline, J. Mhuireach, G. Bennett, S. Sun, Wind & Light: architectural design strategies (3rd ed)[M]. New Jersey: John Wiley & Sons, Inc.2014, p146-211.
[5] Zhang Rui, Tao Jing, Wei Jian, Rong Ma Yan, Li Ting, Wang Chunmei, Wang Chen. Study on indoor air PM2.5 pollution level and its distribution characteristics[J]. *Journal of environment and health*. 2014. 31 (12): 1082-1084(in Chinses).
[6] Wang Lu, Wang Xingjun. Preliminary study on City haze governance based on the concept of green building design strategies [J]. *Green technology*. 2015 (12): 261-262(in Chinses).
[7] Ni Yun, Bao Weizhe, Hu Zhenyu. Build green indoor environment against haze[J]. *Residential technology*. 2014. (9): 31-33(in Chinses).

IOP Conf. Series: Earth and Environmental Science **146** (2018) 012032 doi:10.1088/1755-1315/146/1/012032

Planning and Design of Urban Eco-agriculture Landscape with Health Care Benefits: A Case Study on Future Homeland Project

Xiaoxi Zhang[1], Xiangnan Zheng[2]

[1] School of urban design of Wuhan University
[2] Luojia Pleasant Landscape Planning and Design, Wuhan Ltd

1523338349@qq.com

Abstract. In modern urban life, people pay more and more attention to health. Eco-agriculture landscape with health care benefits has got academic attention. On the basis of experiences of developed countries, combing with regional characteristics & policy support of Wuhan, this paper summed up the main elements and planning methods of the urban eco-agriculture landscape, and it also draws up the comprehensive experience mode of the eco-agriculture landscape with health care benefits in view of the case study on future homeland project. These creative ideas for the planning and design could guide the practice and development of agricultural landscape in the future.

1. Agricultural landscape with ecological functions and its development opportunity

Ecological environment nowadays is increasingly being disrupted. People generally live in a complex, changing, disorder, high-pressure urban environment, leading to the impetuous tendency in society and multiplied sub-health population while the national economic growth gradually slowed down. Eco-agriculture landscape planning with health care benefits cannot only be a slogan and new landscape planning and design cannot focus only on the aesthetic research of visual appearance. Mood regulation and emotional release in high efficiency and scientific methods are the urgent tasks that landscape planning and design face in modern time[1].

Urban agricultural landscape with health and well-being benefits involves the agriculture of metropolitan area and its surrounding areas. It is a kind of regional economy, which is located around the urbanized area and the clearance between the area. It mainly relies on and serves the urban and rural people in the surrounding areas. Its main mission is to provide high quality agricultural and sideline products as well as healthy and beautiful ecological environment, which is mainly based on Non-staple food production project, environmental-ecological engineering, bioengineering and information engineering. On behalf of the extent of market economy development and the level of technological content, it is characterized by multi-level, multi-functional, and multi-directional opening, which is a sustainable development of the modern regional agricultural landscape[2]. Since the 1960s, the population has gradually increased, and the process of urbanization has been accelerated. Pollution-free, unexplored arable land is scarce while traditional agriculture is under great pressure. The world has focused on the research of new kind of agriculture[3]. The boundary between cities and countryside is gradually blurred. Urban ecological agriculture with health benefits in the modern city is the inevitable result of cultural, economic, social and technological development. This new agricultural system model takes the initiative to adapt to urban development, which is the link between urban and rural areas, and broaden the development of urban and rural space. It is a new agricultural landscape with marketization, intensification, industrialization. It has the functions of ensuring the basic living standards, developing the economy, balancing ecology, artistic appreciation, tourism & leisure, exporting, education demonstration and increasing farmers' income[4].

"To speed up the reform of ecological civilization system and build a beautiful China, promote green development, focus on solving environmental problems, intensify the protection of ecosystems, and reform the ecological environment supervision system," was stressed at the 19th National Congress of the Communist Party of China[5].Construction of ecological civilization is the Millennium Development Plan for the sustainable development of the Chinese nation, including that increasing the pilot units of ecological agriculture, improving the ecological civilization system with information technology integrated with the market, promoting green production and consumption and establishing a socialist ecological civilization. Pains for the moment, gains for the millennia[6]. Comprehensive three-dimensional analysis of the elements of agricultural landscape with health care benefits and its positive impact on human health demand immediate attention.

This paper analyses the basic characteristics of future homeland project, the features and needs of its clients, and lays the theoretical foundation for the later construction of agriculture landscape with health care benefits.

2. Eco-agriculture landscape technology and concept of health care benefits

The ecological technology adopted in the eco-agriculture landscape focuses on systematic research to reflect the main elements of health and art perception, such as ecology, function, volume, natural characteristic & diversity, pollution rate, cleanliness and other factors. The merits of planning are mainly related to geographical, terrain and human factor[7]. There are many kinds of applications of eco-agriculture landscape technology, but the most effective and convenient applications are mainly reflected in water resources, planting, materials, construction, processing and demonstration.

2.1. Ecological technology of landscape water resources

Water purification and water saving technology. Constructed wetlands and biological floating island are commonly utilized to purify the water. The constructed wetland mainly contains vertical flow constructed a wetland, subsurface flow constructed wetland and surface flow constructed a wetland, which is mainly used for sewage treatment[8]. Biological floating islands absorb and degrade the nitrogen, phosphorus and other elements in water through the growth process of plants, thereby effectively improving water quality. In the meantime, plant photosynthesis and transpiration regulate micro-climate of the water surface[9]. The method of water-saving mainly includes transdermal enhancement technique, aerial drainage technique, reclaimed water technique and drip irrigation technique, which are mainly implemented with raw materials of the garden and the planning and design of terrain slope and grassed swales.

2.2. Ecological technology of landscape planting

Crop rotation, intercropping and three-dimensional planting could balance the land nutrients, improve the ability of resistance to pests and diseases with the use of plant growth characteristics and their differences, resulting in a colorful and characteristic landscape. The widespread use of ecological pesticides and artificial release of natural enemies are beneficial to killing pests, balancing farmland ecological environment.

2.3. Ecological technology of landscape materials

Ceramic tiles are laid with non-slip landscape materials that are made from a reasonable ratio mix of industrial waste materials and other materials with adding adhesive, resulting in high porosity, good permeability materials after high-temperature firing. The permeable concrete is mostly composed of aggregate, cement, water, redispersible polymer powder and latex. It has the advantages of permeable, breathable, light-weight, antifreeze, high bearing capacity, beautiful and low maintenance cost. Plastic-wood composite material is a kind of new material extracted from waste sawdust and plastic with the features of corrosion-resistant, flexible and simple[10]. Solar photovoltaic glass involved in the photovoltaic process through the light, condenser, scattering, conductive and other properties, demonstrating high-strength, corrosion resistance, anti-aging advantages[11].

2.4. Ecological technology of landscape architecture
The roof garden is a garden on the roof of a building with the decorative benefit, which saves the building's energy consumption and adjusts the building's micro-climate[12]. Vertical green wall place plants on the indoor and outdoor surface of the building. It is fixed by the high-intensity grid, and actively adjusts the micro-climate for the indoor and outdoor environment[13]. The vertical farm makes reasonable and efficient, flexible use of indoor three-dimensional planting space and renewable energy & greenhouse technology, to simulate growth environment of crops, achieving efficient agricultural production and effectively reducing the cost of pesticide use and logistics[14].

2.5. Ecological technology of landscape processing
Agricultural internet of things is an information exchange system offering agricultural landscape product traceability service. It involves the cultivation, production, logistics, warehousing, wholesale, design, manufacture, purification, retail, sale, experience, feedback and other links and coordination of the entire information flow is a requirement of realizing the traceability service of the agricultural landscape. Timely optimization of agricultural landscape quality could increase agricultural income[15].

2.6. Ecological technology of landscape demonstration
IFLD system, a full-body motion sensor based Immersive Forest Landscape Display, integrates multiple Kinect somatosensory controllers and multi-channel seamless fusion projection systems and is applied to the agricultural landscape demonstration. It provides an agricultural landscape environment that visitors completely immersed in, and enable free dialogue between participants and the environment with LED discoloration lights and music, creating a new landscape interactive experience for visitors[16].

The future homeland project is located at Tianpu village and Penghua village, Jiangxia District of Wuhan city in Hubei province with Ezhou city in the east, Liangzi Lake in the west. It sites closer to the central districts of Wuhan, covering and attracting various industry. The number of university teachers and students around is stable with new students registering every year, thus the demand for the popular science of ecological agriculture is obvious, and the scholarly groups are generally easy to accept the lifestyle of landscape with health benefits. Eco-agriculture landscape technology is centrally used in future homeland project, including ecological technology of landscape water resources, such as artificial wetlands and floating islands, automatic irrigation system, which is efficient and beautiful; the whole set of ecological technology of landscape planting, such as the landscape art formed by crop intercropping while using a variety of ecological materials to save energy and regulate microclimate; the whole set of ecological technology of landscape architecture, such as agricultural networking system, biological purification, waste regeneration. The ecological technology displayed in the landscape can bring and improve fun and interactive experience for the visitors.

Rational use of ecological technology enhances the comprehensive appearance of the landscape. In the macro aspect, it creates forward-looking health care concepts and in the meso aspect, it put forward a comprehensive evaluation system and its application model while in the micro aspect, it promotes health concepts and improves the health and happiness index of the affected population[17](Figure 1-2).

IOP Conf. Series: Earth and Environmental Science **146** (2018) 012032 doi:10.1088/1755-1315/146/1/012032

Figure 1. The architecture and landscape, Aerial View

Figure 2. The future home landscape, Square perspective

3. The pattern and design of the eco-agriculture landscape.

On planning and construction of industrial parks of urban eco-agriculture landscape, in addition to emphasizing landscape art and agricultural products which represents geographical characteristics, it still need to analyse the successful experiences of ecological agriculture in western developed countries, including the aspects of ecological services, landscape values, nature conservation and cultural heritage and other multi-functional cross-integration. The agricultural system created in this way can greatly improve the quality of the agricultural landscape[18].

3.1. Pattern analysis of urban eco-agriculture landscape in western countries

France is the first European agricultural country emphasizing ecological agriculture planning in the first place. The concept of ecology and environmental protection is widely accepted among France. With stabilized eco-agricultural products market, it developed and implemented the Eco-agriculture development plan. Its main measures include the establishment of reasonable, flexible and detailed agricultural laws and regulations; to guide and strengthen consumption concept of eco-agricultural products. Germany is a big country of industrial production. In order to clear the pollution caused by industrial manufacturing, the mode of planting industrial crops is developed, and ecological agriculture is expanded vigorously too. Bioenergy crops are used to produce mineral energy and chemical raw material alternatives. A biological raw materials and bio-energy research center was established, which deeply studies the cultivation and new technologies of industrial crops and coordinates its promotion, thus providing sufficient new raw materials for the chemical and paper industry. On the basis of organic agriculture, the United States emphasized the new farming system of sustainable agriculture and put forward the model of sustainable eco-agriculture development. It was a kind of ecological recycling agriculture offering a comprehensive benefits package. Most of the agricultural systems emphasized the combination of agriculture and animal husbandry. it balanced the relationship between the aquaculture and planting industry in the feed, fertilizer and other aspects, and made them make use of each other[19]. Sweden attached importance to the research and development of high-quality food and its derivatives production; the Netherlands placed more emphasis on the eco-agriculture model of transformation of nature due to its natural conditions[20].

3.2. Comprehensive experience mode of urban eco-agriculture landscape

Modularized landscape infrastructure is highlighted in the pastoral farming system which utilizes the high-quality ecological technology to make the agricultural production become industrialized and stabilized with high-yield, and not is affected by the climate and the season. It innovates the design of agricultural and horticultural tourism and improves land use efficiency. Comprehensive experience pattern is mainly achieved by carrying out sightseeing tourism, science exhibition, shopping and other projects with the design of nursery stock garden, flower garden, Chinese herbal medicine garden, vegetable garden, fruits garden and so on. In the same time, it should make full use of universities and research institutions, cooperate with high-tech development zone and agricultural research units, and introduce and develop new technologies and varieties to provide technical support for modern agriculture or provide supporting facilities for the study of high technology. Community farms and public farms are constructed while professional tourism agriculture and experience marketing agriculture are developed as well. With the increase of sites for studying agronomy knowledge, eco-agriculture landscape with interactive experience is fully developed. To build a comprehensive agricultural landscape museum, the materials of the distinctive agricultural production, life, allusions, ancient villages and other materials with regional characteristics should be collected and displayed for visitors to facilitate their study and research[21].

3.3. Future development ideas of urban eco-agriculture landscape

The classic cultural characteristics in the planned landscape are deeply studied and classified. Taking ecological civilization construction on low-carbon as the main tone and emphasizing the artistic creative thinking, the urban eco-agriculture landscape is constructed with interdisciplinary theories to create several types of charms and positive & far-reaching significance[22]. Combining with modern new technologies, new ideas, new species, new materials, the new trend, it brings new ideas and fantastic imagination with a strong sense of the future to the ecological landscape by introducing the space theme, cartoon theme, science fiction theme, popular star theme, art theme, etc or the introduction of famous paintings and contemporary famous artists, and creates new ideas through the use of multi-styles. This model mainly highlights the artistic impression, and effectively enhances the region's fashion sense and advertising effect, resulting in good economic benefits and the development of new areas[23].

3.4. The application of future homeland mode and design

Based on the development of eco-agriculture landscape, future homeland project follows a harmonious and interactive development path with scientific agriculture, ecological farm, health resort , experience culture with the green technology as the driving force, ecological agriculture as support, tourism industry as agglomeration and cultural and artistic industry as soft power, forming an industrial system lead by four major industries, including agricultural cultivation, deep processing of agricultural products, tourism of ecological landscape and health resort, and cultural industry. Turning the development mode from agricultural development led by the primary and secondary industries into a new development path with primary industry, secondary industry and tertiary industry well-coordinated, a future homeland system has been built, interacted the advantages of agricultural science and technology, agricultural planting, deep processing of agricultural products, agricultural sightseeing, crops picking, tourism exhibition, garden culture, green food and health resort, etc. to cultivate a healthy and green industry chain, creating an ecological, scientific and exquisite industrial system with Chinese characteristics, contributing to the formation of model base of ecological agriculture for future homeland in Wuhan suburbs.

Design concepts include ecological harmony, maintaining the integrity of the park's natural ecosystem, respecting the natural properties of landscape vegetation, configuring the edible plants with the fine species of Wuhan city; circular economy, treating the resource flow of the whole park as a whole, establishing the eco-type recycling economy system based on resource recovery and recycling; green health, promoting green and healthy life with the establishment of ecology-friendly production system of edible landscape and the production and processing of agricultural products without chemical pesticides, chemical fertilizers, chemical preservatives and other synthetic substances or genetically engineered organisms and their products; intelligent technology, making full use of high-tech ecological garden technology with green soil and fertilizer, artificial fog, water purification, sewage purification system and the application of modern information technology, providing precision planting, visual management, intelligent decision-making on the grounds of emerging Internet, cloud computing and Internet of Things technology; low-carbon environmental protection, constructing the park with environment-friendly materials and technology to create low-energy, low consumption, low expenditure and other conservation-oriented production and lifestyle[24](Figure 3-4).

Figure3. Future Hall of Future homeland

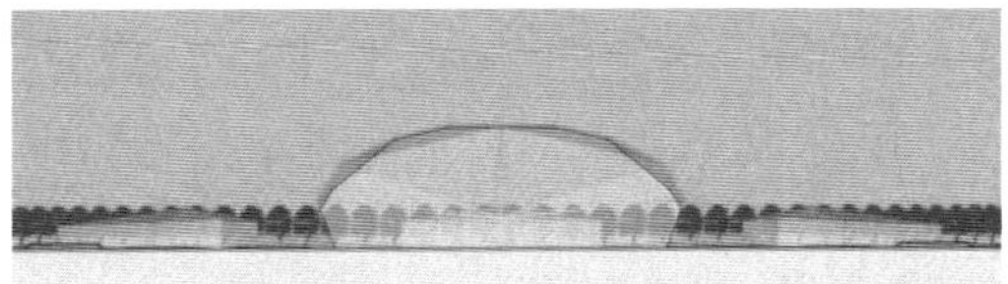

Figure4. Future museum profile

4.Conclusion

Eliel Saarinen once said: "The city is an opened book, from which you can see its ambition. Let me have a look at your city, I can find out the cultural pursuit of city residents."If we also treat the city as a book, then the eco-agriculture landscape is the preface of the book, representing expectations of various disciplinaries to the construction of most healthy and comfortable landscape. Urban residents are eager to have a pleasant space in public agricultural landscape. In this context, we must strictly follow the basic principles and ideas of ecological civilization and highlights the health care benefits with the most advanced technology available, thereby enhancing the happiness of the public[25].

References

[1] Yun X and Chen Z 2010 *Journal of Hunan Agricultural University(Natural Sciences)* Summary on the design of healing garden **S2** 83-85.
[2] Shuqin C 2005 *Journal of Anqing Teachers College(Social Science Edition)* Urban ecological agriculture and its development in China **1**104-106.
[3] Qingsheng Z 2002 *Bulletin of Agricultural Science and Technology* Urban agriculture——The development direction of Chinese urban agriculture in the 21st century **04** 4.
[4] Junzhe M 2004 *Journal of Beijing Agricultural Vocation College* Urban agriculture and urban agricultural industrial structure adjustment in Beijing suburbs **1**13-17.
[5] Http://Cpc.People.Com.Cn/19th/N1/2017/1019/C414305-29595273.Html.
[6] Http://Www.Sohu.Com/a/198804611_675552?Qq-Pf-to=Pcqq.C2c.
[7] Ying P, He X and Zhenrong Y 2009 *Chinese Journal of Applied Ecology* Beijing eco-agricultural landscape and aesthetic quality space evaluation **10** 2455-2460.
[8] Qing Z 2011 *WETLAND SCIENCE* Construction and application of artificial wetland **04** 373-379.
[9] Shanghai Garden Group 2009 *Garden* The application of "ecological floating island" technology in the post-beach park of landscape water system **12** 36-37.
[10] Zhongli M. 2016 The application of ecological techniques in modern agricultural landscape. Wuhan University
[11] Meijun X 2011 *GLASS&ENAMEL* Development of Solar Battery Glass and Film and Their Application **01** 44-47.
[12] Shizhang W 2007 Research on the roof garden design Huazhong University of Science and Technology
[13] Chunshui Z, Jian L and Chunlei T 2008 *City* Vertical Greening of Road Landscape - Application of "Green Wall" **10** 44-47.
[14] Jinghua W and Jingdun J 2013 *Journal of Agricultural Science and Technology* Pattern of Vertical Farm in Chicago and its Revelation **05** 75-79.
[15] Qing H, Chaoyuan C and Yun W 2013 *Computer Systems & Applications* Technical Analysis of the Internet of Things Applied to Traceability System for Agricultural Products **01** 44-47.
[16] Meng L 2013 A full-body motion sensor based Immersive Forest Landscape Display Beijing forestry university
[17] Hubei future home high - tech agriculture Co.LTD
[18] Ying P, He X and Zhenrong Y 2009 *Chinese Journal of Applied Ecology* Eco-agricultural landscape and aesthetic quality space evaluation in Beijing **10** 2455-2460.

[19] 2010 *Rual communication* The characteristics of foreign eco-agricultural development model **18** 38-39.
[20] Ruihe B 2014 *Country Agriculture Farmers(A)* The characteristics of foreign ecological agriculture **1** 31-33.
[21] Wenjun C 2005 *Economic Geography* Study on Urban Agriculture Tourism and Urban Sustainable Development - Taking Guangzhou as an Example **06** 915-919.
[22] Bin S 2007 Chinese and Overseas Architecture Superficial study on creative thinking of environmental art design **01** 63-65.
[23] Bingshe W 2006 *Journal of Xi'an University of Arts and Science(Social Sciences Edition)* Artistic Thinking **06** 1-5.
[24] *Wuhan Future Homeland Overall Planning* Wuhan University
[25] Yun O 2016 *The 7th city and landscape U+L new thinking international academic seminar* (Wuhan:Huazhong University of Science and Technology)

Study on Operation Mode of Anti-poverty Photovoltaic Project Based on Altruism Preference

Min Sun[1], Wei Zeng[2], Wenning Zhang[3] and Cunbin Li[4]

[1,2]Institute of Electric Power Science , Jiangxi Electric Power Corporation of State Grid, Jiangxi 330000, China
[3,4]School of Economics and Management, North China Electric Power University, Beijing 102206, China

Corresponding author e-mail: z_wenning@163.com

Abstract. The operation of anti-poverty Photovoltaic project puts forward higher requirements on multi-participates cooperation. The cooperative behavior among different subjects has altruistic motivation, so based on the stakeholder theory, this paper analyzed the stakeholders' rights and interests, establishes the operation mode of multi-participates collaborative optimization under the altruistic preference. This paper discusses and analyzes the influence of altruistic preference on project performance and project participants. Finally, the model is verified.

1. Introduction

With the increasing tension of traditional energy supply and the progress of industrial technology, countries in the world have increased efforts to research and develop sustainable energy such as solar energy, which has brought the vigorous development of photovoltaic market. The research of photovoltaic industry is emerging, such as technical issues, quality issues and management issues in the construction and operation [1], the organizational structure and risk problem of photovoltaic project construction and operation mode[2]. Through the analysis of the main operation mode of PV system at home and abroad, it is found that our country currently lack early financing and operation management.

Photovoltaic project has the characteristics of high economic benefits, good stability, energy saving and emission reduction effect, low cost of operation and maintenance, and has a great advantage in the poor areas[3]. At present, the research on anti-poverty Photovoltaic project in China is still lack. In order to study the collaborative optimization among different subjects, we must have a thorough understanding of the participants in the project. In 1984, freeman, an American economist, argued that stakeholders were broadly defined as "any individuals and groups that can influence the achievement of an enterprise, or that can be affected by the process of achieving its objectives"[4]. Using stakeholders to analyze the power and benefit of the project participants , build the power benefit matrix, can support the research of the operation model.

There are many participants in anti-poverty Photovoltaic project, how to improve the collaborative optimization of multi - agent project is an important problem that project managers need to solve urgently. Many experts work with the project team from the perspective of contract design. Holmstrom first proposed the issue of moral hazard in team cooperation through contract design [5].

IOP Conf. Series: Earth and Environmental Science **146** (2018) 012033 doi:10.1088/1755-1315/146/1/012033

Then many experts have further research on team cooperation on this basis, Yang [6,7] and others has been studied contract mechanism in the team cooperation, and proved the effectiveness of incentives. In recent years, behavioral economics and behavioral game theory developed rapidly, and the research found that the mutual preference between subjects can promote the cooperation of the team to some extent. Many scholars have also made research on the influence of altruism on team collaboration. For example, Mifune, through economic experiments, show that altruistic preferences have a positive impact on the reputation of the team [8]. Han Jiaojie [9] studied the cooperative behavior of multi-participates, and constructed the project decision model considering altruism preference. Altruistic preference is seldom used in the anti-poverty Photovoltaic project of multi-participates. Therefore, this paper combines the characteristics of photovoltaic power generation project and the related research results of altruism, constructs the team cooperation model of project owners and multiple participants, and analyzes the increased effort and the income of each subject under the altruism preference (including direct income and indirect benefit). It provides theoretical support and policy suggestions for how to improve the altruistic preference of the project participants and enhance the team cooperation.

2. Analysis of stakeholders in the construction and operation of anti-poverty Photovoltaic project

2.1. Stakeholder theory

The development of anti-poverty Photovoltaic project involves many stakeholders, and the stakeholders of the project participate in the project in the form of different roles. At the same time, there is a large number of network relationships among the stakeholders in the process of the distributed photovoltaic project construction, the action strategies of stakeholders influence other stakeholders, and the results are affected by other stakeholders. This network relationship and its structure affect the effectiveness of stakeholder management.

Based on stakeholder theory, stakeholders can be divided into four types by analyzing the rights and influence of each stakeholder in the project. according to the stakeholder matrix (figure 1), the key stakeholders of the project are located in the upper right corner of the matrix, it plays a key role in the success of the project. The upper left of the matrix has great power and small interests, they have great power in the decision of the project, but because of the small interests, there is a lack of interest in the development of the project; The stakeholders in the lower right side have large interests and small influence. The development of the project has great influence on their interests. they are also interested in the development of the project, but they do not have enough capacity and resources to promote the development of the project; The stakeholders in the lower left corner of the matrix have little power and interest, and in the process of the project, organizations often pay little attention to such groups, in contrast, their efforts are also the smallest. Due to the different resources of possession, the role of various stakeholders in the project can also be different. Correct understanding of the power and interests of various stakeholders in photovoltaic pro-poor projects is of great significance for the rapid and effective implementation of photovoltaic poverty alleviation projects.

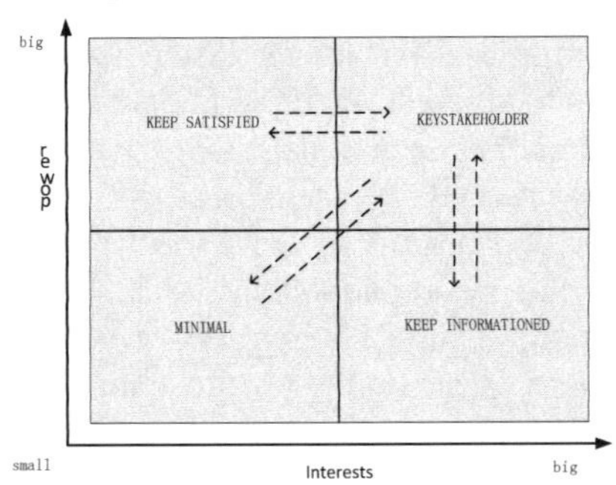

Figure 1. Power benefit matrix

2.2. Analysis of the rights and interests of stakeholders

The construction of anti-poverty Photovoltaic system needs the cooperation of all parties to complete. Speeding up the construction speed of photovoltaic power generation system needs to mobilize the enthusiasm of all parties to ensure the reasonable income of all parties. Professor Ding proposed the identification of project stakeholders through the construction of task dimension, process dimension and role dimension, and established a three-dimensional model. In this paper, through the three-dimensional model, identify the stakeholders. The main stakeholders of the anti-poverty Photovoltaic system are photovoltaic units, grid companies, government, bank lenders, photovoltaic component producers, third-party supervision, third-party certification, roof owners, etc. Each party has its own interests in the project implementation of photovoltaic power generation system, at the same time, the distributed photovoltaic project construction stakeholders also include the construction party, environmental organizations, the public and so on.

Table 1. power benefit analysis of stakeholders

stakeholders	NO	power	interests
project-owner	1	power generation capacity	sell electricity , get income
grid company	2	dispatch and distribution	get subsidies and good reputation
lenders	3	provide loans, supervise	Loan interest income
government	4	policy guarantee	adjust the energy structure, alleviate crisis, get good reputation
construction	5	engineering construction	get construction benefits
component producers	6	provide cells, components	increase sales and revenue
supervision	7	supervise construction	get profit from supervision
certification	8	assess quality of products	get profit
roof owners	9	have the roof, install the photovoltaic equipment	income from rent or electricity
the public	10	use of clean energy	a good living environment
Social group	11	promote use of clean energy	protect environment

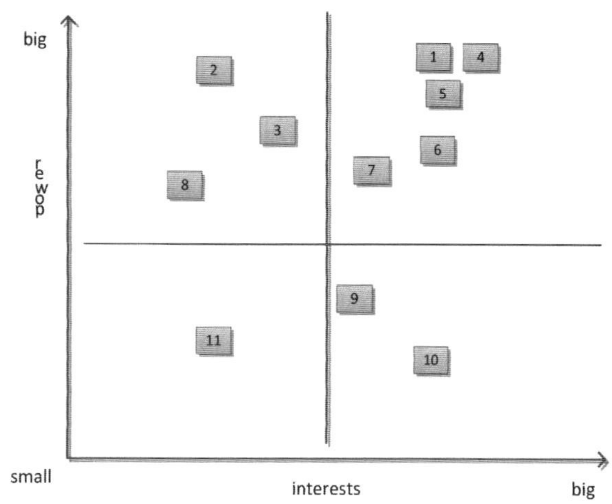

Figure 2. Power interest matrix of stakeholders

In the construction and operation of anti-poverty photovoltaic project, stakeholders want to take different measures to maintain their own interests due to their own resources and location, when these measures are inconsistent, to realize the cooperation optimization of stakeholders cooperation, we must clarify the position of the project stakeholders in the power benefit matrix, clarify roles and status, so as to establish the construction and operation mode of the project. As shown in figure 2 and table1, the main stakeholders of the anti-poverty photovoltaic project and their location in the power benefit matrix are analysed through an investigation of the relevant experts.

3. Multi-agent cooperation model of project based on altruism

3.1. Model hypothesis

Photovoltaic project is generally composed of a project owner and N independent organizations, such as the analysis of the stakeholders, they are independent stakeholders, their influence and interests in the project are different, to maximize their own interests as the goal. Suppose that the i body is called the body i, then $0 \leq i \leq 10$.

In the project, different subjects have different resources and advantages, the completion of the project needs the mutual cooperation among the various subjects, each subject needs to provide help to other stakeholders besides doing his own work.

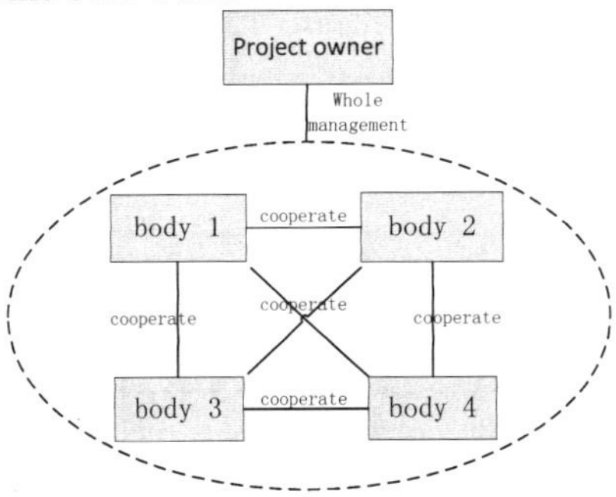

Figure 3. Multi-agent participation in photovoltaic project cooperation mode

Assuming that body i work in the project can be divided into production effort and cooperative effort, respectively by means a and b, a represents the production effort of body i, and $0 \leq a_i \leq 1$,b_{ij}represents the cooperation effort of body j, and $0 \leq b_{ij} \leq 1 (i \neq j)$, each subject also pays the cost, including the production effort cost and the cooperative cost. Suppose the cost function of body i is $C(a_i, b_j)$, and $C'(a_i, b_j) > 0$, $C''(a_i, b_j) > 0$,the cost function of the subject i can be expressed as:

$$C(a_i, b_j) = \frac{1}{2} c_i (a_i^2 + b_i^2) \quad (1)$$

Among them, c_i is the cooperative effort of the cost coefficient[10].

Referring to the collaborative output model [11] between subjects with different power and interests, the output function π of the body i can be expressed as:

$$\pi_i = \theta a_i + (1-\theta) \sum_{j=1, j\neq i}^{N} b_{ji} + \gamma a_i \sum_{j=1, j\neq i}^{N} b_{ji} + \xi_i \quad (2)$$

Among them, θ and $1-\theta$ are the influence factors of the main output efforts and other subjects' cooperation efforts on their output, and $0 < \theta \leq 1$, when $\theta = 1$, indicating that the subjects are independent, and the cooperation efforts of other subjects can not increase the output of the main body. conversely, when$\theta \to 0$, the influence of other subjects has great influence on the main output, and also indicates that the work of the main body has a higher demand for cooperation among other subjects. γ is potential impact of collaboration between subjects on output. ξ_i is a natural environment variable, showing the influence of external environment on the output of body i, $\xi_i \sim (0, \sigma^2)$.

As the contribution of each subject to the output of the project is different, such as the rights and interests of each subject on the project directly affects the contribution rate, the contribution degree of the main body i is k_i, $0 < k_i < 1$, and $\sum_{i=1}^{N} k_i = 1$, so the total output of the project is:

$$\mathrm{\pi} = \sum_{\mathrm{i}=1}^{\mathrm{N}} \mathrm{k}_{\mathrm{i}}\mathrm{\pi}_{\mathrm{i}}. \quad (3)$$

3.2. Model building

1) the utility function of the project owner

This paper assumes that the project owner aims at maximizing the benefits of the project, and its income is affected by two factors: the project output and the main profit sharing. under the risk neutral condition, it can be expressed as

$$V_o = (1 - N\beta)\pi \quad (4)$$

Among them, β is the profit impact coefficient of the team output for body, that is, the profit sharing coefficient, $0 \leq \beta < 1$.

The expected utility is equal to the expected return, so the expected utility of the project owner is

$$\mathrm{EU}(V_P) = (1 - \mathrm{N\beta})\mathrm{E}(\mathrm{\pi}) = (1 - \mathrm{N\beta}) \sum_{i=1}^{N} k_i \, E(\pi_I) \quad (5)$$

2) the utility function of the other body

In this paper, the altruism - utility function has been studied by many scholars at home and abroad, and many scholars have expressed the utility of altruism as a combination of their direct utility and the utility of altruistic behavior, such as Levine[12]and yang chunxue[13], so the altruism can be divided into altruism to their own utility and altruistic behavior, and the latter can directly use altruistic behavior to contribute to the overall benefit of the project team. The direct utility of body i is expressed by μ_i and μ_j.and $\mu_i$is the expected return of body i , it can be expressed as:

$$\mu_i = a_i + \beta E(\pi) - \frac{1}{2} c_i (a_i^2 + b_i^2) \quad (6)$$

$\mu_{ij}$is the cooperative effort of the main body to bring the utility of the other subjects as well as the project team. Body does not know the action selection strategy of the other subjects when the body selects the action strategy, it assumes that the other subjects will pay a greater amount of production effort(a_j=1). thus the utility μ_{ij} produced by the main part I can be expressed as:

$$\mu_{ij} = (1 - \theta) b_i + \gamma b_i \quad (7)$$

Further, the expected utility function EU_i of the body i can be expressed as:

$$EU_i = (1 - \varepsilon_i) u_i + \varepsilon_i u_{ij} \quad (8)$$

For the altruism preference coefficient of the body i, satisfying $0 \leq \varepsilon_i \leq 1$. in particular, when $\varepsilon_i = 1$, body i is completely altruistic; When $\varepsilon_i = 0$, the agent i was completely selfish.

3) multi - agent cooperation model

In the multi - agent project team cooperation, the goal of the project owner is to maximize the individual income, maximize the output efforts and cooperative efforts. The owner and the individual entities in different decision-making levels, have their respective objective function and decision variables, and the project owner decision will directly affect other subjects' behavior selection, so the cooperation model is a double-level programming model. The multi - agent cooperation model of the project team under altruistic conditions can be expressed as:

$$\max_{a,b,\beta} (1 - N\beta) E(\pi) \quad (9)$$

$$(\mathrm{IC})_1 \;\; a_i \in \arg\max_{a_i} EU_i \quad (10)$$

$$(\mathrm{IC})_2 \;\; b_i \in \arg\max_{b_i} EU_i \quad (11)$$

when other relevant conditions such as profit sharing are not changed, the performance of the project team under the altruistic preference conditions is significantly higher than the project team performance under pure selfishness[9].

Therefore, from the above altruistic preference model, we can know that: (1) The increase of altruistic preference increased the cooperative effort of the participants, which increased the input of the project team. (2) The cooperation effort of each subject consumes itself cost, but it also increases the output of other main body, so the output of subject I is related to the assistance and cooperation of other subjects in the project besides the production effort. (3) The increase of the performance output of the participants is greater than the cost of the project team, and then the total output of the project is increased. This shows that the increase of altruistic preference will increase the performance output of the project team. Therefore, for the project owner, incentive and carry forward the altruism behavior of the project participants will benefit to improve the project performance and increase the project output.

3.3. Analysis of an example

In this paper, the various variables of altruism preference are more abstract, and in many cases it is difficult to measure, so the empirical test method chosen in this paper is not to calculate directly the variables involved in 3.2. But taking a small photovoltaic power station as the research object, in a certain period of time, the independent mode, cooperation mode, and cooperation mode with altruistic preference are simulated, and the income of each participant and the project owner are compared in three cases.

Assuming that in the independent operation mode, the small photovoltaic power station, in addition to the project owner of the main involved in the main photovoltaic operators and users, cooperative mode of participation will increase the energy storage operators. The capacity of photovoltaic power station is 200 kw, and the energy storage system is 50KW·h, and the inverter is 65kw.

The load and photovoltaic output of the typical day are shown in fig. 4, the daytime is in the trough, at night is in the peak of electricity, and the photovoltaic output is greater than the load at 9: 00 - 17: 00., there will be a small peak. In the model, the electricity price of power grid is 0.6 yuan/(KW.h), the photovoltaic price is 0.85 yuan/(KW·h), and the photovoltaic subsidy is 0.42 yuan/(KW·h).

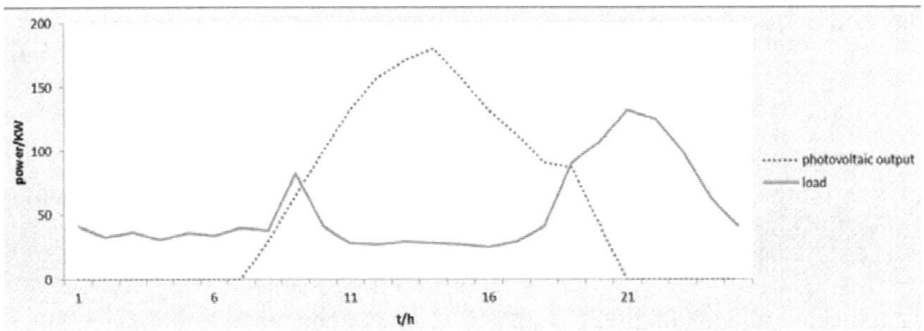

Figure 4. Power curve of photovoltaic output and load

In the independent mode, photovoltaic operators' output is all connected to the net , users directly buy electricity from the grid, energy storage operators used to ensure the reliability of the grid, without direct economic benefits. In the mode of cooperation, photovoltaic output first supply customers, the surplus photovoltaic energy storage equipment connect to the net after the charge; Energy storage operators use off-peak power, send out electricity peak power, energy storage surplus photovoltaic supply users; Users will give priority to photovoltaic output, followed by the energy storage system. In the altruistic cooperative mode, photovoltaic operators and energy storage operators have altruistic preference.

The proceeds of the three subjects in the independent model are[14]:

$$C_{PV.0} = \sum_{t=1}^{T}((V_e P_{PV}(t) + V_{PV} P_{PV}(t)) \quad (12)$$

$$C_{BS.0} = \sum_{t=1}^{T}((V_{i,p}(t) P_{b,f}(t) + V_{i,o}(t) P_{b,c}(t)) \quad (13)$$

$$C_{U.0} = \sum_{t=1}^{T}(-V_i(t) P_d(t)) \quad (14)$$

$C_{PV.0}$,$C_{BS.0}$, $C_{U.0}$ are the typical daily benefits of photovoltaic operators, energy storage operators and users under the independent mode; $P_{PV}(t)$, $P_d(t)$ are photovoltaic output and load power, respectively. $P_{b,f}(t)$,$P_{b,c}$ are energy storage charging and discharge power, respectively. V_{PV}, V_e, $V_i(t)$ are photovoltaic power tariff, grid power price.

The income of the three subjects in the cooperation mode is:

$$C_{PV.1} = \sum_{t=1}^{T}((V_d P_d(t) + V_i P_{d,PV}(t) + V_e P_{b,c,PV}(t) + V_{PV} P_{PV}(t)) \quad (15)$$

$$C_{BS.1} = \sum_{t=1}^{T}((V_i P_{b,dc}(t) - V_e P_{b,c,PV}(t) - V_i P_{b,c,C}(t)) \quad (16)$$

$$C_{U.1} = \sum_{t=1}^{T}(-V_i(t) P_d(t)) \quad (17)$$

$C_{PV.1}$, $C_{BS.1}$, $C_{U.1}$ are the proceeds of the three main bodies of the mode of cooperation. $P_{b,c,PV}(t)$, $P_{b,c,C}(t)$, $P_{b,dc}(t)$ are the charge power of photovoltaic and grid to energy storage equipment. $P_{d,PV}(t)$ is the power of the photovoltaic and energy storage equipment to the load; $P_d(t)$ is the photovoltaic power.

In the model of altruistic cooperation, the three participants of the project are not all altruistic preferences, this paper assumes that the photovoltaic operators and energy storage operators have altruistic preference, the contribution of photovoltaic operators is higher than the energy storage operators, the optimal production efforts and cooperative efforts are: $a^1 = 0.7$, $b^1 = 0.2$, $a^2 = 0.8$,$b^2 = 0.3$, and the contribution of the photovoltaic operators, energy storage operators, users are k_1=0.6, k_2=0.3, k_3=0.1. $\theta = 0.5, \gamma = 0$.

The results of the profit comparison of the independent model and the altruistic cooperation model are shown in table 4.

Table 2. Income comparison of various models

independent mode	cooperation mode	altruistic cooperative mode
297.35	349.36	379.23

As can be seen from table 4, the total income of the three participants in the altruistic cooperative mode is higher than the total income under the independent mode, which is higher than the total income in the cooperative mode.

4. Conclusion

First of all, this paper have the stakeholders analysis of the anti-poverty photovoltaic project and build the stakeholder matrix. Next, this paper introduces the altruism preference to the multi - agent participation in the anti-poverty photovoltaic project, constructs the multi - agent cooperation model of the project owner and other relevant subjects, and analyzes the production effort, cooperative effort, the main output of the altruistic preference team and the output of the whole project. The analysis of the altruism behavior of the participants will improve the project performance and increase the project output. The results are tested by an example. The results show that the increase of altruistic preference will increase the cooperation efforts of the main body and promote the cooperation among team members. The total income of the project under the altruistic cooperation mode is obviously higher than the total project income under the independent model.

Based on the above conclusions, this paper puts forward the following suggestions for the multi - agent cooperation of anti-poverty photovoltaic project:

1) The project owner shall identify and select the main body with altruistic preference to form the team. Because altruistic people will improve their own efforts to improve the output of the project.

2) The spirit of collectivism and team cohesion are the key factors affecting cooperation. Project owners can enhance team cohesion by organizing competitions, competitions and so on, enhance the collective sense of honor of the project team, and improve the social relations among the teams by organizing various amateur activities, enhancing the altruistic preference of team members, enhancing team cooperation and enhancing team performance.

3) In order to promote cooperation better among team members, in addition to material incentive, staff training can reduce the cost of team cooperation, improve the quality of team members, so as to improve the enthusiasm of team members.

4) Altruistic preference may reduce the production efforts of team members, so the client encourages team members to help and cooperate with each other, and to ensure the production level of team members too.

5) Increase the transparency between the main bodies, arrange the action sequence of each subject, and ensure that the latter is mutually beneficial when the main body is not acting simultaneously, so that the efficiency of team production can be improved.

This paper analyzes the impact of the stakeholders and the altruism preferences on the project multi - agent cooperation, and provides theoretical support and suggestions for the project team cooperation.

However, there are a series of deficiencies: how to quantify the theoretical model parameters is still a question, it is difficult to test the model, so the test of the model results only involves three participants, the case study only from the macroscopic angle and not in-depth, so it need follow-up study to further improve.

Acknowledgments
This work was financially supported by national grid corporation science and technology project financing "research and application of improving the absorptive capacity and support technology of photovoltaic power grid in poverty alleviation area"(project no: 5218201700W).

References

[1] G.M. Chen, S.H Ma, Research on quality assurance system of distributed photovoltaic power generation project in China. Solar Energy, 2011 (17):6-9.
[2] X.J.Fan, Research on operation mode of solar photovoltaic grid-connected demonstration project based on contract energy management, Wuhan: Wuhan University of Science and Technology, 2012.
[3] Y, Ma, Z.R. Cao, Innovative photovoltaic pro-poor business model to create green ecological model city, Reform and Open, 2016, 7: 41 - 42.
[4] R.E. Freeman , Strategic management: A stakeholder approach [M]. Boston: Pitman, 1984.
[5] B. Holmstrom, Moral Hazard in Teams, The Bell Journal of Economics, 1982, 13(2): 324–340.
[6] T. Yang, C.W. Chen, An incentive pay system for project management based on responsibility assignment matrix and fuzzy linguistic variables, Expert Systems with Applications, 2009, 36: 12585–12591.
[7] L.D. Qiu, S.Wang, BOT projects : Incentives and efficiency, Journal of Development Economics, 2011, 94(1):127–138.
[8] N. Mifune, H. Hashimoto, T. Yamagishi , Altruism toward in-group members as a reputation mechanism. Evolution and Human Behavior, 2010, 31: 109–117.
[9] J.J. Han, G.H. Zhou, Y.L. Li, Multi - agent cooperative behaviour of project team based on altruistic preference, System Engineering Theory and Practice, 2013, 33 (11): 2776 - 2886.
[10] B. Holmstrom, P. Milgrom, Aggregation and linearity in the provision of intertemporal incentives, Econometrica,1987, 55: 303–328.
[11] T. Kretschmer, P. Puranam, Integration through incentives within differentiated organizations, Organization Science, 2008, 19(6): 860–875.
[12] D.K.Levine,Modeling altruism and spitefulness in experiments, Review of Economic Dynamics, 1998, 1(3):593–622.
[13] C.X. Tang,The pursuit of altruistic economists, Economic Research, 2001 (4): 82 – 90.
[14] N. Zhou, Multi - agent cooperative operation mode and benefit allocation , Grid and Clean Energy, 2016, 5 (32): 133 - 141.

IOP Conf. Series: Earth and Environmental Science **146** (2018) 012034 doi:10.1088/1755-1315/146/1/012034

Assessment of Wind Power Integrated to the Power Systems Using Probabilistic Load Flow Based on PFV with LS

E. Mohammed[1], J. Yu[1], S. Wang[1] and L. Peng[1]

[1]School of Electrical Engineering and Automation, Harbin Institute of Technology, Harbin 150001, China

moheissa183@gmail.com

Abstract. Power flow analysis (PFA) is one of the most important issues of power system analysis and design. This paper aims at developing and applying a high-speed sampling method to produce large-capacity sample set based on pattern feature vector with layer structure (PFV with LS) of IEEE 14-bus system data analytical methods. In this study, sampling based pattern vectors are constructed in three layers. The first and second layer vectors are very low dimensional, and the sampling of the third layer is in parallel for each local area. Samples generated by this sampling method can be applied to the calculation of probabilistic load flow (PLF) and probabilistic static security assessment. Simulation results show that the proposed method can improve efficiency of PLF analysis and probabilistic static security assessment. The advantages of using PLF based on PFV with LS as the power flow limit can minimize the complicated mathematical equations. Furthermore, the algorithm is very simple and accurate, especially when the system connected to the wind power.

1. Introduction

In recent years, power system operation faces a new challenge due to deregulation and restructuring of electricity markets. The PFA is necessary for planning, operation, economic scheduling and exchange of power between utilities. This paper extends the application of the security assessment index. It aims to show the detailed mathematical model of the uncertainties for the grid connected wind farms. Power grid contains other power generation and wind power uncertainties. However, the effects of the uncertainty can be reduced by accurate calculation.

Power flow (PF) is the very important tool for the analysis power systems, and it is used in operational and planning. The objective of PF is calculating unspecified bus voltage angles and magnitudes, active and reactive powers, as well as line loadings and their associated real and reactive losses for certain generation and load conditions. Different mathematical techniques have been used to examine PF analysis. These techniques are Newton-Raphson (NR), Fast-Decoupled (FD), and Gauss-Seidel (GS) methods.

One of the best PF methods is NR method. The NR process usually converges faster than other methods, but it takes longer computational time per iteration. However, the proposed method in this paper is a concern in accounting for any kind of limit generation, not only line and curve but also the combination of line and curve with discontinuities.

The load flow (LF) studies are performed for power system planning, operation, and control. LF studies data is also used for contingency analysis, outage security assessment, as well as for optimal dispatching and stability. The LF problem has received more attention than all the other power system problems combined [1]. The LF studies are used to ensure that electric power transfer from generators to consumers through the grid system is stable, reliable and economic.

LF calculation is a standard tool used to analyze the operation and planning of power systems. Normally, a LF calculation solves system states and power flows for a given fixed value of power generations and load demands. This type of calculation is referred to as deterministic LF (DLF). A DLF calculation does not consider uncertain phenomena in power systems, such as generator outages and load variations. Furthermore, modern power systems are commonly integrated with renewable energy based DG units. These renewable generations, such as wind power and photovoltaic generation, are intermittent and not fully controllable due to the stochastic nature of their prime sources.

The PLF analysis technique based on the static security assessment is considered as important to the operation and planning of power systems. PLF technique is an important probabilistic analysis tool for power system, historically suggested by the Borkowska in 1974 [2]. PLF can be used to quantitatively evaluate the impacts of various uncertain factors on PF when the system connected to the wind power is applied. Considerable work related to PLF has been carried out.

Probabilistic reliability techniques are required to model the impacts of wind power on system's reliability and adequacy [3]. In fact, the correlation between wind power and load directly influences the reliability of system, and for the installation of a new wind farm, the minutely correlation between wind power and load plays a key role in supplying peak load [4]. The static method is widely analyzed for voltage stability assessment when large demand and limited network expansion occur. The relationship between transmitted power and receiving end voltage (PV) curve is one of the powerful tools for stability analysis [5-6]. The probability density functions of loads at different buses show a number of variations and cannot be represented by any specific distribution [7].

Voltage stability is one of the most important and interesting topics for transmission networks. It is defined as the ability of a power system to maintain load voltage magnitudes after disturbances or uncertainty [8]. The increased uncertainty caused by wind power and intermittent energy behavior, it is hard to predict the system operational status. The PV stability analysis should consider how to accurately express the system's risks. Several researchers have studied the risk-assess based voltage stability problems using probabilistic methods [9].

The basic task of PLF analysis is to obtain the statistic or probability-distributions of the PF responses from the probability-distributions of stochastic input variables such as loads, PV, and wind power generations. The PLF methods can be classified into two types, first one Monte Carlo simulation (MC), and the second analytical methods. On the other hand, the MC technique is regarded as a benchmark to assessing the accuracy and effectiveness of other approximate PLF analytical methods [10-11]

Connecting wind farm to the power network increases the variability and uncertainty of power grid. Therefore, applying accurate calculation techniques, such as probabilistic static security assessment, static security analysis method of a power system with large scale of wind power, and stochastic response surface method, is essential to assure system stability [12-14]. To deal with the correlations between the stochastic input variables in the PLF based on PFV with LS analysis. Therefore, the main contributions of this paper, which are not considered in previously published works, can be listed as follows.

- A new PLF method based on PFV with LS is proposed. It can model stochastic input variables following both normal distributions as loads and non-normal distributions as wind power.
- The correlations between all the stochastic input variables in the PLF based on PFV with LS analysis are considered.
- The correctness and effectiveness of the proposed method are demonstrated by comparing the probabilistic static security assessment power flow results for the IEEE 14-bus systems obtained using the proposed method, the amplitude of the bus voltage combined with PLF by traditional load flow method.

2. Steady state analysis

2.1. Overview of Newton-Raphson load flow equations

In this paper, we used base LF equations for the process of calculation, and the obtained results are compared with those of the proposed method, the exponential model for representation the dependence of active power, and reactive power, on the bus voltage magnitude at a load bus in an electric power network. In conventional LF studies the general equations normally solved are:

$$P_i = f_i(\delta, V) = V_i \sum_{j=1}^{n} V_j (G_{ij} \cos(\delta_i - \delta_j) + B_{ij} \sin(\delta_i - \delta_j)) \tag{1}$$

$$Q_i = g_i(\delta, V) = V_i \sum_{j=1}^{n} V_j (G_{ij} \sin(\delta_i - \delta_j) - B_{ij} \cos(\delta_i - \delta_j)) \tag{2}$$

- General power balance equations as:

$$P_i = \sum_{j=1}^{n} |V_i||V_j|(G_{ij} \cos\delta_{ij} + B_{ij} \sin\delta_{ij}) = P_{Gi} - P_{Di} \tag{3}$$

$$Q_i = \sum_{i=1}^{n} |V_i||V_j|(G_{ij} \sin\delta_{ij} - B_{ij} \cos\delta_{ij}) = Q_{Gi} - Q_{Di} \tag{4}$$

- Numerical iteration as:

The set of LF equation (1) & (2) are non-linear and solved by the NR iterative method, which requires finding a Jacobin matrix to update the current estimates of improved solutions. The general forms of the iteration for the load flow analysis are as follows:

Update values = old value + iteration matrix * error

$$p^{(k+1)} = p^{(k)} + A^{-1}[y - f(p^{(k)})] \tag{5}$$

$$\Delta p_i^{(k)} = \Delta c^k \Big/ \left(\frac{df}{dp}\right)^k \tag{6}$$

The variables that need to optimize from equation (1) & (2) are δ and |V| so the general forms of the iterations equation can be written as follows:

$$\delta_i^{(k+1)} = \delta_i^{(k)} + \Delta\delta_i^{(k)} \tag{7}$$

$$\left|V_i^{(k+1)}\right| = \left|V_i^{(k)}\right| + \Delta\left|V_i^{(k)}\right| \tag{8}$$

Then the above equations can be written as:

$$\begin{bmatrix} \Delta\delta \\ \Delta|V| \end{bmatrix} = \begin{bmatrix} J_1 & J_2 \\ J_3 & J_4 \end{bmatrix}^{-1} \begin{bmatrix} \Delta P \\ \Delta Q \end{bmatrix} \tag{9}$$

$$\Delta P_i^{(k)} = P_i^{sch} - P_i^{(k)} \tag{10}$$

$$\Delta Q_i^{(k)} = Q_i^{sch} - Q_i^{(k)} \tag{11}$$

The J_1, J_2, J_3 and J_4 actually are related to kind of buses, for PV-buses the variable need to optimize are δ and Q. Figure 1 shows the flow chart of the NR based PLF. The LF calculations by NR method are embedded within a cycle that simulates many load scenarios for different realizations.

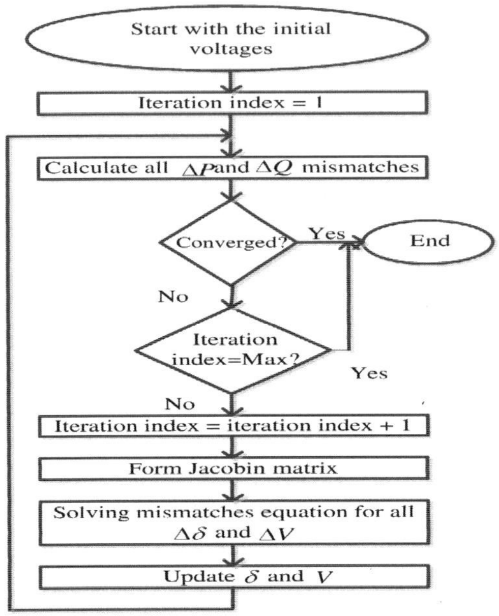

Figure 1. Flowchart of the Newton-Raphson based load flow calculation.

2.2. Probabilistic load flow analysis

This section contains two parts about the PLF analysis. The first one focuses on describes the techniques for PLF calculations. First, a short review of PLF methods is provided. Then, the MC based PLF method is discussed, followed by the convolution based and cumulates based PLF methods. The IEEE 14-bus system is used to demonstrate the different methods. The second part of the section investigates the influence of temporal correlation in stochastic generation and load on the accuracy of PLF based on PFV with LS results.

The solution to a PLF calculation can be obtained in three ways: a numerical approach, an analytical approach and a combined approach. The main advantage of the numerical approach is that nonlinear LF equations can be directly used in the PLF analysis, whereas the disadvantage is that a MC simulation is usually very-time consuming. On the other hand, the numerical approach is much faster in computation speed. However, the main issue is that it involves complex mathematical computation, which usually requires linearization of the LF equations.

2.3. PLF based on the PFV with LS method

The PLF analysis method based on the PFV with LS is proposed in this paper. First, stochastic input variables including loads P_L, *PV* power outputs P_P, and the outputs of wind power P_W are expressed by standard random variables. Second, traditional load flow method is used to represent the PF responses using the standard random variables. Third, the unknown coefficients in the traditional LF are determined by several deterministic PF. Finally, stochastic samples, statistics and probability distributions of the PF responses are established.

The input variables such as: PV generations, wind power, and loads follow different probability distributions. Gaussian, and normal distributions are usually used to model the randomness of P_P, P_W, and P_L, respectively. The inputs in the PLF based on PFV with LS must be standard random variables.

3. Methodology

In this section, we will describe the detailed procedure of the new approach for pattern feature vector with layer structure (PFV with LS), probabilistic load flow (PLF) distribution of power that considers load and wind power generation uncertainties. The simulation data used are the IEEE data test 14 buses. To show the advantages of using PLF based on PFV with LS, the load data each bus is increased.

3.1. Procedure of the proposed method

Based on the above subsections, the overall procedure of the proposed method for PLF based on the PFV with LS can be summarized as follows:

- Firstly, the PFV is composed of total load power of the system, so its dimension is very low.
- Secondly, the PFV dimension equals to the number of targeted sub-systems, and its value is relatively low; the PFVs for the second layer and the third layer are respectively composed of the power ratio coefficients of the second layer to the first layer and of the third layer to the second layer.
- In addition, the PFV for the final can be processed in parallel by decomposing the sub-systems of the second layer into multiple sub-vectors. Thus, the low-dimension of vector and parallel processing of sub-vectors enable high sampling efficiency.

The proposed method is tested by simulation on IEEE 14-bus corrected system, and problem of probabilistic static security assessment of the power system with wind power integration was analyzed in terms of bus voltage, branch current values, CDF, etc. The framework of the proposed assessment based PLF method is shown in figure 2, which shows the process of classification and steps of security assessment from PLF. The 1st and 2nd steps are data processing; the 3rd and 4th steps are using approaches; the 5th to 8th steps assess static security, and the 9th step displays the results.

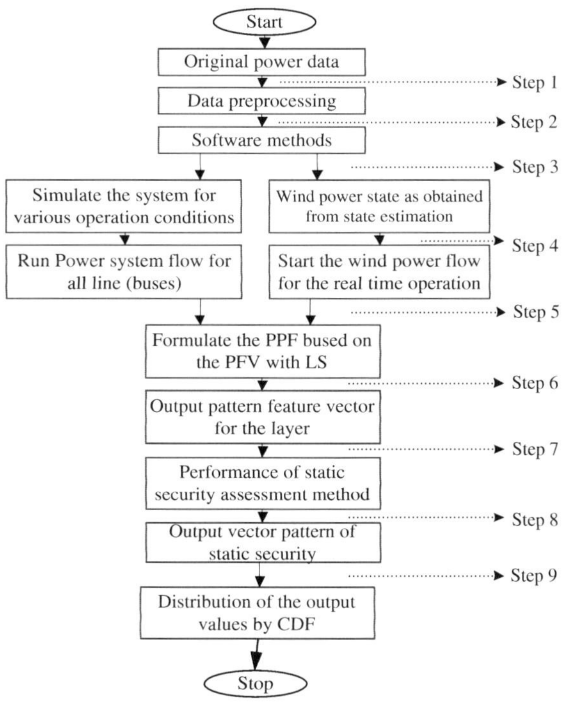

Figure 2. Framework of the PLF method process of security assessment.

4. Case study

The verification of the proposed method in this paper includes three parts. The first proposed method, then the PPF by traditional load flow, and finally the base cases are applied to the PLF based on PFV with LS of the IEEE 14-bus test system. The simulation results of the three methods are compared to demonstrate the accuracy and efficiency of the proposed method.

4.1. Comparisons between proposed method and PPF by traditional load flow method

Accuracy of the proposed method in this section: The modified IEEE 14-bus test system given in figure 3 is composed of 14 buses and 19 branches. The performance is evaluated on the IEEE 14-bus test system using three indexes: power generation, inter-area flow and total load for the system.

Assume that the case study system contains three local areas, the first area containing three load buses L1, L2 and L3, then the second area containing two load buses L4 and L5, and the last area containing three load buses L6, L7 and L8. For simplifying the analysis, power factors of all loads and wind farms are assumed to be constant. Suppose also the bus loads in the system are assumed to follow normal distributions with the mean values being original bus loads and the standard deviation values being 5% of the mean values, the original system is modified which includes two generators of wind farms and eight load buses. 24.65 & 11.70 MW PV generators are added at buses 6 and 7, respectively.

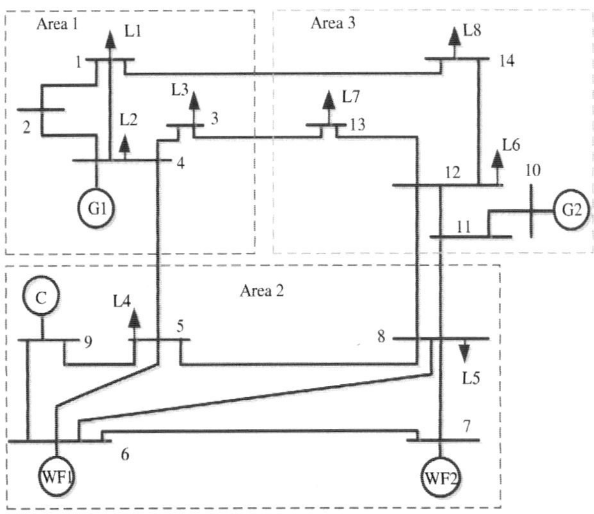

Figure 3. Single-line diagram of the modified IEEE 14-bus system.

In PFV with LS, taking the calculate of load as example, assume that pattern feature vector has 3-layer structure.

The 1st layer is system total load, denoted as P_1

$$X_1 = (P_1)^T \tag{12}$$

The 2nd layer includes loads of three local areas, denoted as $P_{2(1)}$, $P_{2(2)}$, $P_{2(3)}$

$$X_2 = (P_{2(1)}/P_1, P_{2(2)}/P_1, P_{2(3)}/P_1)^T \tag{13}$$

The 3rd layer includes loads of eight buses, denoted as $P_{3(1\text{-}1)}$, $P_{3(1\text{-}2)}$, $P_{3(1\text{-}3)}$, $P_{3(2\text{-}1)}$, $P_{3(2\text{-}2)}$, $P_{3(3\text{-}1)}$, $P_{3(3\text{-}2)}$, $P_{3(3\text{-}3)}$

$$X3 = (P_{3(1-1)}/P_{2(2)}, P_{3(1-2)}/P_{2(1)}, P_{3(1-3)}/P_{2(1)}, P_{3(2-1)}/P_{2(2)}, P_{3(2-2)}/P_{2(2)}, P_{3(3-1)}/P_{2(3)}, P_{3(3-2)}/P_{2(3)}, P_{3(3-3)}/P_{2(3)})^T \tag{14}$$

As shown in figure 4, the P-V curve is obtained based on the continuation PF, plotted by the change of (p.u) power injection on buses on the horizontal axis and the voltage of node on the vertical.

The simulation results of proposed method are usually over optimistic because it takes the uncertainties of load and wind power into consideration. Get the change of PV curves before (base case), and after traditional and proposed methods) connecting wind farm to power grid, at the bus 6 are shown in figure 4. The results show the voltage increment due to injection of wind power. The voltage steady-state stability of the system is more than the case without wind power.

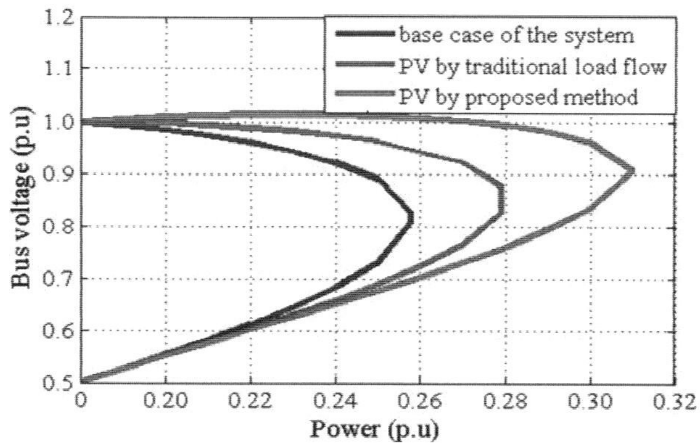

Figure 4. The P-V curve of load bus without and with wind power, with different methods.

LF problem has been solved for the case study considering the wind farms and the bus voltages have been obtained. Figure 5 shows the voltage profile of the network buses without wind farm (base case), with wind farm connected to the system by traditional method, and proposed method. From figure 5 shows that the proposed method gives a better solution for voltage improvement. In this figure, the blue line corresponds to the PLF by traditional LF, and the green line corresponds to the base case.

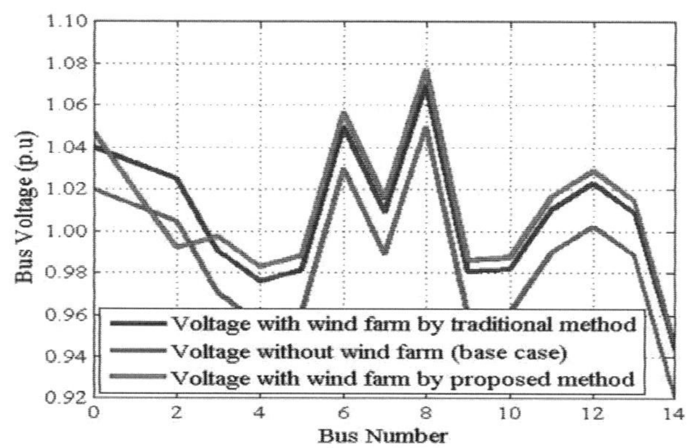

Figure 5. Voltage profile of 14-bus system at the different methods.

In figure 6, the voltage stability margin is obtained by proposed, traditional and base case sequence with the different cases, and cumulative distribution function (CDF) curve of voltage stability margin are plotted. Figure 6 shows the CDF of the voltage at bus 6 in a modified IEEE 14-bus system obtained by using three different methods: base case simulation, traditional method and proposed method. For comparison, the CDF of the voltage obtained from the proposed method match the voltage from traditional method simulation very well. The traditional method is slightly worse than the proposed method, but the result also resembles the proposed results adequately.

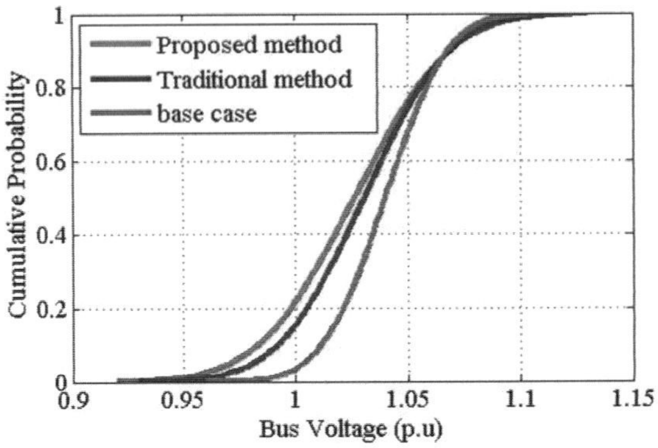

Figure 6. The voltage CDF of bus 6 before and after wind power integration.

4.2. Comparisons of numerical values with different methods

The proposed method (denoted as the PLF based on PFV with LS in the tables and figures in this section), the PFV with LS, and the traditional method are applied to solve the PLF of the IEEE 14-bus test system. The wind farms are located at the buses 6 and 7; simulations are carried out for the three cases. For easy comparison the results are summarized and given in table 1. The means of the bus voltage magnitudes were calculated and are given in table I.

Table 1. Means of the bus voltage magnitudes (p.u) in the IEEE 14-bus modified test system.

Bus No	Calculate mean voltage values at: Base case	Traditional	Proposed	Bus No	Wind injected at bus 6, 7 by Base case	Traditional	Proposed
1	1.04000	1.06000	1.06700	8	1.07000	1.09000	1.09700
2	1.02500	1.04500	1.01200	9	0.97994	1.00070	1.00613
3	0.99000	1.01000	1.01700	10	0.98120	1.00190	1.00750
4	0.97587	0.99601	1.00273	11	1.00970	1.03010	1.03640
5	0.98110	1.00120	1.00800	12	1.02200	1.04300	1.04870
6	1.05000	1.07000	1.07700	13	1.00840	1.02890	1.03480
7	1.00900	1.02940	1.03560	14	0.94146	0.96393	0.96597

When the power generated from the RES is supplied at the load bus locally, the overall demand on the system is significantly reduced. This improves the voltage profile and minimizes the power loss of power grid. Table1 describe the numerical results of the bus voltages for the IEEE 14-bus system at two cases: without and with wind power. From the table it can be inferred that the voltage profile at all the load buses is improved. The means of the branch voltage magnitudes were calculated and are given in table 2.

Table 2. Means of the branch voltage magnitudes (p.u) in the IEEE 14-bus modified test system.

Branch No	Branch voltage values at: Base case	Traditional	Proposed	Branch No	Wind injected at bus 6, 7 by Base case	Traditional	Proposed
1-2	0.330680	0.332681	0.337681	7-8	0.220465	0.223468	0.224893
2-4	0.438033	0.440056	0.445011	1-14	0.391069	0.393069	0.398069
3-4	0.355585	0.357585	0.362585	3-13	0.408001	0.410011	0.415022
4-5	0.364624	0.366624	0.371624	12-13	0.413213	0.415203	0.420218
5-6	0.075854	0.076851	0.079894	12-14	0.418905	0.420000	0.425011
5-9	0.372630	0.374631	0.379632	11-12	0.378136	0.380136	0.385136
6-9	0.423022	0.425011	0.430210	10-11	0.428001	0.430956	0.435051
6-7	0.166247	0.166330	0.167049	8-12	0.433011	0.435012	0.440055
6-8	0.197949	0.200705	0.206138	8-11	0.345760	0.347762	0.352762
5-8	0.388358	0.390358	0.395358				

As can be seen from table 1 and table 2, the calculated value of the PLF result by using the traditional method has been barely affected, but the deviation is much bigger than that using the proposed method. This shows that the relevance of node power injection increases the fluctuation of the node voltage and the branch current values. Therefore, when the system is close to the limit running state, the result of node voltage amplitude and branch, current flows are more likely to go beyond the range by using the proposed method. This is because when the relevance of node power injection is high, the probability of the output of each wind farm or the load of each node increasing or decreasing at the same time is higher. The traditional probability method does not take this into account, so it is easy to underestimate the risk of system operation.

5. Conclusion

This paper presents a method for assessing power flow for wind power integrated into the power systems with the consideration of wind and load-power generation correlations. Using PLF based on PFV with LS for calculations, the possibility load flow can be computed and distribution functions of the voltage magnitude. The effectiveness of the proposed method is verified via the modified IEEE 14-bus system. Simulation numerical results show that the proposed method is appropriate for security assessing load flow of wind-integrated power systems. It is observed that the types, and the parameters of uncertain wind power distribution would affect the security assessment indices of load flow. Thus, appropriate shape and scale parameters of Gaussian distribution for wind power should be carefully tuned. In addition, wind resource and load power generation correlations would also impact the security assessment indices of load flow, and needed to be considered when assessing the load flow security. Security assessment indices with the consideration of random variable correlation would more accurately reflect the system security and provide more effective information for the decision-maker.

Acknowledgment

This study was supported by the National Natural Science Foundation of China under Grant 51377035.

References

[1] Singh S 2012 *J. Int. J. Computer Technology & Applications.* **3(3)** 887-894
[2] Borkowska B 1974 *J. IEEE Trans. Power App. Syst.* **PAS-93, no. 3** 752–759
[3] Hamon C, Perninge M and Soder L 2016 *J. Electric Power Systems Research.* **131** 11-18
[4] Boussion C G F and Van Bussel G J W 2013 *J. Structural Health Monitoring.* **1 and 2** 2423-2430
[5] Almeida A B, De Lorenci E V, et al. 2013 *J. IET Renewable Power Generation.* **7(5)** 521-530
[6] Wan H, McCalley J D and Vittal V 2000 *J. IEEE Trans. on Power Systems.* **15(4)** 1247-1254
[7] Singh R, Pal B C and Jabr R A 2010 *J. IEEE Trans. on Power Systems.* **25(1)** 29-37
[8] Abe S, Fukunaga Y, Isono A and Kondo B 1982 *J. IEEE Trans. Power Apparatus and Systems.* **PAS-101** 3830-3840
[9] Rodrigues A B, Prada R B and da Silva M D G 2010 *J. IEEE Trans. on Power Systems.* **25** 1575-1588
[10] Miao F, Vittal V, Heydt G T and Ayyanar R 2012 *J. IEEE Trans. Power Syst.* **27, no. 4** 2251–2261
[11] Gupta N, Pant V and Das B 2014 *J, Elect. Power Syst. Res.* **106** 203–213
[12] Zhu X, Huang Y, Zhang J, Fei S, Liu J and Li H 2014 *J. Automation of Electric Power Systems.* **38** 46-53
[13] Wang C, Zheng H, Xie Y and Chen K 2005 *J. Automation of Electric Power Systems.* **29(24)** 39-44
[14] Isukapalli S S, Roy A and Georgopoulos P G 1998 *J. Risk Anal.* **18, no. 3** 351–363

IOP Conf. Series: Earth and Environmental Science **146** (2018) 012035 doi:10.1088/1755-1315/146/1/012035

Study on the surface stress and the surrounding dust deposition of solar photovoltaic module

Mingzhi Zhao[1],Hao Sun[1],Yiming Miao[1],Dan Zhang[1] and Shuai Wang[1]

1.School of Energy and Power Engineering, Inner Mongolia University of Technology, Hohhot, China , 010051.

zhaomingzhi2020@163.com; sunhhhh@qq.com

Abstract: Northwest China is rich in solar energy resources and vast land resources, this unique advantage is the reason for the development of solar photovoltaic industry here. However, windy and sandy climate characteristics of northwest are brought hidden trouble to run efficiently and safety for solar photovoltaic modules. This topic through the numerical simulation method to simulate wind load condition of S275 type polycrystalline silicon photovoltaic modules in net wind field, dust particle are added on the basic, to analyze the various factors influencing the settlement of sand and dust around the photovoltaic modules, as well as provides a reference to the design standards for solar photovoltaic system suitable for northwest China climate and landscape environment.

1. Introduction

The solar photovoltaic component toward the development of lightweight, anti wind performance put forward a very high demand, at the same time, the rate of conversion of solar photovoltaic modules to high quantitative development, in order to ensure solar power plants in the desert dust in the environment of efficient and stable operation, analysis of dust in the desert environment of sand movement in gas-solid flow and settlement mechanism and the conditions of solar photovoltaic module surface is necessary[1]. With the rapid development of solar energy industry, more and more researches on the wind resistance performance of solar PV modules are presented. Raddu et al.[2] Studied the effect of wind on the solar flat plate collector. It is concluded that the drag coefficient is the largest when the wind blows from below. Dou Zhenzhen[3] of Chang'an University through the analysis method of wind tunnel test and numerical simulation on comparison of the solar photovoltaic component of wind load, wind angle, inclination, height, spacing and location factors of wind load, obtains the photovoltaic array of solar photovoltaic power plants in the mutual occlusion of the wind load produces certain effect.In the numerical simulation of wind blown sand flow, Hsu[4] studied the characteristics of wind sand flow under low wind speed. Li Yintang[5] calculated by Fluent software, the dust events occurred during the initial stage of the surface layer in different wind speed, particle size, volume fraction of space dust distribution, the results show that the numerical simulation and wind tunnel test is very similar. Based on the above research, this paper adopts the numerical simulation method of S275 type polycrystalline silicon solar PV module and wind load were analyzed in the gas-solid two-phase flow in the sand dust around the settlement, according to the simulation results, analysis of influence factors and the sedimentation of wind load.

2. Physical Model

Solar photovoltaic array is to meet the needs of practical application in high power and a plurality of solar photovoltaic modules connected together constitute the monomer in the array, this selection is shown in Table 1 will be six pieces of polysilicon components S-275D connection in the solar photovoltaic array unit together, simulation model is built according to the actual size 5952 * 1956 * 50mm.

The simulation of solar photovoltaic components due to different conditions, choose the conditions of mirror perpendicular to the surface of the calculation domain settings, can calculate the occlusion rate was 1.79% in the table size calculation and calculation of the domain model, meet the requirements.

Table 1. Size of calculation model

Calculation model height (m)	Calculation model width (m)	Calculation model windward length (m)	Calculation model of leeward length (m)	Blocking rate (%)
13.692	47.616	13.667	39.095	1.79

In the solar photovoltaic installation, PV module under the ground along the distance should be 0.3 ~ 0.5m, so as not to be weed shelter and winter snow buried, this paper choose along the height from 0.5m, this distance is reflected in the calculation model of photovoltaic module and the lower edge of the bottom surface of the spacing of flow field.

3. Calculation Model

3.1 Modeling and Meshing

The ANSYS software is used to model and build the mesh. First set up a small area near the photovoltaic component model, the calculation model for cutting 5 plane through a small domain, the method is cut into 18 geometry, the area using tetrahedral mesh encryption to adapt to complex flow field of photovoltaic module. The external area is the use of mesh in Body sizing--Body of Influence function to generate the positive rate of growth from small area of hexahedral mesh body grid size to 1.05 gradient, can shorten the calculation time reduced the number of grid. The surface of the PV module needs to be encrypted locally, and the overall grid is shown in Figure 1, and the total number of grids is 2 million 60 thousand.

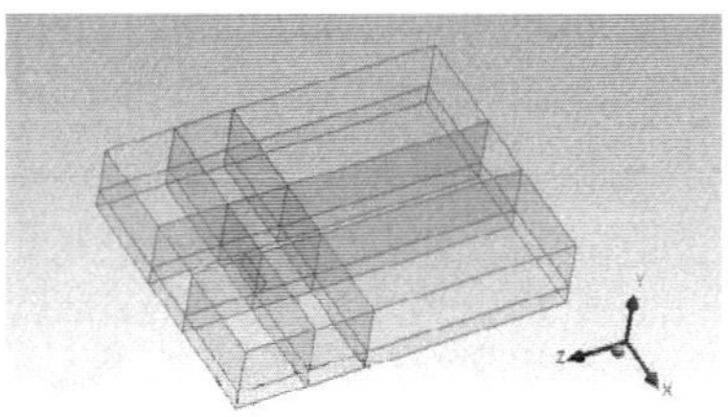

Figure 1. Calculation model after mesh

The calculation process is solved by Fluent pressure basis and the standard $k-\varepsilon$ model is adopted[6]. Using the Euler Euler method to establish the basic equations of wind sand flow, the mixed model numerical simulation model, follow the mass conservation equation, momentum conservation equation, energy conservation equation, discretized by the finite volume method calculation area.

3.2 Boundary Condition Transitivity

3.2.1 Net Wind Field Simulation Boundary Condition Setting

In the numerical simulation assuming fluid motion for incompressible unsteady flow, velocity boundary conditions using the entrance entrance conditions, fully developed flow boundary conditions with open boundary conditions, the computational domain top and sides and a photovoltaic component on surface adopt wall noodles no slip wall, the standard wall function is selected the function simulation of near wall flow.

3.2.2 The Boundary Conditions of the Simulation of the two Phase Flow

The dust particles added in the net wind field numerical simulation has been completed, and the air together constitute the sand two-phase flow, Euler -Mixtrue model is applied to simulate and analyze the above selected process of settlement of dust in the solar photovoltaic module and peripheral surface. The characteristics of the given boundary between Bagnold gravel, sand and dust were 1mm and 0.01 mm in dust deposition, widely used in sport research, in the process of simulation of dust particle size selection as a reference, the sand density is 2650 kg/m^3, it was assumed that the following conditions are established:

(1) if the air is an incompressible fluid, the density is always constant;

(2) assume that dust particles for uniform quality, uniform particle size and sphere, without collision and other reactions, dust particles in the field of entrance along the height of the uniform distribution, the volume fraction of the initial conditions set the average volume fraction of the dust particles.

The 3D double precision should be calculated in the process of pressure based implicit solver; in sand two-phase flow need to set the Y -9.81 in the direction of acceleration of gravity; flow field solution using SIMPLE algorithm, two order upwind scheme use high precision; other parameters besides the settings are in accordance with the default value.

4.Simulation Results and Analysis

4.1 The Influence of Inclination Angle of Wind Farm on Distribution of Velocity Field

The installation of photovoltaic modules in actual project settings and the ground has certain angle, different installation angles will affect the photovoltaic component around the velocity field and surface pressure distribution, which affects the safety stability of PV module structure, so the analysis is very important for the installation angle of the PV modules.

The flow velocity is 10m / s, and the distribution of the velocity field on the symmetry plane of the PV module is 15 °, 30 °, 45 °, 60 °, 75 ° and 90 °, respectively.Comparison of field distribution can see installation speed when the different angle, flow and blunt obstacle of solar PV flow is similar to that of air on the surface structure of the windward side will have a block, the airflow from the outward appearance of radial expansion, blocking above will be on the rise and cross structure the following points, blocking will flow down to the ground, and the general flow around bluff body is different, the solar photovoltaic component along the ground and there is a certain distance (500mm), so the air separation will occur in the lower part, from below through air to roll up to form a small vortex.

Comparison of installation angle of 15 degrees to 90 degrees around the velocity field of PV modules, can be seen with the change of installation angle, the wind PV area is more and more big, the blocking effect of air is more and more big, performance in the PV downstream flow width becomes longer, the upstream PV module cyclotron eddy current with the increase of the angle end from the near to flow more and move up to 90 degrees maximum, basically symmetrical; air separation along the length is increased with the increase of PV installation angle decreases.

4.2 The Factors Affecting the Deposition of Dust around Solar PV Modules

In the foundation of calculation simulation has been completed in accordance with the net wind field, wind blown sand flow setting conditions of dust particles into phase, then the dust in the settlement around the solar photovoltaic component simulation, considering the effect of gravity in the simulation process, the dust particle density is set to 2650kg/m3.

IOP Conf. Series: Earth and Environmental Science **146** (2018) 012035 doi:10.1088/1755-1315/146/1/012035

4.2.1 Effect of Installation Inclination on Dust Deposition

The distribution of the velocity field around the PV module is affected by the inclination of the solar PV module, which affects the sedimentation of the sand particles. At the same time, due to the blocking of the surface of the PV module to the dust particles, the particles collide with the surface of the PV module and consume a part of kinetic energy.

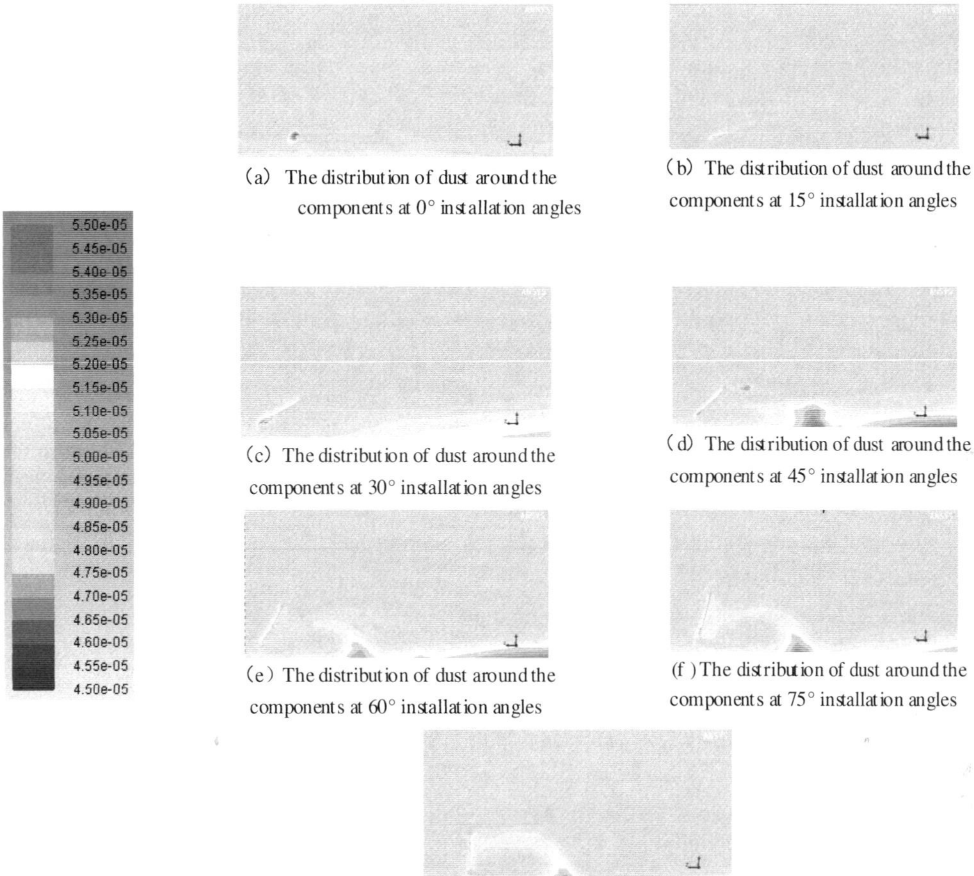

(a) The distribution of dust around the components at 0° installation angles

(b) The distribution of dust around the components at 15° installation angles

(c) The distribution of dust around the components at 30° installation angles

(d) The distribution of dust around the components at 45° installation angles

(e) The distribution of dust around the components at 60° installation angles

(f) The distribution of dust around the components at 75° installation angles

(g)The distribution of dust around the components at 90° installation angles

Figure 2. Volume fraction distribution around the photovoltaic modules at different installation angle

Figure 2 shows the fixed wind speed 10m/s, dust particle size 10um, entrance volume fraction 5e-05%, PV installation angle of 15°, 30°, 45°, 60°, 75° and 90°of PV components symmetrical surface dust volume fraction distribution. It can be seen from the figure, installation angle of PV modules around the sand volume fraction distribution, and volume fraction of large area, it can be considered a dust prone to subsidence of the area, which shows that the inclination angle is very significant influence on the sedimentation.

Comparison of the installation angle of photovoltaic modules around the dust distribution, because of the self gravity of PV modules and dust barrier role in the PV wind to ground will appear different degrees of settlement. 0° of dust only in the thickness of the photovoltaic components by very small obstacles, so in front of the PV module basically no settlement, 90°of PV module maximum of blocking effect of sand flow, the dust is not easy to be blocked by the photovoltaic component dust along the surface of the most obvious decline, settlement.

Because of the PV module below through air will roll up the formation of a small vortex, the airflow will drive the wind direction is hindered by the dust through settlement in the PV back to the

IOP Conf. Series: Earth and Environmental Science **146** (2018) 012035 doi:10.1088/1755-1315/146/1/012035

ground subsidence area will always appear dust, from 0°to 90°to the dorsal subsidence area from the photovoltaic component is getting closer. At 0°, the aeolian sand flows through a small perturbation of the PV module and then continues to flow in the horizontal direction until it is deposited in the far back due to gravity. At 90°, the wind direction to block the dust, the lower the direction of the settlement is also faster.

5.Conclusions

Wind load simulation results show that the wind farm:

With the increase of the inclination angle, the wind PV area is more and more big, the blocking effect of air is more and more big, performance in the PV downstream flow width becomes longer, the upstream PV module cyclotron eddy current with the increase of the angle from the end near the flow more and move up to 90°maximum, basically symmetrical; air separation along the length is increased with the increase of PV installation angle decreases.

Simulation results of wind sand two phase flow by Euler Mixture:

With the increase of installation angle, appear in the PV wind to the ground sedimentation area is more and more obvious, the photovoltaic component subsidence area from the back to the PV module is getting closer, dip the instructions to install more easily in the PV surrounding settlement.

Reference

[1] Zhao Mingzhi,Song Shijin,Zhang Xiaoming 2013 Method research of parabolic trough solar thermal power plant site selection *Renewable Energy Resources* **3**(31):18-22.

[2] Raddu A, Axinte E. Wind forces on structures supporting solar collectors[J]. Journal of Wind Engineering & Industrial Aerodynamics, 1989, V32(1):93-100

[3] Dou Zhenzhen, The Study of The Influence F actor s on PhotovoltaicModules Wind Tunnel Test And Numerical Simulation, *Chang'an University*,2014

[4] Hsu T J, Liu L F. On two-phase sediment transport: Sheet flow of massive particles[J]. Proceedings of the Royal Society A, 2004, V460(2048):2223-2250

[5] Li Yintang,Yu Yuanyuan,Meng Haiyan, Numerical Simulation of the Near-surface Particle Concentration at the Early Stage of Blown Sand Emission, *Journal of Desert Research*, 2010, 30(3):492-497

[6] Bhumralkar C M, Slemmons A J, Nitz K C 1981 Numerical study of local regional tmospheric changes caused by a large solar central receiver power plant *Journal of Applied Meteorology* **20**: 660-77.

Base on DPM model to simulation Sand erosion on PV modules surface

Zhao Mingzhi[1], Miao Yiming[1] and Kang Xiaobo [2]

1. School of Energy and Power Engineering, Inner Mongolia University of Technology, Hohhot, China, 010051.
2. China General Certification Center, Beijing,China, 100000.

zhaomingzhi2020@163.com; 373034206@qq.com ; 402712046@qq.com

Abstract: The erosion wear behavior of dust on the surface of PV modules in desert aeolian gas-solid two-phase flow was analyzed by discrete phase model.This paper investigated the effect of the tilt angle、the wind speed、the particle size、the mass flow to erosion on PV modules surface.The conclution showed that the tilt angle and the wind speed are the major factors to effect the maximum and average erosion rate on PV modules surface. In especially,the effect of wind speed on the surface erosion rate of PV modules varies exponentially.

1. Introduction

The country's most abundant resource of solar radiation distributing in most areas of western China, including Inner Mongolia, Gansu, Qinghai, etc.[1].The rich solar energy resources and the cheap land have provided a convenient condition for the large-scale development of the photovoltaic industry. However,these places are located in Central Asia sandstorm area which is one of the most active regions of dust activity.These areas are mostly wasteland, sand, gobi, desert and where is a frequent area of sandstorms.The dust particle impacted on the surface of PV modules to reduce the power generation efficiency of PV modules and reduce their service life, will have a significant impact on the operation of PV solar power plants, especially for the PV modules. According to the above,to extended the service life and guarantee the long-term efficient operation of solar photovoltaic power generation system,the durability of the components is of major importance.Many scholars have studied the influencing factors of surface erosion. Dong Xiaofeng[2] Based on the natural environment of Xinjiang, the paper analyzes the erosion and wear of wind turbine blades in poor environmental conditions. The effects of impact speed, impact angle, particle size and other factors on erosion and wear of wind turbine blade material were analyzed. Lin Nan[3] used the numerical simulation method to study the influence of the elbow size and the incident angle of the solid particles on the erosion and wear, and the angle of incidence was the important factor affecting the erosion and erosion of the pipe. The experiment also got the same result. Numerical simulation was carried out. This study was to investigate the effect of the different tilt angle、the different wind speed、the different dust particle size and the different mass flow to the erosion phenomenon on PV modules surface.

2. Calculation model And methods

2.1 Theoretical basis

2.1.1 *Continuity equation and momentum equation*

(1) .Continuity Equation of Sand Particle Phase:

$$\frac{\partial(\varphi_p \rho_p u_x)}{\partial x} + \frac{\partial(\varphi_p \rho_p u_y)}{\partial y} = 0 \quad (1)$$

(2) .Momentum Equation of Sand Particle Phase:

$$\frac{\partial}{\partial t}(\varphi_p \rho_p U_p) + \nabla(\varphi_p \rho_p U_p U_p) = -\varphi_p \nabla p - \nabla p_p + \nabla \tau_p + \varphi_p \rho_p g + f_{fp} \quad (2)$$

In formula (1) (2), φ_p and ρ_p represent the volume fraction of the particulate phase density, in units of kg/m^3; U_p is the velocity of the dust particles; p_p is the dust pressure of the dust particles; τ_p is the stress of the dust particles; g is the gravitational acceleration, where g = 9.8 m/s^2; f_{fp} is the interphase force between the gas and the dust particles, which is ignored in this paper.

2.1.2 Erosion theory In calculation of Erosion wear, according to MCLAURYBS proposed wear model to calculate,the equation is as follows:

$$ER = CF_S U_p^{1.73} f(\theta) \quad (3)$$

C is constant, is related to the selected material; U_p is the velocity at which the particles collide with the wall; $f(\theta)$ is the impact angle function, and θ is the angle of the particle collision.

In the numerical simulation, the following formula is used to calculate the wear, and the simulation is carried out using Fluent:

$$R_{erosion} = \sum_{i=1}^{N_{particle}} \frac{m_p C(d_p) f(\theta) U_p^n}{A_{face}} \quad (4)$$

$R_{erosion}$ is the area of the wall of the wear rate of the unit, the unit kg/(m^2·s), also known as wear rate. A_{face} is the area of the material wall calculation unit. $N_{psrticle}$ is the number of particles colliding on the surface of the component per unit area. m_p is the mass flow rate of particles in the process of wall collision. n is the relative exponent of the velocity of the particles relative to the velocity of the gas. θ is the incident angle of impacted when the particles collide PV modules.

$f(\theta)$ is an impact angle function, which is defined as the following equation:

$$f(\theta) = \begin{cases} 22.7\theta - 38.4\theta^2 & \theta \le 0.267 rad \\ 2.0 + 6.80\theta - 7.50\theta^2 + 2.25\theta^3 & \theta > 0.267 rad \end{cases} \quad (5)$$

$C(d_p)$ is a function related to particle diameter, and its expression is :

$$C(d_p) = 1.559 \times 10^{-6} B^{-0.59} F_S \quad (6)$$

B is Brinell hardness of the be eroded material, which material in this paper is solar photovoltaic glass, a value of 6.5; F_S is the particle shape factor, when the particles are spherical to take 0.2, hemispherical when the 0.5, more sharp when taken 1[4,5],Dust particles are herein assumed rule spherical particles, so a value of 0.2.

2.2 Model establishment
Based on the size of standard solar PV modules, The dimensions of the monolithic components are 1580mm×808mm×40mm.The final model has eight PV modules, up and down two rows of installation, each row of four,the size is 3232mm×3160mm×40mm,and the model is 500mm from the ground. The hybrid model of tetrahedron and hexahedron were chosen to mesh the established model,the eventually get the number of grid is about 4.34 million and the number of nodes is about 3.25 million,as shown in Figure 1.

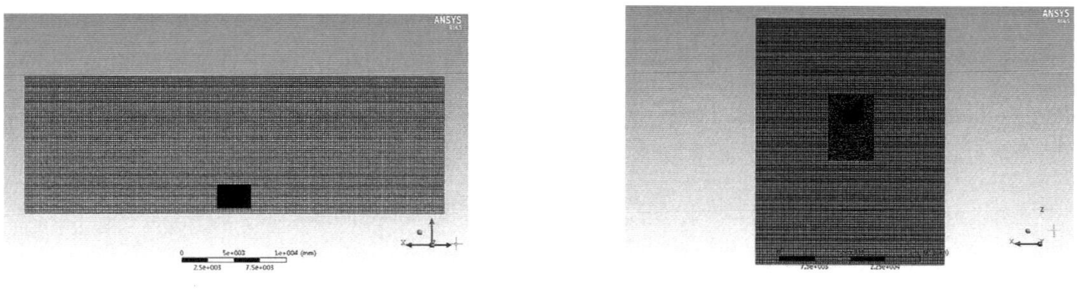

(a).Grid model front view (b).Grid model top view

Figure.1 The model geometry and meshing

2.3 Solution setting
In the DPM model, Assume that the discrete phase is very sparse,thus, The effect of particle-particle interaction and particle volume fraction on the continuous phase were neglected.In this model, First, the flow field of the continuous phase was solved,and then add the discrete phase of the dust particles in this paper phase, set the continuous phase to calculate the iterative calculation of a discrete phase for every 5 steps. The parameter settings are shown in Table 1.

Table 1. Parameter setting in erosion simulation

Boundary Conditions	
Import Speed	5m/s, 10 m/s, 15 m/s, 20 m/s
Gas Density	1.225kg/m^3
Particle Density	2650 kg/m^3
Particle Diameter	10μm~60μm
Mass Flow Rate	0.05kg/s~0.4kg/s
The Sand Incidence Mode	surface

3. Simulation results and analysis
Erosion wear is a type of wear caused by damage to the surface of the material due to the impact of tiny and loose motion particles [6].The factors affecting the erosion are impact angle, abrasive grain velocity, abrasive grain shape, ambient temperature, material structure and so on. In this paper, only the abrasive grain velocity, impact angle, abrasive grain size and abrasive grain mass flow, In the simulation process, the ambient temperature, sand shape, PV modules surface materials, etc. do not change.

By simulation, the maximum erosion rate of the component surface often occurs at the bottom edge

IOP Conf. Series: Earth and Environmental Science **146** (2018) 012036 doi:10.1088/1755-1315/146/1/012036

of the assembly, and there is also a distribution of erosion rates in other areas of the module. Take the inclination angle of 45 °, wind speed 10m/s and particle size 10μm as an example,the surface of the erosion rate distribution shown in Figure 2,the proportions of the various erosion rates on the surface of the module are shown in Table 2.

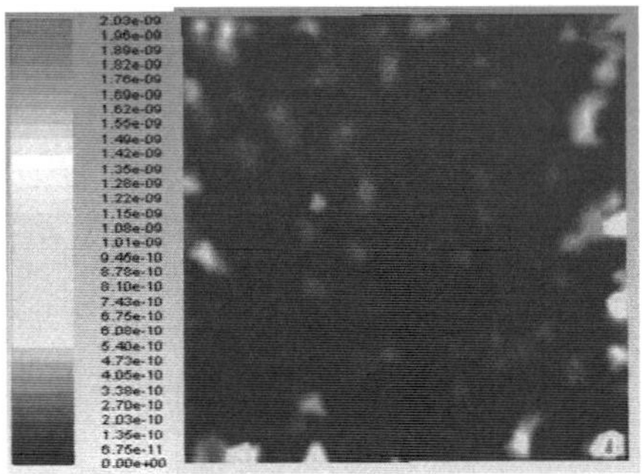

Figure 2. The surface of the module erosion rate distribution

The erosion rate refers to the mass loss of the surface of module material per unit area per unit time in kg/(m^2·s).The erosion rate is calculated by calculating the cumulative damage to the surface of each particle.

Table.2 Percentage of erosion rate

erosion rate（kg/m^{2}s）	percentage（%）
0~5.3917e-10	96.5225
5.3917e-10~1.0783e-09	2.2703
1.0783e-09~1.6175e-09	0.5766
1.6175e-09~2.1567e-09	0.2883
2.1567e-09~2.6958e-09	0.1441
2.6958e-09~3.2350e-09	0.0180
3.2350e-09~3.7742e-09	0.0180
3.7742e-09~4.3133e-09	0.0180
4.3133e-09~4.8525e-09	0.0541
4.8525e-09~5.3917e-09	0.0541
5.3917e-09 or more	0

According to Table 2, it can be seen that the main erosion rate of the surface of the module is low,therefore, only the maximum erosion rate of the surface of the module can not objectively reflect the influence of different factors on the surface erosion rate of the module,can only explain the erosion of the largest erosion site,want to know the overall erosion of the components, but also need to use the average erosion rate to describe.Figure 3 shows the comparison between the maximum erosion rate and the average erosion rate of the module surface under various conditions. The unit is kg/(m^2·s).

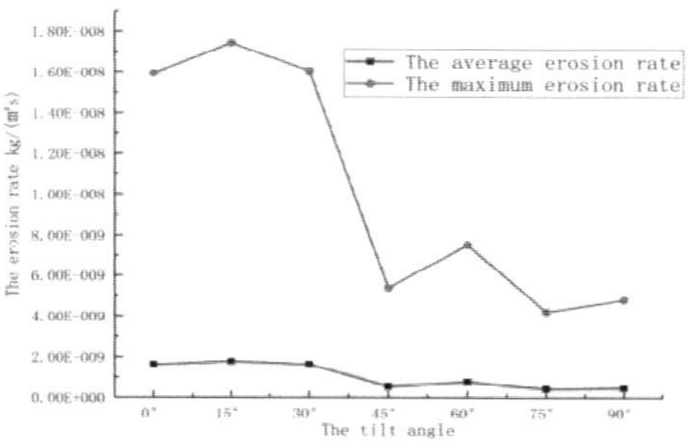

(a). different tilt angle

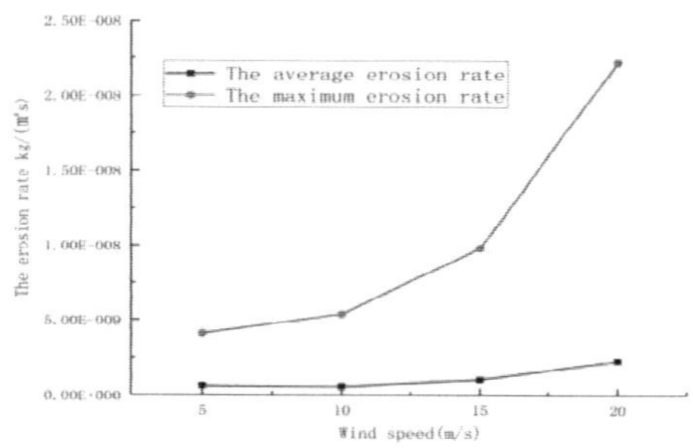

(b). different wind speed

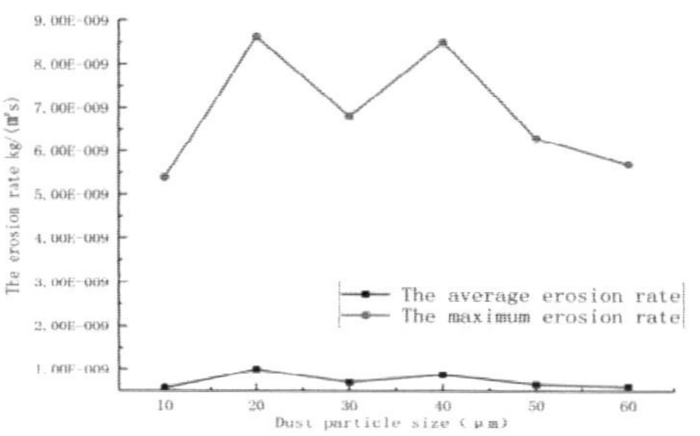

(c). different particle sizes

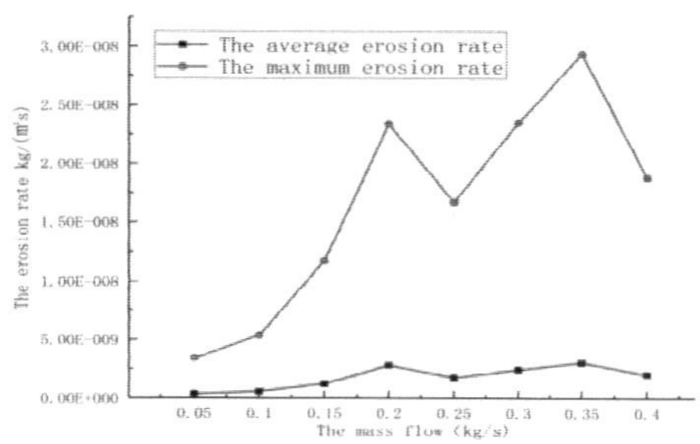

(d). different mass flow

Figure.3 The maximum erosion rate and average erosion rate of the surface of PV modules under different conditions

From the maximum erosion rate curve in the graph, the maximum erosion rate increases exponentially with the increase of wind speed when the other conditions are constant. This result is very close to the model results in Finard's theory of erosion. Indicating that particle velocity is one of the important factors affecting erosion.

When the particle size increases, the maximum erosion rate changes little, basically maintained at the same order of magnitude, the results and Huang Si [7] using DPM model simulation centrifugal pump unsteady solid-liquid two-phase flow wear similar results , Indicating that the change in particle size has little effect on the erosion rate.

When the other conditions are constant, the maximum erosion rate and the average erosion rate are decreasing as a whole as the installation angle increases.The installation angle of the module changes, which is equivalent to changing the impact angle of the sand, According to the erosion theory, the impact angle is another important factor affecting the erosion. When the installation angle of the module is small, The angle between the incident velocity of the sand and the plane of the module is also small,from Figure 3 (a) can be seen when the installation angle of 0°~30°, that is the impact angle of 0°~30°, the component surface erosion rate of the largest.In the erosion theory , The typical ductile material has the largest erosion rate at an impact angle of 20° to 30°, indicating that the module surface material exhibits the properties of the ductile material.

As shown in Figure.3 (d), with the increase of sand mass flow rate, the maximum erosion rate increases first, and then the curve change is fluctuating, When the mass flow rate of sand is less than 0.2kg/s, the maximum erosion rate increases exponentially with the increase of sand mass flow rate, When the mass flow rate is more than 0.2kg/s, the maximum erosion rate fluctuate with increase of sand mass flow rate of soil, and reaches the maximum value of 2.9355e-08 when the mass flow rate is 0.35kg/s. According to the influence of particle mass flow on erosion rate in erosion theory, when the mass flow reaches a certain value,due to the existence of the flow, the collision and rebound between the sand and the sand will affect the erosion rate of the material, which is also the reason why the module surface erosion rate fluctuates when the mass flow rate is greater than 0.2kg/s.

4. Conclusion

(1) With the increase of tilt angle, the maximum erosion rate and the average erosion rate showed a decreasing trend.The change in the tilt angle can be seen as changing the impact angle of the sand.This mean that when the impact angle is 0°—30°, the erosion rate of the PV module surface reaches its maximum value.The material of the PV module exhibits the properties of the Plastic material.So, with the change of impact angle, erosion rate change is more obvious, and that is meaning the impcat angle of sand is one of the important factors to the erosion rate of PV surface.

(2) With the increase of wind speed, the maximum erosion rate and the average erosion rate all increase exponentially, and the maximum erosion rate increases greatly, indicating that wind speed is one of the main factors affecting erosion.

(3) With the increase of sand particle size, the change trend of erosion rate on the surface of PV module was not obvious, and the maximum change ratio between each particle size is only 37.5%, which indicates that sand particle size had little effect on the erosion rate of PV module.

(4) When the sand mass flow rate is less than 0.2kg/s, with the increase of sand mass flow rate, the maximum erosion rate and the average erosion rate increase exponentially; when the sand mass flow rate is more than 0.2kg/s, with the increase of mass flow rate, the maximum erosion rate and the average erosion rate had little effect.

Reference

[1] Zhao Mingzhi,Song Shijin,Zhang Xiaoming 2013 Method research of parabolic trough solar thermal power plant site selection *Renewable Energy Resources* **3**(31):18-22.

[2] Dong Xiao-feng.Study on erosion behavior of wind turbine blade material under gas-solid two-phase flow. 2013 *Xinjiang University.*

[3] Lin Nan, Lan Huiqing, Cui Yue, Zhao Chao.Influence of Erosion Angle and Elbow Size on Erosion Wear. *Science Technology and Engineering* 2013 (18): 5135-40.

[4] J.K.Edwards, B.S.Mclaury, S.A.Shirazi.Supplementing a CFD Code with Erosion Prediction Capabilities. *ASME 1998Fluids Engineering Division Summer Meeting* 1998 Washington DC.

[5] Ma Guangfei.The effect of gate position on gas-solid two-phase flow and wear in gate valve 2016 *Zhejiang Sci-Tech University.*

[6] Ma Ying, Ren Jun, Li Yuandong, Chen Tijun, Li Bing.Study on the Research of Erosion Wear. *Journal of Lanzhou University of Science andTechnology*.2005,31: 21-5.

[7] Huang Si, Zou Wenlang, Zhou Jinju, He Dongping, Peng Tianyang.Calculation of Unsteady Solid -liquid Two -phase Flow and Wear Calculation of Centrifugal Pump Based on DPM Model. 2016 *China Rural Water and Hydropower.* (07):103-6.

Study on the Distribution Law of Sand Dust Particles in Solar Power Stations

Wang Shuai[1] and Zhao Mingzhi[1]

[1]School of Energy and Power Engineering, Inner Mongolia University of Technology, Hohhot, China, 010051.
[1]China General Certification Center, Beijing,China, 100000

zhaomingzhi2020@163.com; 1351587945@qq.com

Abstract: Dust is an important part of the particulate matter in the desert atmosphere environment . It moves to the nearby solar panels by suspension and saltation, which affects solar power. Collecting sand dust with sand sampler and compare with the dust distribution characteristics and parameters on the earth's surface to find out the transport law of dust and variation of dust weight with height in sand sampler. The results showed that the distribution of dust particle size obeyed the exponential distribution law. The dust particle size of surface dust is larger than that of sand sampler and the dust is in poor sorting which was caused by the selection of external factors such as wind. There is a good correlation between the dust in the sand sampler and the specific gravity of dust particle size in the earth's surface. The dust weight increases with height in sand sampler obey the law of power-decreasing. The heigher the solar panel, the surface dust deposition has less influence on it. The results will provide a theoretical basis for the impact of desert dust on solar power station.

1.Introduction

The desert in the Northwest China region is an important source of sand dust. Hobq Desert, as one of the largest deserts in China, it is possible to have a rough understanding of the dust composition in the northwest in researching its composition of dust particle size law. Many of the solar power plants are built in the desert, and most of the dust is transported at a close range, which provides the conditions for the deposition of dust on the batterypanels, the power efficiency of the battery panels have been seriously affected.When the sand falls on the batterypanels, the light can not be shining on the batterypanels through the sand and the area of the light received by the battery board will be affected.When thebattery panel starts to work, the sheltered part rises quickly, it may cause damage to the inside of the battery boardmay.

Sand dust falls on the battery board, not only with the wind speed, dust particle size and the physical properties of the battery itself, but also with the structure of the sand flow. Sand dust falls on the battery board, Ma Shiwei[1] and others are studied in the vertical height of the sand flow and found that the sediment transport decreases exponentially with the increase of height. Zhang Jinchun[2]and others analyzed aerosol concentration inMinqin area. As the desert gradually close to the oasis, the concentration gradually decreased, the concentration decreased with the increase of altitude in the form of power law function. Sand dust in sand source area was more easily affected by underlying surface in near ground within 40m. Kong Dan[3]and others analyzed dust particle size with laser particle size analyzer in Taklimakan Desert, the average particle size of sand dust is about 100μm, the higher the altitude is, the better the sorting, the particle size curve is positive.

This paper analyzes the dust particle size at the source of desert and obtains the law of particle size distribution of sand dust. Designing the shunt-hedging sand sampler and explains the working principle and structure. Arriving that to using index to describe reasonable distribution law in the dust particle size at the Hobq Desert solar power base. The analysis of the dust particle size can provide a better understanding of the movement of sand dust, provide more theoretical basis for the settlement of sand dust and it's helpful to increase the efficiency of solar power generation

2.Experimental Method

Particle size analysis has a unique advantage in analyzing the particle size distribution of dust. Particle size is not a real particle size. It is not easy or even impossible to get the dust particle size, dust presents an irregular state in reality. So in the analysis, we are equivalent to the particle size and the equivalent particle size, that means to think of dust particles as an equivalent sphere of specific diameter. Different sources of dust particles have different particle size characteristics, we explore their differences from the point of view of mathematical statisticswe. Average grain diameter, medium diameter and the frequency distribution curve are the main contents of our research.

The averagee grain diameter is the average level of the particle size distribution, the median diameter is also commonly used particle size index, that is, the particle size of the total content is 50% , the median diameter is not affected by the larger diameter,so the sensitivity is lower than the average diameter. In the dust study, the average particle size is related to the average kinetic energy of dust. If the average particle size is calculated by graph method, first, cumulative frequency distribution curve of dust, and work out the particle size of 16%, 50% and 84%. Then, figure out the average particle size by the formula (1). If use the moment method, first find the percentage of the median and the range of granularity in each particle size range,then substitution formula (2) to calculate the average particle size. When the dust particle size was normal distribution, it is equal in averagee grain diameter and median diameter.

$$M_z = (F_{16} + F_{50} + F_{84})/3 \tag{1}$$

$$M_z = \sum (f_i m_i)/\sum f_i \tag{2}$$

In the formula, $f_i$is the percentage of overall ranges of size range i, m_i is the median of i size range. F_{16}、F_{50}、F_{84} are the corresponding value on the cumulative distribution curve.

The separation factor is related to the degree of sorting of dust, and the graphic method is calculated of ratio of grain size with cumulative content of 25% and 75%, sometimes expressed in terms of aroot mean square. The separation factor is related to thedynamic conditions of dust handling which reflects the stability of transportingdust. Generalized graphic method for calculating standard deviation can be a good reflection of empennage in sorting property, which has been widely used and calculated by formula (3). Standard deviation of moment method was calculated by formula (4).

$$s_1 = (F_{84} - F_{16})/4 + (F_{95} - F_5)/6.6 \quad (3)$$

$$d = [\sum f_i (m_i - M_z)^2 / \sum f_i]^{1/2} \quad (4)$$

In the formula,$f_i$is the percentage of overall ranges of size range i, m_i is the median of i size range. $M_z$is the average particle size of sand dust.

The skewness is related to the symmetry of the particle size distribution and represents the relativeity between the mean particle size and the median particle size. The skewness formula of generalized graphic calculation is (5), and the skewness of moment method is (6). When the skewness is zero,the curve is symmetrical. When it is greater than zero, it is positive skewness and the coarse part is relatively concentrated. When it is less than zero, it is negative skewness and the fine particle size part is more concentrated.

$$S_k = \frac{F_{84}+F_{16}-2*F_{50}}{2(F_{84}-F_{16})} + \frac{F_{95}+F_5-2*F_{50}}{2(F_{95}-F_5)} \quad (5)$$

$$S_k = \sum f_i (m_i - M_z)^3 / \sum f_i d^3 \quad (6)$$

In the formula,f_i is the percentage of overall ranges of size range i, m_i is the median of i size range. M_z is the average particle size of sand dust. d is standard deviation.

The kurtosis is a measure of the distribution curve center section and shape area of tali, and intuitively indicates the degree of the distribution curve. The larger the peak state, the more obvious the peak, the narrow the peak is,the better is the middle part than the tail part. The formula for calculating the peak state of the graph is (7) and the kurtosis formulaof the moment method is (8).

$$K_G = \frac{F_{95}-F_5}{2.44(F_{75}-F_{25})} \quad (7)$$

$$K_G = \sum f_i (m_i - M_z)^4 / \sum f_i d^4 \quad (8)$$

The sieving method usually obtains the percentage of mass fraction of dust particles in an interval. We take the right end endpoint value of each granularity range as the particle size of each granularity range, thus normalizing the particle size within each particle size range to create conditions for data analysis.

3. Dust Sample Collection

The dust samples used are randomly which taken from the solar power station in the heart of the Hobq Desert, there are no factories around, away from roads and towns, and animals are rarely infested and are not affected by human factors. Standard sieve selected was produced from Zhejiang Shangyu City Road Zhang Xing yarn sieve plant, the aperture of 0.01mm ~ 0.09mm, 0.1mm ~ 1mm. JY series of multi-functional electronic balance by the Shanghai Pu Chun Measurement Instrument Co. Ltd. production, the accuracy of 0.01g[4-5].Sift the dry sand particles through 30min in the airless indoor environmentand then then weigh it on the paper that has been measured. The weight of the paper is removed from the measured weight to give the weight of the dust in the different particle size ranges.

IOP Conf. Series: Earth and Environmental Science **146** (2018) 012037 doi:10.1088/1755-1315/146/1/012037

Sand sampler is located in the center of the solar power station of the Hobq Desert. In order to facilitate transport and demolition, sand sampler is generally bolted, fixed on the ground, the sand separator is fixed on the support of the sand sampler through bolts, so the air inlet on the sand separator can rotated in different directions. The height of the inlet of sand separator is 65,110,150,192,235,272 cm, and the number of layers is 6. In order to avoid the influence of wind direction and wind speed on the sediment concentration, except for the height of the sand separator, the conditions of sand separator size, exit direction and entrance direction are the same[6].

4. Results and Discussion

4.1. Sand particle size analysis

The analytical method of sediment grain size (d is sand particle size, mm)is a conventional method for sediment particle size analysis.

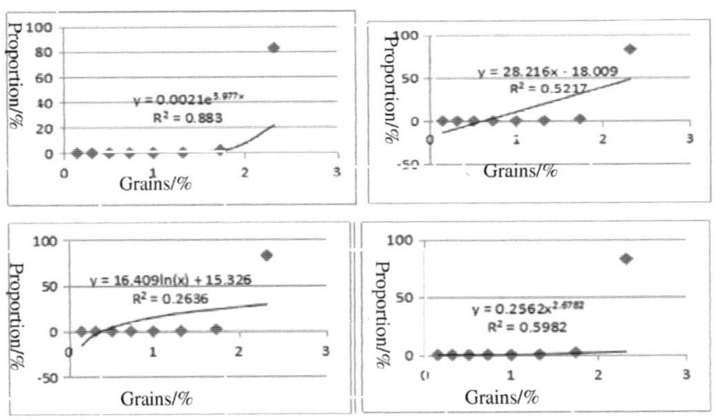

Figure 1. Relationship between Gravity and Sand Grain Size (0-0.1mm)

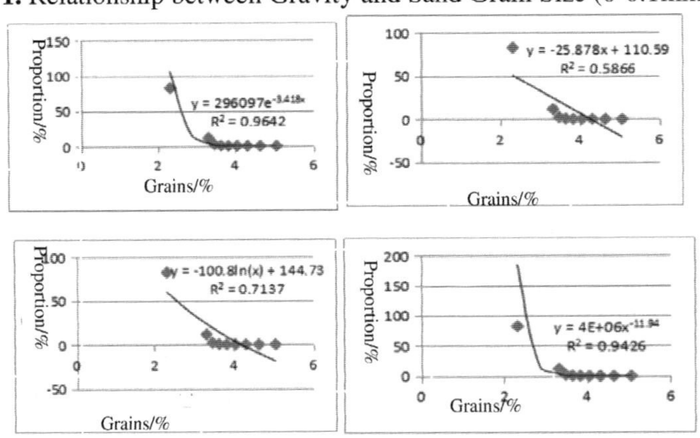

Figure 2 Relationship between Gravity and SandParticle Size (0-0.1mm)

According to Figue 1 and 2,we can see from the correlation coefficient, regardless of the dust particle size is less than 0.1mm, or greater than 0.1mm, the best relationship between specific gravity and dust particle size is exponential relationship. When the dust is less than 0.1mm, the proportion gradually increased, and then gradually reduced until zero. The reason, when dust is collected, there are tiny dust particles that are easily carried by the wind because of levitation, dust particles with larger grain sizes produce saltation energy by impacting.

4.2. Distribution characteristics, Parameter comparison and Correlation analysis between Sand sampler (High 65cm)and Dust particle size on the ground surface

4.2.1. Comparison of distribution characteristics between sand sampler and dust particle size on the ground surface

Table 1. The main grain size parameters of sand dust on the ground in sand sampler

Dust	Median/mm	mean value/mm	Separation coefficient	SkewnessS_K	KurtosisK_G	< 0.01 mm	0.01 ~ 0.02 mm	0.02 ~ 0.03mm	0.03 ~ 0.04 mm
Near ground	0.112	0.124	0.042	0.340	0.797	0	0	0	0
Ground	0.142	0.142	0.037	0.019	0.724	0	0	0.01%	0.04%

Dust	0.04~ 0.05 mm	0.05~ 0.06 mm	0.06~ 0.07 mm	0.07~ 0.08 mm	0.08~ 0.09 mm	0.09~ 0.1 mm	0.1~ 0.2 mm	0.2~ 0.3 mm	0.3~ 0.4 mm
Near ground	0	2.17%	4.35%	6.52%	6.52%	23.92%	54.35%	2.17%	0
Ground	0.13%	0.23%	0.33%	0.68%	2.23%	11.67%	82.51%	1.92%	0.15%

Dust	0.4~ 0.5 mm	0.5~ 0.6 mm	0.6~ 0.7 mm	0.7~ 0.8 mm	0.8~ 0.9 mm	0.9~ 1 mm
Near ground	0	0	0	0	0	0
Ground	0.02%	0.02%	0.02%	0.02%	0.01%	0.01%

It can be seen from Table 1 that the dust in the surface is dominated by fine sand (0.1 ~ 0.25mm), accounting for 83.47%,dust (0 ~ 0.1mm) accounted for 15.32%, coarse sand accounted for 1.21%, which medium sand (0.25 ~ 0.5mm) accounting for 1.13%,coarse sand (0.5 ~ 1mm) accounted for 0.08%. The dust separation coefficient of the surface is 0.037, the sorting is excellent, the skewness is 0.019, almost symmetric, kurtosis is about 0.724, which belongs to the narrow kurtosis type. The dust separation coefficient in sand sampler (high 65cm) is 0.042, good in sorting and skewness is 0.340, positive skewness, kurtosis is about 0.797, which belongs to the narrow kurtosis type. In addition to the skewness, dust particle size parameters in sand sampler and the earth's surface are relatively close to each other. The dust particles on the earth's surface are larger and poor sorting than the dust particles in the sand sampler which is caused by the selection of external factors such as wind.

4.2.2. Correlation analysis of dust particle size distribution on the earth's surface and sand sampler (high 65cm)

Table 2. Correlation between sand sampler and dust particle size distribution on the earth's surface

		Specific gravityofsurface dust	Sand dust ratioof sandsampler
Specific gravity ofsurface dust	PearsonRelevance	1	.953**
	Significance (bilateral)		.000
	N	19	19
Sand dust ratio of sand sampler	PearsonRelevance	.953**	1
	Significance (bilateral)	.000	
	N	19	19

**. Significantly correlated at .01 level (bilateral)

It can be seen from Table 2 that Available from the table, the Pearson correlation coefficient between the two variables is 0.953, the two-tailed test probability p is 0.000 <0.05,and the variables are significantly correlated. The dust particle size distribution in the sand sampler is similar to that in the earth's surface. The dust particle size analysis of earth's surface can well reflect the dust distribution characteristics in the sand sampler and it is of great practical significance to analyze the dust particle size on the earth's surface. It also shows that near ground wind carries dust mainly from the vicinity, only a small part is transported from a long distance.

4.3. Variation law of sediment discharge rate with height

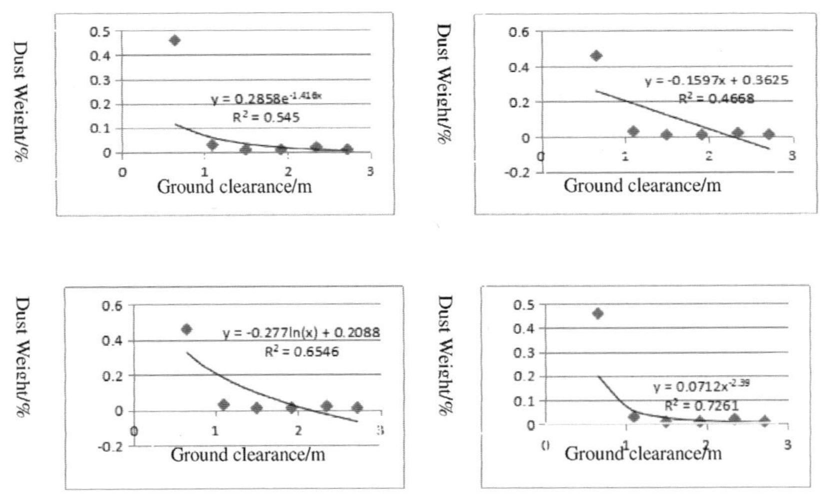

Figure 3. Relationship between dust weight and height

It can be seen from Figure 3 that the sand dust weight decreases with height in the sand sampler, and the best relationship is the power function,which is also the relationship between the sediment transport rate and the height.Dust in the air movement is mainly determined by their own gravity and air resistance, if there is not enough updraft, dust can only rise to a certain height and fall. The weight collected by the sand sampler at the entrance height of 1.1m is much different than that collected by the sand sampler of 0.65m height which shows that most of the dust has fallen below the height of 1.1m. This is consistent with the theory of aeolian sand boundary layer. From here it can be concluded that if the height of solar panels is higher than 1.1m, the amount of dust will be relatively small, the impact of solar power will be small.

5. Conclusion

(1) It is reasonable to describe the dust particle size distribution of the solar power station in Hobq Desert and it is not easy to control the suspended dust drift of dust due to the acquisition and subdivision, some experimental errors may occur.

(2) The surface dust is dominated by fine sand, followed by dust and coarse sand; the dust particles in the surface and sand dust particles in the sand sampler are close to each other. The dust particles in the sand sampler are smaller than the dust particles on the earth's surface, it is concluded that dust particles with smaller size are more likely to take off from the ground surface; sand sampler (65cm) has a great correlation with dust particle size on the ground surface, indicating that most of the sand dust is transported at close range.

(3) The dust weight increases with height in sand sampler obey the law of decreasing power function, indicating the existence of wind-sand boundary layer. The higher the solar panel height, the smaller the impact of dust settling on it.

References

[1] Ma Shiwei.Study on the structure of wind and sand flow [J]. *China Desert*, 1988, 8 (3): 8-22.

[2] Institute of Environmental Science and Engineering, Central South University, Changsha 410083, China); Study on aerosol concentration variation of dust storms in Minqin near ground [J] .*Chinese Journal of Environmental Science*, 2009, 29 (3): 496-504.

[3] Dong Zhibao, Sun Hongyi, Zhao Aiguo. WITSEG sandstorm instrument: multi-channel sandstorm in wind tunnel [J]. *China Desert*, 2003, 23 (6): 714-720.

[4] ZHAO Man-quan, WANG Jin-lian, LIU Han-tao, et al.Effects of structural parameters of sand-bearing apparatus on sand-collecting efficiency [J].*Journal of Engineering*, 2010, 26 (3): 140-145.

[5] Li Ying, Shi Yongji, Jiang Fu Qiang. Development of full wind direction gradient sand sampler [J]. Technical supervision of railway, 2012, 40 (2): 41-43.

[6] Li Zhenshan, Ni Jinren, Liu Xianwan.Study on sand storage efficiency of vertical lattice sand collecting sand instrument [J]. *Sediment Research*, 2003, 1: 24-31.

IOP Conf. Series: Earth and Environmental Science **146** (2018) 012038 doi:10.1088/1755-1315/146/1/012038

Experimental Research on Influencing Factors of Surface Friction Coefficient of Solar PV Panels

Zhang Dan[1] ,Zhao Mingzhi[1]

[1]School of Energy and Power Engineering, Inner Mongolia University of Technology, Hohhot,China 010051.
[2]China General Certification Center,Beijing,China 100000

[1]1078745247@qq.com ,[2]zhaomingzhi2020@163.com

Abstract: This paper analyzes the friction force model of dust on the solar panel. By using the weight of the solar panel on the solar panel, we can find out the angle that can best adapt to the sliding of dust by changing the angle of the solar panel. Coefficient of friction between the panels. The results show that the coefficient of sliding friction decreases with the increase of dust particle size.
Key words: dust; solar panels; coefficient of friction

1.Introduction

As countries pay more and more attention to the development of energy and environment, the proportion of traditional energy in energy consumption due to high pollution and high energy consumption will be smaller and smaller. As a renewable, non-polluting energy source, solar energy will grow rapidly. However, there will be some problems in the development of PV power plants, such as the problem of sand accumulation on PV panels. Its impact on the panel has gradually become a hot spot. As the dust slides down the panel, the sliding friction on the surface creates a drag. It can be seen that the friction coefficient of sand and dust affects the research of sediment accumulation to a certain extent on PV panels.

Many scholars at home and abroad to analyze the friction. Based on a swash plate slider in a hydraulic axial piston pump, W Shi [1] developed a rotational model using molecular dynamics to study the frictional properties of atoms. C Chen et al. [2] used molecular dynamics simulations to determine the effects of different temperatures on the dynamic friction properties of Cu substrates in single crystal and polycrystalline structures. The domestic research on the friction between dust and solar panels is relatively rare. Hu Lin et al. [3] studied the friction caused by the interaction between the rod and the particulate matter.DH Cho [4] and so on nanoscale friction and wear were studied.Wang Yazhen [5] studied the influence of the surface morphology on the friction force and established the friction model of the nano rough surface and the atomic smooth surface.

Although there are many researches on friction at home and abroad, there are few studies on the dust on solar panels. This paper focuses on the analysis of friction. Through the experiment, the variation law of the coefficient of sliding friction was obtained, and the main influencing factors were found out. By comparing the coefficient of friction between the dry and wet solar panels, the correlation between them was found out.

IOP Conf. Series: Earth and Environmental Science **146** (2018) 012038 doi:10.1088/1755-1315/146/1/012038

2.Friction Analysis of Sand Particles on Horizontal Panels

2.1 Sliding friction adhesion friction theory

*2.1.1The friction surface is plastic contact.*The actual contact area is only a very small part. When the load acts on the contact, the stress on the contact point reaches the yield limit of compression and plastic deformation will occur. Thereafter, the stress will not change and the load will continue to increase by increasing the contact area.
*2.1.2Sliding friction is the process of jumping.*The contact points are in a plastic flow state, and when the contact points slide, the contact points instantaneously generate high temperature, thus causing the two objects to stick together. The adhesion points have the adhesive force. Under the action of the friction force, the shear point at the adhesion point slides and the sliding friction Is accompanied by the formation of sticky point and shear process.
2.1.3.Friction is the sum of the resistance caused by the sticking effect and furrow effect
F friction consists of two parts:

$$F = T + P_e = A\tau_b + Sp_e \quad (1)$$

In the formula, T is the shear force, f is the furrow force, A is the adhesion area, ie the actual contact area, the shear strength at the adhesion point, S is the furrow area, and is the furrow force per unit area. Experiments show that the size is related to the sliding speed and lubrication state and is close to the shear strength limit of the soft material, which indicates that the shear occurs inside the soft material and causes the phenomenon of material migration.It is proportional to the yield point of the soft material and not to the lubrication state, and the depth of embedment decreases as the yield point of the soft material increases.

2.2 Sand particles in the horizontal panel on the friction force model

In order to predict the sliding resistance of dust on solar panels, the dust is first reduced to spherical particles, the cross-section of the ball over the center of the sphere is a cylinder with a width of 1, which is convenient for stress analysis. Figure 1 shows a simplified sand particle-solar panel interaction model. Assuming that the force of the solar panels on the dust particle size is in the radial direction, the resultant force in the horizontal and vertical directions can be written as:

$$R = \int_0^{\theta_0} \sigma r \sin\theta \, d\theta$$

$$W = \int_0^{\theta_0} \sigma r \cos\theta \, d\theta \quad (2)$$

In the above formula, R is the sliding friction; W is the vertical load, that is, the wind load; σ is the normal stress; r is the radius of the wheel; is the contact angle;

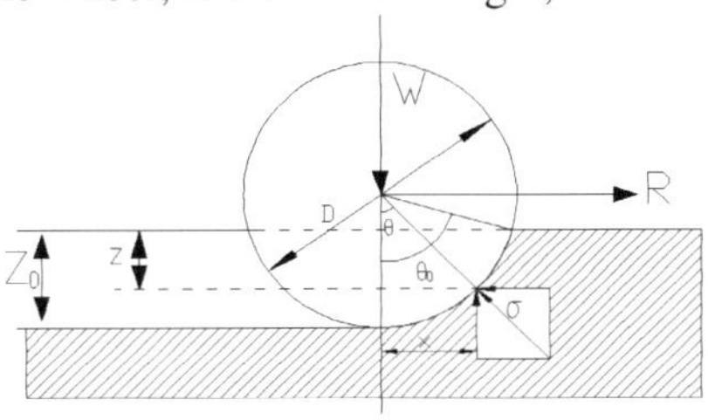

Figure 1. Simplified dust particle-solar panel interaction model

Assuming that the radial pressure σ acting on the dust edge is the normal pressure-settlement relationship under the tempered glass plane of the solar panel at the same depthz [6]

$$\sigma = p = (k_\varepsilon + k_\phi) z^n \quad (3)$$

IOP Conf. Series: Earth and Environmental Science **146** (2018) 012038 doi:10.1088/1755-1315/146/1/012038

In the formula, and are respectively tempered glass cohesion and frictional deformation modulus; n is the subsidence modulus.

The formula (3), and into (2) the first formula and points were:

$$R = \left(\frac{z_0^{n+1}}{n+1}\right)\left(k_\varepsilon + k_\phi\right) \tag{4}$$

Available from Figure 2.9 $x^2 = [D - (z_0 - z)](z_0 - z)$ (5)

When the amount of subsidence is small$x^2 \approx D(z_0 - z)$Existing$2x\,dx = -D\,dz$.From (2) the second formula is:

$$W = \left(k_a + k_\phi\right)\int_0^{z_0} \frac{z^n\sqrt{D}}{2\sqrt{z_0 - z}}\,dz \tag{6}$$

other $z_0 - z = t^2$,then $dz = -2t\,dt$,into（6）Got it

$$w = \left(k_a + k_\phi\right)\sqrt{D}\int_0^{\sqrt{z_0}}(z_0 - t^2)^n\,dt \tag{7}$$

Expand the $(z_0 - t^2)^n$, only take the first two items, the amount of solar panels available after settlement

$$z_0 = \left[\frac{3W}{(k_a+k_\phi)\sqrt{D}(3-n)}\right]^{2/(2n+1)} \tag{8}$$

Substituting equation (4) in Eq. (8), we get sliding friction R

$$R = \left[\frac{(\sqrt{D})^{2n+2}}{(3-n)^{2n+2}(n-1)^{2n+1}n(k_a+k_\phi)(3W)^{2n+2}}\right]^{1/(2n+1)} \tag{9}$$

In order to study the force and force of sand under the same area under wind load and different particle sizes under load, a certain amount of pressure, F, between the planks and the solar panels under a certain area of the planks of S is assumed to be Layers of dust particles loosely arranged, dust particles is a regular ball, as shown in Figure2:

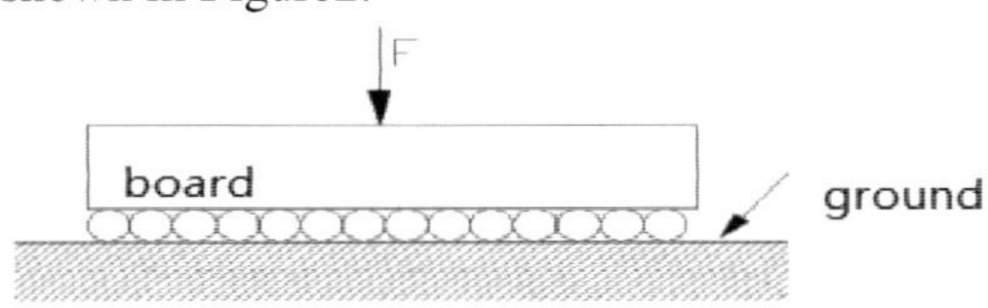

Figure 2. Single layer dust particles placed between the plank and the ground

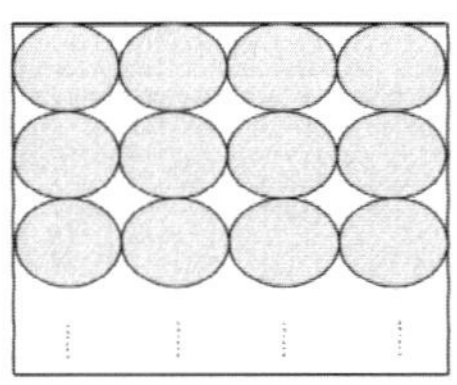

Figure 3. Single-layer dust particles in the solar panel surface arrangement diagram

It is assumed that the dust particles are the most loosely arranged on the solar panel [7], as shown in Figure3. The total number of dust particles is N, then:

$$N = \frac{S}{D^2} \tag{10}$$

The vertical load on each dust particle is:

$$W = \frac{F}{N} = \frac{FD^2}{S} \tag{11}$$

Substituting (11) into (9) gives the sliding friction force R as:

$$R = \left[\frac{S^{2n+2}}{(3-n)^{2n+2}(n-1)^{2n+1}n(k_a+k_\phi)(3F)^{2n+2}D^{3n+3}}\right]^{1/2n+1} \tag{12}$$

By $R = \frac{4}{3}\pi D^3 \rho g f$,Got it:

$$f = \frac{3}{4\pi\rho g D^3} \cdot \left[\frac{S^{2n+2}}{(3-n)^{2n+2}(n-1)^{2n+1}n(k_a+k_\phi)(3F)^{2n+2}D^{3n+3}}\right]^{1/2n+1} \tag{13}$$

In the formula, the density of dust particles. From (13), we can see that when the diameter of dust particles increases, the sliding friction coefficient f decreases.

3.Dust on the sloping panels friction coefficient of variation of sliding law

3.1 Experimental Preparation

Solar panels will be affected by the wind load, meanwhile, the dust on the solar panels will also be affected by the wind load during the sliding process. The wind load equation [8] of the solar panel is as follows:

$$W = 0.5C_w \rho V_{max}^2 S\alpha IJ \tag{14}$$

Where: wind coefficient, take 0.65 +0.009 theta; θ is the angle between the solar panel and the ground to take 90 degrees; for the air density to take 1.225; for the maximum wind speed to take 30; S solar panels to take the area Solar panels are made of wood with an area of 0.0036; I is a factor of 1; α is a height correction factor of 1.631; J is an environment factor of 1.15.W is a maximum of 5.434N, as long as the force applied to the panel is less than this value In line with the experimental conditions, in the following analysis, the pressure generated by the weight of the wood instead of solar panels suffered wind load.

The tilt angle of the solar panel bracket is set to 0°、15°、30°、45°、60°、75°,placed on the solar panels wooden block size of 6 × 6 × 2cm, the dust particles adhere to one side of the block, the adhesion of sand Dust particles face and solar panels close together with a dynamometer slow and uniform pulling wood, and record the dynamometer data, as shown in Figure 4 and Figure 5:

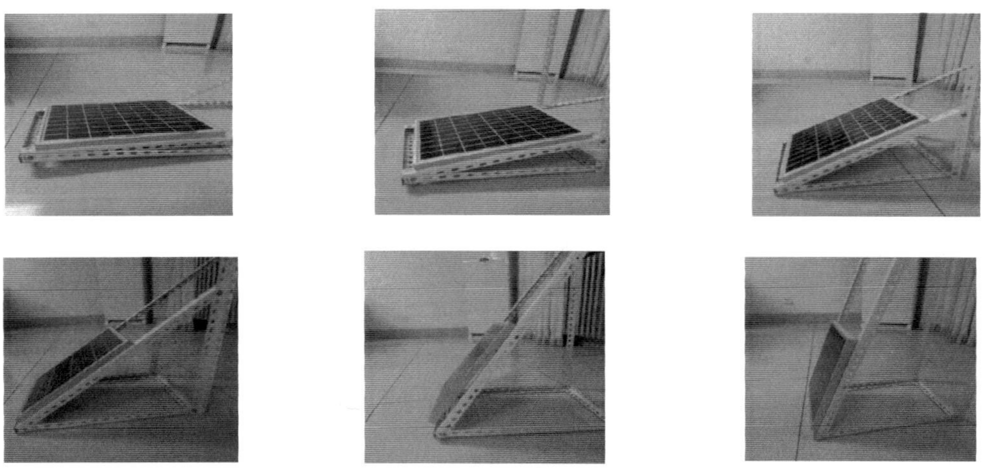

a.Inclination is0° b.Inclination is15° c.Inclination is30°
d.Inclination is45° e.Inclination is60° f.Inclination is75°

Figure 4. Solar panels with different angles of inclination

IOP Conf. Series: Earth and Environmental Science **146** (2018) 012038 doi:10.1088/1755-1315/146/1/012038

Figure 5. Slides on a solar panel

In order to obtain a more accurate coefficient of friction, measured four times at each angle, the first dynamometer pulls only one block of wood, after each test to add a block of wood, up to four until each sliding friction coefficient , And then take the average is the sliding friction coefficient of this angle.

As dust and wood stick together, the force analysis can be regarded as a whole, the weight of the wood along the slope of the solar panel component of the size of the force of friction with the sand the same size, wood Gravity perpendicular to the slope of the solar panel component of the size of the wind load the same size. Wind load here is mainly referring to the pressure resistance, friction resistance is much smaller than the pressure resistance, will not be considered. Thus, the formula for calculating the coefficient of sliding friction between sand and solar panels is obtained as follows:

$$f = \frac{F - m\ g\ \sin\theta}{m\ g\ \cos\theta} \tag{15}$$

Where f is the coefficient of sliding friction between sand and solar panels; F is the reading of the tensioner, that is, the tension; m is the mass of the block; θ is the inclination of the solar panel; g is the acceleration of gravity,take $9.8 m/s^2$.

3.2 Dust on the dry solar panels coefficient of variation of friction

In order to obtain the variation law of the friction coefficient of dust particles on the solar panel, the range of 0.07 ~ 0.08, 0.08 ~ 0.09, 0.09 ~ 0.1, 0.1 ~ 0.2, 0.2 ~ 0.3 was selected in the dust, In the experiment, the sliding friction coefficients measured by different numbers of wood blocks are averaged, and then the sliding friction coefficients obtained from different six angles are averaged. The friction coefficient varies with the particle size of dust Figure 6 shows:

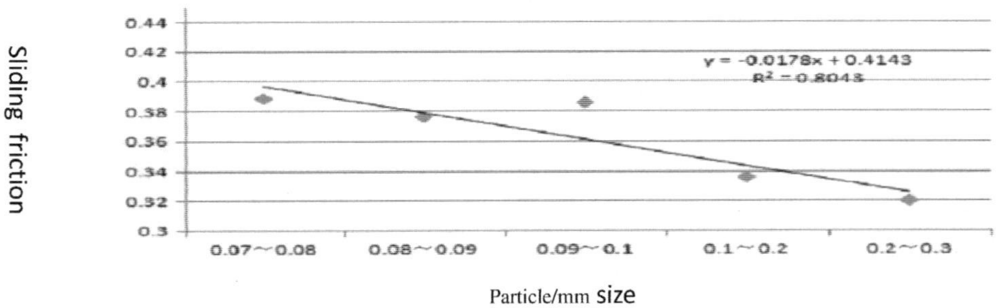

Figure 6. Sliding friction coefficient with the size of dust particle size changes

As can be seen from the figure, the sliding friction coefficient tends to decrease as the particle size becomes larger, and the correlation coefficient between the sliding friction coefficient and the size of the particles is obtained. The correlation coefficient between them is 0.804, indicating that the particle size The size of the coefficient of sliding friction has a certain ability to explain. In less than 0.1mm, the change is relatively stable, basically unchanged; more than 0.1mm, the change is relatively larger. This is part of the reason why the coefficient of friction changes slowly and slowly.

When the dip angle of solar panels changes, each dip angle averages the friction coefficient for different dust particle ranges and loads, and the sliding friction coefficient between dust and solar panels also changes, as shown in Fig.7Show:

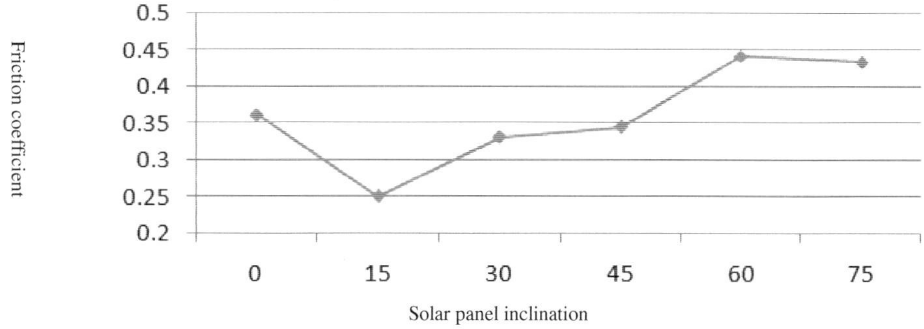

Figure 7. Sliding Friction Varies with Solar Panel Tilt

The coefficient of sliding friction between sand and solar panels is minimized at this angle of inclination as the angle of inclination of the solar panel closely follows the slip of dust particles. It can be seen from Figure 4.10 that the solar panel has a dip angle of about 15 °. For the dust in different places, the physical and chemical properties of its surface are different, the surface topography is also different. Different solar panels, the surface is also different, so this dip is random and accidental.

4 Conclusion

Through experiments, changing the angle of solar panels and the wind load, we can get the rule that the sliding friction coefficient decreases with the dust particle size increasing, which is mainly caused by the decrease of the adhesion effect. From the perspective of changing solar panels, when the dust particle size is relatively small, the friction coefficient tends to increase with the increase of load. It can not be determined which role plays a leading role. When the particle size increases, the coefficient of sliding friction increases with the increase of load decreasing trend. Through the solar panel inclination changes, the coefficient of sliding friction coefficient at the minimum when the inclination of about 15 °, at this angle most can adapt to the sliding of dust.

References

[1] Shi W, Luo X, Zhang Z, et al.Influence of external load on the frictional characteristics of rotary model using a molecular dynamics approach[J].Computational Materials Science, 2016,122:201-209

[2] Chien C, Wang C, Tsai C, et al. Temperature effect on kinetic friction characteristics of Cu substrate composed by single crystal and polycrystalline structures[J].Computational Materials Science, 2016,117:412-421

[3] HU Lin, YANG Ping, XU Ting et al. Static Friction on Round Rods in Particulate Matter [J] .Acta Physica Sinica, 2003, 52 (4): 879-882

[4] Cho D. H, Bhushan B, Nanofriction and nanowear of polypropylene, polyethylene terephthalate,and high-density polyethylene during sliding[J]. Wear, 2016,s352-353:18-23

[5] WANG Ya-Zhen, HUANG Ping.Calculation of Microscopic Sliding Friction on Nanoscale Stochastic Rough Surface [J] .Acta Physica Sinica, 2013,62 (10): 106801-106801

[6] Huang Zuyong. Principles of ground vehicles [M]. Beijing: Mechanical Industry Press, 1985

[7] Li Guangxin. Soil mechanics [M]. Beijing: Tsinghua University Press, 2013

[8] Zhang Qizhu, Liu Zhizhang, Qi Xiaohui, et al.Analysis of wind load on solar PV panels [J]. Energy Technology, 2010, 31 (2): 93-95

IOP Conf. Series: Earth and Environmental Science **146** (2018) 012039 doi:10.1088/1755-1315/146/1/012039

Study on Treatment of Preserved Wastewater by Combined Ultrafiltration-Nanofiltration

Mengxiang Zeng[1] ,Qian YE [1] ,Wanjia LU [2] ,Renjie Ma [1]

[1]Xiamen University of Technology, Xiamen,361024, China
[2]Nanxian No.High School,Yiyang,413200,China

zengmx@xmut.edu.cn

Abstract. The treatment process of preserving soaking waste water by combined ultrafiltration-nanofiltration process was investigated. In the nanofiltration experiment,measuring the effects of different enrichment times on the flux, COD, total phosphorus and ammonia nitrogen to analyze the characteristics of the nanofiltration process and membrane fouling problems during nanofiltration.The results indicated that when the enrichment multiple was 4 times,the removal rate of COD was up to 87%, the removal efficiency of total phosphorus was above 90% and the removal rate of ammonia nitrogen was up to 63%, besides the membrane flux could be maintained at a better level.With prolonged operation, membrane flux decreased by 71%.After being soaked and cleaned with sodium hydroxide solution, membrane flux can be restored to more than 73% of the original. The concentrated waste water soaked by nanofiltration can be treated with advanced oxidation process,and the filtrate can reach the standard for reuse water.

1. Introduction

Candied soaking wastewater has low pH value, high salinity and high molecular weight of organic pollutants[1], and these are the key factors that affect the operation and stability of the wastewater treatment system.In this paper, on the basis of characteristics of preservative soaking wastewater , combined with the decontamination advantages of ultrafiltration and nanofiltration to research treatment process of preservative soaking wastewater with “combined ultrafiltration-nanofiltration”.Ultrafiltration membrane has been widely used in wastewater treatment,and it has a very high removal rate of suspended particulate matter, microorganisms and turbidity in water,but the ultrafiltration membrane itself belongs to mesoporous membrane[2]It’s removal effect on dissolved salts,dissolved organic matter and some small organic molecules in the water is not good, while organic pollutants part of preservative soaking wastewater are small molecules[1].Nanofiltration can not only effectively remove most of the organic matter in water, but also has a certain removal effect on some monovalent ions[3]. In this paper,through the actual wastewater operation test,we optimized the concentration multiples and membrane cleaning program to obtain efficient and stable treatment process, so as to provide some references and basis for solving the pollution of candied fruit soaking wastewater.

2. Experimental system and water quality

2.1. Experimental system

Figure 1 is experimental system diagram of the combined ultrafiltration - nanofiltration, preservative soaking wastewater firstly got into the ultrafiltration membrane,the effluent from the ultrafiltration membrane enters the nanofiltration membrane for concentration treatment.The transmembrane pressure of the ultrafiltration membrane is 0.2-0.4 MPa and the transmembrane pressure of the nanofiltration membrane is dominated as 2.0 MPa[4].Before the experiment, ultrafiltration membrane and nanofiltration membrane were rinsed with pure water to reduce the impact of membrane fouling on the experiment.

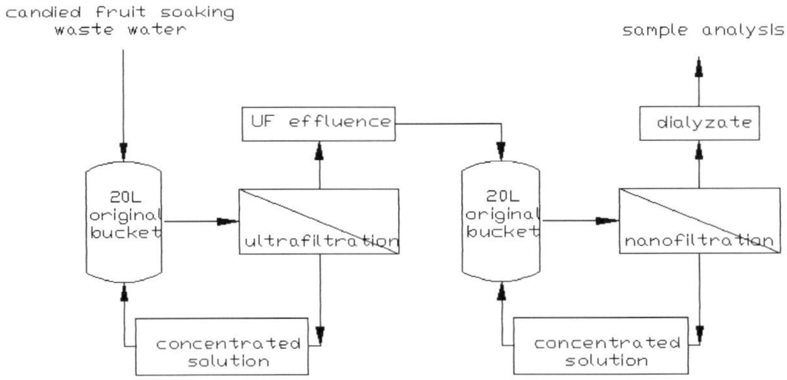

Figure 1. UF-NF system

Ultrafiltration system:The ultrafiltration membrane selected by the ultrafiltration system is 181 tube ultrafiltration ceramic membrane,the operating pressure of equipment is 0.35Mpa. First of all filling the waste tank with clear water, connect the electric source,turn on the machine,after adjusting the parameters, waiting for the machine to run for a period of time,test the water membrane flux of ceramic membrane. Turn off the machine, drain the water and pour the preservative soaked waste water into the two-thirds of the original liquid tank, re-open the machine to run under the specified parameters for 5mins, taking the filtrate to determine its parameters.

Nanofiltration system: 250 composite membrane is selected by nanofiltration system,the operating pressure of equipment is 1.5Mpa. First of all filling the waste tank with clear water, connect the electric source, turn on the machine, after adjusting the parameters, waiting for the machine to run for a period of time, test the water membrane flux of ceramic membrane.Turn off the machine,drain the water and pour the preservative soaked waste water into the two-thirds of the original liquid tank, re-open the machine to run under the specified parameters for 5mins, taking the filtrate to determine its parameters.

Cleaning scheme of membrane: During actual membrane treatment, the permeate flux will decreasewith prolonged operation[5],and membrane fouling problems will emerge.The experiment used different chemical reagents:such as sodium hypochlorite, sodium hydroxide and so on to clean the contaminated film,according to the cleaning effect of membrane, so as to determine the best cleaning method.

3. Experimental drugs and equipment

3.1. Materials

Sulfuric acid and sodium hydroxide,analytical reagent,Xilong Chemical Co., Ltd., ultrafiltration membrane material is the 181 tube ceramic membrane, molecular weight cutoff is 5 × 104U;nanofiltration membrane is GE osmosis membrane, molecular weight cutoff is 3 × 102U.

Ceramic/tubular film small test equipment,CeraMem-0100,Xiamen World Tat Film Technology Co., Ltd.small roll membrane equipment, RNF-0460, Xiamen Fumei Technology Co., Ltd.

3.2. Experimental water quality

The experimental wastewater was collected from a salt immersion pond of a preserves plant in Yongtai County. The water samples showed a light yellow color and a large viscosity.Contents of COD, chloride and organic acid are high,the water quality index of experimental wastewater is shown in Table 1.

Table 1. Experiment waste water quality

COD(mg/L)	NH3-N(mg/L)	TP(mg/L)	pH	Cl-(mg/L)
4800	112	25.2	4.6	8750

4. Results and discussion

4.1. Ultrafiltration of preservative soaking wastewater

The experiment selected the ceramic membrane for ultrafiltration pretreatment, the device is simple and easy to operate.Firstly pouring waste water into the original liquid tank, the operating equipment began to filter, until all the wastewater was filtered out,taking the supernatant to measure its parameters,the results are shown in Table 2:

Table 2. Water quality after UF

COD(mg/L)	NH3-N(mg/L)	TP(mg/L)	pH	Cl-(mg/L)
3650	102	23	4.6	8550

From the experimental results, it can be seen that the removal efficiency of COD can reach 24% after the preservative soaking wastewater was ultrafiltrated,with little change in other parameters.However, the viscosity of water reduced from 2.1CP to 1.2CP, the effluent is more clear,which can effectively avoid membrane fouling during subsequent film process.

4.2. Effect of concentration ratio on the water quality of preservative wastewater during the process of nanofiltration

In this experiment, the ultrafiltration permeate was subjected to nanofiltration concentration experiment under the conditions of operation pressure 15 bar and the temperature 25-40 ° C.The operation mode of all retentate recycle was adopted.The volume of each batch of feed liquid is 20L, and the membrane is not rinsed during the concentration of the feed liquid.In the nanofiltration filtration process, the concentration multiplier has a great influence on the treatment effect.This article investigated different concentrations of dialyzate influenced on the chemical oxygen demand COD, total phosphorus TP, ammonia nitrogen [NH 3 -N] Through flux and so on.

*4.2.1. Influence of concentration ratio on COD.*This experiment use potassium dichromate titration method to measure the COD of dialyzate with different concentration times.

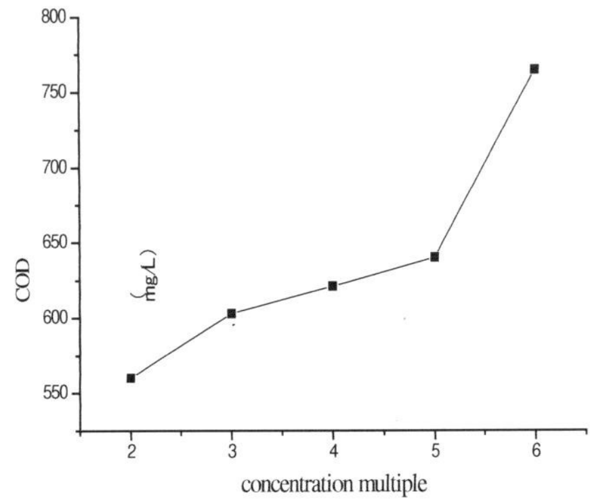

Figure 2. Influence curve between concentration multiple and COD

As shown in Figure 2,during the process of nanofiltration,the COD value of dialyzate gradually increased with the increase of concentration times.The COD value was 562mg /Lwhen the dialyzate was concentrated to 2 times, while the COD value increased slowly when the dialyzate was concentrated from 2 times to 4 times.When the concentration times of dialysate reaches 5, the COD value reached 645mg/L.The upward trend of COD is more obvious when it is concentrated up to 6 times. This is mainly because the nanofiltration is a recycling process. The concentration times of organic pollutants in preservative soaking wastewater increases with the increase of operating time.The increase of concentration of concentrated side liquid leads to the continuous increase of COD concentration on the dialysis side.When the concentration times was 4 times, interception rate of COD was 87% or more,which can ensure better effluent quality and improve processing efficiency.

4.2.2. Effect of concentration times on total phosphorus

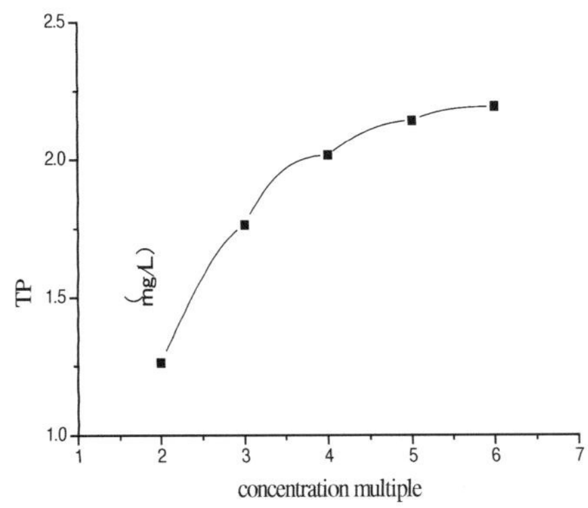

Figure 3. Influence curve between concentration multiple and TP

The effect of nanofiltration concentration times on total phosphorus is shown in Figure 3. After ultrafiltration treatment, total phosphorus concentration of preservative soaking wastewater was 23mg / L.During the process of nanofiltration, the concentration of total phosphorus decreased obviously after treatment by nanofiltration membrane. With the increase of concentration times,the total phosphorus concentration of dialyzate increased gradually, but it was basically below 2.5mg /L and the removal efficiency was over 90% .

*4.2.3. Effect of concentration times on ammonia nitrogen.*The experiment uses ultraviolet spectrophotometry to determine the concentration of ammonia in different dialyzate.Due to the high concentration of ammonia nitrogen in the dialyzate, this experiment was measured by the dilution method,ammonia concentrations of dialysate with different concentration times are shown in Figure 4.

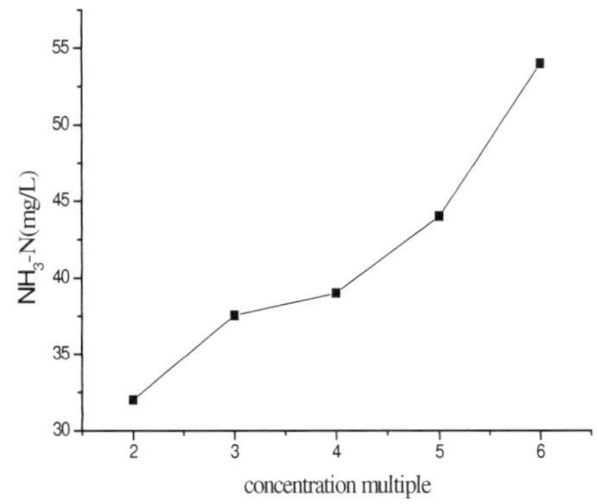

Figure 4. Influence curve between concentration multiple and NH3-N

The results showed that the initial concentration of ammonia nitrogen in the ultrafiltrate effluent of preservative soaking wastewater was 102.0mg / L.After treatment through nanofiltration membrane, the concentration of ammonia nitrogen in the dialyzate was reduced to 23mg / L ~ 60mg / L.During the nanofiltration concentration , With the increase of concentration, dialyzate ammonia concentration gradually increased. When nanofiltration is concentrated at about 4 times,ammonia nitrogen removal rate of nanofiltration effluent was 63%.Thus, Nanofiltration membrane has high efficiency of ammonia nitrogen treatment,with less energy consumption and stable treatment results.

*4.2.4. Influence of concentration ratio on permeation flux.*In this experiment, under the condition of temperature 25 ℃ with the operating pressure was fixed to 2.0MPa at the same time,then all the intercepted liquid was used with the operation mode of internal circulation to conduct the nanofiltration concentration experiment on preservative soaking wastewater.The volume of each batch of feed is 20L, and the membrane is not rinsed during a complete feed concentration. The change of nanofiltration membrane permeation flux with concentration factor is shown in Figure 5.

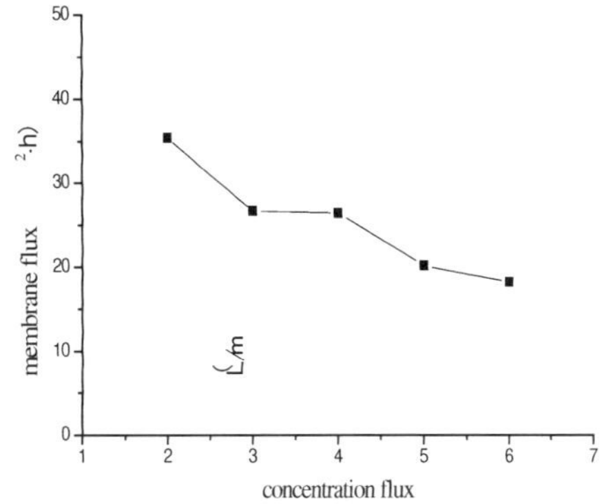

Figure 5. Influence curve between concentration multiple and NF membrane permeate flux

At the beginning of nanofiltration, the membrane permeation flux was larger,which was 41.67 L / (m^2·h). As can be seen from Figure 5, with the increase of the concentration factor,the permeation flux decreases rapidly, which is due to the feed concentration gradually increased, the viscosity also increased,then the membrane permeation flux decreased owing to intensification of concentration polarization.Concentration polarization will increase the osmotic pressure inside the membrane, reducing the effective operating pressure. In addition, the phenomenon of concentration polarization can cause the solute to form a resistance layer on the membrane surface and also prevent the solvent from passing through.That is, the concentration polarization can reduce the permeation flux of the nanofiltration membrane.The effect of concentration polarization layer on the solute permeability is relatively complex. The concentration of some solutes in the concentration polarization layer is higher than that in the bulk liquid material,increasing the driving force of these solutes when they through the nanofiltration membrane,thus the rate of interception is reduced. At the same time, due to the existence of the concentration polarization layer, diffusion resistance of some solutes increase, so that the permeability of these solutes is reduced.

When concentrated to 3 to 4 times, the nanofiltration membrane flux gradually stabilized. The main reason is that as the concentration proceeds, due to the formation of the solute-gel layer on the surface of the membrane, the rate of reverse diffusion of solute tends to be balanced with the concentration polarization, and the secondary resistance of the gel is basically unchanged, so that the permeation flux of the membrane Gradually stable.

After concentration to 5 to 6 times, the permeation flux of the membrane decreased, which is due to membrane fouling caused by the decrease of membrane flux.Membrane fouling is reversible (or irreversible) deposition(Including plugging, adsorption, the formation of filter cake, precipitation, etc.),in the membrane surface or in the membrane pores due to mechanical or physicochemical interactions existed among trapped micelles, particles, suspensions, milk filtrates, salts and macromolecules,resulting in smaller or blocked membrane pore size to make an irreversible change in membrane separation and permeation flux.

4.3. Membrane fouling and cleaning

After the end of the experiment, then clear water permeation flux of nanofiltration membrane was measured through the water with the experiment. It is calculated that through the clear water permeation flux J was 40.91L/(m^2·h), the membrane flux is reduced by 71%.The data show that during the nanofiltration concentration process with preservative soaking wastewater, the original nanofiltration membrane is easily contaminated by some pollutants in the wastewater, which leads to the decrease of separation performance and the increase of operation cost,and the pollutants include organic pollutants, microorganisms Pollutants, inorganic contaminants, etc[6].

Chemical cleaning is an important way to reduce pollution,and the concentration and cleaning time of chemical reagent depends on the chemical resistance of nanofiltration membrane.Common chemical cleaning methods include chelant cleaning, acid-base cleaning, surfactant cleaning and oxidant cleaning , Enzyme cleaning and salt washing[7]. In membrane separation operations, contamination begins when the feed is in contact with the membrane. When the membrane module is operated for a period of time, the permeation flux and the separation performance of the membrane are both decreased.Membrane cleaning can make the membrane performance parameters to restore as much as possible and prolong the service life of the membrane [8].

The two stage cleaning method is used in this experiment.That is, firstly the nanofiltration membrane is immersed in 0.1% sodium hypochlorite for 6 hours, and the sodium hypochlorite of the same concentration is exchanged once among these six hours to measure the permeation flux of the nanofiltration membrane;Then making the nanofiltration membrane immersed in clean water for 6 hours one more time,change the clear water every 2 hours to measure the permeation flux of the nanofiltration membrane;Finally,making the nanofiltration membrane immersed in sodium hydroxide solution for 6 hours to measure the permeation flux of the nanofiltration membrane. After 6 hours, the membrane was rinsed with deionized water to remove organics from the membrane surface and residual sodium hypochlorite,which is as shown in Figure 6.

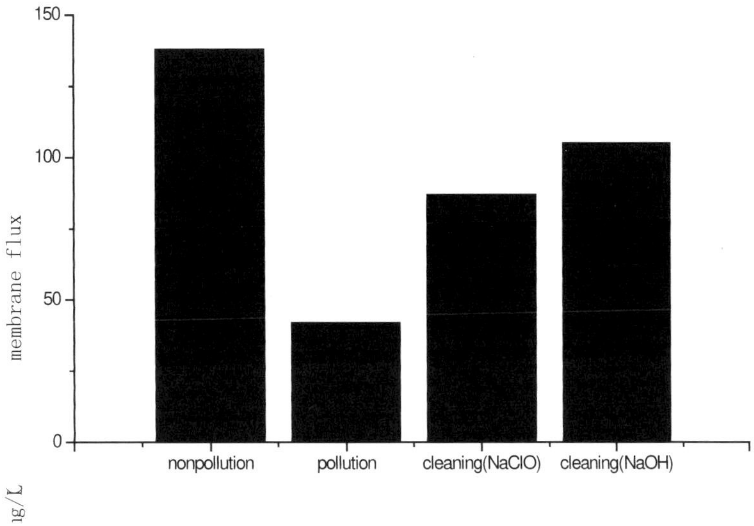

Figure 6. Chart of permeate flux before and after NF membrane pollution

It can be seen from Figure 6 that after using sodium hypochlorite solution to wash, the permeation flux of the nanofiltration membrane was recovered to 90.64L / ($m^2 \cdot h$); after it was soaked in clear water, using sodium hydroxide solution to clean it,so that the permeation flux of the nanofiltration membrane was restored to 102.85 L /($m^2 \cdot h$), corresponding to 73.76% of the new membrane. The analysis suggests that the decrease of membrane permeation flux is mainly caused by organic matter and microorganisms blocking membrane pores and accumulating on the membrane surface to form cake layer.Simple physical cleaning can only remove the loose physical adsorption adsorbed on the membrane surface; sodium hypochlorite and sodium hydroxide are commonly cleaning agents, which can effectively remove the formation of pollution formed by the adsorption and adhesion of organic layer and microorganisms in the membrane surface.However, there is still some membrane permeation flux can not be restored,and the irreversible pollution is a number of small molecular weight organic matter and microorganisms plug the membrane pores,so that the membrane pores become smaller and the membrane resistance increase,both of which make the water flux decrease, this part of the pollution is difficult to remove by physical and chemical cleaning.

5. Conclusion

In this study, the preservative soaking wastewater was treated by combined ultrafiltration-nanofiltration,and analyse the influence of different concentration times of dialysates on on the COD of dialyzate, total phosphorus,ammonia nitrogen and membrane permeation flux after the wastewater was concentrated by nanofiltration,besides the influence of operating conditions on the performance of nanofiltration membrane was investigated. The main conclusions are as follows:

Through chemical analysis,some components in preservative soaking wastewater were discussed.Through the combined ultrafiltration -nanofiltration, the membrane flux can be kept at a good level when the concentration ratio is 4 times, the COD removal rate can reach 87%, the total phosphorus removal rate can reach 90%, and the ammonia nitrogen removal rate can reach 63%.With prolonged operation, membrane permeation flux will decrease.After soaked and cleaned with sodium hydroxide solution, membrane permeation flux can recover to 73.76%.

Fund Project: Fujian Provincial Department of Education JK Class: Study on treatment of high salinity preserved wastewater (Project No.: JAT160343) Fujian Provincial Scienceand Technology Project (No. 2014N0031)

References

[1] Li Shuzhan. Flotation - anaerobic - aerobic combination process of dried fruit dried bean curd wastewater [J]. Industrial Water Treatment, 2013,33 (12): 59.
[2] Wen Xiaoxian, Zhou Zheng, Wei Gang, et al.Experimental study on the deep treatment of coking wastewater by ultrafiltration nanofiltration [J]. Water Treatment Technology, 2010,36 (3): 93-95.
[3] Li Xiang, Zhang Zhongguo, Ren Xiaojing, et al.Research progress on nanofiltration membrane materials [J]. Progress in Chemistry, 2014,33 (5): 1210-1211.
[4] Lu Yonghong, Li Dongliang, Hou Xueyan. Nanofiltration membrane separation technology research [J]. The second session of the National Membrane Separation Technology in the Food Industry Symposium on Proceedings, 2006.
[5] Pei Yuxin, Xu Youyi. Nanofiltration membrane separation technology and its research progress [J]. Membrane Science and Technology, 1999,19 (5): 10.
[6] Yan Zhongsen, Qu Fangshu, Liang Heng et al .. Advances in ultrafiltration membrane fouling and pre-membrane technology [J]. Membrane Science and Technology, 2014,34 (4): 108-109.
[7] Huan Guolan, Zhang Yufeng, Du Qiyun. Mo pollution analysis of extreme control [J]. Water Treatment Technology, 2003,29 (1): 1-3.
[8] Xue Gang, Gao Pin, Dai Xuemin, et al. Micro-flocculent fiber ball filtration - ultrafiltration - Nanofiltration combined bathing wastewater treatment of hotel [J]. Water Treatment Technology 2008.34 (3): 57-58.

Numerical simulation of the spread of bacteria in water flow considering the process of reproductions

Xiaoxi Zhang[1, 2], Chao Zhao[1, 2], Namei Guo[1, 2] and Qiuhua Chen[3,*]

[1]School of Environmental Science and Engineering, Xiamen University of Technology, Xiamen 361024, PR China
[2]Key Laboratory of Environmental biotechnology，Fujian Province University, Xiamen 361024, PR China
[3]School of Civil Engineering and Architecture, Xiamen University of Technology, Xiamen 361024, PR China

[*]Correspongding author: chenqiuhua@xmut.edu.cn

Abstract: In this paper, a simple approach for modelling the reproduction of bacteria in water flow is presented. Firstly, the basic mathematical model for simulating the advection-diffusion of pollutions in water flow is reviewed and validated. Then, a solution strategy of doubling the concentration at regular intervals by the basic model is proposed. This approach is easy to be implemented in the existing flow field solver and is proved to be feasible by simulating a standard example. Finally, the spread of the bacterial pollution in a pool is simulated. The result shows that the water quality variation in the pool can be attributed to both the flow patterns of the water flow and the reproduction characteristics of the bacteria. The recirculation region limits the spread of the bacteria, which induces the continuous deterioration of water quality when considering the reproduction of the bacteria.

1. Introduction

Water has tremendous impact on life health, but its quality is from bad to worse with the evolution of the modern society, which has become the global problems. In the last few decades, more advanced theories and methods have been developing not only to analyse the water quality, but also to predict its developing trends[1].

Numerical simulation, which uses to solve problems numerically, has been widely employed in the field of water environment. With the help of high-speed computers, this method can give comprehensive and visual results of complex problems, which is more economical than the traditional experimental analysis and especially suitable for studying the harmful conditions. The basic mathematical models and numerical methods for modelling water environment can be found in the publication of Peng et al.[2]. In recent years, researches in this area are growing very fast both in theories and applications. Chen et al.[3] presented a new numerical approach by coupling the one-dimensional and two-dimensional shallow water models to accurately predict the water environment in large watersheds. Xu et al.[4] proposed a new hybrid approach of genetic algorithm and fuzzy simulation to solve the water quality management problem. Wang et al.[5] developed a new method for evaluating the water quality in the watershed with limited data and studied the water pollution of a watershed by the Hydrological Simulation Program-Fortran model. Ding et al.[6] provided an approach to simulate the dynamic diffusion and migration process of pollutions and apply it to a practical project. These researches offer us various options to solve different problems about water quality.

However, the existing work for simulating the water quality just considered the physical and chemical factors, such as the formation and diffusion of pollutions, while ignoring the biological factors, such as the population growth of microbes. In ideal conditions, bacteria can increase their population by a factor of two in dozens of minutes. That means a small number of bacteria can easily grow to millions in several hours. Therefore, the increase in population must be considered when the simulation involves microbe contaminates.

In this paper, a simple approach for modelling the reproduction of bacteria in the simulation of water quality and its application is presented. This approach can be easily implemented in the flow solver by the help of an external program, which provides an effective tool to solve these problems.

2. Numerical method

The first step of numerical simulation is to establish the mathematical models among different physical variables. Before the discussion of bacteria, the convective-diffusion model of the general pollution in water flow should be mentioned, which includes the mass conservation, momentum conservation equations and the transport equation of the pollution are need. These equations can be written as follows[2]:

Equation of mass conservation:

$$\frac{\partial \rho}{\partial t} + div\left(\rho \vec{u}\right) = 0 \tag{1}$$

Equation of momentum conservation:

$$\left.\begin{aligned} \frac{\partial(\rho u)}{\partial t} + div\left(\rho u \vec{u}\right) &= -\frac{\partial p}{\partial x} + div\left(\mu \; grad \; u\right) + S_u \\ \frac{\partial(\rho v)}{\partial t} + div\left(\rho v \vec{u}\right) &= -\frac{\partial p}{\partial x} + div\left(\mu \; grad \; v\right) + S_v \\ \frac{\partial(\rho w)}{\partial t} + div\left(\rho w \vec{u}\right) &= -\frac{\partial p}{\partial x} + div\left(\mu \; grad \; w\right) + S_w \end{aligned}\right\} \tag{2}$$

Transport equation of the pollution:

$$\frac{\partial(\rho C)}{\partial t} + div\left(\rho \vec{u} C\right) = div\left(\Gamma grad \; C\right) + S_C \tag{3}$$

where p is the pressure, u, v and w are the velocity components, C means the concentration of the pollution, Γ is the diffusion coefficient of the pollution, ρ is the water density, t is the time, x, y and z is the coordinate component, S indicates the source term. In this study, these equations are solved by the

IOP Conf. Series: Earth and Environmental Science **146** (2018) 012040 doi:10.1088/1755-1315/146/1/012040

computational fluid dynamic solver Ansys Fluent 13.0, more details about the theories of this solver can be found in reference [7].

To validate the above models in simulating the convective-diffusion process of the pollution, a two-dimensional computational domain is built, as shown in figure 1. In this case, a rectangular region, which is 5m wide and 15 m long, is defined as the source of pollution in a channel. The initial concentration of the pollution is defined as 450 g/m^3, the diffusion coefficient is set to 0.64 m^2/s, and the inlet velocity is 0.5 m/s.

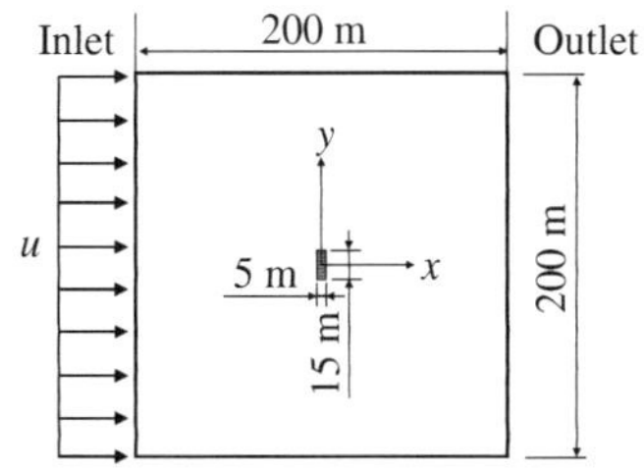

Figure 1. The computational domain of the two-dimension convective-diffusion model.

The simulated concentration distributions at different times are compared with the analytical solutions in figure 2. Note that the analytical solutions are obtained by the formula which is presented in reference [2] and has been validated by several methods. It is obvious that the two results agree well with each other, indicating that the models above are valid and accurate.

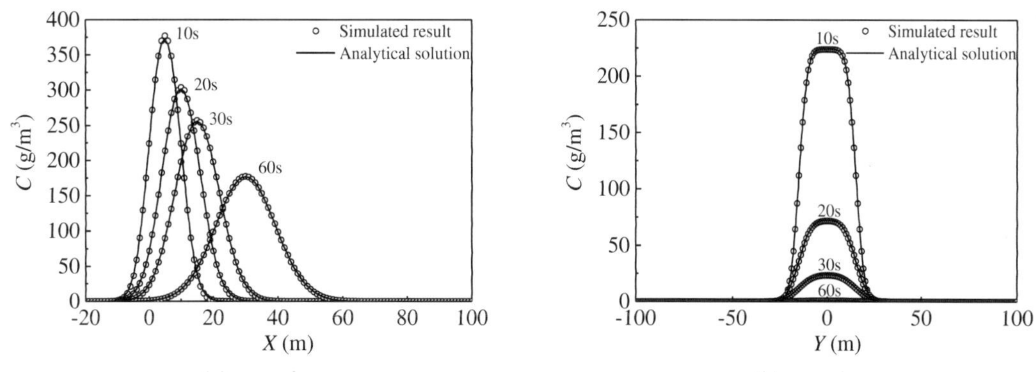

Figure 2. Comparisons between the simulated concentration distributions and the analytical solutions.

From the viewpoint of numerical modelling, bacteria can be treated as a kind of pollution which doubles its concentration at regular time intervals. To implement this, we just need to modify the existing flow field solver which is available for simulating the convective-diffusion process of the general pollution, and the detailed solution procedure is demonstrated in Figure 3. By applying this approach, the case above is simulated again and the doubling time interval is set as 20s. The results are illustrated in figure 4 and it can be seen that the doubling of the concentration of the pollution in the process of advection-diffusion is executed accurately, which shows the feasibility and practicality of this approach.

IOP Conf. Series: Earth and Environmental Science **146** (2018) 012040 doi:10.1088/1755-1315/146/1/012040

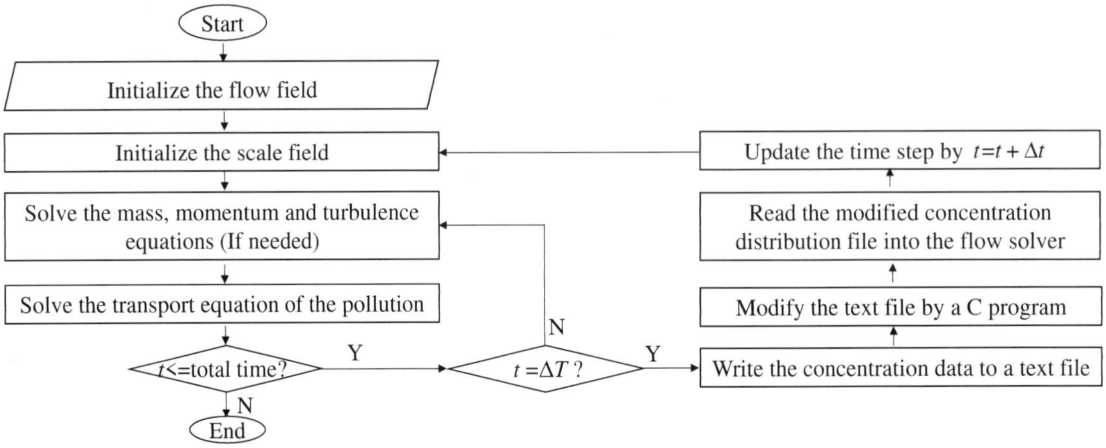

Figure 3. Flow chart of the simulation procedure considering the doubling of the concentration (In this chart t is the time, Δt means the time step size and ΔT indicates the time interval of doubling).

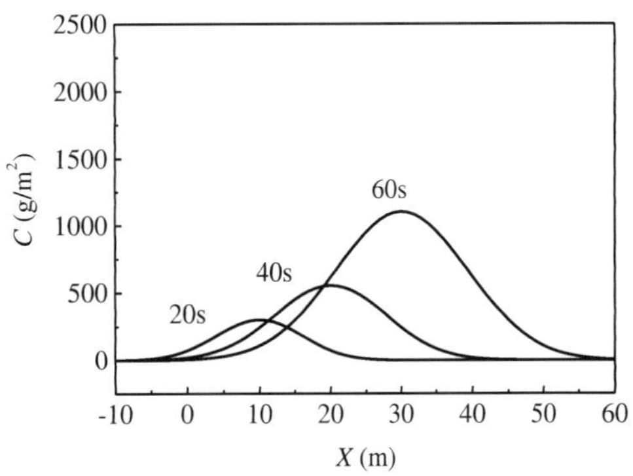

(a) x-axis (Before doubling the concentration)

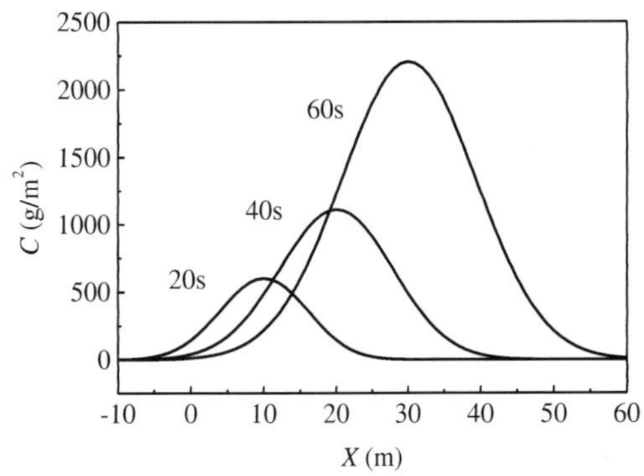

(b) x-axis (After doubling the concentration)

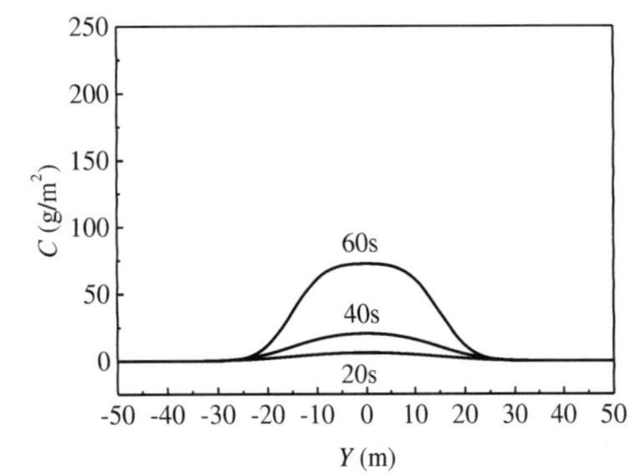

(c) y-axis (Before doubling the concentration)

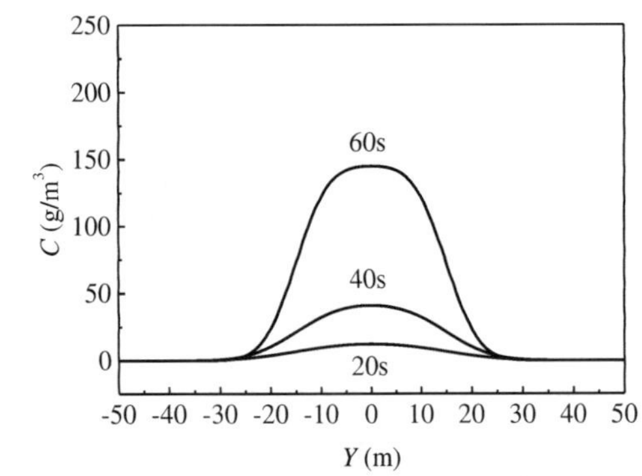

(d) y-axis (After doubling the concentration)

Figure 4. The simulated concentration distributions considering the doubling of the concentration.

3. Application

In this section, the numerical treatment proposed above are employed to simulate the convective, diffusion and reproduction process of the bacteria in a pool. The computational domain of the pool is depicted in figure 5. In the normal operation conditions, water enter into the pool at the inlet with a constant velocity of 0.05m/s and is drained out of the pool at the outlet.

Firstly, water flow without bacteria is simulated to obtain the preliminary hydraulic characteristics in the pool, the velocity vectors of the water are shown in figure 6. The flow can be divided into two patterns: the main flow and the secondary flow. The main flow, which runs along a narrow region, connects the inlet and outlet of the pool, while the secondary flow circulates around the centre of the pool. Obviously, the secondary flow region, especially near the centre of the pool, is not conducive to the spread of pollutants.

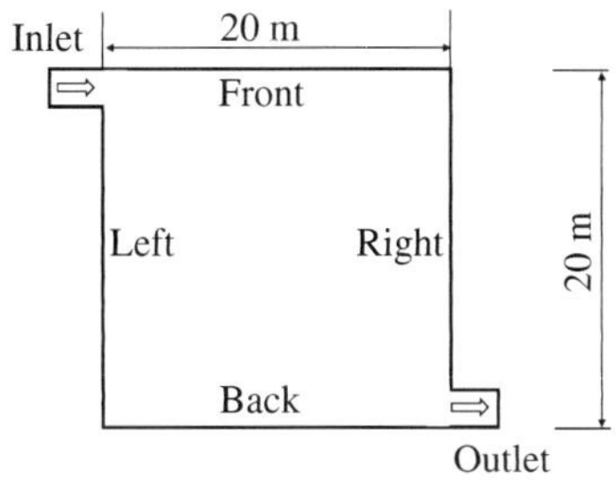

Figure 5. The computational domain of the pool.

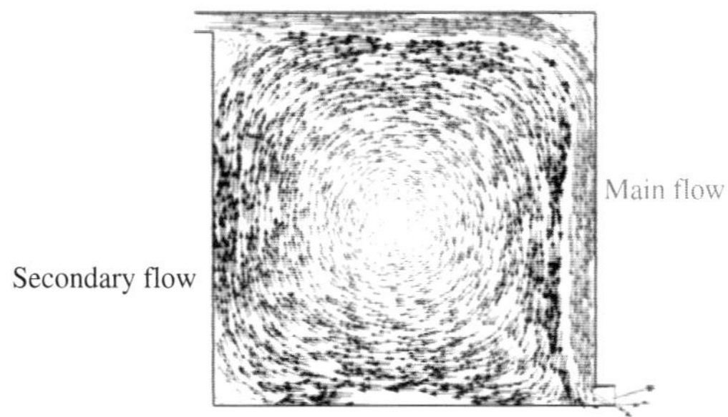

Figure 6. The velocity vectors of the water flow.

Then the bacteria, which are put into the centre of the pool as a point source, are taken into consideration. The initial concentration of the bacteria in the point source is 20,000 per square metre of water and the diffusion coefficient is set as 8.3×10^{-6} cm^2 / s. To investigate the influence of reproduction on the spread of the bacteria, two cases are simulated. In case A, the bacteria is considered as a pollution which cannot reproduce, while in case B the generation time of the bacteria is 20 min. The developments of bacteria distributions over time are shown in figure 7, in which the colours represent the concentration of the bacteria. Figure 8 depicts the trends of both the bacteria concentration in the outflow and the relative area of water which is contaminated over time in case B. Note that the contaminated water is defined as the region where the concentration of bacteria is over 20,000 per square meter. In case A, although the bacteria are limited in the central region of the pool, their concentration decreases quickly and almost reaches to zero after about an hour. Additionally, there are hardly any bacteria in the outflow in the whole process, which means that the water quality is nearly unaffected. In case B, both the area of water which is contaminated by bacteria and the maximal bacterial concentration keeps increasing. After 5 hours, more than half of the pool is full filled with bacteria and about 99% of the water is contaminated after 14 hours. Moreover, the concentration of the bacteria in the outflow is also continue to increase, which means that the water quality is deteriorating irreversibly.

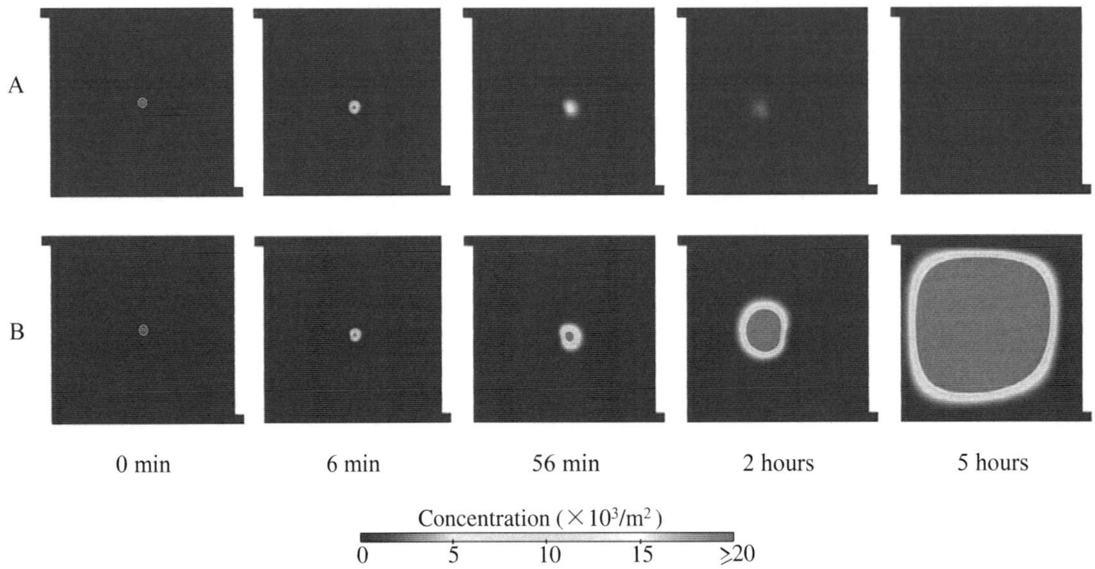

Figure 7. The concentration distributions of the bacteria at different times in Case A and B.

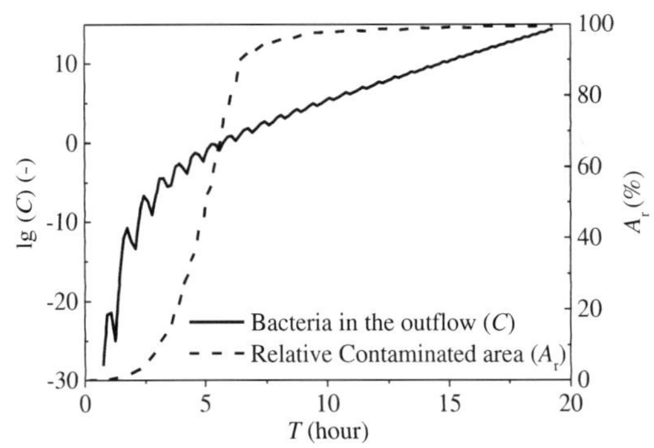

Figure 8. The trend of the bacteria concentration and the relative contaminated area over time in Case B.

The development of bacterial pollution in the pool can be attributed to both the flow patterns and the reproduction of the bacteria. The bacteria source is located in the secondary flow region, so the bacteria circulate in the central area, see figure 6, and hardly flow out of the pool. As a result, in case A, the bacteria still gather around the centre of the pool in a few hours, but the concentration keeps decrease for the convective diffusion effect. In case B, the spreading characteristics of the bacteria is the same as that in the case A, but their concentration keeps increase, for the reproduction of the bacteria. Therefore, the pool is full of bacteria for some time and the quality of outflow water continues to deteriorate.

4. Conclusions

In this paper, a simple approach for modelling the reproduction of bacteria in water flow is presented and is employed to simulate the development of the bacterial pollution in a pool. This approach is easy to implement in the existing flow field solver and is proved to be feasible by simulating a standard example. The application of this approach shows that the evolving characteristics of bacterial pollution in a pool can be attributed to both the flow patterns of the water flow and the reproduction of the bacteria. The recirculation region limits the spread of the bacteria and provides enough time for their constant reproductions, which induces the continuous deterioration of water quality in the pool.

Additionally, the approach in this study should be improved further, for bacteria will not maintain their exponential reproduction endlessly in reality. In addition, further studies should also be emphasized on the validation of this approach by experiments.

Acknowledgements
This work is financially supported by the Education Department of Fujian Province (JA15384, JAT170410), the Natural Science Foundation of Fujian Province (2017J01721), the Xiamen Science and Technology Planning Project of China (3502Z20150051), and the Scientific Research Foundation of Xiamen University of Technology (YKJ15042R, YKJ14025R).

Reference

[1] Sheryl M 2015 Water pollution: a global concern (New York, USA: Callisto Reference)
[2] Peng Z, Yang T, Yang X 2006 Mathematical modelling of water environment and its applications (Beijing, China: Chemical Industry Press) (In Chinese)
[3] Chen Y, Wang S, Wang Z, Liu Z and Zhu D 2012 *J. Hydraul. Eng.* **138** 122
[4] Xu T Y and Qin X S 2013 *J. Environ. Inform.* **22** 39
[5] Wang H, Wu Z, Hu C and Du X 2015 *Environ. Sci. Pollut. R.* **22** 14095
[6] Ding X, Wang S, Jiang G and Huang G 2017 *J. Clean. Prod.* **167** 326
[7] ANSYS, Inc. 2010 Ansys Fluent 13.0 Theory Guide (Pennsylvania, USA: ANSYS, Inc.)

IOP Conf. Series: Earth and Environmental Science **146** (2018) 012041 doi:10.1088/1755-1315/146/1/012041

Gasification of Furniture Waste Sawdust in a Cyclone Gasifier

Adisa Vongsvarnrungruang[1], Duangduen Atong[2] and Viboon Sricharoenchaikul[1,3]

[1]Department of Environmental Engineering, Chulalongkorn University, Bangkok, Thailand
[2]National Metal and Materials Technology Center, Thailand Science Park, Pathumthani, Thailand
[3]Energy Research Institute, Chulalongkorn University, Bangkok, Thailand

Viboon.sr@chula.ac.th

Abstract. Sawdust is one of the major biomass that was produced over 1 million ton per year in Thailand. A large amount of sawdust can be used as renewable energy. Gasification is one of the high-efficiency processes to convert biomass to the gaseous product. Sawdust is suitable to be converted to the gaseous product due to its high proportion of volatile (73%). In this work, sawdust thermal conversion was investigated by using a stairmand type cyclone gasifier with 100 mm diameter. The operation of the cyclone with short residence time and high efficiency of particulate separation from the gaseous product is suitable for a small particle like sawdust. The influences of equivalence ratio (0.15-0.60) and reactor temperature (700°C, 800°C and 900°C) on the waste conversion performance were determined. The gas composition (CO, CO_2, CH_4, H_2, and C_nH_m), carbon and hydrogen conversions, lower heating value and cold gasification efficiency were reported. The data analysis indicated the optimum equivalence ratio is 0.45. Excessively lower or higher equivalence ratio negatively affected the gas composition and heating value of gaseous product. The optimum of reactor temperature is 900°C. The lower temperature also adversely affected the gaseous product, conversion of carbon and hydrogen. The lower heating value of product gas is range from 0.95-3.45 MJ/m^3 and the cold gasification efficiency is range from 28-68%. The optimal condition of sawdust gasification in cyclone gasifier is equivalence ratio of 0.45 and temperature of 900°C at which total carbon and hydrogen conversionsare 90% and 60%, respectively.

1. Introduction

Several types of agricultural waste, forestry products and industry residue in Thailand are used as feedstock for gasification process. Thai government of supports this technology for heat and power purposes [1]. Gasification is the thermal conversion process that is suitable for producing gaseous product from carbonaceous raw materials. Here, the input material should be small or long reaction time is needed to ensure effective conversion to fuel gas. Sawdust from furniture industry is of small particle size and has appropriate characteristics to be converted to the fuel gas products because of its high volatile matter (around 70%) and carbon content (around 50%) with low moisture content (around 6%). From previous researches, sawdust was gasified in several reactor types such as spout-fluid bed reactor [2] and entrained-flow reactor [3, 4]. Though problems of small particle being carried away or short-circuited through the reactor before complete reaction resulted in contamination of the product gas. Furthermore, some alkalis were found to exit the gas stream by attaching to the surface of those small particles [2]. Thus, in order to mitigate these problems, a cyclone reactor was selected due to its potential of efficient separation of small particle from the gaseous product as well as its simple operation.

The purpose of the present work is to investigate the gasification parameters during the conversion of sawdust to fuel gas product by cyclone gasification process. The performance was evaluated in terms of gas composition (CO, CO_2, CH_4, H_2 and C_nH_m), carbon and hydrogen conversions, lower heating value and cold gasification efficiency.

2. Material and Methods

The biomass feedstock in this study is sawdust which comes from a furniture plant located in an eastern province of Thailand. The particle size between 150 - 250 µm was used and kept at 105 °C in an oven for 24 hours prior to any experiments to vaporize the moisture. The feedstock was collected in one batch to minimize the characteristic variation.

The characteristics of biomass feedstock, proximate analysis, ultimate analysis and lower heating value are shown in Table 1. The main characteristics of feedstock consists of high carbon content (45%), high volatile matter (73%), low moisture content (6%) and low ash content (6%). Note that the ultimate analysis was determined by following ASTM-D5373 using C-H-N analyzer except the oxygen composition was assumed by difference and the proximate analysis was determined by following ASTM-D5142-02. Bomb calorimeter was used to analyze lower heating value that resulted in the value of 19.2 MJ/kg which is typical for lignocellulosic materials.

As shown in Figure 1, the major particle size distribution of this feedstock is in range of 150-250 µm. This size of feedstock is optimum for gasifying particles according to Moghiman et al. [5] who reported the satisfactory range for wood particle size in the cyclone gasification of wood powder of 150-600 µm because the particle with small size of less than 150 µm may leave the cyclone and carry away the alkali compound with the producer gas. The bigger particle size of more than 600 µm may leave the cyclone before the complete gasification. Another reason is heat transfer, the smaller particles could be heated more rapidly and evenly with minimized temperature gradient which leads to higher rate of gasification reactions.

The Stairmand type cyclone design yielded a reactor with 100 mm diameter and 400 mm height. This design was selected because it is suitable for small particle that required short residence time and high efficiency of particulate separation from the gaseous product. The gasifier was made from stainless steel grade 310S which can withstand high temperature and corrosive nature of expected by-products.

The experimental set up is shown in Figure 2 which consists of six major parts: carrier gas preheating system, biomass feeding system, cyclone reactor, electric heater, tar condenser and sampling gas train for analysis. To begin the experiment, carrier gas, nitrogen and oxygen, was heated up to 200°C prior to entering and flowing through the feeding tube to the cyclone reactor. The biomass feeding system which located perpendicular to the feeding tube fed the sawdust continuously to the feeding tube by screw conveyor. Then, the sawdust was mixed with the carrier gas and entrained to the cyclone reactor where it came into contact with the hot cyclone wall and rapidly heated up. After that, the producer gas left the reactor through the upper section while the char product was separated to char box at the bottom of the cyclone. The producer gas moved into the tar condensation system to extract tar content and to decrease its temperature. Then the producer gas was sampled by peristaltic pump for analysis of the gas compositions (CO, CO_2, CH_4, C_nH_m and H_2) before released to the atmosphere.

Table 1. Proximate, Ultimate Analysis and Lower heating of feedstock.

Ultimate analysis (% dry basis)	C	H	N	O^a
	44.99	6.31	0.32	48.38
Proximate analysis (% wet basis)	Moisture	Volatile matter	Fixed carbon	Ash content
	6.12	72.7	14.8	6.38
Lower heating value (MJ/kg)	19.2			

[a] by difference

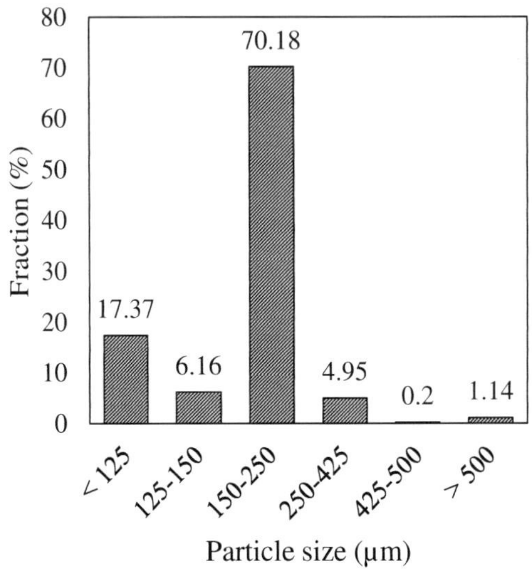

Figure 1. The particle size distribution of biomass feedstock.

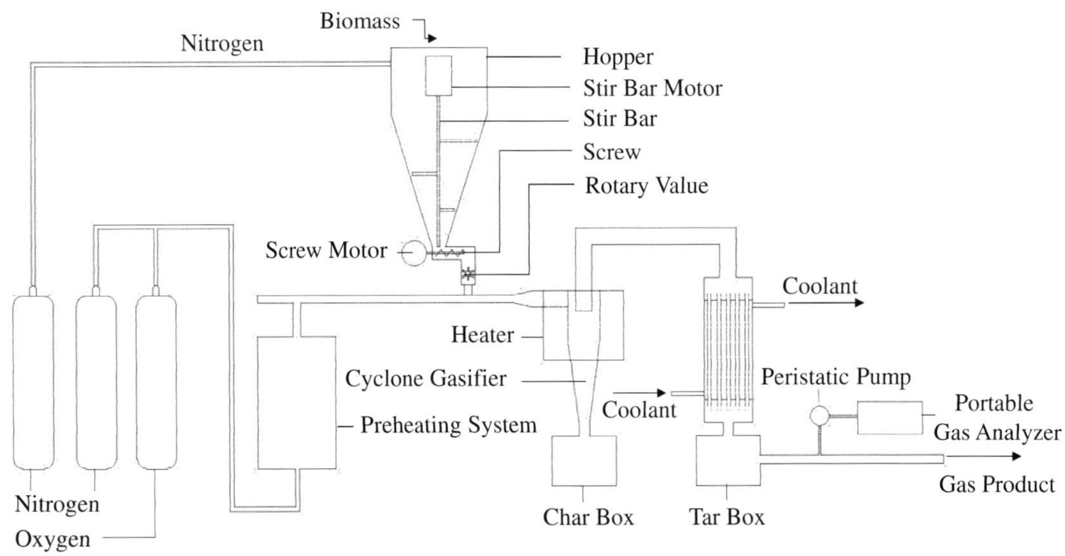

Figure 2. Diagram of the cyclone gasification system.

This study is focused on gasification temperature and equivalence ratio that may affect the gas product distribution. The gasification temperature was varied from 700°C, 800°C and 900°C and the equivalence ratio (ER) was varied from 0.15, 0.30, 0.45 and 0.60. The carrier gas consists of nitrogen and pure oxygen that acted as the gasification agent. The biomass feed rate was determined at 0.32-0.47 kg/h.

3. Result and Discussion

The varied experimental conditions are temperature and ER in which their effect on the main gasification parameters would be discussed here. The sawdust was fed to the feeding tube. Then combined with the carrier gas and move to the hot cyclone wall by centrifugal force. The hot cyclone wall provided the thermal energy to sawdust within short residence time resulting in the gasification process rapidly occurred which was summarized as several following reactions.

Oxidation reaction

$C + ½ O_2 \rightarrow CO$ $\Delta H = - 111$ kJ/mol (1)

$C + O_2 \rightarrow CO_2$ $\Delta H = - 394$ kJ/mol (2)

$CO + ½ O_2 \rightarrow CO_2$ $\Delta H = - 284$ kJ/mol (3)

$CH_4 + 2O_2 \rightarrow CO_2 + 2H_2O$ $\Delta H = - 803$ kJ/mol (4)

Boudouard reaction

$C + CO_2 \rightarrow 2CO$ $\Delta H = + 172.5$ kJ/mol (5)

Water-gas reaction

$C + H_2O \rightarrow CO + H_2$ $\Delta H = + 131.3$ kJ/mol (6)

Water-gas shift reaction

$CO + H_2O \rightarrow CO_2 + H_2$ $\Delta H = - 41.2$ kJ/mol (7)

Methanation reaction

$C + 2H_2 \rightarrow CH_4$ $\Delta H = - 74.9$ kJ/mol (8)

Steam reforming reaction

$CH_4 + H_2O \rightarrow CO_2 + 3H_2$ $\Delta H = + 206.2$ kJ/mol (9)

3.1. Effect of the ER

In this study, ER was varied from 0.15 to 0.60. The ER is the ratio of actual mole of oxygen input to the stoichiometric mole of oxygen for particular oxidation processes which depends on the biomass feeding rate.This means that the higher of ER, the higher amount of oxygen.

Oxygen content plays an important role in oxidation reaction (equation(1-4)) which produces CO_2 as the main product. Higher oxygen availability led to increased CO_2 generation as shown in Figure 3. At 700°C, 800°C, and 900°C, the ER increased, the CO_2 conversion increased from 6.78% to 28.27%, 11.54% to 21.16 and 11.41% to 27.59, respectively. In contrast, the CH_4 composition was slightly decreased because it can react with oxygen to produce CO_2 and H_2O according to the oxidation reaction (equation(4)). However, when the ER was increased up to the optimum level, the CO conversion decreased because the CO was used in the interaction (equation(3)) in which it would combine with O_2 and convert to CO_2. As a result, the optimum ER of reactor temperature 700°C, 800°C and 900°C is 0.30, 0.45 and 0.45, respectively. Hence the increase rate of oxidation reaction led to greater sawdust degradation which decrease amount of char products as shown in Table 2.

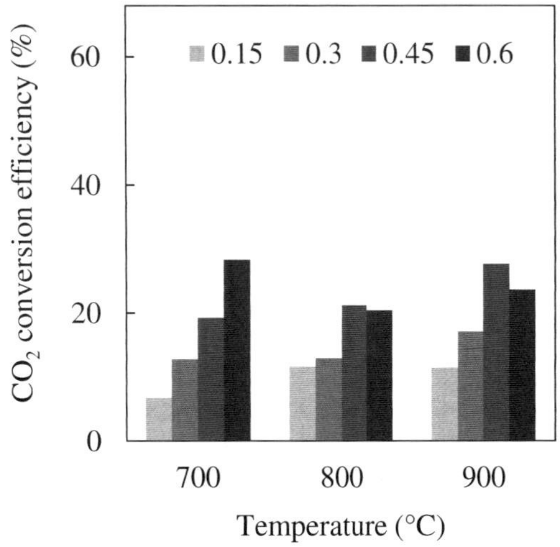

Figure 3. Effective of equivalence ratio on carbon conversion to CO_2 at 700°C, 800°C and 900°C.

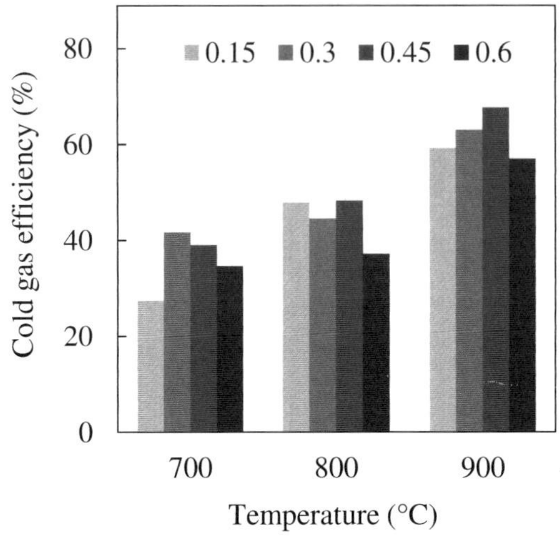

Figure 4. Effective of equivalence ratio on cold gas efficiency at 700°C, 800°C and 900°C.

The cold gas efficiency depends on the composition of producer gas, CO, CH_4 and H_2 and hence the producer gas at optimum ER cause the highest value of the cold gas efficiency at 700°C, 800°C and 900°C of 38.96%, 48.33% and 67.66%, respectively. Figure 4 shows cold gas efficiency data from the experiments.

A similar trend of CO, CO_2 and H_2 was also reported by Yijun on the experiment with rice husk using a cyclone reactor [6]. The result indicated that when the ER increased to 0.3, greater CO, CO_2 and H_2 could be observed. On the other hand, the carbon conversion of sawdust (33.30-60.17%) was higher than that of rice husk (27.1 – 39.7%) at 700 °C in cyclone reactor which resulted from oxidation reactions (equation(1-4)).

At 800°C, the results displayed similar trend to that of Shaozeng who studied on cyclone air gasification of wood powder [7]. The results showed that H_2 and CO_2 increased with high ER. The gasification in an entrained-flow reactor [3] of sawdust has a similar trend of carbon conversion when compared with the gasification of sawdust in cyclone reactor. At high temperature (800°C – 900°C), the ER had little effect on carbon conversion.

3.2. Effect of reaction temperature

The increasing of reactor temperature is the option to enhance the energy input to the system which leads to greater chemical reaction rate of gasification reactions, particularly, the endothermic ones, Boudouard and water-gas reactions. The Boudouard and water-gas reactions have a faster reaction rate with the high temperature. The high effectiveness of these reactions can be observed via the CO and H_2 composition increased as shown in Figure 5. At greater ERs from 0.15 to 0.60, the higher temperature resulted in increased CO and H_2 conversions from 20% to 50% and 3.7% to 26%, respectively. Furthermore, the oxidation reaction (equation(1)) and steam reforming reaction (equation(9)) were also increased the CO and H_2 composition.

The synthesis gas, CO and H_2, greatly influence the LHV and cold gas efficiency. High synthesis gas generation leads to increased LHV and cold gas efficiency as shown in Figure 6. At ER 0.45, the LHV and cold gas efficiency increase from 1.46 to 3.45 MJ/m^{3}and 38.96% to 67.66%, respectively. If oxygen is readily available, CO_2 composition would increase with temperature but at 800°C and 900°C the composition of CO_2 was relatively steady. This probably due to the reaction rate of the equation(1-4) which may already reached the maximum.

At high temperature, the methanation reaction (equation(8)) produces more CH_4 but the steam reforming reaction (equation(9)) also consumes CH_4 as the raw material in the reaction. This may explain the relatively stable CH_4 composition in the produce gas with increasing temperature.

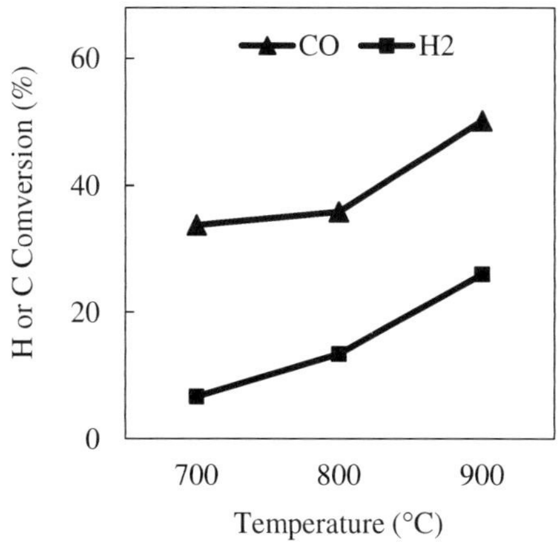

Figure 5. Effect of temperature on carbon conversionto CO_2 and hydrogen to H_2 at ER 0.45.

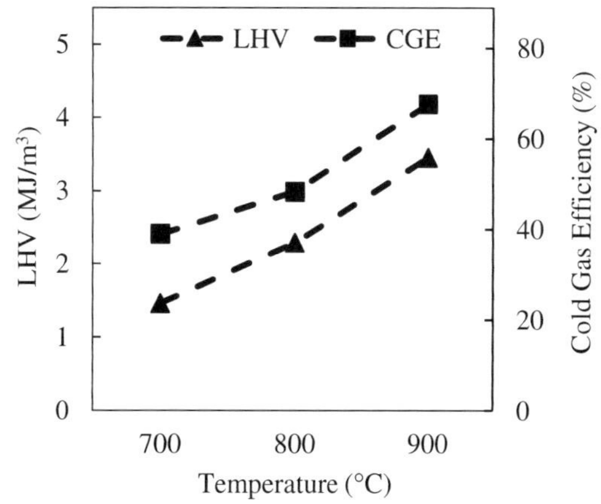

Figure 6. Effect of temperature on LHV and cold gas efficiency at ER 0.45.

Table 2. The gasification parameters for ER 0.15, 0.30, 0.45 and 0.60 at different temperature.

Temperature	700°C				800°C				900°C			
ER	0.15	0.30	0.45	0.60	0.15	0.30	0.45	0.60	0.15	0.30	0.45	0.60
Feeding rate (kg/h)	0.36	0.41	0.39	0.39	0.42	0.38	0.46	0.34	0.41	0.44	0.47	0.32
Carbon conversion (%)												
CO	20.01	33.23	33.74	28.58	34.44	30.53	35.83	28.67	41.45	46.19	50.15	44.49
CO_2	6.78	12.70	19.18	28.27	11.54	12.83	21.16	20.40	11.41	17.05	27.59	23.64
CH_4	6.51	9.42	7.24	7.76	10.60	9.56	9.94	6.32	13.28	12.53	12.56	10.19
Total	33.30	55.35	60.16	64.61	56.58	52.92	66.93	55.39	66.14	75.77	90.30	78.32
Hydrogen conversion (%)												
CH_4	17.29	25.03	19.24	20.63	28.18	25.40	26.41	16.79	35.29	33.29	33.38	27.09
H_2	3.68	7.62	6.73	6.11	10.62	14.19	13.39	12.89	14.69	20.52	25.97	20.70
C_nH_m	0.55	0.58	0.36	0.29	0.78	0.52	0.47	0.85	0.85	0.52	0.24	0.08
Total	21.52	33.23	26.33	27.03	39.58	40.11	40.27	30.53	50.83	54.33	59.59	47.87
LHV (MJ/m3)	0.95	1.65	1.46	1.30	2.06	1.75	2.29	1.27	2.66	3.02	3.45	1.97
Cold gas Efficiency (%)	27.52	41.70	38.96	34.61	47.87	44.55	48.33	37.12	59.29	63.05	67.66	57.10
Char product (%)	9.52	8.62	5.58	4.03	8.19	5.61	4.02	3.65	6.40	3.83	3.76	5.17

4. Conclusions

Investigation of the performance of cyclone gasification of furniture waste sawdust was carried out in this work. The parameters used to define the process efficiency were carbon and hydrogen conversions, product gas heating value and cold gas efficiency. The operating parameters investigated was temperature and equivalence ratio. Generally, the increasing of ER has an effect to the oxidation reactions which increase carbon conversion to favourable CO gas product. Meanwhile the increasing of reaction temperature significantly effects the endothermic reactions which enhanced carbon to CO and hydrogen to H_2. Thus the optimum conditions that produce high performance synthesis gas is 900°C at ER 0.45when the highest carbon to CO, hydrogen to H_2, LHV and cold gas efficiency can be achieved.

Acknowledgments

The work cited in this paper was supported by Department of Environmental Engineering, Faculty of Engineering,Chulalongkorn University and National Metal and Materials Technology Center (MTEC).

IOP Conf. Series: Earth and Environmental Science **146** (2018) 012041 doi:10.1088/1755-1315/146/1/012041

References

[1] Laohalidanond K, Chaiyawong P and Kerdsuwan S 2015 *Energy Procedia* **79** 385
[2] Thamavithya M, Jarungthammachote S, Dutta A and Basu P 2012 *Int. J. Energ. Res.* **36** 204
[3] Zhao Y, Sun S, Zhou H, Sun R, Tian H, Luan J and Qian J 2010 *Fuel. Process. Technol.* **91** 910
[4] Weiland F, Hedman H, Marklund M, Wiinikka H, Ohrman O and Gebart R 2013 *Energy Fuels* **27** 932
[5] Moghiman M, Hashemi T, Zahmatkesh I and Daghighi Y 2007 *J. Energy. Inst.* **80** (1) 29
[6] Zhao Y, Sun S, Che H, Guo Y and Gao C 2012 *Int. J. Hydrogen. Energ.* **37** 16962
[7] Sun S, Zhao Y, Tian H, Ling F and Su F 2009 *Bioresource. Technol.* **100** 4047

IOP Conf. Series: Earth and Environmental Science **146** (2018) 012042 doi:10.1088/1755-1315/146/1/012042

Experimental Research on the Refrigeration Performance of PVT Solar Heat Pump in Summer Night

Jianquan Chen, Jili Zhang*, Mingsheng Liu, Shixiang Lu

Institute of Building Energy, Dalian University of Technology, Dalian, China 116024

axcjq188@163.com, *corresponding email: zjldlut@dlut.edu.cn

Abstract. A method of refrigeration utilizing photovoltaic-thermal (PVT) solar heat pump is proposed, and long-wave radiation cooling is the key point. The PVT solar heat pump system combined cooling, heating and power (CCHP) is designed and constructed. The refrigeration performance of PVT solar heat pump during summer night is experimentally tested, including chilled water making and ice making. The refrigeration performance difference between clear sky night and cloudy sky night is compared and analyzed. The environmental condition more suitable for refrigeration is given according to the comparison result.

1. Introduction

PVT solar heat pump (SHP) is a comprehensive utilization of solar energy. Part of the solar radiation absorbed by PVT module in daytime is used for power generation by PV cell, while the other part is converted into thermal through heat pump. The temperature of PV cell will reduce through the thermoelectric synergy, which improves the photoelectric conversion efficiency at the same time [1]. However, most conventional SHP is in a shutdown state for most of the time, including nighttime, due to the less heating demand in summer, which cause a low system utilization. Considering the large cooling demand in summer, if utilizing SHP for refrigerating during the non-heating period can be realized, the comprehensive system utilization will be improved and cooling demand will be satisfied. PVT SHP has the PVT module with a large surface area, which is especially suitable for dissipating heat to low-temperature sky through long-wave radiation at night. Therefore, PVT module is an ideal heat-dissipating carrier and can be used as the condenser when utilizing heat pump for refrigerating.

The sky of night is a natural cold source and effective sky temperature of clear sky in summer nighttime can even be lower than -10℃ [2]. Therefore, scholars have done much research on the long-wave radiation cooling. The heat dissipation performance of roofs in different structures was tested, which shows that the roof temperature is significantly lower than the ambient temperature during nighttime[3]. Long-wave radiation cooling was applied in the air-conditioning system of teaching building to reduce energy consumption[4]. The heat dissipation performance of the flat radiant panel and the PVT module with water as cooling medium during summer night was tested, which can reach 90 W/m^2 and 120W/m^2 respectively[5]. According to the existing research, the long-wave radiation cooling power is relatively low compared with the cooling load of building, so the cooling demand of building can not be fully satisfied. Taking the initiative to increase the temperature of heat dissipation object can enhance the long-wave radiation cooling power from the viewpoint of heat transfer principle. By using refrigerant as working fluid, heat pump can provide a higher heat dissipation temperature (condensation temperature), while making low-temperature chilled water.

In view of the advantages of PVT module in long-wave radiation cooling and the shortcomings of conventional PVT SHP mentioned above, a method of utilizing PVT SHP for refrigerating is proposed.

The PVT SHP runs refrigeration mode at summer night and utilizes PVT module as the condenser, while dissipating heat to sky and surroundings by long-wave radiation and convection through PVT module.

2. Experimental system design and testing principle

2.1 *Experimental system design*

The new PVT SHP needs to add refrigerating function based on the existing power generation and heating function. From the point of view in cost and integration, the components used for refrigerating and heating should be shared as much as possible. From the perspective of operational, the system should be able to switch between refrigerating and heating mode easily. From the view of energy consumption, refrigeration operates at night while mainly cooling load occurs during daytime, so cold-storage device is essential. Based on these design thoughts, the PVT SHP system combined cooling, heating and power is designed, as illustrated in figure 1. The system mainly consists of compressor, four-way reversing valve, cold-storage tank, thermal-storage tank, throttle valve, PVT module and photovoltaic inverter. Compressor, throttle valve, PVT module and the main pipeline is the shared part of refrigerating and heating. Four-way valve is used for refrigerating and heating mode switching. Cold-storage tank and thermal-storage tank are energy storage devices. PVT module can act as evaporator or condenser depends on the operation mode.

Figure 2 shows the refrigerating mode. The innovative point is that the PVT module, which absorbs solar radiation when running heating mode, acts as the condenser to release heat to sky and surroundings by long-wave radiation and convection. Ice-storage tank acts as the evaporator of the system, both contains cold-storage function.

The experimental system equips with 4 PVT module is constructed after concrete design based upon the schematic diagram. The copper coil that built-in ice-storage tank transfers heat with the water in the tank directly, which can realize chill water making and direct evaporative ice making. Figure 3, figure 4 and figure 5 show part of the physical components of the experimental system.

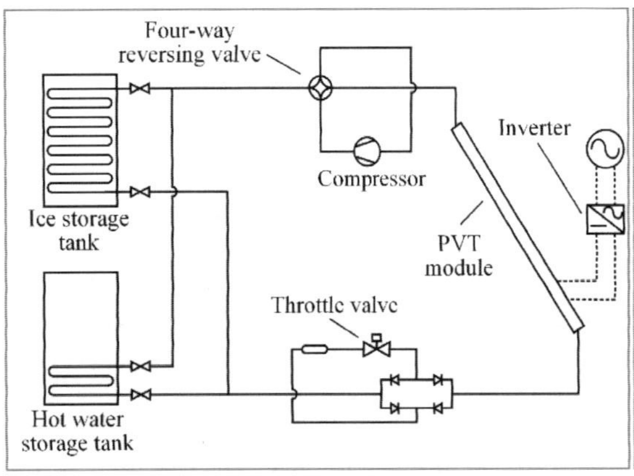

Figure 1. System schematic diagram.

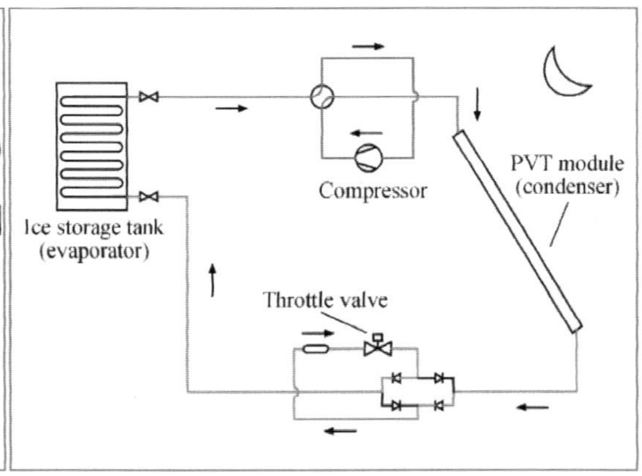

Figure 2. Refrigerating mode.

Figure 3. PVT module.

Figure 4. Ice-storage tank/hot-water tank.

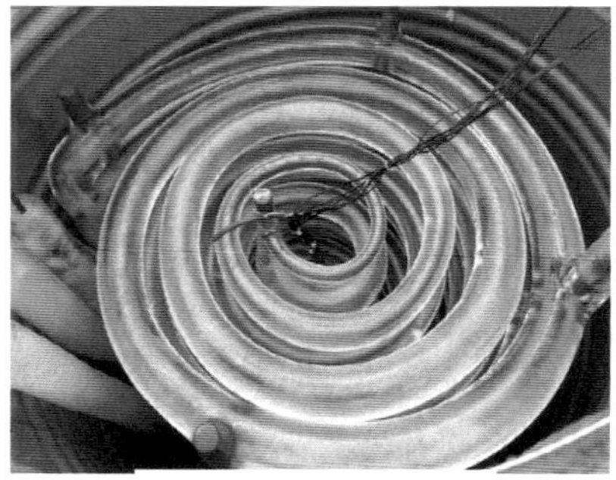

Figure 5. Copper coil

2.2 *Evaluation method of refrigeration performance*

The performance index of refrigeration cycle is usually expressed as coefficient of performance (COP), representing the amount of cold energy that can be achieved upon unit's power consumption.

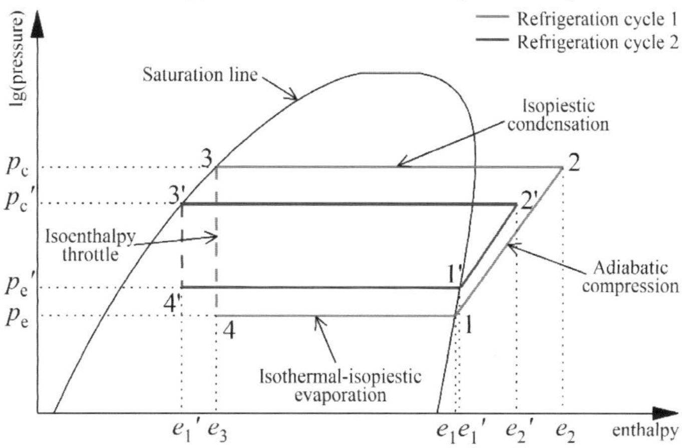

Figure 6. Refrigeration cycle.

The $1 \rightarrow 2 \rightarrow 3 \rightarrow 4 \rightarrow 1$ shown in figure 6 is a theoretical refrigeration cycle, including the following four basic processes: compression, condensation, throttling and evaporation. The theoretical COP of refrigeration cycle can be expressed as:

$$COP_{\text{th}} = \frac{Q_{\text{th}}}{P_{\text{th}}} = \frac{e_1 - e_3}{e_2 - e_1} \tag{1}$$

In practical refrigeration cycle of PVT SHP system, the Q_{th} in formula (1) refers to the actual refrigeration capacity Q_{real}, while the P_{th} refers to the actual power P_{real}. The Q_{real} can be calculated using the following formula:

$$Q_{\text{real}} = m_{\text{ref}} \left(e_{\text{ref-out}} - e_{\text{ref-in}} \right) = A\overline{K}\left(\overline{T}_{\text{w}} - \overline{T}_{\text{ref}} \right) \tag{2}$$

Where, m_{ref} is the refrigerant mass flow, $e_{\text{ref-in}}$ and $e_{\text{ref-out}}$ is the enthalpy of refrigerant at the inlet and outlet of the evaporator, A is the heat transfer surface area betweencopper coil and water in the evaporator, $\overline{K}$ is the average heat transfer coefficient between refrigerant and water, $\overline{T}_{\text{w}}$ is the average water temperature, $\overline{T}_{\mathrm{ref}}$ is the average refrigerant temperature which generally equate to evaporation temperature.

When the condensation temperature decreases orevaporator temperature increases, the refrigeration cycle becomes $1' \rightarrow 2' \rightarrow 3' \rightarrow 4' \rightarrow 1'$, as shown in figure 6, and theoretical COP becomes that shown in equation (3):

$$COP_{\mathrm{th}}' = \frac{e_1' - e_3'}{e_2' - e_1'} > \frac{e_1 - e_3}{e_2 - e_1} = COP_{\mathrm{th}} \tag{3}$$

Equation (3) points out lower condensation temperature and higher evaporation can achieve higher COP, which is also applicableinpractical refrigeration cycle. For PVT SHP system, lower condensation temperature means lower surface temperatures of PVT module and better heat dissipation performance. Evaporation temperature related to the state of heat-exchange medium in the evaporator.

3. Experimental results analysis

A large number of PVTSHP refrigeration experiments were carried out during the summer nighttime from August to September 2017 in Dalian City. The refrigeration mode contains chilled water making and ice making, and environmental condition includes clear sky night and cloudy sky night.

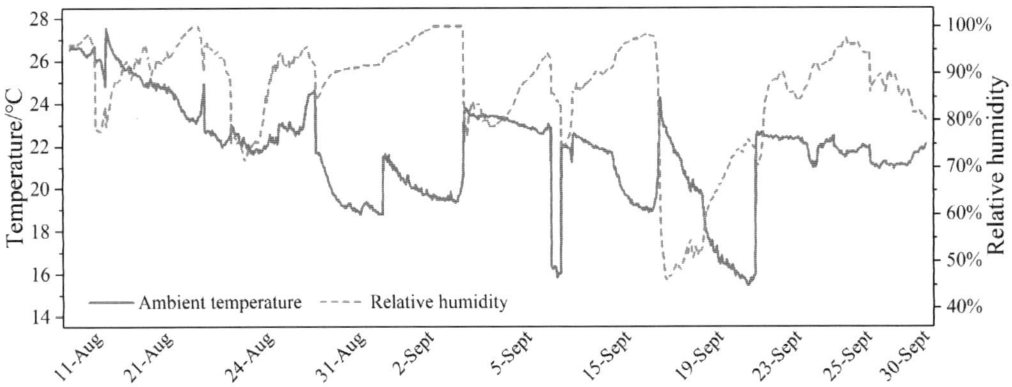

Figure 7. Outdoor meteorological parameters.

The meteorological parameters of the environment during these experiments is shown in figure 7. The ambient temperature range from 16℃ and 28℃ and the relative humidity from 45% to 100%, which is consistent with the summer night meteorological features in Dalian City.

According to the chilled water temperature requirement (<12℃) in air conditioning system, the experimental data of ice making and 5℃~12℃chilled water making were chosen to process and analyze, and these stages also called as effective refrigeration stage.

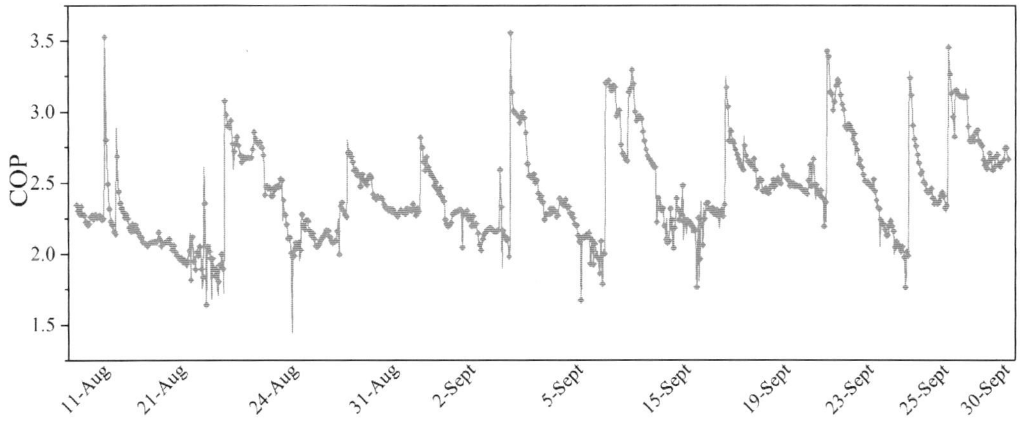

Figure 8. COP in effective refrigeration stage.

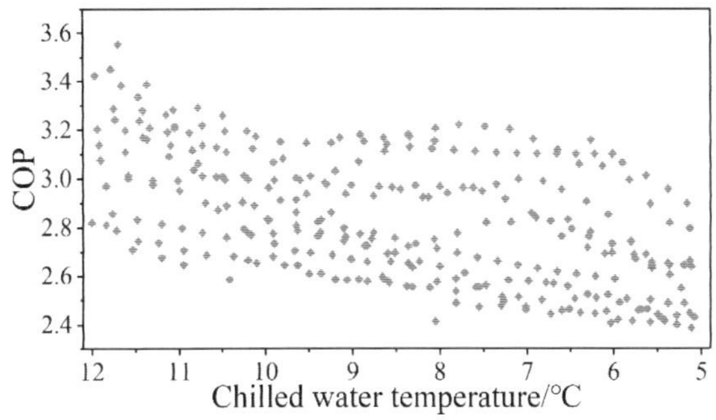

Figure 9. Chilled-water-making COP.

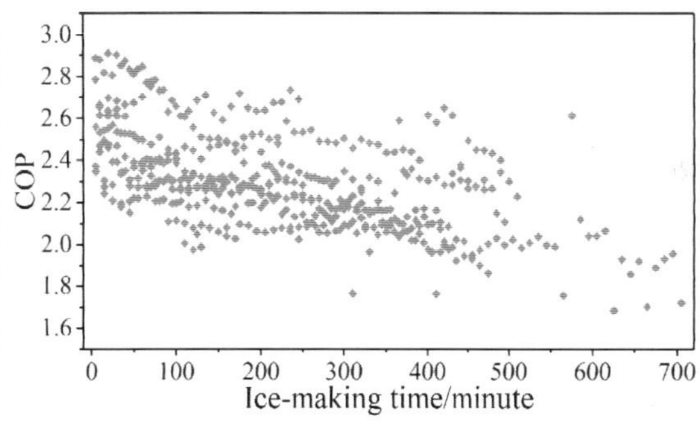

Figure 10. Ice-making COP.

The COP of effective refrigeration stage is shown in figure 8. As can be seen that the COP ranges from 1.7 to 3.5, and the average level is 2.4. Figure 9 and figure 10 illustrate a huge number of experimental results in chilled water making and ice making. The COP of chilled water making varies from 2.4 to 3.5 with an average level of 2.8, and presents overall downtrend with the chilled water temperature declining, as is shown in figure 9. Figure 10 shows the average COP of ice making is 2.3, with a range of 1.8 ~ 3.5,and the COP also shows a descending trend as the ice making time increases.

From the formula (2), it can be found that for the chilled water making, when water temperature decreases, the temperature difference between water and refrigerant will reduce accordingly, causing refrigeration capacitydecline. For ice making, the ice thickness increases with ice making time, so heat transfer coefficient will decrease accordingly, also causing refrigeration capacity decline. In both cases, the control system will increasethe temperature difference of heat transfer by reducing evaporation temperature to ensure sufficientrefrigeration capacity. From equation (3), it can be seen that reducingevaporation temperature will cause COP dropping, which is the reason why COP presents overall downtrend both in chilled water making and ice making. Due to the ice layer on the surface of the coil, as shown in figure 5, the thermal resistance of ice making is higher than chilled water making, causing a lower evaporator temperature to ensure refrigeration capacity, so the average COP of ice making is 19.4% lower than chilled water making. However, the volume of cold storage equipment for ice storage is smaller than that for chilled water storage in the case of same cold storage capacity, and ice can also providea lower cooling temperature, which means ice storage is better than chilled water storage in terms of cooling quality and system miniaturization.

From the experimental results above, it can be concluded that it is feasible to utilize PVT SHP for refrigerating in summer nighttime. The average COP of chilled water making is 2.8, and ice makingis 2.3. The energy efficiency of chilled water making can reach the level 4 (2.8~2.99) in Chinese air conditioning energy efficiency rating standards.

4. Refrigeration performance in different environmental conditions

Any changes of the outdoor environmental conditions will affect the refrigeration performance of the system since PVT module radiate heat to the surroundings and sky directly. Therefore, figuring out the impact of environmental condition on the refrigerating performance is significant for improving the system performance and optimizing the operation strategy. The summer nighttime can be classified into clear sky and cloudy sky according to meteorological characteristics, and less cloudiness is the main feature of clear sky compared with cloudy sky. In the following comparison, clear sky refers to the average cloudiness less than 0.1, while cloudy sky refers to more than 0.3.

Figure 11 presents the COP of chilled water making in clear sky and cloudy sky. The COP in clear sky ranges from 2.6 to 3.4 with an average level of 3.0, while that in cloudy sky varies from 2.4 to 3.0 with an average level of 2.7. The average COP in clear sky is 12.4% higher than that in cloudy sky. Figure 12 shows the COP of ice making with a same ice-making time of 300 minutes, which can be seen that the COP in clear sky varies from 1.9 to 2.9 with an average level of 2.4, while that in cloudy sky ranges from 2.0 to 2.6 with an average level of 2.3. The average COP in clear sky is 4.8% higher than that in cloudy sky.

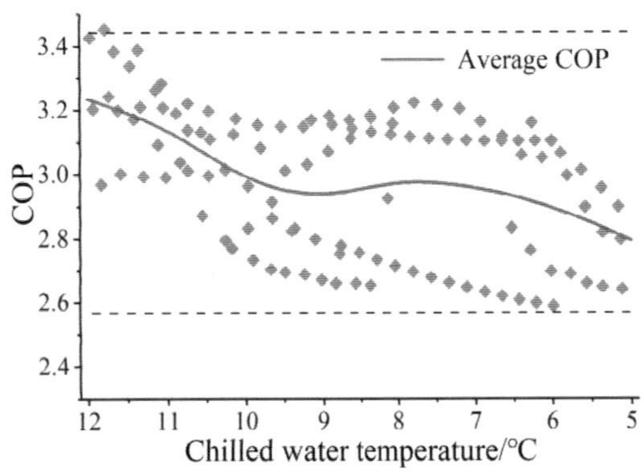

(a) Chilled-water-making COP in clear sky

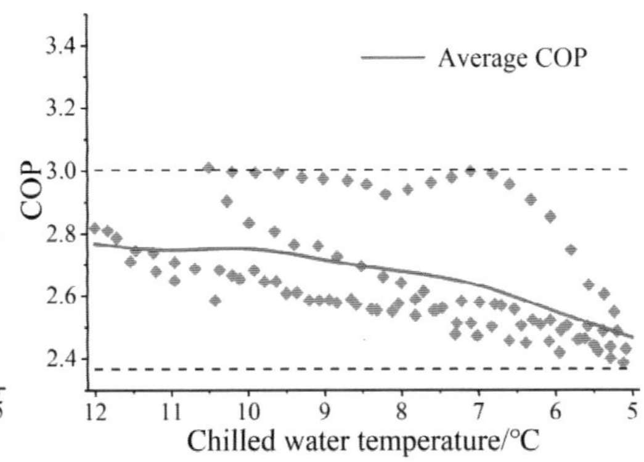

(b) Chilled-water-making COP in cloudy sky

Figure 11. Chilled-water-making COP in clear sky and cloudy sky.

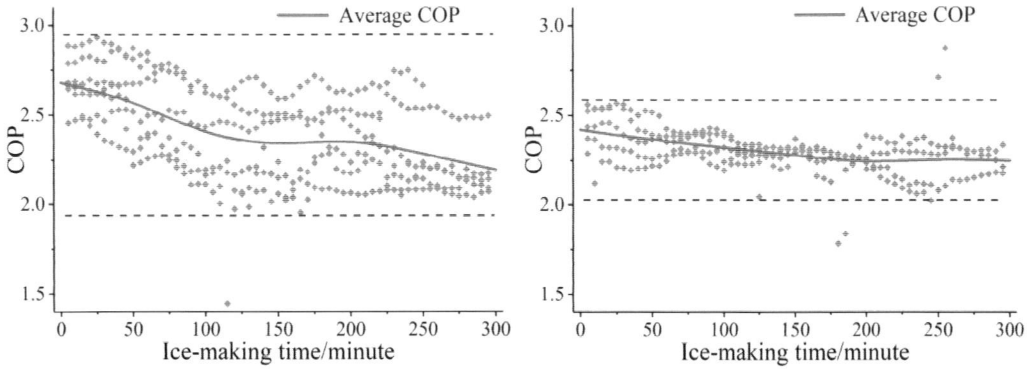

(a) Ice-making COP in clear sky (b) Ice-making COP in cloudy sky

Figure 12. Ice-making COP in clear sky and cloudy sky.

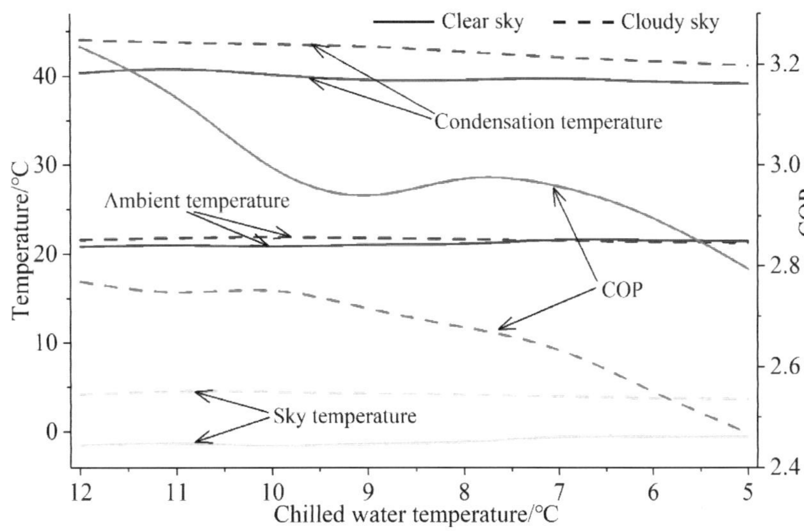

Figure 13. Influencing factor of chilled-water-making COP in clear sky and cloudy sky.

From the experimental result, the COP in clear sky is higher than cloudy sky, either in chilled water making or ice making. From the perspective of radiation heat transfer, less cloudiness in atmosphere causes less long-wave radiation absorbed from the ground, resulting in a lower effective sky temperature, which is conducive to radiative heat dissipation for PVT module. Therefore, the condensation temperature will decrease and the COP will be promoted according to equation (3).Experimental data also supports this conclusion. Take the testing data in chilled water making as an example, as shown in figure 13 (the data in figure13 is the average level of all these experiments). As can be seen from figure 13, the average ambient temperature between clear sky night and cloudy sky night shows little difference, while the effective sky temperature in clear sky is about 5 ℃ lower than that in cloudy sky, causing a lower condensation temperature of 2℃ ~ 4℃. Therefore, the heat dissipation performance of the PVT module is better, anda better COP can be achieved in clear sky. From the comparison result of different environmental conditions, it can be concluded that the average

COP in clear sky is higher than cloudy sky due to the lower sky temperature, so the PVT SHP system is recommended to run refrigerating mode in clear sky.

5. Conclusion

By analyzing the characteristics and the existing problems of PVT SHP and combining with the cooling demand in summer, a method of utilizing PVT SHP for refrigerating during summer nightis proposed. The main innovation is using PVT module as the condenser of heat pump to dissipate heat through long-wave radiation. The PVT SHP combined with cooling, heating and power was designed and a concrete experimental system was set up.

The experimental result shows that utilizing PVT SHP for refrigerating is feasible. The average COP of chilled water making is 2.8 andice making is 2.3. The refrigeration performance of chilled water making is higher than that of ice making, while ice making have advantages in cooling quality and system miniaturization.

From the comparison result of the performance in different environmental conditions, it can be concluded the refrigeration performance in clear sky is better than that in cloudy sky due to the lower effective sky temperature. The averageCOPof chilled water making in clear sky is 3.0, which is 12.4% higher than that in cloudy sky night. The comparison conclusionalso provides a reference for the optimization of the PVT SHP operation strategy, that is utilizing PVT SHP for refrigeratingis recommended tooperatein clear skynight.

Acknowledgment

This work is supported by National Key Research and Development Project of China No.2017YFC0704100 (entitled New Generation Intelligent Building Platform Techniques).

References

[1] Liang XU, Shuhong LI, Xiaoshong ZHANG, et al. Comparative Study on Different Types of PV/T System[J]. Building Energy & Environment, 2017(9):1-7.
[2] Duffie J A, Beckman W A, Mcgowan J. Solar Engineering of Thermal Processes[J]. Journal of Solar Energy Engineering, 1994, 116(1):549.
[3] Khedari J, Waewsak J, Thepa S, et al. Field investigation of night radiation cooling under tropical climate[J]. Renewable Energy, 2000, 20(2):183-193.
[4] Simonetti M, Fracastoro G V, Perino M. CFD TRANSIENT ANALYSIS OF NIGHT COOLING STRATEGY APPLIED TO SCHOOL BUILDING[J]. 2007.
[5] Erell E, Etzion Y. Radiative cooling of buildings with flat-plate solar collectors[J]. Building & Environment, 2000, 35(4):297-305.
[6] Eicker U, Dalibard A. Photovoltaic–thermal collectors for night radiative cooling of buildings[J]. Solar Energy, 2011, 85(7):1322-1335.

Effect of Flue Gas Recirculation on Reheated Steam Temperature of a 1000MW Ultra-supercritical Double Reheat Boiler

K Ma[1], C F Li[1], W P Yan[1], J W Sun[2], P Cai[2] and Q L Huang[2]

[1]MOE's Key Lab of Condition Monitoring and Control for Power Plant Equipment, School of Energy and Power Engineering, North China Electric Power University, Baoding, PR China
[2] NanjingInspectionCenterofQualityPerformanceforElectricEquipment, Nanjing, PR China

kai_ma@126.com

Abstract. In view of the problem that the reheat steam temperature is lower than the design value in the operation of a 1000MW ultra-supercritical double reheat boiler in a power plant, this paper puts forward the method of flue gas recirculation (FGR) to improve it. Two schemes are put forward:the extraction of flue gas from the economizer or the draft fan to the bottom burner of the furnace. Indifferentload conditions,using the standard method of boiler thermodynamic calculation,the influence of different FGR schemes on boiler operation parameters is calculated. The results show that above FGR schemescan obviously improve thereheat steam temperature; with the increase of the amount of recirculating gas, the rise of the reheat steam temperature increases; with the decrease of the load, the influence of FGR on steam temperature increase; the scheme of extracting recirculating flue gas from the economizer outlet has little effect on the boiler efficiency, which is more suitable for the boiler, and the reasonableFGR rate is about 10%.

1. Introduction

In response to the requirements of the current national exploration and promotion of efficient clean coal and electricity technology, the ultra-supercritical doublereheat power generation technology is being paid more and more attention [1-3]. The units with the ultra-supercritical double reheat technology can increase the thermal efficiency by about 2% more than the single reheat unit [4].But the reheater series and the reheat steam heat absorption of the double reheat boiler have both increased, Yan et al. [5] pointed out that because of the doublereheat technology, the heat absorption ratio of 1000WM capacity unit's superheated steam decreased and the reheat steam's heat absorption proportion increased. That makes the structure of the thermal system of the double reheat unit more complex, and the coordinated control of the reheat steam becomes more difficult. And it may lead to the phenomenon that the reheat steam temperature is lower than the design value when the unit is running, which reduces the operation economy of the unit.

In view of the difficulty of the problem of the reheat steam temperature of the double reheat unit below the design value. Many scholars have carried on the related research.Dang et al. [6] has studied the problem of low reheat steam temperature of the 660MW ultra-supercritical boiler by remoulding the heating area of reheater.Zhang et al. [7] has studied the effect of FGR on reheat steam temperature and set a 600MW single unit as an example to calculate. In summary, the transformation of reheater heating area and the use of FGR can effectively solve the problem of low temperature of reheat steam, but the method of remolding the heating area of reheater is influenced by the structure and layout of the boiler.

Therefore, based on the original temperature regulation methods of reheat steam,flue gas recirculation is proposed to increase reheat steam temperature in this paper. Two FGR schemes are developed for the different extraction points of recirculating flue gas and will be adopted to calculate thermodynamic parameters of boiler, and then the effect of different recirculation rate will be analyzed to obtain a reasonable scheme.

2. Overview of the boiler and FGR schemes

This is a 1000MW ultra-supercritical double reheat tower boiler. The temperature regulation methods of reheat steam are swing nozzle and flue gas baffle. The main design parameters of the boiler are shown in table 1.

Table 1. Main design parameters of the boiler. SUP: superheated steam outlet pressure; SUT: superheated steam outlet temperature; SIP: single-reheated steam outlet pressure; SIT: single-reheated steam outlet temperature; SEP: secondary reheated steam outlet pressure; SET: secondary reheated steam outlet temperature; BMCR: boiler maximum continuous rating; THA: turbine heat-acceptance.

Load	SUP (MPa)	SUT (℃)	SIP (MPa)	SIT (℃)	SEP (MPa)	SET (℃)
BMCR	33.03	605	11.17	613	3.30	613
75%THA	23.12	605	7.55	613	2.24	613

2.1. Problems in operation

The temperature of single-reheated and secondary reheated steam are lower than the design value under the actual working conditions. The specific parameters are shown in table 2.

Table 2. The steam temperature of the operation condition. DV: design value; RV: runtime value.

Load	SIT		SET	
	DV(℃)	RV(℃)	DV(℃)	RV(℃)
BMCR	613	605.8	613	613
75%THA	613	595.3	608.7	594.5

2.2. The principle of FGR

The principle of the FGR technology of steam temperature regulation: The gas recirculation fan extracts a part of low temperature flue gas from the outlet of the economizer or induced draft fan and sends into the furnace. Thereby, it reduces the level of flue gas temperature inside the furnace and increases the amount of the flue gas in the furnace. Finally, the reheater and other heat exchangers can be enhanced and the reheat steam temperature is improved.

2.3.Concrete scheme of flue gas recirculation

According to the structure of the boiler and the arrangement of the heating surface, the following FGR schemesare formulated:

Scheme 1: the recirculating flue gas is extracted from the exit of coal economizer and sent to the bottom of the burner at the bottom of the hearth.

Scheme 2: the recirculating flue gas is extracted from the outlet of the induced draft fan and sent to the bottom of the burner at the bottom of the hearth.

The different FGR extraction locations correspond to different gas temperatures. The gas temperature from the economizer outlet is about 379 degrees, and thegas temperature from the outlet of the draught fan is about 117 degrees. TheFGRschemes are shown in Figure 1.

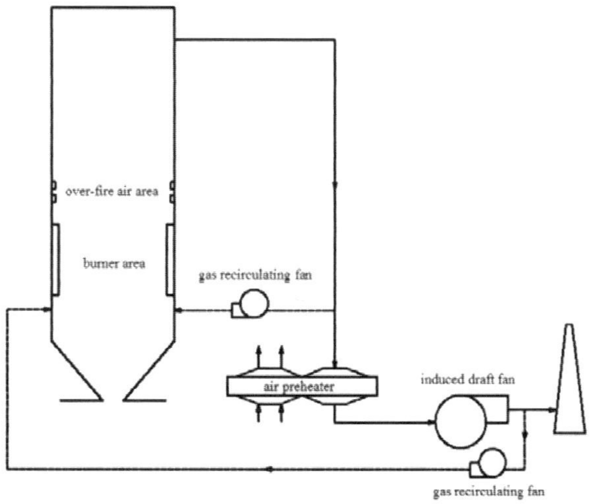

Figure 1. Schematic diagram of the recirculating flue gas extraction positions.

In view of above FGR schemes, under the BMCR and 75%THA load conditions, this paper calculated and analyzed the influence of different schemes on boiler parameters when there was no recirculation, 5%, 10% and 15% FGR rate.

3. The thermodynamic calculation method for boiler with FGR

According to the standard method of thermal calculation of boiler unit, the thermal calculation of boiler unit is realized through VB software programming. Because of the use of FGR, the flue gas volume, composition and gas enthalpy from the return point to the extraction place have changed. When the boiler thermodynamic calculation is carried out, it is necessary to recalculate the gas characteristic parameters and the enthalpy of the flue gas.

The volume and enthalpy of mixed flue between the return point and the exit point are:

$$V_r = V + r * V_c \tag{1}$$

$$I_r = I + r * I_c \tag{2}$$

Where, V is the flue gas volume (m^3kg^{-1}); r is the FGR rate (%) and I is the flue gas enthalpy ($kJkg^{-1}$).

Then, the flue gas temperature after mixing is solved:

$$\theta_r = I_r / (VC)_r \tag{3}$$

Where, θ_r is the mixed flue gas temperature (℃); $(VC)_r$ is the average thermal capacity of the mixed flue gas (kJ (kg℃)$^{-1}$).

It is necessary to pay attention to the amount, the flow rate and the characteristics of flue gas and the parameters of the enthalpy meter should be based on the actual flow of flue gas after mixing the heated surface.

4. Results and discussions

4.1. Comparison of the results of flue gas recirculation schemes under the BMCR load condition

The influence of the ratio of different recirculation flue gas on the theoretical combustion temperature, the flue gas temperature of the furnace outlet, the temperature of the single-reheated and secondary reheated steam are shown in the following figures.

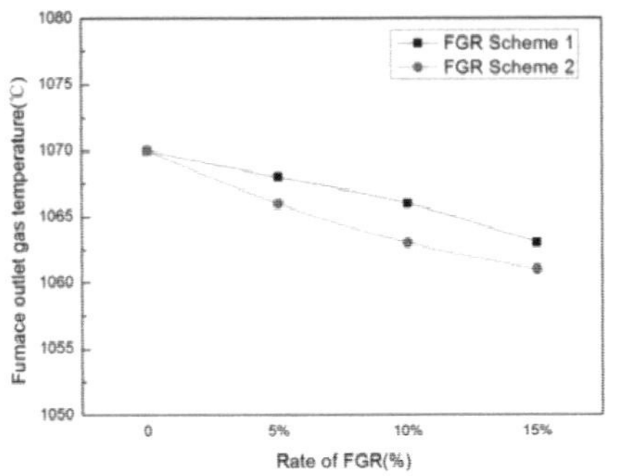

Figure 2. Furnace outlet gas temperatures under the BMCR load.

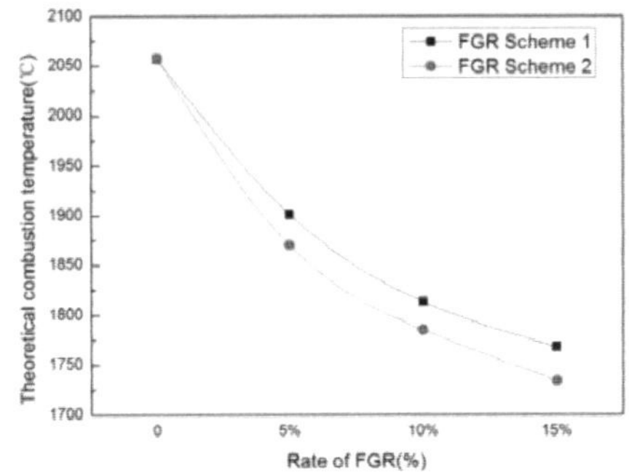

Figure 3. Theoretical combustion temperatures under the BMCR load.

As shown in the figures, the two schemes put the low temperature flue gas into the bottom area of the furnace, which will have an impact on the combustion and heat transfer in the boiler furnace. And the final result is that the furnace outlet gas temperature will decrease. Compared with the two schemes, the flue gas extraction point of scheme2 is the induced draft fan, which results in a lower gas temperature. For every 1% increase in recirculation rate, the furnace outlet of scheme 1 is reduced by 0.4℃ and the theoretical combustion temperature reduced by 25℃, while scheme 2 reduced by 0.6℃ and 28℃.

Due to the increase of flue gas volume and flue gas flow velocity on the convection heating surface, the flue gas heat release coefficient and convective heat transfer increase, which eventually leads to the rise of the temperature of the outlet working medium.

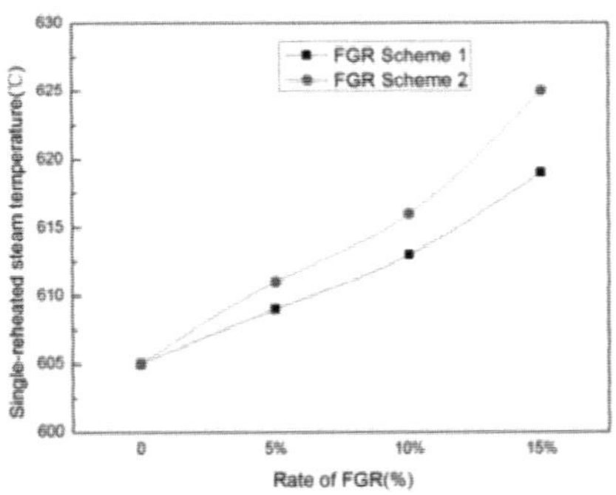

Figure 4. Single-reheated steam temperatures under the BMCR load.

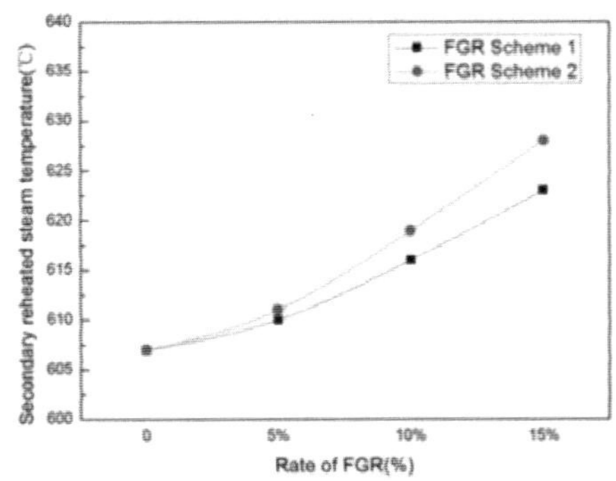

Figure 5. Secondary reheated steam temperatures under the BMCR load.

Figure 4 and 5 show that for every 1% increase in recirculation rate, the single-reheated steam temperature of scheme 1 increased by 0.8℃;the secondary reheated steam temperature increased by 1℃, while scheme 2 increased by 1.2℃ and 1.4℃.

4.2. Comparison of the results of flue gas recirculation schemes under the 75%THA load condition

The influence of the ratio of different recirculation flue gas on the above parameters under 75%THA load condition is shown in the following figures.

IOP Conf. Series: Earth and Environmental Science **146** (2018) 012043 doi:10.1088/1755-1315/146/1/012043

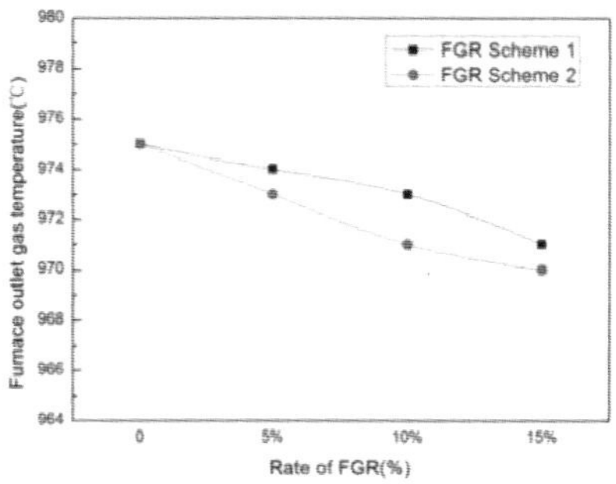

Figure 6. Furnace outlet gas temperatures under the 75%THA load.

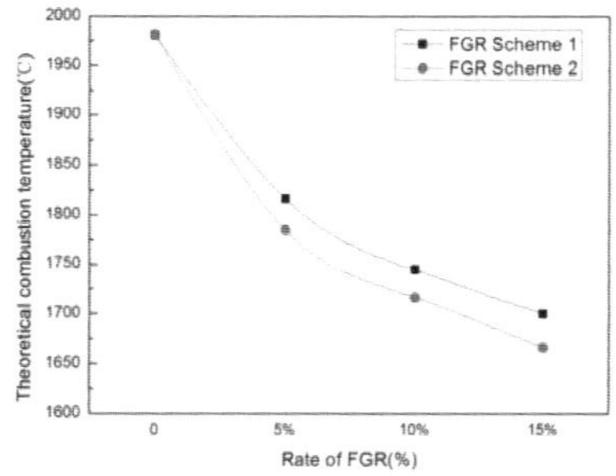

Figure 7. Theoretical combustion temperatures under the 75%THA load.

Figure 6 and 7 show that under the 75%THA load, for every 1% increase in recirculation rate,the furnace outlet gas temperature of the scheme 1decreased by 0.2℃ and the theoretical combustion temperature decreased by 23℃, while scheme 2 decreased by 0.4℃ and 26℃.

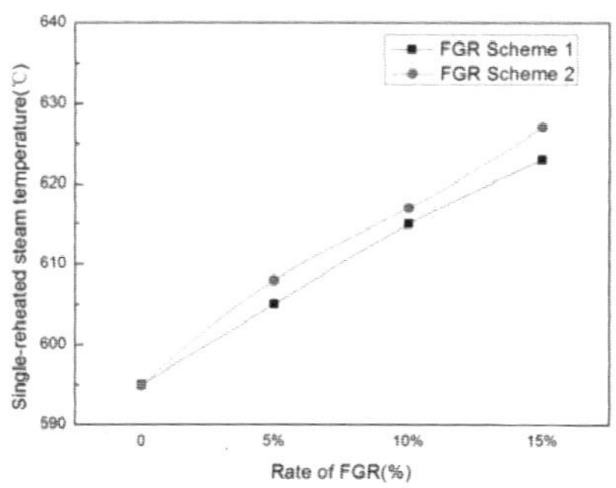

Figure 8. Single-reheated steam temperatures under the 75%THA load.

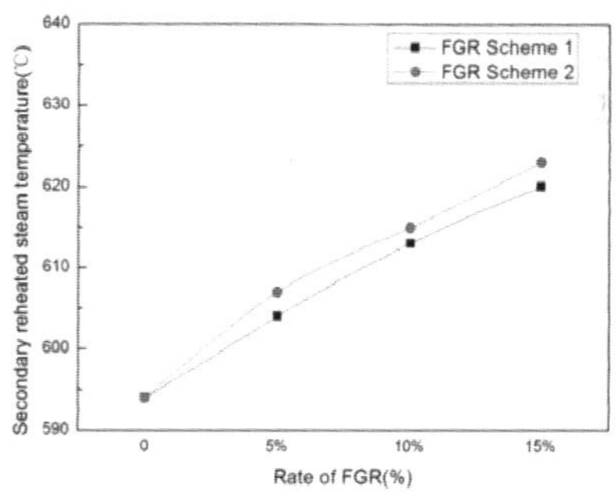

Figure 9. Secondary reheated steam temperatures under the 75%THA load.

Figure 8 and 9 show that under 75%THA load, for every 1% increase in recirculation rate, the single-reheat steam temperature of scheme 1 increased by 2℃ and the secondary reheated steam temperature increased by 1.8℃, while scheme 2 increased by 2.2℃ and 2℃.

The above results show that under the condition of BMCR and 75%THA load, when the FGR rate is about 10%, scheme 1 and 2 can make the steam temperature of single-reheated and secondary reheated reach the design value. With the increase of recirculation ratios, the drop of the furnace outlet gas temperature and the rise of reheated steam temperature are increased. And at low load conditions, the influence of the recirculation of the flue gas on the reheated steam temperature is increased. Compared with the scheme 1, the exhaust gas temperature of scheme 2 increased, and the boiler efficiency decreased. Therefore, the scheme 1 is more suitable for this boiler's FGR scheme.

5. Conclusion

Through the above calculation and analysis, this paper draws the following conclusions:

The reheat steam temperature of the double reheat boiler can be significantly improved by the FGR scheme of sending the recirculating flue gas into the bottom of the furnace. With the increase of the amount of recirculating flue gas, the rise of reheated steam increases and with the decrease of load, the influence of recirculating flue gas on reheated steam temperature increases. The scheme of extracting flue gas into the bottom of the furnace from the economizer is the suitable FGR scheme for the boiler, and the reasonable FGR ratio is about 10%, which can provide a reference value for solving the low boiler reheat steam temperature and the operation optimization of the unit.

References

[1]Yan W P, Zhao Y M, Li H X and Liu L H 2014 Analysis of the design of a thermal system for a 1350 MW secondary reheat power generator unit Journal of Engineering for Thermal Energy & Power vol 29:35-40.

[2] Xu G, Zhou L Y, Zhao S F, Liang F, Xu C and Yang Y P 2015 Optimum superheat utilization of extraction steam in double reheat ultra-supercritical power plants Applied Energyvol 160:863-872.

[3] Chen L, Zhou X and Zhao H 2016 Modeling and Simulation of 1000MW Double Reheat Ultra-supercritical Unit based on Particle Swarm Optimization International Conference on Information Engineering for Mechanics and Materials.

[4] Zhao Y M, Yan W P and Liu L H 2013 Influence Analysis of Design Parameters of an Ultra-Supercritical Double Reheat Unit Turbine Technology.

[5] Zhang X L, Yan W P, Liu L H and Zhu Y F 2016Optimization design of steam turbine with double-reheat in a 1350 MW ultra-supercritical unit Thermal Power Generation vol 45:110-116 [in Chinese].

[6] Dang L J, Yang H, Ying W Z, Wang Z H, Fan S D, Liu C H and Liu M Q 2017 Analysis and Technical Retrofit on Low Reheat Steam Temperature of a 660 MW Ultra-supercritical Boiler Journal of Chinese Society of Power Engineering vol37:261-266 [in Chinese].

[7] Zhang G, Xu W, Wang X and Yang Y 2015 Analysis and optimization of a coal-fired power plant under a proposed flue gas recirculation mode Energy Conversion & Management vol 102:161-168.

IOP Conf. Series: Earth and Environmental Science **146** (2018) 012044 doi:10.1088/1755-1315/146/1/012044

Light Affects Electrochemical Characterization of Electrogenic Microalgae *Desmodesmus* sp. A8

Y C Wu[1], Y Xiao[2], HY Fu[1] and F Zhao[2]

[1]College of Environmental Science and Engineering, Xiamen University of Technology, Xiamen 361024, China
[2]Institute of Urban Environment, Chinese Academy of Sciences, Xiamen 361021, China

fzhao@iue.ac.cn

Abstract. Electrogenic microalgae attached on cathode could enhance the generation of current in microbial fuel cells. For understand the electrochemical characterizations of microalgae, *Desmodesmus* sp. A8, as a model electrogenic microalgae, was investigated by cyclic voltammetry analysis under illumination and dark. *Desmodesmus* sp. A8 showed higher current response under incandescent (4.46 μ A) than that under fluorescent (2.16 μ A) at -0.44 V vs. Ag/AgCl, while the current decreased significantly to 0.14 μ A in dark. With the increasing illumination time, current generation enhanced rapidly and reach to a certain value remained stable, finally. From the light source and working time, we demonstrated that light profoundly impacted the electricity generation of electrogenic microalgae*Desmodesmus* sp. A8.

1. Introduction

Microbial fuel cells (MFCs) provide new ideas for new energy production and water pollution control with the aid of electrochemically active microorganisms [1]. MFCs inoculatedphotosynthetic microorganisms can generate electricity from respiratory andphotochemical activities [2,3]. Photosynthetic microorganisms such as *Chlorella vulgaris* [4] and *Rhodopseudomonaspalustris* DX-1[5] have been applied to expand the functions of MFCs.

Microalgae produce approximately half of the atmospheric oxygen, simultaneously consuming the greenhouse gas of carbon dioxide [6]. The photosynthesis process involves in the production of reactive oxygen species and oxygen which are important electron acceptors for electricity generation in MFCs [7,8]. Besides, an additional benefit of the application of microalgae in MFCs is that carbon dioxide can be effectively reduced by algae in cathode chamber [9]. Other functions of microalgae such as N, P abortion [4], biodiesel production [10], algal biomass supply [11] and self-regeneration ability [12] increase the potential application of MFCs in renewable energy generation and wastewater treatment.

Light is one of the most important factors affecting the photosynthesis efficiency of microalgae [13]. In the present study, the effect of light on electrochemical characterizations of electrogenic microalgae *Desmodesmus* sp. A8 were investigated.

2. Methods and materials

2.1 Microalgae

The *Desmodesmus* sp. A8, previously isolated from Jimei wastewater plant (Xiamen, China) [13]. The microalga was cultured with BG11 medium under illumination [14].

2.2 *Electrochemical analysis*

Cyclic voltammetry (CV) was performed in a three-electrode system withplatinum wire as counter electrode, 3 mm diameterglassy carbon electrodeas working electrode, and Ag/AgCl (KCl, sat.) as reference electrode. During the tests, the working electrode was covered with *Desmodesmus* sp. A8. CV was conducted in the potential range of −1.0 to 0.2 V at a scan rate of 20 mV/s using an electrochemical workstation (Autolab 302N, Netherlands).

3. Results and discussion

3.1Morphological characterization

The morphological characteristics of *Desmodesmus* sp. A8 are shown in Figure 1, the dents at the pole of the coenobia and the ribs on the cell surface are characters of *Desmodesmus* [15]. In cellular suspensions, two-cell coenobia were the most frequently.

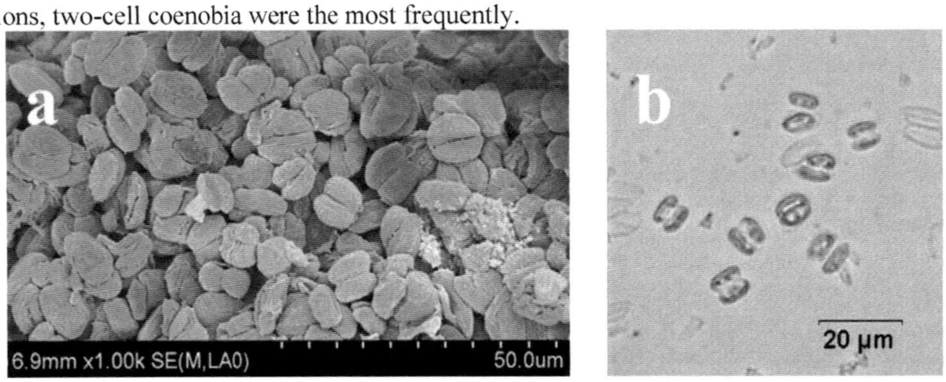

Figure 1. Scanning electron microscope (a) and light microscope (b, 400×) micrographs of the cells of strain A8.

3.2 Effect of light source on electrochemical characterizations

Light source is a significant factor affecting the photosynthesis of microalga mainly due to the difference in wavelength range coverage [16]. In this study, fluorescent and incandescent lamps were used to investigate the effects of light source on electrochemical characterizations ofstrainA8. As shown in Figure 2, peak current of strain A8 under the incandescent and fluorescent lamps were 4.46 and 2.16μ A at -0.44 V vs. Ag/AgCl respectively, which decreased to 0.14 μ A in dark.

Light distribution of the incandescent lamp was similar to the adsorption spectrum of *Desmodesmus* sp. A8 in the range of 660-690 nm [17]. Therefore, *Desmodesmus* sp. A8 under fluorescent light with lower current generation should be ascribed to a greater wavelength range of fluorescent light which are more unsuitable for the activation of chlorophyll pigments enabling A8 to produce less oxygen as the electron acceptor.

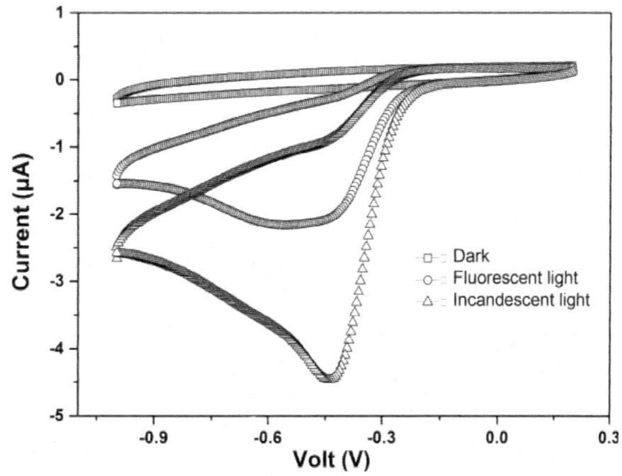

Figure 2. Effect of light sources on electrogenic characterization of *Desmodesmus* sp. A8.

3.3 Illumination affected electrochemical characterization

To investigate the effect of illumination on electrochemical characterization of strain A8, CVs of A8 were carried out under illumination (1500 lux) and dark. As shown in Figure 3, a reduction peak was observed at potential -0.4 V. The peak current rapidly increased from 0.013 (dark) to 2.23 μ A (illumination) and continuously increased to 4.26μ A after illumination for 10 hours. The current peak stayed constant as illumination time increased to 14 hours.

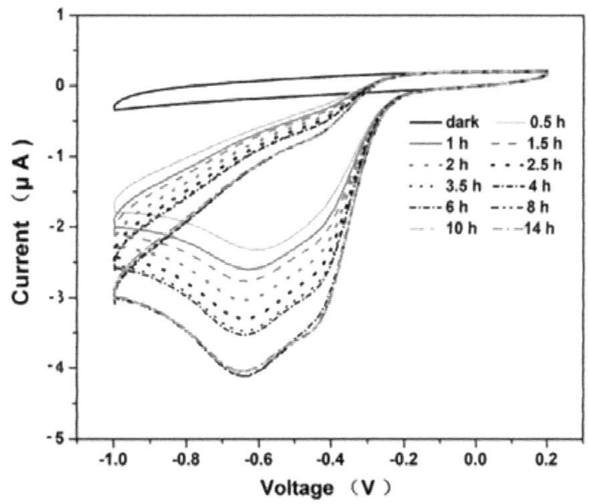

Figure 3. Cyclic voltammetry analysis of *Desmodesmus* sp. A8 under illumination.

As shown in Figure 4, the peak current at potential -0.4 V rapidly decreased to 0.012μ A when the A8 culture condition was changed to dark, and current output stayed constant as culture time increased to 14 hours in dark.

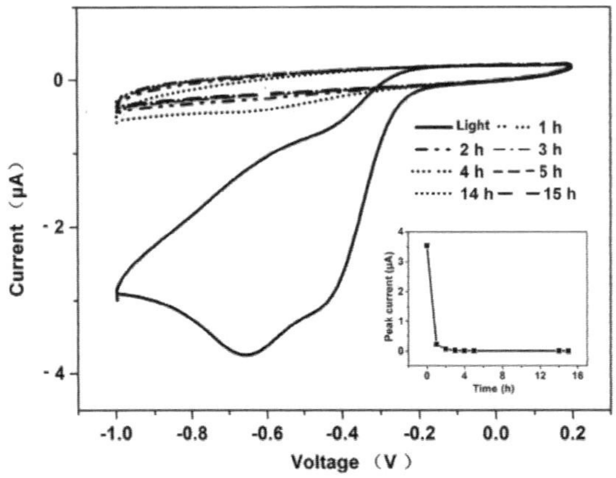

Figure 4. Cyclic voltammetry analysis of *Desmodesmus* sp. A8 under dark.

4. Conclusions

This study demonstrated that illumination significantly enhanced electricity generation of *Desmodesmus* sp. A8. Peak current of microalgae was significantly promoted when A8 was illuminated with incandescent lamp compared with fluorescent lamp. This study presented an insight into improving the performance of MFCs inoculated with microalga.

Acknowledgements

This work was supported by High Level Talent Program of Xiamen University of Technology (E2015028), the Young Teacher Project of Fujian Province (JA15372) and Xiamen Science and Technology Project(3502Z20179029).

References

[1] Bond DR, Lovley DR. Electricity production by *Geobactersulfurreducens*attached to electrodes. 2003*Appl Environ Microb***69**1548-55.

[2] Rosenbaum M, He Z, AngenentL, Light energy to bioelectricity: photosynthetic microbial fuel cells. 2010*CurrOpinBiotec***21**259–64.

[3] Alistair J. McCormick ,Bombelli P, Scott AM., Alexander J., Alison G, Adrian C and Christopher J. Photosynthetic biofilms in pure culture harness solar energy in a mediatorless bio-photovoltaic cell (BPV) system. 2011*Energy Environ Sci***4** 4699-709

[4] Wang X, Feng Y, Liu J, Lee H, Li C, Li N and Ren N. Sequestration of CO_2 discharged from anode by algal cathode in microbial carbon capture cells (MCCs). 2010*Biosensors Bioelectronics***25**2639-43.

[5] Xing D, Zuo Y, Cheng S, Regan JM and Logan BE. Electricity generation by *Rhodopseudomonaspalustris* DX-1.2008*EnvironlSciTechnol*, **42**4146-51.

[6] Schenk P, Thomas-Hall S, Stephens E, Marx U, Mussgnug J, Posten C,Kruse O and Hankamer B. Second generation biofuels: Highefficiency microalgae for biodiesel production. 2008*BioEnergy Res***1**20–43.

[7] Wu Y, Wang Z, Zheng Y, Xiao Y and Zhao F.Light intensity affects the performance of photo microbial fuel cells with *Desmodesmus*sp. A8 as cathodic microorganism. 2014*ApplEnerg***116**86-90.

[8] Cai P, Xiao X, He Y, Li W, Zang G, Sheng G. Reactive oxygen species (ROS) generated by cyanobacteria act as an electron acceptor in the biocathode of a bio-electrochemical system. 2013*BiosensBioelectron***39** 306-10.

[9] Li X, Gan K and Sun YX. Effects of differentnitrogen and phosphorusconcentrations on the growth, nutrient uptake, and lipid accumulation of a freshwater microalgae Scenedesmus sp. 2010*BioresourceTechnol***101**5494–500.

[10] Sathish A, Sims RC. Biodiesel from mixed culture algae via a wet lipid extraction procedure.2012*BioresourceTechnol***118**: 643-7.

[11] Chen R, Li R, Deitz L, Liu Y, Stevenson RJ and Liao W. Freshwater algal cultivation with animal waste for nutrient removal and biomass production. 2012*Biomass Bioenerg* **39** 128-38.

[12] Lan JCW, Raman K, Huang CM and Chang CM. The impact of monochromatic blue and red LED light uponperformance of photo microbial fuel cells (PMFCs) using *Chlamydomonasreinhardtii*transformation F5 as biocatalyst. 2013*BiochemEng J*, **78** 39-43

[13] Wu Y, Guan K, Wang Z, Xu B and Zhao F. Isolation, identification and characterization of an electrogenic microalgae strain. 2013*PLoS ONE***8**: e73442

[14] Yoon J H, Sim SJ, Kim MS and Park TH. High cell density culture of *Anabaena variabilis* using repeated injections of carbon dioxide for the production of hydrogen.2002*Int J Hydrogen En***27**1265-70.

[15] Lürling M. Phenotypic plasticity in the green algae *Desmodesmus* and *Scenedesmus*with special reference to the induction of defensive morphology.2003*Ann Limnol***39**85-101.

[16] Terry K L. Photosynthesis in modulated light: quantitative dependence of photosynthetic enhancement on flashing rate. 1986*Biotechnology Bioen***28**988-95

[17] Wu Y, Xiao Y, Wang Z and Zhao F. Performance of bioelectrochemical systems inoculated with *Desmodesmus* sp. A8 under different light sources, 2016*Bioremediat J***20**233-9.

IOP Conf. Series: Earth and Environmental Science **146** (2018) 012045 doi:10.1088/1755-1315/146/1/012045

Analysis of Energy Return on Investment of China's Oil and Gas Production

Zhaoyang Kong[1], Xiucheng Dong[2,*, a] and Qingzhe Jiang[2,*, b]

[1]School of Environment and State Key Joint Laboratory of Environment Simulation and Pollution Control, Tsinghua University, Beijing 100084, P.R. China;
[2]School of International Trade and Economics, University of International Business and Economics, Beijing 100029, China.

*Corresponding authors. E-mail: [a]dongxiucheng@cup.edu.cn; [b]jiangqz@uibe.edu.cn

Abstract. China ranks No.1 and No.3 respectively in global oil and gas consumption. Examination of viability of oil and gas resources could provide useful information in gauging economic vulnerability of future oil and gas supply in China. Energy Return on Investment (EROI) is an important index to characterize the viability of a natural resource from an energy viewpoint. This paper calculates EROI for oil and gas exploration ($EROI_{OGE}$) and EROI for light oil products ($EROI_{LOP}$) in China. The results show that the $EROI_{OGE}$ decreased from approximately 16.4 in 1985 to 8.4 in 2003, and then increased to 12.2 in 2012. The $EROI_{OGE}$ in recent years are due to the increasing of gas production. As a trade-off between the decrease of oil extraction efficiency and the increase of oil processing efficiency, the $EROI_{LOP}$ fluctuated around 4. The results suggest that China should develop the natural gas industry and improve the oil processing efficiency vigorously.

1. Introduction

Nearly 60% of the world's energy consumption is meet by oil and gas, and their availability has a critical impact on economies of many countries [1-2]. China is the largest energy consumer in the world, and its energy consumption has increased from 5.7×10^8 tce (tonnes of coal equivalent) in 1978 to 43×10^8 tce in 2015 [3]. In China, oil and gas provide approximately 24% of its total energy consumption [4]. As a result of its limited domestic production capacity, however, China has to import more and more oil and gas from others counties. In the past 20 years, China's oil-import dependency has increased from 9.8% in 1996 to 60.6 in 2015. In addition, China's gas-import dependency grows also rapidly and is projected to reach over 40% by 2030 [5]. The high imported oil and gas dependency poses a risk to energy security. If China seek to induce import dependency, it must develop domestic oil and gas resources vigorously. Therefore, it is necessary and important to calculate the viability of China's domestic oil and gas resources. EROI is a useful approach for estimating the viability of an energy source from an energy viewpoint [6]. This paper calculates EROI for oil and gas exploration ($EROI_{OGE}$) and EROI for light oil products ($EROI_{LOP}$) in China.

2. EROI Methodology

The general equation for EROI is as follows:

$$\mathrm{EROI} = \frac{\text{Energy returned (outputs)}}{\text{Energy required (inputs)}} \tag{1}$$

The numerator is the sum of all energy produced, and the denominator is the sum of the energy inputs. Before calculating the EROI, it is necessary to choose the suitable system boundary, which is perhaps the most important decision in an EROI analysis [7]. The system boundary of this paper is shown in Figure 1. The energy outputs of $EROI_{OGE}$ is oil and natural gas. The energy inputs include the energy consumption in extraction. Here, the energy outputs of $EROI_{LOP}$ are light oil products including gasoline, kerosene, and diesel. To calculate the $EROI_{LOP}$, we must consider not only the energy consumed in oil extraction but also the energy required for oil transportation and processing.

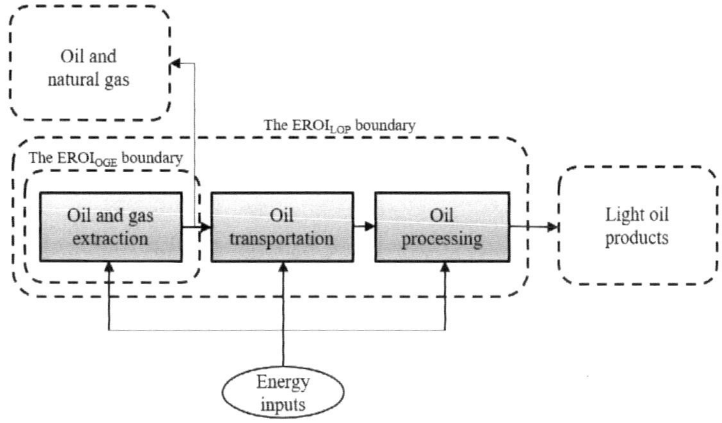

Figure 1. The system boundaries for $EROI_{OGE}$ and $EROI_{LOP}$

The formula for $EROI_{OGE}$ is as follows:

$$EROI_{OGE} = \frac{E_{OGE}}{E_{e,O\&G}} \tag{2}$$

Where E_{OGE} refers to total energy outputs of oil and gas extraction.

$EROI_{LOP}$ is expressed as follows:

$$EROI_{LOP} = \frac{E_{LOP}}{E_{e,oil} + E_{t,oil} + E_{p,oil}} \tag{3}$$

Where, $E_{e,oil}$, $E_{t,oil}$, and $E_{p,oil}$ refer to the total energy input of oil extraction, transportation and processing, respectively.

In calculating the $EROI_{LOP}$, we find that the amount of oil extracted are not equal to the amount of oil processed, as will be shown in Section 3.2. However, to calculate the $EROI_{LOP}$, we must take into account equal volumes. Safronov and Sokolov (2014) [4] also encountered this problem in calculating the $EROI_{LOP}$ for Russian oil companies. Thus, Safronov and Sokolov (2014) proceeded as follows (Figure 2):

(1) Equalize mounts by notionally increasing or decreasing oil extraction;

(2) Proportionally change the energy inputs for oil extraction;

(3) Calculate the $EROI_{LOP}$.

In this scheme, the authors make an important assumption, that is, the average energy inputs for oil extraction (or oil processing) did not change as the production scale changed. However, this assumption is usually not true, and in fact, the average energy inputs for oil extraction (or oil processing) do change with a changed production scale. In oil processing, for example, changes in scale have important implications for energy efficiency [8-10]. In fact, based on the method of Safronov and Sokolov (2014) [4], we could calculate the $EROI_{LOP}$ by adjusting Equation (3) to exclude the assumption. First, we simultaneously divided the numerator and denominator by the amount of oil processing (M_p):

$$EROI_{LOP} = \frac{E_{LOP} / M_p}{E_{e,oil} / M_p + E_{t,oil} / M_p + E_{p,oil} / M_p} \quad (4)$$

The calculation of the $EROI_{LOP}$ requires M_p to be equal to the volumes of oil extraction (M_e) and oil transportation (M_t), so we obtain the following equation:

$$EROI_{LOP} = \frac{E_{LOP} / M_p}{E_{e,oil} / M_e + E_{t,oil} / M_t + E_{p,oil} / M_p} = \frac{E_{LOP,per}}{E_{e,oil,per} + E_{t,oil,per} + E_{p,oil,per}} \quad (5)$$

Where, $E_{LOP,per}$ refers to the light oil production per tonne of oil processing, and $E_{e,oil,per}$, $E_{t,oil,per}$, and $E_{p,oil,per}$ refer to the energy input per tonne of oil extracted, oil transportation and oil processing, respectively.

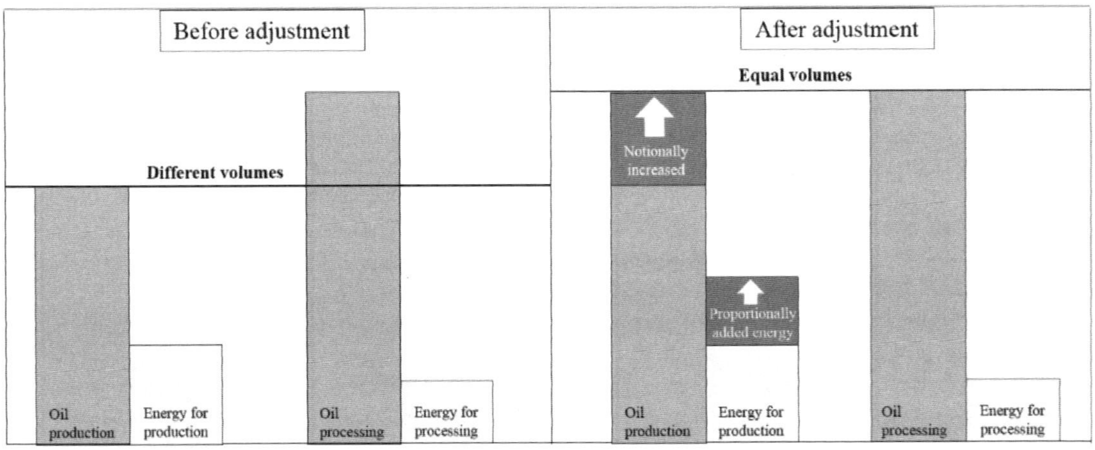

Figure 2. Scheme for calculating the $EROI_{LOP}$

The energy input of oil extraction is part of the energy input of light oil production. However, the cost of oil and gas is mixed together, so we have to assume that $E_{e,oil,per}$ is equal to $E_{e,O\&Gl,per}$ (the energy input per tonne of oil and gas extracted). Thus, Equation (5) is adjusted as follows:

$$EROI_{LOP} = \frac{E_{LOP,per}}{E_{e,O\&G,per} + E_{t,oil,per} + E_{p,oil,per}} \quad (6)$$

As $EROI_{OGE} = 1/E_{e,O\&G,per}$, Equation (6) is modified as follows:

$$EROI_{LOP} = \frac{E_{LOP,per}}{1 / EROI_{OGE} + E_{t,oil,per} + E_{p,oil,per}} \quad (7)$$

Thus, it is much easier for us to calculate the $EROI_{LOP}$ using Equation (6) or (7).

The energy outputs and inputs for China's oil and gas extraction are derived from National Bureau of Statistics of China [3]. In China, oil transportation relies on pipelines, and we assume that the average distance from oilfield to oil processing plant is approximately 1000 km. The unit energy consumption by oil pipeline transport is approximately 0.3 MJ/ton-km [11]. Because it is unavailable to obtain the accurate data of refining energy consumption in different refineries, we use the average refining energy consumption of Sinopec for substitution [12-13].

3. Results and discussion

The $EROI_{OGE}$ and $EROI_{LOP}$ in China are shown in Figure 3. It is clear that, as a result of the depletion of oil reserves, oil production needs more energy inputs. In addition, the growth in energy inputs in turn leads to a decrease in the $EROI_{OGE}$ in 1985-2003.However, after 2003, the $EROI_{OGE}$ increased from 8.4 to 12.2, which may result from the increasing gas production with relatively high EROI. As a result of the interaction between the decrease of oil extraction efficiency and the increase of oil processing efficiency, the $EROI_{LOP}$ fluctuated about 4.

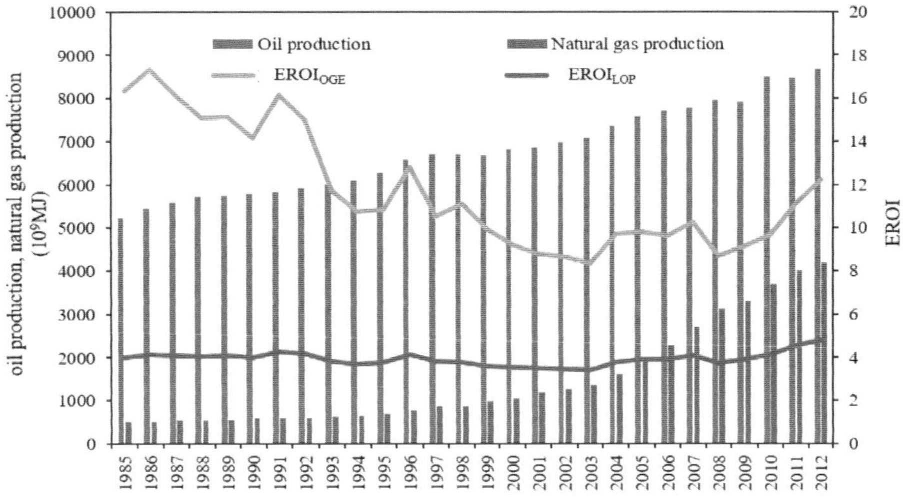

Figure 3. The $EROI_{OGE}$ and $EROI_{LOP}$ in China

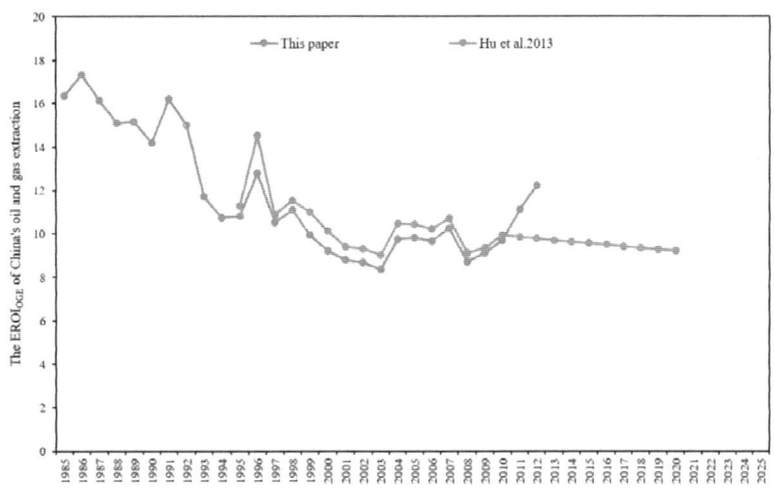

Figure 4. The $EROI_{OGE}$ values of this study and Hu *et al.*'s study

Hu et al. (2013) [14] showed that China's $EROI_{OGE}$ fluctuated from 12 to 14:1 in the mid-1990s and decreased to 10:1 in the period from 2007-2010 (Figure 4). The EROI results of this paper are similar to those of Hu et al. Hu et al. (2013) [14] further predicted that China's $EROI_{OGE}$ will continue to decline in 2011-2020 and will drop to 9.6: 1 by 2020. The result of this prediction is different from the result of this paper. According to the actual data, the $EROI_{OGE}$ calculated in this paper rises in 2011-2013. We argue that China's $EROI_{OGE}$ will continue to rise in the short term. The reason is that China's natural gas development is relatively late and its EROI is currently in the rising stage. EROI trends of an energy resource are impacted by two main factors [15]. One is technological component. With the progress of energy production, the mining technology will be gradually mature and the energy inputs used in the extraction process will decrease. However, technological progress has theoretical limits (Figure 5a). The other is physical resource component. The energy resource that offer the best returns is exploited first. Attention then turns to resources with lower returns as production continues. This mode of exploitation leads to a gradual increase in energy input per unit output (Figure 5b). Given the above two components, the EROI of an energy source will first increase and then decrease (Figure 5c). Assuming that the EROI trends for oil and natural gas are shown in Figures 6a and 6b, respectively, and the natural gas development stage is later than oil, the EROI of oil and gas may be shown in Figure 6c. Its trend is to rise first, then fall, then rise again, and finally fall. Now, China's oil and gas development is in accordance with stage 4 (t_3 to t_4) in Figure 6c, which is an ascending stage rather than the descending stage proposed by Hu et al.

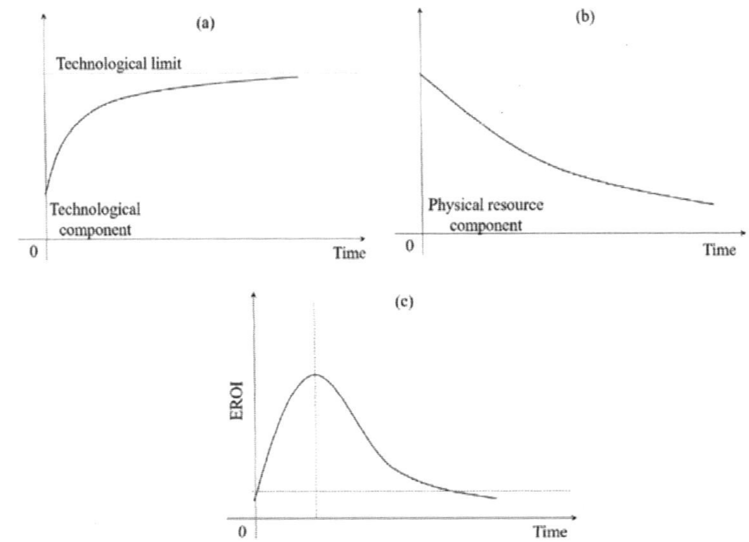

Figure 5. EROI trends in technical constraints and resource constraints

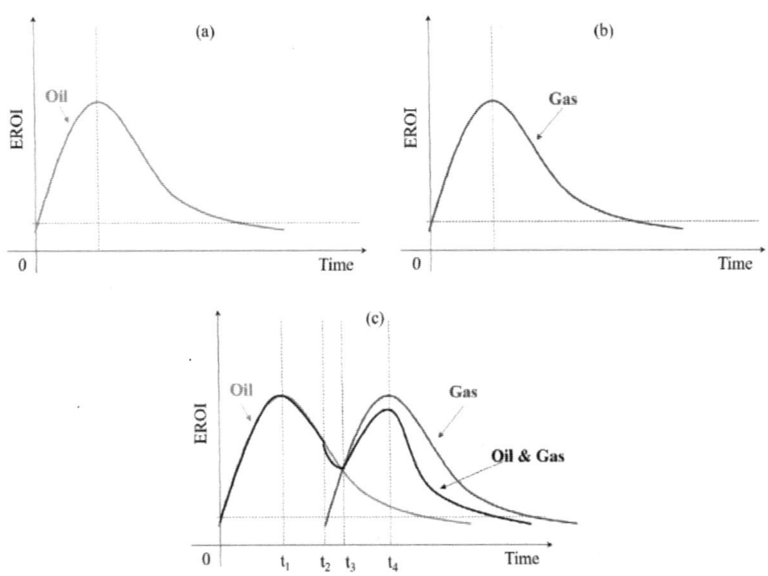

Figure 6. Illustration of EROI trends for oil and gas

4. Conclusion
This paper calculated the EROI for oil and gas exploration ($EROI_{OGE}$) and EROI for light oil products ($EROI_{LOP}$) in China. The results show that the $EROI_{OGE}$ in China decreased from 19.1 in 1986 to 9.6 in 2003, and recovered to 15.7 in 2013. The increase of $EROI_{OGE}$ is due the increasing of gas production with relatively high energy return. According to the analysis in Section 4, in the short term, the increase of gas production will further improve the $EROI_{OGE}$. In the future, some measures should be taken by the government to increase gas production such as rationalizing the domestic gas pricing mechanisms. The $EROI_{LOP}$ is much lower than $EROI_{OGE}$ and fluctuated around 4. To increase the $EROI_{LOP}$, some measures could be taken to improve the oil processing efficiency such as increasing the R&D investment.

Acknowledgements
This work was supported by the National Natural Science Foundation of China (71722003, 71273277, No. 71690244,), the Philosophy and Social Sciences Major Research Project of the Ministry of Education (No.11JZD048), the National Key R&D Program (2016YFC0208901), and China Postdoctoral Science Foundation (2017M620809).

References
[1] BP Statistical Review of World Energy 2017. Available online: http://www.bp.com/en/global/corporate/energy-economics/statistical-review-of-world-energy.html (accessed 21 January 2017).
[2] Munasinghe, M. The sustainomics trans-disciplinary meta-framework for making development more sustainable. Int. J. Sustain. Dev. 2002, 5, 125-182.
[3] National Bureau of Statistics of China. China Statistic Yearbook 1989-2013. China Statistics Press: Beijing, China, 2014. [In Chinese].
[4] Safronov, A.; Sokolov, A. Preliminary calculation of the EROI for the production of crude oil and light oil products in Russia. Sustainability 2014, 6, 5801-5819.
[5] Kong, Z.; Dong, X.; Zhou, Z. Seasonal imbalances in natural gas imports in major northeast Asian countries: variations, reasons, outlooks and countermeasures. Sustainability 2015, 7, 1690-1711.
[6] Gagnon, N.; Hall, C.A.S.; Brinker, L. A preliminary investigation of energy return on energy investment for global oil and gas production. Energies 2009, 2, 490-503.
[7] Cleveland, C.J.; O'Connor, P.A. Energy return on investment (EROI) of oil shale. Sustainability 2011, 3, 2307-2322.
[8] Huang, Z.H.; Liu, L.L. Sacle economy and cost analysis: from the perspective of the relationship between refining size and cost. Chemical Techno-Economics 2002, 3, 28-31. [In Chinese]
[9] Zhang, G.S. Study on economy of scale of refinery and process units. Petrol. Refinery Eng. 2007, 37, 58-62. [In Chinese].
[10] Lin, B.; Xie, X. Energy conservation potential in China's petroleum refining industry: evidence and policy implications. Energy Convers. Manag. 2015, 91, 377-386
[11] Ou, X.; Xiaoyu, Y.; Zhang, X. Life-cycle energy consumption and greenhouse gas emissions for electricity generation and supply in China. Appl. Energy 2011, 88, 289-297
[12] Sinopec. Sinopec Statistic Yearbook 2001-2013. China Petrochemical Press: Beijing, 2014
[13] Sinopec, Annual Reports. Available online: http://www.sinopecgroup.com/group/gsjs/gsbg/ (accessed 20 August 2014).
[14] Hu, Y.; Hall, C.A.S.; Wang, J.; Feng, L.; Poisson, A. Energy return on investment (EROI) of China's conventional fossil fuels: historical and future trends. Energy 2013, 54, 352-364
[15] Dale M, Krumdieck S, Bodger P. A Dynamic Function for Energy Return on Investment. Sustainability 2011, 3, 1972-1985.

IOP Conf. Series: Earth and Environmental Science **146** (2018) 012046 doi:10.1088/1755-1315/146/1/012046

Ship Target Segmentation in SAR Images Using A Modified Nonlocal Active Contour

Xiaoqiang Zhang[1,*], Boli Xiong[1] and Gangyao Kuang[1]

[1]National University of Defense Technology, Electronic Science College, Changsha, Hunan, China

*Corresponding author. Email: zxqdark@163.com

Abstract. In ocean surveillance, ship targets are the focus of attention. The ship target segmentation of SAR images is still challenging due to not only speckle noise but also the variations of backscattering intensities. A new segmentation method for ship targets is proposed in this paper, based on the nonlocal processing principle. The level set is implemented by using nonlocal comparison of patches. In the procedure of pairwise interaction, only the local homogeneity of image features is utilized, and this is important in optimizing regions with homogeneous features. In order to make the method applicable for SAR images, a distribution metric is incorporated into the energy function to suppress the multiplicative noise. Experimental results based on TerraSAR data illustrate the effective performance of the proposed method on the ship target segmentation: It can overcome the speckle noise and intensity variations.

1. Introduction

Synthetic Aperture Radar (SAR) has been widely used in ocean surveillance due to its unique advantage of independence of both time and weather conditions. In many maritime SAR images, ship targets are the focus of attention. Ship target segmentation is the fundamental work of image interpretation, which is still challenging. The multiplicative noise effects are the traditional problem in SAR image processing. It may lead to intensity inhomogeneity and fuzzy edges, which frustrate the segmentation performance. For this problem, tremendous research efforts have been devoted to eliminating the speckle effects[1]. Efficient stochastic models of speckle noise, focusing on the multiplicative nature, are proposed[2]. Suitable distributions that fit SAR images better (e.g. K, Gamma and G^0) are employed to eliminate the speckle effects[3]. By contrast, the work that aims at variations of backscattering coefficients is very limited. As known, the grey level of a pixel in the SAR image represents the proportion of microwave backscattered from the corresponding region. When we implement target segmentation form the background, the intensities of targets and background are usually supposed to be homogeneous. However, in SAR systems, multiplicative speckle noise and other factors usually make the regions of interest inhomogeneous. Furthermore, as for ships, there may be obvious intensity variations over the ship target due to the complicated structure of the ship's hull, especially in high resolution images.In Figure 1, although the ship is a strong scattering-object, some dark regions familiar with the background are embedded in it, caused by shadows or flat surfaces. This kind of variations of local statistics bring much trouble to extract the ship from the background completely and precisely.

IOP Conf. Series: Earth and Environmental Science **146** (2018) 012046 doi:10.1088/1755-1315/146/1/012046

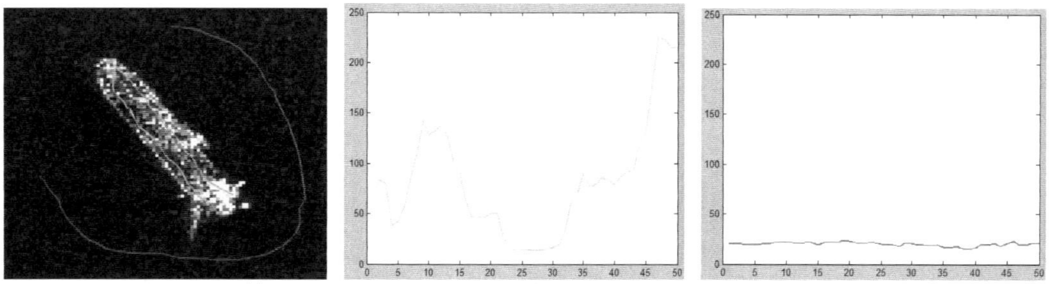

(a) (b) (c)

Figure 1. The difficulty of ship segmentation in SAR images. (a) shows a SAR chip containing a ship target with strokes in various colors. (b) and (c) display profiles of the SAR chip along the green and the red strokes.

The active contour model (ACM) has been widely used in image segmentation. It guides the evolution of contours toward a target's boundary by means of minimization of an energy function. There are two kinds in ACMs: edge-based models and region-based models[4]. The snake and the geodesic active contour (GAC) models[5] are the classic edge-based models. They use some edge detector and guide the active contour towards sharp gradients of pixel intensity. These models are usually sensitive to noise. To solve this problem, region-based active contour models are proposed incorporating global information to optimize the homogeneity of regions inside and outside the contour. Region-based models (i.e. Chan–Vese (CV)[6], local and global intensity fitting (LGIF) [7], and local Gaussian distribution fitting (LGDF) models[8] have better anti-noise performance in optical and medical images. Besides, Some hybrid ACMs are proposed such as Lankton's hybrid models[9] and selective binary and Gaussian filtering regularized level set (SBGFRLS) [10]. They absorb the advantages of those two kinds and have been demonstrated the great performance in optical and medical images. Due to the good performance, ACM has also been widely studied in SAR image processing. Many improved works are based on the classic ACMs. Germain et al. introduced a likelihood-ratio edge detector into the snake model to localize edges[11]. Shuai presented an improved level set algorithm with stationary global minimum using Gamma distributions[3]. Hu et al. utilized the Kullback–Leibler (KL) distance of Edgeworth to improve segmentation[12]. What's more, Tu replaced the Euclidean distance by a new ratio distance and proposed the modified CV model and the modified LGIF model[13].

In the literature, most of these methods are based on the hypothesis that the regions inside and outside the contour are homogeneous or follow the same distribution. That's to say, they may not deal with the intensity variations of ship targets. Focusing on this problem, a level set algorithm based on the nonlocal interactions between pairs of patches inside and outside the segmented regions is proposed in this paper. To improve the performance, a ratio distance, which is inspired by Tu's idea in [13], is integrated in nonlocal comparisons of patches.

2. Nonlocal active contour model

Nonlocal image processing means nonlocal comparisons of patches that are extracted from the image. It is firstly proposed by Buades et al, and has been applied in denoising, inverse problems and classification[14]. Jung et al use the nonlocal energy to drive the active contour and optimize the homogeneity of the segmented region. It is called nonlocal active contour model[15].

In the image f, a patch around a pixel $x \in I$ can be defined as $p_x(t) = f(x+t), t \in \left[-\frac{\tau}{2}, \frac{\tau}{2}\right]^2$ and the size is $(\tau+1)^2$. The nonlocal active contour model is implemented by minimizing the following energy function,

$$E(\Omega) = E_{\mathrm{NL}}(\Omega) + \lambda E_{\mathrm{R}}(\Omega) \tag{1}$$

Where $R(\Omega)$ is a smoothing term regularizing the contour of the region, usually defined as the length of the boundary, and the parameter λ is a weight controlling the regularity of the contour. More importantly, $E_{\mathrm{NL}}(\Omega)$ is the nonlocal energy term describing the dissimilarity inside and outside Region Ω, and it is defined as

$$\begin{aligned} E_{\mathrm{NL}}(\Omega) &= \bar{E}_{\mathrm{NL}}(\Omega)+\bar{E}_{\mathrm{NL}}(\Omega^{c}), \\ \bar{E}_{\mathrm{NL}}(\Omega) &= \int_{\Omega\times\Omega} G_{\sigma}(x,y)d(p_x,p_y)\,\mathrm{d}x\,\mathrm{d}y \end{aligned} \tag{2}$$

Where Ω^{c} is the complementary region of Ω, and $G_{\sigma}(x,y)$, a Gaussian kernel of scale σ, is used as a decaying function of $\|x-y\|$. $d(p_x,p_y)$ evaluates the distance between patches p_x and p_y. A specific metric can be used to deal with some images. This will be discussed in the next section.

For numerical implementation, the contour of the target can be evolved by a level set function φ. Then, the energy function in (X) can be rewritten as

$$E_{\mathrm{NL}}(\varphi)=\int_{I\times I}\left[1-\left|H(\varphi(x))-H(\varphi(y))\right|\right]G_{\sigma}(x,y)d(p_x,p_y)\,\mathrm{d}x\,\mathrm{d}y \tag{3}$$

The regularization term can be defined as

$$E_{\mathrm{R}}(\varphi)=\int_{I}\left|\nabla H(\varphi(x))\right|\mathrm{d}x \tag{4}$$

The contour evolution can be computed numerically by minimization of the energy functions.

3. Patch comparison in SAR images

As described above, the patch comparison metric $d(p_x,p_y)$ plays a crucial role in nonlocal image processing. To utilize pixel values, the orientation of textures and local statistical features, several metrics are introduced in patch comparison in , such as L^2 distance, L^2 distance based on Gabor filters and sliced Wasserstein distance. Although achieve success in optical and synthetic images, they are not suitable for measuring the distance between SAR image patches due to the speckle noise. On the one hand, the multiplicative nature of the speckle noise has been investigated by Feng et al, and it can be seen that the ratio distance is a useful metric for patch comparison in SAR images. On the other hand, according to the research of Deledalle et al, probabilities characteristics have satisfactory performance in SAR image processing. These two points of view has been further supported by the work of Tu et al.. They incorporate the ratio distance into the distribution metric, and modify the classic CV model by replacing the Euclidean distance. The energy function of the modified CV (MCV) model is defined as

$$\begin{aligned} E_{MCV}(\Omega) &= \lambda_1\bar{E}_{MCV}(\Omega)+\lambda_1\bar{E}_{MCV}(\Omega^{C}) \\ \bar{E}_{MCV}(\Omega) &= \int_{\Omega}\left(\varepsilon\left\{\log^{2}\left[\frac{p(x,\Omega)}{p_{\mathrm{img}}}\right]\right\}-\varepsilon\left\{\log\left[\frac{p(x,\Omega)}{p_{\mathrm{img}}}\right]\right\}^{2}\right)^{1/2}\mathrm{d}x \end{aligned} \tag{5}$$

Where p represents the probability density function of the region, p_{img} represents the probability density function of the raw image, and $\varepsilon(\bullet)$ denotes the expectation.

The efficiency of the MCV model in image segmentation has been demonstrated based on the MASTAR data set. As previously mentioned, however, the classic scheme of the CV model has limited ability to deal with intensity variations.

Inspired by the preceding works, we introduce the distribution metric of the ratio distance into patch comparison to compute the nonlocal energy. Therefore, the dissimilarity measure can be rewrote as:

$$E_{\mathrm{NL}}(\Omega)=\int_{\Omega\times\Omega}G_\sigma(x,y)\left(\varepsilon\left\{\log^2\left[\frac{P_x}{P_y}\right]\right\}-\varepsilon\left\{\log\left[\frac{P_x}{P_y}\right]\right\}^2\right)^{1/2}\mathrm{d}x\,\mathrm{d}y$$
$$+\int_{\Omega^C\times\Omega^C}G_\sigma(x,y)\left(\varepsilon\left\{\log^2\left[\frac{P_x}{P_y}\right]\right\}-\varepsilon\left\{\log\left[\frac{P_x}{P_y}\right]\right\}^2\right)^{1/2}\mathrm{d}x\,\mathrm{d}y \quad (6)$$

Where P_x and P_y denote the probability density functions of the patches p_x and p_y. The distance can be considered as the standard deviation between the log-likelihood of two distributions, P_x and P_y. When the energy function is minimized, the minimum variances will be obtained. During this process, the contour will be driven close to the target boundaries so that the homogeneity of the regions inside and outside the contour will be optimized.

4. Experimental results and analysis

TerraSAR X-band images with 1m resolution are used in the experiment. To demonstrate the performance of the proposed segmentation method, we utilize 4 chips with apparent intensity variations, which are shown in Figure 2. What's more, some traditional models, such as RSF, CV and LGIF, are added to comparison, and their results are displayed in Figure 3, figure 4 and figure 5. It can be seen that these method cannot deal with intensity variations and they fail to extract the ship from the background completely.

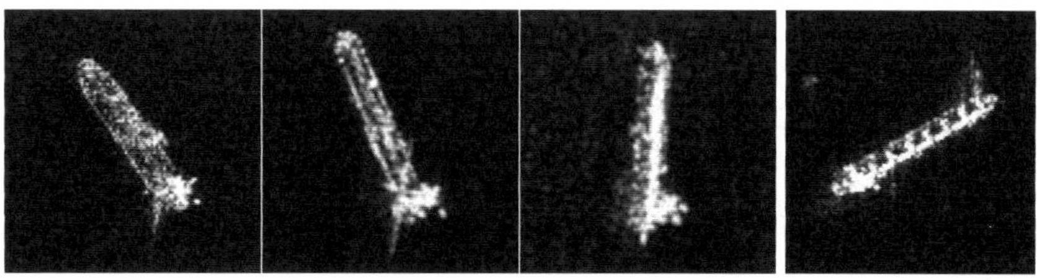

Figure 2. Chips from SAR images

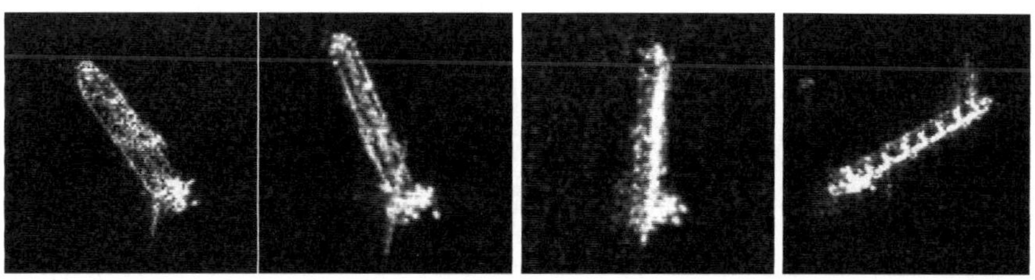

Figure 3. RSF

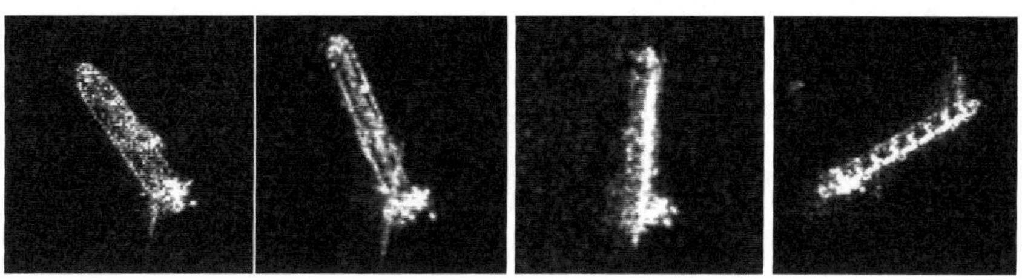

Figure 4. CV

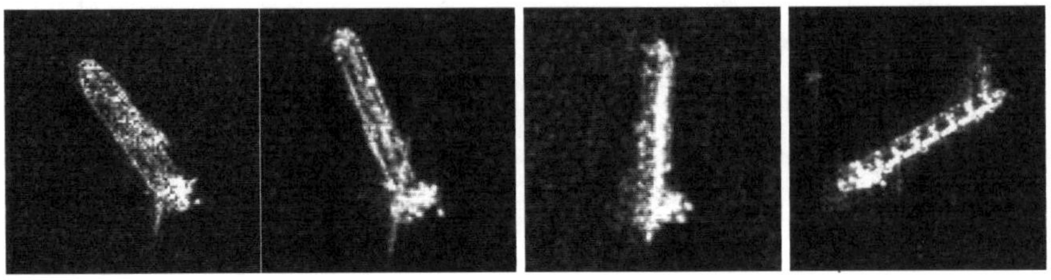

Figure 5. LGIF

Figure 6 shows the results of the proposed method with the parameters $\lambda=10, \sigma=10, \tau=5$. Each target is completely segmented from the background.

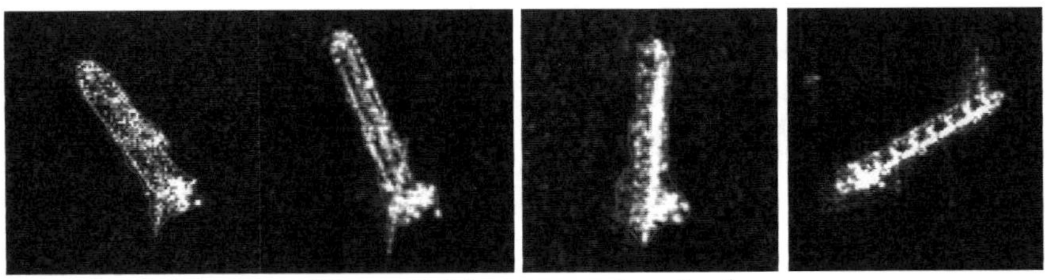

Figure 6. The proposed method

Thanks to the advantage of level set, the proposed method can be used in multi target segmentation. In Figure 7, there are 3 ships in the sub-image of the TerraSAR X-band data. All of them are completely extracted by the proposed method.

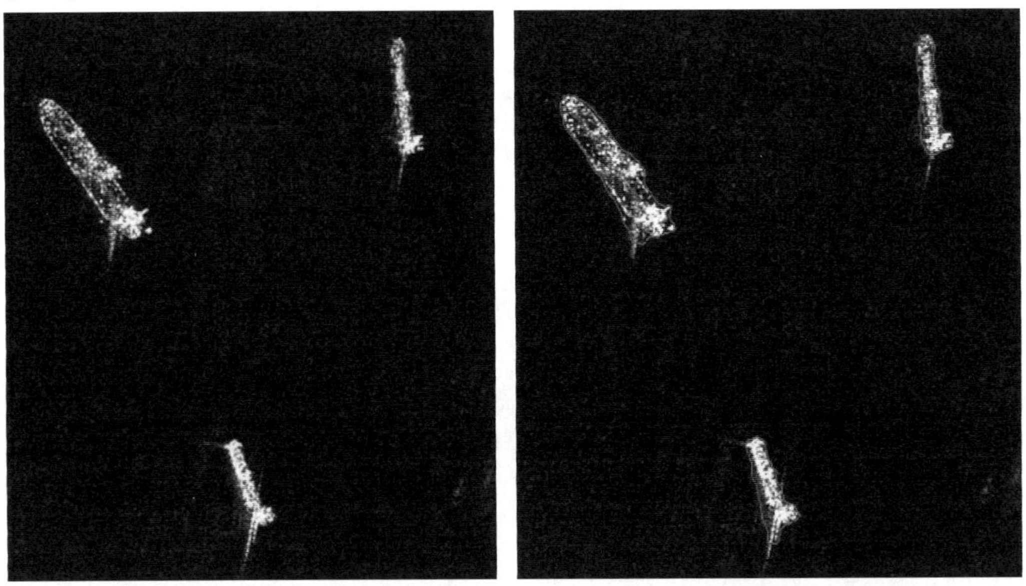

(a) (b)

Figure 7. Image of several ships. (a) is the original image, and (b) is the result of our method

5. Conclusion

A modified nonlocal active contour model for SAR image segmentation is proposed in this paper. On the one hand, the distribution metric of the ratio distance is used to overcome the multiplicative speckle noise. On the other hand, the patch comparison of the nonlocal scheme can deal with the intensity variations in SAR images, so that the proposed method can extract ship targets from the background completely. The real SAR images from TerraSAR X-band data with 1m resolution are used in the experiments to demonstrate the performance of the proposed method. Compared with some popular ACMs, the proposed method achieves apparent development in accuracy and robustness of ship segmentation.

References

[1] H Maitre, *Processing of Synthetic Aperture Radar Images*. Hoboken, NJ, USA: Wiley, 2010.
[2] G Gao, Statistical modeling of SAR images: A survey, *Sensors*, vol. 10, no. 1, pp. 775–795, Jan.2010.
[3] Y Shuai, H Sun, and G Xu, SAR image segmentation based on level set with stationary global minimum, *IEEE Geosci. Remote Sens. Lett.*, vol. 5, no. 4, pp. 644–648, Oct. 2008.
[4] M Kass, A Witkin, and D Terzopoulos, Snakes: Active contour models, *Int. J. Comput. Vis.*, vol. 1, no. 4, pp. 321–331, Jan. 1988.
[5] Caselles V, Kimmel R, Sapiro G, 1997, Geodesic active contours, *International Journal of Computer Vision,* 22**,** pp. 61–79.
[6] Chan T F, Vese L A, 2001, Active contours without edges, *IEEE Transactions on Image Processing*, 10, pp. 266-277.
[7] Wang L, Li C, Sun Q, Xia D, Kao C, 2009, Active contours driven by local and global intensity fitting energy with application to brain MR image segmentation, *Computerized Medical Imaging and Graphics*, 33, pp. 520-531
[8] Wang L, He L, Mishra L, Li C, 2009,Active contours driven by local Gaussian distribution fitting energy, *Signal Process*, 89, pp. 2435-2447.
[9] Lankton S, Nain D, Yezzi A, Tannenbaum A, 2007, Hybrid geodesic region-based curve evolutions for image segmentation, *Proc. SPIE,* vol. 6510, pp. 2-10.
[10] Zhang K, Zhang L, Song H, Zhou W, 2010, Active contours with selective local or global segmentation: A new formulation and level set method, *Image Vis. Comput.*, 28, pp. 668-676.
[11] Germain and P.Réfrégier, Edge location in SAR images: Performance of the likelihood ratio filter and accuracy improvement with an active contour approach, *IEEE Trans. Image Process.*, vol. 10, no. 1, pp. 72–78, Jan. 2001.
[12] L Hu, Y Ji, Y Li, and F Gao, SAR image segmentation based on Kullback–Leibler distance of Edgeworth, *Proc. PCM Adv. Multimedia Inf. Process.*, 2010, pp. 549–557.
[13] Song Tu, Yi Su, Yu Li. Convex active contour model for target detection in synthetic aperture radar images. *Journal of Applied Remote Sensing*. 2015, 9 (1) :095084
[14] A Buades, B Coll, and J M Morel, A non-local algorithm for image denoising, *Proc. IEEE Comput. Vis. Pattern Recog.*, Jun. 2005, vol. 2, pp. 60–65
[15] M Jung, G Peyre, and L D Cohen, Non-local active contours, *SIAM J. Imag. Sci.*, vol. 5, no. 3, pp. 1022–1054, 2012.

IOP Conf. Series: Earth and Environmental Science **146** (2018) 012047 doi:10.1088/1755-1315/146/1/012047

Allelopathic Effects of Two Organic Acids on *Microcystis Aeruginosa*

T Chai[1,2*], H D Zhu [1], H Z Yan[1], D Zhao [1], X Y Liu[3] and H Y Fu[1,2]

[1]College of Environmental Science and Engineering, Xiamen University of Technology, Xiamen 361024, China
[2]Key Laboratory of Environmental Biotechnology(XMUT)，Fujian Province University
[3]Shenyang Environmental Monitoring Central Station, Shenyang Liaoning Province 110016

*Corresponding Author. Phone:0592-6291079, e-mail:chait@xmut.edu.cn

Abstract: We selected two organic acids, heptanoic acid and benzoic acid, with the same carbon length but with different molecular structures to study their mechanisms of algal inhibition and to examine their combined algal inhibition ability. The results showed that the algal inhibition ability of benzoic acid was greater than that of heptanoic acid. The median lethal concentrations of benzoic acid and heptanoic acid were 55.02 mg/L and 83.90 mg/L, respectively. Heptanoic acid and benzoic acid could decrease the Total Superoxide Dismutase (T-SOD) and Catalase(CAT) activities of *Microcystis aeruginosa* and cause increased Malondialdehyde(MDA) production. The T-SOD and CAT activities initially increased and then decreased. The MDA content quickly decreased after reaching a maximum value. Benzoic acid had a greater effect on the antioxidant enzyme system of *Microcystis aeruginosa* than heptanoic acid. A hybrid evaluation was performed on the combined algal inhibition effect of heptanoic acid and benzoic acid. The results indicated that benzoic acid and heptanoic acid showed a synergistic effect and that their compound algal inhibition effect was superior to the separate effects of the two organic acids acting independently.

1. Introduction

Due to the rapid development of the economy and the swift growth of the population, water eutrophication is gradually increasing [1] [2]. It is becoming more difficult to manage increasingly frequent algal blooms using traditional physical and chemical algae removal methods [3] [4]. Therefore, some novel, high-efficiency and safe algae removal technologies are in rapid development. Based on plant allelopathy, the direct application of allelochemicals to inhibit algal propagation has become a new and important method of efficiently treating eutrophic waters [5]. As an important type of secondary metabolite, with the most species secreted by aquatic plants in natural waters, organic acid allelochemicals play an important role in the study of algal inhibition.

According to general research, the inhibition mechanism of broad-spectrum allelochemicals, such as organic acids, on all types of algae mainly consists of disrupting the operation and photosynthesis capability of the cell membrane structures and antioxidant enzyme systems of algal cells [6] [7]. Additionally, organic acids may affect the synthesis of microcystic toxins. Ma Ruixia [8] found 3 organic acid allelochemicals (ferulic acid, benzoic acid and p-hydroxybenzoic acid) produced from the decay process of wheat straw that had an inhibitory effect on the activity of nitrate reductase at a specific concentration and pH value and might interfere with the synthesis of microcystic toxins. Liu et al. [9] found that high concentrations of organic acid components of *Canna indica* could reduce the chlorophyll content in *Microcystis aeruginosa* and produce oxidative stress. The algal inhibition mechanism of organic acids might also be closely related to their chemical structures. Wu et al. [10] found that fatty acids changed the permeability of cell membranes, which caused the leakage of cell contents, further damaged the membrane structures and thus affected the synthesis of microcystic toxins. Jiang Wenxin et al. [11] studied the inhibitory effect of palmitic acid and stearic acid on the growth of *Microcystis aeruginosa*[12] and compared the structure-activity relationship of algal inhibition of long-chain saturated fatty acids. Their results showed that the algal inhibition ability of a long-chain fatty acid was related to the carbon chain length and the degree of unsaturation of the fatty acid. Although many types of organic acid allelochemicals exist, the mechanisms of action of many organic acids remain unknown. Heptanoic acid and benzoic acid, which were selected for this research, are two organic acid allelochemicals with the same carbon length but with different molecular structures. Their impact on algal blooms is seldom reported. Our study used the common algae of algal blooms, *Microcystis aeruginosa*, for algal inhibition experiments, utilized various physiological indexes of algal liquid to measure algal inhibition, compared the differences in allelopathy of the two organic acids, examined the algal inhibition ability of the combination of both organic acids, supplemented the existing research on the allelopathic algal inhibition mechanism of organic acids and provided a theoretical reference for the application of organic acids for algal inhibition.

2. Experimental materials

2.1 Selection of experimental materials

Heptanoic acid and benzoic acid are organic acids that have been isolated and identified from plants with allelopathic properties. They are fatty acids and aromatic acids with the same carbon length but with different molecular structures and chemical properties. Heptanoic acid (98%) was purchased from Aladdin Chemistry Co., Ltd. Benzoic acid was purchased from Sinopharm Chemical Reagent Co., Ltd. as a type of analytical reagent.

Microcystis aeruginosa was obtained from the Freshwater Algae Culture Collection at the Institute of Hydrobiology (serial number FACHB-905).

2.2 Preparation of experimental materials

(1) *Microcystis aeruginosa* in the logarithmic phase of growth

The algae filament of *Microcystis aeruginosa* was inoculated in sterilized BG11 culture medium, and amplification culture was performed in a light incubator. The light intensity of the incubator was set to 3000 lx, and the temperature was approximately 25±1.0°C. *Microcystis aeruginosa* was used for the experiment after reaching the logarithmic phase of growth.

(2) Preparation of the stock solutions of the two organic acids

A sterilized BG11 culture solution was used for the ultrasonic dissolution of heptanoic acid and benzoic acid. A 1.0 $g \cdot L^{-1}$ stock solution was prepared for storage at a temperature of 4°C after filtration by a 0.22 μm microfiltration membrane.

(3) Preparation of the mixed organic acids

Heptanoic acid and benzoic acid were mixed at a ratio of 1:1. A sterilized BG11 culture solution was used for the ultrasonic dissolution. A 1.0 $g \cdot L^{-1}$ stock solution was prepared for storage at 4°C.

3. Experimental contents and methods

3.1 Experimental contents

Moderate amounts of *Microcystis aeruginosa* in the logarithmic phase of growth were used to inoculate sterilized 250 mL triangular flasks. Stock solutions of heptanoic acid and benzoic acid and the mixed organic acids were added. Then, 150 mL of sterilized BG11 culture medium was added to each flask to achieve final total concentrations of the two organic acids and the mixed organic acids of 5 $mg \cdot L^{-1}$, 10 $mg \cdot L^{-1}$, 20 $mg \cdot L^{-1}$, 50 $mg \cdot L^{-1}$ and 100 $mg \cdot L^{-1}$. The initial OD_{680} value of the algal culture was 0.06±0.005. The control group did not contain any allelochemicals. Each group contained 3 parallel samples that were placed in a light concussion incubator. The temperature was set at 28±1.0°C, and the light intensity was set at 3500 lx. The OD_{680} of the algal liquid was measured every 24 h. The Total Superoxide Dismutase (T-SOD) activity, Catalase(CAT) activity and malondialdehyde (MDA) content were measured every 48 h.

3.2 Measurement methods

(1) Measurement of the number of algal cells

In the experiment, the algal density was represented by the OD_{680}, which was measured by an ultraviolet-visible spectrophotometer.

(2) Measurement of the T-SOD activity

The T-SOD activity in the algal cells was calculated using the following formula:

T-SOD activity ($U.cell^{-1}$) = (control OD_{550}-measurement OD_{550})/control OD_{550} /50% x the total volume of reaction liquid (mL)/sampling amount (mL) /algal density ($cell.mL^{-1}$)

(3) Measurement of the CAT activity

The CAT activity of the algal cells was calculated by measuring the OD_{405} value.

CAT activity ($U.cell^{-1}$) = (control OD_{405}-measurement OD_{405}) x 271 x 1/60 x sampling amount / algal density ($cell.mL^{-1}$)

(4) Measurement of the MDA content

MDA content (nmol/algae) = measurement OD_{532} - control OD_{532}/standard OD_{532} - blank OD_{532} x 10 nmol/mL/algal density (algae/mL)

(6) Calculation of the half maximal inhibitory concentration

The half maximal inhibitory concentration (EC_{50}) of the allelochemicals was calculated by the Bliss method using R procedural programming [13].

(7) Evaluation of the combined effect of the mixed acids

Combined index methods were adopted to evaluate the combined algal inhibition effect of benzoic acid and heptanoic acid, which mainly included the toxic unit method [14], additive index method [15] and similarity index method [16].

4. Experimental results and analysis

4.1 Effects of the two organic acids on the algal density of Microcystis aeruginosa

As shown in Figure 1 and Figure 2, benzoic acid and heptanoic acid both inhibited the growth of *Microcystis aeruginosa*, but their inhibitory abilities were different. The half maximal inhibitory concentration of heptanoic acid (83.90 mg/L) was greater than that of benzoic acid (55.02 mg/L). Therefore, heptanoic acid exhibits a weak algal inhibition ability. When the concentration of heptanoic acid was less than 20 mg/L, the growth of *Microcystis aeruginosa* was slightly accelerated. The algal inhibition rate of 100 mg/L of heptanoic acid was 48.63% on the 7th day. The algal inhibition effect of the heptanoic acid was consistent with the "hormesis effect" of some allelochemicals described in existing research. The algal inhibition rate of 100 mg/L of benzoic acid reached 99.92% on the 7th day after the experiment. However, the algal inhibition rate of low concentrations of benzoic acid was very low. When the concentration of benzoic acid was less than 20 mg/L, the algal inhibition rate of was less than 14% on the 7th day.

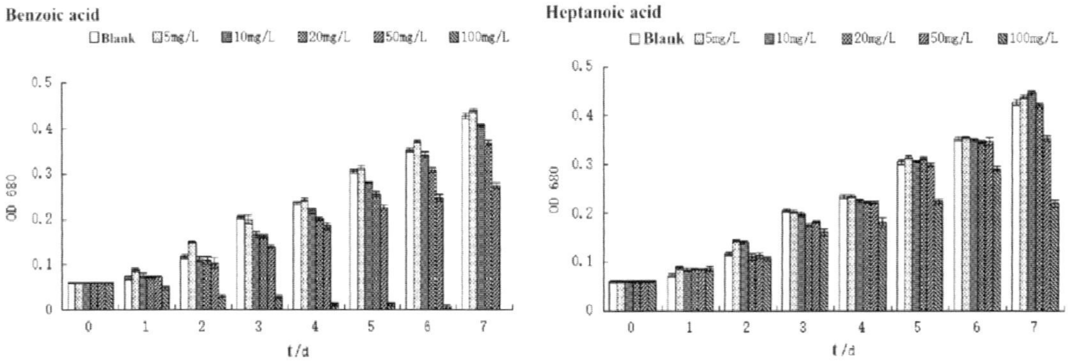

Figure 1. Effects of the two organic acids on the cell density of *Microcystis aeruginosa.*

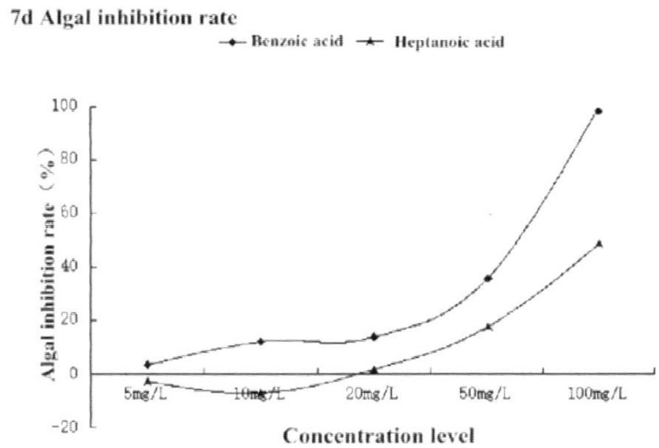

Figure 2. Algal inhibition rates of different concentrations of organic acids on the 7th day of the experiment.

4.2 Effects of the two organic acids on the antioxidant enzyme activity of Microcystis aeruginosa

As important constitutive enzymes in an enzymatic defense system, SOD and CAT supplement each other and maintain the dynamic steady state of the Reactive oxygen species(ROS) level in algal cells. When *Microcystis aeruginosa* is under severe stress, the ROS level can increase and damage the algal cells. When ROS cannot be cleared in a timely manner, the activities of SOD and CAT are inhibited.

(1) Effects of the two organic acids on the T-SOD activity of *Microcystis aeruginosa*

As displayed in Figure 3, the T-SOD activities of the control group and the treatment groups with 5 mg/L and 10 mg/L of the two organic acids were stable during the experiment, and heptanoic acid had little effect on the T-SOD activity of *Microcystis aeruginosa*, which might be responsible for its weak allelopathic inhibition of *Microcystis aeruginosa*. The T-SOD activities of the treatment groups with medium and high concentrations of benzoic acid increased to varying degrees on the 1st day of the experiment. Except for the group treated with 100 mg/L of benzoic acid, the T-SOD activities of the other groups treated with benzoic acid continued to increase on the 3rd day of the experiment. Meanwhile, the T-SOD activity of the group treated with 100 mg/L of benzoic acid began to sharply decrease on the 3rd day, which probably occurred because the high concentration of benzoic acid caused the ROS level in *Microcystis aeruginosa* to exceed a certain threshold. The SOD activity reached a specific limit, and the ROS balance could not be maintained, which damaged the algal cell membranes and biological macromolecules and obviously reduced the SOD activity.

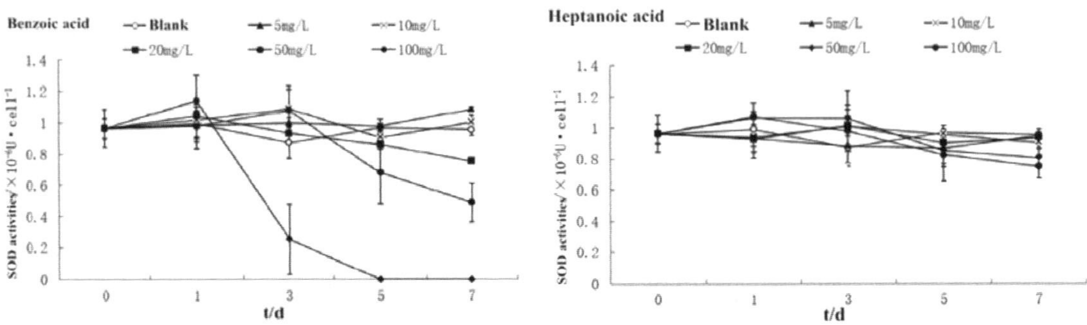

Figure 3. Effects of the two organic acids on the T-SOD activity of *Microcystis aeruginosa.*

(2) Effects of the two organic acids on the CAT activity of *Microcystis aeruginosa*

As shown in Figure 4, the CAT activity of *Microcystis aeruginosa* showed different responses to various concentrations of the two organic acids. As displayed in Figure 4, heptanoic acid had a relatively small effect on the CAT activity of *Microcystis aeruginosa.* The CAT activity of the group treated with 100 mg/L of heptanoic acid was obviously decreased on the 7th day of the experiment; however the CAT activities of the other treatment groups exhibited small fluctuations during the experiment. The CAT activities of the groups treated with benzoic acid increased on the 1st day and started to decrease on the 3rd day of the experiment. In addition, the rate of decrease was directly proportional to the amount of organic acids added. The results showed that the CAT activity of *Microcystis aeruginosa* might be stimulated by benzoic acid. However, the CAT system of *Microcystis aeruginosa* was severely damaged and exhibited reduced activity with continued stress, which increased the difficulty of maintaining the hydrogen peroxide balance in algal cells. The accumulation of hydrogen peroxide continued to damage structures such as the cell membranes and various enzyme systems, which thus formed a vicious cycle and affected the metabolism of the algal cells. CAT and SOD are important components of the antioxidant enzyme system in algal cells. According to the experimental results, the two enzymes had similar responses to benzoic acid, possibly because the two enzymes are isozymes. The ROS level was maintained in the body of the algal cells, possibly because the two enzymes exhibit similar stress responses to adverse factors in the environment [17].

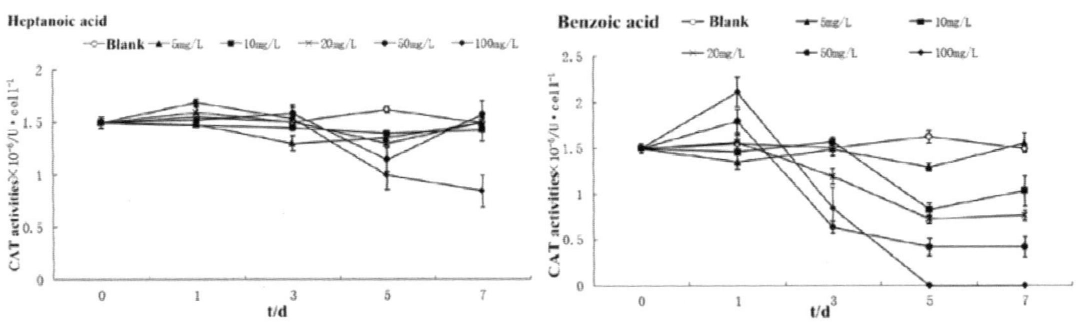

Figure 4. Effects of the two organic acids on the CAT activity of *Microcystis aeruginosa.*

4.3 Effects of the two organic acids on the MDA content of Microcystis aeruginosa

MDA is produced from the peroxidatic reaction of *Microcystis aeruginosa* in response to environmental stress. Under normal circumstances, changes in the MDA content of the algal cells reflect the degree of damage of algal cells or membrane structures [18]. As shown in Figure 5, the two organic acids affected the MDA content of *Microcystis aeruginosa* to varying degrees. The MDA content of the groups treated with a high concentration of the two acids quickly increased, reaching a

IOP Conf. Series: Earth and Environmental Science **146** (2018) 012047 doi:10.1088/1755-1315/146/1/012047

maximum value on the 1st day and sharply decreasing on the 3rd day of the experiment, which indicated that the algal cells were seriously damaged. At this time, the algal density of *Microcystis aeruginosa* exhibited a state of slow growth. The cell membranes were damaged. However, many of the *Microcystis aeruginosa* cells did not immediately die; rather, the widespread death of algal cells occurred slowly during prolonged stress, which also indicated that damage to the cell membrane structures and the accumulation of MDA resulted in the death of *Microcystis aeruginosa*.

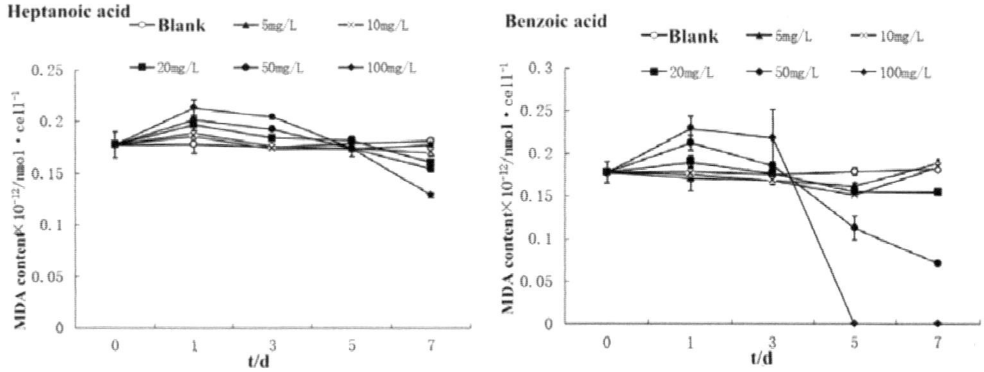

Figure 5. Effects of two organic acids on the MDA content of *Microcystis aeruginosa.*

4.4 Evaluation of the algal inhibition effect of the combined organic acids

Figure 6 shows the combined effect of heptanoic acid and benzoic acid, mixed in a ratio of 1:1, on the growth of *Microcystis aeruginosa*. Compared to the control group, the various treatment groups showed slow growth or death of algal cells. The algal density of the various treatment groups was lower than that of the control group on the 7th day of the experiment. Various treatment groups showed an attenuated increase in algal density compared to that in the control group starting on the 2nd day. Over time, the gap between the algal densities of the various treatment groups and the control group became increasingly wider, which indicates that the combination of the two organic acids had a strong inhibitory effect on *Microcystis aeruginosa*. The algal inhibition rate gradually increased with increasing concentrations of the organic acids. When the combined algal inhibition abilities of the two organic acids were compared to the separate effects of each acid, treatment with both heptanoic acid and benzoic acid showed a stronger algal inhibition ability than treatment with heptanoic acid and benzoic acid separately, which indicates that the combination of the two acids might have a synergistic effect on algal inhibition. As shown in Table 3-1, the half maximal inhibitory concentrations were calculated on the 7th day of the experiment: the concentration of benzoic acid was 55.02 mg/L; the concentration of heptanoic acid was 83.90 mg/L; the concentration of the benzoic acid and heptanoic acid mixture in equal proportions was 48.11 mg/L. The data further showed that the compound algal inhibition effect of benzoic acid and heptanoic acid was superior to the separate effects of the two organic acids.

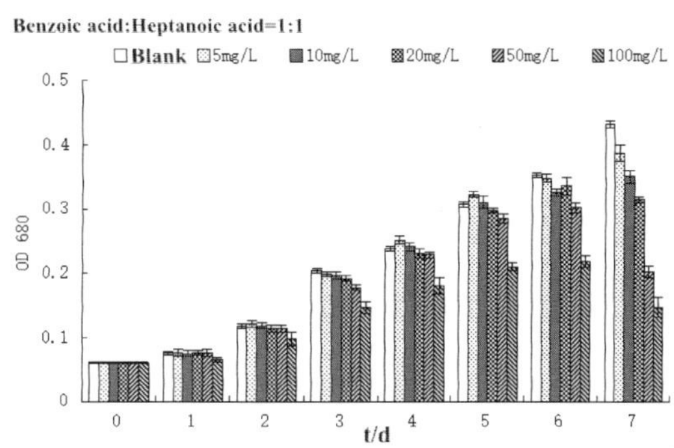

Figure 6. Effect of the two organic acids, mixed in a ratio of 1:1, on the algal density of *Microcystis aeruginosa.*

The toxic unit method, additive index method and mixture toxic index method were adopted to evaluate the combined effect of the mixture of the two organic acids. The results are shown in Table 3-2. Benzoic acid and heptanoic acid showed a synergistic effect. The combined effect of the two organic acids indicates that the mixture of different organic acids might produce algal inhibition characteristics different from those in the case of the single organic acids, which could provide a basis for the development of a mixed algistat in the future.

Table 3-1 Half maximal inhibitory concentrations of the separate and combined organic acids

Allelochemicals	Half maximal inhibitory concentration/mg·L^{-1}
Benzoic acid	55.02
Heptanoic acid	83.90
Benzoic acid + heptanoic acid	48.11

Table 3-2 Evaluation parameters of the combined effect of the two organic acids

Allelochemical	Method	Parameter							Evaluation
		TU_1	TU_2	TU_3	M	M_0	AI	M_{T1}	
Benzoic acid + Heptanoic acid	TU	0.44	0.29		0.73	1.66	-	-	Synergy
	AI	0.44	0.29		0.73	1.66	0.37	-	Synergy
	M_{T1}	0.44	0.29		0.73	1.66	-	1.62	Synergy

5. Conclusions

(1) Heptanoic acid and benzoic acid had an inhibitory effect on *Microcystis aeruginosa*. Benzoic acid exhibited a stronger algal inhibition ability than heptanoic acid. The median lethal concentration of benzoic acid was 55.02 mg/L, while the median lethal concentration of heptanoic acid was 83.90 mg/L.
(2) The two organic acids decreased the SOD and CAT activities of *Microcystis aeruginosa* and increased the MDA content. Benzoic acid had a greater effect on the antioxidant enzyme system of *Microcystis aeruginosa* than heptanoic acid.
(3) The two organic acids were mixed at a 1:1 ratio to evaluate their combined algal inhibition effect. Benzoic acid and heptanoic acid had a synergistic effect, and their compound algal inhibition effect was superior to the separate effects of the two organic acids.

Acknowledgements
This work was financially supported by National Natural Science Foundation of China(51109181), Guiding Project of Fujian Province of China (2018Y01010025).

Reference

[1] Nne M. Hansen, Claudia Hernández-Martínez, Axel Falcón-Rojas, Evaluation of Eutrophication Control Through Hypolimnetic Oxygenation, In Procedia Earth and Planetary Science, Volume 17, 2017, Pages 598-601.

[2] Shouliang Huo, Chunzi Ma, Beidou Xi, Yali Zhang, Fengchang Wu, Hongliang Liu, Development of methods for establishing nutrient criteria in lakes and reservoirs: A review, In Journal of Environmental Sciences, 2017, https://doi.org/10.1016/j.jes.2017.07.013.
[3] Mohammad Hadi Dehghani, Removal of cyanobacterial and algal cells from water by ultrasonic waves — A review, In Journal of Molecular Liquids, Volume 222, 2016, Pages 1109-1114.
[4] Guo L G. Algae removal and control technologies[J]. China Water Resources. 2006, 17: 34-36.
[5] Wang J, Zhu J Y, Liu S P, et al. Generation of reactive oxygen species in cyanobacteria and green algae induced by allelochemicals of submerged macropHytes[J]. ChemospHere, 2011, 85(6): 977-982.
[6] Inderjit, Duke S O.EcopHysiological aspects of allelopathy[J]. Planta, 2003, 217(4):529-539.
[7] Hong Y, Hu H Y, Xie X, et al. Gramine-induced growth inhibition, oxidative damage and antioxidant responses in freshwater cyanobacterium Microcystis aeruginosa[J]. Aquatic Toxicology, 2009, 91(3): 262-269.
[8] Ma R X. Study on Influence of Allelochemicals on Activity of Nitrate Reductases[J]. Environmental Science, 1999, 20(1):80-83.
[9] Liu X Y, Fu H Y, Huang G H, Chai T, Gao P F, Wu Y C. Allelopathic effect of organic acids from Canna indica on Microcystis aeruginosa[J]. Environmental Engineering, 2015, 9(12): 5769-5774.
[10] Wu J T, Chiang Y R, Huang W Y. Cytotoxic effects of free fatty acids on phytoplankton algae and Cyanobacteria[J].Aquatic Toxicology, 2006, 80(4): 213-218.
[11] JIANG W X, JIA Y, WANG C Y, WANG Q, TIAN X J. The inhibitory effects of palmitic acid and stearic acid on Microcystis aeruginosa[J]. Ecology and Environment Sciences, 2010, 19(2):291-295.
[12] Jin X C, Li Z C, Zheng S F. Studies on the Growth Characteristics of Microcystis aeruginosa[J]. Research of Environmental Sciences, 2004, S1: 52-54.
[13] Jiang X D, Yan Z L, Zhou G F.Application of R Program in Calculating Median Lethal Dosage with Bliss Algorithm[J]. Practical Preventive Medicine, 2011, 3(18):431-432.
[14] Marking L L, Dawson V K. Method for assessment of toxicity or efficacy of mixtures of chemicals[J]. US Fish Wildlife Service Invest fish Control,1975, 67:1-8.
[15] Marking L L. Method for assessing additive toxicity of chemical mixtures[J]. Aquatic toxicology and hazard evaluation ,1977, 634:99-108.
[16] Christensen E R, Chen C Y. Modeling of combined toxic effects of chemicals[J]. Hazard assessment of chemicals, 1989, 6:125-186.
[17] WU Z B, Deng P, Wu X H, et al. Allelopathic effects of submerged macropHyte Potamogeton malaianus on Seenedesmus obliquus[J]. Hydrobiologia, 2007, 592:465-474.
[18] Allen R D. Dissection of oxidative stress tolerance using transgenic plants[J]. Plant pHysiology, 1995, 107:1049-1054.

IOP Conf. Series: Earth and Environmental Science **146** (2018) 012048 doi:10.1088/1755-1315/146/1/012048

Enhancing Advanced Oxidation Process by Microbubbles Technology and the Analysis of Its Degradation Process

Jin Wang[1] and Dengxin Li[1,*]

[1]College of Environmental Science and Engineering, State Environmental Protection Engineering Centre for Pollution Treatment and Control in Textile Industry, Donghua University, Shanghai, China, 201620

*lidengxin@mail.dhu.edu.cn

Abstract. The simulated dyestuff containing Indigo Carmine(IC), a highly toxic indigoid dye used as a textile coloring agent, was studied using hydrogen peroxide as oxidant enhanced by microbubbles. The decolorization rate, chemical oxygen demand (COD) and total organic carbon (TOC) removal extent were dramatically increased by the presence of microbubbles. The discoloration was observed in this system up to 94.27%, compared with about 49.6% without hydrogen peroxide in 60 minutes, and COD and TOC removal rate can achieve to 72% and 53.07%, respectively. In addition, the main degradation pathway was analyzed by Fluorescence (FLS), Ultraviolet Visible (UV-Vis) and Ultra High Liquid Chromatography and Time of Flight Mass Spectrometry (UPLC-Q-TOF-MS). The results showed that hydroxyl radicals (•OH) induced strong oxidizing effects in the target solution and destroyed the chromophoric groups of the molecular structure. The application of microbubbles technology in the advanced oxidization process may provide an effective and low-cost approach for treating dye wastewater.

1. Introduction

Printing and dyeing wastewater effluent discharges of textile industrial wastewater often contains a large amount of non-biodegradable dye[1]. There will be harmful to the environment if these dyes were discharged into the environment without any treatment. Various organic colorant used dyes in the dye industries, indigo carmine is one of them. Besides it is also used as an additive in pharmaceutical tablets and capsules for medical diagnostic purposes[2]. Physical and chemical methods have been studied to remove indigo carmine in the waste water. Physical treatments such as adsorption, membrane filtration and coagulation flocculation[3]. However, further treatment still required with physical treatments due to these treatments only convert the pollutants from the liquid phase to a solid phase, beyond that it is also producing massive sludge, causing membrane fouling and phase change of the pollutants. Chemical treatment for instance bacterial and fungal cultures relied biological processes, but the limitation of those is the poor biodegradability characteristic of the dyeing wastewater[4]. For these reasons, it is necessary to find economical and efficient method to degrade indigo carmine[5].

Advanced oxidation processes (AOPs) were reported as one of a promising way to treat wastewater[6]. AOPs such as ozonation, hydrogen peroxide, ultrasonic, ultraviolet (UV) radiation and photocatalytic treatments has been utilised to decolorize textile wastewater[7]. However, finding a cost-effective treatment that insures complete elimination of dye wastewater is still in demand.

Table 1. Application examples of advanced oxidation process

AOPs	processing object	treatment efficiency
UV/O_2	ethenol	TOC 18.0 → 1.5 mg/l
	chlorobenzene	40 → 20.6 ug/l
	dioxin	6500 → 5000 pg/l
	PVA	TOC removal 96.6%
H_2O_2/O_2	humic acid	TOC removal 84.3%
	LAS	TOC removal 72.6%
UV/TiO_2	dye wastewater	TOC removal40.6%
$UV/H_2O_2/O_2$	BOD=0mg/l	BOD/COD 0→0.4
H_2O_2/O_2	chlorobutane	chlorobutane removal 93%

In recent years, microbubbles technologies have drawn increasing attention due to their widely applications in various fields in our society, such as biomedical engineering, water treatment and nanomaterials. The definition of microbubbles is the bubbles which diameters are less than 50 μm, the characteristic of microbubbles in water is small size, high adhesive efficiency and low floating tendency[8]. The microbubbles generator can produce a milky and high intensity microbubbles solution. The surface of bubbles immersed in a liquid phase would be charged due to the ions are adsorbed at the gas - liquid interface. The spatial distribution of ions would be influenced by these charged bubbles surface to form an electric double layer, the zeta (ζ) potential would be used to evaluate strength of this electric double layer. Meanwhile, free radicals can also be generated during the collapse of air microbubbles, even without dynamic stimulus[9]. It is previous reported that by changing the circumstance of the adsorbed ions around the gas–liquid interface of the microbubbles would occur some interesting phenomena[10]. However, It is a very limited amount of total •OH come from the collapse of air micro-bubbles[11]. Therefore, in our study, hydrogen peroxide was added as oxidation to enhance this technology in the practical applications.

In this paper, some critical parameters were studied, such as pH, hydrogen peroxide dosage and gas-flow rate. An investigation of these parameters during the treatment is vital for understanding the mechanism of degradation. Besides, the FLS, UV-Vis and MS were used to analysis degradation process and products of the degradation pathway.

2. Experimental methods and materials

2.1 Experimental apparatus

The microbubbles generator was made by Xiazhichun Environmental Tech (China). Figure 1 shows an experimental set-up which contains a recycling pump, an efficient gas-water mixer and a microbubble nozzle. The simulated dyestuff wastewater prepared in a 5-L beaker and then circulated by the microbubbles generator. In this experiment: the gas and water were sucked in the pump and then mixed by the turbulent flow, next the gas was dissolved in the water at a pressure of 0.4 MPa in the gas-water mixer, the pressurized water was then delivered to a tank at atmospheric pressure via a specially-designed nozzle which comprised of a bottom valve.

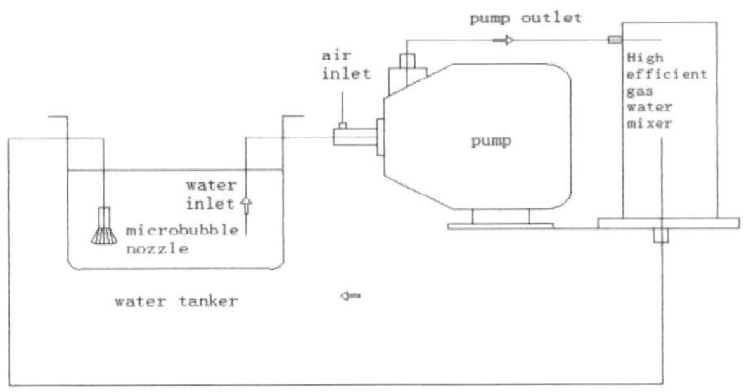

Figure 1. Schematic diagrams of the experimental apparatus

2.2 Materials

A commercial dye, indigo carmine (figure 2.) with the chemical formula of $C_{16}H_8N_2Na_2O_8S_2$ (molecular weight of 466.37) was used in this study. The simulate dye wastewater was prepared by dissolving the indigo carmine in deionized water. Indigo carmine, hydrogen peroxide (30%), sodium hydroxide, sulphuric acid purchased from Sinopharm Chemical Reagent.

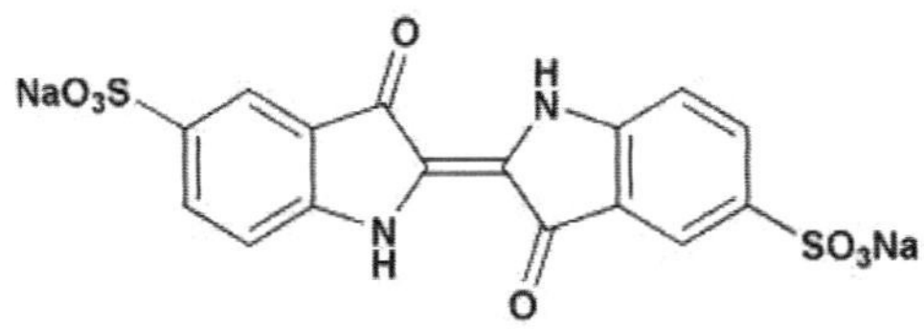

Figure 2. Chemical structure of indigo carmine dye (IC)

2.3 Analysis methods

Free-radicals produced by the collapsed microbubbles were detected by the fluoro-spectro-photometer (QM/TM, Unite States PTI), a method for detecting the hydroxyl radical (•OH), Ce^{3+} can produce its characteristic fluorescence with is excitation and emission wavelength at 280nm and 360nm in diluted acid, respectively. But the fluorescence will be quenched if Ce^{3+} is oxidized to Ce^{4+} by oxidizing agent such as hydroxyl radical (•OH). Therefore, we can indirectly determine the concentration of hydroxyl radicals (•OH) by testing the changes of fluorescence intensity [12].

For testing the effective of removing dye in the wastewater, indigo carmine was chosen as the target dye to determine the generation of •OH from microbubbles collapsing in three different systems. Indigo carmine dye wastewater was prepared in a 10 L reactor, microbubbles was continuously produced in the solution, the total organic carbon (TOC) content of the water samples was measured by a TOC analyser (Shimadzu, Japan) to quantize the level of indigo carmine removal efficiency. Chemical oxygen demand (COD) removal rate was also measured by the potassium dichromate standard method, the decolorization extent measurement was carried out with a UV-2300 UV-Vis (Avantes, Netherlands) spectrophotometer. COD, TOC and color removal rate (R) were calculated as follows:

$$R = \frac{Co - Ci}{Co} \times 100\% \ldots\ldots \qquad \ldots\ldots\ldots\ldots(1)$$

Where Co and Ci are the initial and final concentration value of dye , TOC, or COD respectively[13].

The pathway analysis was analyzed by UV-Vis and mass spectroscopy techniques. Quantitaive analysis of IC was done by UV visible spectrophotometer in a spectronic UV/Vis spectrophotometer set at 608 nm, the intermediates generated during the degradation were analysed by a UPLC & Q-TOF MS (Waters, USA) equipped with Mass Spectra which has a duo spray ion source, the data were recorded in positive ion mode, the ion source parameters: capillary voltage=5.5kV; temperature=550°C and scan range (m/z) =100-400 amu.

3. Results and discussion

3.1 Oxidation Enhancement

In order to investigate the enhancement of the oxidation by microbubbles, the performance of the microbubble with hydrogen peroxide system, a conventional mechanical agitation machine with hydrogen peroxide system, and microbubble-only system were compared in the indigo carmine degradation process, the initial dye concentration used in the experiments was about 100mg/l, the oxidation efficiency was represented by the colour, COD and TOC removal rate.

With the introduction of microbubbles, the oxidation activity in the wastewater increased rapidly. The rate of the oxidation was obviously faster in the microbubble with hydrogen peroxide system. Discoloration was observed in this system up to 94.27% compared with about 49.6% for the mechanical agitation with hydrogen peroxide system in 60 minutes, and COD and TOC removal rate can achieve to 72% and 53.07%, respectively. As shown from the figure 3. The molecule of the dye stuff in the waste water starts to be oxidized to mineral ions results in the decreasing of the concentration of COD[14]. In the indigo carmine degradation process, some aromatic compounds will be transferred into aliphatic compounds by ring-opening reaction. The total organic carbon concentration is one the index could indicate this reaction[15].

Oxidation efficiency initially increased sharply and then increased slightly, COD and TOC have the same tendency, it can be explained by the reason of the large driving forces for the mass transfer of the microbubbles and then reached saturated. The total mass transfer coefficient in the hydrogen peroxide-microbubble system was much larger than that in the mechanical agitaion system by maximizing the surface area-to-volume ratio[2, 8]. Furthermore, the formation of the •OH could be accelerated by the microbubbles during the oxidation process[16].

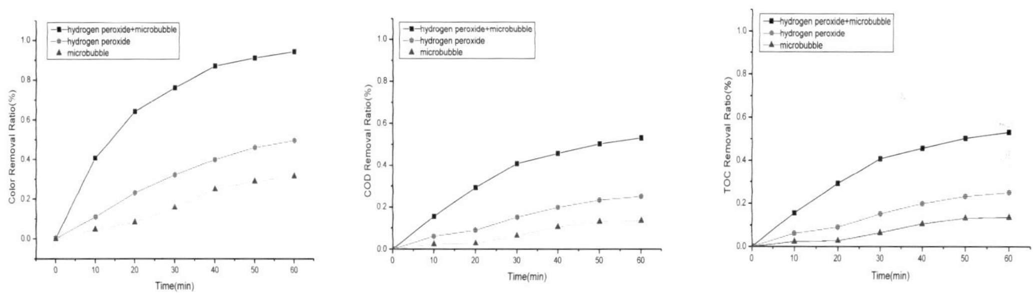

(a) (b) (c)

Figure 3. Different system of removal extents of colour (a), COD(b) and TOC (c) of IC

3.2 Influence of the reaction variables

3.2.1 Effect of the hydrogen peroxide dosage. The impacts of the dosage of hydrogen peroxide was investigated varying the initial hydrogen peroxide concentration introduced into the reactor and the rest of the operational variable parameters kept constant. Figure 4. shows the difference in the experiment carried out from 0 ml hydrogen peroxide to 30 ml hydrogen peroxide. The degradation of the indigo carmine with the microbubbles and hydrogen peroxide in 60 minutes: the color removal extent of the indigo carmine was almost 100%, TOC decrease 72.7%, the result showed the higher the hydrogen peroxide dosage, the higher the degradation extent. The interaction between microbubbles and hydrogen peroxide could be explained as follows: from previous studies, the concentration of free-radicals involved reaction mechanism keeps increasing as the introduction of hydrogen peroxide, so this would eventually enhance the organic compounds degradation[11].

IOP Conf. Series: Earth and Environmental Science **146** (2018) 012048 doi:10.1088/1755-1315/146/1/012048

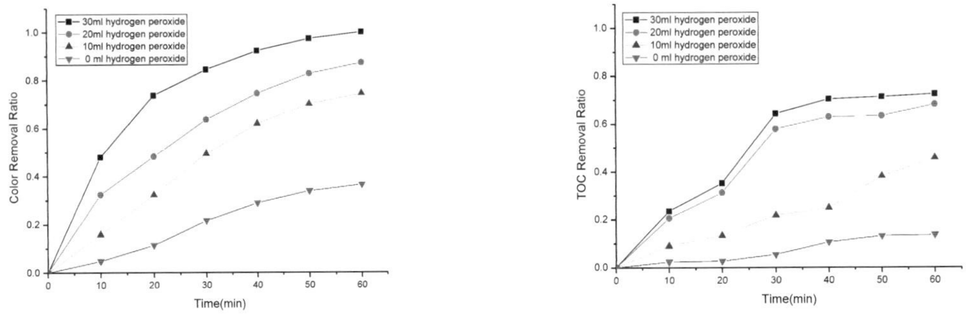

(a) (b)

Figure 4. Effect of the hydrogen peroxide dosage on the decolorization and TOC removal rate of IC Dye concentration=100mg/L, pH7.0, Reaction pressure=0.4 MP.

3.2.2 Effect of solution pH. Another important factor is the influence of pH in the wastewater, pH plays a vital role to remove colour and COD in the degradation process of the dye. For purpose of finding the optimum operational conditions, the experiments of the solution varied in pH was carried out with the range from strongly alkaline to strongly acidic by adjusting by sodium hydroxide and sulphuric acid. The changes in indigo carmine concentration and TOC removal are shown in figure 5 (a) and (b), respectively. The dye decomposition efficiency is greatly dependent on the pH of the solution and the pH would affect the amount of hydroxyl radicals produced by hydrogen peroxide-microbubbles system. The optimal pH value found in this study was 2 and a higher mineralization rate and colour removal efficiencies could be yielded under this pH. Dissolved oxygen and hydrogen peroxide react with OH^- was reported to produce HO_2• and O_2•$^-$ free radicals[1]. OH^- ions can accelerate the production of O_2•$^-$ free radicals while H^+ ions can promote the formation of HO_3• free radicals, both ions could enhance •OH free hydroxyl radical generation. But in our study, we found under the strongly acidic condition, it would be more efficient to remove the colour and TOC[17]. Therefore, in the hydrogen peroxide-microbubbles system, more hydroxyl radicals are formed at lower pH.

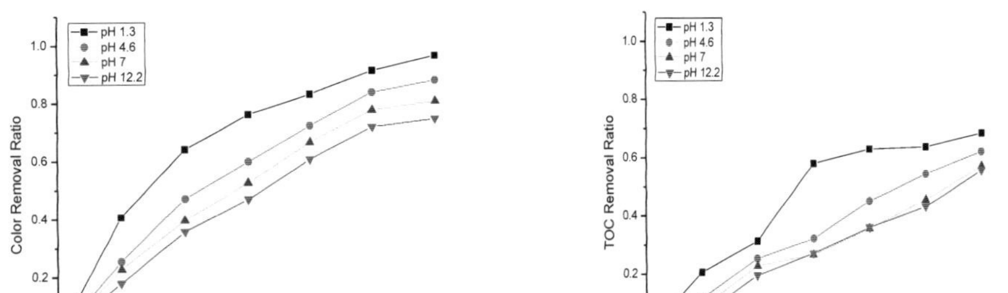

(a) (b)

Figure 5. Effect of the solution pH on the decolorization (a) and TOC removal rate (b) of IC Dye concentration=100mg/L, hydrogen peroxide dosage=30ml, Reaction pressure=0.4 MP.

3.2.3 Effect of reaction pressure. Experiments of indigo carmine were also performed at different reaction pressure in this study to reveal the potential index that reaction pressure played in the degradation process. As shown in figure.6. The solution with reaction pressure of 0.4 MP indicates the highest oxidation rate a hiher reaction pressure would increase the efficiency. It has been reported that the formation of free radicals is dependent on the concentration of dissolved oxygen[18]. It presents that oxygen microbubbles would increase the dissolved oxygen which could enhance the decolorization ability of indigo carmine.

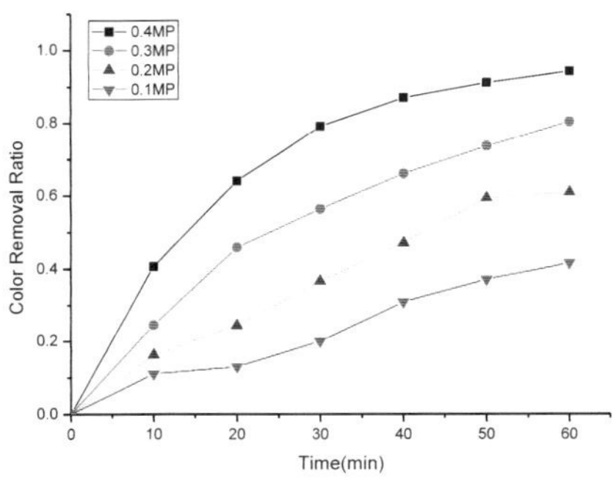

(a)

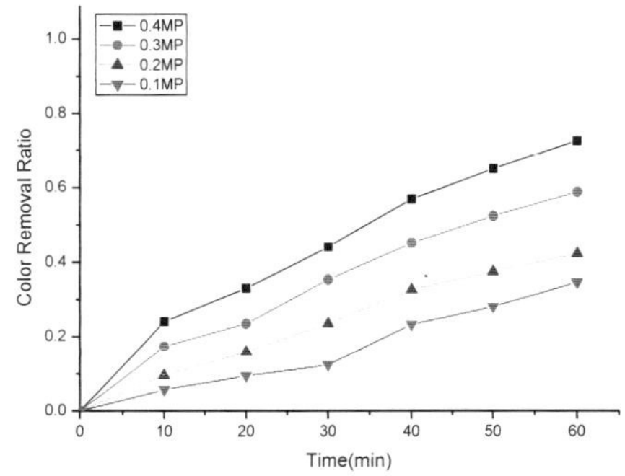

(b)

Figure 6. Effect of the reaction pressure on the decolorization and TOC removal rate of IC Dye concentration=100mg/L, hydrogen peroxide dosage=30ml, pH7.0

3.3. Mechanism of oxidation degradation

3.3.1 Exploration of degradation

Fluorimetry method is used to determine the mechanism of the indigo carmine in figure 7.[12], suggesting that •OH radicals were generated by the collapse of oxygen microbubbles.

Three systems in terms of decolorizations were compared during the study. TOC and COD removal efficiency (Dye concentration=100mg/L, hydrogen peroxide dosage=30ml, pH=2, Reaction pressure=0.4 MP). The results are showed in the figure 7. Results found that all the samples in the three systems increased with the continuous micro-bubbles produced into the solution and then reached steady state. The fluorescence intensity of the samples from the hydrogen peroxide-microbubble system was much higher than that from the other two systems. But it is clearly implied that the amount of hydroxyl radicals was higher in the hydrogen peroxide-microbubble system which would contribute to the degradation of the dye molecules.

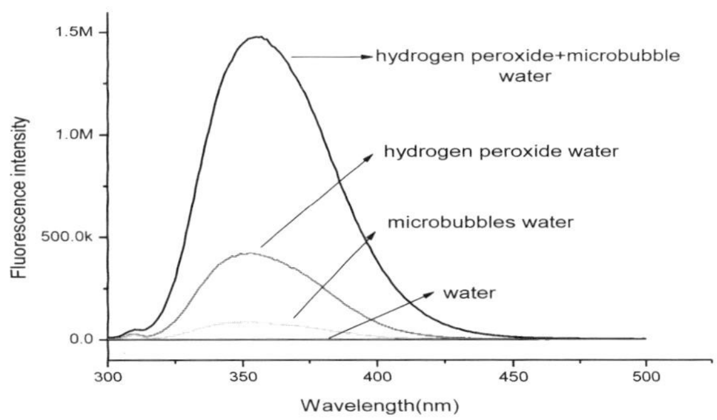

Figure 7. Fluorescence detection of hydroxyl radical formed in microbubble water under different conditions

To understand the degradation mechanism of the indigo carmine in degradation process and determine the chemical reaction during the process of the change of colour. COD and TOC. UV-Vis and UPLC-Q-TOF MS were used to analysis in the oxidation process.UV-Vis absorption spectra of the degradation in whole degradation and at some typical time were investigated. Figure 8. showed the variation of absorbance intensity at two specific wavelengths: 608 nm and 275 nm. The peak of appeared at 275 nm in the visible region indicates the benzene rings in the indigo molecule. This is due to the maximum absorption peak of aromatic compounds is mainly at 210-320nm[19]. The peak in the visible region at 608 nm which corresponding to the blue colour of the solution and it is because of the n→ π group orbital of the double bond (C=C) system[20]. The decreasing of absorbance at 608nm close to zero indicated the degradation of indigo. During the process, the phanic trending was: the residual absorbency at visible light region decreased fast that suggesting the initial accumulation and the intermediates might be further degraded to mineral compounds[4, 5, 21].

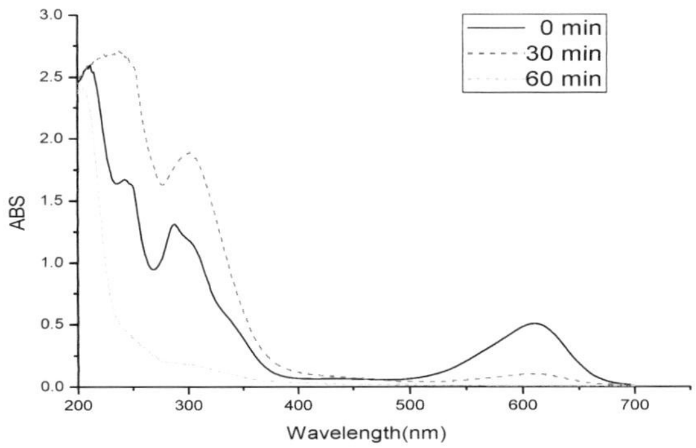

Figure 8. UV-Vis spectra of indigo carmine during degradation

Moreover, Ultra Performance Liquid Chromatography – Quadrupole – Time of Flight Mass Spectrometer (UPLC-Q-TOF MS) gives an accurate and specific detection with reliable identification of the analysis in a precise instrument. It provides a qualitative analytical data for the characterization of the oxidation products in the process of degradation of the dye ingredients. Four intermediate compounds were found by LC-MS technique, as shown in figure 9. We detected that $C_{16}H_{10}O_8N_2S_2$ (A), $C_8H_5O_5NS$ (B), $C_8H_6O_6NS$ (C) and $C_8H_70_5NS$ (D) are appeared, it can be shown in the figure 10. The reaction process could be considered that the free radical first attack the double bond (C=C), in the compound A which oxidized to the B occurred with the decolorization of indigo carmine, but the product B peaked at about 5 min and then gradually decreased, due to the transformation to C. C=O groups of carboxyl acids in the C were oxidized to the hydroxyl groups. Based on the results of UV-Vis and UPLC & Q-TOF MS analysis, the possible degradation process of indigo carmine is concluded here. In the first step, lower molecular weight compounds were formatted by the olefinic bond cleavage. Then the second step is the further oxidation of lower molecular weight organic matter.

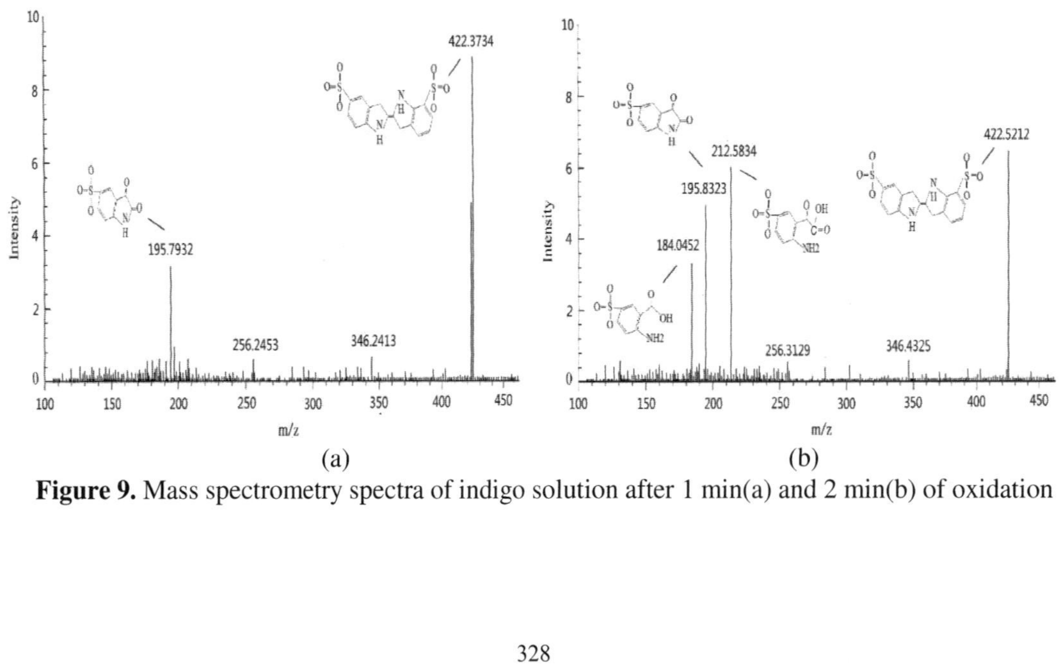

(a) (b)

Figure 9. Mass spectrometry spectra of indigo solution after 1 min(a) and 2 min(b) of oxidation

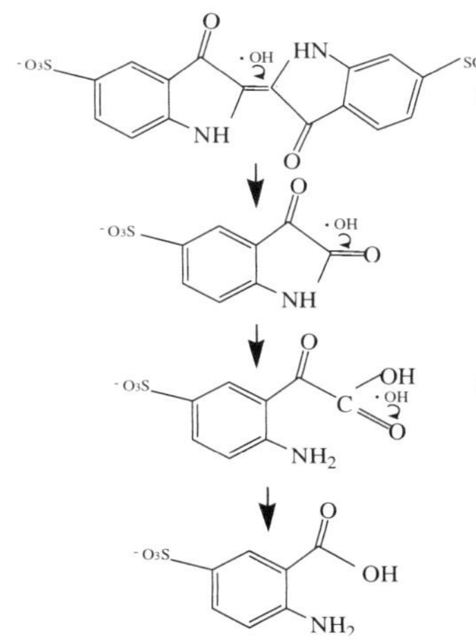

Figure 10. Degradation course of indigo carmine

4. Conclusions

Microbubble technology applied to increase the mass transfer rate reaction as well as to enhance the oxidation of indigo carmine gained increasing attention in the recently work. It is clearly exhibited by the color disappearance, COD and TOC decreased in indigo carmine solutions in this study. The removal rate of color, COD and TOC were 94.27%, 72% and 53.07%, respectively in 1 hour. Fluorimetry method was used to determine the mechanism of the indigo carmine, showing that hydroxyl radical (•OH) radicals were generated by the collapse of oxygen microbubbles. The analysis of UV-Vis, MS spectra showed that the double bond and the groups of carboxylic acids were destroyed at the first stage of the degradation.

Although, research on microbubble is still required to indicate the mechanism for the microbubbles induced oxidation, the high decolorization achievements of this technology presents a promising way to treat dye wastewater with indigo carmine.

Acknowledgments

This work was financially supported by the National Natural Science Foundation of China (U1660107), Textile Light Application Basic Research of China (J201503), the Fundamental Research Funds for the Central Universities (CUSF-DH-D-2018076) and Specialized Research Fund for the Doctoral Program of Senior Education of China (Grant no. 20130075110006).

Reference

[1]. Bossmann, S.H., et al., New Evidence against Hydroxyl Radicals as Reactive Intermediates in the Thermal and Photochemically Enhanced Fenton Reactions. The Journal of Physical Chemistry A, 1998. 102(28): p. 5542-5550.

[2]. Chu, L.B., et al., Enhanced ozonation of simulated dyestuff wastewater by microbubbles. Chemosphere, 2007. 68(10): p. 1854-60.

[3]. Gulyas, H., et al., Combining activated carbon adsorption with heterogeneous photocatalytic oxidation: Lack of synergy for biologically treated greywater and tetraethylene glycol dimethyl ether. Environmental Technology, 2013. 34(11): p. 1393-1403.

[4]. Flox, C., et al., Electro-Fenton and photoelectro-Fenton degradation of indigo carmine in acidic aqueous medium. Applied Catalysis B: Environmental, 2006. 67(1-2): p. 93-104.

[5]. Gemeay, A.H., et al., Kinetics and mechanism of the heterogeneous catalyzed oxidative degradation of indigo carmine. Journal of Molecular Catalysis A: Chemical, 2003. 193(1-2): p. 109-120.

[6]. Stasinakis, A.S., Use of selected advanced oxidation processes (AOPs) for wastewater treatment–a mini review. Global Nest Journal, 2008. 10(3): p. 376-385.

[7]. Ballesteros Martín, M.M., et al., Degradation of a four-pesticide mixture by combined photo-Fenton and biological oxidation. Water Research, 2009. 43(3): p. 653-660.
[8]. Chu, L.-B., et al., Enhanced treatment of practical textile wastewater by microbubble ozonation. Process Safety and Environmental Protection, 2008. 86(5): p. 389-393.
[9]. Takahashi, M., Zeta potential of microbubbles in aqueous solutions: electrical properties of the gas-water interface. The journal of physical chemistry. B, 2005. 109(46): p. 21858-21864.
[10]. Takahashi, M., K. Chiba, and P. Li, Formation of hydroxyl radicals by collapsing ozone microbubbles under strongly acidic conditions. The journal of physical chemistry. B, 2007. 111(39): p. 11443-11446.
[11]. Lorençon, E., et al., Generation of reactive oxygen species in titanates nanotubes induced by hydrogen peroxide and their application in catalytic degradation of methylene blue dye. Journal of Molecular Catalysis A: Chemical, 2014. 394: p. 316-323.
[12]. Xiang，R.X, W. Wenhua, and L. Huabing, Fluorescence method for the determination of hydroxyl radicals produced by Fenton reaction. analytical chemistry, 1998(12): p. 1460-1463.
[13]. Hua, L., H. Ma, and L. Zhang, Degradation process analysis of the azo dyes by catalytic wet air oxidation with catalyst CuO/gamma-Al2O3. Chemosphere, 2013. 90(2): p. 143-9.
[14]. Koch, M., et al., Ozonation of hydrolyzed azo dye reactive yellow 84 (CI). Chemosphere, 2002. 46(1): p. 109-113.
[15]. Tasaki, T., et al., Degradation of methyl orange using short-wavelength UV irradiation with oxygen microbubbles. J Hazard Mater, 2009. 162(2-3): p. 1103-10.
[16]. Li, P., M. Takahashi, and K. Chiba, Enhanced free-radical generation by shrinking microbubbles using a copper catalyst. Chemosphere, 2009. 77(8): p. 1157-60.
[17]. Shu, H.Y. and C.R. Huang, Degradation of commercial azo dyes in water using ozonation and UV enhanced ozonation process. Chemosphere, 1995. 31(8): p. 3813-3825.
[18]. Levec, J. and A. Pintar, Catalytic wet-air oxidation processes: A review. Catalysis Today, 2007. 124(3–4): p. 172-184.
[19]. Li, H.-X., et al., Reductive decolorization of indigo carmine dye with Bacillus sp. MZS10. International Biodeterioration & Biodegradation, 2015. 103: p. 30-37.
[20]. Mahmoodi, N.M., et al., Decolorization and aromatic ring degradation kinetics of Direct Red 80 by UV oxidation in the presence of hydrogen peroxide utilizing TiO2 as a photocatalyst. Chemical Engineering Journal, 2005. 112(1-3): p. 191-196.
[21]. Zaied, M., et al., Spontaneous oxidative degradation of indigo carmine by thin films of birnessite electrodeposited onto SnO2. Applied Catalysis B: Environmental, 2011. 107(1-2): p. 42-51.

Effects of Important Factors on Effluent Trading: A Bayesian Estimation-Based Inexact Two-Stage Stochastic Programming Method

Junlong Zhang[1], Yongping Li[2,]* and Jing Liu[3]

[1] Research Scientist, College of Environmental Science and Engineering, Qingdao University, Qingdao, Shandong 266071, China
[2*] Research Scientist, School of Environment, Beijing Normal University, Beijing 100875, China
[3] Department of Environmental Engineering, Key Laboratory of Environmental Biotechnology (XMUT), Fujian Engineering and Research Center of Rural Sewage Treatment and Water Safety, Xiamen University of Technology, Xiamen 361024, China

*E-mail: yongping.li@iseis.org

Abstract: We proposed a Bayesian estimation-based inexact two-stage stochastic programming (BITSP) for identifying factors' effect on effluent trading. BITSP incorporates nutrient fate modeling with Bayesian estimation and inexact two-stage stochastic programming (ITSP). Based on the water quality protocols, Bayesian estimation is used to analyze parameter uncertainty of nutrient modeling as well as provide the random inputs for the optimization process. ITSP can then be used for dealing with multiple uncertainties associated with randomness and intervals. A case study for water management in the Xiangxihe watershed is conducted. Results reveal that strict environmental limits increase the desire for permit trading program. The results also reveal that treatment rate have an obvious effect on effluent trading through changing the buying and selling behavior of point sources

1. Introduction

A number of research efforts have been conducted for recognizing optimal factors, such as treatment rate of dischargers, environmental limit, the assign of river assessment points, and market scale [1-5]. Doyle et al. [6] proposed a coupled hydrologic-economic modeling approach for optimizing the scale of markets for water quality trading. Chen et al. [7] developed a water environmental functional zone-based effluent trading systems framework for analyzing the effect of background water quality, the location of river assessment points, and tradable discharge permits. Environmental limit and treatment rate are central regulatory features for the permit trading program. Environmental limit regulates the pollutant emission within the watershed. Environmental penalties would be paid when it is exceeded. Treatment rate denotes the rate of pollutants that are treated by the point sources. The treatment can generate discharge permit for store or sale. Previously, few research works were conducted on effluent trading models to analyze such central regulatory features.

Therefore, this study aims to develop a Bayesian estimation-based inexact two-stage stochastic programming (BITSP) for identifying important factors' effect. BITSP incorporates nutrient fate modeling with Bayesian estimation, and inexact two-stage stochastic programming (ITSP). We conduct a real case study for the application of BITSP in the Xiangxihe watershed. The modeling approach will (i) address multiple uncertainties expressed as intervals and probability distribution functions; (ii) disclose the effect of important factors in effluent trading.

2. Case study

The Xiangxihe River is located in Hubei province, China (30°57'N to 31°34'N, and 110°25'E to 111°06'E). The point and nonpoint sources pollute the river. Firstly, nutrients from farmlands that are distributed on the hillside can adhere to eroded soil and be dissolved in surface runoff, then transport into the river. This can cause serious agricultural nonpoint source pollution (NPS) pollution. Secondly, phosphorus minings distribute densely within the watershed. Water can percolate from the phosphate waste rocks; this can lead to point pollution [8]. Thirdly, many phosphorus-related industries are situated near the river, with 1.97 million ton of wastewater sluiced into the Xiangxihe River annually, containing large amounts of phosphoric chemicals. In this study, total phosphorus (TP) is selected as the main water quality indicator. Nonflood season (i.e. November to May of the following year) and flood season (i.e. June to October) are considered as the two planning periods. Multiple main crops are also taken into consideration associated with different irrigation water, fertilizer application, and agricultural benefit. The effluent trading design problem can be solved through the BITSP method. The BITSP model incorporates SWAT, Bayesian estimation, and ITSP within a framework. The formulation of BITSP is presented in the "Appendix".

3. Results and discussion

In this study, six levels of environmental limit for the whole watershed are examined through varying the multiplier (ε) for the limits. Figure 1 illustrates the detailed trading process for agricultural zones, chemical plants and phosphorus mining companies. Blue bars represent the purchasing amount; while red bars represent the selling amount. From the results, the trading process would not change when ε takes 1, 0.95 and 0.9. When ε is less than 0.85, the purchasing and selling amount would be increased as the multiplier is reduced. For example, chemical plants would sell 26.04 ton, 28.57 ton and 47.62 ton discharge permit when ε takes 0.85, 0.8 and 0.75, respectively. The total trading amount would increase from 56.17 ton ($\varepsilon = 0.85$) to 134.96 ton ($\varepsilon = 0.75$) during nonflood period. The reduced multiplier represents the more strict environmental allowance. The results can be due to the reason that strict environmental limits lead to high level of environmental penalties; this increase the desire for permit trading program.

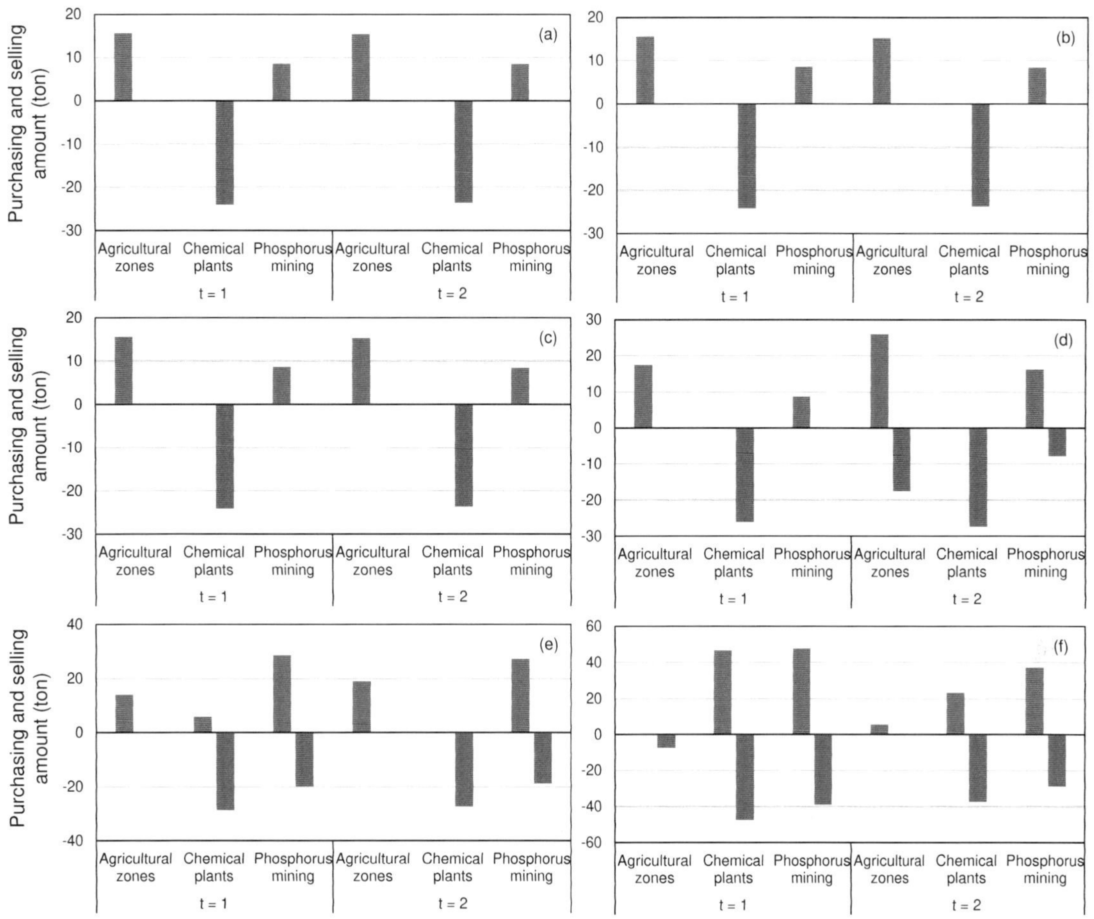

Figure 1. The detailed trading process for agricultural zones, chemical plants and phosphorus mining companies [(a) $\varepsilon = 1$, (b) $\varepsilon = 0.95$, (c) $\varepsilon = 0.9$, (d) $\varepsilon = 0.85$, (e) $\varepsilon = 0.8$, (f) $\varepsilon = 0.75$]

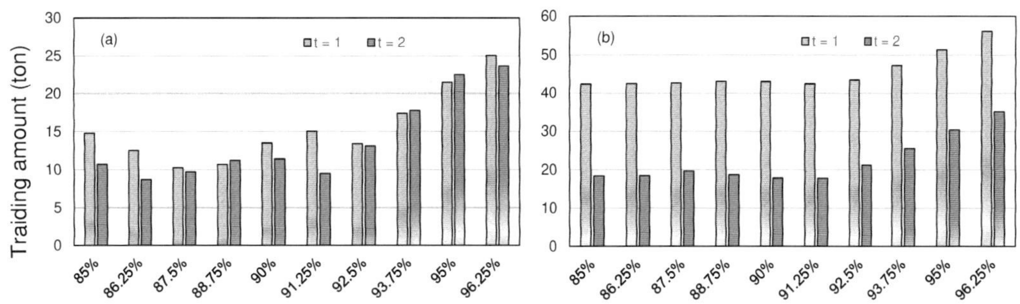

Figure 2. Total trading amount under different treatment rate [(a) $\varepsilon = 1$, (b) $\varepsilon = 0.85$]

Figure 2 illustrates the total trading amount when waste water generated from chemical plants is handled with different treatment rates when environmental limit is at the two levels ($\varepsilon = 1$ and $\varepsilon = 0.85$). From the results, the trading amount would fluctuates as treatment rate is raised. For example, in nonflood season, the trading amount would decrease from 14.79 ton ($\eta = 85\%$) to 10.23 ton ($\eta = 87.5\%$), then increase to 13.40 ton ($\eta = 92.5\%$), and rise to 15.04 ton ($\eta = 91.25\%$). As treatment rate rise, the plants would purchase less permits because higher treatment rate can bring about more permits for the plants. This can account for the decrease of total trading amount. When treatment rate is further raised, more permits would be generated and the plants can thus sell more permits to other pollution sources to obtain benefit. This can explain the rising trend of total trading amount when treatment rate is larger than 92.75%. In addition, similar changing rules can be found when environmental limit is more strict ($\varepsilon = 0.85$); while the trading amount is less fluctuant.

4. Conclusions

In this study, a Bayesian estimation-based inexact two-stage stochastic programming (BITSP) method is proposed for identifying the effect of important trading factors. The BITSP model incorporates SWAT, Bayesian estimation, and ITSP within a framework. BITSP can help (i) address multiple uncertainties presented as intervals and probability distribution functions; and (ii) disclose the effect of important factors in effluent trading. We conduct a real case study for the application of BITSP in the Xiangxihe watershed. Results reveal that strict environmental limits increase the desire for permit trading program. The results also reveal that treatment rate have an obvious effect on effluent trading through changing the buying and selling behavior of point sources.

Acknowledgement

This research was supported by the National Key Research Project (2016YFC0207800), Natural Science Foundation of China (51779008 and 51608422), and the National Key R&D Program of China (2016YFC0502800).

Appendix

The formulation of the BITSP:

$$\begin{aligned}\text{Max } f^{\pm} = & \sum_{i=1}^{4}\sum_{j=1}^{9}\sum_{t=1}^{2} AP_{ijt}^{\pm} AB_{ij}^{\pm}\left(x_{ij}^{-} + \Delta x_{ij} o_{ijt}\right) \\ & - \sum_{i=1}^{4}\sum_{k=1}^{3}\sum_{t=1}^{2} p_k CA_t^{\pm} D_{ikt}^{\pm} + \sum_{s=1}^{5}\sum_{t=1}^{2} CB_{st}^{\pm}\left(Y_{st}^{-} + \Delta Y_{st} z_{st}\right) \\ & - \sum_{s=1}^{5}\sum_{t=1}^{2}\left[k_{1s}^{\pm} + k_{2s}^{\pm}\left(\eta_{Cs}^{\pm}\right)^{k_{3s}^{\pm}}\right]\left[WG_s^{\pm}\right]^{k_{4s}^{\pm}} - \sum_{t=1}^{2} SP_t^{\pm} SPM_t \\ & - \sum_{s=1}^{5}\sum_{t=1}^{2} W_{st}^{\pm} CEH_t^{\pm} + \sum_{m=1}^{6}\sum_{t=1}^{2} PB_{mt}^{\pm}\left(V_{mt}^{-} + \Delta V_{mt} b_{mt}\right) - \sum_{m=1}^{6}\sum_{t=1}^{2} F_{mt}^{\pm} CEK_t^{\pm}\end{aligned} \tag{1a}$$

Subject to:

$$\left\{\begin{aligned} & \sum_{s=1}^{5}\sum_{i=1}^{4}\left[PAP_t\,(s,\, i) + PAS_t\,(i,\, s)\right] + \sum_{m=1}^{6}\sum_{i=1}^{4}\left[PAP_t\,(m,\, i) + PAS_t\,(i,\, m)\right] \\ & + \sum_{m=1}^{6}\sum_{s=1}^{5}\left[PCP_t\,(m,\, s) + PCS_t\,(s,\, m)\right] + \sum_{n=1}^{3}\sum_{i=1}^{4} PAP_t\,(n,\, i) \\ & + \sum_{y=1}^{4}\sum_{s=1}^{5} PCP_t\,(y,\, s) + \sum_{f=1}^{5}\sum_{m=1}^{6} PMP_t\,(f,\, m)\end{aligned}\right\} = SPM_t \tag{1b}$$

IOP Conf. Series: Earth and Environmental Science **146** (2018) 012049 doi:10.1088/1755-1315/146/1/012049

$$\begin{aligned} ACE_{it} &= AA_{it} + \sum_{n=1}^{3} PAP_t\left(n,\ i\right)/T_A\left(n,\ i\right) - \sum_{n=1}^{3} PAS_t\left(i,\ n\right) \\ &+ \sum_{s=1}^{5} PAP_t\left(s,\ i\right)/T_A\left(s,\ i\right) - \sum_{s=1}^{5} PAS_t\left(i,\ s\right) \\ &+ \sum_{m=1}^{6} PAP_t\left(m,\ i\right)/T_A\left(m,\ i\right) - \sum_{m=1}^{6} PAS_t\left(i,\ m\right) \end{aligned} \tag{1c}$$

$$\begin{aligned} CCE_{st} &= AC_{st} + \sum_{y=1}^{4} PCP_t\left(y,\ s\right)/T_C\left(y,\ s\right) - \sum_{y=1}^{4} PCS_t\left(s,\ y\right) \\ &+ \sum_{i=1}^{4} PCP_t\left(i,\ s\right)/T_C\left(i,\ s\right) - \sum_{i=1}^{4} PCS_t\left(s,\ i\right) \\ &+ \sum_{m=1}^{6} PCP_t\left(m,\ s\right)/T_C\left(m,\ s\right) - \sum_{m=1}^{6} PCS_t\left(s,\ m\right) \end{aligned} \tag{1d}$$

$$\begin{aligned} MCE_{mt} &= AM_{mt} + \sum_{f=1}^{5} PMP_t\left(f,\ m\right)/T_M\left(f,\ m\right) - \sum_{f=1}^{5} PMS_t\left(m,\ f\right) \\ &+ \sum_{i=1}^{4} PMP_t\left(i,\ m\right)/T_M\left(i,\ m\right) - \sum_{i=1}^{4} PMS_t\left(m,\ i\right) \\ &\sum_{s=1}^{5} PMP_t\left(s,\ m\right)/T_M\left(s,\ m\right) - \sum_{s=1}^{5} PMS_t\left(m,\ s\right) \end{aligned} \tag{1e}$$

$$\begin{aligned} &AA_{it} + \sum_{n=1}^{3} PAP_t\left(n,\ i\right)/T_A\left(n,\ i\right) + \sum_{s=1}^{5} PAP_t\left(s,\ i\right)/T_A\left(s,\ i\right) \\ &+ \sum_{m=1}^{6} PAP_t\left(m,\ i\right)/T_A\left(m,\ i\right) \geq \sum_{n=1}^{3} PAS_t\left(i,\ n\right) + \sum_{s=1}^{5} PAS_t\left(i,\ s\right) + \sum_{m=1}^{6} PAS_t\left(i,\ m\right) \end{aligned} \tag{1f}$$

$$\begin{aligned} &AC_{st} + \sum_{y=1}^{4} PCP_t\left(y,\ s\right)/T_C\left(y,\ s\right) + \sum_{i=1}^{4} PCP_t\left(i,\ s\right)/T_C\left(i,\ s\right) \\ &+ \sum_{m=1}^{6} PCP_t\left(m,\ s\right)/T_C\left(m,\ s\right) \geq \sum_{y=1}^{4} PCS_t\left(s,\ y\right) + \sum_{i=1}^{4} PCS_t\left(s,\ i\right) + \sum_{m=1}^{6} PCS_t\left(s,\ m\right) \end{aligned} \tag{1g}$$

$$\begin{aligned} &AM_{mt} + \sum_{f=1}^{5} PMP_t\left(f,\ m\right)/T_M\left(f,\ m\right) + \sum_{i=1}^{4} PMP_t\left(i,\ m\right)/T_M\left(i,\ m\right) \\ &+ \sum_{s=1}^{5} PMP_t\left(s,\ m\right)/T_M\left(s,\ m\right) \geq \sum_{f=1}^{5} PMS_t\left(m,\ f\right) + \sum_{i=1}^{4} PMS_t\left(m,\ i\right) + \sum_{s=1}^{5} PMS_t\left(m,\ s\right) \end{aligned} \tag{1h}$$

$$\left(V_{mt}^{-} + \Delta V_{mt} b_{mt}\right) MG_{mt}^{\pm} - F_{mt}^{\pm} \leq MCE_{mt} \quad \rho C_s^{\pm}\left(Y_{st}^{-} + \Delta Y_{st} z_{st}\right)\left(1 - \eta_{Cs}^{\pm}\right) - W_{st}^{\pm} \leq CCE_{st} \tag{1i}$$

$$e_k^{\pm}\left(\sum_{j=1}^{9}\left(x_{ij}^{-} + \Delta x_{ij} o_{ijt}\right)\right) - D_{ikt}^{\pm} \leq ACE_{it} \quad ACE_{it} \leq TA_t \quad CCE_{st} \leq TC_t \quad MCE_{mt} \leq TM_t \tag{1j}$$

$$\sum_{i=1}^{4} ACE_{it} + \sum_{s=1}^{5} CCE_{st} + \sum_{m=1}^{6} MCE_{mt} \leq T_t \qquad ACE_{2t} + CCE_{2t} \leq T_{\mathrm{I},\ t} \quad ACE_{3t} + \sum_{s=3}^{5} CCE_{st} \leq T_{\mathrm{III},\ t} \tag{1k}$$

$$ACE_{1t} + CCE_{1t} + \sum_{m=1}^{3} MCE_{mt} \leq T_{\mathrm{II},\ t} \qquad \sum_{m=4}^{6} MCE_{st} \leq T_{\mathrm{IV},\ t} \quad ACE_{4t} \leq T_{\mathrm{V},\ t} \tag{1l}$$

$$PAP_t\left(s,\ i\right) = PAS_t\left(i,\ s\right) \geq 0 \qquad PAS_t\left(s,\ i\right) = PAP_t\left(i,\ s\right) \geq 0 \tag{1m}$$

$$PAP_t\left(m,\ i\right) = PAS_t\left(i,\ m\right) \geq 0\,(1) \qquad PAS_t\left(m,\ i\right) = PAP_t\left(i,\ m\right) \geq 0 \tag{1n}$$

$$PCP_t\left(m,\ s\right)=PCS_t\left(s,\ m\right)\geq 0 \quad PCS_t\left(m,\ s\right)=PCP_t\left(s,\ m\right)\geq 0 \tag{1o}$$

$$PAP_t\left(n,\ i\right)=PAS_t\left(i,\ n\right)\geq 0 \quad PCP_t\left(y,\ s\right)=PCS_t\left(s,\ y\right)\geq 0 \tag{1p}$$

$$PMP_t\left(f,\ m\right)=PMS_t\left(m,\ f\right)\geq 0 \quad 0\leq D_{ikt}^{\pm}\leq e_k^{\pm}\left(\sum_j\left(x_{ij}^{-}+\Delta x_{ij}o_{ijt}\right)\right) \tag{1q}$$

$$0\leq o_{ijt}\leq 1 \quad 0\leq z_{st}\leq 1 \quad 0\leq b_{mt}\leq 1 \quad 0\leq W_{st}^{\pm}\leq \rho C_s^{\pm}\left(Y_{st}^{-}+\Delta Y_{st}z_{st}\right)\left(1-\eta_{Cs}^{\pm}\right) \tag{1r}$$

$$0\leq F_{mt}^{\pm}\leq\left(V_{mt}^{-}+\Delta V_{mt}b_{mt}\right)MG_{mt}^{\pm} \tag{1s}$$

The detailed nomenclatures for the variables and parameters can be found in Zhang et al. [9]. The values of $e_k^{\pm}$ are obtained based on analysis of the propagation of parameter uncertainty by using MCMC. Model (1) can be transformed into two submodels corresponding to lower and upper bounds of the objective function values, as recommended in Huang and Loucks [10].

References

[1] Lankoski J, Lichtenberg E, Ollikainen M, 2008. Point/nonpoint effluent trading with spatial heterogeneity. *American Journal of Agricultural Economics* **90** 1044-1058

[2] Mesbah S M, Kerachian R, Torabian A, 2010. Trading pollutant discharge permits in rivers using fuzzy nonlinear cost functions. *Desalination* **250** 313–317

[3] Lee M, Douglas-Mankin K R, 2011. An environmental trading ratio for water quality trading. *Transactions of the ASABE* **54** 1599-1614

[4] Zhang J L., Li Y P, Huang G H, 2014. A robust simulation-optimization modeling system for effluent trading-a case study of nonpoint source pollution control. *Environmental Science and Pollution Research* **21** 5036-5053

[5] Mahjoobi E, Sarang A, Ardestani M, 2016. Management of unregulated agricultural nonpoint sources through water quality trading market. *Water Science and Technology* **74** 2162-2176

[6] Doyle M W, Patterson L A, Chen Y Y, Schnier K E, Yates A J, 2014. Optimizing the scale of markets for water quality trading. *Water Resources Research* **50** 7231-7244

[7] Chen L, Han Z X, Li S, Shen Z Y, 2016. Framework design and influencing factor analysis of a water environmental functional zone-based effluent trading system. *Environmental Management* **58** 645-654

[8] Jiang L G, Liang B, Xue Q, Yin C W, 2016. Characterization of phosphorus leaching from phosphate waste rock in the Xiangxi River watershed, Three Gorges Reservoir, China. *Chemosphere* **150** 130-138

[9] Zhang J L, Li Y P, Huang G H, Baetz B W, Liu J, 2017. Uncertainty analysis for effluent trading planning using a Bayesian estimation-based simulation-optimization modeling approach. *Water Research* **116** 159-181

[10] Huang G H, Loucks D P, 2000. An inexact two-stage stochastic programming model for water resources management under uncertainty. *Civil Engineering and Environmental Systems* **17** 95-118

Ethernet Based Embedded Precision Acceleration with Temperature and Humidity Acquisition System

Jin Han[1], Biyao Wang[2, *] and Nuonuo Xue[2]

[1]Key Laboratory of Contemporary Design and Integrated Manufacturing Technology, Ministry of Education, Northwestern Polytechnical University, Xi'an, China,710072
[2]School of Automobile, Chang'an University, Xi'an, China,710064

*biyaoss@126.com

Abstract. This paper proposes an efficient precision acceleration with temperature and humidity acquisition system based on 24-bit A/D Converter, IIC temperature and humidity sensor, Ethernet controller and microcontroller. This paper expounds in detail the realization of acquisition hardware, UDP protocol in microcontroller, the acquisition software as well as many other important aspects of the system. On the basis of the foregoing, build the entire acquisition system and apply it to the road roughness measuring instrument. Through the practice of the vehicle-mounted environment inspection, experimental results show that this system can stable operation for a long time and have a good application prospect.

1. Aims and background

The High-speed High-accuracy and long-time data acquisition of accelerometer signal is an important aspect of the road roughness measurement system [1, 2]. In practical engineering application, the traditional method to digital accelerometer output signal uses voltage to frequency converter ADC (i.e. AD650)[3]. But this method has many disadvantages: the conversion rate is not fast enough, exists blind area for weak signal, exists temperature drift, system volume is larger, and is difficult to improve the measurement precision etc.

With the performance improvement of the technology, the accelerometer output signal is more and more sensitive and weak. Acceleration acquisition system is increasingly demanding more precision ADC. Traditional voltage to frequency methods have been difficult to meet the needs of the measurement, whereas the current fastest growing 24-bit Δ-Σ A/D converter is a kind of very good solution [4]. This paper from the practical point of view, using the 24-bit Δ-Σ A/D converter LTC2440 combined with Ethernet controller RTL8019AS, IIC temperature and humidity sensor HTU21D and enhanced 8051 series microcontroller W77E58, design and realize the 10 Mbit/s Ethernet interfaced acceleration and temperature and humidity data collection scheme, and apply it in vehicle-mounted road roughness measuring system [5, 6].

2. Hardware system

Hardware system uses two LTC2440 chips to separately collect two acceleration sensor output signal. Microcontroller reads the digitized data from LTC2440, then drive Ethernet controller RTL8019AS to send data to the computer via UDP protocol. System data flow is shown in Figure 1.

IOP Conf. Series: Earth and Environmental Science **146** (2018) 012050 doi:10.1088/1755-1315/146/1/012050

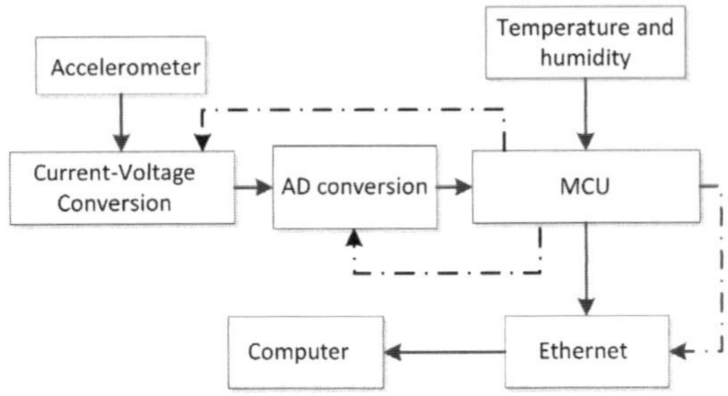

Figure 1. System data flow.

2.1. System structure
Two pieces of LTC2440 are respectively connected to P1.0 ~ P1.2 and P1.3 ~ P1.5 of the microcontroller via Serial Peripheral Interface (SPI). The microcontroller starts two LTC2440's A/D conversion process and collects the resulting data at a fixed frequency. Then use connected to the microcontroller bus network chip RTL8019AS to send the data to the host computer via UDP network protocol. Furthermore, in order to improve the stability of the system, the system also uses a watchdog chip MAX813L and feeds the watch dog pulse to P1.6 of the microcontroller. For temperature and humidity data, use P1.7 and INT1 connect with IIC temperature and Humidity sensor SHT25 to measure the temperature and Humidity of the environment, the system structure is shown in figure 2.

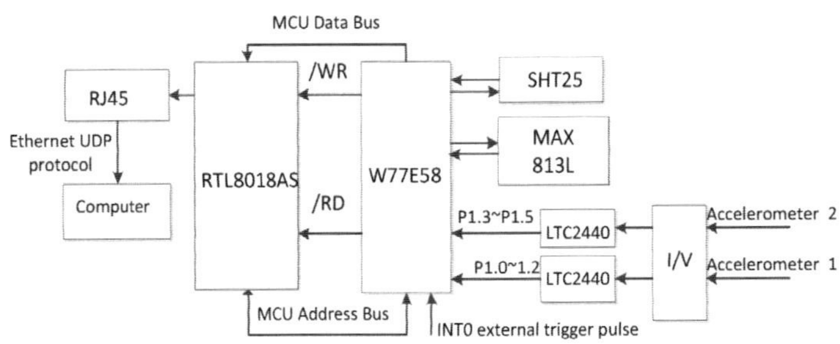

Figure 2. System structure.

2.2. Current to voltage conversion
Acceleration sensor signal current to voltage conversion can be simply achieved by connecting a precision resistor between the output signal and GND. But the resistor will take a greater impact on the acceleration signal, leading to the presence of a large deviation between the measured value and the true value, so this method is not desirable. In this paper, using the operational amplifier built a current to voltage conversion circuit, its principle is shown in figure 3. The op-amp working on negative feedback status, so the op-amp output voltage is as as in equation (1).

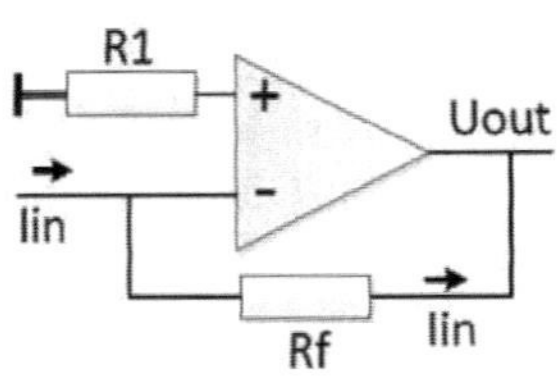

Figure 3. Current-voltage Conversion.

$$U_{out} = -I_{in} \times R_f \quad (1)$$

Therefore we can adjust the gain of the circuit by changing the value of resistor R_f.

IOP Conf. Series: Earth and Environmental Science **146** (2018) 012050 doi:10.1088/1755-1315/146/1/012050

2.3. A/D conversion

The core of high-precision data acquisition is the selection of A/D chip. In this application, the selection of A/D chip is mainly based on input polarity, effective resolution and conversion time to determine.

Input polarity. Due to the application of the accelerometer output bipolar signals, thus require A/D chip can support dual polarity of the input signal.

The resolution. A/D chip can measure the minimum voltage variation. This can be calculated by using equation (2).

$$\frac{U_{FSR}}{U_{\min}} = \frac{2^N - 1}{1} = 2^N - 1 \qquad N \geq \log_2\left(1 + \frac{U_{FSR}}{U_{\min}}\right) \tag{2}$$

In the formula, U_{FSR} is full range of chip. $U_{\min}$ is the A/D chip minimum detect voltage. N is the A/D chip output length of bits. Then available:

The input range of acceleration of this application system is -10g~+10g, the minimum resolution is 0.0001g. So as in equation (3).

$$\frac{U_{FSR}}{U_{\min}} = \frac{10-(-10)}{0.0001} = 2\times10^5 \qquad N \geq \log_2\left(1+2\times10^5\right) \approx 18 \quad \text{(round up)} \tag{3}$$

So the system must ensure that the final measurement result has 18 bits precision. Because the system supply voltage drift, temperature drift, other conversion error. Therefore the A / D conversion process should reach at least 20 actual conversion accuracy.

Conversion time. Because the system requirements for accelerometer measurement update rate is output once every 2.5ms, i.e. 400Hz. At the same time considering every 5 ~ 6 time of collection can perform the arithmetic average filter to the data, so the A/D chip conversion rate should be above 2400 Hz.

At present, 24 bits A/D conversion chip mainly include two companies (ADI and TLC) products, their typical models are AD7710 and LTC2440 respectively. The comparison of two kinds of chips are shown in Table 1.

Table 1. AD7710 and LTC2440[7,8].

Devices	Input Range	Bits	Resolution	Bits (100Hz)	Resolution (100Hz)	Maximum sampling frequency
AD7710	-5~5V	24	0.596 uV	18	38.1 uV	1KHz
LTC2440	-2.5~2.5V	24	0.298 uV	21	2.3 uV	3.5KHz

Visibley, in terms of high frequency acquisition LTC2440 has greater advantages than AD7710. At 100Hz frequency acquisition LTC2440 can still achieve an effective resolution of 21 bits, for the present application is a better choice. LTC2440 through 3 line digital interface communication with microcomputer, contain the serial data output (SDO), the serial clock (SCK) and chip enable (CS). According to SPI protocol MCU read 32-bits data from LTC2440, the 3rd as a sign bit, 4 to 27 for the 24 bits of A/D conversion (bit 4 is high, bit 27 is low).So total data of one LTC2440 are 25 bits. LTC2440 operation timing is shown in Figure 4.

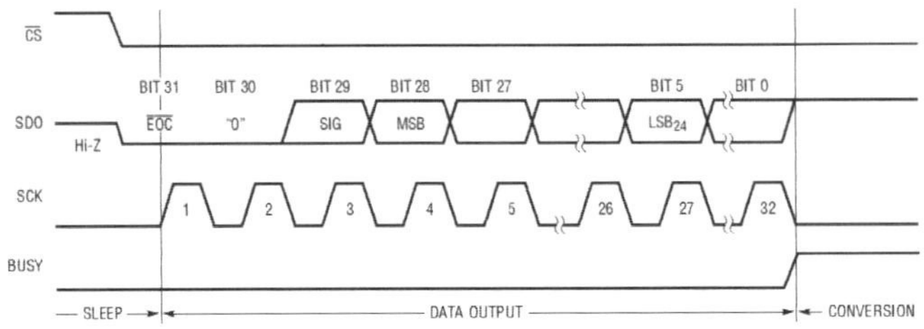

Figure 4. LTC2440 operation timing.

Using high-precision A/D chip for data acquisition can give a more bits code, and convert the raw sensor signals into more bits digital data. Then using MCU simple arithmetic average processing can further improve the accuracy of the system and reduce noise.

Temperature and Humidity measurement. This system uses the integrated temperature and humidity sensors to measure temperature and humidity of the environment. Acquisition temperature to modify the collected acceleration data and acquisition environment humidity to maintain the reliability of the detection instrument. Nowadays, there are many temperature and humidity sensor produced by many different companies. The Table 2 shows the temperature and Humidity sensor chips that are popular at present.

Table 2. Temperature and Humidity sensor chips [9,10,11,12].

Chip	Temperature Range	Temperature Accuracy	Humidity range	Humidity Accuracy
SI7021	-10~85℃	0.4℃	0~100%RH	3%
SHT25	-40~120℃	0.2℃	0~100%RH	1.8%
HIH9131	-40~125℃	0.6℃	0~100%RH	1.7%
HTU21P	-40~125℃	0.3℃	0~100%RH	2%

As can be seen from the table, the comprehensive performance of SHT25 is better than the others, so the SHT25 is selected for the measurement of temperature and humidity in this project. SHT25 temperature measurement range is -40℃~120℃, its accuracy can up to 0.2℃ of temperature and 1.8% of humidity. In this application, every time INT0 interrupt signal is received, the microcontroller collected once the current temperature and humidity using I2C Protocols, every time can gather 26 bits of data（RH：12bits，T：14bits），the data is sent along with the A/D conversion data of acceleration to the host computer. PC software receives, calibrates and stores data [13].

Ethernet data communication. MCU synchronously acquisites the data of two A/D chips, and also reads the state of external interrupt pin INT0. When the INT0 signal arrives, the microcontroller performs arithmetic average filter to recent five groups acquisition data, and collects the current temperature and humidity(26 bits), then adds 4 bits as packet ID, consisting of an 10-byte packets. As is shown in Figure 5.

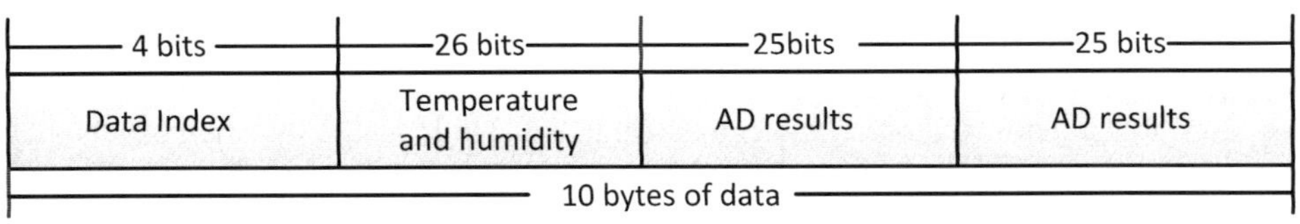

Figure 5. Packet Content.

The system uses a 10Mb/s Ethernet chip RTL8019AS to send the collected data to the computer, the circuit connection is shown in Figure 6.

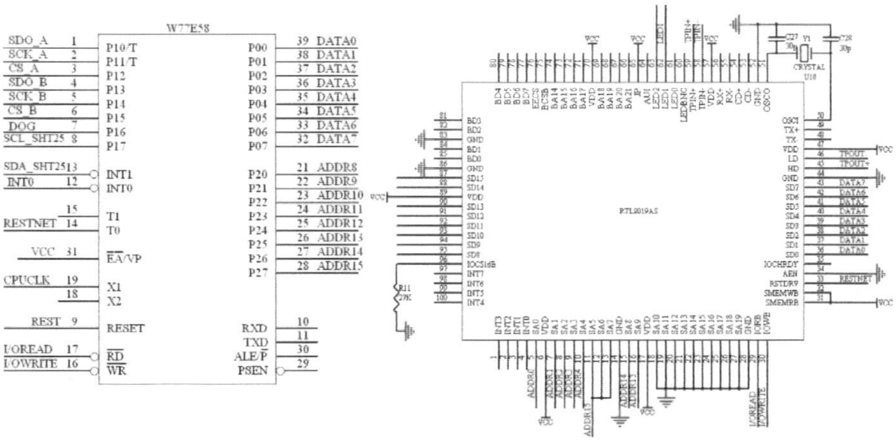

Figure 6. MCU and RTL8019AS connection.

Collect and transmit data flow is: MCU collects multiple sets of data into the RAM buffer, and group packages based according to UDP protocol, while setting each UDP package can only have a maximum of 50 groups of collected data (10 bytes for each group, 50*10= 500 bytes), Package format is shown in Figure 7, the MCU controls RTL8019AS through remote DMA operations to write the package data to RTL8019AS internal SRAM, and then sent to the Ethernet via RJ45 interface. The computer uses UDP protocol to receive and store data.

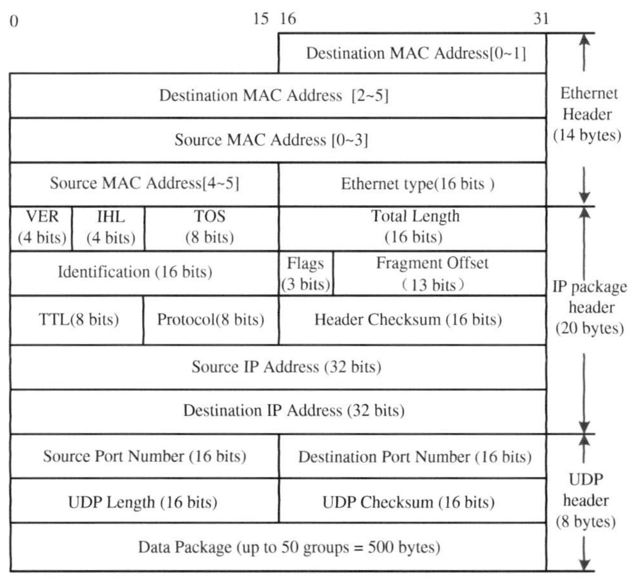

Figure 7. UDP Packet.

As can be seen from Figure 7, using UDP protocol to establish data package needs to add additional 42 bytes Ethernet header, which includes IP header and UDP header (14 + 20 + 8 = 42). The front 40 bytes are fixed number, only the last two bytes (16-bit UDP checksum) needs to be calculated based on the collected data. Therefore, the speed of the microcontroller group package is fast when using UDP protocol, which can increase the communication speed.

MCU program design. The main parts of the microcontroller software include: read LTC2440 data, INT0 interrupt process, arithmetic average filter, read temperature and humidity, group UDP package, UDP package transmission and other parts. The main program flow is shown in Figure 8.

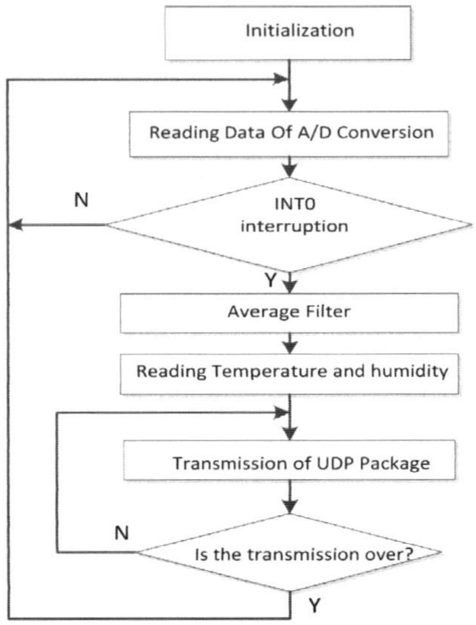

Figure 8. MCU Software Flow.

3. Software Develop

The text of your paper should be formatted as follows:

MCU collects and sends data to the computer via the UDP protocol, so we need to develop the software systems in the computer to receive and store the collected data, the focus of software work includes data correction and data storage.

3.1. Data correction

MCU collected voltage is 25-bits signed binary number. It can be assumed that at a fixed temperature, there is a linear relationship between the acquired voltage M and the measured acceleration A . i.e.

$$A = kM - b \tag{4}$$

Wherein k is the slope of the calibration model, M is the measured digital voltage, b is the intercept of the calibration model. When the temperature changes, the intercept b can be considered constant, only slope k is changed, and k varies quadratic relationship with temperature. That is as in equation (5).

$$A = \left(cT + dT^2 + e\right)M - b \tag{5}$$

By the way of experiments, we can get many group datas of A, T, M . Based on these datas, we can set up equations and solve to get c, e, d, b . So we can obtain at a temperature T of the environment and acquire value of M corresponding to the acceleration A .

3.2. RESULTS

By using the VC++.net 2010, a software development platform, we developed a data collection and storage system [14]. In the system, UPD protocol is used to communicate with computer, so as to collect data transmitted by accelerometer real-timely, then the data are corrected by conversion Eq.(5). The system can operate stably in specific instrument for detecting roughness of road. What's more, the detection accuracy can reach 98% and the data errors repeatedly measured in different temperature are below 2%. The interface of data collection software is shown in figure 9 and the road roughness measurement vehicle is show in figure 10.

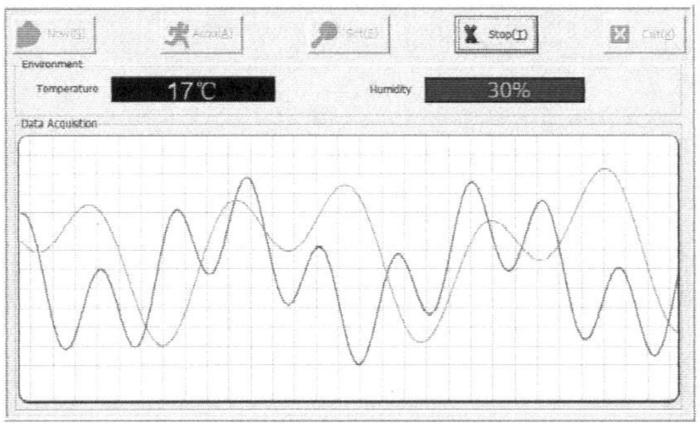

Figure 9. Acquisition Software.

Figure 10. Road Roughness Measurement Vehicle.

4. Conclusions

This paper designs and implements the accelerometer with temperature and humidity acquisition program with an Ethernet interface, very suitable for the accelerometer signal acquisition correction and storage in road roughness measurement system. Through the practice of the vehicle-mounted environment inspection, experimental results show that this system can stable operation for a long time and have a good application prospect.

References

[1] Philippe Nitsche, Rainer Stütz, Michael Kammer, Peter Maurer: Comparison of Machine Learning Methods for Evaluating Pavement Roughness Based on Vehicle Response. Journal of Computing in Civil Engineering,28(4), 04014015.
[2] Yuchuan Du, Chenglong Liu, Difei Wu: Measurement of International Roughness Index by Using Z-Axis Accelerometers and GPS. Mathematical Problems in Engineering.2014.
[3] Voltage-to-Frequency and Frequency-to-Voltage Converter AD650 Data Sheet. Analog Devices, Inc.2013.
[4] Xiaowei Liu, Honglin Xu, Liang Yin, Zhiqiang Gao, Mingyuan Ren:A low noise sigma-delta microaccelerometer interface circuit. IEICE Electronic Express,11(12), 1-6(2014).
[5] Samia Boubaker, Férid Rehimi, Adel Kalboussi: Impact of intersection type and a vehicular fleet's hybridization level on energy consumption and emissions. Journal of Traffic and Transportation Engineering (English Edition), 3(3),253-261(2016).
[6] Stefan Torbruegge, Burkhard Wies: Characterization of pavement texture by means of height difference correlation and relation to wet skid resistance. Journal of Traffic and Transportation Engineering (English Edition), 2(2),59-67(2015).
[7] Signal Conditioning ADC AD7710. Analog Devices, Inc.2004.
[8] LTC2440 24-Bit High Speed Differential ADC with Selectable Speed/Resolution[R]. Linear Technology Corporation.2002.
[9] Datasheet SHT25. Sensirion.2014.
[10] I2C HUMIDITY AND TEMPERATURE SENSOR Si7021-A20 Datasheet. Silicon Laboratories.2015.
[11] Honeywell Humidlcon Digital Humidity/Temperature Sensors. Honeywell.2013.
[12] HTU21P(F) RH/T Sensor IC – Miniature Relative Humidity and Temperature. Measurement Specialties.2015.
[13] YaPing Yang, Bo Wang, YongTao Liu: Real-time Image Acquisition and Display System Based on Field Programmable Gate Array. Journal of the Balkan Tribological Association, 22(2), 1085-1091(2016)
[14] Jianguang Zhao, Mangui Liang, Jingjing Fan: Traffic Adaptive Routing Algorithm For Wireless Sensor Network. Journal of the Balkan Tribological Association,22(2), 1098-1107(2016)

The influence of migration on the reconstruction area of soil contaminated by complex heavy metals

NAN LU[1, 2, 3, 4,*], GANG LI[1, 2, 3, 4] and YANG WEI[1, 2, 3, 4]

[1]Shaanxi Provincial Land Engineering Construction Group Co., Ltd., Xi'an 710075, China
[2]Key Laboratory of Degraded and Unused Land Consolidation Engineering, the Ministry of Land and Resources, Xi'an 710075, China
[3]Shaanxi Provincial Land Consolidation Engineering Technology Research Center, Xi'an 710075, China
[4]Institute of Land Engineering and Technology, Shaanxi Provincial Land Engineering Construction Group Co., Ltd., Xi'an 710075, China

[*]Corresponding author e-mail: lunan8836@126.com

Abstract. In the heavy metal compound pollution area, the isolation and repair technology was used to seal the fixed contaminants. And then, the land organic reconstruction carried out. In order to test the isolation effectiveness of reconstructed land, two experiments have been carried out. There were the migration status of heavy metals in reconstruction soil profile and its effects on crops, respectivly. The contents of the heavy metal pollutants such as Hg, Cu, Cd, Zn, Pb, As, Cr and Ni were studied. The results showed that Hg, Cu, Cd, Zn and Pb were the main heavy metals that caused pollution in the study area. Besides, the concentration of Pb was very serious. After two years of planting, the content of heavy metals in each layer of the reconstructed and profile were lower than that of the soil environmental quality standard (GB 15618-1995). The migration and accumulation of heavy metals on the various soil profiles were not found. And the edible part of corn and peanuts planted which were planted in the reconfiguration soil profile conld conform the food safety requirements.

1. Introduction

China was rich in mineral resources, and the heavy metal pollution caused by the exploitation of mineral resources was very common in the mining process [1]. Heavy metal polluted soil and high risk area had emerged in the mining area and the surrounding area. Jiao. et al. investigated the pollution of farmland soil in the northern Guangdong mining area, indicated that the Cd pollution was serious and the Ni contamination existed to varying degrees. In addition to Cu and Zn in the brown rice, other heavy metals with different concentrations were also detected [2]. On the basis of investigating heavy metal pollution and plant accumulation in farmland soil in great Baoshan mining area, Zhang. et al. found that the contents of Cu, Pb and Cd in soil samples were much more than the grade II standard of soil environmental quality [3].

Tongguan, located at the E110°09′32″~110°25′27″ and the N34°23′33″~34°39′01″. The north of Tongguan was near the Yellow River, and its south near Qinling Mountains, containing Au, Ag, Pb and so on. It was famous for rich reserves of gold and long history of mining. As the pillar industry of Tongguan, the gold industry contributed more than 70% of the county's financial revenue. While the frequent mining activities were bringing huge economic benefits, and at the same time, left many problems such as heavy metal pollution and so on.

In the heavy metals compound polluted area, the pollutants were buried and fixed by isolation and repair technology. The structure of land profile was reconstructed on the isolation layer to achieve the purpose and requirements of isolating pollutants, protecting the structure and function of the isolation layer, and meeting the cultivation of the reconstructed profile. Under the conditions of isolation, the release and migration of heavy metals was more complex than in natural conditions. The types of material, heavy metal content, physical and chemical properties, and the thickness of section layer were related to the reconstruction of soil profile, and it was also closely related to the properties of the isolation layer. On the one hand, the profile reconfiguration leaded to the change of temperature, humidity, pH and other environmental conditions. On the other hand, the composition of the isolation layer and the tailings containing heavy metals might be neutralized, adsorbed, and ion exchange, and so on. These chemical reactions have an impact on the original mineral components and their occurrence state in pollutants, and inhibited the diffusion and migration of heavy metals. It was great significance to study the effect of isolation and repair technology by means of monitoring and evaluating the heavy metals contents of each layer in the reconstructed soil profile regularly.

2. Materials and Methods

2.1. Experimental method

In order to fully simulate the migration of heavy metals in the reconstructed profile of composite contaminated soil, the in situ field experiment was carried out. Firstly, the compound heavy metals pollutants was compacted, and then covered the isolation layer. Specifically, the pure soil layer and the farming soil layer were spread out respectivly. The pure soil layer thickness was 40 cm, and its compaction coefficient should not be less than 0.83. The thickness of the farming soil layer should be 30 cm with packing naturally. The reconstituted soil profile configuration as shown in Figure 1. In order to compare the difference of land quality between the application of isolation technology and the uncontaminated. Set the reconstituted soil profile as the experimental group, and the contrast was treated as an unpolluted ordinary farmland, which was recorded as CK. The plane size of each cell was 3 m×3 m. According to the site conditions of the community, the uniformity of light, micro terrain and other factors were considered.

The crops were planted for two years in the natural rainfall situation. During this period, soil samples were collected regularly to determine the content of heavy metals in the farming and pure soil layers of the reconstructed soil. The migration of the heavy metals in the soil profile and the accumulation in the plant samples were investigated and evaluated.

2.2. Sample collection

The peanuts were planted in the first year, with planted the corn in the next year. After the crop harvest, the soil and plant samples were collected. According to the requirements of the technical specification for soil environmental monitoring, the sampling points of surface soil samples or heavy metal pollutants were mainly in one place. As a fixed-point sampling location, the "plum blossom sampling" was used within the range of 5 m, and the 2~3 subsample combined into a mixed sample, retained the sample weight of 1 kg. The sampling depth was 0 cm~30 cm, 30 cm~50 cm, 50 cm~70 cm from top to bottom successively. The location of the sampling point should be recorded accurately.

2.3. Sample processing and determination

2.3.1 Treatment and determination of soil samples. The soil and heavy metal pollutants samples were dried by natural wind, removed sundries and grind, sieved the 0.149 mm nylon mesh, determined the following indicators: pH and Hg, Pb, Cd, As, Cr, Zn, Cu, Ni. And the samples were determed in accordance with the relevant standards, such as GB/T 17141-1997, HJ 491-2009, GB/T 17139-1997,

GB/T 17138-1997, GB/T 22105.1-2008, GB/T 22105.2-2008 and so on.

2.3.2 Treatment and determination of plant samples. Washed the crop seeds with tap water, and rinsed and dried with deionized water. Seeds were placed at 105 °C to kill for 30 min, and then adjusted the temperature to 75 °C to dry to the constant weight, which was recorded dry weight (g). Samples were pulverized with the agate mortar, sieved the 0.25 mm nylon mesh. Determined the following indicators: Hg, Ni, Cd, As, Pb. The samples were determed in accordance with the relevant standards, such as GB 5009.17-2017, GB 5009.123-2014, GB 5009.138-2017, GB 5009.11-2014, GB 5009.15-2014, GB 5009.12-2017.

2.4. Evaluation method of heavy metal pollution

The soil environmental quality standard included three grades, of which the grade II was the soil limitted value for the protection of agricultural production and the maintenance of human health [4]. GB15618-1995 was used as a basis for evaluation of heavy metal pollutants and the accumulation of heavy metals in the reconstructed profile. The soil samples of heavy metal pollutants and regular monitoring were evaluated using the method of "accumulative pollution index".

The contents of heavy metals in the seeds of the harvested crops were compared with those in the normal cropland, and the effectiveness of the isolation treatment technology was evaluated. The safety of edible parts were evaluated by comparing the limit of heavy metals contents in GB 2715-2016 and GB 2762-2017.

3. Discussion

3.1. Type and content of heavy metals in the samples

According to the standard method, 30 samples of heavy metal pollutants were collected. And the content of 8 common heavy metals were measured, such as Hg, Cr, Ni, Cu, Cd, Zn, As and Pb. Evaluation of the results of heavy metals in samples by using the "accumulative pollution index". Descriptive statistics of heavy metal pollutants were shown in Table 1. Table 1 showed that in the construction area, the content of As in heavy metal pollutants was lower than the background value of soil elements in Shaanxi. The contents of Cr, Ni, Cu, Zn, Cd and Pb changed limitedly, and the coefficient of variation were less than 10%. The change of Hg content was more than 50%, with most of the high content of points. It showed that the Hg was greatly influenced by human activity. The average contents of Hg, Cu, Zn, Cd and Pb were all higher than the background values of soil elements in Shaanxi. The average contents of Hg, Cu, Cd and Pb were higher than the grade II of GB15618-1995(pH>7.5), with 1.29~8.6 times more than the standard limit, respectively. The contents of Cd and Pb exceeded the standard significantly. Analysis of the excess scale from the sampling point, Hg, Cd and Pb had the higher standard exceeding ratios of 68.3%, 46.1% and 16.9%, respectively.

Table 1. Descriptive statistics of heavy metal pollutants.

Element	Range	Mean	Standard deviation	Variation Coefficient (%)	Shaanxi soil element background value	Soil environmental quality standard grade II（pH>7.5）
Hg	0.69~7.82	4.65	2.96	63.66	0.06	1.0
Cr	51.01~58.68	53.46	3.70	0.07	55.00	250
Ni	7.06~10.77	8.60	19.40	2.26	24.00	60
Cu	110.56~145.63	129.01	14.40	0.116	17.90	100
Zn	47.13~492.08	197.74	208.15	1.06	45.30	300

As	1.62~2.42	2.02	0.327	0.17	9.96	25
Cd	0.47~5.34	2.11	2.28	1.08	0.793	0.6
Pb	2506.9~3369.9	3000.1	362.97	0.12	11.50	350

The contents of heavy metal elements in the soil of Guanzhong area of Shaanxi Province were used as the background value [5]. The application of the "ground cumulative pollution index method" to the heavy metal content of pollutants, the results of the evaluation were shown in Figure 2. According to the figure 2, it was known that the cumulative index of the 8 kinds of heavy metals from large to small was Pb(7.52)>Cu(2.35)>Zn(2.20)>Hg(2.01)>Cd(1.49)>Cr(-0.56)>Ni(-1.90)>As(-2.76)in turn. Inside, Cd was enriched with partial moderate pollution. The concentration of Hg, Cu and Zn were all moderate pollution, and the concentration of Pb in this area was extremely serious. There were no enrichment of Cr, Ni and As at 100% sampling points. The ratio of Hg, Cd, Cu, Zn and Pb to moderate concentration were 32.4%, 23.1%, 30.8%, 30.8% and 23.1%, respectively. The Cd of the 15.3% sampling points were partial to moderate enrichment. For Cu, 23.1% of the sampling points were heavily enriched. And for Pb, 53.8% sampling points were seriously enriched. According to data analysis, the cumulative effect of Hg, Pb, Cu, Zn and Cd in heavy metal pollutants were obvious. Many reasons accounting for this were as follows: first, the Gold deposits belong to rock gold deposits. The main associated elements were Pb, Cu, Pb and Zn, due to their metallogenic characteristics and associated metal mineral contents. Second, a variety of flotation agents were added to, which resulted in the high content of heavy metal pollutants in the mineral processing. Lastly, in Tongguan area, amalgamation, cyanidation and other mineral processing technology and gold extraction process were applied.

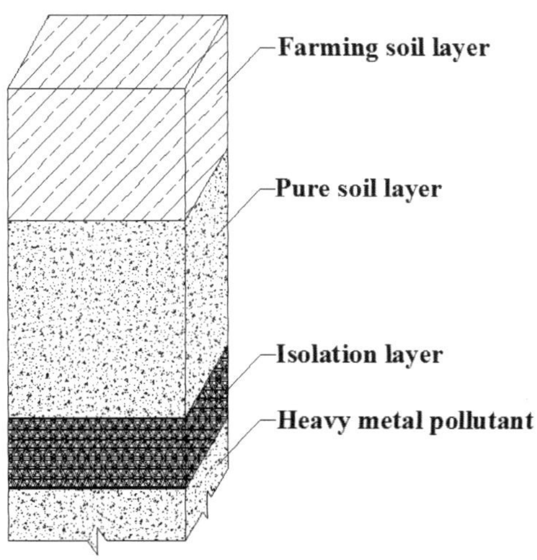

Figure.1. The reconstituted soil profile configuration diagram

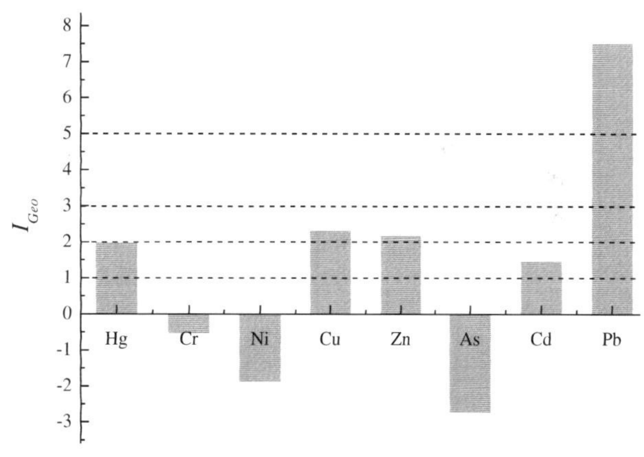

Figure.2. The average I_{Geo} of heavy metals

3.2. The change of heavy metal content in the reconstructed soil profile

Soil was complex and spatiotemporal variability due to the influence of soil matrix, climate, and biological factors [6]. Two aspects of variation in the plane distribution and variation in the vertical profile of the soil. The distance between the land and pollution sources, the way of using land and the utilization time determined the distribution and pollution degree of the heavy metal elements in the soil [7]. Under the influence of point source pollution, the impact of solid pollution sources was the highest, followed by liquid, and the smallest gas. The concentration of heavy metals in soils was high, with the influence of pollution decreasing gradually with increasing distance. Linear pollution sources such as rivers and irrigation channels, soil pollution levels of heavy metals were generally decreasing from center to both sides. Non-point source pollution was mainly influenced by wind direction, industrial layout and other factors. Heavy metals in agricultural land were often affected by the combined effects of single or multiple sources of pollution, such as industry, transportation and irrigation. It was beneficial to control the pollution of heavy metal pollutants and restore the land ecology by using the method of reconstruction for the composite contaminated soil. While eliminating point source pollution, pollution caused by traffic or irrigation should be prevented. Taked the planting mode, planting time and the distance between the section and the heavy metal contaminants into consideration, if the heavy metal contaminants could be strictly isolated, there were no adverse effects on the reconstituted soil profile. However, regular monitoring and testing were still needed.

Different land use patterns led to different contents of heavy metal elements in soils. Heavy metals in soil were mainly concentrated in the 20~60 cm soil layer, and there were some differences in the vertical distribution of heavy metals content. Therefore, it was very important to sample and monitor the depth of the soil layer of the reconstructed profile. According to the experimental design, 30 samples of soil samples were collected to determine the content of 8 heavy metal elements in soil samples. The median value of heavy metal content were shown in Table 2.

Table 2. Heavy metal content in reconstructed profile of compound polluted soil.

Treatment	Profile depth (cm)	The median heavy metal content (mg kg^{-1})							
		Hg	Cr	Ni	Cu	Zn	As	Cd	Pb
CK	0~30	0.22	51.4	29.0	31.8	53.7	3.3	0.19	31.6
	30~50	0.38	46.2	30.4	54.3	43.2	2.7	0.16	51.1
	50~70	0.32	55.7	31.3	45.2	49.1	4.9	0.19	49.0
	Mean	0.29	51.1	30.1	42.1	49.4	3.6	0.18	42.1
Isolation treatment	0~30	0.29	38.1	32.8	36.5	31.9	2.8	0.12	41.3
	30~50	0.29	43.6	27.8	41.0	37.4	3.3	0.15	42.6
	50~70	0.32	50.6	33.6	44.6	34.5	3.3	0.17	46.4
	Mean	0.30	43.2	31.6	40.1	34.2	3.1	0.14	43.1
Soil environmental quality grade II		1.0	350	60	100	300	20	0.6	350

After two years of cultivation, the contents of heavy metals in reconstructed soil profiles were compared with those of soil environmental quality standard grade II and the contents of heavy metals in corresponding soil profiles in CK. The results showed that the contents of heavy metals in the layers of reconstructed soil profile were all lower than GB 15618-1995, and accumulation of heavy metals on the soil profiles was not observed. The average value of heavy metal content at per unit thickness of soil were calculated, and compared the reconstructed soil profile with those of the control soil profile. The results showed that the contents of Cr, Cu, Zn, As and Cd in different profile depth were all lower than the CK. The contents of Hg, Ni and Pb were slightly higher than CK, increased by 2.5%, 5.1% and 2.3%, respectively. Combined with the results that Hg, Cd, Cu, Zn and Pb in pollutants were the main heavy metals that cause pollution, we can see that most of the heavy metal elements were effectively isolated. However, the contents of Hg and Pb were not significant compared with CK. There was no obvious change in the content of heavy metals in the sections of the reconstituted profile, which indicated that the effect of heavy metal blocking was good. However, in order to prevent the diffusion and migration of heavy metals, the contents of heavy metals in all layers of reconstructed soil profiles should be paid close attention to.

3.3. Heavy metal content in the seeds

In order to evaluate the effectiveness of isolation and reconstruction techniques, the accumulation of heavy metals in crops grown in soil were monitored, and a total of 20 samples of peanut and maize were collected. The average content of heavy metals were shown in Table 3.

Table 3. Content and limit of heavy metals in edible part of cropping plants.

Treatment	Sample	Average value of heavy metal content (mg kg-1)				
		Hg	Ni	As	Cd	Pb
CK	Corn [a]	ND	0.42	0.02	0.03	0.25
	Peanut [b]	ND	0.55	0.14	0.06	0.32
Isolation treatment	Corn [a]	ND	0.42	0.03	0.02	0.25
	Peanut [b]	ND	0.52	0.14	0.07	0.24
Food safety limits		0.02	1.0	0.5	0.1	0.5

[a] The element content was extremely low, exceeding the detection limit of 0.003 mg kg^{-1}.
[b] The element content was extremely low, exceeding the detection limit of 0.003 mg kg^{-1}.

The concentration of heavy metals in plants had a certain correlation with the content of heavy metals in the soil grown by plant roots. By monitoring the content of heavy metals in plants, it could indirectly reflect the heavy metals contents of soils. Isolation treatment compared with the CK of planted crops. There were no obvious changes in the content of Hg, Ni and Pb in corn grains. While, the content of As was slightly higher than CK, Cd content slightly lower. However, the contents of Hg, As, Ni and Pb in peanut grains were at a same level or lower than CK, Cd content slightly higher. By means of comparison, it was found that there was no significant difference in the contents of heavy metals in the planted crops between the reconstructed soil profiles and the CK. The contents of heavy metals in the seeds of corn and peanut under all treatment conditions were significantly lower than the content limit of GB 2715-2016 and GB 2762-2017.

4. Conclusion

1) The following conclusions were drawn by the analysis of heavy metal pollutants in the study area by the method of "accumulation pollution index": As was not a typical polluted metal in the study area. Hg, Cu, Cd, Zn and Pb were the main heavy metals that caused pollution in the study area. Inside, the concentration of Pb was very serious, and Pb was the typical pollution element in the study area.
2) After two years of cultivation, the contents of heavy metals in the layers of reconstructed soil profiles were all lower than the grade II of the "Soil Environmental Quality Standard" (GB 15618-1995). The migration and accumulation of heavy metals on the various soil profiles were not found. There was no significant change in the content of heavy metals in different depth of soil profile.
3) The content of heavy metals in the edible parts of corn and peanut planted in the compound pollution reconstructed soil profile were lower than the food safety limits. In other words, the crops planted in the soil profile reconfiguration were in accordance with the food safety requirements.

References

[1] TIAN K, YANG J, SUN Z J, et al. 2017, Preparation of Soil Certified Reference Materials for Heavy Metals in Contaminated Sites [J]. *Rock and Mineral Analysis*, **36**(1): 82-88.

[2] JIAO H P, ZHU H R, JIANG H Y, et al. 2017, Distribution Characteristics and Correlations of Heavy Metals in Contaminated Farmland Soils in Yuebei Mining Area [J]. *Chinese Agricultural Science Bulletin*, **33**(19):93-100.

[3] ZHANG H, JIN Q W, HUANG R L, et al. 2017, Characteristics of Heavy Metal Pollution in Agricultural Soils and Bioaccumulation in Plants of Dabaoshan Mine [J]. *Soils*, **49**(1):141-149.

[4] ZHANG M. 2012. Study on the changes of soil quality after reclamation of coal mining subsidence area-A case study of reclaimed area filling with fly ash in Jiawang, Xuzhou, China [D]. *Chinese academy of agricultural sciences*.

[5] Xue C Z, Xiao L, Wu Q F, et al. 1986. Studies of background values of ten chemical elements in major agricultural soils in Shaanxi Province [J]. *Acta universities agriculturalis boreali-occidentalis (NATURAL SCIENCE EDITION)*, (3):30-53.

[6] DUAN H M. 2012. The study on heavy metal pollution of urban soils in YongKang [D]. *Zhejiang Normal University*.

[7] LIU J F, GU N, ZHANG K H. 2012. Progress and prospect of soil heavy metal spatial differentiation and migration [J]. *Geography and Geo-Information Science*, **28**(2):99-103.

Occurrence and Distribution of Arsenic in Water and Soil at Inland-Arid/Semi-Arid Basin

Yihui Dong [1,2], **Jiale Li** [1,*] **and Jinjing Zan** [1]

[1] School of Water Resources and Environmental Engineering, East China University of Technology
[2] School of Environmental Studies, China University of Geosciences

*Corresponding author's e-mail: lijiale39@126.com

Abstract. Hetao Basin, an inland-arid/semi-arid basin, is well known for the widely distribution of high arsenic groundwater and the serious local arseniasis causing by drinking high arsenic groundwater. Shallow groundwater in the basin generally occurs in the Quaternary alluvial, alluvial-pluvial, and fluvial-lacustrine aquifers, which are unconfined or leaky-confined. In this study, soil samples, surface water and groundwater samples were taken from Hetao Basin to analyse the occurrence and the distribution of arsenic in water and soil environments in this inland-arid/semi-arid basin. This study will provide a scientific basis for protect the safety of water supply in high arsenic groundwater areas.

1. Introduction

Hetao Basin (HB) is located in the west of Inner Mongolia with an area of about 13,000 km^2, and it extends north to the foot of Yin Mountains, south to the scarp in the northern edge of Ordos Loop, west to Ulanbuhe Desert, and east to the piedmont hills of Manhan and Lingel Mountains (Figure 1). It covers 500 km from east to west and 20-90 km from south to north, with an area of gently from southwest to northeast. Several polymetallic (Fe, Cu and Zn) sulfide ores are distributed along Yin Mountains. Among them, two large-scale deposits Tanyaokou and Dongshengmiao in the north of Hangjinhouqi have been mined since the 1970s. HB has a long history of using Yellow River as the water resource for irrigation purposes since the Qin Dynasty.

The climate is continental arid and semi-arid, with an average annual precipitation of 150-400 mm decreasing from east to west and evaporation of 2,000-2,600 mm. In this area, the precipitation from June to August accounts for 60-80% of annual precipitation. The Yellow River, flowing through HB with a length of approximately 350 km and an average annual runoff of 20-35 cubic meters, has been used as an water source for the irrigation purpose since Qin Dynasty at 221 BC and as an important agricultural food production base [1].

Shallow groundwater at HB generally occurs in the unconfined or leaky-confined Quaternary alluvial, alluvial-pluvial, and fluvial-lacustrine aquifers [2]. The aquifer sediments are mainly composed of alluvial arenaceour mud and sand, and lacustrine muddy clay and arenaceour mud which are rich in organic matter.

Groundwater is recharged by vertically infiltrating meteoric water in the plain, the lateral flow of fracture water from bedrock along the front of the Yin Mountains, irrigation return flow, surface water runoff and leakage from the Yellow River. Groundwater flow is generally slow in this area and is affected by geological structure, the gentle surface topography with a low hydraulic gradient, and the low hydraulic conductivity of fine-grained sediments. In general, two flow regimes can be detected in HB: (1) flow in the alluvial apron in front of the Yin Mountains from north to south and (2) flow inside the plain along the southwest-northeast direction, flowing finally to the east. Shallow groundwater is discharged mainly via evapotranspiration and artificial extraction, finally to Wuliangsuhai Lake in the east of HB.

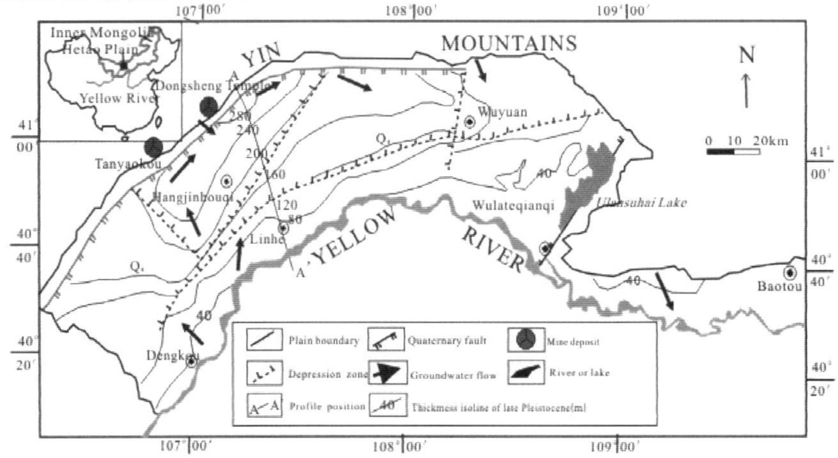

Figure 1. Location of HB, Inner Mongolia (modified from [3])

2. Materials and methods

2.1. Site description

Hangjinhouqi County, located in the western part of the HB, is selected as the study area. It has been concerned for over two decades because of the most serious endemic arsenicosis in Inner Mongolia province, as a representative for northern China. Groundwater with high levels of arsenic (As), TDS or high levels of F is the only source of potable water for about 50% of local residents in rural areas in the last several years. To date, the population of arsenicosis patients has exceeded 1100 people. At the margin of the Yin Mountains in the north of Hangjinhouqi county, there are two large-scale polymetallic (S, Pb, Zn) sulfide ore deposis, which have been mined since 1970s. Hangjinhouqi county is also a part of Hetao Irrigation Area. The vast majority of agricultural irrigation water in this study area is supplied by these four diversion channels: the main irrigation channel, the Ula Rriver, the Yangjia river, and the Huangji channel. There are also four drainage channels: the main drainage channel and drainage channels labeled simply No.1, No.2 and, No.3 drainage channels. The main drainage channel is located in the north of Hangjinhouqi County, which is the lowest place in the county. Thus, all the irrigation return flows, meteoric water, floods from the Yin Mountains front and residual water, the precipitation, the piedmont flood, the seepage, the industrial and domestic wastewater flow into the main drainage channel.

2.2. Sampling and analysis

In order to get a general knowledge of As occurrence and distribution in environment at Hangjinhouqi county, soil samples, borehole sediment samples, surface water samples and groundwater samples were collected for chemical analysis. Subsoil samples were collected from abandoned irrigation channels, abandoned drainage channels and farmlands, with different depth intervals (0-20, 20-40, 40-60, 60-80, 80-100 cm). River sediment samples were collected from irrigation channels and drainage channels. Two boreholes (TYM and TYS) were dug in low-As areas and the other two (HF and SH) were in high-As areas. And data of boreholes TYS, HF and SH were cited from Deng's publication [4].

All soil and sediment samples were taken in clean aluminum boxes and sealed by airtight strips and were stored in a refrigerator until processed within 7 days.Sediment samples and soil samples were air-dried in the dark at room temperature, crushed in an agate mortar, and the passed through a 100-gauge mesh. 0.5 g of dry samples was initially digested with 9 ml of concentrated HNO_3, 2 ml of $HClO_4$ and 3 ml of HF at 180℃ for 24 hours. Then the solution was evaporated to dryness and 50 ml of 2% (V/V) HNO_3 was added prior to analysis. The As contents were measured by hydride generation atomic fluorescence spectrometry (HG-AFS) (AFS-820; Titan Instruments, Beijing, China).

Groundwater samples were collected after wells were pumped for approximately around 10 minutes. Samples were firstly filtered (<0.45μ m) and stored in 50 ml acid-washed HDPE bottles and acidified by ultra-pure HCl for total dissolved As analysis. All samples were kept at 4℃ and total As concentrations measured by HG-AFS within 7 days. As speciation separation was conducted in the field. Water samples were firstly filtered using 0.45 μm membrane filters (Whatman) to to separate particulate As and soluble As, then passed through silica-based strong anion-exchange cartridges (Supelco) to separate arsenite and arsenate species.

3. Results and discussion

3.1. As in soils and sediments

Table 1 shows the As, TOC and TP contents in soils and sediments and pH of samples. The national soil background level of As is 11.2 μg/g, while the background level in HB is 7.5 μg/g. As contents in abandoned irrigation channels range from 7.57 μg/g to 19.86 μg/g with an average of 10.26 μg/g, in abandoned drainage channels from 6.58 μg/g to 20.12 μg/g with an average of 13.06 μg/g, in farmland from 8.83 μg/g to 22.81 μg/g with an average of 15.06 μg/g. As in most of the subsoil samples is above the background value. The highest average content of As occurs in farmland, while the lowest value occurs in abandoned irrigation channel. As contents in river sediment from drainage channel are higher than from irrigation channel. Compare with low-As area where As contents ranging between 3.04 μg/g and 26.43 μg/g with an average of 12.95 μg/g, borehole sediments in high-As area have higher As contents between 6.80 μg/g and 58.53 μg/g with an average of 16.45 μg/g.

The correlation coefficients between As contents and other elements in soils and sediments were analysed. Lithogenic metals such as Ca and Co show a high correlation with As. The strong positive correlation between As and Fe, Mn, TOC at level of 0.01, which is also observed in sediments from SH borehole [4], suggests that these components are of importance for enrichment of As in soils and sediments. The good positive correlation between As and Si indicates that As is possible bound to silica colloids which was also evidenced using Nuclear Magnetic Resonance (NMR), X-ray Absorption (XAS) and micro X-ray Fluorescence (μXRF) [5].

A total of 26 soil and sediment samples were tested with sequential chemical extractions to differentiate pools of solid phase associated As and adsorbed and coprecipitated As (Figure 2). Subsoil samples contains three groups: the profile S1, S5, and S9 were dug from abandoned Laoxi irrigation channel, from an abandoned tributary of the second drainage channel, and from a farmland in high-As area, respectively, to get subsoil samples with a depth interval of 20 cm under the ground surface. River sediment sample of R2 were from an irrigation channel, while R6, R7, and R8 were from drainage channels. Borehole sediments of TYM were from low-As area. For the all 26 samples, residual and Fe/Mn oxide-bound As are the dominated fractions, accounting for over 50% of the total As. Water-soluble As contents ranges from 0.045 mg/kg to 1.326 mg/kg, and have a tiny percentage of total As. Exchangeable As contents are generally closed to carbonate-bound As contents in most samples, with exception in S5-6, S9-1, and river sediments. High As contents generally corresponded to high Fe/Mn oxide-bound As and organic matter/sulfide-bound As, indicating the affinity of As to Fe/Mn minerals, organic matter, and sulfide. Fe/Mn oxide-bound As and organic matter/sulfide-bound As are the major active As form in these samples.

3.2. As in surface water

IOP Conf. Series: Earth and Environmental Science **146** (2018) 012052 doi:10.1088/1755-1315/146/1/012052

As concentrations in irrigation channel water range from 6.2 μg/L to 17.1 μg/L with an average of 8.5 μg/L, while in drainage channel water from 5.9 μg/L to 25.3 μg/L with an average of 15.5 μg/L. The major irrigation channel, Yangjiahe River and the middle reach of Huangjin irrigation channel have As concentrations below 10 μg/L, except for the low reach of Huangjin irrigation channel in Shanba town containing 17.1 μg/L of As.

Table 1. Summary of As, TOC and TP contents in soils and sediments and pH of samples

	As (μg/g)		pH		TOC (%)		TP (μg/g)	
	Range	Mean	Range	Mean	Range	Mean	Range	Mean
Subsoil								
Abandoned irrigation channel	7.57-19.86	10.26	8.17-8.67	8.45	0.08-0.30	0.2	335.68-641.84	453.79
Abandoned drainage channel	6.58-20.12	13.06	7.84-8.83	8.35	0.08-1.47	0.33	355.71-756.85	512.92
Farmland	8.83-22.81	15.06	8.49-8.80	8.65	0.08-0.52	0.26	251.89-658.25	455.64
River sediment								
Irrigation channel	7.58-7.69	7.64	8.70-9.18	8.94	0.04-0.10	0.07	363.24-490.70	426.97
Drainage channel	14.97-27.50	19.10	ND		ND		ND	
Borehole sediment								
Low-As area	3.04-26.43	12.95	8.41-9.73	8.89	BDL-2.00	1.25	122.77-889.26	501.97
High-As area	6.80-58.53	16.45	ND		ND		ND	

BLD: Below detection limit
ND: No detected

As concentrations in drainage channel water range from 5.9 μg/L to25.3 μg/L with an average of 15.5 μg/L. The highest concentration occurs at the third drainage channel, while the lowest concentration occurs at the major drainage channel near Hongfeng village.

One surface runoff sample collected from the Yin mountains front contains 11.2 μg/L of As. The rainwater collected in August 2011 contains 8.8 μg/L of As. As concentration in rainwater at HB is 0.28 μg/L in July 2009 [2]. Based on the results of 35 rainwater samples in Asia, As is found below 10 μg/L.

3.3. As in groundwater

In Hangjinhouqi county, dissolved As concentrations in groundwater are in the range of 0.5 μg/L to 720.7 μg/L with an average of 105.9 μg/L. 66.88% of (103) groundwater samples exceed the standard for allowable As concentration 10 μg/L in drinking water.

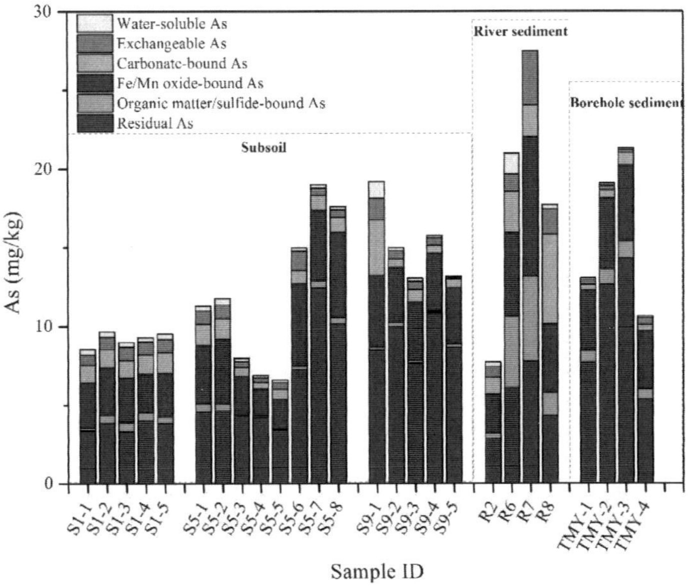

Figure 2. As forms (including water-soluble As, exchangeable As, carbonate-bound As, Fe/Mnoxide-bound As, organic matter/sulfide-bound As, and residual As) in subsoil from different depth intervals (0-20, 20-40, 40-60, 60-80, 80-100, 100-120, 120-140, and 140-160 cm), river sediment, and low-As borehole sediment from Hangjinhouqi county

All the samples were taken from wells with depths between 3 m and 80 m. The vertical profile of As clearly shows that high As concentrations were mainly observed from depth between 10 m and 40 m (Figure 3). The shallow groundwater samples from wells with depths less than 10 m had low As concentrations (less than 10 μg/L or closed to 10 μg/L). Those groundwater samples from wells with depths more than 40 m had As concentration less than 50 μg/L, except for one sample collected from a 80-meter-deep well with a As concentration of 140.2 μg/L. The highest As concentration in shallow groundwater is 720.7 μg/L with a well depth of 27 m.

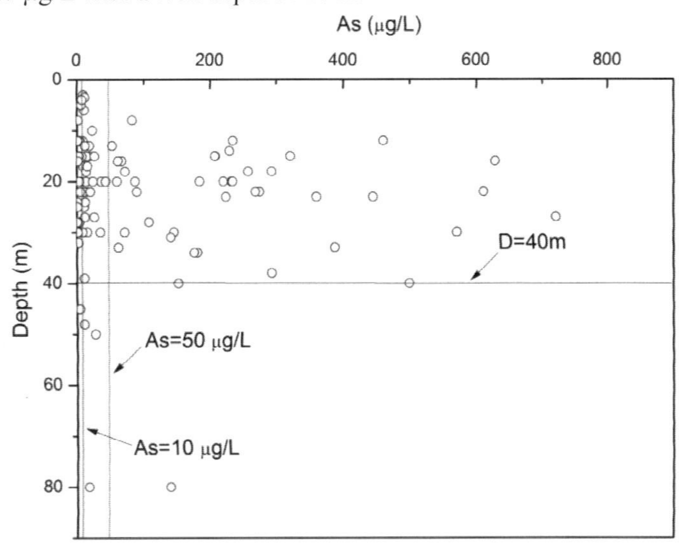

Figure 3. Depth distribution of As concentration in groundwater from groundwater at Hangjinhouqi county

As concentrations in groundwater varied in different districts of the study area. In the southern part of Hangjinhouqi county where groundwater mainly receives recharge from the Yellow River and has good runoff conditions, groundwater had low As concentrations <10 μg/L, or between 10 μg/L and 50 μg/L. In the center district, groundwater generally had As concentrations below 100 μg/L. High-As groundwater is distributed in the north of Hanjinghouqi county along the Yin Mountains where the groundwater flow is sluggish. This high-As district, is also a mosaic transition zone between The polymetallic sulfide mining area and the irrigation area. Groundwater in this district receives the recharge from the Yin Mountains front, as well as the seasonal recharge from the irrigation water and irrigation return in the main drainage channel.

4. Conclusion

As in most of the subsoil samples is above the background value. The highest average content of As occurs in farmland, while the lowest value occurs in abandoned irrigation channel. Compare with low-As area where As contents ranging between 3.04 μg/g and 26.43 μg/g with an average of 12.95 μg/g, borehole sediments in high-As area have higher As contents between 6.80 μg/g and 58.53 μg/g with an average of 16.45 μg/g. As concentrations in irrigation channel water range from 6.2 μg/L to 17.1 μg/L with an average of 8.5 μg/L, while in drainage channel water from 5.9 μg/L to 25.3 μg/L with an average of 15.5 μg/L. The major irrigation channel, Yangjiahe River and the middle reach of Huangjin irrigation channel have As concentrations below 10 μg/L, except for the low reach of Huangjin irrigation channel in Shanba town containing 17.1 μg/L of As. In Hangjinhouqi county, dissolved As concentrations in groundwater are in the range of 0.5 μg/L to 720.7 μg/L with an average of 105.9 μg/L. 66.88% of (103) groundwater samples exceed the standard for allowable As concentration 10 μg/L in drinking water. The vertical profile of As clearly shows that high As concentrations were mainly observed from depth between 10 m and 40 m. The shallow groundwater samples from wells with depths less than 10 m had low As concentrations (less than 10 μg/L or closed to 10 μg/L). Those groundwater samples from wells with depths more than 40 m had As concentration less than 50 μg/L, except for one sample collected from a 80-meter-deep well with a As concentration of 140.2 μg/L. The highest As concentration in shallow groundwater is 720.7 μg/L with a well depth of 27 m.

Acknowledgments

Authors wish to acknowledge the financial support by International Atomic Energy Agency Coordinated Research Project （IAEA No. 21122）, Open Fund of Jiangxi Key Laboratory for Mass Spectrometry and Instrumentation, East China University of Technology (No. JXMS2017016) and Water Resources and Water Environment Research Center Fund of Water Resources Ministry in China (No. ZXKT201709).

References

[1] T, L., H. S, C. JL, T. HX and J. CY 2012. Comparison of arsenic geochemical evolution in the Datong Basin (Shanxi) and Hetao Basin (Inner Mongolia), China. *Applied Geochemistry* **27** 2315-2323

[2] Guo, H., B. Zhang and Y. Zhang 2011. Control of organic and iron colloids on arsenic partition and transport in high arsenic groundwaters in the Hetao basin, Inner Mongolia. *Applied Geochemistry* 26 360-370

[3] Deng, Y., Y. Wang and T. Ma 2009. Isotope and minor element geochemistry of high arsenic groundwater from Hangjinhouqi, the Hetao Plain, Inner Mongolia. *Applied Geochemistry* **24** 587-599

[4] YM, D., W. YX, M. T, Y. H and H. J 2011. Arsenic associations in sediments from shallow aquifers of northwestern Hetao Basin, Inner Mongolia. *Environmental Earth Sciences* **64** 2001-2011

[5] Masion, A., J. Rose, F. Ziarelli and J. Y. Bottero 2006. NMR evidence of arsenic binding to silica colloids. *Goldschmidt conference abstracts* 398.

IOP Conf. Series: Earth and Environmental Science **146** (2018) 012053 doi:10.1088/1755-1315/146/1/012053

Analysis of Carbon Reduction Strategy on Distributed Combined Cooling, Heating and Power (CCHP) Systems Design in Rural Area

Y Zhu [1], B Shao [1], Q L Tong [1], Y F Liu [1], Y X Li [2], X X Yan [1], N Wu[1] and T Du[3]

[1] State Key Laboratory of Green Building in Western China, School of Environment and Municipal Engineering, Xi'an University of Architecture and Technology, Xi'an 710055, China
[2] Ankang Environmental Engineering Design Limited Company, Ankang 725000, China
[3] Environmental monitoring center of Shaanxi province, Xi'an 710055, China

*E-mail: xinyingcg@163.com

Abstract. An optimal design model of distributed CCHP systems is established based on the minimum system cost. Renewable energy is added as auxiliary energy. By proposing carbon emission reduction constraints, the influence for optimal design of the system could be analysed, which generated by the changes in carbon emission reduction rate. Some reasonable results would be achieved, such as: 1. A trade-off among system cost, building energy consumption and carbon emission reduction can be received, and optimization schemes under different carbon emission reduction strategies are obtained. 2. System lowest carbon reduction rate is 20%, and the highest is 50%; systems lowest primary energy saving rate is 4.91%, and the highest is 19.09%. 3. With the change of carbon emission reduction strategy, energy consumption and costs of the distributed CCHP systems show diversity differences at the same time. This article suggests the carbon emission reduction rate of CCHP systems is 40% in Ankang rural.

1. Introduction

In 2015, the Paris climate change conference adopted the Paris agreement to address global climate change after 2020[1]. Due to the large consumption of Chinese building energy, it accounts for about 1/3 of the total energy consumption of the society[2,3]. Zeng et al. proposed that increase the proportion of renewable energy in building energy consumption can effectively reduce carbon emission[4]. Distributed combined cooling, heating, and power (CCHP) system is a form of distributed energy system, which can simultaneously use primary energy and a variety of renewable energy sources to supply cold energy, heat energy, and electric power[5,6]. Therefore, how to balance the relationship between system cost, building energy consumption and carbon emission reduction is particularly important. Many scholars have studied the environmental benefits of distributed CCHP[7-11]. As proportion of renewable energy consumption continues to decline because of the lack of technical support[12,13,14]. Therefore, this paper will build the optimization model of distributed CCHP systems for rural residential buildings, taking Qiyan community, Dazhuyuan town, Hanbin district, Ankang city, Shaanxi Province, China as an example to verify the practicability of the model.

2. Statement of the Problems

Ankang municipality in the new rural community construction is fully rolled out [15,16]. This paper takes Community Qiyan, Dazhuyuan, Hanbin as the case study area, which is a rural community reconstructed after the disaster and plans to resettle 567 households[17](Figure 1). The authors used EnergyPlus8.6 to generate typical daily load(Figure 2). In this study, the peak power of PV panels is 0.17 kW and each PV panel covers an area of 1.3 m^2. The area available for mounting PV panels is 3900 m^2. The maximum installed PV panels capacity is 510 kW. Solar radiation is based on relevant meteorological data[18]. Total biomass available for 518,400 tons[19]. The price of natural gas is 2.5 yuan/m^3 in Ankang, and heat value is 10.6 kWh/m^3. Efficiency of internal combustion engine generator is 0.35, the fuel price per unit of electricity generated is 0.671 yuan/kWh. The carbon emission factor of natural gas under the standard conditions is 2.1622 kg CO_2/m^3, carbon emission per unit of electricity generated by this device is 0.583 kgCO_2/kWh. The residential electricity price is 0.4983 yuan/kWh in Shaanxi Province, the carbon emission factor of power grid is 0.89 kgCO_2/kWh[20]. Based on early distributed CCHP systems[20,21], this study developed a variety of renewable energy and energy storage technologies for the system, where L, R and D corresponding means cold, heat and electricity three forms of energy [22, 23].

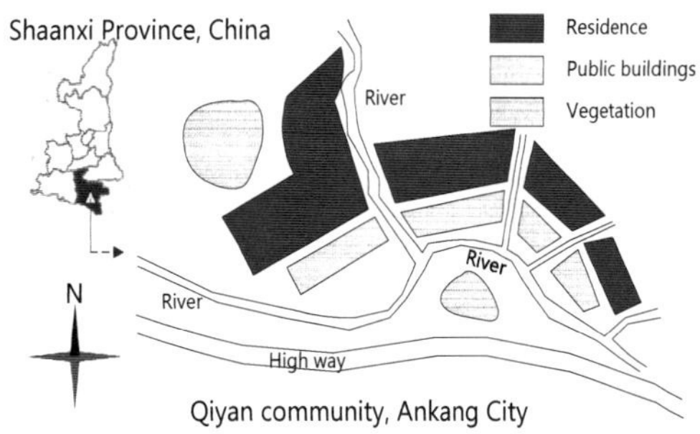

Figure 1. Community map.

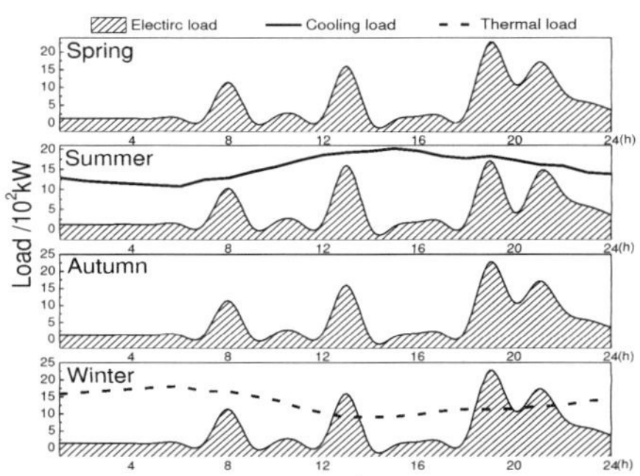

Figure 2. Typical daily load.

3. Modelling formulation

3.1 Objective function

The model is the target function of the total cost of CCHP system. This daily economic cost, given in Equation (1) involves the following terms: installation cost (C_{INV}), operational and maintenance cost (C_{OM}), purchasing electricity cost (C_{ELEC}), fuel cost (C_{FUEL}), and government subsidies (C_{SUB}).

$$\min\left(C_{\mathrm{INV}}+C_{\mathrm{OM}}+C_{\mathrm{ELEC}}+C_{\mathrm{FUEL}}-C_{\mathrm{SUB}}\right) \tag{1}$$

Equipment system can be divided into 4 categories: a. use of primary energy production equipment; b. first-level energy conversion equipment; c. secondary energy conversion equipment; and d. storage of energy storage devices.

$$C_{\mathrm{INV}}=4/365*\left(\begin{array}{l}\sum_k\sum_n CAG_{k,n}*UG_k*COG_{k,n}*IR\Big/\left(1-1/(1+IR)^{LG_k}\right)+\sum_n CAH_n*UH*COH_n*IR\Big/\left(1-1/(1+IR)^{LH}\right)\\+\sum_n CAC_n*UC_n*COC_n*IR\Big/\left(1-1/(1+IR)^{LC_n}\right)+\sum_n CAS_n*US_n*COS_n*IR\Big/\left(1-1/(1+IR)^{LS_n}\right)\end{array}\right) \tag{2}$$

Where $CAG_{k,n}$, CAH_n, CAC_n and CAS_n respectively is design capacity of class a, b, c and d equipment, kW; $UG_{k,n}$, UH_n, UC_n and US_n respectively is number of class a, b, c and d equipment; $COG_{k,n}$, COH_n, COC_n and COS_n respectively is capacity cost of class a, b, c and d equipment, yuan/kW; IR is the depreciation rate, specified at 10% in this study; LG_k, LH, LC_n and LS_n respectively is lifetime of class a, b, c and d equipment, year.

$$C_{\text{OM}} = \sum_s \sum_h \sum_k \sum_n OMG * G_{s,h,k,n} + \sum_s \sum_h \sum_n OMC_n * CG_{s,h,n} + \sum_s \sum_h \sum_n OMH * HG_{s,h,n} + 4/365 \times \sum_n OMS_n * CAS_n \tag{3}$$

Where $OMG_{k,n}$, OMH_n, OMC_n and OMS_u respectively is the running cost of class a, b, c and d equipment, yuan/kWh; $G_{s,h,k,n}$, $HG_{s,h,n}$ and $CG_{s,h,n}$ respectively is energy production of class a, b and c equipment, kWh; s stand for season; and h represents hour.

$$C_{\text{ELEC}} = \sum_s \sum_h PE_h * EP_{s,h} \tag{4}$$

Where PE_h is hourly electricity price, yuan/kWh; $EP_{s,h}$ is electricity form the grid, kWh.

$$C_{\text{FUEL}} = \sum_s \sum_h \sum_k \sum_n PF_{k,n} * G_{s,h,k,n} / \eta_{k,n} \tag{5}$$

Where $PF_{k,n}$ is price of fuel, yuan/kWh; $\eta_{k,n}$ is capacity efficiency of production equipment.

$$C_{\text{SUB}} = 4/365 * \sum_k \sum_n SUB_{k,n} * CAG_{k,n} / LG_k + \sum_s \sum_h \sum_k \sum_n SUBV_{k,n} * G_{s,h,k,n} \tag{6}$$

Where $SUB_{k,n}$ is one-time investment subsidy for equipment, yuan/kW; $SUBV_{k,n}$ show device running subsidy, yuan/kWh.

3.2 Constrains

The energy produced during the operation of the device cannot exceed its capacity, given in Equation (7), (8), (9).

$$G_{s,h,k,n} \le UG_k * CAG_{k,n} \tag{7}$$

$$CG_{s,h,n} \le UC_k * CAC_{k,n} \tag{8}$$

$$HG_{s,h,n} \le UH_k * CAH_{k,n} \tag{9}$$

Electricity generated by the internal combustion engine is fixed with the residual heat generated, given in Equation (10). Where ζ is ratio of power generation and surplus heat of gas combustion engine; ζ is specified at 0.7 in this study.

$$\zeta * G_{s,h,k=\text{NG},n=\text{R}} = G_{s,h,k=\text{NG},n=\text{D}} \tag{10}$$

Solar PV power generation is limited to equipment capacity and solar radiation intensity, given in Equation (11). Where RS is the surface area of the PV panel, m^2; $SI_{s,h}$ is direct solar radiation, kW/m^2.

$$G_{s,h,k=\text{PV},n=\text{D}} \le UG_{k=\text{PV}} * RS * SI_{s,h} * \eta_{k=\text{PV},n} \tag{11}$$

The secondary energy conversion equipment needs to convert the heat energy into the cold energy during the summer, which is converted into heat by the heat exchanger during the spring and autumn period. The constrains of energy conversion is given in Equation (12), (13). Where θ_n is the efficiency of energy conversion equipment.

$$\theta_{n=\text{L}} * \sum_k G_{s=\text{sum},h,k,n=\text{R}} = CG_{s=\text{sum},h,n=\text{L}} \tag{12}$$

$$\theta_n * \sum_k G_{s=\text{spr/aut/win},h,k,n} = CG_{s=\text{spr/aut/win},h,n} \tag{13}$$

The energy storage device at h=1, the storage energy is 0, there may be energy input ,energy output is 0, given in Equation (14), (15), (16). Where $EST_{s,h,n}$ is energy in energy storage devices, kWh; $IST_{s,h,n}$ is energy to enter the energy storage devices, kWh; $OST_{s,h,n}$ is energy flowing out of the energy storage devices, kWh.

$$EST_{s,h=1,n} = 0 \tag{14}$$

$$IST_{s,h=1,n} \ge 0 \tag{15}$$

$$OST_{s,h=1,n} = 0 \tag{16}$$

When the energy storage device is at h > 1, its operation can be represented by the Equation (17), (18), (19). Where β_n is the efficiency of energy storage.

$$EST_{s,h,n} = EST_{s,h-1,n} + \beta_n * IST_{s,h-1,n} - OST_{s,h-1,n} \tag{17}$$

$$0 \leq EST_{s,h,n} + \beta_n * IST_{s,h,n} - OST_{s,h,n} \leq CAS_n \tag{18}$$

$$0 \leq EST_{s,h,n} \leq CAS_n \tag{19}$$

The system energy supply of each moment must meet the needs of the user load, the consumption of electricity the absorption chiller and the ground source heat pump (20), (21). Where $ED_{s,h,n}$ is user loads, kWh; σ is the power consumption coefficient of absorption chiller; $\varepsilon_{s,n}$ is the power consumption coefficient of ground source heat pump.

$$CG_{s,h,n=\mathrm{D}} + EP_{s,h} + OST_{s,h,n=\mathrm{D}} \geq IST_{s,h,n=\mathrm{D}} + ED_{s,h,n=\mathrm{D}} + \sigma * CG_{s,h,n=\mathrm{L}} + \sum_n HG_{s,h,n} * \varepsilon_{s,n} \tag{20}$$

$$CG_{s,h,n=R/L} + HG_{s,h,n=R/L} + OST_{s,h,n=R/L} \geq IST_{s,h,n=R/L} + ED_{s,h,n=R/L} \tag{21}$$

The carbon emission reduction rate of the system is given in Equation (22). Where CEG is electricity carbon emissions, $kgCO_2/kWh$; $CEF_{k,n}$ is the carbon emissions of system equipment, $kgCO_2/kWh$; $CENG$ is natural gas carbon emissions, $kgCO_2/kWh$; ϕ is the efficiency of electric chiller, specified at 5[24]; RT is the carbon emission reduction rate, set by the decision-maker.

$$\begin{aligned} &\sum_s \sum_h EP_{s,h} * CEG + \sum_s \sum_h \sum_k \sum_n G_{s,h,k,n} / \eta_{k,n} * CEF_{k,n} \\ &\leq (1-RT)\left(\sum_s \sum_h ED_{s,h,n=\mathrm{D}} * CEG + \sum_s \sum_h ED_{s,h,n=\mathrm{L}} / \phi * CEG + \sum_s \sum_h ED_{s,h,n=\mathrm{R}} / \psi \times CENG\right) \end{aligned} \tag{22}$$

4. Results and Discussions

Based on the above-mentioned, this paper sets six scenarios as follow: the carbon emission reduction rate of 0, 10%, 20%, 30%, 40%, 50%. Comparing the Scenarios, weighing the relationship between cost of distributed CCHP systems, residential building energy consumption and carbon emission reduction, obtaining the optimal configuration of the distributed CCHP systems under different carbon reduction rate. The following will analyze the change rule of optimal configuration, energy consumption, energy saving rate of primary energy and cost under different carbon emission reduction rate.

4.1 Change rule of optimal configuration

Optimal configuration under different carbon emission reduction rate is shown in Table 1. According to the table, when set of carbon emission reduction rate lower than 20%, the system configuration does not change, that is, the value of the minimum system carbon reduction rate by calculation, set of carbon emission reduction rate exceeds 50%, the model has no solution, namely the value for the system to achieve the highest rate of carbon emissions. Because of the renewable energy equipment does not produce carbon emissions, with the increase of carbon reduction rate, the demand will increase, but the utilization of biomass and solar power is limited, so the biomass boiler and the equipment capacity of solar photovoltaic panels have no change. On the whole, the capacity of the ground heat pump is on the rise. There are unpredictable changes in the capacity of all other technologies, indicating that the optimal configuration of the system is extremely complex.

System cost under different carbon emission reduction rate described in Figure 3. With the increase of the consumption of natural gas, the amount of waste heat generated by the internal combustion engine is gradually increasing, so the consumption of biomass energy is gradually reduced. PV panels power generation has neither fuel cost nor carbon emission, resulting in the device running at full load, so the utilization of solar resources is unchanged under all scenarios. The utilization of geothermal energy is only related to the capacity of ground source heat pump. Combined with Table 2, consumption of geothermal energy varies with the change of the capacity of ground source heat pump. The reference object of energy saving rate of primary energy is the traditional energy system. Lowest energy saving rate of primary energy of the system is 4.91%, and the highest energy saving rate of primary energy of the system is 19.09%. Cost under different carbon emission reduction rate is shown in Figure 4. Because the optimal configuration does not change under the carbon emission reduction rate of 20%, the cost is invariable. In this system, only PV panels are subsidized, and its capacity and operation mode do not vary with the change of carbon emission rate, so the subsidy cost will not change.

Table 1. Optimal configuration under different carbon emission reduction rate.

Technologies	Carbon emission reduction rate, Capacity /kW					
	0	10%	20%	30%	40%	50%
NG ICG	781	781	781	766	598	1043
PV	510	510	510	510	510	510
BB	450	450	450	450	450	450
GP	154/171	154/171	154/171	173/192	385/427	351/390
AC	1713	1713	1713	1658	1346	1589
HE	1535	1535	1535	1514	1279	1901
IS	382	382	382	553	2177	198
HS	725	725	725	725	340	901
BA	404	404	404	456	873	2170

5. Conclusions

Based on the early distributed CCHP systems, conclusions generated as follow: 1. Lowest carbon emission reduction rate was 20% and highest was 50%. The lowest energy saving rate of primary energy was 4.91% and the highest was 19.09%. 2. As carbon emission reduction rate increased from 20% to 50%, natural gas consumption increased 47,441 kWh, grid power purchase reduced 23838 kWh, solar energy consumption remained unchanged, biomass energy consumption decreased by 3322 kWh, and energy consumption of geothermal showed an irregular upward trend. 3. As carbon emission reduction rate increased from 20% to 50%, total cost increased 13,200 yuan, of which C_{INV}, C_{OM}, C_{FUEL} increased, C_{ELEC} reduced, C_{SUB} remain unchanged.

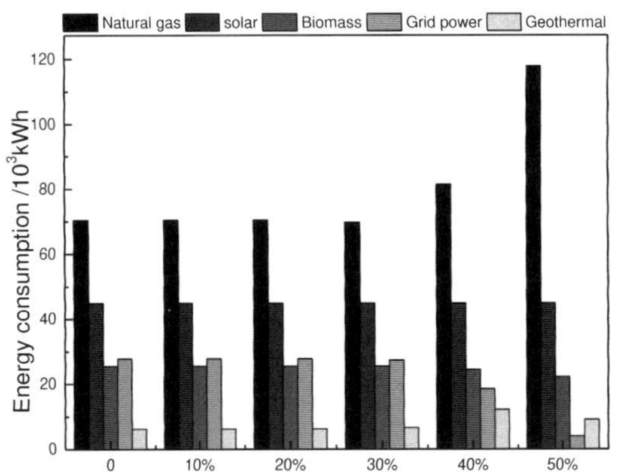

Figure 3. System energy consumption under different carbon emission reduction rate.

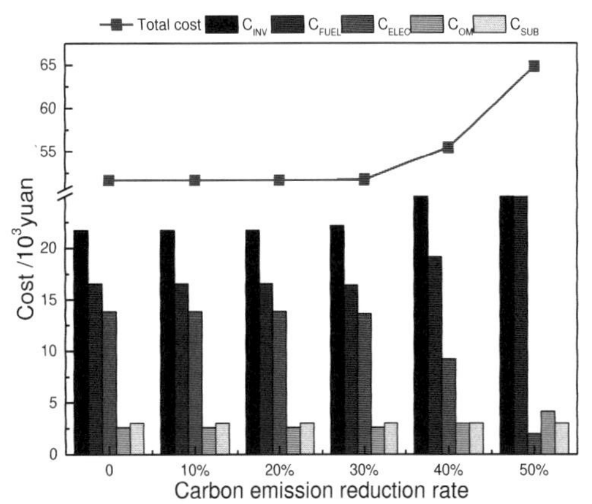

Figure 4. Costs under different carbon emission reduction rate.

Acknowledgments

This research was supported by the National Key Research Projects (2016YFC0207800), National

IOP Conf. Series: Earth and Environmental Science **146** (2018) 012053 doi:10.1088/1755-1315/146/1/012053

Natural Science Foundation of China (51608422), and Undergraduate Training Program for Innovation and Entrepreneurship of Shaanxi Province (1244).

References

[1] Gao X and Teng F 2016 *Energy of China*. **(2)** 29-32

[2] Jiang X S 2014 Modelling and operation optimization of a multi-energy complementary distributed heating and cooling system. *South China University of Technology.*

[3] Zhu Y 2014 Integrated full-infinite programming methods for energy systems management. *North China Electric Power University.*

[4] Zeng M, Zhu X L, Xue S, et al. 2012 Interactive Influence between Carbon Emission Reduction Policies and Renewable Energy Promotion Polices. *East China Electric Power*, **(7)** 1130-33

[5] Zhou R J, Ran X H, Mao F L, et al. 2012 Energy-Saving Coordinated Optimal Dispatch of Distributed Combined Cool, Heat and Power Supply, *Power System Technology*, **(6)** 8-14.

[6] Zhang C J 2016 Modeling, Optimization and Control of Distributed Energy System in the CCHP System, *Shandong Jianzhu University*.

[7] Bracco S, Dentici G and Siri S 2013 *Energy*. **55(55)** 1014-24

[8] Capuder T and Mancarella P 2014 *Energy*. **71(7)** 516-33

[9] Descateaux P, Astudillo M F and Amor M B 2016 Assessing the life cycle environmental benefits of renewable distributed generation in a context of carbon taxes: The case of the Northeastern American market, *Renewable & Sustainable Energy Reviews*. **53** 1178-89

[10] Li L, Mu H, Li N, et al 2016 *Energy.* **109** 947-60

[11] Yousefi H, Ghodusinejad M H and Noorollahi Y 2017 *Energy & Buildings*. **138** 309-17

[12] Tian Y S 2014 Status and trend of rural energy development of China in 2013, *Energy of China* **36(8)** 10-14

[13] Chen Y and Zhu Y L 2011 Evaluation of Carbon Emissions From Renewable Energy Consumption by Rural Residents in China, *China Population Resources and Environment.* **133(9)** 88-92

[14] Qiu H G, Yan J B, Jiang Y, et al. 2015 Renewable Energy Consumption in Rural China:Current Situation and Major Driven Factors, *Journal of Beijing Institute of Technology (Social Sciences Edition).* **17(3)** 10-15

[15] Ding Y Z 2012 Strategies for Environmental Protection in Building the New Type of Countryside, *Modern Economics*. **(9)** 110-12

[16] Chen J 2011 Energy-saving Design Research of Urban Residential Building in Southern Shanxi, *Xi`an University of Architecture and Technology*.

[17] Du N N 2014 Reconstruction of mountain type new rural community planning practice in Qiyan Community, Dazhuyuan Town, Hanbin District, Ankang ,*Development of Small Cities & Towns*.

[18] Zhang Q Y, Yang H X, et al. 2012 *Typical Meteorological Database Handbook for Buildings*. (Beijing: China Architecture & Building Press) pp 290-300

[19] The Statistics Bureau of Shaanxi Province 2015 *Shaanxi Statistical Yearbook 2015*. (Beijing: *China Statistics Press*)

[20] Zhang C X 2012 Energy saving analysis of CCHP system, *North China Electric Power University*.

[21] Zhu C B 2016 Optimization Research on Combined Cooling, Heating, and Power System, *North China Electric Power University.*

[22] Zhou Z, Liu P, Li Z, et al. 2013 *Applied Thermal Engineering*. **53(2)** 387-96

[23] Zhang T, Zhu T, Gao N P, et al. 2015 Optimization Design and Multi-criteria Comprehensive Evaluation Method of Combined Cooling Heating and Power System, *Proceedings of the CSEE*. **35(14)** 3706-13

[24] Xiong Y 2014 Efficiency Calculation and Optimization Analysis of Distributed Energy System, *Wuhan University of Science and Technology*.

IOP Conf. Series: Earth and Environmental Science **146** (2018) 012054 doi:10.1088/1755-1315/146/1/012054

A Rapid Active Power Regulating Technique for A Virtual Small-hydro Cluster

Fang Jialiang[1], Zha Weiqiang1, Li Zhi[1], Zhang Xinghai[1], Li Shichun[2]*, Tang Hongyan[2] and Lei Xiaolin[2]

[1] State Grid Ningbo Power Supply Company Ningbo, Zhejiang Province, China
[2] College of Electrical Engineering and Renewable Energy, China Three Gorges University,Yichang, Hubei Province, China

Li Shichun's e-mail:lschunu_023@126.com

Abstract. A rapid active power regulating technique for a virtual small-hydro cluster is studied in this paper. The virtual small-hydro cluster consists of several decentralized small-hydro power stations, which quickly responds to the active power target instructions of real-time dispatch and AGC. Furthermore, the power fluctuation of the delivery cross-section caused by intermittent power source in short-term scale can be efficiently smoothed. When any tie line transmission power reaches its thermal stability limit, the active power target instructions is allocated according to the sensitivity priority of small-hydro power plants to threshold-crossing lines among the small-hydro cluster. When all the tie lines transmission powers are within their thermal stability limits, the active power target instruction is allocated on the basis of the priority of the summation of product of tie line power variations and sensitivity of small-hydro power stations to tie lines. Under the constraint of fulfilling the active power target instructions, the simulation results show that the adopted control strategy and allocation algorithm can effectively decrease the power fluctuation of each tie line.

1. Introduction

Large-scale intermittent power generation integration increases the active power fluctuation and frequency excursion of power grid, however, we can still employ the adjustment ability of distributed generators to smooth the power fluctuation caused by themselves. In distribution networks, the typical distributed generators involve wind turbine, photovoltaic(PV) power generation and small-hydro power station basically, etc.

At present, the synthetic inertial control technique and primary frequency auxiliary control technique with respect to wind turbine have been raised already, but they are not popularized and utilized in the power grid yet, the MPPT strategy is adopted in most wind farms in the domestic and overseas[1-4]. The coordination control between PV and energy storage device to achieve active power smoothing control normally, however, the investment cost will increase visibly for the large-scale application of this technology[5-7]. In some areas of China, distribution network incorporates small hydro-power with high penetrations, and also the active power fluctuation can be restrained applying small hydro-power stations. As a matter of fact, the small hydro-power station contains run-off-station and non-run-off-station. For run-off-station, the output is stochastic, the storage capacity is deficient normally and the active power control ability is limited. Therefore, the active power adjustment in the region where contains a large number of run-off-stations is difficult[8, 9]. While the non-run-off-station possesses a certain storage capacity and active power adjustment ability on hours level time scale, thus as mall-hydro cluster that consists of multiple non-run-off-stations will definitely provide with assistance on active power adjustment of power grid.

Therefore, a virtual small-hydro cluster consists of several decentralized non-run-off-stations in this study, which quickly responds to the active power target instructions of real-time dispatch and AGC. Because almost all the non-run-off-stations are not utilized to participate in power grid dispatching, the raised technique is highly valued in popularization.

2. The control strategy and control algorithm of virtual small-hydro cluster

2.1. The control strategy of virtual small-hydro cluster

The proposed control strategy is illustrated in Figure 1. A virtual small-hydro cluster consists of several small-hydro stations possessing a little adjustment ability responds to regional real-time dispatch and Automatic Generation Control (AGC) collectively. The real-time dispatch delivers basic point active power through regional dispatch center and AGC delivers active power regulation instruction to the small-hydro cluster. The virtual small-hydro cluster control center calculates the total active power adjustment target value combining the actual operating status of each small-hydro station and allocates it among the small-hydro stations according to the proposed controlalgorithm subsequently. Meanwhile, the regulation of small-hydro cluster should be effective in suppressing the active power fluctuation of cross-section tie lines resulting from intermittent power.

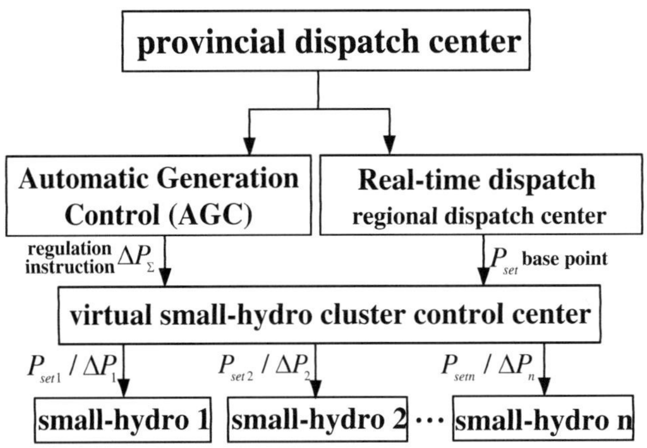

Figure 1. The control strategy of virtual small-hydro cluster.

2.2. The mathematical model of small-hydro station

The output power of hydropower unit is given by[10]:

$$P_G = 9.81\eta_h\eta_G(Q - Q_0)H \tag{1}$$

where η_h is the efficiencycoefficient of hydraulicturbine,η_G is theefficiencycoefficient of generator unit, Q is the flow of hydraulicturbine, Q_0 is the no-loadflow of hydraulic turbine, Histhe operating head of hydraulic turbine. Q is given by[10]:

$$Q = k_u G\sqrt{H} \tag{2}$$

where $k_u$isthe proportionalitycoefficient, G is the ideal opening.

According to (1) and (2), the rated output power P_{GN}, the rated flow Q_N and the no load flow Q_0 are described as follows:

$$P_{GN} = 9.81\eta_h\eta_G(Q_N - Q_0)H_N \tag{3}$$

$$Q_N = k_u G_N\sqrt{H_N} \tag{4}$$

$$Q_0 = k_u G_0 \sqrt{H_0} \tag{5}$$

where H_N is the rated head, H_0 is the no load head. The mathematical equations in pu unit corresponding to (1) through dividing (1) by P_{GN} and dividing (2), (5) by Q_N is given by:

$$P_G^* = \left(\frac{G\sqrt{H^*} - Q_0^*}{1 - Q_0^*} \right) H^* \tag{6}$$

According to (6), the total active power in real-time of a small-hydro station i containing m units is given by:

$$P_{hi} = \sum_{j=1}^{m} \left[\left(\frac{\sqrt{H_{ji}^*} - Q_0^*}{1 - Q_0^*} \right) H_{ji}^* P_{GNji} \right] \tag{7}$$

where Q_0^* equals 0.05.

2.3. *The active power allocation algorithm of virtual small-hydro cluster*

2.3.1. The calculation of regulation limits. The active power regulation upperlimits andlower limits of the small-hydro stations keep changed alone with operating status. Consequently, virtual small-hydro control center should upload the upper limits and lower limits to dispatching center for real-time, which contributes to assign instructions reasonably for the regional dispatch center and AGC.

Universally,ΔP_i^+ andΔP_i^-denotes the active power regulation upperlimit andlowerlimit of the small-hydro station i respectively. For the unit$j$in the small-hydro station$i$, the active power regulation upperlimit andlower limit should be determined combining the operating headH_{ji}. When $H_{ji}>H_N$, the active power regulation upperlimit of unit$j$ is rated valueP_{GNji}. Hence, ΔP_i^+ is given by:

$$\Delta P_i^+ = mP_{GNji} - P_{hi} \tag{8}$$

When $H_{ji}<H_N$, the active power regulation upperlimit of unit is the output power that the guide vane opening attains the maximum value. Hence, ΔP_i^+ is given by:

$$\Delta P_i^+ = \sum_{j=1}^{m} \left[\left(\frac{\sqrt{H_{ji}^*} - Q_0^*}{1 - Q_0^*} \right) H_{ji}^* P_{GNji} \right] - P_{hi} \tag{9}$$

In any case, ΔP_i^- is given by:

$$\Delta P_i^- = -P_{hi} \tag{10}$$

According to (8) and (9), the total active power regulation upperlimit of the virtual small-hydro cluster containing n units is given by:

$$\Delta P_\Sigma^+ = \sum_{i=1}^{n} \Delta P_i^+ \tag{11}$$

According to (10), the total active power regulation lowerlimit of the virtual small-hydro cluster is given by:

$$\Delta P_\Sigma^- = \sum_{i=1}^{n} \Delta P_i^- \tag{12}$$

$\Delta P_i^+$and $\Delta P_i^-$are uploaded every 5s in the practicalapplication.

2.3.2. The calculation of total active power adjustment instruction. The virtual small-hydro cluster responds to the dispatchcontrol in accordance with 2.1. Regional real-time dispatch sends base active power instruction and AGC system sends regulating instruction to the small-hydro cluster. Thus, the total active power adjustment target value ΔP_Σ must be calculated firstly and then can be allocated among small-hydro stations. The total active power target value of small-hydro cluster$P_\Sigma$can be obtained easily whenP_{set}andΔP_{AGC} are received from dispatch control center:

$$P_\Sigma = \Delta P_{AGC} + P_{set} \tag{13}$$

The active power(P_{hi}) of each small-hydro station in real time is sampled, thenΔP_Σ is given by:

$$\Delta P_\Sigma = P_\Sigma - \sum_{i=1}^{n} P_{hi} \tag{14}$$

2.3.3. The allocation algorithm aiming at inhibiting active power fluctuation of cross-section

2.3.3.1. Calculation of the active power allocation priority p. The active power should be allocated among the small-hydro cluster when$\Delta P_\Sigma \neq 0$.

Sampling the real-timetransmission active power$P_k(n)$ and the last value $P_k(n\text{-}1)$ for the cross-section tie line k, one obtains the power fluctuation $\Delta P_k = P_k(n)\text{-}P_k(n\text{-}1)$. Then the ratio of$\Delta P_k$ to the total powerfluctuation of all the cross-section tie line can be obtained:

$$\Delta P_k\% = \left|\Delta P_k\right| / \sum_{k=1}^{n} \left|\Delta P_k\right| \tag{15}$$

Based on that, one can calculates the index:

$$\zeta = \sum_{i=1}^{n} S_{ki} \Delta P_k\% \tag{16}$$

where S_{ki} is the sensitivity of the output active power of the small-hydro station i relative to the transmission active power of the cross-section tie line k. The priority p for active power allocation among small-hydro stations is determined by ζ, and the p is lower for the greater ζ.

2.3.3.2.Upward regulations of small-hydro stations. In general, for the small-hydro station i withpriorityp, if $\Delta P_\Sigma > 0$ and $H_{ji} > H_N$,ΔP_i^+ is given by (8), and if $H_{ji} < H_N$,ΔP_i^+ is given by (9). Considering the primary frequency of the small-hydro station, the actual adjustable active power upper limit for the small-hydro stationiis given by:

$$\Delta P_{i\max} = \Delta P_i^+ - \Delta f_i / R_i \tag{17}$$

where R_i is the adjustmentcoefficient of small hydropower unit, Δf_i is the frequency deviation forthe grid-connected point of the small-hydro station$i$.

If$\Delta P_{i\max} < \Delta P_\Sigma$, the active power allocation instruction for the small-hydro station$i$ is$\Delta P_{hi} = \Delta P_{i\max}$, the remaining total active power regulation instruction is updated to $\Delta P_\Sigma\text{-}\Delta P_{hi}$ and it will be allocated sequentiallyby the small-hydro station with priority $p=i+1$, the distribution process remains unchanged among subsequent small-hydro stations until the ΔP_Σ is allocated altogether or $p=n$. Instead, if $\Delta P_{i\max} > \Delta P_\Sigma$, the active power allocation instruction for the small-hydro station$i$ is$\Delta P_{hi} = \Delta P_\Sigma$, the total active power regulation instruction is allocated by the small-hydro station $i$with the highest priority completely.

2.3.3.3.Downward regulations of small-hydro stations. For the small-hydro station i withpriority p, if$\Delta P_\Sigma < 0$, equally, considering the primary frequency of the small-hydro station, the actual adjustable active power lower limit for the small-hydro stationiis given by:

$$\Delta P_{i\min} = P_{hi} - \Delta f_i / R_i \tag{18}$$

If$\Delta P_{i\min} < \text{-}\Delta P_\Sigma$, the active power allocation instruction for the small-hydro station$i$ is$\Delta P_{hi} = \text{-}\Delta P_{i\min}$, the remaining total active power regulation instruction is updated to $\Delta P_\Sigma\text{-}\Delta P_{hi}$ and it will be allocated sequentiallyby the small-hydro station with priority $p=i+1$, the distribution process remains unchanged among subsequent small-hydro stations until the ΔP_Σ is allocated altogether or $p=n$. If$\Delta P_{i\min} > \text{-}\Delta P_\Sigma$, the active power allocation instruction for the small-hydro station i is $\Delta P_{hi} = \Delta P_\Sigma$, the total active power regulation instruction is allocated by the small-hydro station $i$with the highest priority completely.

2.3.4.The allocation algorithm considering security constraintof cross-section. The active power regulation ofsmall-hydro stations is achieved through changing their active power, as a result, the transmission active power of the cross-section tie lines may exceed the stability constraint. Consequently, the transmissionpowers of the cross-section tie lines must be restricted within their stabilityconstraint, namely satisfying:

$$P_k \le (1-\gamma_k)P_{k\_\max} \tag{19}$$

where $P_{k\text{-max}}$and γ_k are separately the stability constraint and the stability allowance of the cross-section tie line k.

If$\Delta P_\Sigma \neq 0$ and any cross-section tie line hits its stability limit, namely satisfying$P_k \ge (1-\gamma_k)P_{k\text{-max}}$, the active power allocation priority$p$is determined byS_{ki}, the specific allocation algorithm is the same as 2.3.3).

3. Simulation verification

3.1. The simulation system

A regional power gridis utilized to validate the rapid active power regulating technique for a virtual small-hydro cluster, as shown in Figure 2. In this power grid, the generation units contain four thermal power plants(as shown in Figure), a number of small-hydro power stations(five small-hydro station), and wind farms(including a wind power base and a wind farm).The incorporated capacity of wind power is about 800MW,the power transmission between the regional power grid and the neighbour power grid is achieved via 4 tie lines(tie line 1, 2, 3, 4), 5 dispersesmall-hydro stations are exploited to build the virtual small-hydro cluster control center,whose target is to receive control instruction from dispatch center and allocate the active power instruction among the 5 small-hydro stations. In this regional power grid, ,the transmission powers of the cross-section tie lines fluctuate obviously influenced by the intermittent power,consequently, the neighbour power grid is surely impacted by the transmission power fluctuation. In this section, the proposed control strategy and allocation algorithm in section II are adopted to perform the effect on suppressing the active power fluctuation of cross-section tie lines resulting from intermittent power. The simulation model is built using PSD-BPA.

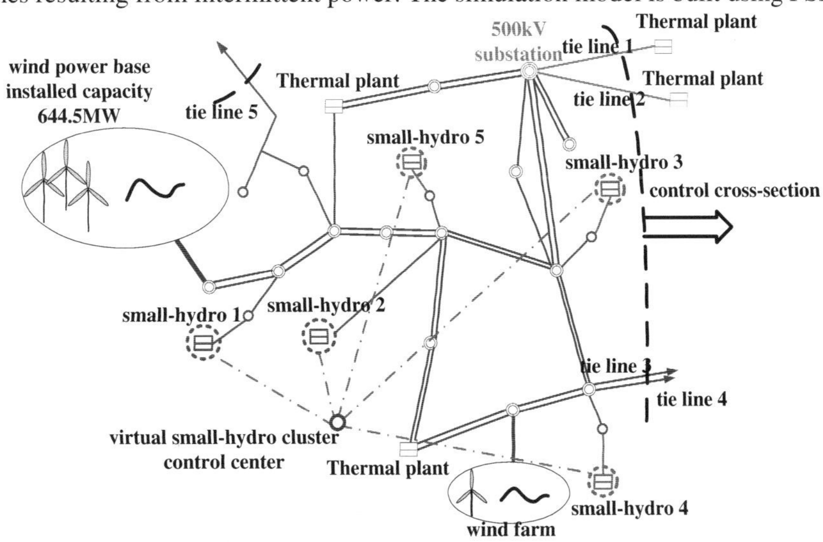

Figure 2. The regional power grid.

3.2. The verification of control strategy and allocation algorithm

At t=0s, the wind power decreases 40.5 MW in sixty-fifth seconds in the regional power grid, which leads to the transmission power fluctuation of the 4 tie lines. The balck curves display the transmission power fluctuation without control strategy and allocation algorithn, as illustrated in Figure 3 (a-d). Meanwhile, when the control strategy of virtual small-hydro cluster is employed equally, the 3 allocation algorithms, including average allocation, positive sequence allocation, inverse sequence allocation are compared to show the performance on inhibiting power fluctuation of every tie lines. To be clear: the average allocation algorithm indicates ΔP_{Σ} is assigned averagely among 5 small-hydro stations, the positive sequence allocation algorithm indicates ΔP_{Σ} is assigned according to the priority p determined by ζ, which is used in this study, and the inverse sequence allocation algorithm indicates ΔP_{Σ} is assigned according to the inverse priority p.

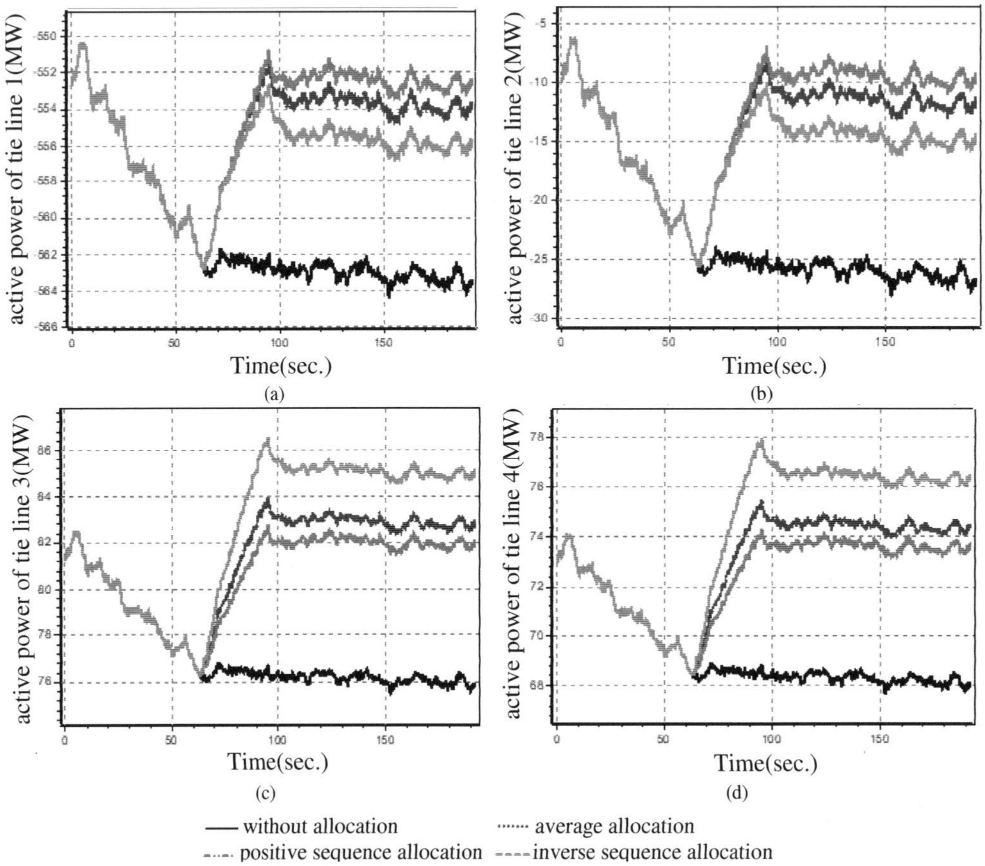

Figure 3. The transmission active power of the cross-section tie line 1-4 for different controlalgorithm.

When the wind power decreases 40.5MW, the total active power instructionΔP_{Σ} from the dispatch center is +40.5MW. According to the comparative curves of the 3 different allocation algorithms as illustrated in Figure 3 (a, b, c, d), the transmission power fluctuations of tie lines are inhibited rapidly from sixty-fifth seconds to ninety-fifth seconds, which demonstrates the rapid regulation characteristic of the constructed virtual small-hydro cluster. In addition, it can befound that the power fluctuations arerestrained most smoothly when the positive sequence allocationis executed for any a tie line, the smooth effect takes second place adopting the average allocation algorithm ,while the worst smooth effect is arisen using the inverse sequence allocation algorithm. Therefore, it can beseen that the positive sequence allocation algorithm is effective on smoothing the transmission power fluctuations of the cross-section tie lines resulting from the intermittent windpower.

4. Conclusion

In this paper, a rapid active power regulating technique for a virtual small-hydro cluster is proposed to reduce the power fluctuation resulting from the intermittent power within regional power grid. The control stratedgy and allocation algorithm are specially valid to smooth the transmission power fluctuations of all tie lines, which is favourable to weaken the disturbance and influence on neighbour power grid. This regulating technique can be popularized by the regional power grid with massive small-hydro power and intermittent power.

Acknowledgments
This work was support by the innovation consulting project of state grid ningbo power supply company (FC11), high level introduction of talent research start-up fund of China Three Gorges University(1910103) and Hubei provincial research center on microgrid engineering technology, China Three Gorges University(2016KDW01).

5. References

[1] Fischer M, Engelken S, Mihov N, et al. Operational experiences with inertial response provided by type 4 wind turbines[J]. Renewable Power Generation Iet, 2016, 10(1):17-24.
[2] Asmine M, Langlois C. Field measurements for the assessment of inertial response for wind power plants based on Hydro-Québec TransÉnergie requirements[J]. Iet Renewable Power Generation, 2016, 10(1):25-32.
[3] MORREN J, PIERIK J, DE HAAN S W H. Inertial response of variable speed wind turbines[J]. Electric power systems research, 2006, 76(11): 980-987.
[4] Sun Y Z, Lin J, Li G J, et al. A review on the integration of wind farms with variable speed wind turbine systems into power systems[C]// Sustainable Power Generation and Supply, 2009. SUPERGEN '09. International Conference on. 2009:1-6.
[5] Sang B, Wang D, Bo Y, et al. Optimal Allocation of Energy Storage System for Smoothing the Output Fluctuations of New Energy[J]. Proceedings of the Csee, 2014, 34(22):3700-3706.
[6] Yang X, Jie R, Xiangjun L I, et al. Slope Control Method with Variable Coefficients of Battery Energy Storage System for Smoothing Photovoltaic Power Fluctuation[J]. Automation of Electric Power Systems, 2016.
[7] Han T, Lu J P, Qiao L, et al. Optimized scheme of energy-storage capacity for grid-connected large-scale wind farm[J]. Power System Technology, 2010, 34(1):169-173.
[8] S. Zhou, X. Zhang, J. H. Liu, "The trend of small hydropower development in China," Renewable Energy, vol.34, pp. 1078-1083, 2009.
[9] B. Dursun, C. Gokcol, "The role of hydroelectric power and contribution of small hydropower plants for sustainable development in Turkey," Renewable Energy, vol. 36, pp. 1227-1235, 2011.
[10] Shu Z, Deng C, Huang W, et al. An Optimal Strategy For Under-frequency Pump Shedding of Pumped Power Storage Stations Based on Large-scale Wind Power Integration[J]. Proceedings of the Csee, 2013, 33(13):76-84.

Conceptual design of an integrated PV catchment storage system for rainwater harvesting

M E Ya'acob[1,2], **M M Ideris**[3], **N A Latiff**[1] and **A. Noor Iskandar**[1]

[1] Department of Process & Food Engineering, Faculty of Engineering, Universiti Putra Malaysia, Serdang, 43400, Selangor, Malaysia
[2] Centre for Advanced Power and Energy Research (CAPER), Universiti Putra Malaysia, Serdang, 43400, Selangor, Malaysia
[3] National Hydraulic Research Institute of Malaysia (NAHRIM)

Corresponding Email: fendyupm@gmail.com

Abstract. The issue of freshwater resources being continuously contaminated and polluted deemed to create water supply shortages. In urban areas where working population dominates, demands for freshwater supply increases rapidly over the years which lowers the government's capability to supply and keeping in pace with the basic need in the near future. The RWHPV system are proposed to be located at the Faculty of Engineering Universiti Putra Malaysia PV in Serdang district which has a good rainfall collection throughout the year. The objective of this work is to feasibly identify and design a suitable rainwater catchment using PV array combinations, to calculate, design and install underground storage tanks with rainwater feasibility study and lastly to suggest the system as Clean Development Mechanism (CDM) for urban flood mitigation for public awareness and research purposes.

1, Introduction

Malaysia as a tropical climate country has enormously received 3,000 mm average/annum of rainfall, with an estimated 900 billion m^3 of water supply runoff dispersed into the river system [1]. However, due to the increase in population, higher demand from industry and domestic/agriculture user, delay of treatment plant and most importantly lack of maintenance creates water crisis in urban area. The statement by Chief Minister of Selangor recently *"Selangor will experience severe water supply issues due to higher demand and delay of the Langat 2 water treatment plant"* [2], proves that preventive and realistic measures has to be table out in order to rectify the crisis.

Rainwater harvesting approach are generally define as a method for inducing, collecting, storing and conserving local surface runoff for agriculture production, domestic uses, and groundwater recharge [3], [4]. The technique provides a cheaper and environmentally means for providing water resources at the same time reduce storm water discharges and urban floods. This practice was used throughout Asia, Africa and Middle East region (mostly in off-grid area) which serve as the main source of drinking supply [5], [6]. Comprehensive review by Basinger et al. [7] provides guideline for catchment area or storage tank capacity with a reliability factor based on water demand. The concept is further supported by urban environment improvement using green strategies [8], [9].

This work presents a conceptual approach for rainwater harvesting system using PV catchment concept with an underground water storage tanks. The objective of this research is to identify and design a suitable rainwater catchment using PV array combinations. To calculate, design and install underground storage tanks with rainwater feasibility study. Roof catchment area size and storage tank capacity typically are the main target system parameters for the rainwater harvesting.

2, Conceptual design and layout

Research collaboration between Universiti Putra Malaysia (UPM) and National Hydraulic Research Institute of Malaysia (NAHRIM) has been mutually bound to initiate the project showcase involving design and performance, installation and operational cost. The site plan for this project is approximately1000 mtsq comprising green and wet area as shown in Figure 1.

Figure 1. Site overview for PV Catchment mechanism and storm water retention basin

Common method used in the design of rainwater harvesting is the storage area of rainwater sub-surface called as cistern. Figure 2 illustrate the rainfall data for the nearest NAHRIM Weather Station and Figure 3 shows the location and conceptual layout of the system in Universiti Putra Malaysia.

KAJANG rain fall data from 1986 to 2006

Average Monthly Rainfall

No.	Month	Total (mm)
1	January	135.6
2	February	148.8
3	March	237.7
4	April	251.6
5	May	170.1
6	June	103.5
7	July	159.1
8	August	180.5
9	September	189.2
10	October	200.6
11	November	281.9
12	December	194.8

Annual Rainfall

No.	Year	Total (mm)
1	1986	1757.0
2	1987	2070.0
3	1988	2308.5
4	1989	1078.0
5	1990	1469.0
6	1991	1206.3
7	1992	2028.0
8	1993	2719.0
9	1994	2187.0
10	1995	2933.5
11	1996	2373.5
12	1997	2471.5

Average annual rainfall 2253.38

WILAYAH PERSEKUTUAN rain fall data from 1986 to 2006

Average Monthly Rainfall

No.	Month	Total (mm)
1	January	147.5
2	February	195.9
3	March	238.8
4	April	296.6
5	May	224.5
6	June	162.7
7	July	143.2
8	August	172.3
9	September	243.5
10	October	264.8
11	November	337.8
12	December	190.0

Annual Rainfall

No.	Year	Total (mm)
1	1986	1891.0
2	1987	2605.5
3	1988	3265.5
4	1989	1976.5
5	1990	1756.5
6	1991	2608.0
7	1992	2469.0
8	1993	2551.5
9	1994	2970.0
10	1995	2559.5
11	1996	2908.0
12	1997	2858.5

Close

Average annual rainfall 2617.63

Figure 2. Rainfall trend based on nearest Weather Station. Data from http://www.nahrim.gov.my/

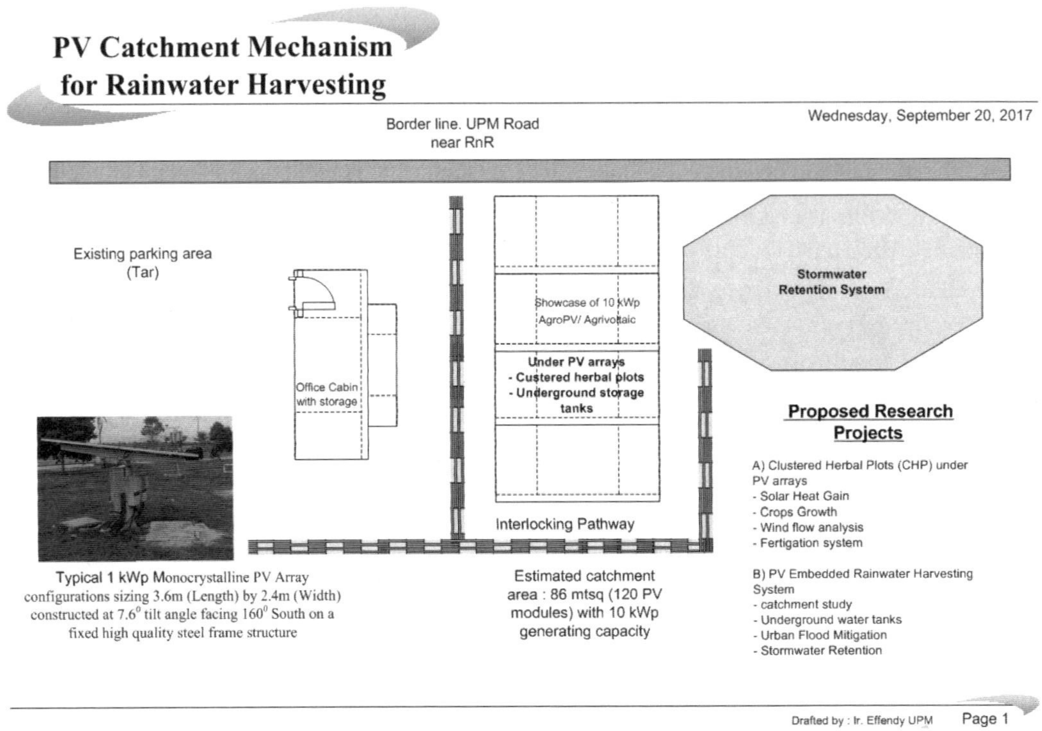

Figure 3. Proposed overall layout for PV Catchment Mechanism at UPM site

A simple modification of PV skirting system, water storage and silicon bonding between each PV module as the method of catchment mechanism are shown in Figure 4.

Figure 4. Development of RWHPV using PV skirting, water storage and silicon bonding for collecting rainwater.

3, Result and discussion

The assessment are based on the Tangki NAHRIM software http://www.nahrim.gov.my/en/products/418-perisiantangkinahrim.html to determine and calculate the suitable size of tank depending on catchment area. The software provides some important parameter i.e Tank size (m^3), Rainwater captured and Delivered (m^3), Reliability (%), Coefficient of Rainwater

Utilization (%), Storage Efficiency (%) and Time Tank Empty (%). The basic equation implied in the software for rainwater catchment system can be calculated using hydrological model as shown below

$$Q = C \times I \times A,$$

Where: Q = Discharge m^3/time,
C = Coefficient of Runoff (depends material used),
I = Total annual rainfall in m,
A = area of catchment in m^2.

Table 1 shows the tank size (m^3) for the roof catchment area of 86 m^2 and the runoff coefficient of 0.9. The water demand is estimated at 200 lt/day. The reference data are from Rainwater Station in Kajang, Selangor (Data: 1986 – 2006) and Sungai Balak, Selangor (Data: 2009 – 2016).

Table 1. Estimated 200 liter/day water demand in UPM for agricultural use (Rain Data: Kajang, Selangor).

Tank Size (m^3)	Rainwater Captured (m^3)	Rainwater Delivered (m^3)	Reliability (%)	Coefficient of Rainwater Utilization (%)	Storage Efficiency (%)	Time Tank Empty (%)
10	3254.3	1521.38	99.24	46.75	47.40	0.78
20	3254.3	1533.0	100.00	47.11	47.74	0.00
40	3254.3	1533.0	100.00	47.11	47.74	0.00
80	3254.3	1533.0	100.00	47.11	47.74	0.00

It is found that a 10 m^3 rainwater tank can meet the water demand with reliability of 99.24% and increases to 100% reliability if 80 m^3 tank size are used. Based on storage efficiency most of the tank size are similar with average 47.74% with the same capacity of rainwater captured. The appropriate design of storage tank can prevent overspill with regards to flooding. Oversize tank should also be avoided to reduce cost of construction where in this case, 20 m^3 tank size are suggested for optimum capacity storage.

4, Conclusion

This works shares some conceptual approach for flood mitigation by means of rainwater harvesting using RWHPV system. The proposed model suggested a suitable tank size for rainwater storage using Tangki NAHRIM software with some limitation of Rainfall data. The cleaner production concept are reflected based on the Nexus Integration of Rainwater, Green Energy and Land resources.

Appendix

Supplementary data associated with this article can be found, in the online version, at http://www.nahrim.gov.my/en/products/418-perisiantangkinahrim.html

Acknowledgement

The author would like to delegate our thanks to the Research Management Centre (RMC), Universiti Putra Malaysia for the approval of research funding under the UPM Putra IPM Grants Scheme (Vote no: 9459400)

Reference

[1] Abdul Rahman H.2014 Water shortage in Malaysia: Again? *Malaysian Consumer Law Journal*. **1** 115-128
[2] Selangor to face critical water shortage from 2017-2019, MB says [Internet]. [cited 2016 Mar 25].Available from: http://www.themalaymailonline.com/malaysia/article/selangor-to-face critical-water-shortage-from-2017-2019-mb-says#JrYBSJoPlHc2ErrG.99
[3] Sonbol MA 2006 Sustainable systems of water warvesting in arid regions, A case study : Sinai Peninsula – Egypt
[4] Hajani E, Rahman A 2014 Rainwater utilization from roof catchments in arid regions: A case study for Australia. J Arid Environ [Internet]. **111** 35–41. Available from: http://dx.doi.org/10.1016/j.jaridenv
[5] Nicholson N, Clark SE, Ph D, Wre D, Long B V, Spicher J, et al. 2009 Rainwater harvesting for non potable use in gardens : A comparison of runoff water quality from green vs . traditional roofs. *World Environ Water Resour Congr* Gt Rivers 1478–87
[6] Behzadian K, Kapelan Z, Morley MS 2014 Resilience-based performance assessment of water recycling schemes in urban water systems. In: *Procedia Engineering* p. 719–26
[7] Basinger M, Montalto F, Lall U 2010 A rainwater harvesting system reliability model based on nonparametric stochastic rainfall generator. *J Hydrol*. **392**(3–4) 105–18
[8]Chuxiong Y, Environment U. Yunnan Chuxiong Urban Environment Improvement Project (RRP PRC 45507) 1–7.
[9] Zhang Z, Szota C, Fletcher TD, Williams NSG, Werdin J, Farrell C 2018 Influence of plant composition and water use strategies on green roof stormwater retention. *Sci Total Environ*. **625** 775–81.

Research on Nuclear Facilities Decommissioning Funds for China

Qu Yun-huan, Li Guang-hui, Li Xiao-ding and Dong Yi-man*

Nuclear and Radiation Safety Center, Ministry of Environmental Protection, Beijing 100082, China

Abstract. Nuclear facility decommissioning technology is difficult, long processing time, requires considerable capital investment, the impact of nuclear power in the future sustainable development is enormous. However, China has not yet established a sound economic policy and system of decommissioning nuclear facilities, the world can learn from less experience, the issues of nuclear facility decommissioning funds need in-depth study. This paper analyzes the management status of decommissioning fund, retirement fund collection and management mode in major nuclear power countries around the world, and studies the management status of retirement funds in China, summarizes the outstanding problems in decommissioning funds in China, and puts forward relevant suggestions, the establishment of civilian nuclear facilities decommissioning fund, as soon as possible to develop civilian nuclear facilities decommissioning fund management approach, accelerate the estimation of the minimum guaranteed value of special funds for retirement, to promote the establishment of early public scientific research nuclear facilities retirement funds.

Keywords: Nuclear Facility; Decommissioning Funds; Research

1. Introduction

Decommissioning of nuclear facilities has a huge impact on the sustainable development of nuclear power in the future. It is a huge systematic project and requires considerable capital investment [1,2]. The decommissioning cost of nuclear facilities is a long time after the closure of related facilities. Therefore, it is the first choice for all countries in the world to do the retirement fund reserve ahead of schedule. The [3] (Chairman Decree No. sixth) of the law of the People's Republic of China on the prevention and control of radioactive pollution proposes that the decommissioning expenses of nuclear facilities should be withheld and included in the investment estimate or the production cost. The measures for the extraction and management of the decommissioning expenses of nuclear facilities shall be stipulated by the financial department of the State Council and the competent department of pricing, together with the administrative department of environmental protection and the competent department of nuclear facilities of the state council. As China's nuclear energy construction and operation have formed the enterprise operation, according to "who pollutes who governance" the basic principle, nuclear facilities decommissioning funds should be undertaken by nuclear power enterprises. But the economic management in our country at present is not perfect, the lack of effective financial system, it is difficult to avoid business risk and financial risk of nuclear power enterprises in the future, need to reference to international experience, in-depth study of the nuclear power plant decommissioning fund management mode, establish and perfect the policy system and the decommissioning of nuclear facilities.

2. Overview of Decommissioning Management of Foreign Nuclear Power Plants

Almost all of the nuclear power plants in France belong to state-owned enterprises or owned by the majority of the state owned companies. Decommissioning of nuclear power plants is under the supervision of the French Atomic Energy Commission (CEA) and the national radioactive waste management agency (ANDRA). Unlike other countries, France does not have a dedicated funding agency, which is funded by waste producers. French electric power company (EDF) has established a relatively complete decommissioning reserve for nuclear power plants. For PWR nuclear power plant EDF, according to the initial investment of 15% from the retired fees, part of the funds retained by EDF in its interior, and is responsible for the management of funds, but should be examined and approved by the Ministry of Finance on behalf of the. After 2002, France introduced the provisions of the present value of the reserve fund to reassess the decommissioning reserve for the nuclear power plant. The present value of the reserve fund is used to estimate the present value of the reserve fund in the calculation, so as to ensure that the provision can be used at the time of provision. The present value conversion is not mandatory in France, but since 2005, the present value conversion must be carried out, and the present value is converted according to the present value rate of 2.5% (according to the actual situation will fluctuate).

British military nuclear facilities decommissioning is managed by the Department of Defense (MOD), the decommissioning of civilian nuclear facilities is governed by the nuclear decommissioning authority (NDA), and the entire decommissioning process is supervised by the health and Safety Agency (HSE) and the environmental protection agency. NDA is a public institution, not affiliated with government departments, funded by the UK Department of Commerce, enterprise and system reform. Its strategy, plan and budget are approved by the Ministry of Commerce, enterprise and system reform. NDA through the management and operation contract signed with the owner of the company, to fulfill the retirement responsibility. The owner company implements the retirement work according to the work plan formulated jointly with NDA, and the actual cost of the work is paid by the NDA, and the owner uses his own employees or subcontractors to complete the decommissioning work. There are two sources of NDA costs, part funded by the UK Department of Commerce, enterprise and system reform, and part of the business revenue, including operating income and fuel processing revenues paid by owner firms.

The United States Department of energy (DOE) is responsible for the decommissioning of military nuclear facilities in the United States, the decommissioning of civilian nuclear facilities is managed by the American Nuclear Safety Agency (NRC), and the entire decommissioning process is supervised by the Department of environmental protection (EPA). The United States does not have a decommissioning fund for nuclear power plants, but requires nuclear power plant operators to reserve deposits that can be used for retirement.

3. Management Model of Decommissioning Fund for Nuclear Power Plant

According to preliminary estimates, decommissioning of nuclear power plants has increased the cost of nuclear power, with an increase of about 2% to 5% [4]. Due to the differences of economic system, accounting system and traditional practices, the management mode and mode of decommissioning fund for nuclear power plants adopted by different countries are different. Some countries set up a system to collect retired capital, some countries require the electric power company to set up the retirement fund. But no matter what kind of situation, the country needs to introduce specific retirement fund management measures.

3.1. Ways of Collecting Funds

There are four ways to collect decommissioning funds for nuclear power plants: (1) Prepayment. Before the operation of power plants began to pay the expenses for retirement funds in a separate account, the provisions shall not be used for other purposes. (2) External payment method. The payment from customers' electricity charges to extract a certain percentage, after years of accumulation, set up a fund to invest in or into a free power company controlled by the special commission of decommissioning funds dominated trust fund. (3) Internal reserve mode. Similar to external payments, electricity is extracted from electricity bills, but electricity companies can use this fund to borrow money for investment or for other purposes. (4) Margin, letter of credit or insurance. It is purchased by the electric power company, and when the electric power company fails to repay the debt, the payment of the retirement fee is guaranteed.

3.2. Management Style

On the whole, the retirement funds in the world can be divided into two levels and three types in the management style. Two levels of fund management. One is whether it is incorporated into the national budget. The two is whether it is relatively independent of enterprise finance. Where the management of funds is not independent of the enterprise's finance, it belongs to the reserve provided by the enterprise in order to fulfill its future obligations. Where the management of funds is relatively independent of enterprise finance, even if it is not in the category of national budget management, it is often regarded as the name of "fund".

Three types of fund management [5]. The first is internal reserves. So far, the French EDF's funds required for decommissioning of its nuclear power plants are managed by the enterprise's internal reserves, and the relevant responsibilities and capital delivery are dealt with by commercial contracts. According to the interpretation of the French financial sector, because the French nuclear size, if the huge retirement funds into national fund management, is bound to set up special investment company, the French administrative authorities on balance, gave up this plan. The second is to set up a separate account that is independent of the assets of an enterprise. The most typical representative is the United States nuclear power plant decommissioning fund management system. The federal regulations in the United States have regulations on the amount of retirement funds, the types of effective financial instruments, the independence of funds, the principle of capital investment, and the use of funds. According to the US system, the American Nuclear Regulatory Commission (NRC) oversees the decommissioning fund for nuclear power plants, but the tariff rates are regulated by the Federal Energy Regulatory commission. The third is the establishment of a national nuclear waste fund that does not fall into the national budget category. The typical representatives are the national nuclear waste fund of Sweden and Finland. The fund defines the ownership of capital very clearly. Key features include the fund management system operation: prediction of different charges based on four nuclear plants according to the different rates of pay, and have different share of the fund's assets; nuclear power plants have the statutory right to return the funds from the fund for future retirement expenditure. Clear legal ownership of the fund is inherent characteristic of Swedish fund management system [6].

3.3. Maintain Value and Avoid Financial Risk

The common problems faced by different types of funds are to increase the value of investment and avoid the risk of investment. Nuclear power plant decommissioning fund management, on the one hand to prevent risks from the management system, in order to protect the nuclear power plants have sufficient funds to fulfill future nuclear responsibility; on the other hand to allow these funds from their situation to find suitable investment channels, in order to reduce the cost and improve the competitiveness of nuclear power nuclear power.

The Swedish retired do not total funds, funds were initially placed in the "risk of bank" (the Swedish central bank) account, then put them from the internal to the state reserve fund as a financial guarantee mechanism by government bonds, because this new funding system can ensure the lower unit electricity fund extraction amount. The new nuclear waste fund committee can invest the fund's money through the Treasury office, so it can get a fixed return at a market rate for a fairly long period of time. At the same time, the commission can also use the opportunity of interest liberalization to increase returns through the reinvestment of different interest periods.

The decommissioning fund of the United States nuclear power plant is managed by the enterprise fund, but the federal regulations stipulate that the funds must be separated from the holder's assets and trust the third party to manage the money. In the standard review plan, the Moodie rating of the trust company is required to be at least "BBB"". The Bank of New York in May 2006, a survey report "to prepare for change"-closed nuclear decommissioning trust investment method pointed out in the [7], U.S. nuclear decommissioning trust investment strategy has been stressed from the review of each investment financial qualifications, and emphasized the diversity of investment, and specifically pointed out that the power companies pursue investment returns for the typical tax 6-7%.

The decommissioning reserve of the French nuclear power plant is managed by the internal fund of the enterprise [8]. From the EDF financial annual report, the annual extraction amount of the reserves allocated to the generation cost is calculated according to the nominal discount rate of 5%, that is, EDF takes all the reserves from the extraction to the use of the nominal annual return of 5% to increase the value of the obligations, in order to reduce the burden on consumers, or to improve their market competitiveness. The reserve for decommissioning of nuclear facilities in France is the capital which is not used at present and will be used for the future, and its total amount increases year by year. In the past EDF reserve does not establish a special account, but use the money to buy foreign power companies and expand foreign market share, EDF has reached a considerable scale in English, German, Spanish, Italian, Brazilian, Argentine and other countries of the mergers and acquisitions and market share, which on the one hand by the European Parliament external pressure, another also that the French government departments to worry about safety investment. Therefore, the new law passed by France in 2006 requires EDF to establish a special account, and gradually establish "mortgage" special assets". According to the Bank of Paris disclosed that the current EDF "special assets" composition is: 45% stocks, 55% bonds. That is to say, the French government requires EDF to gradually change the form of reserve investment from overseas self-expansion to the form of investment in financial markets. The new trend that France requires nuclear power plants to invest in value added through financial markets is, to some extent, similar to that of the United States, but there are differences. The United States requires that the retirement fund be independent of the enterprise finance and entrust the third party (trust) to manage the money; France does not require the three reserve to be independent of the enterprise finance, nor does it require the entrusting third party to manage the money.

3.4. And the Relationship of Enterprise Financial System

Under the mechanism of external fund management, the enterprise provides the necessary funds for the external retirement fund according to the regulations, and the enterprise does not reflect its retirement nuclear responsibility in the accounts. In retirement funds according to the mechanism of internal fund management, according to international accounting standards, the responsibility to the future retirement provision form reflected in the balance sheet, in the calculation of the allowance while allowing the use of "net present value method". Time in the nuclear power plant is completed, all the discounted future retirement responsibility as a provision in the balance sheet to confirm (the debt side); at the same time, the future retirement funds as the book value of nuclear power plant (capital asset side).

4. Nuclear Facilities Decommissioning Fund Management Status in China

The initial purpose of China's nuclear energy development is for national defense and military needs. Therefore, the state undertakes the responsibility of all decommissioning of nuclear facilities. This kind of funds is provided by the central government in full support. With the rapid development of society and economy, the construction and operation of nuclear energy have been operated in an enterprise, and the civil nuclear fuel cycle facilities and public welfare research facilities are also increasing.

According to statistics, China has three nuclear power generating units and power generation in 1990s, Qinshan a 300 thousand kilowatts of nuclear power units has been running for nearly 20 years, the design life of 30 years, the Dayawan nuclear power station two sets of 1 million kilowatts of nuclear power units has been running for nearly 18 years, the other in the operation of nuclear power plants are concentrated in the grid time after 2002 design life is 40 years. If the life extension of nuclear power plant is not considered, Qinshan Phase I and Dayawan nuclear power station will retire in 2021 and 2033, and about 10 nuclear power units will be retired in about 2042~2050 years. It is foreseeable that the decommissioning of nuclear power plants will be a long and arduous project in the future.

Decommissioning of nuclear power plants not only lasts a long time, but also consumes a lot of money. The International Atomic Energy Agency (IAEA) stipulates that the three stage of decommissioning is recognized by the international community. It takes about 40~60 years, or even hundreds of years. According to preliminary estimates, the cost of decommissioning of large nuclear power plant is about 10%~20% of its infrastructure investment, according to the current cost (the two generation, three generation), a million kilowatt pressurized water reactor nuclear power plant decommissioning need at least 3 billion Yuan, if not immediately retired, in addition to considering retirement itself, but also to ensure the safety of maintenance during storage service facilities and equipment operation and maintenance costs.

At present, in the decommissioning of nuclear facilities, all civil nuclear facilities have no corresponding economic policy and management measures. Financing is an unavoidable problem in retirement plan, and it is also a key issue in decommissioning. There must be sufficient funds for retirement in the future. The nuclear power plants in China have different technical routes and many reactors, and the decommissioning costs of different reactors are different. Therefore, the specific extraction methods should be studied in detail. In addition, the relevant issues such as the division of responsibilities during decommissioning of nuclear power plants are not clear, so it is necessary to establish decommissioning fund for nuclear power plants and study the relevant management methods.

In addition, the construction and management of public research facilities scattered in the nuclear industry, education, science and technology, health and other government departments or local, in the early construction did not fully consider the responsibility and funding problems in decommissioning of nuclear facilities, resulting in a number of civilian nuclear facilities has reached early life cannot implement funding channels, unable to implement nuclear facilities decommissioning, increase the risk of nuclear safety[9,10].

The future, nuclear power will be more widely used in all walks of life, retired economic policy and management means, if not as soon as possible to establish and perfect, will not be conducive to the healthy and steady development of nuclear energy, the country is difficult to circumvent the financial risk, the risk of nuclear safety, environmental risk and social risk.

5. Thoughts and Suggestions on the Management of Retired Funds in China

Establishment of decommissioning fund for civilian nuclear facilities. In view of the fact that decommissioning of nuclear facilities is closely related to national nuclear safety and environmental safety, considering the lag, long-term and arduous nature of decommissioning of nuclear facilities, it is suggested that the decommissioning fund for civilian nuclear facilities be established as soon as possible, all civil nuclear facilities are included in the management of decommissioning funds.

Administrative measures for the decommissioning of civilian nuclear facilities as soon as possible. According to the 2003 enactment of the "People's Republic of China radioactive pollution prevention and control law" put forward to extract and manage the expenses for retirement of nuclear facilities, by the competent departments of Finance and the State Council jointly with the competent price departments, nuclear facilities administrative regulations for environmental protection under the State Council set. Therefore, the relevant departments should implement the responsibility as soon as possible and formulate the provision, management and use of the decommissioning fund for civilian nuclear facilities.

A Study on the Estimation of the Minimum Guaranteed Value of Special Funds for Retirement. It is recommended that the Ministry of Finance, in conjunction with the competent authorities of nuclear industry, nuclear power authorities and nuclear safety supervision departments, and with the participation of enterprises, learn from the international experience and carry out the research work of various types of decommissioning cost estimation, and constantly updated and improved.

To Promote the Establishment of Early Public Scientific Research Funding for Nuclear Facilities Decommissioning. The construction and management of the early public scientific research nuclear facilities shall be the national behavior. The decommissioning funds of such facilities shall be borne by the state. Considering that the early public scientific research nuclear facilities scattered in the nuclear industry, education, science and technology, health and other government departments or places, And the retirement of nuclear facilities should not be scattered management, it is recommended that the state in the central government to set up such facilities for the retirement of special funds, special funds by the Ministry of Finance, nuclear safety supervision departments in conjunction with the relevant departments to manage and implement.

6. References

[1] U.S.NRC. Decommissioning Nuclear Power Plants [EB/OL]. http://www.nrc.gov/reading-rm/doc-collections/fact-sheets/decommissioning.html,2013-03-13.

[2] Wang Hai-liang. Nuclear power plant decommissioning fund management [J]. Chinese nuclear power, 2013 (6) 2:168-172.

[3] The People's Republic of China Radioactive Pollution Prevention Law. 2003.10.1.

[4] Luo Shang-geng, et al. Retirement of nuclear facilities and radiation facilities [M]. Beijing: China Environmental Science Press. 2010: 55

[5] Lu Yan. The retirement regulations and capital management of the world nuclear power plant [J]. China nuclear industry, 2017 10:21-25.

[6] F.TCHAPGA. Topical Session on Funding Issues in Connection with Decommissioning of Nuclear Power Plants[C]. NEA/RWM/WPDD, 2005.4.

[7] Bank of New York. Preparing for Shutdown-Shifting Attitudes Toward Nuclear Decommissioning Trusts[R]. New York: Bank of New York, 2006.6.

[8] CONTE D, et al.. French Approach to Regulatory Review: lessons learned from the experience of the last 20 years, IAEA Proceedings of an International[C]. Conference, Athens, 2006.

[9] Dong Yi-man, Qu Yun-huan, et al. Study on spent fuel disposal and nuclear facilities decommissioning funds [J]. Environmental pollution and prevention, 2013 (35) 10:99-103.

[10] Liu Jian-qin, An Hong-xiang, et al. Study on Civilian Nuclear Facilities Decommissioning Funds [J]. Environmental science and management, 2017 (42) 3:20-25.

Effects of Streptomycin Wastewater on the Activity of GSH-PX in Zebrafish before and after Aerobic Biological Treatment

Yi Nan Liu[1] , Hong Yan Shen[1]

[1]College of Environmental Science and Engineering, Pharmaceutical Molecular Chemistry Laboratory of Hebei Province, Hebei University of Science and Technology, Shijiazhuang, China.

Email: 1178611805@qq.com

Abstract. Most wastewater treatment plants now use aerobic biological treatment to dispose wastewater, but the toxicity of influent and effluent is unclear. In this study, subacute toxicity test was used to set four exposure groups and one control group to measure GSH-PX activity when zebrafish was exposed for 3,6,9,12,15 days at the indicated volume ratio, which is 10%, 30%, 50%, 70% respectively for influent and 10%, 20%, 30%, 40% for effluent. For influent, because of the response to the oxidative stress, the GSH-PX activity was activated in all other exposure groups except for the 70% volume group. The GSH-PX activity in 70% volume was significantly lower than the control group indicates that because of the sustaining increase of volume ratio, antioxidant system in zebrafish was destroyed and antioxidant enzyme activity was reduced. But for effluent, GSH-PX activity in each group was less than that in influent but still slightly induced, which means that after aerobic biological treatment, the sewage can reduce oxidative damage to zebrafish, but still have induction effect. For both influent and effluent, maximum of GSH-PX activity was both at the 30% exposure volume (18.7523U/mgprot for influent, 10.8023 U/mgprot for influent) which can provide the data basis for future research. All in all, GSH-PX activity in zebrafish muscle tissue can response to the oxidative stress and the changes can be used as an indicator of oxidative damage. And it is necessary to use the anaerobic method for the secondary treatment.

1.Introduction

Antibiotic wastewater is a kind of high-concentration organic wastewater containing refractory materials and biological toxic substances [1]. In China, over 700 kinds of antibiotics are produced and used, with an annual output of approximately 210,000 tons of raw materials, as well as 30,000 tons for export in 2016 [2]. Shijiazhuang, the provincial capital of HeiBei, as the most productive place for antibiotics, it's production accounts for one sixth of the country [3]. Streptomycin is one of the common antibiotics, it extracted from the culture medium of streptomyces lividans [4]. Nowadays, there are still many technical difficulties in the production of antibiotics, resulting in many problems such as low utilization rate of raw materials, low purification purity and high residual antibiotic content in wastewater [5]. Discharge of wastewater is large. Most of sewage treatment plants now use aerobic biological treatment to dispose wastewater with considerable economic, environmental and social benefits. However, the toxicity of the influent and effluent is unclear [6].

The discharge of antibiotic wastewater released into the water can consume the oxygen in water and then cause the oxygen deficiency and destroy the self-purification function of water. It may lead to the eutrophication and have certain impacts on the aquatic organisms. Zebrafish is one of the model organisms, which has the characteristics of easy feeding, rapid development, annual spawning, large fecundity and easy collection of eggs [7]. In 1996, the Organization for Economic Cooperation and Development (OECD) included zebrafish in its testing standards for health toxicology and environmental toxicity [8]. GSH-PX activity in muscle tissue of zebrafish is one of the most important antioxidant enzymes in the fish body. The change of its activity directly affects oxygen free radical in the intracellular and its final product-lipid peroxides (LPO) [9].

This study selected a sewage treatment plant's influent and effluent as the water samples which deals with streptomycin wastewater primarily and using zebrafish as the toxicity test organism to evaluate the effect of streptomycin wastewater on GSH-PX activity in muscle tissue of zebrafish.

2. Materials and methods

2.1. Experimental Materials

2.1.1. Instruments and reagents. UV-visible light spectrophotometer (UV-2550); high speed centrifuge (TG16-WS); hypothermia refrigerator (BBC-226STV); high purity water distiller (SYZ-A). Protein kits, GSH-PX were purchased from a biological engineering research institute.

2.1.2. Test animals. Zebrafish were obtained from aquafarm in Shijiazhuang. In this experiment, mean length of fish used in this experiment was 34.0±0.3mm and mean weight of fish used in this experiment was 0.24±0.03g. Before the experiment, disinfection treatment employ 5% salt water, and then domesticate a week in 48h of aeration tap water. Feeding was halted 7d prior to testing. Healthy individuals of fish with similar size and weight were employed for test.

2.2. Experimental Methods

2.2.1. Preliminary experiments. Health zebrafish were put into 100% streptomycin wastewater (water volume concentration percentage), and no death within 96 h. For the influent, subacute experiment set up four experimental groups, volume ratio is 10%, 30%, 50%, and 70% respectively. For the effluent, subacute experimental set up four experimental groups, volume ratio is 10%, 20%, 30%, 40%, respectively. The effects of streptomycin wastewater with different volume fraction on antioxidant capacity of zebrafish muscle tissue are determined.

2.2.2. Exposure experiment. The subacute toxicity experiments lasted 15d. There are one control group and four experimental groups in the subacute experiments. There are three parallel experiments in each group, and put in 20 fish randomly. The experiment uses the static displacement method and changes the water every day. Besides, experiment assay is carried on 3, 6, 9, 12, and 15d.

2.3. Assay Methods

Zebrafish was dissected. 0.20g muscle tissue was obtained and rinsed in physiological saline. Muscle tissue was blotted with water and weighed. Samples was added to the mass ratio of 1:9 saline and grinded in the ice-water bath for 6-8min. 10% of the tissue samples were removed for experimentation and centrifugated for 10 minutes at speed of 3500 $r \cdot min^{-1}$ to get supernatant for future determination. The supernatant was collected for the experiment.

2.4. Statistical Analysis

All statistical analyses were expressed as mean ± standard deviation (SD). One-way Anova of all data were checked using SPSS 17.0. Significant difference analysis uses the method of least significant difference (LSD). It analyzes the data in the groups on the same day. $P<0.05$ was considered statistically different, $0.01<P<0.05$ statistically significantly different, and $P<0.01$ statistically very significantly different.

3. Results and Discussion

GSH-PX is found in almost all tissues of animals and has anti-oxidant functions to protect cell membranes from the irritation and damage of peroxides. It can reduce the incidence of cell mutations and clear out H_2O_2 that produced in tissues ability [10]. If the GSH-PX activity is suppressed, it may affect the body's antioxidant function and cause tissue damage [11]. This study selected a sewage treatment plant's influent and effluent as the water samples. Subacute toxicity method was used to set four exposure groups and one control group to measure GSH-PX activity when exposed for 3,6,9,12,15 days at the indicated volume ratio, which is 10%, 30%, 50%, 70% respectively for the influent and 10%, 20%, 30%, 40% for the effluent. Almost all exposure groups had higher GSH-PX activity than the controls either in the influent or in the effluent, except for the exposure to 70% influent water.

3.1. Effects of Streptomycin Influent and Effluent on GSH-PX Activity in Zebrafish

3.1.1. Effects of Streptomycin Influent on GSH-PX Activity in Zebrafish. The activity of GSH-PX in zebrafish muscle with different concentrations of streptomycin influent in different exposure times is shown in Figure.1(a). For influent, when treated with 10% exposure volume, the activity of GSH-PX was significantly induced ($P<0.01$) from the ninth day. When treated with 30% exposure volume, the activity of GSH-PX was activated at all days of exposure, especially when exposed to the 6th, 9th and 12th days. When the number of exposure days was identical, the maximum enzyme activity all appeared in the 30% exposure group. The remarkable increase of GSH-PX activity indicates that when exposed to a certain concentration of streptomycin wastewater, zebrafish body may increase the intracellular enzyme activity and improve the body's antioxidant protection mechanism, so that H_2O_2 and oxygen free radicals can be promptly cleared and non-accumulation in cells. Avoid cell lipid peroxidation caused by cell damage simultaneously. Due to the reduction of lipid peroxidation products and the continual improvement of cellular anti-oxidant functions, oxidative stress damage in tissues can be minimized [12]. Li et al studied the effects of diethylstilbestrol on GSH-PX of Cadmium-contaminated Grass Carp. They found the maximum GSH-PX activity appeared at the mid-value of exposure concentration (1mg/L) and proposed that grass carp had the strongest ability to scavenge oxygen free radicals when exposed to the concentration of 1mg/L [9], which is consistent with the results of this study.

When treated with 50% exposure volume, the GSH-PX activity was significantly induced ($P<0.01$). However, the maximum was reduced by 21.40% compared to the 30% exposure group (When exposed to the 12th day). With the continuous increase of exposure concentration, the activity of enzyme gradually decreased, which may lead to the inadequate antioxidant capacity and the damage of antioxidant system [13]. This phenomenon was even more evident when exposed to 70% exposure group. Except for this group which exposed for 6 days, the enzyme activity of other groups was significantly lower than that of the control group. Especially, on the 15th day of exposure, the enzyme activity was only 27.40% of the control group. An assumption that the increase of concentration and prolongation of exposure time, the scavenging capacity of zebrafish's own cells is not enough to eliminate the free radicals, even damage the cells irreversibly, which hinders the synthesis of GSH-PX. So that enzyme activity was significantly lower than the control group. We have tried to compare our predictions with available conclusions made from another study. Liu et al studied the effect of ammonia nitrogen stress on activities of SOD and GSH-PX in muscle tissue of loach. According to the concentration of NH3-N, four experimental groups and one control group were set up. Concentration ratio were 50 mg/L, 150 mg/L, 250 mg/L, 450 mg/L. Muscle tissue was grinded for measure the GSH-PX activity at 2, 7, 14 and 21 days. They found when the concentration was 250mg/L, compared with the 150mg/L concentration group, GSH-PX activity was seriously decreased. GSH-PX activity was 57.46% of that in the control group when exposed to 450 mg/L for 21 days. They proposed that the activity of GSH-PX in muscle is affected by the exposure time and concentration of ammonia nitrogen. With the extension of exposure time, the antioxidant capacity of muscle tissue is inadequate, which will accumulate free radicals and decline the enzyme activity. Therefore, the enzyme activity of the high concentration group showed a significant decrease at 21d [14]. That can provide the basis for this research's conjecture.

3.1.2. Effects of Streptomycin Effluent on GSH-PX Activity in Zebrafish. For effluent, streptomycin effluent within 15d of GSH-PX activity in zebrafish muscle was shown in Figure.1(b). GSH-PX activity in each group was less than that in influent and the change trend was stable. Indicating the oxidative damage to zebrafish was reduced after aerobic biological treatment. However, the zebrafish muscle tissue was still slightly induced (0.01<P<0.05) compared with the control group.

It can be seen in the Figure.1(b), under the same exposure conditions in different groups, the trend in the activity of GSH-PX changed variously with the increase of exposure time. The results of this experiment are keeping with those of Yao et al, who proposed that GSH-PX clears lipid peroxides induced by reactive oxygen species and hydroxyl groups in order to protect the integrity of cell membrane structure and function [15]. And changes in intracellular lipid peroxides are also affected by the cell's own metabolites [16].

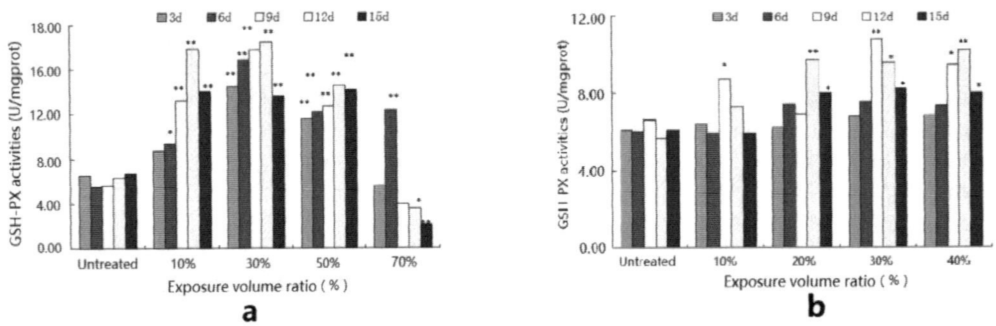

Figure 1. Effect of streptomycin wastewater (a for influent, b for effluent) on GSH-PX activities in zebrafish muscle tissue.

(The same day all the data are compared with control group, *p <0.05, ** p<0.01.)

3.2. Comparison of GSH-PX activity with influent and effluent

When treated with the same exposure volume (10%,30%) of influent and effluent, different results can be observed. When treated with 10% exposure volume of influent, the activity of GSH-PX increased at first and then decreased with the extension of exposure time, which accord with "bell-shaped" distribution. However, GSH-PX activity changes are not regular when treated with the effluent (10%), GSH-PX activity decreased on 6th day, then increased on 9th day, and decreased again on 12th day and 15th day. For both influent and effluent, the maximum of GSH-PX activity was both at the 30% exposure volume (18.7523 U/mgprot in influent, 10.8023 U/mgprot in effluent) indicating that when zebrafish exposed to 30% volume of wastewater, body's enzyme activity and antioxidant capacity are the largest, which can be used as a threshold for future research.

On the other hand, for the influent, under the same exposure time, the activity of GSH-PX also showed a "bell-shaped" distribution as the exposure concentration increased. Shen et al studied the effect of streptomycin in influent on SOD activity and MDA content in zebrafish muscle. They found the SOD activity in zebrafish muscle was significantly induced on the 3rd day, and the SOD activity in the 50% volume fraction reached 30.56U/mg. 70% of the volume fraction of SOD activity in the experimental group gradually decreased with the extension of exposure time which SOD and GSH-PX can indicate the degree of oxidative damage in zebrafish [17], which is similar to the results of this study. Heleen Van Acker proposed that different concentrations of antibiotic wastewater produced reactive oxy-gen species (ROS) at a rate greater than the clearance rate, unbalancing the oxidation and antioxidant systems of zebrafish muscle tissue, causing tissue damage eventually[18]. But it can be seen in Figure.1(b), no similarities were observed in the effluent. Zhao et al discussed the effect of biochemical method on streptomycin effluent on the activity of SOD in zebrafish muscle. They found that when exposed for 9~15 days, 10%, 20%, 40% of the experiment group had no significant difference from the control group [19]. These experimental results indicate that the zebrafish antioxidant defense system eventually reached a new balance and the fish antioxidant defense system has a function of self-regulating [20].

4. Conclusion

In this study, streptomycin Wastewater was used for subacute toxicity test by static laboratory exposure method, and experiments determined the zebrafish muscle GSH-PX activity. For influent, except for the 70%, the GSH-PX activity was induced in all other exposure groups because of the response to the oxidative stress. Owing to the continuous increase of volume ratio, antioxidant system in zebrafish was destroyed and antioxidant enzyme activity was reduced. Therefore, the GSH-PX activity in 70% volume was significantly lower than the control group. For effluent, GSH-PX activity in each group was less than that in influent but still slightly induced indicates that after the aerobic biological treatment, the sewage can reduce oxidative damage to zebrafish, but still have induced effect. It is essential to use the anaerobic method for the secondary treatment. For both influent and effluent, maximum of GSH-PX activity was both at the 30% exposure volume, which can be used as a threshold for future research. GSH-PX activity in zebrafish muscle tissue can response to the oxidative stress and the changes can be used as an indicator of whether oxidative damage occurs.

5. Acknowledgement

This study was supported by the National Natural Science Foundation of China (41373096), National Environmental Protection Project (201509041-05), Hebei Province Natural Science Fundation (B2014208068), Hebei Province Pharmaceutical Molecular Chemistry Laboratory Open Foundation, Hebei Province Environmental Protection Public Interest Project, and Hebei Province Key Disciplines Construction Fund Project.

6. References

[1] Shi Ruiming, Wang Feng, Yang Yuping, et al. Status and Research Progress of Antibiotic Wastewater Treatment [J]. Journal of Shan Dong Chemical Industry, 2007,36(9):10-14.
[2] Cheng Xianwei, Liang Yinxiu, Zhu Hui, et al. Research Progress on Constructed Wetlands for Treatment of Antibiotics in Water [J]. Journal of Wetland Science, 2017,15(1):125-132.
[3] Xu Yan, Zhang Yuan, Guo Changsheng, et al. Characteristics of Antibiotics, Resistant Bacteria and Resistance Genes in WangYangGou, Shijiazhuang [J]. Journal of Agro-Environment Science, 2014,33(6):1174-1182.
[4] Su Jianjiang, Huang Fuyi, Zhu Yongguan, et al. Research Progress on Environmental Antibiotic Resistance Genes [J]. Journal of Biodiversity Science, 2013,21(4):481-487.
[5] Zeng Linxuan, Zhang Qiuyun, Liu Peihong, et al. Progress of Antibiotic Pharmaceutical Wastewater Treatment Technology [J]. Journal of Safety and Environmental Engineering, 2005,12(4): 62-68.
[6] Shao Yiru, Xi Beidou, Cao Jinling, et al. Distribution and Removal of Antibiotics in Urban Sewage Systems [J]. Journal of Environmental Science and Technology, 2013,36(7):85-94.
[7] Alexander J, Stainier DY. A Molecular Pathway Leading to Endoderm Formation in Zebrafish [J]. Journal of Current Biology, 1999,9(20):1147-1157.
[8] Wang Jiajia , Xu Chao, Tu Yunjie, et al. Experimental Research and Application of Zebrafish and Its Embryo in Toxicology [J]. Asian Journal of Ecotoxicology, 2007,2(2):123-136.
[9] Li Yuanling, Lan Xin, SiWantong, et al. Effects of Diethylstilbestrol on GSH-PX, SOD and MDA of Cadmium-contaminated Grass Carp [J]. Journal of Agro-Environment Science, 2008,27(1):0350-0353.
[10] Yao Xin, Qian Yuanshu. Effects of Fluoroquinolones on Activities of Extracellular GST, GR and GSH-PX in Extrahepatic Tissue of Rats [J]. Journal of Chinese Antibiotics, 2005,30(7):423-427.
[11] Schauer R, Sommer U, K rüger D, et al. The Terminal Enzymes of Sialic Acid Metabolism [J]: Journal of Biosci Rep,1999,19(5):373.

[12] Ding Limin, Huang Xiaomin, Zhang Zhuoyi. Effects of Procyanidin on The Activity of SOD and GSH-PX Induced by Hydrogen Peroxide in Endothelial Cells [J]. Journal of Chinese Medicine Emergency, 2013,22(5):714-717.
[13] Yin Jin, Hu Yixiu, Hu Yuming, et al. Effect of Grape Seed Proanthocyanidins Extract on MDA, SOD and GSH-PX in Mice [J]. Journal of China Tropical medicine,2007,8(7):1285-1287.
[14] Liu Yang, Leng Yufei, Yu Lianyang. Effect of Ammonia Nitrogen Stress on Activities of SOD and GSH-PX in Different Tissues of Loach [J]. Journal of Anhui Agriculture, 2011,39(2):1069-1072.
[15] Yao Xin. Effects of Fluoroquinolones on Extrahepatic Drug Metabolizing Enzymes and GR, GSH-PX Activities [D]. Chongqing Medical University, 2003.
[16] P.A. Olsvik, T. Kristensen, R. Waagbø, et al. mRNA Expression of Antioxidant Enzymes and Lipid Peroxidative Stress in Liver of Atlantic Salmon Exposed to Hyperoxic Water During Smoltification [J]. Journal of Comparative Biochemistry and Physiology, 2005, PartC:314-323.
[17] Shen Hongyan, Wang Linxin, Yang Jindi, et al. The Effect of Streptomycin in Influent on SOD Activity and MDA Content in Zebrafish Muscle [J]. Journal of Hebei University of Science and Technology, 2014,35(3):303-309.
[18] Heleen Van Acker, Tom Coenye. The Role of Reactive Oxygen Species in Antibiotic-Mediated Killing of Bacteria [J]. Journal of Trends in Microbiology, 2017,25(6):456-466.
[19] Zhao Yue, Cao Zhihui, Wang Bing, et al. Effect of Streptomycin Wastewater Effluent Treated with Biochemical Method on SOD Activity and MDA Content in Zebrafish [J]. Journal of Jiangsu Agricultural Sciences, 2015,43(8):348-351.
[20] Li Min, Zhao Yue, Cao Zhihui, et al. Effect of Streptomycin Wastewater on POD Activity and GSH Content in Zebrafish [J]. Journal of Safety and Environmental Engineering, 2015,22(3):1-11.

IOP Conf. Series: Earth and Environmental Science **146** (2018) 012058 doi:10.1088/1755-1315/146/1/012058

Taking the Red Bay as an Example to Study the Numerical Simulation Method for the Thermal Drainage along the East Coast of Guangdong Province

Aizhen Liu, YingWang, Yuxin Zhu, XinXu and Boqun Liu

Laboratory of Environmental Protection in Water Transport Engineering, Tianjin Research Institute of Water Transport Engineering, Tanggu, Tianjin, 300456, China

147491256@qq.com

Abstract. In view of the characteristics of the current in the east coast of Guangdong Province, based on the hydrological observation in the summer and winter in red bay, a two-dimensional ocean current mathematical model based on triangular meshes is established. Using the prediction results of the South China Sea regional circulation as the boundary conditions, the numerical simulation of the ocean current and the temperature drainage is carried out in the Red Bay. Because of the seasonal difference in current and the heat exchange parameters, the current and thermal drainage in summer and winter is predicted. The influence range and influence degree of summer and winter thermal drainage are compared and analysed. This method is more suitable for the prediction of current and temperature drainage along the east coast of Guangdong Province.

1. Introduction

In recent years, the construction of coastal power plants is increasing. In the Red Bay, the Taiping Ling nuclear power plant and Haifeng thermal power plant are to be built. Red Bay is in the north of the South China Sea, which is a semi closed bay, located on the east coast of Guangdong. As we all know, the sea current in the South China Sea is more complex [1], which is very different from other waters dominated by the tidal current.

The full tide hydrology test of the four seasons was carried out in the Red Bay Area in 2015[2]. The observed results show that the characteristics of the current flow are in accordance with the characteristics of the coastal current in East Guangdong. The flow direction of the East Guangdong coastal flow changes with the monsoon. The general characteristics are: the southwest monsoon is prevalent in summer, and the flow direction is drifting in the northeast, and the northeast monsoon is prevailing in winter, and the flow is changed to the southwest [3]. Its main driving mechanism is the strong wind stress formed by the wind field and the coastal terrain [4].

In view of the above features, based on the current mathematical model of summer and winter in the Red Bay, the prediction of thermal drainage in summer and winter is carried out. The influence range and influence degree of summer and winter thermal drainage are compared and analysed.

2.Numerical simulation method

2.1.Two dimensional hydrodynamic and water quality model

The paper use the method of MIKE21 Flow Model hydrodynamic numerical simulation which isa general numerical modelling system for the simulation of water levelsand flows in estuaries, bays and coastal areas. It simulates unsteady two dimensionalflows in one layer (vertically homogeneous) fluids and hasbeen applied in a large number of studies.

2.2Model setting

(1)Settings of computational domain and grid

The computational range of the model is about 81km*66km. The calculation range and water depth of the model are shown in Figure 1. Figure 2 is grid diagram which have 51978 nodes in the area. The nodes are divided into 26653 non overlapping triangular units. The minimum space step length is 20m, the maximum space step is about 2km, and the time step is automatically adjusted from 0.01-20s.

Water depth and offshore boundary, according to the charts produced by the Ministry of People's Liberation Army Navy, is in consistent with water depth data of " Tianwei Jiao to Great Kunshan".

(2)Settings of boundary condition

The open boundary of this study is extracted from the prediction results of the regional circulation of the South China Sea [5]. Because of the flow in the Red Bay is complex. In addition to tidal influence factors, but also by the circulation and coastal current effects induced by the local topography. Therefore, the model only considering the power flow is not consistent with the dynamic characteristics of the sea area. The South China Sea Regional Circulation prediction model takes account of the comprehensive factors such as monsoon, temperature and salinity. As an open boundary, it can accurately simulate the current in the Red Bay Sea area.

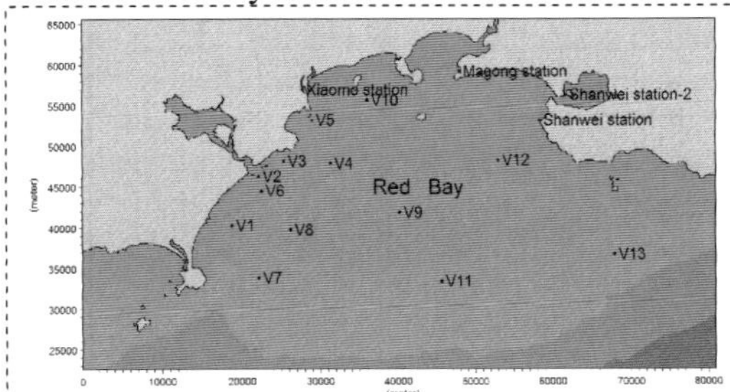

Figure 1. Calculation range and verifying point position

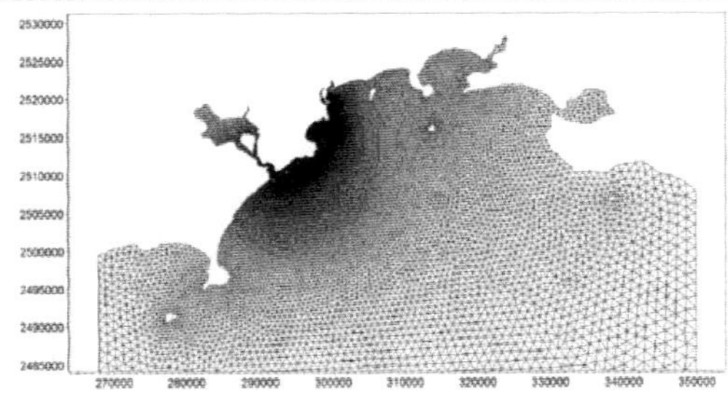

Figure 2. Model computing grid

(3)Parameter setting of temperature model

This model uses the data of temperature, humidity, wind speed, wind direction, rainfall and evaporation measured by Shanwei meteorological station in July and December.

3.Model verification and analysis of the results of ocean current calculation

3.1 Model verification

To test the reliability of the model, according to the above information and given conditions, compare the measured results and calculation results of the tidal level, flow velocity and flow direction for each measuredstation. As the article length is limited, we omitted the verifying curve of model validation[5].

The verification of the two-dimensional model is precise. The calculation results of the model are in agreement with the change trend of the actual measurement. The current model is a good reflection of the flow in the Red Bay.

3.2 Simulating analysis of the ocean current

Figure 3 to Figure 4 is the flood tide in summer. During the flood tide at spring tide of summer, the current flows from the north to the west of the Red Bay. The current flows westward along the coastline and finally from the west side of the bay to the west. During the ebb tide at spring tide of summer, the current flows from the southwest to the Red Bay. The flow runout of the Red Bay along the east coast line. The flow direction of calculation is consistent with the measurement. It can be seen from the velocity distribution, the velocity distribution is the same during the flood tide and the neap tide in summer. The velocity in the Red Bay is less than the outside of the Bay. The eastern velocity is slightly larger than the West. The calculated results are also in accordance with the distribution trend of the measured values. During the flood tide at neap tide of summer, the current flows from the west to the Red Bay. The current flows eastward along the coastline and finally from the east side of the bay to the east. The flow direction of the flood tide is the same as the ebb tide. The velocity distribution is the same. The velocity in the Red Bay is less than the outside of the Bay. The eastern velocity is slightly larger than the West. The calculated results are also in accordance with the distribution trend of the measured values.

Figure 5 to Figure 6 is the flood tide in winter. During the spring tide, the current flows to the northwest from the east side of the bay. The water flows southwest to the outside of the Bay. The currents in the bay present anti-clockwise rotation. During the neap tide, the current flows to the northeast from the west side of the bay. The water flows southeast to the outside of the Bay. The currents in the bay present clockwise rotation.

From the above calculation results, the Red Bay flow field calculated by the simulation is consistent with the measured results. The calculation method of this model is reasonable and feasible.

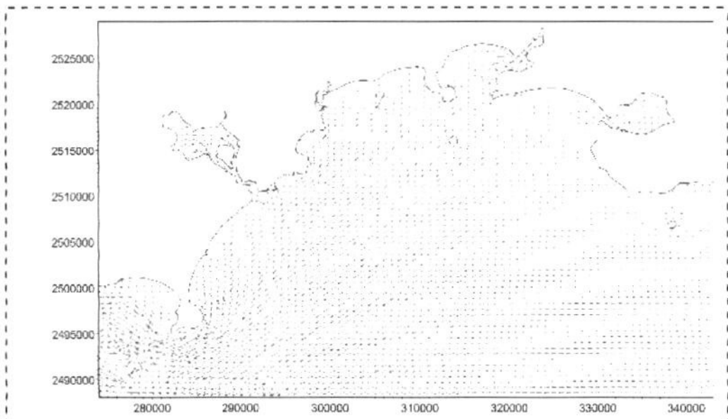

Figure 3. Velocity vector of flood tide at spring tide in summer

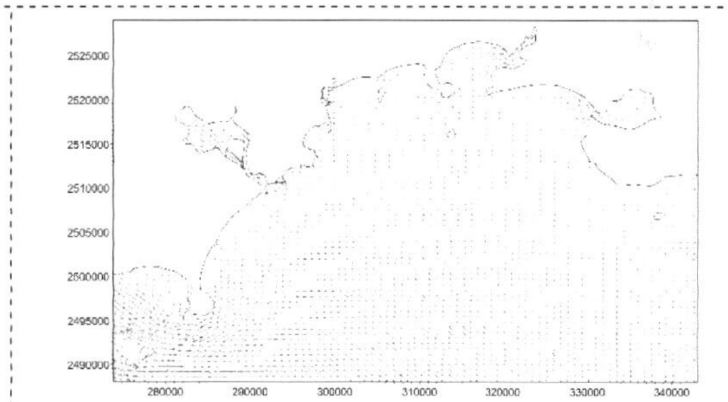

Figure 4. Velocity vector of ebb tide at spring tide in summer

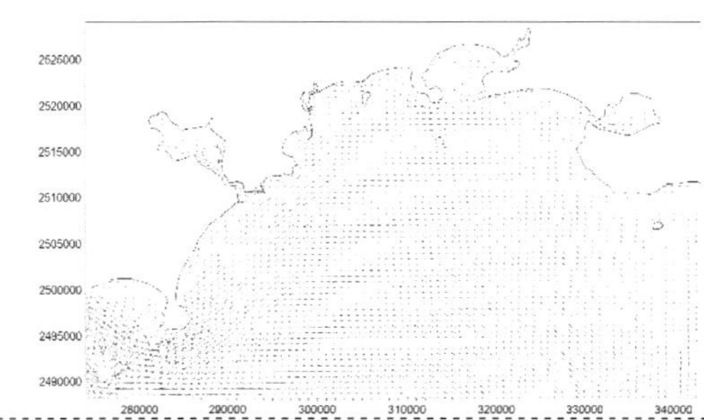

Figure 5. Velocity vector of flood tide at spring tide in winter

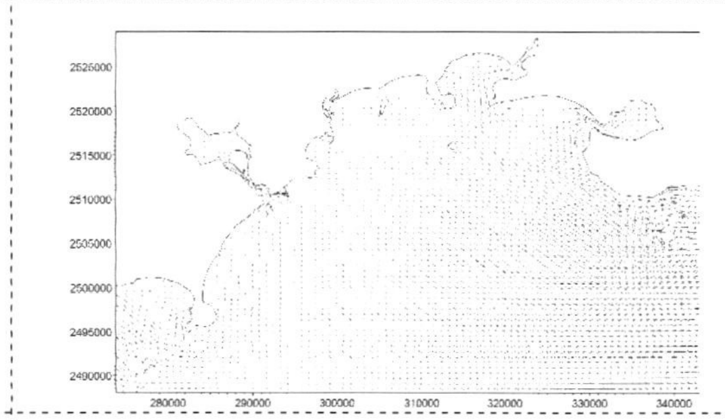

Figure 6. Velocity vector of ebb tide at spring tide in winter

4. Numerical simulation and comparative analysis of thermal drainage in summer and winter

4.1 Construction of power plant in Red Bay

The Guangdong Taiping Ling nuclear power plant and the Haifeng power plant are to be built in the Red Bay. The Guangdong Taiping Ling nuclear power project plans to build 2 million kilowatt class nuclear power units. The circulating water system of the power plant uses a direct current water supply system. The power plant uses the sea water around the site as a cooling water source. The circulating cooling water of the power plant is 130 m^3/s, and the temperature of the extraction and drainage is 7.83 degrees centigrade.

At the northeast direction of the Taiping Ling nuclear power plant, about 7km is the site of the Haifeng power plant. The Haifeng power plant plans to build 2 million kilowatt units, with a cooling water amount of $72m^3/s$ and a 7.5 centigrade temperature of the water intake and drainage. The geographical location and sensitive target map of the project is shown in Figure 7 and Figure8.

4.2 Analysis of temperature rise in summer and its influence on the marine environment

Figure 7 is the maximum temperature rise envelope of the whole tide in summer, and table 1 is the largest distribution of the influence of summer temperature rise. It can be seen from the chart and table that the thermal drainage of the Taiping Ling power plant in Guangdong mainly spread along the northeast and southwest. The range of temperature rise is mainly around the water distribution near the drain. The maximum temperature rise above 1℃ in summer is about 77.6ha.

The main influence of temperature drainage in Haifeng thermal power plant is near the shore, and the direction of diffusion is South-North. The range of temperature rise is mainly around the water distribution near the drain. The maximum temperature rise above 1℃ in summer is about 335.2ha. The maximum temperature rise in the ecological red line area is 0.76℃.

Table 1. Thermal drainage discharge environmental impact analysis on water environment in summer.

The spatial distribution of temperature rise	The maximum possible impact area of different temperature rise (ha)			
	≥4.0℃	≥3.0℃	≥2.0℃	≥1.0℃
Taiping Ling nuclear power plant	——	——	6.2	77.6
Haifeng thermal power plant	1.4	12.9	170.4	335.2

4.3 Analysis of temperature rise in winter and its influence on the marine environment

Figure 8 is the maximum temperature rise envelope of the whole tide in winter, and table 2 is the largest distribution of the influence of winter temperature rise. It can be seen from the chart and table that the thermal drainage of the Taiping Ling power plant in Guangdong mainly spread along the northeast and southwest. The range of temperature rise is mainly around the water distribution near the drain. The maximum temperature rise above 2℃ in winter is about 27.0ha. The maximum temperature rise above 3℃ in winter is about 1.4ha, and the maximum temperature rise above 4℃ is about 0.4ha.

The main influence of temperature drainage in Haifeng thermal power plant is near the shore, and the direction of diffusion is South-North. The range of temperature rise is mainly around the water distribution near the drain. The maximum temperature rise above 2℃ in winter is about455.9ha. The maximum temperature rise in the ecological red line area is 0.76℃. The maximum temperature rise above 3℃ in winter is about 173.1ha, and the maximum temperature rise above 4℃ is about 128.6ha.

Table 2. Thermal drainage discharge environmental impact analysis on water environment in winter.

The spatial distribution of temperature rise	The maximum possible impact area of different temperature rise (ha)		
	≥4.0℃	≥3.0℃	≥2.0℃
Taiping Ling nuclear power plant	0.4	1.4	27.0
Haifeng thermal power plant	128.6	173.1	455.9

IOP Conf. Series: Earth and Environmental Science **146** (2018) 012058 doi:10.1088/1755-1315/146/1/012058

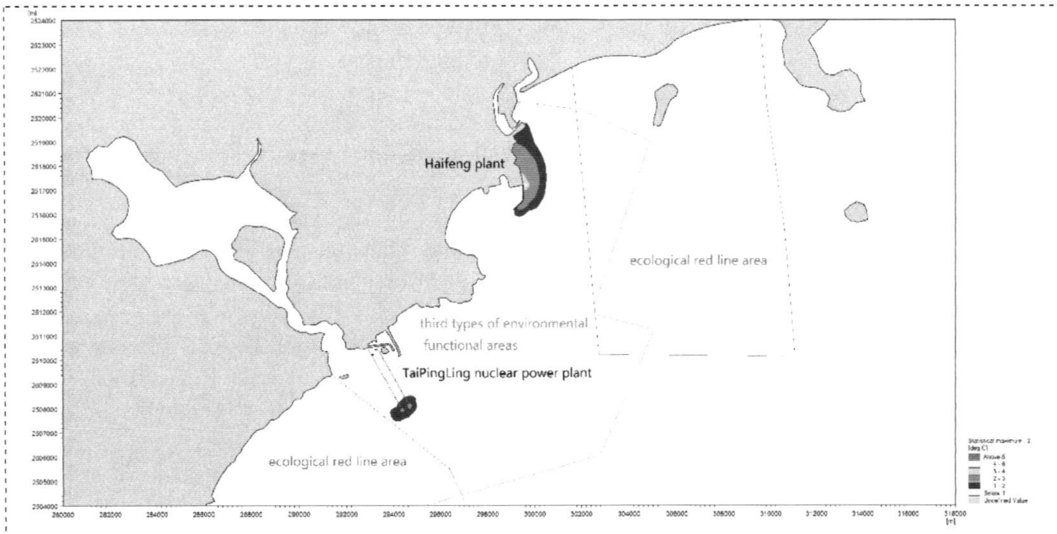

Figure 7. The influence range of the maximum temperature rise of the whole tide in summer

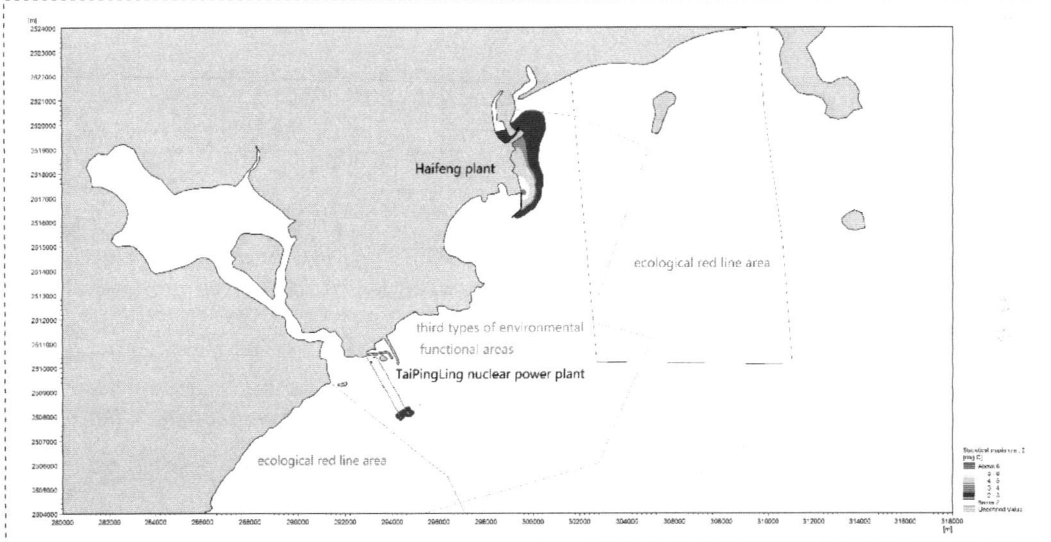

Figure 8. The influence range of the maximum temperature rise of the whole tide in winter

4.4 Comparison and analysis of the effects of thermal drainage in winter and summer

To sum up, the thermal drainage of the power plant in the Red Bay is mainly influenced by the wind and the current. The diffusion region and morphology of the thermal drainage are basically consistent with the characteristics of the sea current.

Because of the large difference of sea current in summer and winter and the different parameters of heat diffusion, the difference of temperature rise range in summer and winter is also great. The temperature rise range of the power plant in the Red Bay is larger than that in summer, and the difference of the diffusion direction is little. According to the standard of Water Quality of Sea Water (GB 3097-1997), the temperature rise in summer is controlled by 1℃ in summer, and the temperature

IOP Conf. Series: Earth and Environmental Science **146** (2018) 012058 doi:10.1088/1755-1315/146/1/012058

rise in winter is controlled by 2℃. The environmental impact of thermal emission from power plants should be taken into consideration in summer and winter. We should evaluate its impact on the marine environment in accordance with the corresponding evaluation criteria.

5. Summary

Through the numerical simulation of the sea current and thermal drainage in Red Bay, the following conclusions are obtained:

- The flow patterns in the Red Bay are very different in summer and winter. We also found this rule in the process of research in Daya Bay and DapengBay[6][7]. Measurements and calculations should be taken into consideration in different seasons (at least in summer and winter) and in different tidal types.
- The open boundary of the two-dimensional numerical simulation of the Red Bay is reasonable, which is provided by the results of the prediction of the South China Sea Regional Circulation. This method is also suitable for other regions along the east coast of Guangdong Province, such as Daya Bay, Dapeng Bay and Jieshi Bay etc.
- The diffusion region and morphology of the thermal drainage are basically consistent with the characteristics of the sea current.
- The environmental impact of thermal emission from power plants should be taken into consideration in summer and winter. We should evaluate its impact on the marine environment in accordance with the corresponding evaluation criteria.

6. References

[1] He Qi.et al 2012 *Study on the sea currents in the northern shelf and slope of the South China Sea based on the observation(Acta Oceanologica Sinica Vol.34)*No.1 pp 17-28

[2] Hu Xiaomin*The typical full tide observation of the four seasons of the first phase of the Taiping Mountain nuclear power plant in Guangdong(The First Institute Of Oceanography,Soa)*pp 126-320

[3] Huang Qizhou.et al 1992.9*General situations of the current and eddy in the South China Sea* (Advance in earth sciences *Vol.7)*No.5 pp 1-9

[4] Su Jilan2005.*Overview of the South China Sea circulation and its dynamics (Acta Oceanologica Sinica Vol.27)*No.6pp1-8

[5] Liu Aizhen*Optimization Research on the head of CNPDC Taiping Mountain nuclear power plant drainage in Guangdong (Tianjin Research Institute for Water Transport Engineering)*pp 42-69

[6] Zhu Yuxin*Report on the demonstration of the use of the Shenzhen LNG emergency Peak Station (Tianjin Research Institute for Water Transport Engineering)*pp 63-89

[7] Wang Xinhai*Research Report on the test analysis and scheme design of submarine pipeline diffuser in Shell petrochemical project*pp 16-35

IOP Conf. Series: Earth and Environmental Science **146** (2018) 012059 doi:10.1088/1755-1315/146/1/012059

Phytoplankton Community Structure and Water Quality Assessment in the Northwest of Liaoning Province Water Supply Project Waters

Xin Xu[1], Zhaohui Ni[2], Ziwei Shen[2], Mingjing Tian[1] and Boqun Liu[1]

[1] Laboratory of Environmental Protection in Water Transport Engineering, Tianjin Research Institute of Water Transport Engineering, Tianjin, China
[2] Yangtze River Fisheries Research Institute, Chinese Academy of Fishery Sciences, Wuhan, China

562611006@qq.com

Abstract. The current species composition, distribution and standing crop on the phytoplankton were investigated and studied in the waters of the Northwest of Liaoning Province water supply project, during May to June in 2010. The water supply project consists of the water supply (the middle and lower reaches of Yalu River) and water benefited areas (Liaohe and Dalinghe River). There were 241 species of phytoplankton, which belonged to 91 genera of 8 phyla. The overall evaluation suggested the water supply area could be poor and medium nourishment and the water was polluted slightly. However, the water benefited area could be medium-eutrophic, which was medium-polluted, and the eutrophication was more serious in some regional water.

1. Introduction

Phytoplankton is the primary producer of water and the main source of dissolved oxygen in the water body. The community composition, dominant species, pollution indicator species and diversity, and standing crops of phytoplankton are the important indicators of water pollution and nutritional levels [1].

This paper was mainly based on the aquatic ecological survey in the early stage of the northwest of Liaoning Province water supply project. The northwest of Liaoning Province water supply project was a project for the sake of solving the urban and rural life, production and ecological water shortage problems in the northwest of Liaoning Province and Liaohe River area. It was a large-scale inter basin water transfer project of transferring water from the abundant water area of Yalu River, the eastern part of Liaoning province. The northwest of Liaoning Province water supply project mainly involves three areas: the middle and lower reaches of Yalu River (water-supply area), Liaohe River Basin and Dalinghe River Basin (intake area).

So far, the aquatic ecological surveys about the three major basins mainly focus on the fish resources, plankton, periphytic algae, benthic animal, bacteria and water quality monitoring, etc, which belonged to several sections of rivers and the related reservoirs[2]. The complete survey of phytoplankton in the above three water areas were relatively fewer. This paper aimed at a comprehensive survey of phytoplankton community structure situation in northwest of Liaoning Province and related water areas， and discussed the population characteristics and ecological significance of phytoplankton in the three water areas. On the one hand, it synthetically evaluated the water quality in combination with physical and chemical indexes, on the other hand, it provided references for the environmental planning of water supply projects.

2. Materials and methods

2.1. Sampling point setting

In May and June of 2010, according to the main water system component of the water supply project areas (Yalu River, Liaohe River, and Dalinghe River areas), phytoplankton sampling points were arranged. There were 7 water systems and 32 sampling sections (Figure 1).

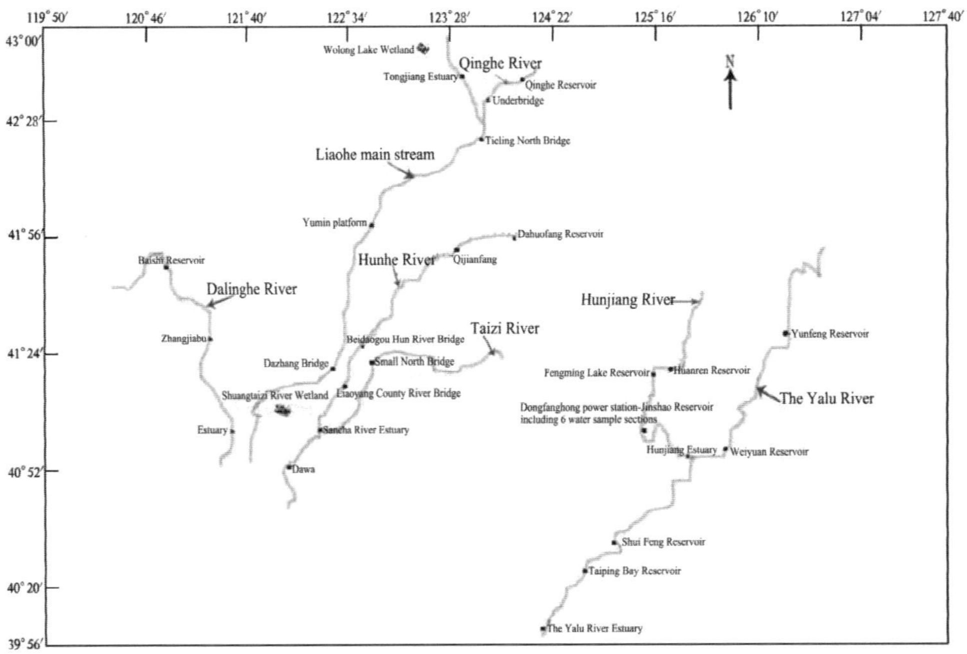

Figure 1. Sampling sites

2.2. Sample collection, analysis and identification

Phytoplankton qualitative samples:phytoplankton is collected on the surface of the water by using 25# plankton net (mesh diameter 0.064mm) and fixed with 4% formalin solution, and then identified by indoor microscopy. Phytoplankton is identified as a genus or species[3].

Quantitative sample:1L water samples are taken by conventional methods and fixed by 10mL Lugol's solution,and then preserved by concentration to 30mL after the settlement of 48h.Before counting, shake the sample to take 0.lmL and count 100 fields of vision under 400 times the mirror, then convert it into the number of algae in the water sample per liter.

2.3. Data analysis

The number of phytoplankton is calculated according to N=[V0/V1]×n, where N is the number of phytoplankton in 1 liters of water (/L); n is the number of phytoplankton counted; V0 is the volume of 1 liters water sample after precipitation and concentration (mL); V1 is the water volume (mL) of counting samples. The calculation of phytoplankton biomass refers to the calculation of wet weight of related species, that is, the biomass conversion formula of phytoplankton is biomass (mg/L) = individual (cell) mean wet weight (mg/ind) × ind/L.

The diversity index [4] is calculated with two general-purpose indicators, the Shannon-Wiener index (H ') and the Pielou index (J). Shannon-Wiener index is highly dependent on the number of species and species distribution uniformity. Pielou index can well reflect species evenness. The calculation formula is as follows:

$$H' = \sum_{i=1}^{s} p_i \log_2 p_i \text{ ,Pi=ni/N}$$

$$J = H'/\log_2 S$$

In the formula, S is the total number of species, N is the total number of individuals, and Ni is the number of the i species in the community.

3. Results and analysis

3.1. The composition and distribution of phytoplankton species

A total of 241 species of phytoplankton belonging to 91 genera and 8 phyla had been identified. Among them, the most species were baoillariophyta and chlorophyta, 119 species and 74 species respectively, followed by 22 species of cyanophyta. There were 8 phyla, 68 genera, and 180 species of phytoplankton in the water-supply area of the middle and lower reaches of Yalu River, including 104 species of baoillariophyta, 43 species of chlorophyta, 14 species of cyanophyta, and others were distributed but less. There were 7 phyla, 65 genera, and 143 species of phytoplankton in the benefit area of Liaohe River, including 66 species of baoillariophyta, 50 species of chlorophyta, 14 species of cyanophyta.There were 6 phyla, 36 genera, and 60 species of phytoplankton in the Dalinghe River, including 24species of baoillariophyta, 26 species of chlorophyta, 4 species of cyanophyta and 4 species of euglenophyta.

In the view of the species composition of phytoplankton, the water area above was a diatom-green algae type on the whole, which was consistent with the algal structure in the northeast waters of China summarized by Hongsong [5]. The waters of middle and lower reaches of Yalu River and Liaohe River both belonged to diatom-green algae type, while the waters of the Daling River belonged to green algae-diatom type. The distribution of species composition in the three major waters was obvious. The Yalu River waters were mainly composed of river water bodies, while the Liaohe River and Daling River waters were gradually transformed into lake water bodies. The dominant species of the phytoplankton in the Yalu River waters were diatoms, including Asterionella formosa, Fragilaria capucina, etc. The dominant species of the phytoplankton in the Liaohe River waters were mainly green algae, including Chlorella vulgaris, Chodatella quadriseta, etc. The dominant species of the phytoplankton in the Dalinghe River waters were Synedra acus, Ankistrodesmus, etc. The proportion of indicator species of pollution in the three major waters was mostly medium eutrophication and eutrophication.

3.2. Phytoplankton cell abundance

The distribution of phytoplankton cell abundance at different stations in the three large basins was shown in Figure 2. The phytoplankton cell abundance in the Yalu River basin of the water supply area was between 22.95 × 104ind/L~621 × 10^4ind/L, with an average of 184.14 × 10^4ind/L. There were 7 categories of phytoplankton, and the number of diatoms was the most, with an average value of 105.14 × 10^4ind/L, accounting for 57.1% of the total phytoplankton cell abundance, and secondly the average of green algae was 42.57 × 10^4ind/L (accounting for 23.1%),and then the average of hidden algae and cyanobacteria were 19.68 × 10^4ind/L (accounting for 8.2%) and 15.05 × 10^4ind/L (accounting for 10.7%), but euglenophyta, chrysophytes and dinoflagellates were distributed on individual sites, but were less.

The water-supply area of Liaohe River phytoplankton cell abundance ranged from 55.8×10^4ind/L ~2597.1×10^4ind/L, and the cell abundance in Liaoyang County Hun River Bridge was to a maximum of 2597.1 × 10^4ind/L,with an average value of 885.49 × 10^4ind/L.There were 5 categories of phytoplankton. The number of green algae was most, and the average value of that was 618.7× 10^4ind/L, accounting for 69.9% of the total cell abundance, and next the average of diatoms and cyanobacteria were 170.32×10^4ind/L (accounting for 19.2%) and73.83×10^4ind/L (accounting for 8.3%),and then cryptophyta distributed on individual sites but less. Euglenophyta cell abundance in Hunhe River in Shenyang Qijianfang was the greatest, up to 162×10^4ind/L, and other waters less.

The water-supply area of Daling River phytoplankton cell abundance ranged from 310.5~1335× 10^4ind/L, with an average value of 757.17 × 10^4ind/L. Daling River phytoplankton contained 6 categories. The number of diatoms was the largest, reaching 360.83×10^4ind/L, accounting for 47.7% of the total cell abundances, and secondly the average of green algae and cyanobacteria were 241.08× 10^4ind/L (accounting for 31.8%) and 152.43×10^4ind/L (accounting for 20.1%) ,and then hidden algae, euglenophyta and chrysophytes cell abundance were less.

From the abundance of phytoplankton cell, the Yalu River basin was mainly in poor and medium nutrition type by reference to the evaluation standard. The proportion of the biomass of cryptoalgae was up to 37%, because the cryptoalgae biomass in Huanren Reservoir in the upper reaches of the Hun River reached as high as 4.87mg/L. Liaohe River Basin was subordinated to medium trophic type. Shenyang Qijianfang waters of the Hun River was euglena-chlorophyta type, which was highly nutritious, seriously polluted. It might be related to the pollution of industrial and domestic sewage from the riverbank town in the city of Fushun[6]. Shuangtaizi River wetland was the middle eutrophication type mainly, and Wolong Lake wetland was the poor nutrition type. Daling River was the main type of nutrition, and the estuary of Daling River was the middle eutrophication type. According to the comprehensive evaluation results of water quality, the watet quality in the Yalu River basin was better than that of Liaohe and Dalinghe River basin.

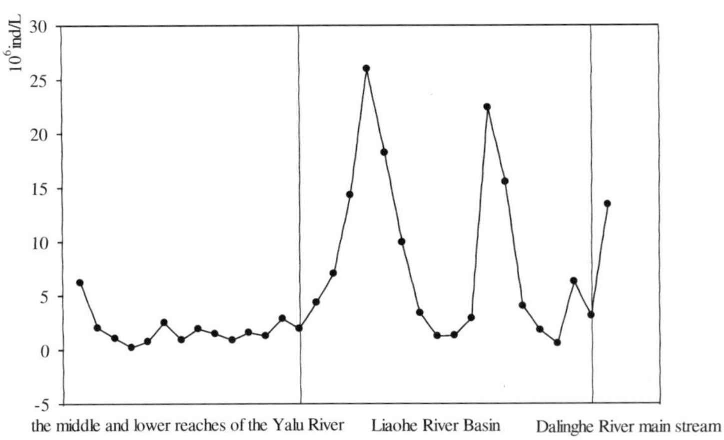

Figure 2. Curve: Horizontal distribution of phytoplankton abundance

3.3. Phytoplankton diversity index

The Shannon-Wiener index and Pielou index of phytoplankton distribute at 32 different sites in the three big watersheds, as shown in Figure 3. Referring to the evaluation standard, Shannon-Wiener index and Pielou index are used to evaluate the water quality of the three big watersheds. The Shannon-Wiener index and Pielou index were as follows. The indexes of the main stream of the middle and lower reaches of Yalu River to Hunhe River waters are respectively 1.28~3.82, 0.34~0.81;the indexes of Liaohe River waters were respectively 0.17~4.02, 0.066~0.84; the indexes of the main stream of the Dalinghe River were 2.28~2.69, 0.59~0.64. The comprehensive evaluation results were as follows: the Yalu River waters was clean-less polluted- β mesosaprobic type, in which the middle and lower reaches of Yalu River waters to Hunhe River waters was mainly β mesosaprobic type, the water quality of the middle and lower reaches of Hunhe River was better and was mainly clean-less pollution type. The Liaohe River watershed was clean- α -medium pollution type, in which the Shuangtaizi River and Wolong Lake wetland were the mainly clean-less pollution type; the Qinghe River was mainly β -medium pollution- α -medium pollution type, with poor water quality; The main streams of the Hunhe River, Taizihe River and Liaohe River were mainly clean - less pollution - β mesosaprobic type. The Dalinghe River was β mesosaprobic type.

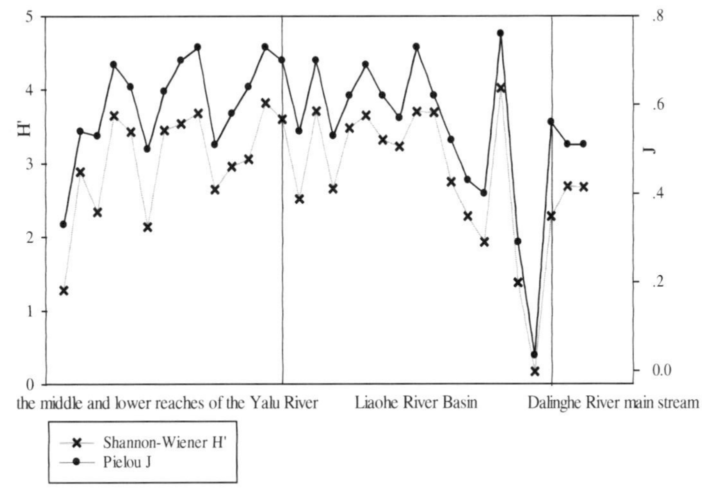

Figure 3. Curve: Horizontal distribution of phytoplankton diversity index

3.4. Water quality parameters

Water quality parameters of each section were as shown in Table 1. According to the national surface water quality standard, the data of the the Yalu River River Basin showed that the water quality evaluation of the the Yalu River River ranged from III to IV, and the N/P range was 25.2~91.4, which was more than 20 for phosphorus limitation. So the possibility of algal blooms was relatively small. The water quality evaluation of the Liaohe River Basin and Daling River Basin ranged from IV to V. The N/P range was 8.0~17.6 in the middle and lower reaches of Liaohe River, where the cell abundance were up to 2597.1 $\times 10^4$/L, and obvious tendency to eutrophication were demonstrated. In general, the water quality in the water supply area should be better than that in the water receiving area.

IOP Conf. Series: Earth and Environmental Science **146** (2018) 012059 doi:10.1088/1755-1315/146/1/012059

Table 1. A table with the mass concentration of water nutrients.

River system	The Yalu River Basin		Liaohe River Basin		Dalinghe River Basin
	The Yalu River main stream	Hunjiang River	Hunhe River-Daliaohe River-Taizi River	Liaohe River main stream-Qinghe River	
TN	0.87-1.97	0.83-1.97	2.09-4.05	1.87-2.47	2.0-2.72
average	1.55	1.41	3.206	2.07	2.65
TP	0.02-0.077	0.005-0.0174	0.036-0.354	0.03-0.16	0.0137-0.143
average	0.0355	0.0129	0.179	0.092	0.132
N/P	25.2-91.4	63.8-372	8.0-70	11.7-62.4	16-146.5
average	52.1	127.3	28.8	33.6	37.2

4. Concluion

According to the comprehensive analysis of phytoplankton species composition, dominant species, cell abundance, diversity index and related physical and chemical indexes, the community structure of the phytoplankton in the middle and lower reaches of Yalu River was mainly diatoms - green algae type, which was the nutrient type in the poor, and the individual water bodies were on the trend of eutrophication. The community structure of the phytoplankton in Liaohe River water area and main stream of Dalinghe River water area were chlorophyta-diatom type, which was medium eutrophication and lake type water body.

All in all, the water quality of the middle and lower reaches of Yalu River was generally better than that of the Liaohe River and Daling River areas. Therefore, on the point of alleviating drought and water shortage, and improving the water quality of the northwest of Liaoning Province, the northwest of Liaoning Province water supply project was necessary. However, in the long run, in order to solve the problems of drought and water shortage, deterioration of water environment and wetland restoration in the northwest of Liaoning Province, we should strengthen the measures to protect and control the water environment in water-supply area or water receiving area, so as to prevent pollution from the source.

5. Acknowledgments

This work was supported by National Nonprofit Institute Research Grant of TIWTE (TKS170204).

6. References

[1] Lepistö, L. Rosenström, U. *The most typical phytoplankton taxa in four types of boreal lakes. Hydrobiologia,* 1998, 369/370: 89-97.

[2] Liu L,Zhang M.*Phytoplankton community structure and water quality in lower reaches of Yalu River. Journal of Eastern Liaoning University*（Natural Science Edition）, 2007, 35(2): 497-498.

[3] Han MS. *Atlas of freshwater plankton. Beijing: Agricultural Press*, 1980.

[4] Sun J,Liu DY.*The application of diversity indices in marine phytoplankton studies. ACTA OCEANOLOGICA SINICA*,2004, 26(1): 62-75.

[5] Hong S,Chen JS.STRUCTURE CHARACTERISTICS OF AQUATIC COMMUNITY FROM THE MAIN RIVERS IN CHINA[J]. ACTA HYDROBIOLOGICA SINICA,,2002, 26(3): 295-305.

[6] Zhang HL,Sun LN,Luo Q,etc.*Seasonal variations and sources of pollutants in surface water of Hunhe River. CHINESE JOURNAL OF ECOLOGY*,2011, 30(1): 119-125.

Towards Thermodynamic Architecture: Research on Systems-based Design Oriented by Renewable Energy

LI Linxue[1] **and TAO Simin**[1]

[1]College of Architecture and Urban Planning Tongji University, Shanghai, 200092, China

land_well@hotmail.com

Abstract. The research presented in this paper, "Thermodynamic Architectural Prototype", hopes to put forward a new way from systems-oriented view to interpret renewable energy issue in buildings. It also helps to develop the concept of energy flow in architectural field in both theoretical and practical aspects, through a series of specific design studies. From prototype to paradigm, those design studies aims to provide an effective method for systems-based design through four phases: case studies, complex knowledge production, prototype generation and implementation, energy formation.

Preface
"An alternate agenda for energy—one that is more specific to architecture—becomes methodologically possible once the principles of these energy systems and hierarchies are more fully understood."—Kiel Moe, 2013

1. Context and Challenge
In 21st Century some pioneers of architectural education and practice begin to pay close attention to the role that energy and thermodynamics play in buildings. By introducing thermodynamics into architecture, it emphasizes two urgent tasks: one is to know what kind of architectural form can be used as a system to effectively organize the relationship between architecture and environment, the other is to learn how to channel the energy flow among the subsystems of a building[1].

1.1. Big new task
As eco-crisis and climate change have become the ubiquitous agenda within the contemporary urbanization in China, the development of "green buildings" characterized by energy saving and renewable energy utilization has turn to be a research tendency in architecture. "Thermodynamic Architecture" has tried to redefine the relationship of matter, energy, space, material and system[2]. It takes maximum energy conservation and sustainable energy supply as the principle to guide the form generation.

1.2. Integrated / Systems perspective / Multi-disciplinarity
The energy issues of contemporary architecture require complex scientific perspectives. The task of "Thermodynamic Architecture" is to study what kind of architectural form would efficiently organize the relationship between buildings and climate, as well as the energy flow in a whole building system

and its internal subsystems. In this process, it needs a shift in thought patterns to remedy the alienation between technology and design methods.

1.3. Climatic responsive design
The development of modern artificial environment technology has made buildings losing their characteristic which caused by regional climate. We believe that the architectural form as an interpretation of climate and environment, can feedback the thermodynamic performance of the site by special shape, material and spatial organization. How to continually utilize the renewable energy on the site and reconnect the link between climate and design is a critical topic in this study.

2. Research by Design
Since 2014, the research team led by Professor Li Linxue from Tongji University has been studied in the renewable energy agenda with the integration of interdisciplinary specialists. They are trying to deal with three dilemmas: how can thermodynamic science drive buildings as a system from different scales? How can interdisciplinary tools be integrated on the same platform within a site's condition? How, as John. T. Lyle said, "*shape the forms to show the processes*"?

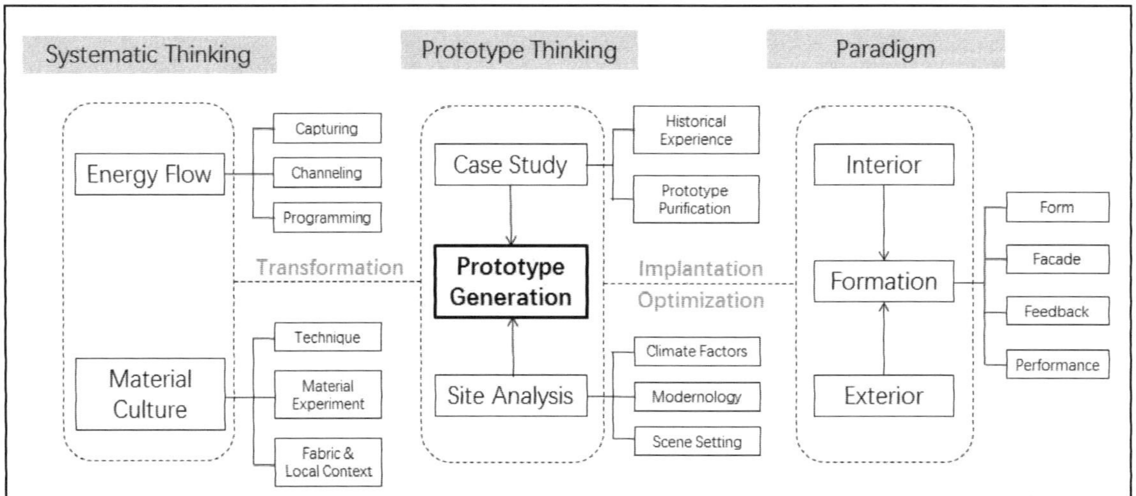

Figure 1. The general framework of "*Thermodynamic Architectural Prototype*" (Source: Tao Simin)

Therefore, the research group took "Thermodynamic Architectural Prototype" as the topic, and selected three independent keywords - Wind, Light and Heat - as the main targets of renewable energy utilization to develop design projects in three semesters. Through abstract or figural methods such as monographic study, prototype establishment and software simulation, they chose a specific site in China and proceeded from architectural design(a large-scale public building such as a commercial complex or a museum)to explore the relationship between forms and energy(figure.1). By focusing on the utilization and transformation of natural energy sources in a specific environment, a new epistemology of thermodynamic architecture is put forward and the systematic research on the design methodology is carried out.

3. PHASE 1-2-3-4 :Prototype to Paradigm
The research process includes four phases, and each phase requires corresponding modelling study and diagram analysis.

3.1. PHASE 1: Case study: reading, observing, analyzing and illuminating

"Prototype" represents the underlying and historical experience, it indicates a fundamental contribution[3]. The case studies expect to recall the energy clues in those buildings and reveal the typical inherent architectural laws by prototypes extraction. It may include fragments and the entireties, small and large scales, single building and settlement. And it may also concern with technics, materials, spaces, surfaces, devices and organizations[4]. The aim is to refine their significant "prototypes" by analysing the strategies in the selected cases, about how they capture, channel or program energy. The step may provide references for the subsequent design process and expand them to a wider field.

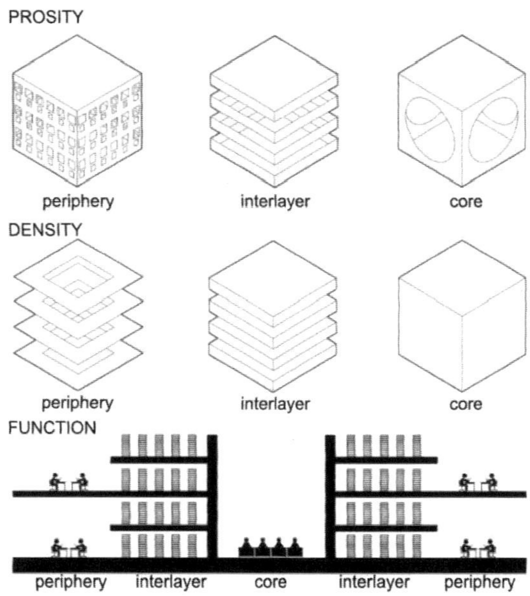

Figure 2. "Peel" the Phillips Exeter Academy Library for prototypes (Source: Zheng Xin, Zheng Siyao, Lyu Xinxin. Tutors: Li Linxue, Zhou Jianjia)

Take "light" as the topic, the research group selected Phillips Exeter Academy Library for a case study in this phase. They peeled the whole library into three sections: periphery, interlayer and core (figure.2). The periphery provides a homogeneous facade, the two sizes of windows guarantee the ample sunshine for the reading place; the interlayer as a relation between hall and reading place ensures suitable light intensity; the core is not just a simple hall, in fact it is a centrality, a symbolism, meanwhile it can be the reflection of light. After that they exacted the prototype in each section with three keywords - porosity, density and function - to answer the question according to the local climatic characteristics: how the case represents significance in the architectural energy agenda?

3.2. PHASE 2: Interdisciplinary and trans-disciplinary approach: a complex production of knowledge

Through the construction of new knowledge systems, research in energy and thermodynamic architecture will provide an opportunity to restore an autonomous instrumentality in a wide range of discourses of sustainability. In this phase, the researchers concern about all the natural elements in a built environment. Within the interdisciplinary knowledge system of architecture, environment, energy and material science, the research group is trying to do a discourse re-construction. For example, a set of quantitative analyses were carried out by a group member in order to take the middle-east wind tower as his prototype, and some experiments were conducted to discuss the influence of the air flow angle and the opening form of a wind tower on ventilation performance (figure.3).

In addition, we also need to pay attention to "material culture" which closely related to the climate and historical context on site. Material interacts with the environment in the form of energy flow. The redefinition of material pedigree, tectonic and perception is actually the aggregation of visible matter and invisible energy[5]. The research group once took a series of drain ability test: since it rains frequently in Shanghai, the facade drainage is quite important. Clay can be gradually eroded and stained if immersed in water for a long time. Different types of texture are tested under water to optimize the drain ability. The V shape texture, though very simple, works efficiently. Base on the texture, the technique for this facade was also explored (figure.4).

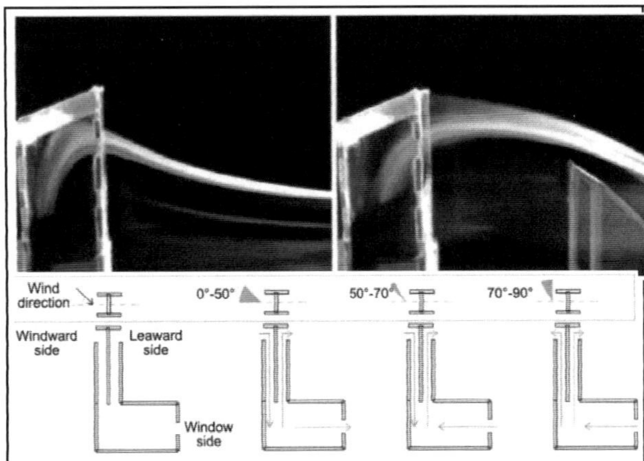

Figure 3. Analysis for Wind tower (Source: Zhang Runze. Tutors: Li Linxue, Zhou Jianjia)

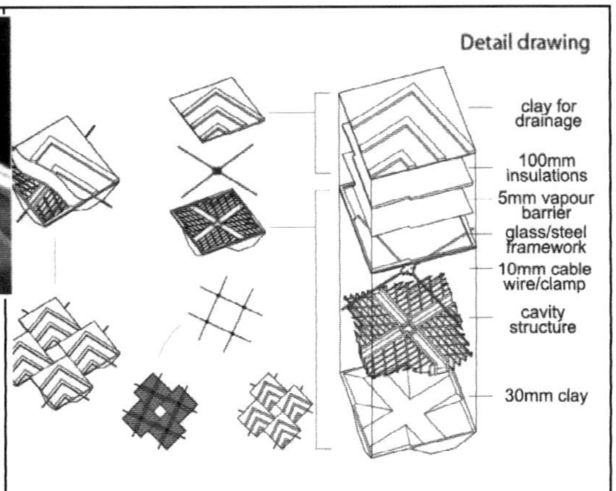

Figure 4. Material experiment (Source: Liang Qianhui, Lin Jingzhi, Wang Jingkai. Tutors: Li Linxue, Zhou Jianjia)

3.3. PHASE3: Prototype generation and implementation

Looking for original energy clues in buildings from the perspective of prototype is to reveal the most typical thermodynamic rules in its design. Base on the experience of thermodynamic abstraction from case studies, the research group embeds the initial prototype into the climate, environment and context of the site. By the process of angling, segments, addition, merging, overlapping or distortion, the prototype can be evolved and evaluated to fit a specific environmental mechanism.

As mention earlier, a group member took the middle-east wind tower as the original prototype. In this scheme he focused on the sustainable utilization of wind energy on the site. According to the wind direction and speed of the site, prototype transformation steps like geographic control, volume division and contour test for 3D facade were taken by software simulation for better wind energy utilization in the local environment (figure.5).

The "light" scheme came from photosynthesis. The prototype generation took a composition of several factors into account: porosity, layers and channel. In order to test and optimize the performance of the prototypes in the light environment of site, we tried to produce many variations of the original prototype by different dimension, thickness, height, direction and form, and used the prototype models to simulate a whole day's natural light condition in the light laboratory (figure.6). The most reasonable prototype could be selected in this process and applied to the next phase.

In the "heat" study our research group was inspired by ant colony. We started with the inner cavity section analysis, tried to explore the combination modes of different channel branches, heights and cross section forms, and considered the optimization method by simulation. The final prototype generation combines the climate on site and the required functional elements (in this case it is an industrial museum) and it represents a better heat regulation (figure.7).

3.4. PHASE 4: Energy formation: systems-oriented design

Kiel Moe put forward the concept of "energy formation" in his book *Convergence: an Architectural Agenda for Energy*, and he explicated materials, energy systems and amortization as three factors that should be converged through design to maximize the architectural and ecological power of buildings[6]. In this view, the basic way of generating architectural form is the capture and guidance of energy. The process from prototype to paradigm can be seen as a systems-based integral translation. Therefore the resulting form may be called "energy materialization".

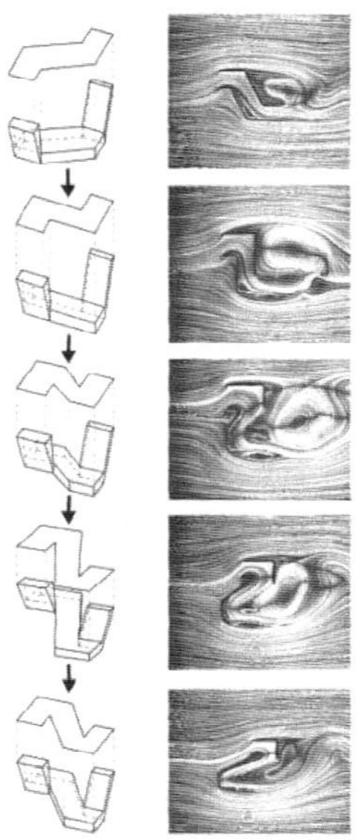

Figure 5. Prototype developing process oriented by wind (Source: Zhang Runze. Tutors: Li Linxue, Zhou Jianjia)

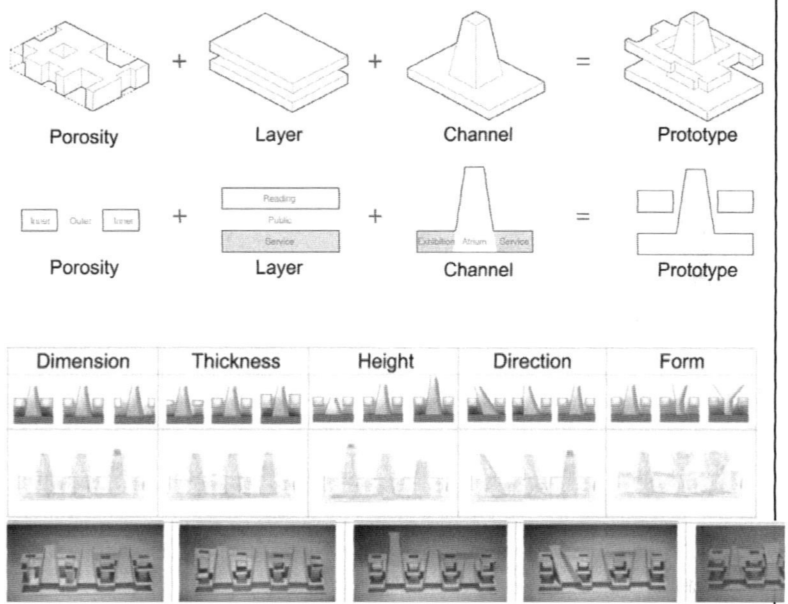

Figure 6. Prototypes test in "Photosynthesis" (Source: Zheng Xin, Zheng Siyao, Lyu Xinxin. Tutors: Li Linxue, Zhou Jianjia)

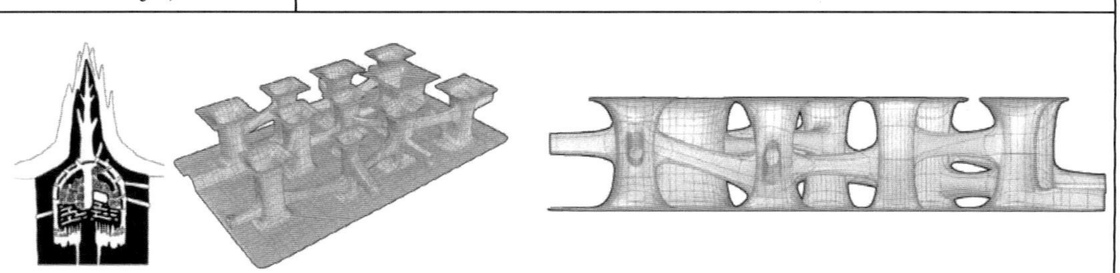

Figure 7. Ant colony and heat formation (Source: Li Hao, Gan Chongyu. Tutors: Li Linxue)

From prototypes to final design, it is a process of morphological and systematic transformation. First is a scale conversion on the prototype. Then taking into account the function, infrastructure, traffic organization, structure and the specific on-site climatic which need to be confront after prototype scaling, the energy formation begins to take shape, the interior and exterior space are naturally generated[7]. Finally, combining the adjustment by software simulation and detail structure models, the way to make use of on-site renewable energy can be expressed through the building form. The form appears to be open and able to fit in the local climate and context. Still take "Photosynthesis" as an example, the main objective of the design is to better guide and organize the natural light to meet the daily use of the building. In the final scheme, we can see the "chimneys" of varying size come from "channels" in the prototype, same as the spaces coming from "porosity", functions coming from "layers". They are massing into a whole building (figure.8).The building should be put into the site for environmental performance simulation, and the model also needs to be used for testing the feasibility of construction (figure.9).

4. Conclusion: New paradigm for thermodynamic architecture

This study focuses on the thermodynamic principles of natural energy flow in the architecture and the utilization of the natural energy sources in specific climate environment. This project employs an innovative academic vision, integrates inter-disciplinary team, re-examines the guiding role of renewable energy in the initial design phase from a systematic perspective. And with the help of ecological simulation and digital design technology, the frontier method of "*Thermodynamic Architecture Prototype*" is built to provide a deeper theoretical support for the development of green buildings.

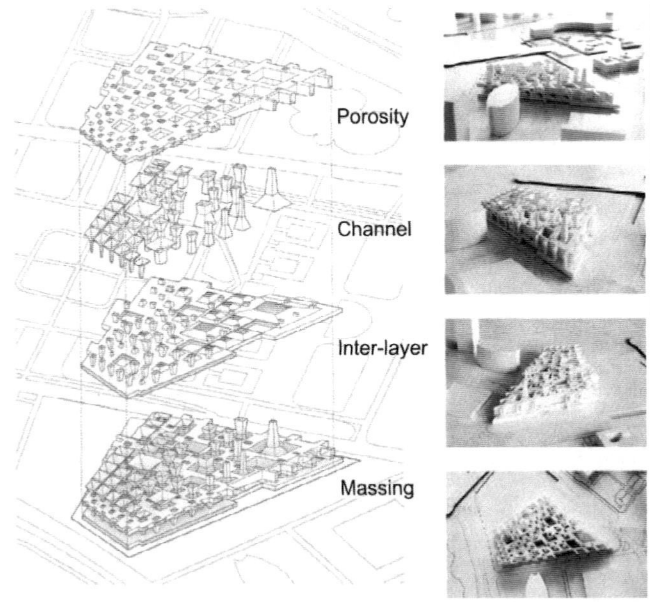

Figure 8. Prototype transformation and implantation (Source: Zheng Xin, Zheng Siyao, Lyu Xinxin. Tutors: Li Linxue, Zhou Jianjia)

Figure 9. Performance and details (Source: Zheng Xin, Zheng Siyao, Lyu Xinxin. Tutors: Li Linxue, Zhou Jianjia)

Acknowledgments

The paper is supported by Key Laboratory of Ecology and Energy-saving Study of Dense Habitat (Tongji University), Ministry of Education of China.

References

[1] Li L 2015 *Time+Architecture* **10** 142
[2] Li L and Tao S 2016 *South Architecture* **27** 173
[3] Moneo R 1978 *Oppositions* **13** 40
[4] Banham R 1969 *The Architecture of Well-tempered Environment* (Chicago: The University of Chicago Press)pp 89-92
[5] Hensal M 2013 *Performance-Oriented Architecture: Rethinking Architecture Design and the Built Environment* (UK: John Wiley & Sons Ltd.)p 121
[6] Moe K 2013 *Convergence: An Architectural Agenda for Energy* (London &New York: Routledge) pp 22-25
[7] Braham W W 2015 *Architecture and Systems Ecology: Thermodynamic Principles of Environmental Building Design* (London &New York: Routledge)pp 23-28

IOP Conf. Series: Earth and Environmental Science **146** (2018) 012061 doi:10.1088/1755-1315/146/1/012061

Building digital topography model of the intertidal zone between Caofeidian Nanpu town and Tanggu district using multi-period remote sensing data

Zhang Z G[1], Wu F[1], Li T Q[1], Li Q[1], Gao Z H[1] and Jin D J[1]

[1]China Aero Geophysical Survey & Remote Sensing Center for Land and Resources, 31 Xueyuan Road, Hiadian District, 100083 Beijing, P.R.C

E-mail:zhang_zonggui@qq.com

Abstract. The intertidal zone, a change zone of land and sea interaction, has abundant land reserves, tourism and other resources, which has a location advantage of great engineering construction in China's coastal areas and thus is very important to investigate and monitor the topography of the tidal zone and also keep track of its topographic changes. In this study, the new section of the Bohai sea coastal zone from Nanpu to Tianjin Bohai sea area has been chosen as an example. Based on the terrain feature of the information expressed by the multi-period remote sensing image information and the tidal water level information as well as the tidal data acquisition time in the remote sensing image data, we have been calculated the water line (or edge point) digital elevation topography inversion analysis, and also constructed the digital terrain elevation model, and most importantly have provided a kind of quantitative method to get tidal zone (tide beach) terrain information.

1. Introduction

China's18,000 kilometers of coastal zone is rich in nature resources, and the location advantage of coastal cities are obvious, such as tidal flat land resources, port resources, salt resources, fishery resources, oil resources, natural gas resources, tourism resources and placer resources, etc. [1]. Whereas there also exists fragile ecological environment and frequent disasters problems, including the population growth and urbanization, sea level rise and coastal erosion, shortage of fresh water resources and deterioration of water environment and degradation of fishery resources, and so on [2].

The intertidal zone of Bohai Sea mainly contains clay, sandy, rocky, artificial and plant, and among them the clay intertidal zone or tidal flat is the main intertidal zone of the Bohai Sea coast, and is also a typical silty tidal flat in China. Its stratum is basically covered by quaternary strata, and it is mainly composed of clay powder, sandy silt, sandy clay, silty clay, and other fine particles. Affected by the tidal water, the intertidal zone of Bohai Sea combines with the two sides of a certain range of land and shallow sea transition strip have rich land resources, tourism resources, etc., having the location advantages of great engineering construction in China's coastal areas. Thus, it is very important to investigate and monitor the topography of the tidal zone of Bohai Sea and also keep track of its topographic changes.

The traditional topographic measurement is limited by manpower, material resources and financial resources, and cannot conduct large-scale dynamic monitoring and is difficult to meet the monitoring requirements. Affected by the tide and silt, the people do not go, and the ship does not come, so the

Muddy intertidal zone in the terrain mapping is in the blank. With the development of remote sensing technology, remote sensing monitoring method has the characteristics of rapid large-area monitoring. Remote sensing technology can quickly obtain images that reflect the instantaneous state of the ground.

By this method, for tidal flats, it is possible to obtain the instantaneous water boundary information acted by tides and so on, furthermore, to calculate the elevation of the water line (or edge point) according to the tide data at the time of remote sensing image data acquisition. The tide terrain inversion based on remote sensing water line is becoming more and more realistic.

In this study, the new section of the Bohai sea coastal zone from Nanpu to Tianjin Bohai sea area has been chosen as an example. Based on the terrain feature of the information expressed by the multi period remote sensing image information, combined with the addition of the tidal water level information, we have made the digital elevation topography inversion analysis, and also constructed the digital terrain elevation model, and importantly have provided a kind of quantitative method to get tidal zone (tide beach) terrain information.

2. Survey of intertidal zone

According to the aerial geophysical remote sensing survey and application project area of Bohai coastal zone in 2016, combining with the situation of Bohai intertidal zone and the field reconnaissance, the tidal flat inversion test area chosen in this work is located in Bohai bay, which is between Tangshan South Town and Tianjin Binhai New Area Tanggu tide station, as shown in Figure 1. The tidal flat consists of silt and sandy, and the sandy tidal flat is full of various shapes of gravel and shells, covered with wormwood on its shore. Beach width (to the highest water level) is about 13m, 0-5m slope of about 10°, 5-13m slope of about 15 °. The length, width and slope of the selected tidal flat are more suitable as a topographic inversion test area for tidal flats.

Figure 1. Sketch map(orange box) of topographic inversion test area of tidal flat.

3. Tidal flat topography construction method

3.1. Technique flow

The OLI (Operational Land Imager) Remote sensing image data was obtained by collecting multiple temporal phases LandSat8 and treated by the procedure of band extraction, atmospheric correction, image cutting, etc. The instantaneous water margin lines at different periods were then extracted by human-computer interaction to get the serialized water line data set. Combined with high tide level grid data generated by tide level data, the elevation of water level point was calculated. Followed by space interpolation, a highly reliable digital tidal flat elevation model DTFEM (Digital Tidal Flat

Elevation Model) was built. And finally use the aerial photography or unmanned aerial vehicle remote sensing data to verify.

3.2. Remote sensing water line extraction

The method of remote sensing water line extraction was to select the water line of image similar to the tide (high tide). The water line extracted from satellite images reflected the state of a tide moment, assuming that the water line was a contour line, and then the elevation of water level line can be calculated on the basis of the tide level data of nearby tide station. And then to inversion the tide level of tide gauges (tidal high base). A series of contour lines can be formed by obtaining a series of contour lines under different tide conditions. The digital elevation model DEM was established by digital elevation model construction technique to obtain the approximate terrain of tidal flat.

Using TM images, Kevin et al. [3] extracted the water margin of the Nile delta coast and monitored the changes in its coastline, whereas Ryu et al. [4] extracted the water channel line from the Gomso bay and discussed the discussed the relationship between the accuracy of the data in different bands and the fluctuation. Zheng Z S et al. [5] and [6] extracted the water line, using TM/ETM images for many years combined with different tidal ground spectral characteristics and the decision tree method as well as the regional growth algorithm, and simulated the satellite transit time water level by hydrodynamic model, in order to solve the problem of tide gauge shortage in the method of tide zoning correct. Hu W et al. [7] used MODIS /TM images to carry out the Multi time scale tidal flat elevation inversion of Dongsha shoal on the central coast of Jiangsu province, and they obtained the tidal terrain model with good precision through the fusion of MODIS images and TM images.

3.3. Remote sensing data selection

Considering the feasibility and cost and the incomplete high score data, LandSat8 and OLI images are collected in this study as the main remote sensing data source for analysis. The basis for selection is based on the appropriate the spatial resolution, the temporal resolution and spectral resolution.

In 2015, LandSat8 acquired a total of 23 scenes of remote sensing images covering the research area, excluding cloud and fog-stained images. And 11 scenes of OLI images were selected in this study, the transit time and of the corresponding tide height of each scene and are shown in table 1.

Table 1. General survey of remote sensing data.

Scenes number	Date	Sensor type	Transit time	Tide value （cm）
2015019	2015/1/19	OLI	10:48	141.77
2015035	2015/2/4	OLI	10:48	116.75
2015083	2015/3/24	OLI	10:47	151.84
2015115	2015/4/25	OLI	10:47	189.03
2015163	2015/6/12	OLI	10:47	201.91
2015195	2015/7/14	OLI	10:47	126.93
2015227	2015/8/15	OLI	10:47	67.62
2015259	2015/9/16	OLI	10:48	97.69
2015275	2015/10/2	OLI	10:48	118.05
2015307	2015/11/3	OLI	10:48	149.99
2015339	2015/12/5	OLI	10:48	113.09

IOP Conf. Series: Earth and Environmental Science **146** (2018) 012061 doi:10.1088/1755-1315/146/1/012061

3.4. Water level extraction and elevation of waterside line

In this study, the water body index method (NDWI, MNDWI) [8] is used to extract the water line, and then the remote sensing water line is further constructed according to the characteristics such as colour, texture and direction of the remote sensing image. The polynomial is used to fit the instantaneous tide position of the imaging time, and then this instantaneous tidal value is assigned to the corresponding water line, and the approximate elevation line of the tidal flat is obtained.

3.4.1. Water line extraction. The water line extraction technology flow is based on remote sensing image data. The water index NDWI and MNDWI [8] and [9] can be calculated by using formula (1) and (2), and then to get the imaging binary map by selecting the water index with the chosen threshold, finally, to extract the water line in the binary map.]

Because the water has the strong absorption characteristics of near-infrared and middle-wave infrared, combined with the experimental data, this study selects the two commonly used water bodies to extract the water line, that is, the normalized water index NDWI and the improved normalized water index (MNDWI):

$$NDWI = (Green - NIR) / (Green + NIR) \quad (1)$$

$$MNDWI = (Green - MIR) / (Green + MIR) \quad (2)$$

Here, Green is the Green band, NIR is the Near InfraRed band, MIR is the Middle InfraRed band.

The first step is to calculate the water index of the 11 scenes of OLI images using the above formula to obtain the NDWI and MNDWI data. The second step is to use the real colour image of each stage to visually select the water and land boundary points in the image to determine the threshold of the water body index. The third step is to compare the position of the water and land boundary points of the different periods under different thresholds, and select the applicable water index for each period as shown in Table 2.

Table 2. Water index and threshold.

Scenes number	Water index	Threshold	Scenes number	Water index	Threshold	Scenes number	Water index	Threshold
2015019	MNDWI	0.2	2015083	NDWI	0.1	2015227	NDWI	0.3
2015035	MNDWI	0.2	2015115	NDWI	0.1	2015259	NDWI	0.1
2015339	MNDWI	0.3	2015163	NDWI	0.1	2015275	NDWI	0.1
			2015195	NDWI	0.1	2015307	NDWI	0.2

3.4.2. Water line screening and painting. Because of the existence of long trenches on the tidal flat, there are cross or overlapping phenomena in the multi-temporal water lines extracted from the images, which make it difficult for the mathematical simulation to make the terrain singular. In order to eliminate the abnormalities of mathematical simulation, appropriate artificial editing can reduce the impact of the tide. In addition, it should screen the water line, and remove the impact of the high sediment concentration of seawater of the low tide, and also extract the obvious anomalies of the water line, as well as sea ice affect the water line, and the results are displayed in Figures 2, 3.

Figure 2. Extracts access to all water margin lines.

Figure 3. Painted and screened water line.

3.4.3. Waterside line high tide acquisition. In this study, we select the measured data of Nanpu tide station to calculate the corresponding tide value of each water line (Figure 1). The tidal high data of the Nanpu tide station records the tidal level of 3 to 4 moments in each day, that is, the maximum or minimum tide value of the period. According to the dynamic characteristics of the tide of the sea, the polynomial function is chosen to fit the change of daily tide level, and the tide level of the water line during the satellite transit is calculated, which is used as the water line tide value in Table 1.

4. Results and analysis of DEM construction in tidal flat

4.1. Tidal flat DEM

The modified water lines are given the corresponding instantaneous tide height values (the high tide datum level), and the tidal flat contour data in the research experimental zone is generated. The tidal beach DEM is constructed by ArcGIS. As the interpolation method of ArcGIS is mainly based on the point vector interpolation, it needs re-sample the high line data of the tide level and select the appropriate interpolation method. In the interpolation, the distribution of the sampling points has a direct influence on the interpolation results, and the sampling point overdrive will reduce the interpolation precision and lose the detailed information, while the distribution is easy to show the terrain singularity.

In this study, we study the resampling of line vector to point vector by using five kind of accuracy of 30m, 60m, 90m, 120m and 150m, respectively. Based on the above five kinds of resampling precision point vector data, the interpolation methods such as Kriging, IDWI and Irregular Triangulation Network are compared with different parameters. The results show that the irregular triangles are connected to the tidal level contour vector data. And it is not sensitive to the change of the resampling precision. Therefore, it is necessary to preserve the details of the tide contour data as much as possible. Finally, the tide contour data is finally resampled by 30m and generated the tidal flat digital elevation model by using the Irregular Triangulation Network is shown in Figure 4.

Figure 4. Approximate topography build of tidal flat using tide level contour and TIN in test area.

4.2. Construction result analysis

The DEM data of the experimental area measured by aerial photogram in 2016 were used as the validation data for the DEM retrieval. In order to facilitate the error statistics, the aerial photogrammetric DEM data is resampled to the same 30m×30m resolution grid size as the DEM retrieval. Then, to select randomly several topographic data of two DEM overlapped regions, the retrieved DEM is subtracted from the error of the DEM obtained by aerial photogrammetry. From the range of values of the difference, the maximum and minimum values of DEM elevation are retrieved locally, and its error is between -5m~2m, whereas the individual differences are -8.81m and 5.71m, according to the statistics data. It should be noted that this model is only approximate terrain inversion, and the aerial photogrammetric obtain of DEM is not low tide level acquisition, and thus DEM inversion results are only relative assessment, do not strictly assess the accuracy.

5. Conclusion

In this study, the extraction of water line has been developed by using the remote sensing data. Although the inversion of tidal flat topography is only a kind of approximate topography, it is still an effective way to keep track of the terrain data in the absence of measured terrain data.

Moreover, the remote sensing image data of the test area in this study is only 30 meters, but the intertidal zone terrain can meet the precision requirement of 1: 10000 ~ 1: 50000 scale remote sensing mapping & engineering drawing. Based on such technology, topographic mapping of muddy tidal flats that is difficult to reach (no man can go, and no ship can come) in traditional surveying and mapping can be carried out quickly and effectively with the use of multi-temporal satellite remote sensing data. The technical method is not only the energy saving cost, but also can fill the intertidal zone with long and wide muddy beach mapping blank.

If the domestic high resolution satellite remote sensing image data is complete, the accuracy can be improved to meet the accuracy requirements of 1:5000~1:25000 scale mapping, by using the high resolution GF-1/GF-2 optical remote sensing image data. With the use of high resolution domestic satellite remote sensing data, the precision of the built intertidal terrain model is improved. A high precision digital elevation terrain model (DEM) built by inversion in intertidal zone will be effectively applied to the coastal zone storm surge dam project construction, tidal flat reserve land resources survey and survey, muddy tide to bring the country to carry out major construction projects (nuclear power plant construction, cross-sea channel project, large-scale petrochemical refining construction, coastal cities Group construction, etc.) site selection, coastal comprehensive geological survey, coastal zone resource development and land and sea co-ordination of sea and land terrain seamless stitching.

Acknowledgments
We thank Prof. Zhang Yongjun (China Aero Geophysical Survey & Remote Sensing Center for Land and Resources) for fruitful discussions and help. Supports from the Sun Xiaoming (Tianjin Geological Survey Center of China Geological survey) and Yin Ping (Qingdao Institute of Marine Geology, China Geological Survey) are also gratefully acknowledged.

References
[1] Wang X D, Cheng F, Kang F, Xie H L, Liu F T and Meng L 2014 *Marine Science Bulletin* **33** 559
[2] Zhang D D, Zhang Z G, Wu F, Jin D J, Liu C L, Xiao C L, Li Q 2016 *Proc. Of The 20th China Remote Sensing Conf.*(Shenzhen:The 20th China Remote Sensing Conference)
[3] Kevin W and El Asmar H M 1999 *Geomorphology* **29** 93
[4] Ryu J H, Won J S and Min K D 2002 *Remote Sensing of Environment* **83** 442
[5] Zheng Z S, Zhou, Y X, Shen F, Jiang X Z and Tian B 2007 *Remote Sensing for Land & Resources* **2** 57
[6] Zheng Z S, Zhou Y X, Jiang X Z and Shen F 2007 *Remote Sensing Technology and Application* **22** 35
[7] Hu W 2012 *Study on remote sensing inversion method of inconstant tidal flats DEM* (Nanjing:Nanjing University)
[8] Wu D, Sun J, Huang W Q, Guo J and Wei S 2016 *Journal of Surveying and Mapping* **4** 8
[9] Zhang Z D, Zang R G 2009 *CHINESE JOURNAL OF PLANT ECOLOGY* **33** 833

Analysis on the operating characteristics of improved UASB to treat the thermal dehydration wastewater of sludge

Li Huili[1], Pei Yuanmei[1] and Long Zhiqiang[1,2]

[1] School of Civil Engineering, Lanzhou University of Technology, Lanzhou730050
[2]Key Lab of Urban Environment and Health, Institute of Urban Environment, Chinese Academy of Sciences, Xiamen 361021

379063403@qq.com; 3422177453@qq.com; 969370482@qq.com.

Abstract In order to investigate COD removal efficiency and production capacity of energy, the anaerobic sludge bed reactor with high diameter ratio of 10 was used to treat the thermal dehydration wastewater (TDW) of sludge. The experiment was divided into 7 stage, the results showed that on the condition of the acclimated sludge, the reactor would be stable after 25 days; then the maximum removal rate of COD was up to 68%, the maximum amount of Nissan CH_4 was 575.53 ml/L, and methane production ratio was 104.05ml/ (g COD·d). The concentration of ammonia nitrogen (NH4+-N) was the main factor affected the stable operation of the reactor. When the concentration of NH4+-N reached 1.5g/L, the inhibition phenomenon to methanogenic bacteria appeared. And when it was up to 2.4g/L, the methanogenic bacteria were seriously inhibited. The optimal sludge loading of the experiment was 0.10~0.24 kg COD/ (kg Sludge·d), and the optimal volume load was 5~13 kg COD/(m^3·d).

1. Introduction

With the improvement of emission standards and industrial development level in China, the Sewage quantity of the urban sewage treatment plant has increased greatly, and the byproduct- the surplus sludge, has also increased. Main components in excess sludge are persistent organic pollutants including protein, polysaccharide, fat and humic substances, polycyclic aromatic hydrocarbons (PAHs), polychlorinated biphenyls (PCBs), and two (PCDD/Fs), and so on[1-2]. In 2015, the sludge production in Chinese urban sewage is about 35 million tons. Therefore, sludge reduction, recycling and harmless are particularly important for protecting environment, sludge dewatering is an important technical method to achieve those objectives. The technology of sludge hydrothermal carbonization treatment may effectively reduce the sludge, at the same time, the residual sludge can be produced used to activated carbon [3-4]. The process of sludge hydrothermal carbonization treatment will produce a large number of sludge thermal dehydration liquid, which is the production of the processes of a hydrothermal carbonization treatment (130～250℃) and high pressure mechanical filter dehydration. The composition of the TDW is complex, containing various kinds of organic matters, they are high in concentration ,especially high nitrogen. Therefore, it is difficult to degrade.

At present, there are few researches on the treatment of the TDW of the sludge (TDWS) in China, and the mainly methods was advanced oxidation, dilution treatment and anaerobic SBR. According to the research of Wang D et al. [5], they used the methods of Fenton and UV-Fenton to treat the TDWS, only part of COD was being removal. The experimental conclusion showed the removal rate of COD were 40%, 85%, respectively, but the removal rate of NH4+-N and total nitrogen were very low. In the experiment of Zhu W et al.[6] , the mixed liquid of the simulated domestic sewage and the TDW was been deal with the SBR reactor. The experiment found that when the ratio of the TDW was 0.05% of the simulated domestic sewage, the operation of SBR system would not be impact, and the indexes of the effluent could achieve the A level of GB18918-2002 (China). But, the huge dilution ratio is less operational in practical applications. According to the results of treating the TDW with the sequencing batch anaerobic digestion (ASBR)[7-8], the removal rate of the COD, NH4+-N, chroma and other indicators were more than 90%. This means the anaerobic treatment method was effective on sludge thermal dehydration. However, the process of treating 200mL wastewater was a period of 26 days. The hydraulic load is too low for practical application.

Summarizing of the above research results, in this experience, the TDWS was treated with the anaerobic sludge bed reactor (UASB) on the condition of the temperature of （52.5±2.5）℃, for studying the removal capacity of various pollutants, the influence factors of the reactor operation and production prospects. The result would be provided a reference for the treatment method of theTDW.

2. Experimental equipment and analytical method

2.1. The reactor of experiment

An improved upflow anaerobic sludge blanket reactor (UASB) was used in this study, of which the ratio of height to diameter was 10. This reactor was similar to the expanded granular sludge bed (EGSB) without reflow, in order to improve the rising velocity, promote the formation of granular sludge, and reduce sludge loss in reactor.

The basic dimensions of the reactor were shown in Table 1; the structure of the reactor was shown in figure 1.

Table 1. Sizes of the UASB

	Reaction area	Three-phase separation area	Setting area	The reactor
high(dm)	7	0.77	0.73	--
internal diameter (dm)	0.7	0.5	1.5	--
volume(L)	2.69	0.15	1.29	5

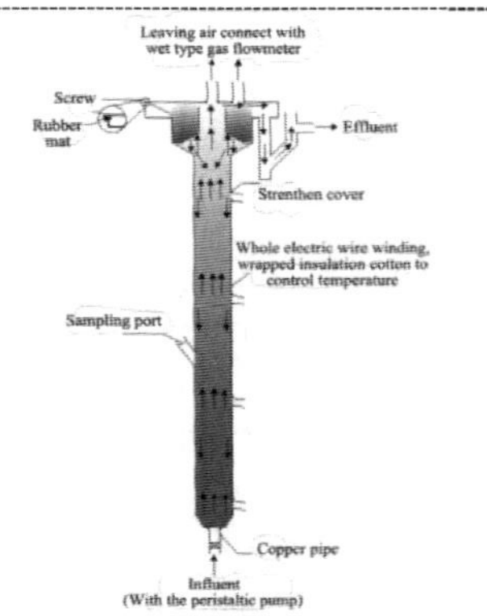

Figure 1. Process sketch map of the UASB

2.2 Experimental wastewater and inoculated sludge

In this experiment, the TDWS was taken from a wastewater treatment plant in Xiamen, their wastewater was from the food industry park wastewater nearby.

As a result, the characteristics of the TDWS had high concentration of organic matter, complex and biodegradable components. The basic parameters of waste water quality were shown in table 2:

Table 2. Original quality of wastewater without diluting

COD	BOD_5	pH	alkalinity	Nitrite nitrogen

(mg/L)	(mg/L)		(mgCaCO$_3$/L)	(mg/L)
50107.5	27175	7.4	8658.65	0.66
NH4+-N (mg/L)	Nitrate nitrogen (mg/L)	total nitrogen (mg/L)	phosphate (mg/L)	total phosphorus (mg/L)
3011.17	176.1	3192.7	115.56	152.26

2.3. Experiment analysis project and method

The main test items were monitored by standard methods, such as: COD, BOD_5, pH, alkalinity, gas production and gas composition, acidity and acid composition, ORP, SS, and VSS, and so on. Testing equipment were such as the fast COD measuring instrument of the Lian Hua 5B-3C, a type BOD apparatus of the LH-BOD601, the gas chromatograph of Zhejiang Fuli 9790-II, the wet gas flowmeter, and the pH meter of the Ohaus ST3100, respectively. The experimental analysis methods of other test items was mainly from "water and wastewater monitoring and analysis methods", China Environmental Science Press, 2002 edition.

2.4. Operating conditions

The temperature of the reactor was controlled at (52.5±2.5) ℃, making the normal operation of UASB in the high temperature range. The pH value was greater than 5 (there was the alkalinity in the TDWS, which can avoid the low pH to the impact of methanogenic activity). To form granular sludge and maintain the high sludge concentration in the reactor, the rising velocity of the anaerobic reactor was 0.5m/h~0.655m/h. Hydraulic retention time (HRT) was 69.5h. The concentrations of the influent COD and other pollutants were controlled by dilution method. The experiment divided into 7 stages. The influent COD concentration was 5000mg/L, 7000mg/L, 1000mg/L, 12000mg/L, 15000mg/L, 20000mg/L and 23000mg/L, respectively.

3. Experimental results and analysis

3.1. Reactor operation analysis

3.1.1. The removal efficiency of COD

As the acclimated pilot test sludge was inoculated with sludge on the start-up of reactor, the influent COD concentration was 5000mg/L. From figure 2, during 1~15th day, the COD concentration of effluent increased gradually, and the COD removal rate decreased continuously. On 16th day, the COD removal rate was only 11.78%, which indicated the microbial activity decreased in the reactor because the living environment changed. The COD removal rate increased gradually from the seventeenth day. On 25th day, under the condition of the same influent COD concentration, the COD removal rate reached 49.69%, which indicated that the microorganism gradually adapted to the environment and the activity was recovered.

During 25~40th days, the removal rate of COD increased slightly to 55%. From 41th day, when the influent COD increased to 7000mg/L, the treatment efficiency of the reactor began to increase slowly. On 66th day, the effluent COD was 2302mg/L, the removal rate was up to 66%, the volume load was 5.22kgCOD/ (m^3.d), and the sludge load was 0.1 kgCOD/ (kg sludge·d), respectively. When the influent COD increased to 12000mg/L, the removal rate of COD shown a large fluctuation, which had a great impact on microorganisms. At this time, the volume load was 13.29kgCOD/ (m3.d), the sludge load was 0.24 kg COD/ (kg sludge·d), and the removal rate of COD was up to 65%. According to the 107 day's operation, with the increase of influent COD concentration (from 5000~12000mg/L), the removal rate of COD showed an upward trend, reached more than 65% after the stable operation.

The removal rate of COD was lower than that of similar wastewater by using batch anaerobic treatment method (ASBR) on the temperature of 35℃[7]. The reason may lie on the wastewater quality and HRT. In the experiment of ASBR, The TDW's quality parameters: COD concentration was 8650 mg/L, pH was 7, chroma was 96 and NH_3-N was 125 mg/L, the COD removal rate was

IOP Conf. Series: Earth and Environmental Science **146** (2018) 012062 doi:10.1088/1755-1315/146/1/012062

more than 89%. The acid dehydration wastewater's COD concentration was 28225 mg/L, pH was 5.1, chroma was 192, and NH_3-N was 286mg/L, the COD removal rate was more than 97.5%. But the HRT of ASBR was 26 days, which is much larger than that of the present experiment (69.5h).

To improve the removal rate of pollutants, a batch of acclimated anaerobic sludge was added in to the reactor on 108th day. The COD removal rate dropped sharply to 12%, which indicated that the ecological environment of the reactor suffered a serious impact. After a week, the COD removal rate was recovered to 50%. From 117th day, when the influent COD concentration increased to around 15000mg/L, the COD removal rate of this stage was 59~68%. When the influent COD concentration was 18000mg/L~23000mg/L, the sludge load was 0.43kgCOD/ (kg sludge·d), and the COD removal rate gradual declined from 58% to 35%. The results of adding sludge showed that COD removal efficiency of the reactor was about 65%, and the influent COD should be controlled below 15000mg/L.

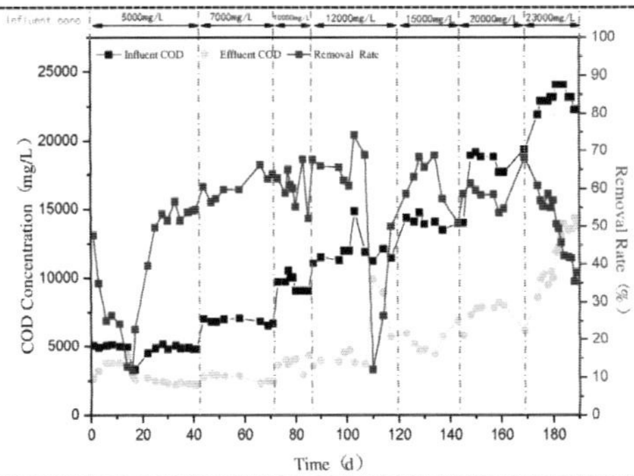

Figure 2. The removal rate of COD, the concentration of the influent and effluent COD

3.1.2. The concentration of Volatile fatty acid (VFA)

In anaerobic fermentation process, variety of organic acids will be produced. The state of the main bacteria in the anaerobic fermentation process, the operation efficiency and the acidification trend of the reactor can be obtained by monitoring the changes of organic acids in the reactor.

From Figure 3, the Graphical of influent VFA concentration was similar with the change of the effluent VFA concentration. At the beginning, volatile acid accumulation and the removal rate of VFA was 0. This meant methanogens did not use the organic acids, the activity was very low. During 17 ~ 30th days, the COD removal rate increased from 12% to 51.5%, the VFA removal rate was unchanged. The results showed bacteria producing acid recovered gradually after a period of 17 days of adaptation, and the adaptation period of methanogens was more than 30 days. During 31~66th days, the removal rate of organic acids was about 30%, and the COD removal rate increased from 56% to 66%. During 67~107th days, the removal rate of organic acids increased gradually, which indicated the activity of methanogenic bacteria was enhanced, and the COD removal rate was above 60%. On 97th day, the removal rate of VFA reached to 98.19%, and the corresponding COD removal rate was about 65.59%, which showed that the COD removal efficiency was directly related to the VFA degradation. When the influent COD concentration was 12000 mg/L, the removal rate of VFA reached more than 85%.

After the addition of new mud on 107th day, the VFA concentration of effluent increased sharply, indicated methanogenic bacteria were been a serious impact. After the adaptation period of 10 days, the removal rate of VFA reached to98.5%, and the removal rate of COD increased to more than 55%.

When the VFA concentration reached 9000mgHAC/L (corresponding to 20000mg/L COD of the influent), the VFA removal rate was fluctuation (max 81%, min 65%), it meant that the pollutants concentration of the influent caused effect on methanogenic activity. However, the COD removal rate was still above 51%, which showed the concentration of pollutants had no significant effect on the bacteria producing acid. After 20 days, the VFA and COD removal rate were all increased. When the VFA concentration of the influent reached 12000mgHAC/L (corresponding to 23000mg/L COD of the

IOP Conf. Series: Earth and Environmental Science **146** (2018) 012062 doi:10.1088/1755-1315/146/1/012062

influent), the VFA removal rate and COD removal rate continued to decline, indicating the pollutant concentration of the influent were causing serious impact on the two main bacteria in the anaerobic reactor. Increasing the influent COD concentration to 24000mg/L, the removal rate of VFA continue to decline. On the 186th day, the VFA concentration of the effluent reached 6243mgHAC/L, the removal rate of VFA and the removal rate of COD were only 48% and 41.89%. At this time, a large number of VFA has been accumulated in the reactor, and the removal rate of COD was decreased and the gas production efficiency was reduced. This state indicated that acidification had occurred in the reactor. Therefore, the influent VFA concentration should be controlled lower than 5000 mgHAC/L of, which was helpful to ensure the stable operation of the reactor.

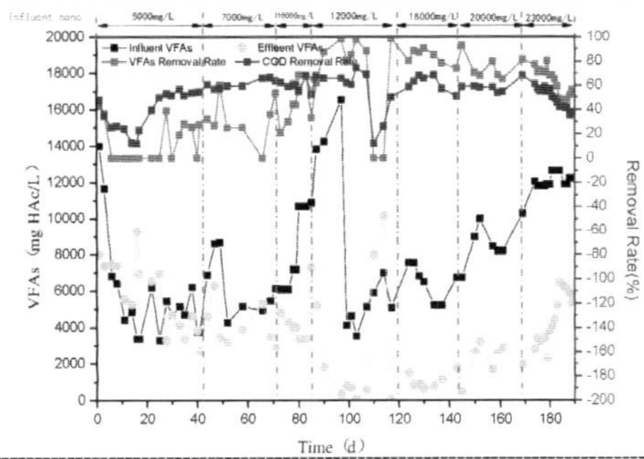

Figure 3. The removal rate of VFA and COD, the VFA concentration of influent and effluent

3.1.3. Gas production and CH_4 production

The output of CH_4 is an important factor to measure the feasibility and efficiency of UASB reactor to treat the TDWS. From figure 4, After 23 days of experiment, the reactor had a relatively stable gas production rate. On 23th day, the gas production rate was 115.2mL/L (the ratio of the volume of gas produced to the effective volume of the reactor in the same day), and the methane production was 14.58 ml/L. With the increase of influent COD concentration, the gas production increased. When the influent COD concentration was 10000mg/L, the daily output of CH_4 and the methane production ratio increased continuously. On 87th days, the daily output of CH_4 reached the maximum of 575.53 mL/L, the methane production ratio was 104.05ml/ (g COD.d). On 88th day, when the influent COD increased to 12000mg/L, the high COD concentration impacted on microorganisms, gas production began to decline. The methane production decreased from 548.58 mL/L to 342.76mL/L, and the methane production rate decreased from 95.32ml/ (g COD.d) to 60.75ml/ (g COD.d). It showed that the activity of methanogenic bacteria decreased, and 12000mg/L COD was the controlling concentration for maintaining a good operation of the reactor. After adding the new mud, gas production was not measured due to equipment commissioning.

During 147~161th days, when the influent concentration of COD was 20000~23000 mg/L, the average daily production of CH_4 was about 550ml/L. However, the gas production efficiency continued to decline, and the methane production rate decreased from 60.61 ml/ (g COD.d) to about 41.22 ml/ (g COD.d). The average methane production ratio was 58.95ml/ (g COD.d). The results showed that the activity of methanogenic bacteria decreased obviously.

IOP Conf. Series: Earth and Environmental Science **146** (2018) 012062 doi:10.1088/1755-1315/146/1/012062

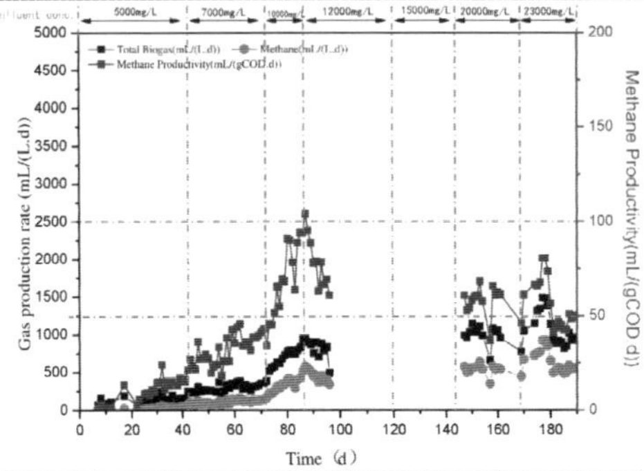

Figure 4. Gas/CH_4 production per day and CH_4 production per COD

3.1.4. The change of ammonia nitrogen (NH_4^+-N) concentration

NH_4^+-N is the one of the nutritional sources of anaerobic microorganisms, suitable concentration of NH_4^+-N can improve the activity of methanogenic; in addition, NH_4^+-N can enhance the alkalinity of the reaction system and increase the ratio of VFAs to ALK, thereby improving the buffer capacity of the system for volatile fatty acids. On the other hand, with the increase of NH_4^+-N concentration, the concentration of free ammonia in the system will increase, which will strongly inhibition the production of methane[9].

From figure 5, NH_4^+-N was monitored from the 40th day. With the increase of influent COD concentration, the influent NH4+-N concentration increased from 684.5mg/L to 2826.5mg/L. As can be seen from figure 5, the overall trend of NH4+-N concentration of the effluent was increased with the influent. However, the removal rate of NH4+-N, the removal rate of COD and the influent and effluent concentration of NH4+-N were not significant relevance. This showed that the change of NH4+-N concentration was not a single factor.

When COD concentration of the influent was 5000~7000 mg/L and the average concentration of the influent NH4+-N was 700 mg/L, the average concentration of the effluent NH4+-N was about 536 mg/L, and the average NH4+-N removal rate was about 22.7%. Due to the demand of anaerobic process for nitrogen, some NH4+-N was consumed. At this time, from the removal rate of NH4+-N, indicated that the dominant bacteria group has not been formed to converse ammonia effectively. When COD concentration of the influent was 10000~12000 mg/L and the average concentration of the influent NH4+-N was 1100 mg/L, the average concentration of the effluent NH4+-N was about 611mg/L, and the average NH4+-N removal rate was about 40.7%. Meanwhile, the removal rate of COD was higher (the average removal rate was 63.4%), and the methane ratio of producing gas was about 50%. It indicated that the concentration of NH4+-N had no inhibition action on the microorganisms in the reactor, and the bacteria group conversing ammonia was formed in the reactor. When COD concentration of the influent was 15000 mg/L and the average concentration of the influent NH4+-N was 1600 mg/L, the average concentration of the effluent NH4+-N was about 1150mg/L, and the average NH4+-N removal rate and the average COD removal rate dropped down to 29% and 60%, respectively. Compared the concentration and removal rate of VFA, the NH4+-N concentration became affected to methanogens; however, NH4+-N had no significant inhibitory action to methanogenesis process. When COD concentration of the influent was 20000 ~ 23000 mg/L and the average concentration of the influent NH4+-N was 2400 mg/L, the average concentration of the effluent NH4+-N was about 1800mg/L, and the average NH4+-N removal rate and the average COD removal rate gradually decreased to 15% and 45%, respectively. In this stage, the VFA in the reactor was greatly accumulated, and the gas daily output decreased from 7.4L to 4.8L. Indicated that methanogens had were seriously affected by the concentration of NH4+-N.

IOP Conf. Series: Earth and Environmental Science **146** (2018) 012062 doi:10.1088/1755-1315/146/1/012062

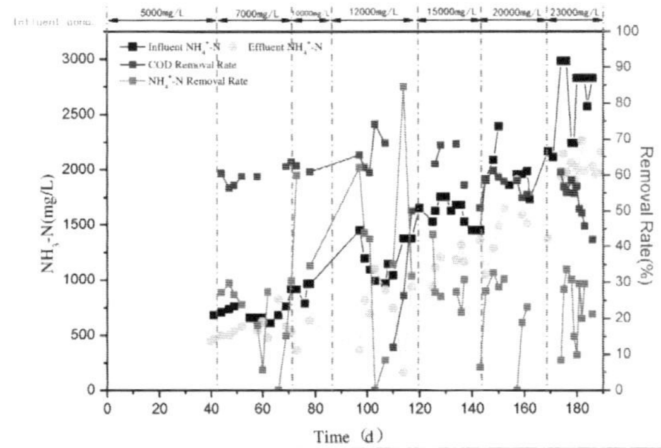

Figure 5. The removal rate of the NH_4^+ -N and COD, the NH_4^+ -N concentration of influent and effluent

In this experiment, the results of NH4+-N restraining methanogenic bacteria was different from that of He SJ et al. [10] and Borja R et al.[11] , Sossa K et al. [12], Gerardi M H et al.[13]. The inhibition concentration of NH4+-N in the experiment was 1.5g/L, which was much larger than that of other experiments (0.8g/L). Analyzing the reasons, an acclimated inoculated sludge was the key, of which microbial adapted to the toxicity of ammonia. The significant inhibitory concentration of NH4+-N was 2.4g/L, which was less than that of He SJ et al. (4g/L)[10], indicated that under the high pH value, the high concentration of NH4+-N was greater toxicity to the methanogens than others.

3.2. The affecting factors of the reactor operation

3.2.1. Alkalinity and pH value

Alkalinity is an important index to monitor the running state of the reactor, and to some extent, it can also reflect the accumulation of organic acids in the reactor. Usually, on the condition of the good operation and non-acidification in the internal reactor, the alkalinity change of influent and effluent wastewater remained the same trend.

From figure 6, when the influent COD concentration was 5000~15000mg/L, the change trend of the alkalinity was the same and the alkalinity of effluent had been greater than influent. It indicated that NH4+-N coming from the decomposition of organic nitrogen may lead to the increase of alkalinity in anaerobic system. When the influent COD concentration was greater than 20000mg/L, the effluent alkalinity was close to the influent alkalinity, which indicated that there were a lot of substances consumed alkalinity.

Comparing with the VFA concentration in the same period, it was found that, VFA in this stage was accumulated in a large amount. This meant that the methanogenic bacteria were not able to metabolize it in time; meanwhile, the alkalinity of the reactor was consumed by organic acids so that the alkalinity decreased. The time period was corresponded with the sudden decline of the COD removal rate due to acidification.

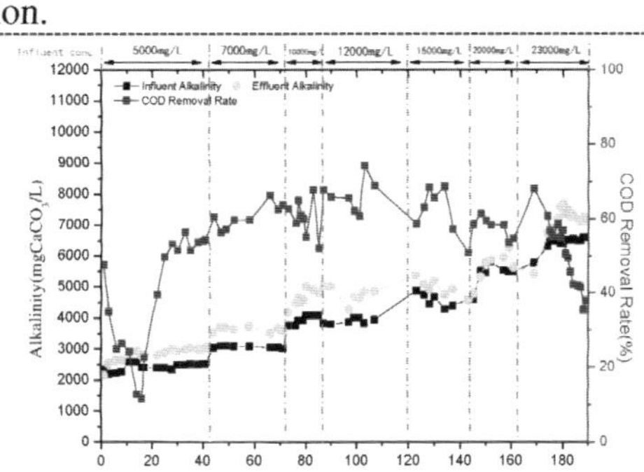

Figure 6. The relationship of influent/effluent alkalinity and COD removal rate in UASB

The optimum growth pH range of the bacteria producing acid and methanogens are 5.00-6.50 and 6.60-7.50 [14]. It can be seen that the pH value can intuitively reflect the survival status of two main bacteria in the reactor. From the figure 7, When the influent COD concentration less than 7000 mg/L, the pH value was higher and the wastewater was not acidified because of the high dilution multiple. At this time, the organic acids of anaerobic digestion in the reactor decreased pH value of effluent.

When the influent COD concentration was 10000~23000mg/L, the partly influent was acidified and the influent pH decreased because of the dilution ratio reduced. After the anaerobic digestion process, the decomposition of nitrogen content and the production of methane would increase the alkalinity, leading to the effluent pH value higher than that of influent. In the whole operation period, the pH value in the reactor was maintained between 7.0~8.0.

Using the sequencing batch anaerobic reactor to treat the TDWS in the research of Dong H et al.[7] , the low concentration of ammonia (286mg/L) did not affect the operation of reactors, when the influent pH value was 6.7~7.2, the effluent pH value was about 7. Compared with the experimental results, the pH value of the effluent in this study was about 7.7 after the stable operation of the reactor. Under the high pH conditions, the toxic effects of NH4+-N would increase.

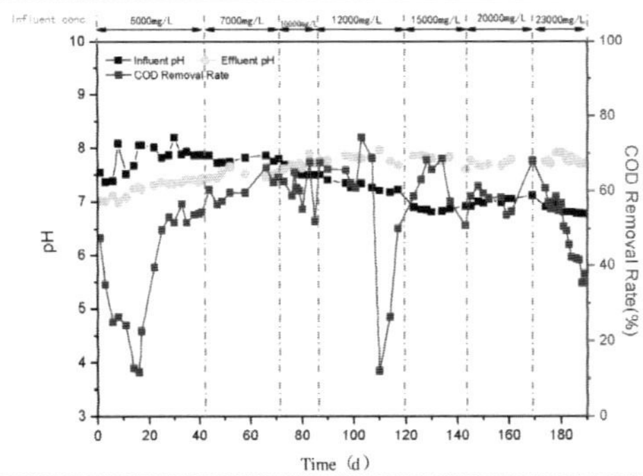

Figure 7. The relationship of influent/effluent pH and COD removal rate in UASB

3.2.2. *ORP*

In the initial stage of the reactor, the value of Oxidation-Reduction Potential in the reactor was relatively volatile. Frome the figure 8, After seventy-first days, the ORP showed a steady downward trend. After stable operation, the ORP value was always less than -320 mV, which could ensure the normal metabolism of methanogens.

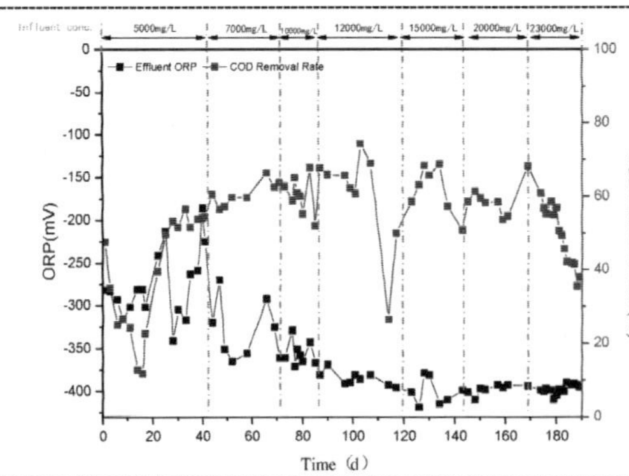

Figure 8. ORP in UASB

3.2.3. Sludge load and volume load

Table.3 Reactor loading and COD removal rate

operation time d	Influent COD mg/L	Sludge load gCOD/(gVSS·d)	volume load kgCOD/(m^3·d)	The COD removal after stable %
28~106	5000~10000	0.1	5.22	51~68
108~142	12000~15000	0.24	13.29	50~68
143~186	20000~23000	0.42	15.08	41~61
187~189	23000	0.35	24.68	35~41

According to the sludge quality in each stage of the experiment, the reactor load in different operation stages was calculated. As shown in the table3, when the sludge load was less than 0.24 g COD/ (g VSS·d) and the volume load was less than 13.29 kg COD/ (m^3·d), the reactor could be stable operation. When the sludge load reached 0.42 g COD/ (g VSS·d), the treatment efficiency of the reactor decreased continuously. Even if the sludge load was reduced in the later stage, the operation state of the reactor could not be recovered, which showed this load exceeded the load requirement of the stable reactor. In contrast to similar experiments, the sludge organic load was 0.9gCOD/ (g VSS·d) ~3.0gCOD/ (g VSS·d) in the experiment of Ohtsuki T et al. [15] to treat acidified sucrose wastewater; The maximum load of mature granular sludge was 3.7g COD/ (g VSS·d) at the temperature of 55℃, which was studied by Syutsubo et al.[16]. In this study, the capacity of the sludge was lower than that of the similar studies, which may be related to the high concentration of NH4+-N in wastewater and the complex and difficult degradation of water quality.

Compared with the volumetric loading, such as the research of Zhang ZJ et al.[17], when the corn alcohol wastewater was deal with the UASB reactor and anaerobic biofilter under the condition of 55(±2)℃, the volume load of the reactor was 3 kg COD/(m^3·d)~20 kg COD/(m^3·d); Under the same conditions to deal with, the actual volume load was more than 10 kg COD/ (m^3·d) in the process of treating the brewery wastewater[18]. The results of previous experiments showed that the volume load used in this study was appropriate.

4. Conclusions

The TDWS was treated by the UASB reactor with the high ratio of height to diameter. After 7 operation stages, the following conclusions were obtained:

1. On the condition of the temperature of 52.5(±2.5)℃, the influent COD concentration of the TDW should be controlled below 15000mg/L for the stable operation of UASB. After the stable operation, the COD removal rate could reach 65%, the daily output of CH_4 was more than 550 ml/L, and the methane production ratio was up to 100 ml/ (g COD·d).

2. Because of the high pH value of the TDW, the ammonia toxicity to methanogens was increased. Therefore, in order to ensure the stable operation and the efficiency of the reactor should control the influent NH4+-N concentration, making it less than 1000 mg/L. When the concentration of influent NH4+-N reached 1500 mg/L, the inhibition of anaerobic digestion began to appear.

3. The optimal sludge loading of the UASB reactor was 0.10~0.24 kg COD/ (kg sludge·d), and the optimal volume load was 5~13 kg COD/ (m^3·d) for the treatment of the TDWS.

Acknowledgements

The authors are pleased to acknowledge Lanzhou University of Technology and Coordination of Institute of Urban Environment, Chinese Academy of Sciences.

Funding

This article was funded by National Natural Science Foundation of China (51568039).

References

[1] Li B, Zhang C, Zhao Y. Characterization and pretreatment of sewage sludge(in Chinese): Metallurgical industry pressMetallurgical industry press; 2010.
[2] Tian D. Chemical speciation analysis of pollutants in municipal sludge (in Chinese): East China Normal University; 2006.
[3] Xue HY, Zhang X, Jiang WT, Song ZW. Process and Mechanism Comparative Analysis of Dewatered Sludge's Thermal Drying and Microwave Drying. Environmental Science & Technology. 2014.
[4] Huang W, Fan T. State key laboratory of metal matrix composites (in Chinese). Materials Review. 2014(s1):131-5.
[5] Wang D, Du S, Zhu W, Cui B, Zhao R, Yang X. Analysis of Processing Dewatering Filtrate of Hydrothermal Conditioning Sludge. Journal of Green Science & Technology. 2012.
[6] Zhu W, Li X, Wang C, Wang D, Xu S. Study on synergistic degradation of sludge heat dehydration filtrate and simulated domestic wastewater in SBR system (in Chinese). Water & Wastewater Engineering. 2014(s1):196-201.
[7] Dong H, Shen Z, Jiang W, Zhou H, Cui W. Experimental Research on Anaerobic Treatment of Sludge Hydrothermal Dewatering Filtrate. Environmental Sanitation Engineering. 2013.
[8] Gao YQ, Peng YZ, Wang SY, Wang JL, Zhang JY. Using Sludge Hydrolysis-acidification Liquor as Carbon Source for Nitrogen and Phosphorus Removal in A~2/O System. China Water & Wastewater. 2009.
[9] Fang-Fang YU, Jian-Dong WU. Toxicity Study of Ammonium on Methanogenic Bacteria in Anaerobic Granular Sludge. Chemistry & Bioengineering. 2008.
[10] Shijun HE, Wang J, Xuan Z. Effect of ammonium concentration on the methanogenic activity of anaerobic granular sludge. Journal of Tsinghua University. 2005;45(9):1294-6.
[11] Borja R, Sánchez E, Weiland P. Influence of ammonia concentration on thermophilic anaerobic digestion of cattle manure in upflow anaerobic sludge blanket (UASB) reactors. Process Biochemistry. 1996;31(5):477-83.
[12] Sossa K, Alarcón M, Aspé E, Urrutia H. Effect of ammonia on the methanogenic activity of methylaminotrophic methane producing Archaea enriched biofilm. Anaerobe. 2004;10(1):13-8.
[13] Gerardi MH. The microbiology of anaerobic digesters. Wiley-Interscience, A John Wiley & Sons. 2003.
[14] Ren N, Wang A. Principle and Application of Anaerobic Biotechnology.
[15] Ohtsuki T, Watanabe M, Miyaji Y. Start up of thermophilic UASB (upflow anaerobic sludge blanket) reactors using microcarrier and mesophilic granular sludge. Waterence & Technology. 1992;26:877-86.
[16] Syutsubo K, Harada H, Ohashi A. Granulation and sludge retainment during start-up of a thermophilic UASB reactor. Water Science & Technology. 1998;38(8–9):349-57.
[17] Zhang ZJ, Zhang RJ, Cheng GU, Wang S, Pan H, Zhang H. Treatment of Corn Ethanol Production Wastewater by Thermophilic UAHB Reactor. China Water & Wastewater. 2000.
[18] Wang J. Engineering example and safety measures of UASB treating high concentration brewing wastewater. China Population, Resources and Environment. 2016(s1):53-7.

IOP Conf. Series: Earth and Environmental Science **146** (2018) 012063 doi:10.1088/1755-1315/146/1/012063

Wind turbine fuzzy logic individual pitch control based on chaotic optimization

Xuan Gong [1] **and Wenyi Li** [2]

[1] College of Energy and Power Engineering, Inner Mongolia University of Technology, Hohhot, China, 010051
[2] College of Electrical Engineering , Inner Mongolia University of Technology, Hohhot, China, 010051

lwyyyll@vip.sina.com

Abstract. Fuzzy-logic-based algorithms applied in the individual pitch control make the effect of the controlled system imperfect. Thus, an adjustable factor algorithm of adaptive fuzzy PID controller based on chaos theory was proposed. To achieve good control effects, the chaos algorithm was applied in designing the fuzzy PID control system to optimize the parameters of the membership functions of controller under the control of the fuzzy PID. Simulation based on matlab/simulink was made to analyze and to compare fuzzy PID individual pitch control system with and without the chaos algorithm respectively. Simulation results show that, the fuzzy PID individual pitch control system based on chaos algorithm was much more smooth and steady and response faster than the system without chaos algorithm within the rated wind speed.

1. Introduction

Individual pitch control system is an important component of the large wind turbines. It has good effect in the improvement of stability and dynamic stability of wind turbine power output[1-2].The combination of fuzzy control and traditional PID control has been applied in pitch control, due to complex nonlinear system of wind turbines[3]-[8].However, the rules of fuzzy controller and parameter setting have the nature of subjectivity and uncertainty. The control effect has been made imperfect because of the long-term accumulation of experience [9]. With the gradual improvement of the complexity of the controlled object and the continuous improvement of the control effect, the parameters of fuzzy controller are particularly important to be set and optimized.

Chaos motion has characteristics of randomness and ergodicity and inherent regularity [10-11]. Chaos motion can get approximate optimal parameters of fuzzy controller according to self regular pattern. In this paper, fuzzy-logic-individual pitch control based on chaos optimization is proposed. Parameters of fuzzy PID controller are optimized using chaos optimization algorithms.

2. Wind turbine individual pitch control strategy

The main strategies of individual pitch control can be divided into two types, which include the one based on blade acceleration signals weight coefficient and the one based on blade azimuth angle weight coefficient [12]. With the consideration of the first strategy being hardly achieved, in this paper, the second strategy will be adopted. The weight coefficient can be determined by the differences of azimuth angle during the wind turbines blades rotate.

Wind turbines operate constantly during the normal working condition. The blade can be located during azimuth angle θ [13]. The force of each blade in different position makes difference, by which weight coefficient can be confirmed [14]. Each weight coefficient K of individual pitch control system can be calculated by equation (1).

$$K_i = \frac{3\left(1+\frac{1}{2}\frac{R}{L}\sin\theta_i\right)^2}{\sum_{i=1}^{3}\left(1+\frac{1}{2}\frac{R}{L}\sin\theta_i\right)^2} \tag{1}$$

Where i shows the number of blade, and θ_i shows the angle of azimuth. θ_i can be gotten by using equation $\theta_i = \omega t + (i-1)\frac{2}{3}\pi$. R (m) is radius of blade, and L(m) is the height of blade shaft relative to ground. ω(rad/s) shows the angular velocity of wind turbine rotation. Each variation of blade can be totally known with the calculation of equation (2).

$$\begin{cases} \Delta\beta_1 = K_1 * \Delta\beta \\ \Delta\beta_2 = K_2 * \Delta\beta \\ \Delta\beta_3 = K_3 * \Delta\beta \end{cases} \tag{2}$$

Where $\Delta\beta$ is the variation of pitch angle. The weight coefficient K_i of each pitch angle distributes the variation of each blade $\Delta\beta_i$, to achieve individual pitch control, which is shown in Figure 1.

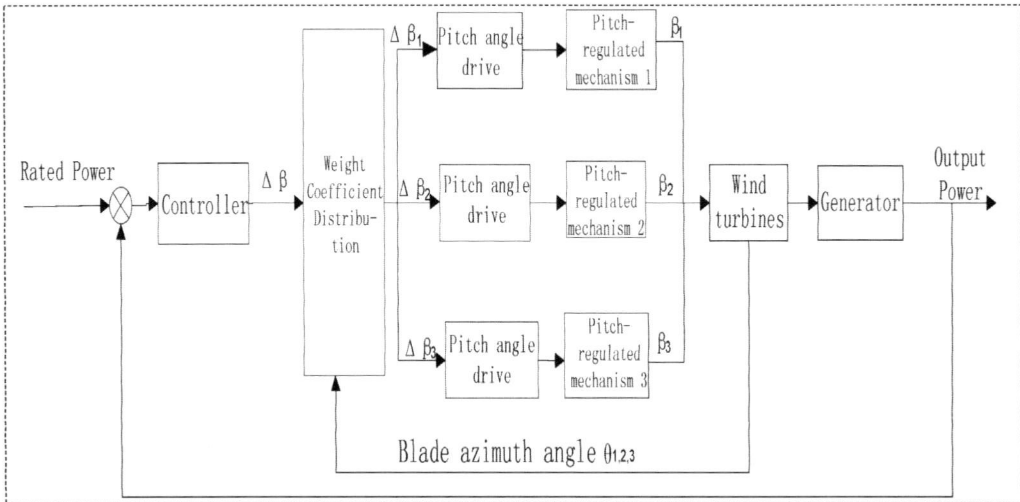

Figure 1. Schematic diagram of individual pitch control weight number distribution

The final control effect is affected by the distribution of K_i and the output of pitch angel $\Delta\beta$. The output of $\Delta\beta$ is usually dominated by the combination of fuzzy control and traditional PID control. By using the relative rules of fuzzy control, the parameters including Kp, Ki, Kd of PID will be adjusted online. The output of systems achieved by the PID algorithms is shown in Figure 2.

The output power of wind turbines is kept around the rated power. Two-dimensional fuzzy controller with dual input and three output is used to obtain the control effect. Where *E* is the power difference and *EC* is the change rate. The parameters *Kp*, *Ki*, and *Kd*. are the output of PID controller. To simplify the algorithm, trigonometric function was adopted to be Double-Input membership function, and Gaussian function was used to be three-output membership function. The final control effect can be achieved by fuzzy inference and defuzzification. The effect of final control will be influenced by the of controller parameters adjustment. In this paper, individual pitch control is improved by optimizing parameters of the fuzzy controller.

3. Fuzzy controller based on chaos optimization

In spite of good robustness and simple application, the final control effect of fuzzy control can be affected by many other factors. The final effect will be influenced by quantification factor, scale factor, membership function and fuzzy control rules[15]. Chaos optimization with chaotic variable search can be an effective method to optimize the membership function of fuzzy controller[10].In this paper, chaos optimization algorithm with randomness and ergodicity is proposed, the fundamental principles of which is shown in Figure 2.

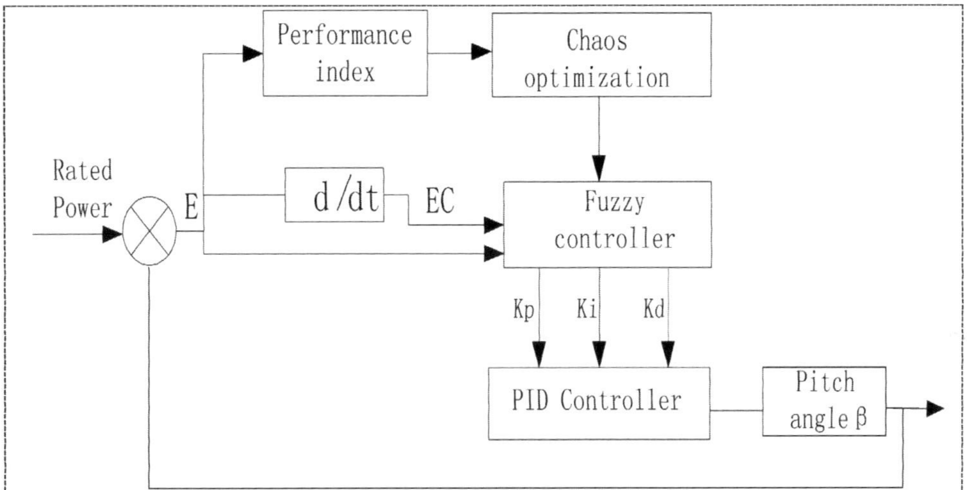

Figure 2. Principle diagram of fuzzy control based on chaos optimization

The membership functions of parameters E and EC in individual pitch controller are adopted in trigonometric function. The eigenvalue can be gotten by the following equation.

$$\{a,b,c\} = CH\left[\sup(x)\right] \tag{3}$$

Where a,b,c is the left, right, and vertex of a triangle respectively.

The membership functions after regularization are shown in equation (4).

$$\mu(x_i) = \begin{cases} (x_i - a_i^l)/(c_i^l - a_i^l), x_i \in \left[a_i^l, c_i^l\right], \\ (c_i^l - x_i)/(b_i^l - c_i^l), x_i \in \left[c_i^l, b_i^l\right], \\ 0, \quad \text{Others} \end{cases} \tag{4}$$

Where i takes values in 1-2, *and* x_1 is input of E *and* x_2 is input of EC. Evaluation function of control results can be shown in equation (5) [16].

$$J = \omega_1 \int_0^1 t\left|e(t)\right| dt / \max(e(t)) + \omega_2 \sigma \tag{5}$$

Where σ can be considered to be the overshoot of centesimal system, besides, $\omega_1 + \omega_2 = 1$.

The chaotic variable can be obtained by means of the Logistic map, as shown in equation (6).

$$X_{n+1} = \mu X_n^* (1 - X_n), \quad 0 < X_0 < 1 \tag{6}$$

Where n is frequency of Chaos optimization, and can take values in 1 to N. The parameters are in the chaotic states when μ takes the value of 4, where X_n is completely ergodic within 0 to 1. Variables to be optimized will be converted into chaotic variables and the global optimum can be found by searching through chaotic variables.

Parameter optimization steps are as follows:

Step.1: x can be considered to be in the state of chaos when μ take the value of 4. The state of chaos will be extremely sensitively affected by the initial value. Three different initial values are taken value within 0 to 1 and plugged into Logistic map separately, where 3 chaotic variables can be obtained.

Step.2: The parameters to be optimized can update according to equation(7):

$$\hat{x}^{*}_{i,n} = x_i^* + z(t)x_n \tag{7}$$

Where i takes value of 1, 2, 3 and $z(t)$ is time-varying parameter. x_i^* is a set of parameters in the optimal performance index at present, and $\hat{x}^{*}_{i,n}$ is a new set of parameters.

Step.3: Chaotic rough search is made according to the performance index. The value of performance index is compared with the performance index in limited optimization times. The value of best performance index can be chosen as sub-optimal value.

Step.4: The optimal value can be found from minimum performance index during iteration.

4. Chaos optimization fuzzy PID Independent variable pitch simulation

Part of relevant parameters of individual pitch fuzzy PID controller can be obtained from table 1. The random wind speed near rated speed has been simulated shown in Figure.3. The control effect of the independent variable pitch fuzzy controller based on chaos optimization has been simulated and analyzed, according to chaos optimization method in this paper.

Table 1. The main parameters of WTGS.

Wind turbine and generator parameters	Value
Rated power	1500 kw
Rated wind speed	11 m/s
Impeller radius	38.5 m
Wind rotary inertia	$1\times10^4\ kg/m^2$
Cut-in wind speed	4 m/s
Cut-out wind speed	25 m/s
Generator pole number	2
Generator rotor inertia	70 kg/m^2
Generator voltage	690 V

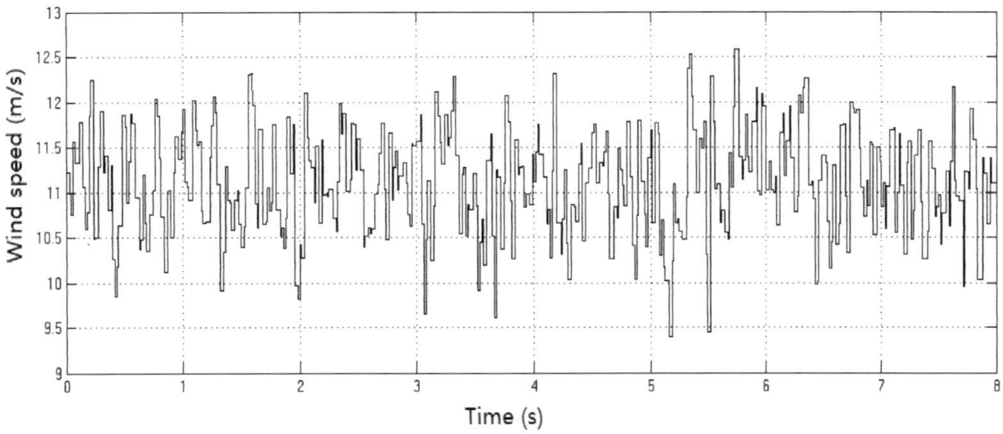

Figure 3. Diagram of random wind speed signal

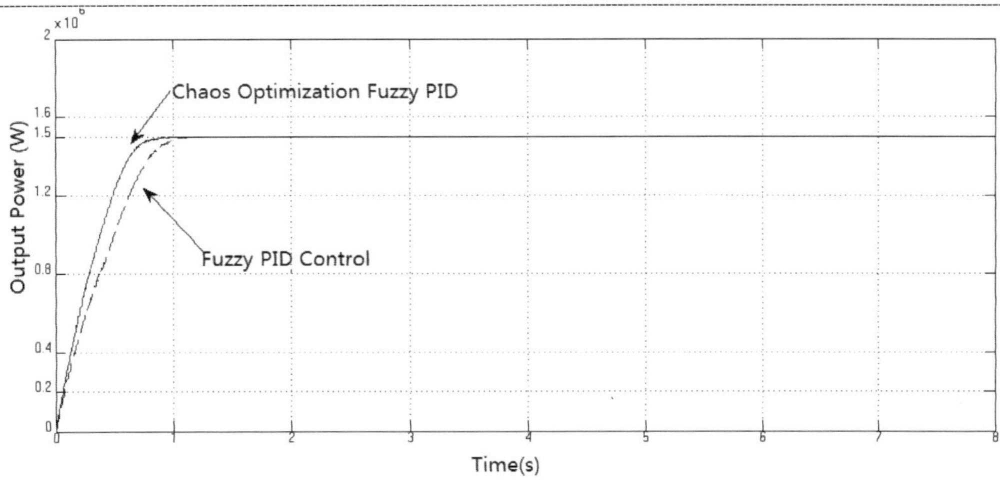

Figure 4. Output power of fuzzy controller based on chaos optimization and without chaos optimization

Figure.4 shows the output power simulation of fuzzy PID control which has been Chaos optimized and not optimized. The simulation results show that under the same conditions, the fuzzy PID controller optimized by chaos algorithm controls better and faster than the one without optimized.

5. Conclusion

The final control effect is affected by the uncertainty of the parameters in the independent pitch fuzzy controller. In this paper, a chaos optimization method for individual pitch fuzzy PID control is proposed, and the control effect of the independent pitch fuzzy controller has been meliorated. Simulation results based on matlab/simulink shows that the independent variable pitch fuzzy controller optimized by chaos algorithm responses faster and controls more accurate than the one without optimized. It can be used as a reference for the improvement of the individual pitch fuzzy controller.

Acknowledgments

The Wind and Solar Energy Utilization Technology Key Labs Co-constructed by Province and the Ministry of Education (No.201611), the Scientific Research Innovation Project by Inner Mongolia. China （No. B20161012805）.

References

[1] Ying You, Yang Fan. Simulation development and field testing of individual pitch control on wind turbines ［J］. Acta Energiae Solaris Sinica, 2013, 34 （5）:882—888.

[2] JIN Xin,WANG Yaming, LI Lang, ed al. Dynamics Loads Optimization Analysis of Wind Turbine Based on LQG Independent Pitch Control ［J］. Proceedings of the CSEE, 2016, 36 （22）: 6164—6170.

[3] Muyeen S M, Takahashi R, Murata T, et al. A variable speed wind turbine control strategy to meet wind farm grid code requirements ［J］. IEEE Transactions on Power Systems, 2010, 25 （1）:31—34.

[4] SONG Yigang,WANG Zhixin. Design and simulation of fuzzy sliding-mode robust controller for large wind generating unit ［J］. Proceedings of the CSEE,2008,28 （14）:136—141.

[5] CHEN Jiawei,CHEN Jie,CHEN Ran,ed al. Adaptive fuzzy logic control technique for variable-speed wind turbines ［J］. Proceedings of the CSEE,2011,31 （21）:93—100.

[6] ZHOU Feihang,LIU Jun,Fuzzy Control Strategy of Direct-Drive Wind Power System Based on State Feedback ［J］. Power System Technology,2016,40 （9）:2758—2763.

[7] WEI Zheng,CHEN Ran,CHEN Jiawei,ed al. Wind turbine-generator unit variable-speed pitch control based on judgment of power changes and fuzzy control ［J］. Proceedings of the CSEE,2011,31 （17）:121—126.

[8] Guo Peng. Variable pitch control of wind turbine generator combined with fuzzy feed forward and fuzzy PID controller ［J］. Proceedings of the CSEE,2010,30 （8）:123—128.

[9] HUO Wenming,ZHOU Zhifeng. Research of DFIG Vector Control System Based on the Optimized Fuzzy PI Control ［J］. Journal of Electric Power,2016,31 （3）:229—235.

[10] RONG Panxiang,DONG Wenbo,YU Lin,ed al. Research and application of turing fuzzy PID controller based on chaos theory ［J］. Journal of Harbin University of Science and Technology,2014,19 （2）:26—30.

[11] WANG Ershen,PANG Tao,QU Pingping. Improved particle filter algorithm based on chaos particle swarm optimization ［J］. Journal of Beijing University of Aeronautics and Astronautics,2016,42 （5）:885—890.

[12] Yao Xingjia,Liu Yue,Guo Qingding. A control method for split range individual pitch based on feed-forward azimuth angle weight number assignment ［J］. Acta Energiae Solaris Sinica,2012,33 （4）:532—539.

[13] PENG Ji-shen,DING Zi-pei,DING Bai- wen. Individual Pitch Control Based on Azimuth Angle Weight Number Assignment ［J］. Control Engineering of China,2014,21 （6）:829—837.

[14] LIU Haoming,TANG Qiaoqiao,ZHANG Zhankui. Study of Individual Pitch Control Based on Azimuth Angle and Load Feedback ［J］. Proceedings of the CSEE,2016,36 （14）:3798—3805.

[15] LI Xiangfei. Researeh on the control system design based on chaos optimization theory ［D］. Changsha:Central South University,2003.

[16] ZHANG Xiaohui,DAI Guanzhong,XUN Aiping. The study on drawing and filtering fuzzy control rules by using genetie algorithms ［J］. Control Theory and Applications,1998,15 （3）: 379—384.

IOP Conf. Series: Earth and Environmental Science **146** (2018) 012064 doi:10.1088/1755-1315/146/1/012064

An Improved Network Partition Method of Voltage Control Considering Coordinated Active and Reactive Power Sources

Sh Zhang[1], Y Huang[2] and B Zh Liu[2]

[1] Electric Power Research Institute, State Grid Ningxia Electric Power Co, Ltd. , Yinchuan, China

[2] School of Electrical and Electronic Engineering, North China Electric Power University, Beijing, China

0zhangsh0nx@163.com, 18010151721@163.com, bzliu@ncepu.edu.cn

Abstract. In order to coordinate distributed generarions and reactive power compensation equipment to improve the voltage level, this paper proposes an improved network partition method of voltage control considering coordinated active and reactive power sources. According to the basic theory of the Newton-Raphson method, the sensitivity expression of the node voltage *vs* the node active power, the node voltage *vs* the node reactive power are deduced from the modified power flow equations. Two kinds of sensitivity matrices are calculated by the perturbation power flow calculation. The calculation process and steps are introduced respectively. Based on these two sensitivity matrices, a kind of comprehensive sensitivity matrix is synthesized. The electrical distance matrix is calculated to complete mapping the load nodes ino the corresponding voltage control region with the help of mapping algorithm. The validity of the proposed method is verified with the simulation results of IEEE 33 nodes.

1. Introduction

In recent years, distributed generarions (DGs) in the distribution network has increased rapidly. The traditional distribution network will gradually evolve into an active distribution network (ADN) with a large number of adjustable and controllable active/reactive power resources. Its operation and management will also encounter many challenges. Especially voltage fluctuation can not be ignored. It is an important research topics of how to coordinate DGs and reactive power compensation equipment to improve the voltage level.

Based on the sensitivity of a small system, it proposed a new method to coordinate the DG active power and reactive power compensation device to adjust the voltage more effectively [1]. The influence of DG output on the voltage of the whole network is analyzed Based on the single-phase power flow model [2]. Existing research that the active distribution network DG output and compensation equipment switching will have great impact on the power grid voltage level, regulation mode, the trend of the distribution. [3][4] Reference [5] proposed a novel overvoltage preventive control method which three phase models are used to overcome the overvoltage problems with high penetration photovoltaic (PV)generation In the active distribution network.

Based on the above analysis, this paper completes the following work: according to the basic theory of the Newton-Raphson method, the sensitivity expression of the node voltage vs the node active power, the node voltage vs the node reactive power are deduced from the modified equations of power flow. Two kinds of sensitivity matrices are calculated by the perturbation power flow, which are synthesized and the electrical distance matrix is calculated. The partitioning of the network is done by mapping algorithm. The validity of the proposed method is verified with the simulation results of IEEE 33 nodes.

2. Sensitivity Matrix of Active and ReactivePower Source Partition

Active and reactive power is decoupled for the transmission network, reactive power and node voltage have a strong correlation. Node reactive power is very important to maintain the node voltage level in the sub-region. While the distribution network is non-decoupled for active and reactive power. Both the node injection active power and reactive power can affect the node voltage, especially in the line resistance parameters and reactance parameters of the almost identical value of the low-voltage network. The effect of the voltage regulation is not obvious and limited only by reactive power adjustment or by active regulation.

In order to ensure every partition region has the adequate and reasonable control power sources, the method proposed in this paper uses the electrical distance to describe the relationship between the node voltage and the node injection power source.

A sensitivity matrix is expressed with the reactive power injection of source node between the voltage of other reactive power nodes. For a certain power grid, the correction equations of Newton-Raphson method can be formulated as

$$\begin{bmatrix} \Delta \boldsymbol{P} \\ \Delta \boldsymbol{Q} \end{bmatrix} = \begin{bmatrix} \boldsymbol{H} & \boldsymbol{N} \\ \boldsymbol{J} & \boldsymbol{L} \end{bmatrix} \begin{bmatrix} \Delta \boldsymbol{\theta} \\ \Delta \boldsymbol{V} \end{bmatrix} \tag{1}$$

In the equations (1) , $\Delta \boldsymbol{P}, \Delta \boldsymbol{Q}$ are the node injection increment of active and reactive power; $\Delta \boldsymbol{\theta}, \Delta \boldsymbol{V}$ are the increments of node voltage amplitude and phase angle; $\boldsymbol{H}, \boldsymbol{N}, \boldsymbol{J}$ and $\boldsymbol{L}$ are the sub-arrays of the Jacobian matrix, respectively. For a normal operation mode, $\boldsymbol{H}, \boldsymbol{N}, \boldsymbol{J}$ and $\boldsymbol{L}$ can be obtained according to a certain power flow calculation results.

To determine the sensitivity of the relationship between node active power and node voltage, we can make $\Delta \boldsymbol{Q} = 0$. The linearized equations (1) can be simplified as

$$\Delta \boldsymbol{V} = \left[\boldsymbol{N} - \boldsymbol{H} \boldsymbol{J}^{-1} \boldsymbol{L} \right]^{-1} \Delta \boldsymbol{P} = \boldsymbol{S}_{VP} \Delta \boldsymbol{P} \tag{2}$$

In the equation (2), the coefficient matrix $\boldsymbol{S}_{VP}$ is the sensitivity matrix between node injection active power and node voltage.

To determine the sensitivity of the relationship between node active power and node voltage, we can make $\Delta \boldsymbol{P} = 0$. The linearized equations (1) can be simplified as

$$\Delta \boldsymbol{V} = \left[\boldsymbol{L} - \boldsymbol{J} \boldsymbol{H}^{-1} \boldsymbol{N} \right]^{-1} \Delta \boldsymbol{Q} = \boldsymbol{S}_{VQ} \Delta \boldsymbol{Q} \tag{3}$$

In the equation (3), the coefficient matrix $\boldsymbol{S}_{VQ}$ is the sensitivity matrix between node injection reactive power and node voltage.

Sensitivity matrix $\boldsymbol{S}_{VP}$ and $\boldsymbol{S}_{VQ}$ show the corresponding electrical distance from two different aspects. According to these two sensitivity matrices, the whole network can be partitioned into several voltage control regions.

For a typical power system operation mode, the results of power flow calculation is the basis for the network partition. Different from the previous scenes, this paper considers the active and reactive coordination to complete the voltage control region partition. Sensitivity matrix $\boldsymbol{S}_{VP}$ and $\boldsymbol{S}_{VQ}$ will be calculated and adapted to improve the previous partition method.

3. Calculation and Analysis of Comprehensive Sensitivity

For a certain network, suppose the number of load node is L, the number of active power sources is N_p, the number of reactive power sources is N_q, the number of reactive power source partitions is M, he number of reactive power sources belonging to the ith volatge control partition is N_{qi} ($i \in M$). The comprehensive sensitivity between M and load node l is calculated as follows.

3.1. Comprehensive Sensitivity Calculation of reactive power

The whole process can be divided into the following steps:

- Input the data necessary for the calculation of the power flow, including: branch information (line parameters R_L, X_L, B_L; transformer parameters R_T, X_T, K; network topology, etc); node information (generator parameters (as needed) P_G, Q_G, V_G, θ_G; distributed generation parameters (as needed) P_{DG}, Q_{DG}, V_{DG}, θ_{DG}; load power data P_D, Q_D; reactive power compensation data Q_C, etc).
- Make the original power flow calculation, obtain the original results.
- Based on the Newton-Raphson method, set the node type before carrying out the power flow calculation. When making the reactive power sensitivity calculation: for the reactive power injection node $i \in M$ which takes part in the primary voltage control, it's node type is set to PV node, otherwise it is set to PQ node. While those reactive source nodes $i \notin M$ are set to PQ nodes [6].
- Let a small perturbation with load reactive power of node l ($l \in L$), including the junction nodes which have no load power. The node type should be reverted back to the its original type, i.e. set the small perturbation to zero, while the comprehensive sensitivity computation is completed.
- Make the reactive perturbation power flow calculation, obtain the perturbation results.
- Base on these two power flow results, the comprehensive sensitivity of reactive power is defined as

$$S_{VPi}^{k} = \frac{\sum_{k \in M} \Delta Q_k}{\Delta V_i} \quad (i \in L) \tag{4}$$

- By using the above steps, the comprehensive sensitivity can be calculated while partitioning all the active and reactive power source.

3.2. Comprehensive Sensitivity Calculation of active power

The whole process can be divided into the following steps:

- When making the active power sensitivity calculation: based on the Newton-Raphson method, all the node type are set as the original type.
- Let a small increasement with active power of generator node g ($g \in N_p$). The node power should be reverted back to the its original value, i.e. set the small perturbation to zero, while the comprehensive sensitivity computation is completed.
- Make the active perturbation power flow calculation, obtain the actinve perturbation results.
- Base on the original results and the actinve perturbation results, the comprehensive sensitivity of active power is defined as

$$S_{VPi}^{k} = \frac{\sum_{k \in M} \Delta V_k}{\Delta P_i} \quad (i \in L) \tag{5}$$

3.3. Comprehensive Sensitivity Calculation of reactive power and active power

IOP Conf. Series: Earth and Environmental Science **146** (2018) 012064 doi:10.1088/1755-1315/146/1/012064

- According to the comprehensive sensitivity Sensitivity matrix $\boldsymbol{S}_{VP}$ and $\boldsymbol{S}_{VQ}$, a comprehensive sensitivity matrix between the and load node and the active/reactive power source partitions is established[8] as equation (5).

$$\boldsymbol{S}_{region} = \boldsymbol{S}_{VQ} + \boldsymbol{S}_{VP} = \begin{bmatrix} S^1_{VQ1} & S^2_{VQ1} & \mathrm{L} & S^M_{VQ1} \\ S^1_{VQ2} & S^2_{VQ2} & \mathrm{L} & S^M_{VQ2} \\ \mathrm{L} & \mathrm{L} & \mathrm{L} & \mathrm{L} \\ S^1_{VQL} & S^2_{VQL} & \mathrm{L} & S^M_{VQL} \end{bmatrix} \tag{6}$$

3.4. Analysis of the comprehensive sensitivity

The method proposed in this paper establishes the comprehensive sensitivity matrix between load nodes and voltage control region from the angle of the normal power system operation mode. Comprehensive sensitivity of active and reactive power in equation (5) reflects the comprehensive power response characteristic of the active/reactive power source partition with load nodes.

The sensitivity value is positive generally. The greater its value of $\boldsymbol{S}_{\text{region}}$, indicating that the accused in the case of the same perturbation, the partition Gj reactive response output is greater than the other partition, it is clear that the node i should be partitioned into this sub-region. Therefore, the definition of a comprehensive sensitivity shows the comprehensive control effect power in the cluster for a controlled load nodes. The analysis results are apparently close to the actual power system operation.

The partition method proposed in this paper impoves the previous algorithm proposed by [6], the main difference lies in the comprehensive sensitivity which is formulated from the perspective of the reactive power and active power coordination. Reference [6] only adapted the scomprehensive sensitivity based on reactive power. So this method proposed in this paper is more universal and versatile, especially suitable for distribution network partition occasions.

4. Mapping the Load Node into the Partition

On the basis of comprehensive sensitivity matrix in formula (6), the electrical distance [7] between two node i and j is shown as follows.

$$\lambda_{ij} = -\lg\left|S_{ij}\right| \tag{7}$$

The whole matrix D, representing the electrical distance, can be formulated between the load node and voltage control partition on the basis of formula (7). The dimension of matrix D is $L \times M$, where L is the number of load nodes, M is the number of number of partitions. A row of a matrix D represents the proportion of the load node to each partition. The whole matrix D can be defined as

$$\boldsymbol{D} = \begin{bmatrix} \lambda_{11} & \lambda_{12} & \mathrm{L} & \lambda_{1M} \\ \lambda_{21} & \lambda_{22} & \mathrm{L} & \lambda_{2M} \\ \mathrm{L} & \mathrm{L} & \mathrm{L} & \mathrm{L} \\ \lambda_{L1} & \lambda_{L2} & \mathrm{L} & \lambda_{LM} \end{bmatrix} \tag{8}$$

Accordiong to the electrical distance matrix D, this paper adapts the mapping partitioning algorithm in [8] adapted "minimum electrical distance" principle to partition the load node into the nearest voltage control partition. This mapping algorithm can reflect the comprehensive control effects of all active and reactive sources to load node. But if two minimum electrical distance" belonging to a certain load node are almost the same value, this load node will be mapped into the corresponding control partitions. Such case will be shown in Section 5.

5. Simulation case

IOP Conf. Series: Earth and Environmental Science **146** (2018) 012064 doi:10.1088/1755-1315/146/1/012064

In order to verify the validity and correctness of the above-mentioned partitioning method, the IEEE33 node testing system is used whose wiring diagram is shown in Figure 1. The testing system has 33 nodes, 37 branches, 5 contact switches (indicated by dashed lines). The initial contact switch is off, so the line number of normal operation is 32. The system has the rated voltage of 12.66kV, the total network load of 3.715MW + j2.300MVar, where node 1 for the power node.

The original system has only one power node (that node 1) supplying power to the whole network. In this paper some appropriate changes is made on the original network in order to achieve power grid partition. Three distributed generations are added as the active and reactive power source.

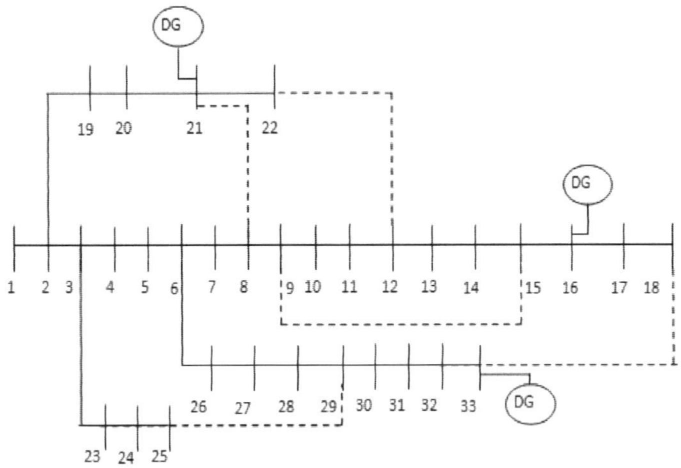

Figure 1. IEEE33 node system with distributed generations

5.1. Simulation Case 1

In this case, three DGs has the enough reactive outputs. The maximum reactive power of the node 16 is 200kvar micro-gas turbine power generation system. The reactive power of node 21 nodes is 400kvar PV power generation system. A doubly fed wind power generation system connecting to node 33 f the reactive capacity of 900kvar.

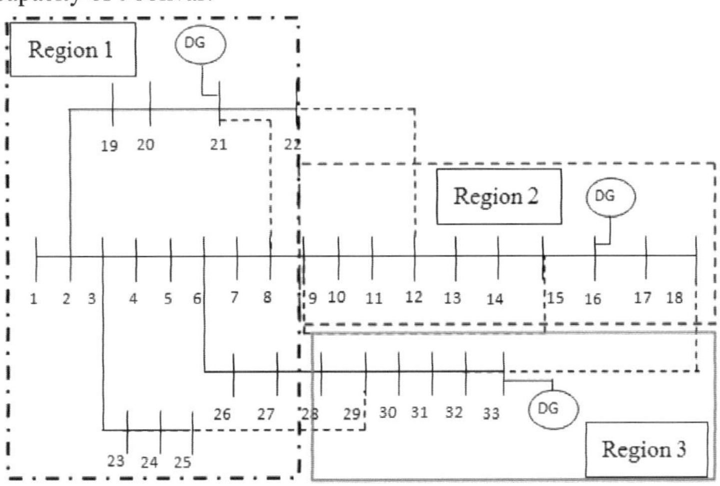

Figure 2. Partition results with enough Var output ability of the three distributed generations

As can be seen from the Figure 2, the four power sources are divided into three voltage control partitions of the entire network. Node 1 and node 21 belong to the same partition. Node 16 nodes and node 33 are distributed in different areas. This partition results ensure that the voltage fluctuations in the region, there are appropriate control methods to ensure that the system voltage level.

5.2. Simulation Case 2

In this case, two DGs has the enough reactive outputs. The maximum reactive power of node 21 nodes is 400kvar PV power generation system. A doubly fed wind power generation system connecting to node 33 f the reactive capacity of 900kvar. As for the node 16, for some reason, its reactive power does not reach the specified value. Its maximum reactive output is only 150kvar. In thess hypothetical scenario the grid partition may have different results which is shown in Figure 3.

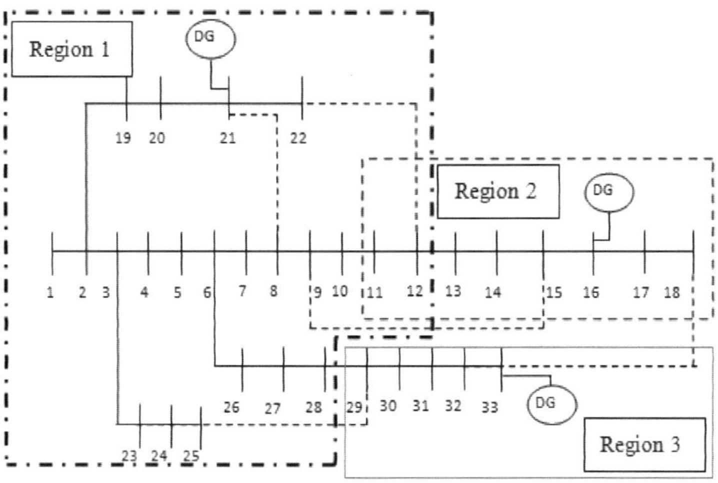

Figure 3. Partition results with different Var output ability of the three distributed generations

Compare Figure 2 and Figure 3, different settings of the reactive power source and active sources have a great impact on the system voltage control partitions. Part of the nodes belonging to the district has changed, these nodes are the node 9, 10, 11 and 12. And specially, the node 11 and node 12 belong to the two different partitions. It means if such a node voltage changes, a certain separate control method cannot improve the voltage level to normal standard. Comprehensive control method must be adapted at the same time to complete the voltage restoring goals.

6. Conclusions

In this paper we proposed an improved network partition method. According to the basic theory of the Newton-Raphson method, the sensitivity expression of the node voltage vs the node active power, the node voltage vs the node reactive power are deduced from the modified equations. Two kinds of sensitivity matrices are calculated by the perturbation power flow, which are synthesized and the electrical distance matrix is calculated. The partitioning of the network is done by mapping algorithm. The validity of the proposed method is verified with the simulation results of IEEE 33 nodes.

References

[1] Ferreir A P D F, Carvalho P M S and Ferreir A P D F M 2013 Distributed energy resources integration challenges in low-voltage networks; voltage control limitations and risk of cascading. *IEEE Trans on Sustainable Energy*. 4(1), 82-88
[2] Wang Z Q, Zhu S Z and Zhou S X 2004 Impacts of distributed generation on distribution system voltage profile. *Automation of Electric Power Systems.* 28(16), 56-60
[3] Thomson M and Infield D G 2007 Network power-flow analysis for a high penetration of distributed generation. IEEE *Trans on Power Sysems*. 22(3), 1157-1162
[4] Tonkoski R, Lopes L A and El-Fouly T H 2011 Coordinated active power curtailment of grid connected PV inverters for overvoltage prevention. *IEEE Trans on Sustainahle Energy*. 2(2),

139-147

[5] Liu Y B, Wu W C, Zhang B M, Ju Y T and Li Z L 2014 Overvoltage Preventive Control Method Based on Active and Reactive Power Coordinated Optimization in Active Distribution Network. *Automation of Electric Power Systems*. 38(9), 184-191

[6] Wang J T, Zhou J, Yuan M, Ye W H and Xu Z H 2014 An Improved Partition Method of Reactive Voltage Control considering Dynamic Reactive Power Sources. China, Advances in Engineering Research. 5(2),820-823

[7] Zhao J Q, Liu F C and Deng Y 2010 Network partitioning for reactive power/voltage control based on a mapping division algorithm. Automation of Electric Power Systems. 34(7),36-39

IOP Conf. Series: Earth and Environmental Science **146** (2018) 012065 doi:10.1088/1755-1315/146/1/012065

The optimal deployment model of energy storage systems considering large-scale wind farms connected

T Yan[1], M X Zhang[1], X F Song[2], B Z Liu[3] and Y Huang[3]

[1]State Key Laboratory of Operation and Control of Renewable Energy & Storage Systems, Haidian District, Beijing, 100192, China
[2]State Grid Xinjiang Electric Power Company, Economic Research Institute, Urumqi, 830011, China
[3]State Key Laboratory of Alternate Electrical Power System with Renewable Energy Sources, North China Electric Power University, Beijing 102206, China

yantao@epri.sgcc.com.cn
sfbrzmx@163.com
sxf024@163.com
bzliu@ncepu.edu.cn
1765506332@qq.com

Abstract. In order to solve the serious wind power curtailment problem caused by the rapid development of large-scale wind power, a multi-point layout energy storage system is proposed in the grid with the large proportion of wind power. Considering the energy storage costs, wind power curtailment costs and the penalty for loss of load, the energy storage planning model is established by introducing the problem of unit commitments with restrictions, which can be solved by using the mature commercial software CPLEX. Finally, taking the standard IEEE 24 node system as an example, different case studies are carried out. It is found that energy storage cost, wind farm location and access power have an important influence on energy storage layout and configuration capacity.

1. Introduction

In recent years, with the rapid growth of wind power installed capacity in the world, wind power has become the main way of human using of clean energy [1]. However, the volatility of wind power output is random and anti-peaking, large-scale wind power curtailment problem is increasing [2-3]. At the same time, due to the restriction of safe and stable operation of power system, the wind farm output is forced to be cut down, and the phenomenon of abandoned wind is generated. The abandonment of wind farm not only causes the waste of energy but also makes the operation of the wind farm difficult to reach the planned utilization hours, which has become a major factor restricting the development of wind power. The energy storage device [4] as the means of transferring electric energy in time, can be introduced into the grid with the large proportion of wind power, which contributes to the large-scale wind power acceptance capacity and reduces wind power curtailment.

Although there are a lot of relevant researches at home and abroad, there are still many problems in energy storage configuration based on improving the new energy acceptance capacity [5-8]. For example, these studies rarely consider the characteristics of energy storage operations. However, the operation characteristics of different energy storage methods vary greatly, and the regulation effect on wind power is different. The operation mode of the energy storage system is flexible, and different operation modes can be adopted to reduce the wind rejection of the power grid in different scenarios. Therefore, in solving the planning problem of energy storage optimal allocation, considering the actual operation of the energy storage system and adding the corresponding operating conditions constraints, the specific charging and discharging process can be optimized, which can make full use of energy storage capacity.

Based on the above analysis, Considering the energy storage costs, wind power curtailment costs and the operation characteristics of energy storage, the energy storage planning model is established, which can be solved by using the mature commercial software CPLEX. Finally, taking the standard IEEE 24 node system as an example, different case studies are carried out.

2. The optimal deployment model of energy storage systems

There are two main ways to configure energy storage in power grid with wind power: one is to configure it on the wind farm side in order to smooth wind power output; another is to configure it on the appropriate nodes in the grid to improve wind power acceptance capacity by means of Peak load regulation and valley filling. In this paper, the energy storage is mainly aimed at reducing the abandoned wind, so the energy storage is configured in the whole network.

2.1. Objective function

The objective function of the model consists of three parts: investment cost of energy storage system f_{ess}, cost of wind power curtailment f_{wind} and the penalty for loss of load f_{loss}.

$$\min f = \delta_s f_{ess} + \delta_w f_{wind} + \delta_d f_{loss} \tag{1}$$

$$f_{ess} = (1 + r_{op} + r_{ma} + r_{sc}) r_{de} \sum_{k \in \Omega_{es}} (c_p p^r_{es,k} + c_e e^r_{es,k}) \tag{2}$$

$$f_{wind} = \sum_{j \in \Omega_w} \sum_{s \in \Gamma} d_s \sum_{t=1}^{T} (p^{\max}_{w,t,s,j} - p_{w,t,s,j}) \tag{3}$$

$$f_{loss} = \sum_{m \in \Omega} \sum_{s \in \Gamma} d_s \sum_{t=1}^{T} (p^{\max}_{load,t,s,m} - p_{load,t,s,m}) \tag{4}$$

In equation (1), δ_s is cost factor of energy storage, δ_w is cost of unit wind power curtailment, δd is the penalty factor for loss of load. In equation (2), r_{op} , r_{ma} and r_{sc} are cost conversion coefficients of the operation, maintenance and disposal of the energy storage system. c_p and c_e are the unit power and unit capacity costs of energy storage system. $P^r_{es,k}$ and $e^r_{es,k}$ are the rated power and rated capacity of the energy storage system allocated at the node k . Ω_{es} is the set of energy storage nodes to be configured. In equation (3), Ω_w is the set of wind farms. The letter 's' stands for a typical day, the number of days of which is d_s. T is the number of periods per typical day, Γ is the set of typical days. $p^{max}_{w,t,s,j}$ and $p_{w,t,s,j}$ are the maximum output and actual output of wind farm j on typical day s within hours t.

2.2. Constraints

The primary constraints of the model include the system DC power flow constraints, conventional power plant operation constraints, wind farm operation constraints, energy storage system operation constraints and system standby constraints.

System power flow constraints

$$\boldsymbol{P}_{g,t,s} + \boldsymbol{P}_{w,t,s} + \boldsymbol{P}_{e,t,s} - \boldsymbol{A}\boldsymbol{P}_{ij,t,s} = \boldsymbol{P}_{load,t,s} \tag{5}$$

$$p_{ij,t,s} - b_{ij} n_{ij} (\theta_{i,t,s} - \theta_{j,t,s}) = 0 \tag{6}$$

$$|p_{ij,t,s}| \le p^{\max}_{ij} \tag{7}$$

In equation (4), $\boldsymbol{P}_{g,t,s}$, $\boldsymbol{P}_{ws,t,s}$, $\boldsymbol{P}_{e,t,s}$ and $\boldsymbol{P}_{load,t,s}$ stand for output vector of conventional power plant, output vector of wind farm, charge discharge power of energy storage and load active demand vector at each node on typical day s within hours t respectively. $\boldsymbol{A}$ stands for node-to-branch incidence matrices of lines. $\boldsymbol{P}_{ij,t,s}$ are branch active power vectors of lines on typical day s within hours t respectively.In equation (5), b_{ij} is admittance of branch ij, n_{ij} is the number of line ij. $\theta_{i,t,s}$, $\theta_{j,t,s}$ and $\theta_{ref,t,s}$ are voltage phase angle of bus i, bus j and slack bus on typical day s within hours t respectively.In equation (6), p_{ij}^{max} is the maximum active power limit of branch ij.

Conventional operation constraints

$$\sigma_t p_{g,i}^{\min} \le p_{g,t,s,i} \le \sigma_t p_{g,i}^{\max} \tag{8}$$

$$\Delta p_{g,i}^{\min} \le p_{g,t,s,i} - p_{g,t-1,s,i} \le \Delta p_{g,i}^{\max} \tag{9}$$

Constraint (7) is the upper and lower bound of the output of conventional thermal power unit. Constraint (8) is climbing and landslide restraint of conventional thermal power unit. σ_t is 0-1 variable standing for the operation state of conventional thermal power unit. Where max and min stand for high/low limit of corresponding variables. $p_{g,t,s,i}$ and $p_{g,t-1,s,i}$ are active power of conventional power plant i on typical day s within hours t and hours t-1 respectively, $\Delta p_{g,i}^{min}$ and $\Delta p_{g,i}^{max}$ stand for equivalent landslide and climbing rate of conventional power plant i.

$$\sigma_t - \sigma_{t-1} = \varsigma_{t-1}^{on} - \varsigma_{t-1}^{off} \tag{10}$$

$$\varsigma_t^{on} + \varsigma_t^{off} \le 1 \tag{11}$$

$$\varsigma_t^{on} + \sum_{k=t+1}^{\min(T,t+T_{\min}^{on}-1)} \varsigma_k^{off} \le 1 \tag{12}$$

$$\varsigma_t^{off} + \sum_{k=t+1}^{\min(T,t+T_{\min}^{off}-1)} \varsigma_k^{on} \le 1 \tag{13}$$

Constraints (10)-(13) are the minimum start-up and shut-down time constraint of thermal power unit. ζ_t^{on} and ζ_t^{off} are 0-1 variables indicating whether to start or stop operation for conventional power plant i at the end of the time interval t. '1' means 'yes', and '0'means 'no'. T_{min}^{on} and T_{min}^{off} are the minimum start-up and shut-down time of thermal power unit.

$$0 \le p_{w,t,s,j} \le p_{w,t,s,j}^{\max} \tag{14}$$

$$0 \le p_{load,t,s,m} \le p_{load,t,s,m}^{max} \tag{15}$$

Where $p_{ws,t,s,j}$ is output of wind-storage combined system j on typical day s within hours t.

Energy storage system operation constraints

$$\begin{cases} 0 \le e_{es,t,s,k} \le e_{es,k}^{r} \\ -p_{es,k}^{r} \le p_{es,t,s,k} \le p_{es,k}^{r} \\ p_{es,k}^{r,\min} \le p_{es,k}^{r} \le p_{es,k}^{r,\max} \end{cases} \tag{16}$$

$$\begin{cases} e_{es,1,s,k} = e_{es,T,s,k} = 0 \\ e_{es,1,s,k} = e_{es,T,s,k} + p_{es,T,s,k}\Delta t \\ e_{es,t,s,k} = e_{es,t-1,s,k} + p_{es,t-1,s,k}\Delta t \end{cases} \tag{17}$$

In equation (16), $e_{es,t,s,k}$ is capacity of energy storage system j on typical day s within hours t. In equation (17), $e_{es,1,s,k}$, $e_{es,T,s,k}$ and $e_{es,t,s,k}$ are storage energy of energy storage system k on typical day s within hours 1, T and t. $p_{es,T,s,k}$ and $p_{es,t-1,s,k}$ are charge-discharge power of energy storage system k on typical day s within hours T and t-1. Δt is time interval,1h.

System standby constraints

$$\begin{cases} \sum_i p_{g,t,s,i}^{\max} + \sum_j p_{w,t,s,j} + \sum_k p_{s,t,s,k} \ge \sum_m p_{load,t,s,m} + R_{u,t}, \\ \sum_i p_{g,t,s,i}^{\max} + \sum_j p_{w,t,s,j} + \sum_k p_{s,t,s,k} \ge \sum_m p_{load,t,s,m} - R_{d,t}. \end{cases} \tag{18}$$

Where $R_{u,t}$ and $R_{d,t}$ are the positive and negative rotating reserve capacity limit within hours t.The objective function consists of equations (1)-(4), and the constraints consists of equations (5)-(18). The case has been implemented on a personal computer, using CPLEX under MATLAB.

3. Case study

3.1. Example system settings

This paper takes the IEEE 24 bus system as an example system [9]. System parameters are available in the literature [9]. Suppose that all the load and power capacity of the system are increased to 3 times of the original value, and the maximum total load is increased to 8550MW. On this basis, 7 new transmission lines are added to increase the transmission capacity of the system. The new lines are respectively : 6-10(1)、7-8(2)、10-12(1)、14-16(1)、16-17(1)、20-23(1). Among them, 6-10 (1) indicates that 1 line is added to the branch 6-10. Assume that nodes 3, 4, 5, 6, 8, 9, 10 are allowable building points for energy storage. The compressed-air system is selected as the energy storage system, and cost parameters [10] of which are shown in Table 1. The time series datas of wind power output and load demand are shown in figure 1.

Table 1. The cost parameters of energy storage system.

r_{op}	r_{de}	c_p(RMB/kW)	c_e(RMB/kWh)
0.01	0.05	200	4000

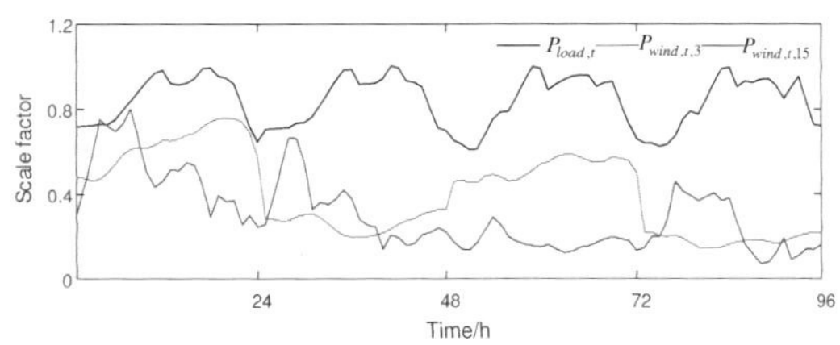

Figure 1. Wind power output and load demand.

3.2. Planning result analysis

3.2.1.Influence of energy storage cost on deployment of energy storage

Two wind farms are connected to nodes 3 and 15 respectively, and the integration power is 1400MW and 300MW respectively. Considering that the energy storage cost has an important influence on the results of deployment of energy storage, this section changes the value of the energy storage cost coefficient δ_s , then solve the model and obtain planning results as shown in table 2 and figure 2. The results show that with the decrease of energy storage cost coefficient δ_s , the energy storage planning power increases gradually and the cost of wind power curtailment gradually becomes smaller. In addition, when the energy storage cost coefficient drops to 0.6, the effect of reducing wind power curtailment is significant.

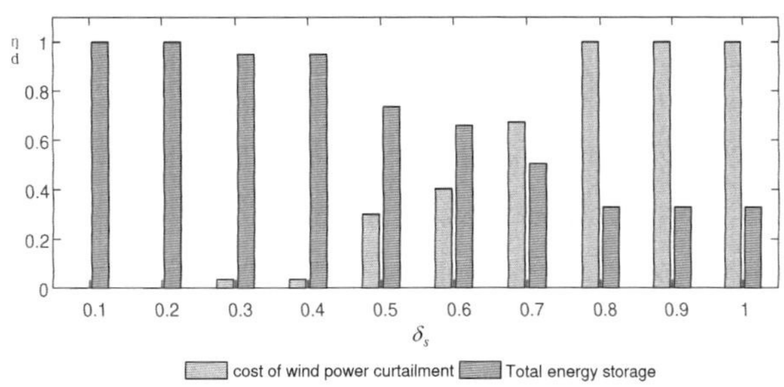

Figure 2. Cost of wind power curtailment and total energy storage under different δ_s.

Table 2. Planning schemes under different δ_s.

δ_s	0	0.2	0.4	0.6	0.8	1
Location/power(MW) of Stored energy	3/120; 4/120; 8/120;9/120	3/120; 4/4; 8/120;9/120	3/120; 8/106; 9/120	3/120; 9/120	3/120	3/120
Total energy storage/MW	480	364	346	240	120	120
f_{wind} /Million RMB	0	0.02	1.21	13.92	34.53	34.53

Influence of wind power integration power on deployment of energy storage In this section, the value of the cost coefficient of energy storage is set to 0.5. In order to analyze the deployment of energy storage under different wind power integration power, two wind farms are connected to nodes 3 and 15 respectively, and the node 15 wind farm power is fixed to 300MW while the node 3 wind farm power is changed, then solve the model and obtain planning results as shown in table 3. The curve of energy storage operation under scenario 1 is shown in figure 3($P_{w,3}$=1400MW). Analysis of Table 3, it can be found that with the increase of wind power integration power, the total power of energy storage gradually increases. However, the cost of wind power curtailment is not a monotonic change. This is because the results of the optimization model are achieving comprehensive optimization between the cost of energy storage and cost of wind power curtailment.

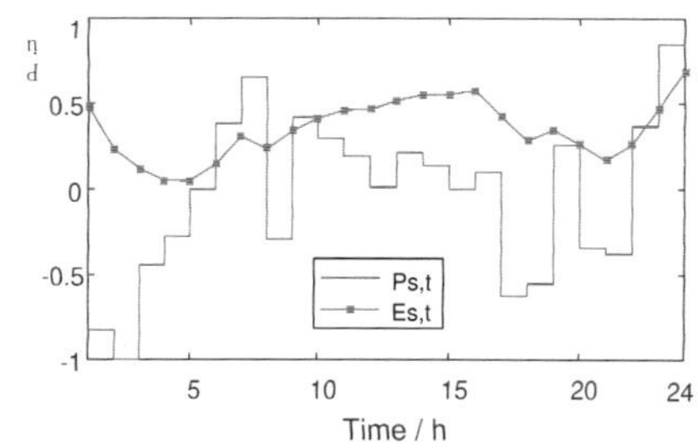

Figure 3. The curve of energy storage operation under scenario 1.

Table 3. Planning schemes under different $P_{w,3}$.

$P_{w,3}$/MW	1100	1200	1300	1400
Location/planning power(MW) of Stored energy	-	3/12	3/110	3/120; 8/28; 9/120
Total energy storage/MW	0	12	110	268
f_{wind} /Million RMB	0	0.07	0	10.3

Influence of wind power integration position on deployment of energy storage In this section, the value of the cost coefficient of energy storage is set to 0.5. The node 15 wind farm integration power is fixed to 300MW, another wind farm integration power is set to 1400MW. Changing its integration position, solve the model and obtain planning results as shown in table 4. The results show that when the total wind power integration power of the system is certain while wind power integration position changing, deployment of energy storage is different, and the cost of wind power curtailment is also different. However, the energy storage systems are the preferred configuration in the wind farm side to enhance the capacity of wind power utilization.

Table 4. Planning schemes when changing wind power integration position.

Wind power integration position/MW	3	6	13
Location/planning power(MW) of Stored energy	3/120; 8/28; 9/120	5/8; 6/120	5/16; 9/84; 10/26
Total power of energy storage/MW	32.48	15.51	15.27
f_{wind} /Million RMB	10.3	12.41	25.05

4. Conclusion

The abandoned wind caused by the wind power output characteristics and power system operation constraints is an important issue for the development of wind power. Based on the characteristics of energy storage operation, the optimal allocation model of energy storage is established to reduce the abandoned wind. The effect of energy storage on reducing the abandoned wind is verified by comparing planning results under different energy storage cost, wind power integration power and position. The results show that as the cost of energy storage decreases, the role of energy storage in reducing abandoned wind will be more pronounced. In addition, with the energy storage to reduce the cost of storage can reduce the abandoned wind effect is more obvious; the storage layout in the whole network is more conducive to play the energy adjustment ability, promote the wind power accommodation. More importantly, the multi-point energy storage layout in the whole network is more conducive to play the role of energy storage regulation.

Acknowledgments

This work is supported by the National Nature Science Foundation of China (51677174) and Science and Technology Project of SGCC (simulation research on large scale energy storage cluster control technology to improve transmission capacity of the renewable energy base, 5230HQ16016U). We would like to express our most sincere appreciation.

References

[1] Huang L H, Song L L, Zhou R W and Li G 2017 Characteristics analysis of wind power fluctuations for large-scale wind farms *Proceedings of the CSEE* **5** 103-106
[2] Shu Y B, Zhang Z G, Guo J B and Zhang Z L 2017 Study on key factors and solution of renewable energy accommodation *Proceedings of the CSEE* **37** 1-8
[3] Zou J, Lai X and Wang N B 2015 Mitigation of wind curtailment by coordinating with pumped storage *Power System Technology* **39** 2472-2477
[4] Ai X and Dong C F 2015 Review on the application of energy storage technology in power system with renewable energy sources *Modern Electric Power* **32** 1-9
[5] Wang S Y, Jiang Q Y and Ge Y F 2017 Optimal Configuration of energy storage considering wind ramping events *Power System Technology* **41** 1-7
[6] Han X N, Li J M, Wen J Y, Ai X M, Li J H and Luo W H 2015 Optimization for robust energy storage allocation in power system with multiple wind farms integrated *Proceedings of the CSEE* **35** 2120-2127
[7] Jia L, Hu Z C, Song Y H and Ding H J 2017 Joint planning of distribution networks with distributed energy storage systems and electric vehicle charging stations *Proceedings of the CSEE* **37** 73-84
[8] Zheng L, Hu W, Lu Q Y, Min Y, Yuan F and Gao Z H 2014 Research on planning and operation model for energy storage system to optimize wind power integration *Proceedings of the CSEE* **34** 2533-2543
[9] Macrae C A G, Ernst A T and Ozlen M 2016 A Benders decomposition approach to transmission expansion planning considering energy storage *Energy* **112** 795-803
[10] Wang K Y 2016 Risk measure based approach for CVPP operation *North China Power University*

IOP Conf. Series: Earth and Environmental Science **146** (2018) 012066 doi:10.1088/1755-1315/146/1/012066

The Research of V2G Technology Real-Time Charge and Discharge Capacity Prediction

E Leugoue [1], J Zhang [1] and S R Dibonji Ndjansse[2]

[1]North China Electric Power University, Changping District, Beijing 102206, China
[2]Hefei University of Technology, 193 Tunxi Road, Hefei, Anhui 230009, China

Corresponding author's e-mail: emiliejolie90@yahoo.fr

Abstract. The problem of spatial randomness and time randomness is very important for the characteristics of electric vehicle access network. Based on monte carlo stochastic simulation method, the daily mileage and the starting time nodes of different types of electric vehicles in the dispatch area were randomly selected and the prediction model of charging capacity of V2G system is presented. Taking the Beijing area as an example, the real-time charge and discharge capacity of V2G system is predicted in 24 hours a day, to provide data reference for power grid dispatch center.

1. Introduction

At present, Three key technologies of battery, motor and electronic control has basically meet the needs of the electric vehicle application, but its charge and the corresponding supporting infrastructure problems limited the popularization of electric vehicles. Emerging V2G (Vehicle to grid)[1-3] technology refers to two-way interaction between electric vehicle as mobile energy storage unit and power grid and its core thought is to connect idle electric cars to the grid in case of time and battery capacity. When the grid load too high, electric vehicle storage energy is dispatched to feed power grid, and when the grid too low, the excess energy in the grid is stored in electric vehicles. Through the corresponding optimization scheduling method, V2G technology can solve the problem of electric car charging and avoid the adverse effects of the charging cluster effect on the power grid. It can also respond to the frequency of the power system, curb the fluctuation of the peak valley and the network of renewable energy, maintain efficient and safe operation of power grid and reduce operating costs. At the same time, electric car[4] users can charge during low electricity prices and feed electricity to the grid during peak electricity consumption , to Lower the cost of electric vehicles.

Based on the analysis of the development status of V2G technology at home and abroad, various factors that affect the real-time charging and discharging capacity of V2G technology were considered, and taking Beijing as example, Monte carlo simulation model[5-7] was established in this paper, and the real-time charge and discharge capacity of V2G technology in the Beijing area was predicted in 2020 and 2024, to promote the development of V2G technology in China's electric vehicle industry.

2. Analysis of the influence factors of V2G technology in real-time charging and discharging capacity

2.1. Electric vehicle classification

According to the characteristics of vehicle use, electric cars can be classified into three types: electric cars, private electric cars, and electric cars that can be dispatched. Among them, utility electric vehicles generally include: bus electric cars, rental electric cars and official electric cars. These cars are all public service and usually have fixed driving characteristics. Their average driving time is relatively long, and the peak of power grid load is still in normal use, so this kind of car is generally not used as a V2G power source. Private electric cars are different from electric cars, and their use has a strong randomness. Private electric cars idle most of the time, and can be charged at night. On the premise of establishing an agreement with the user, some private electric cars can be netted and give back electricity to the grid during peak hours to realize the fluctuation of the power grid, thus to improve the efficiency of power grid and reduce the cost of electric vehicles.

2.2. Scale analysis of electric vehicles

If it is 6 million motor vehicles in 2016 of Beijing, average annual vehicle ownership is expected to grow by 200,000 vehicles per year. The market share of electric vehicles will increase significantly in the next 10 years. At present, it accounts for 5% of motor vehicle ownership The future increases by 5 to 10 percentage points per five years. The prediction results are as table 1.

Table 1. The forecast of electric vehicle ownership in Beijing from 2020 to 2024

year	Number of motor vehicles (10 thousands of cars)	Number of electric cars (10 thousands of cars)	Market share
2020	700	70	10%
2024	800	160	20%

In the initial development phase of electric vehicles, take public transport facilities as the main operating objects, with the development of electric vehicle related technologies and facilities, private electric cars are the main force in the future electric car market. Table 2 shows the percentage forecast of the total number of electric vehicles in the market by public electric vehicles and private electric vehicles in each stage. The assumption is that the market penetration rate of the electric vehicle can be increased continuously. About 20% in 2020 and 30% in 2024.

Table 2. Estimation of the amount of electric vehicles(10 thousands of cars)

year	Number of electric cars	Number of buses	Number of private cars	Number of schedulable vehicles
2020	70	42(60%)	28(40%)	5.6(20%)
2024	160	64(40%)	96(60%)	28.8(30%)

2.3. Charge and discharge time

Different types of electric vehicles can charge and discharge at different times. The charging and discharging time and its duration influence the charging and discharging power of electric vehicle to a great extent. In order to facilitate the real-time prediction and calculation of the charging and discharging capacity of V2G technology, the characteristics and laws of different types of electric vehicles are approximated, as shown in figure 1.

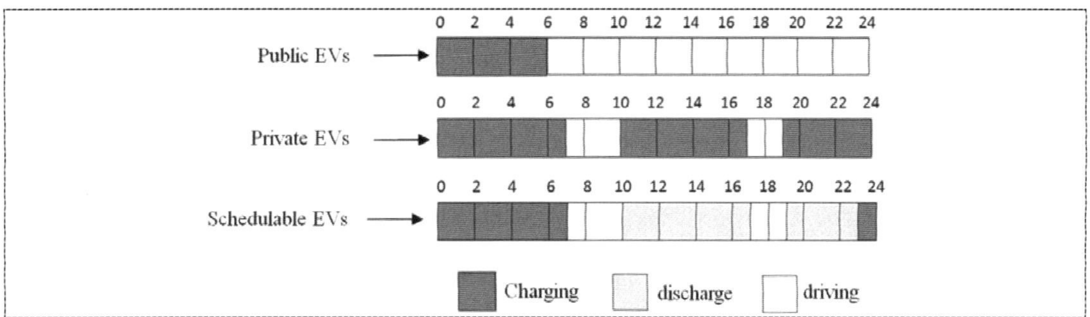

Figure 1. Charge, discharge, and travel hours for three types of EVs.

2.4. Battery system related parameters

Influenced by various factors such as the type, driving conditions and production manufacturers of electric vehicles, the relevant parameters of the battery system are significantly different, which will also affect the charging and discharging power capacity of electric vehicles. In order to simplify the calculation, the parameters of the electric vehicle battery system involved include the rated capacity, rated voltage, charging discharge efficiency, total charging time and power consumption, as shown in table 3.

Table 3. Battery parameters for electric vehicles

Rated capacity (A·h)	Rated voltage (A·h)	Charge/discharge efficiency	Total charging time (h)	Power consumption (kWH/km)
100	320	0.9	5	0.125

3. Prediction of real-time charge discharge capacity of V2G technology

Monte carlo method is also called statistical simulation method, which is used to solve the mathematical or physical problems of non-deterministic (probability or random) mathematics. The charging and discharging behaviors of electric vehicles are highly random, and the monte carlo method can simulate the random behavior well. The simulation process is very close to the actual situation, which is conducive to the accurate prediction of real-time charging and discharging capacity of V2G technology.

First, charging and discharging periods are determined according to different types of electric vehicles; second, during the time of charge and discharge, through the Monte Carlo method, according to that start charge, discharge time obey uniform distribution[8], and the average daily driving range obeys lognormal distribution[9] and normal distribution[10], the starting charge, discharge time and daily mileage of electric vehicles are randomly selected; then, calculate the initial of the corresponding vehicle; finally, the charge and discharge of all electric vehicles are randomly assigned. The total charge power and the total discharge capacity of V2G power supply can be calculated from 24 hours of the day. At the random extraction of start charging time, based on charging power and charging demand (ensure that the battery must be fully charged prior to the end of charging), calculate the required charging time. Narrow the sampling range of the starting charging time when the required length of charging is met. The specific process is shown in figure 2.

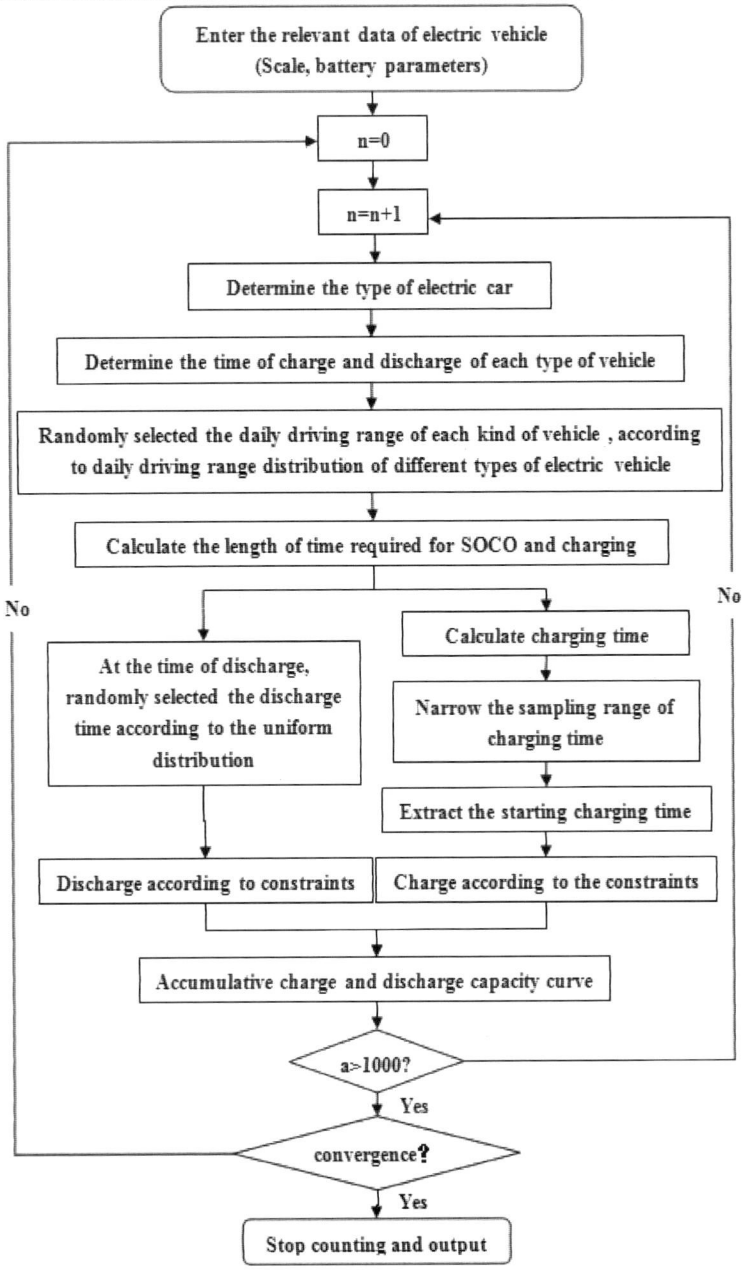

Figure 2. Flow chart of proposed algorithm based on Monte Carlo simulation.

4. Simulation results analysis

Monte carlo simulation runs at least 1000 times, and the variance coefficient is required to be less than 0.05%. The V2G discharge capacity and the charging load power curve are predicted for 24 hours a day using vehicle forecast data of the electric car in 2020 and 2024, as shown in figure 3.

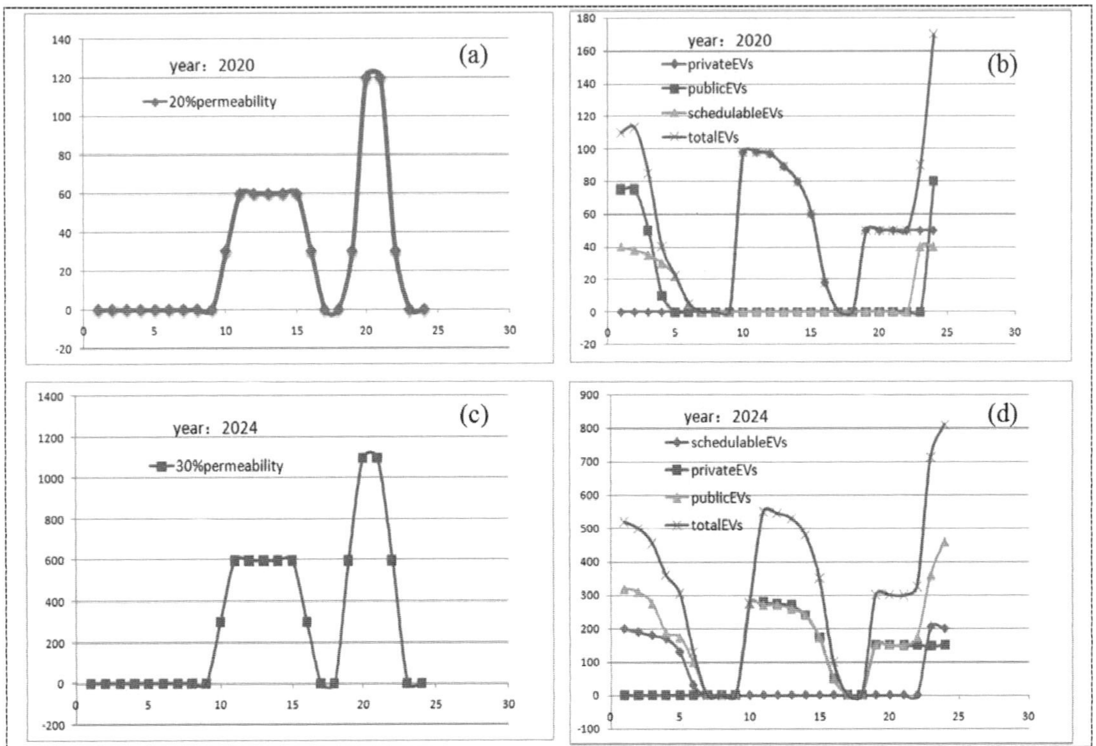

Figure 3. The possible charging and discharging capacity of EVs on each day from 2020 to 2024.

The charging load of electric cars is concentrated between 9:00 and 17:00 during the day, which is the private electric car's centralized charging time. As the number of private electric vehicles increased significantly, the charging load has risen sharply from 2020 to 2024. The evening charge is set at 19:00, which is also the time for private electric car charging. Electric car charging load obviously increases, especially in 2024, and the evening peak load is evident. This result from concentrated charging at night after the private electric cars sunrise. At the same time, as the scale of the available electric vehicles increases, it's a bit of a filling for the charging load increase resulted from evening rush hour of electric vehicles. At the same time, there is a residual capacity to meet the base load of the evening peak. It plays a role in helping the power grid to operate safely and steadily. Electric cars and electric cars that can be scheduled are charged at night time, which play a certain "fill the valley" function. From the simulation results diagram, the large-scale private charging can be a hidden danger to the stability and security of the power grid in future. To meet its charging demand, more units should be added, and thus to increase the cost of grid operation. How to guide its transfer charge more effectively is very beneficial to the safety and economical operation of the power grid. The time-sharing mechanism is used to guide charging of electric vehicles In low-power valley load times.

5. Conclusion

Based on monte carlo stochastic simulation method, monte carlo random sampling method is proposed to predict charging and discharging capacity of electric vehicles, and the simulation results are analyzed. The transformation of the modern transportation industry caused by the widespread application of electric vehicles and its impact on the development of power industry are revolutionary and far-reaching The intelligent charging and discharging management of electric vehicles is effective in dispatching and controlling charging and discharging capacity, which can obviously improve the economic and security of the entire power system operation. How to solve the optimal scheduling and control of electric vehicle charging and discharging problems fundamentally determines whether it can reduce or even avoid into the size of electric vehicles connected to the electricity grid of the negative effects on the electric power system. It also determines whether it can play its best role.

References

[1] Nguyen HNT, Zhang C, Mahmud MA. Optimal Coordination of G2V and V2G to Support Power Grids With High Penetration of Renewable Energy. IEEE Transactions on Transportation Electrification. 2015, 1(2):188-195.
[2] Karfopoulos EL, Panourgias Kostas A, Hatziargyriou ND. Distributed Coordination of Electric Vehicles providing V2G Regulation Services. IEEE Transactions on Power Systems. 2016, 31 (1):329-338.
[3] Ota Y, Taniguchi H, Baba J, Yokoyama A. Implementation of autonomous distributed V2G to electric vehicle and DC charging system. Electric Power Systems Research, 2015, 120:177-183.
[4] Chung SH, Kwon C. Multi-period planning for electric car charging station locations: A case of Korean Expressways. European Journal of Operational Research. 2015, 242 (2):677-687.
[5] Resat H, And HSW, Dixon DA. Probability-Weighted Dynamic Monte Carlo Method for Reaction Kinetics Simulations. Journal of Physical Chemistry B, 2016, 105 (44):11026-11034.
[6] Saariaho M, Subbotin A, Ikkala O, Ten Brinke G. Comb copolymer cylindrical brushes containing semiflexible side chains: a Monte Carlo study. Macromolecular Rapid Communications. 2015, 21 (2):110-115.
[7] Lee YJ, Maeng JY, Lee E. Melting behaviors of icosahedral metal clusters studied by Monte Carlo simulations. Journal of Computational Chemistry. 2015, 21(5):380-387.
[8] Bellingeri M, Kriegel I, Scotognella F. One dimensional disordered photonic structures characterized by uniform distributions of clusters. Optical Materials. 2015, 39:235-238.
[9] Asmussen S, Jensen JL, Rojas-Nandayapa L. On the Laplace Transform of the Lognormal Distribution. Methodology & Computing in Applied Probability. 2016, 18 (2):441-458.
[10] Mukhopadhyay P. Multivariate Normal Distribution and Related Distributions. Weed Research. 2015, 33(1) :79-88.

IOP Conf. Series: Earth and Environmental Science **146** (2018) 012067 doi:10.1088/1755-1315/146/1/012067

Study on the Effects of DOM on the Bioaccumulation of Perfluorinated Acids in *Daphnia magna*

Z N Dai[1,2], A LYang[1,2] and H Y Fu[1,2]

[1]School of Environmental Science and Engineering, Xiamen University of Technology, Fujian, 361024, China
[2]Key Laboratory of Environmental biotechnology of Fujian Province, University, 361024, China

daizn@xmut.edu.cn

Abstract: Perfluoroalkyl acids (PFAs), one kind of emerging contaminants, have attracted great attentions in recent years. However, the study about their bioaccumulation mechanism remains scarce. In this research, the bioaccumulation of 6 kinds of PFAs in water flea*Daphnia magna*was studied in the presence of dissolved organic matter (DOM), and the possible interactions between DOM and PFCs were explored by spectra analysis. The results showed that DOM could decrease the bioaccumulation of PFAs at 10 and 20 mg•L^{-1}, and increase the bioaccumulation at 1 mg•L^{-1}. The decrease of BAF was in the range of 11 to 42% and 23 to 77% for FA and HA, respectively. The increase of BAF was in the range of 12 to 44% and 8 to 73% for FA and HA, respectively. The spectrum analysis indicated that the interaction of PFAs and DOM was not due to the bond formation reaction, but the electrostatic attraction.

1. Introduction

Perfluorinated acids (PFAs) is an organofluorine compound with all hydrogens replaced by fluorine on a carbon chain, but the molecule also contains at least one different atom or functional group. PFAs have a wide range of industrial applications [1]. PFAs such as Perfluorooctanesulfonates(PFOS) andperfluorooctanoic acid(PFOA) have been found in water, air and soil, including remote locations such as the Arctic and the Antarctic[2].PFCs have been found in the aqueous environment ubiquitously in particular.These contaminants have probably been present in the environment and in biota for many years.however, their environmental and biological effects have been realized only recently. PFAs is one kinds of persistent organic pollutants, and they are not known to degrade by any natural processes.

Humic substances are one kind of typical dissolved organic matter (DOM), and it including fulvic and humic acids. According to previous studies,about 40-60% of dissolved organic carbon in natural waters was humic substance [3]. The diversity of functional groups of humic substances defines a broad range of interaction with pollutants[4]. The photolysis rates of organic compounds could be increased, and the concentration of pollutants in water could be reduced with the interaction of relatively hydrophobic organic compounds andhumic substances[5]. Previous studies have shown that humic substances can distinctly influence bioaccumulation and toxicity of contaminants such as pesticidesand chlorinated hydrocarbons in aquatic biota [6].However, the mechanism of the increase of the bioaccumulation is still unclear. Though some hypothesis has been raised that the increase of bioaccumulation is correlate with the content, types and contact time of humic substances[7], they are still not supported by any real evidence and the mechanism need further research.It's important to develop methods to measure the sorption of pollutants to humic substances to predict their fate in environment. Studies dealing with the environment behavior of humic substances indicated that the major light-absorbing of these solute in the range of 200 to 800 nm in many natural waters, and spectroscopic techniques is useful in the study of the interaction of DOM and organic pollutant[8].

Humic and fulvic acids are the most abundant components of DOM and widely believed to be representatives of DOM behavior[9]. The goal of this study is to evaluate the effects of humic substances on the bioaccumulation of PFAs by *D. magna*. The spectral analysis was adopted to study the possible mechanism of the binding between humic substances and PFAs.

2. Materials and Methods

2.1 Chemicals

PFAs including PFOS, PFOA, perfluorononanoic acid (PFNA), perfluorodecanoic acid (PFDA), perfluoroundecanoic acid (PFUnA) and perfluorododecanoic acid (PFDoA) were obtained from Acros Organics (New Jersey, USA). The methanol was purchased from J. T. Baker (chromatography grade, USA). Humic acid (HA) was from Sigma-Aldrich.

2.2 Bioaccumulation experiments

HA and FA was used to study the effects of DOM on the bioaccumulation of PFAs in *D. magna*, and was prepared with the concentration of 1, 10 and 20 $mg \cdot L^{-1}$ in AFW, respectively. The duration of the bioaccumulation was 21 days. The detailed experiment setting is shown in previous study[10, 11].

2.3 Spectral properties of DOM and the complex of DOM and PFAs

2.3.1 Binding of DOM and PFAs Humic acid and fulvic acid were chose as the typical DOM. A series content of DOM stock solution were added into 50 ml PP tubes, and then they were diluted to 1, 10 and 20 $mg \cdot L^{-1}$, respectively. PFOA and PFOS solution were added into these tubes to make sure that each concentration of DOM combined with 0, 1, 5, 10, 20 and 50 $\mu g \cdot L^{-1}$ PFAs, respectively. Each concentration gets 3 replications. The measured pH values for these solutions were range from 8.26 to 8.63, and it thought to be no obvious effects on the spectral analysis. The mixed solutions were shaken at 120 rpm in darkness for 72 h. After equilibrium, these solutions were conducted by spectroscopic analysis.

2.3.2 UV/Vis spectra analysis UV/Vis spectra of DOM and DOM-PFAs samples were recorded on a Hewlett–Packard 8453 spectrophotometerin a 1-cm quartz cuvette and scanned from 200 to 800 nm with the interval of 5 nm.

2.4 Extraction and analysis of PFAs

The extraction and analysis of PFAs in daphnia and water refer to previous studies [10]. The recovery rate of PFAs in HA solution was determined use the same extraction and analysis method. A total of 10 ng PFAs standard solution was added in 20 mg/L HA solution and the results indicated that the recovery rate of PFA was range from 82.1% to 98.3%.

3. Results and Discussion

3.1 Effects of DOM on the bioaccumulation of PFAs

As described in our previous study [11], the BAF of PFAs was increased with the increasing perfluoroalkyl chain length under the effects of HA, FA for the carboxylate, and the BAF of sulfonate is higher than the one of carboxylate with the same or longer perfluoroalkyl chain length.For FA, it was enhanced the bioaccumulation of short-chain PFAs (PFOA, PFNA) in *D. magna* when the FA concentration is below 10 $mg \cdot L^{-1}$ (Figure 1).However, it inhibited the bioaccumulation when the FA concentration reach to 20 $mg \cdot L^{-1}$; the BAF decreased by 27% and 11% for PFOA and PFNA, respectively.

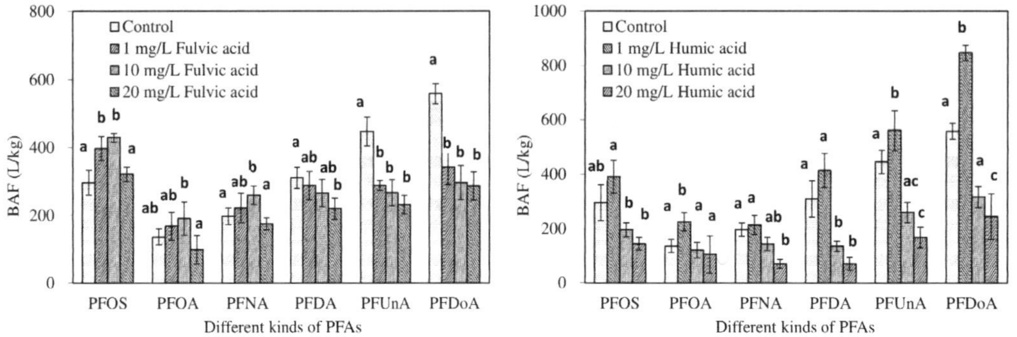

Figure 1. The effects of DOM with different concentrations on the bioaccumulation of PFAs in *D. magna* (different letters means there are significant difference between them ($P< 0.05$) and the groups that have at least one same letter means there are no significant difference between them ($P> 0.05$)).

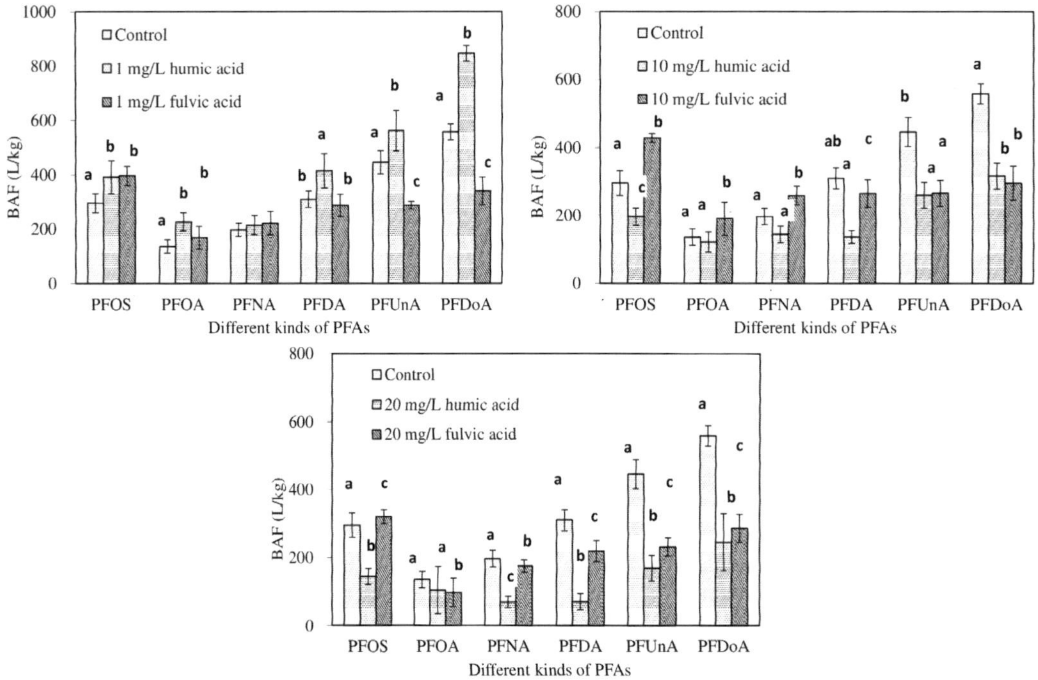

Figure 2. Comparison of BAFs of PFAs between the same concentration of HA and FA (different letters means there are significant difference between them ($P< 0.05$) and the groups that have at least one same letter means there are no significant difference between them ($P> 0.05$)).

It's no doubt that DOM can influence the bioaccumulation of organic chemicals. In commonly, DOM lead to a decrease of the accumulation of these compounds, and it has become generally accepted [12-13]. However, the enhancement of the bioaccumulation of organic compounds due to DOM also has been found in several studies [14].

As shown in Figure 2, at the concentration of 1 mg·L^{-1}, HA and FA enhanced the bioaccumulation for all kinds of PFAs, while FA only enhanced the short-chain PFAs. When the concentration comes to 10 mg·L^{-1}, HA decreased the bioaccumulation; however, FA still enhanced the bioaccumulation for short-chain length PFAs, while the decrease of bioaccumulation for the long-chain PFAs was reinforced. The results indicated that, the quantity and types of DOM should be considered when evaluating the bioaccumulation of PFAs in natural aquatic environment. In addition, the functional group of PFAs also should be considered.

3.2 UV/Vis spectroscopic analysis

The UV/Vis spectrum of HA- PFAs and FA-PFAs solutions at different HA/FA concentrations were described in Figur 3. As shown in this figure, UV spectrum of HA and FA was smooth, broad and featureless though the HA/FA concentration was ranged from 1 to 20 mg·L^{-1}. According toKorshin et, al.,[15], each aromatic chromophore in DOM is expected to have a distinct absorption bands, however, DOM is a complex of many of these chromophore, and for some of them, the absorption bands is at the neighboring wavelength. Thus, the overall spectrum of the DOM represents a superposition of these bands. No new peak was observed after the adding of the different concentration of PFAs (ranged from 5 to 1000 μg·L^{-1}). However, the adding of PFAs to these solutions affect the absorbance of the HA/FA solutions.

For HA, the adding of PFAs caused the red shift of the HA characteristic peak, and this shift was increased with the increasing PFAs concentration. In other words, PFAs can increase the absorbance of HA solution, and it was increased with the increasing of PFAs concentrations. In addition, the increment was enhanced with the increasing of HA concentration. For FA, the change was similar toHA with the adding of PFAs at different concentrations. The PFAs result in the red shift of FA characteristic peak, and this shift was increased with the increasing PFAs concentration. Relevant research suggested that the redshift of the absorption peak means the transition energy of the electron was reduced and the possibly reason is the increase of the conjugation of the molecular system [16]. Thus, the results indicated that PFAs can enhance the conjugation system of HA and FA.

3.3Characteristic UV/Vis spectra parameters

Various absorption wavelengths at 254, 280, 365 and 465 nm as well as ratios like A_{250}/A_{365} (E_2/E_3), A_{465}/A_{665} (E_4/E_6) have been cited in literature for the spectral differentiation of the complex solutions of HA-PFAs and FA-PFAs [17].

It's said that light absorbance at 254 nm (A_{254}) correlates well with the traditional comprehensive index of organic matter, such as total organic carbon [2], and the aromaticity of DOM [18]. For FA, The value was increased with the increasing PFAs concentration, it means the PFAs could increase the aromaticity of the FA solution, and it was in accord with the red shift of the spectrum of FA with the adding of PFAs. For HA, there were no obvious trends for the value of A_{254}. Generally speaking, the value of control group is between the low PFAs concentration group (1 and 5μg·L^{-1}) and the high PFAs concentration group. The reason could be attributed to the complicated composition of HA. The numerous functional groups and chromophores interrupt the effects of PFAs. In addition, the relatively low concentration of PFAs maybe another reason, and it was deduced by the fact that the A_{254} in 50 μg·L^{-1} group is higher than other groups.

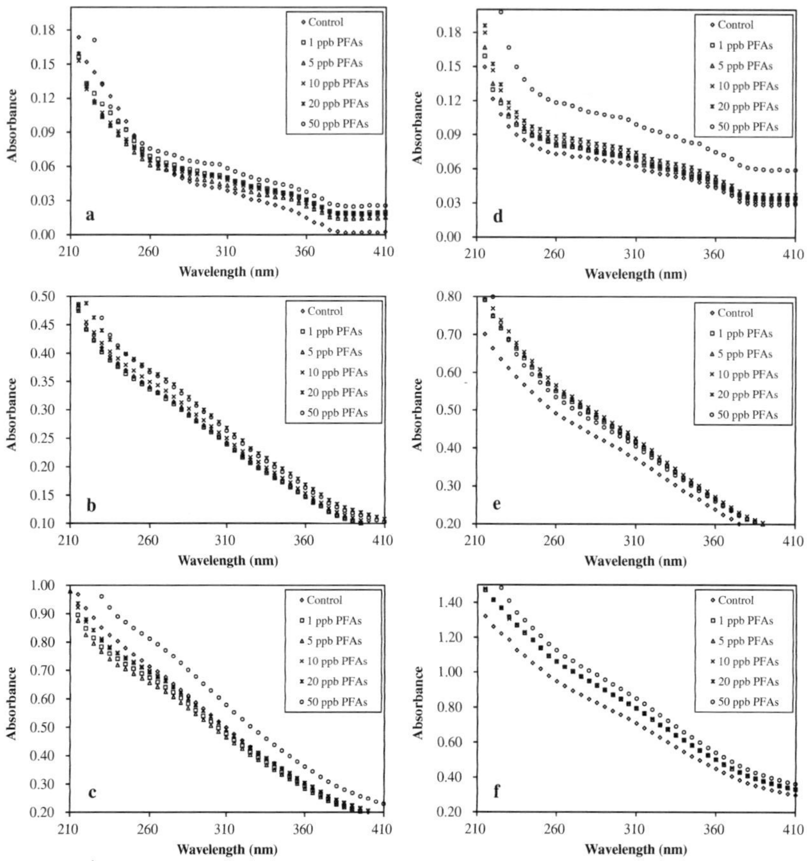

Figure 3. UV/Vis absorptivity of HA- PFAs and FA-PFAs solution in water (the HA concentration is 1, 10 and 20 mg·L^{-1} in graph a, b and c, respectively; the FA concentration is 1, 10 and 20 mg·L^{-1}in graph d, e and f, respectively)

For some kinds of pollutants, such as phenolic substances, aniline derivatives and benzoic acids, theπ-π* electron transitions occur when the wavelength is 280 nm[19, 20]. According toWeishaaret al.,[21], the absorbance at 280 nm is well correlated with the molecular weight of organic matters. Thus, the molar absorbance at 280 nm may reflect the change of DOM before and after the binding of PFAs in molecular levels. As shown in Figure 4, the absorbance of different concentrations of HA and FA at the wavelength of 280 nm were well correlated with the PFAs concentration, and the correlation coefficients were range from 0.65 to 0.98. we can deduced that PFAs was bond to HA and FA, and these kinds of bond enlarged the molecular weight of HA/FA.

E_2/E_3 is the absorbance ratios measured at the wavelength of 250 and 365 nm, and similar to A_{254}, it could correlate with the molecular size and aromaticity[22]. As shown in Table 1, besides the control group, the values were decreased with the increasing PFAs concentrations for each HA/FA concentration group. However, there were no significant differences between them (P> 0.05). For HA, the values of E_2/E_3 in 1 mg·L^{-1} group were higher than10 and 20 mg·L^{-1} group, and they were close to each other for the latter two groups. For FA, the value in 1 mg·L^{-1} group were lower than 10 and 20 mg·L^{-1} group and, like HA, they were close to each other for the latter two groups.

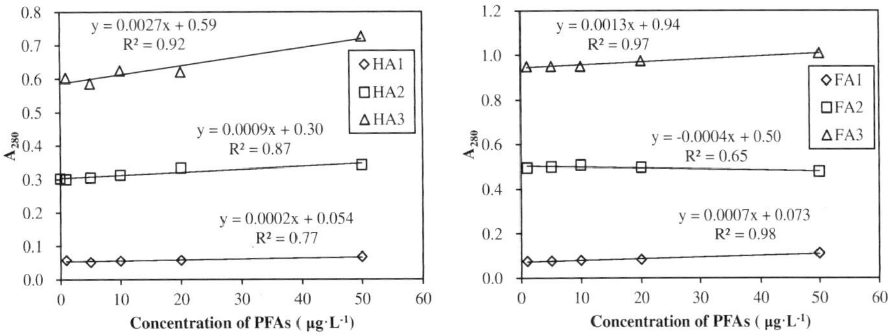

Figure 4. Absorbance of different concentration of humic acid and fulvic acid at the wavelength of 280 nm as the function of PFAs concentration (HA1, HA2 and HA3 represent the humic acid at 1, 10 and 20 mg·L^{-1}, respectively, and the same to the FA1, FA2 and FA3)

4. Conclusions

In conclusion, the results indicated that, the bioaccumulation of PFAs by *D. magna*could significantly affect by HA and FA. They could decrease the bioaccumulation of PFAs at 10 and 20 mg·L^{-1}, and increase the bioaccumulation at 1 mg·L^{-1}. The decrease of BAF was in the range of 11 to 42% and 23 to 77%, while the increase of BAF was in the range of 12 to 44% and 8 to 73%for FA and HA, respectively.The spectrum analysis indicated that the interaction of PFAs and DOM was not due to the bond formation reaction, but the electrostatic attraction.

Acknowledgements

This study is supported by the Fujian Science and Technology Guiding Project (2018Y0079) and Fujian Educational and Scientific Research Project for Young and Middle-aged Teachers(JA15369).

References

[1] Renner, R., Growing concern over perfluorinated chemicals. *Environmental Science & Technology*, 2001. 35(7): p. 154-160.

[2] Stock, N., Furdui, V., Muir, D., Mabury, S., Perfluoroalkyl contaminants in the Canadian Arctic: Evidence of atmospheric transport and local contamination. *Environmental Science & Technology*, 2007. 41(10): p. 3529-3536.

[3] Kannan, K., Koistinen, J., Beckmen, K., Evans, T., Gorzelany, J., Hansen, K., Jones, P., Helle, E., Nyman, M., Giesy, J., Accumulation of perfluorooctane sulfonate in marine mammals. *Environmental Science & Technology*, 2001. 35(8): p. 1593-1598.

[4] Pignatello, J.J. and B. Xing, Mechanisms of slow sorption of organic chemicals to natural particles. *Environmental Science & Technology*, 1995. 30(1): p. 1-11.

[5] Kulikova, N. and I. Perminova, Binding of Atrazine to Humic Substances from Soil, Peat, and Coal Related to Their Structure. *Environmental Science & Technology*, 2002. 36(17): p. 3720-3724.

[6] Song, N., Yang, Z., Zhou, L., Wu, X., Yang, H., Effect of dissolved organic matter on the toxicity of chlorotoluron to Triticum aestivum. *Journal of Environmental sciences*, 2006. 18(1): p. 101-108.

[7] Pan, B., S. Ghosh, and B. Xing, Dissolved organic matter conformation and its interaction with pyrene as affected by water chemistry and concentration. *Environmental Science & Technology*, 2008. 42(5): p. 1594-1599.

[8] Chen, S., Jiao, X.C., Gai, N., Li, X.J., Wang, X.C., Lu, G.H., Piao, H.T., Rao, Z., Yang, Y.L.,, Perfluorinated compounds in soil, surface water, and groundwater from rural areas in eastern China. *Environmental Pollution*, 2015. 211: p. 124-131.

[9] Chai, X.L., Ji, R., Wu, J., Tong, H.H., Zhao, Y.C., Abiotic association of PAEs with humic substances and its influence on the fate of PAEs in landfill leachate. *Chemosphere*, 2010. 78(11): p. 1362-1367.

[10] Dai, Z, Xia, X, Guo, J, Jiang, X, Bioaccumulation and uptake routes of perfluoroalkyl acids in*Daphnia magna*. Chemosphere, 2012. 90(5): p. 1589-1596.
[11] Xia X, Dai.Z.,Andry H. Rabearisoa, Comparing humic substance and protein compound effects on the bioaccumulation of perfluoroalkyl substances by Daphnia magna in water. *Chemosphere*, 2015. 119: p. 978- 986.
[12] McCarthy, J.F., B.D. Jimenez, and T. Barbee, Effect of dissolved humic material on accumulation of polycyclic aromatic hydrocarbons: Structure-activity relationships. *Aquatic toxicology*, 1985. 7(1-2): p. 15-24.
[13] Barron, M.G., Bioconcentration. Will water-borne organic chemicals accumulate in aquatic animals? *Environmental Science & Technology*, 1990. 24(11): p. 1612-1618.
[14] Yang, Y., Shu, L., Wang, X., Xing, B., Tao, S., Effects of Composition and Domain Arrangement of Biopolymer Components of Soil Organic Matter on the Bioavailability of Phenanthrene. *Environmental Science & Technology*, 2010. 44(9): p. 3339-3344.
[15] Korshin, G.V., C.W. Li, and M.M. Benjamin, Monitoring the properties of natural organic matter through UV spectroscopy: a consistent theory. *Water Research*, 1997. 31(7): p. 1787-1795.
[16] Zhang, Q. and H. Li, Study on the interaction of pesticide and soil humic acid by UV/Vis spectroscopy. *Journal of Southwest China Normal University* (Natural Science Edition), 2008. 33(2): p. 87- 92.
[17] Uyguner, C.S. and M. Bekbolet, Evaluation of humic acid photocatalytic degradation by UV-vis and fluorescence spectroscopy. *Catalysis Today*, 2005. 101(3-4): p. 267-274.
[18] Lopes, C., Abreu, S., Válega, M., Duarte, R., Pereira, M., Duarte, A., The Assembling and Application of an Automated Segmented Flow Analyzer for the Determination of Dissolved Organic Carbon Based on UV - Persulphate Oxidation. *Analytical letters*, 2006. 39(9): p. 1979-1992.
[19] Braun, D., A. Floyd, and M. Sainsbury, Organic Spectroscopy. 1988, John Wiley: New York.
[20] Traina, S.J., J. Novak, and N.E. Smeck, An ultraviolet absorbance method of estimating the percent aromatic carbon content of humic acids. *Journal of Environmental Quality*, 1990. 19(1): p. 151-153.
[21] Weishaar, J.L., Aiken, G.R., Bergamaschi, B.A., Fram, M.S., Fujii, R., Mopper, K., Evaluation of specific ultraviolet absorbance as an indicator of the chemical composition and reactivity of dissolved organic carbon. *Environmental Science & Technology*, 2003. 37(20): p. 4702-4708.
[22] Peuravuori, J. and K. Pihlaja, Molecular size distribution and spectroscopic properties of aquatic humic substances. *Analytica Chimica Acta*, 1997. 337(2): p. 133-149.

IOP Conf. Series: Earth and Environmental Science **146** (2018) 012068 doi:10.1088/1755-1315/146/1/012068

Study On Simultaneous Determination of Ni, Pb and Cd by Ion Chromatography

P.F. Gao *, X.W. Zhang, H.Z. Kuang, Q.Q.Li, Y. Li

College of Environmental Science and Engineering, Xiamen University of Technology, Xiamen 361024, China.

gaopf@xmut.edu.cn

Abstract. In this paper, a ion chromatographic method was established for simultaneously determining three toxic heavy metal ions(Ni^{2+}, Pb^{2+}, Cd^{2+}) . the optimal chromatographic conditions are using a conductivity detector, with the Metrosep C_4 column was used, the mixture of 0.07mmol/L nitric acid and 0.5mmol/L dipicolinic acid as eluent, the flow rate was 0.25 mL/min. Three kinds of heavy metals all showed satisfying linear relationships, and the correlation coefficients ranged from 0.9979 to 0.9994. the detection limits were in the range of 0.01-0.05 mg/L. The method is an easy, accurate and reliable way to assess soil heavy metal pollution.

1. Introduction

As one kind of persistent pollutants, the heavy metals are considered as main contaminations in the environment due to their toxicity and ability to accumulate (Khan 2015). The determination of heavy metal content in soil is gradually obtaining attention by people along with the development of economy and industry (Stasinos 2013, Al-Khashman 2018).

The detection mode include traditional chemical analysis methods and instrumental analysis methods. Traditional chemical methods do not require expensive and sophisticated equipment, but their operation procedures are often not convenient and efficient enough. Just like complexometric titration method or spectrophotometry. Some instrumental analytical techniques are suitable for the measurement of heavy metal ion, such as flame atomic absorption spectrometry (Omidi 2015, Leśniewska 2016), inductively coupled plasma atomic emission spectrometry (ICP-AES) (Kos 1996, Wang 2017), inductively coupled plasma mass spectrometry (ICP-MS) (Hussain 2017), laser atomic fluorescence spectrometry and total reflection X-ray fluorescence (TXRF) (Macedomiranda 2009) and so on. These methods have some advantages, such as high sensitivity, lack of chemical interferences, low detection limits, and are suitable for determining trace elements. But these measurement methods require high testing costs, expensive and sophisticated apparatuses and professional technical staff, and primary laboratory can not meet these requirements.

Relative to the two kinds of methods mentioned above, Ion chromatography is a rapid, highly selective and sensitive method, which will be widely used in Environmental Analysis(Moutaz 2015, Bruno 2006, Liu 2010), its hardware cost of equipment devices are moderate. The running costs of this method are also not expensive.

Therefore, the aim of this paper is to develop a rapid and high effective method for the determination of heavy metal ion, and applied in the assessment of soil heavy metal pollution.

2. Methods and materials

2.1. Chemicals

All reagents were analytical grade. Individual pure standards containing 1,000 mg L^{-1} of trace metals in 0.5 normal nitric acid (Ni, Cd, Cr, Pb, Cu, Zn, As, Na, K, Ca, Mg and Mn) , hydrochloric acid and nitric acid were purchased from National drug group chemical reagents Co., Ltd. 2,6-Pyridinedicarboxylic acid was purchased from Sigma-Aldrich. Ultrapure water was got with a Milli-Q system (Millipore , USA).

2.2. Instrumentation

Heavy metal ion were analyzed ion-chromatography(Metrohm, 883 Basic IC plus) with Metrohm Metrosep C4-150/4.0 column and with nitric acid-2,6-Pyridinedicarboxylic acid as eluent.

A closed-vessel microwave digestion method is used for soil samples dissolution with Anton-parr Mutilwave 3000. 0.5 g soil sample was decomposed with HNO_3 acid in a PTFE digestion vessel by using microwave heating for 5 min at 500 W of microwave power. Then the digestion solution was diluted to 25 mL and filtrated by 0.45μm micropore filter.

3. Results and discussion

3.1Optimization of Separation Conditions

The chromatographic behaviour of several heavy metal ions with different eluents was studied. Only the detection peaks of four heavy metal ions were observed, including Ni, Pb, Cd and Mn, when the composition of the mobile phase was 0.07 mM HNO_3 and 0.5mM 2,6-Pyridinedicarboxylic acid. As was show in the Table 1.

Table 1. The conditions of ion peak under the different eluents

HNO_3+2,6-Pyridinedicarboxylic acid	Ni	Pb	Cd	Mn
0.13 mM-0.9mM	-	-	+	+
0.13 mM-0.7mM	-	+	+	+
0.07 mM-0.9mM	-	-	+	+
0.07 mM-0.7mM	-	+	+	-
0.07 mM-0.5mM	+	+	+	+
0.07 mM-0.3mM	+	-	+	-

Note: the plus sign means its peak was observed and negative sign showed no peak.

The effect on the separation result were investigated with different flow rate of eluent. The results was shown in Table 2. Finally, a good separation was obtained at 0.25 mL/min flow-rate, it is shown in Fig. 1. No peak overlap was found. However, the peak time of manganese takes more than 20 min, and it is not selected as the detected object.

Table 2. Retention time varies with changes in the flow rate of the mobile phase

Current Speed	Ni	Pb	Cd	Mn
0.4mL/min	3.6min	5.1min	5.1min	29.2min
0.25mL/min	5.2min	8.2min	8.9min	32.1min

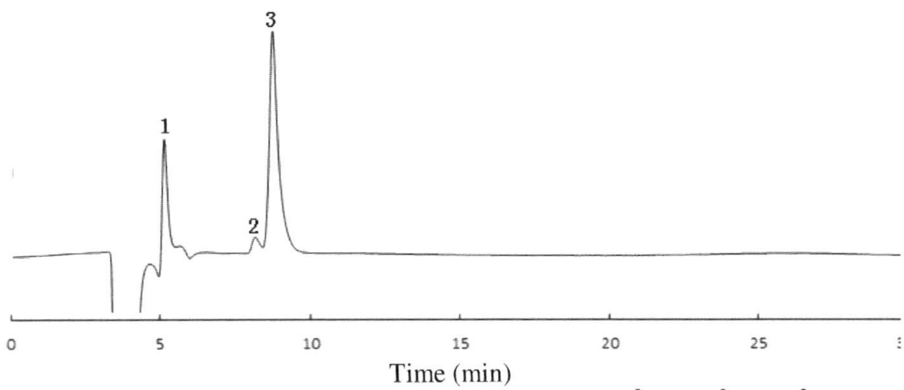

Figure 1. Chromatogram of three metallic ions(1.Ni^{2+}, 2. Pb^{2+}, 3.Cd^{2+})

3.2 *Effects of common alkali metal ions and alkaline earth metal ions in soil*

It is generally known that Na^+ and K^+ are the common alkali metal ions in soil, and Ca^{2+} and Mg^{2+} are the commonly coexist alkaline-earth metal ions. The influence of Na^+, K^+, Ca^{2+} and Mg^{2+} on the heavy metallic ions was investigated. Under the optimized separation conditions, the peak times of Common cation ion(K^+, Na^+, Ca^{2+},Mg^{2+}) more that 30 minutes, and the retention times of Ni^{2+}, Pb^{2+} and Cd^{2+} were not disturbed by the coexist alkali metal ions and alkaline earth metal ions. At the same time, the peak areas of three heavy metallic ions were investigated with four kinds of concentrations Na^+, K^+, Ca^{2+} and Mg^{2+}. The four kinds of concentrations are respectively 5mg/L,10mg/L, 15mg/L, 20mg/L.

As was shown in Fig.2 to Fig.5, the peak areas of Pb^{2+} and Cd^{2+} were not influenced by the alkali and alkaline-earth metals variations, and the peak area of Ni^{2+} was not observed when K+ concentration was more than 10mg/L. The same phenomenons occurred when the coexisting ions were Na^+, Ca^{2+} or Mg^{2+}. In general, the concentrations of K^+, Na^+, Ca^{2+}, Mg^{2+} were 0.1g/kg in the acid soils of southern China(Wang, 2008). their concentrations were less than 2mg/L in the soil sample solutions according the preceding pretreatment method. Therefore, there is no interference in the determination of Ni^{2+} in the practical application with the existence of K^+, Na^+, Ca^{2+} or Mg^{2+}.

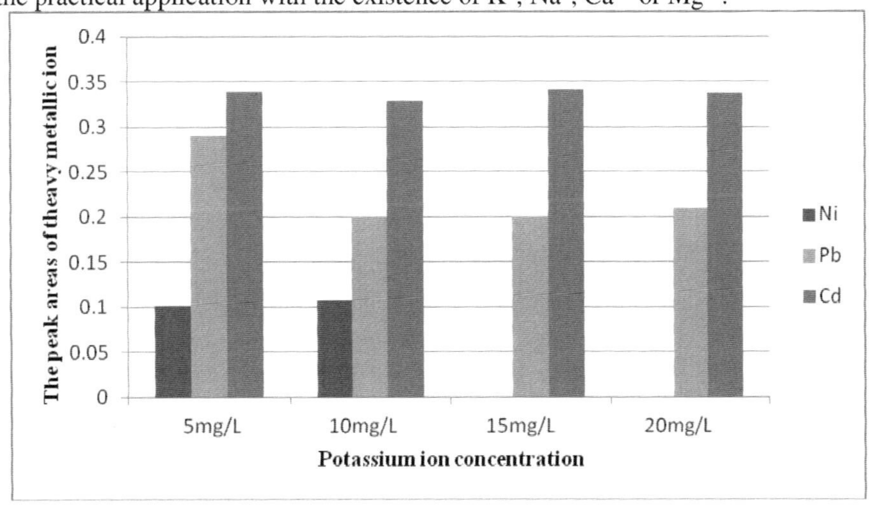

Figure 2. Influence of different K^+ concentrations on the peak area of heavy metallic ions

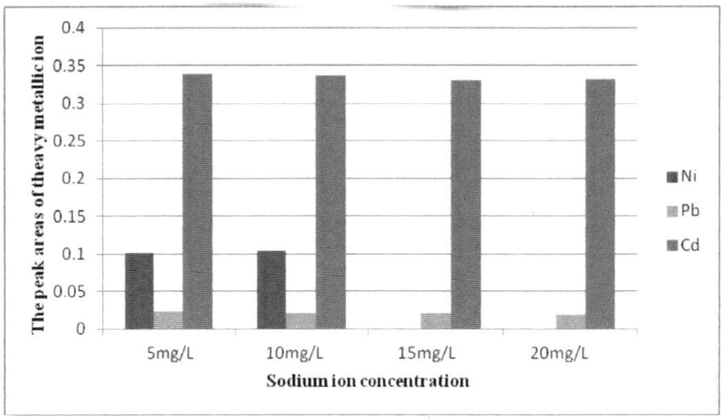

Figure 3. Influence of different Na+ concentrations on the peak area of heavy metallic ions.

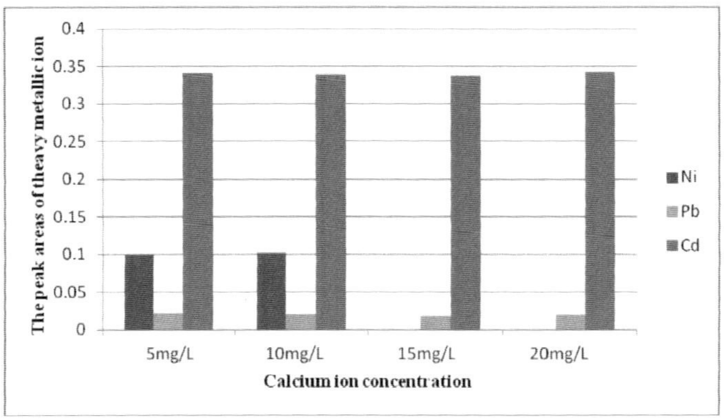

Figure 4. Influence of different Ca^{2+} concentrations on the peak area of heavy metallic ions.

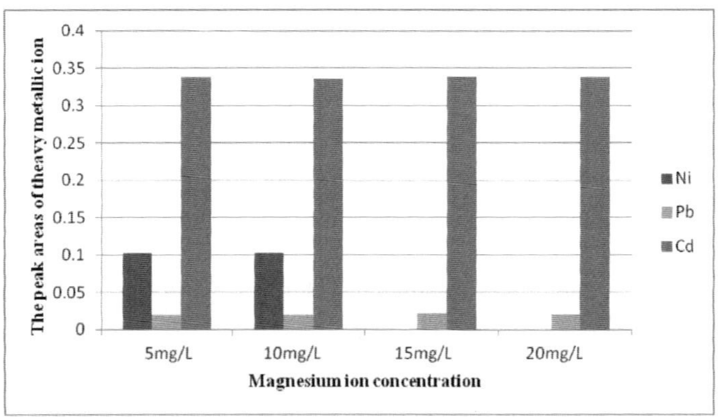

Figure 5. Influence of different Mg^{2+} concentrations on the peak area of heavy metallic ions.

3.3 Linear equation and detection limit

It is studied that the linear equation and detection limit of this determination method for three heavy metal ion by ion chromatography. The results show in Table 3. This method is simple, fast and has the advantage of good reproducibility.

Table 3. linear regression equations and detection limits of three metals elements.

No.	linear equations	R^2	Detection range（mg/L）	Detection limits（mg/L）
Ni	$y = 0.0041x + 0.0193$	0.9983	0.1-10	0.03
Pb	$y = 0.0012x + 0.0015$	0.9994	0.2-20	0.05
Cd	$y = 0.0142x + 0.0101$	0.9979	0.1-10	0.01

Next this method was used to detect the heavy metal in the surface soil of farmland in Xiang'an District. As was shown in Table 4. The soil of the tested area can reach the first grade standard.

Table 4. Contrast between the detection results and Environmental Quality Standard for Soils (mg/Kg).

Element	Sample	Recovery rate	Standard I	Standard II
Ni	23.18	94%	40	50
Pb	55.36	102.3%	90	300
Cd	Not detected	105%	0.2	0.3

4. Conclusions

In this study, a method has been developed for the determination of Ni, Pb and Cd in soils based ion chromatography with a high regression quality of squared correlation coefficient(0.9979-0.9994). The detected limits respectively are 0.03, 0.05 and 0.01. According to the experimental results, this method is rapid , high sensitivity and selectivity and low running cost.

Acknowledgments:

This research was supported by funding from Science and Technology Planning Project of Xiamen City with Grant 3502Z20179042.

References

[1] Khan, A., Khan, S., Khan, M.A., et al. The uptake and bioaccumulation of heavy metals by food plants, their effects on plants nutrients, and associated health risk: A review[J]. Environmental Science and Pollution Research, 2015, 22(18): 13772-13799.

[2] Stasinos, S, Zabetakis I. The uptake of nickel and chromium from irrigation water by potatoes, carrots and onions[J]. Ecotoxicology and Environmental Safety, 2013, 91: 122-128.

[3] Al-Khashman O.A., Jaradat A.Q. Assessment of groundwater quality and its suitability for drinking and agricultural uses in arid environment[J]. Environ Monit Assess, 2018, 190:48-61.

[4] Omidi, F., Behbahani, M., Shahtaheri, S. J., Salimi, S. Trace monitoring of silver ions in food and water samples by flame atomic absorption spectrophotometry after preconcentration with solvent-assisted dispersive solid phase extraction. Environmental Monitoring & Assessment, 2015, 187(6): 361.

[5] Leśniewska, B., Kisielewska, K., Wiater, J. Fast and simple procedure for fractionation of zinc in soil using an ultrasound probe and FAAS detection. validation of the analytical method and evaluation of the uncertainty budget. Environmental Monitoring & Assessment, 2016, 188(1), 29.

[6] Kos V., Budic B., Hudnik V., Lobnik F., Zupan M. Determination of heavy metal concentrations in plants exposed to different degrees of pollution using ICP-AES. Anal. Bioanal. Chem., 1996, 354(5-6): 648-652.

[7] Wang, X., Wu, Y., Wu, C., Wu, Q., Niu, Q. Trace elements characteristic based on ICP-AES and the correlation of flavonoids from sparganii rhizoma. Biological Trace Element Research, 2017: 1-6.

[8] Hussain, S.A., Kareem, M.A., Rasool, S.N., Omar, S.Y.A., Saleh, A., Al-Fwuaires, M.A., et al. Trace element determination and cardioprotection of terminalia pallida, fruit ethanolic extract in

isoproterenol induced myocardial infarcted rats by ICP-MS. Biological Trace Element Research, 2017, (36): 1-10.

[9] Macedomiranda, M.G., Zarazua, G., Mejíazarate, E., Avilapérez, P., Barrientosbecerra, B., Tejeda, S. Simultaneous determination of elemental content in water samples by total reflection x-ray fluorescence and atomic absorption spectrometry. Journal of Radioanalytical & Nuclear Chemistry, 2009, 280(2): 427-430.

[10] Moutaz A., Al-Dabbas & Lamyaa A., Ali & Adnan H. Afaj. Determination of heavy metals and polycyclic aromatic hydrocarbon concentrations in soil and in the leaves of plant (Eucalyptus) of selected locations at Kirkuk-Iraq. Arab J. Geosci., 2015, 8:3743-3753.

[11] Bruno, P., Caselli, M., Gennaro, G.D., Ielpo, P., Ladisa, T., Placentino, C.M. Ion chromatography determination of heavy metals in airborne particulate with preconcentration and large volume direct injection. Chromatographia, 2006, 64(9-10): 537-542.

[12] Liu, S.T., Chen, Z.H., Xie, J.B., Lin, J., Chen, Z.J., Rao, P.F. High-performance ion exchange chromatography of intact bacterial cells in the manner of molecules: 1. establishment of methodology. Analytical Chemistry, 2010, 82(20): 8544-8550.

[13] Wang, J.H., Liu Y.,Meng, H.Y.,Li, G.H., Xu, R.S., Lv, Y.C.,Hu, C.X.,Li, S.Y., Liang, L.H. Accumulation characteristics and seasonal variation of surface soil salts in open vegetable fields under intensive vegetable production. Ecology and Environment, 2008, 17(4): 1624-1629.

IOP Conf. Series: Earth and Environmental Science **146** (2018) 012069 doi:10.1088/1755-1315/146/1/012069

Reduction and Heavy Metals Removal of Excess Sludge by Radio Frequency Discharge Plasma

SHIYI SUN[1,2], SONGLIN ZHANG[1,*], WEIWEI ZHANG[2], JIADONG MENG[2] and LEI WANG[2]

[1]College of Geography & Environment Science, Northwest Normal University, Lanzhou, 730070, P. R. China
[2]College of Environmental Science & Engineering, Xiamen University of Technology, Xiamen 361024, P. R. China

[1]wangl@xmut.edu.cn; [2]13893314212@163.com

Abstract. Excess sludge is an unavoidable by-product in wastewater treatment. The conventional methods(landfill, incineration etc.) may cause secondary pollution. In this study, both reduction and heavy metals removal of excess sludge by radio frequency discharge plasma(RFDP) were investigated. The experimental results showed that RFDP was an effective removal method on total chemical oxygen demand (TCOD) and mixed liquid suspended solids(MLSS). Discharge process could also damage the cell structure of sludge, so that organic matter and heavy metals within the cell structure could enter the liquid phase. At the same time, combined with spectral diagnosis, the radical reactions was discussed.

1. Introduction

Excess sludge without properly disposed will naturally contaminate the water and air to some extents[1]. The traditional methods for treating excess sludge include composting, landfill and incineration. However, apart from organic matters and secondary clay minerals, as a complex system, sludge includes abundant harmful matters of heavy metal and pathogen etc. for Cd, Cr, Cu, Ni, Zn, Hg, Pb and As etc[2]. Direct landfill or composting will make the concentration of heavy metal in the soil seriously exceeded[3]. Heavy metals enter the human body mainly through the food chain, water and breathing. Recently, the disposal of excess sludge has attracted more interest. *Chironomid Larvae* could be used to treat excess sludge, but the shortcomings of biological methods are slow reaction speed and harsh reaction condition[4]. Nina et al. found that the sludge volume index would rise sharply when ultrasonic worked[5]. Therefore, it is necessary to study other optimal methods for treatment.

Radio Frequency Discharge Plasma(RFDP) technology is a new electrochemical method to remove toxic and harmful organic matter in water. When RFDP working, active substances are generated near the electrodes. The solution around platinum electrode vaporized quickly to form a stable steam sheath and it can generate highly active components such as $\cdot H$, H_2O_2 and $\cdot OH$[6]-[7]. These highly active particles are continuously produced and can be easily transferred to the solution, making complex organics in the water completely degraded to simple ones, water and carbon dioxide. It could apply to waste-water disinfection and purification. In conclusion, sludge treatment by radio frequency discharge plasma is innovative and promising.

IOP Conf. Series: Earth and Environmental Science **146** (2018) 012069 doi:10.1088/1755-1315/146/1/012069

2. Experimental

Experimental device was mainly consist of a RF power supply(13.56MHz) and a reactor, which is shown in figure 1. Container of sludge was made of glass, and it was installed with an inter-layer of cooling water. Platinum electrode was connected to the high voltage terminal. Sludge in this experiment was collected from a sewage plant and 200ml sludge was poured into container each time . Container bottom was placed with magnetic stir bar .in order to make reaction evenly and sufficiently. The whole set of device worked stably and power was controlled at 120W. Maya2000 Pro spectrometer (American Ocean Optics Company) was used to measure spectrogram, and TBS 1052B oscilloscope (American Tektronix Company) was used to measure voltage, current wave form.

During the process of reaction, a few samples were taken for data measurement in a while. The total chemical oxygen demand (TCOD) and supernate chemical oxygen demand (SCOD) 5 min after centrifugation (4000r/min) were measured at the usage of potassium dichromate method. A standard method was used to measure mixed liquid suspended solids (MLSS) and mixed liquor volatile suspended solids (MLVSS). Concentration changes of several kinds of heavy metals in supernate were obtained through detection of inductively coupled plasma spectrometer(ICP-OES OPTIMA 7X00DV ,American Perkinelmer Company) In addition, biodegradability of sludge was evaluated by BOD_5/COD . Finally, pH and temperature changes during the process were detected.

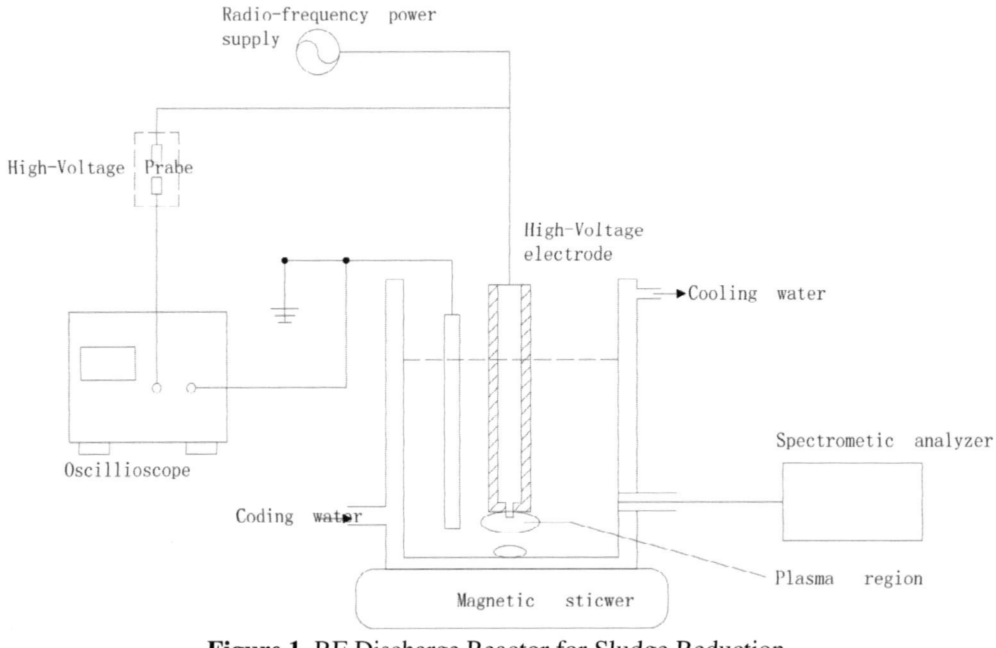

Figure 1. RF Discharge Reactor for Sludge Reduction.

The energetic electrons in the RF plasma region cause a series of complex free radical reactions, as figure 2 shown[8]:

$$H_2O + e \rightarrow \cdot OH + H \quad (1)$$

$$OH + \cdot OH \rightarrow H_2O_2 \quad (2)$$

$$HO_2 \rightarrow \cdot O_2^- + H^+ \quad (3)$$

$$HO_2 + \cdot HO_2 \rightarrow H_2O_2 + O_2 \quad (4)$$

$$OH + H_2 \rightarrow \cdot H + H_2O \quad (5)$$

$$OH + \cdot OH \rightarrow H_2O + \cdot O \quad (6)$$

The spectrum also detected the yellow light emitted by the sodium(to adjust the conductivity) and the black-body radiation of the platinum .

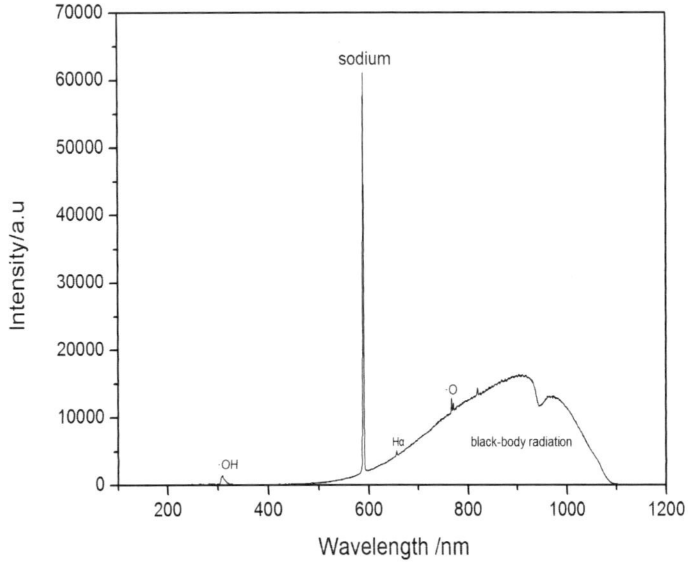

Figure 2. Spectrum of RF Plasma Reactor in Sodium Chloride Solution(conductivity 600us/cm).

3. Results and discussion

3.1. COD variations and suspended solid degradation

Under the common function of generated high temperature (it can reach 6000K nearby electrode), shock wave, ultraviolet light and chemical active particles, cell structure of sludge was damaged. As is shown in Fig.3, intracellular matters were released to liquid phase, which made SCOD continuously increasing. Meanwhile, organic matters in sludge reduce,constantly, which made TCOD presented the down trend. BOD_5/COD change rules during the process are shown in Fig.4, the value increased from 0.16 to 0.62, which showed that RF discharge plasma technology can rapidly degrade organic matters and effectively improve the biodegradability.

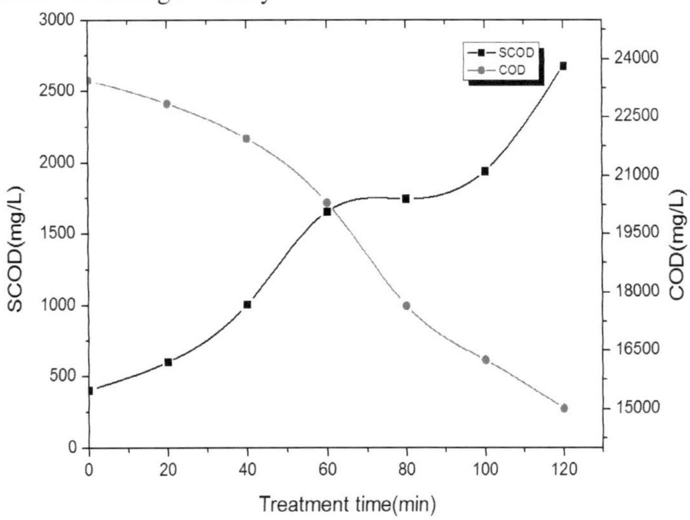

Figure 3. Evolution of COD During the Treatment Time(power, 120w; T, 298K; pH_0, 7.4).

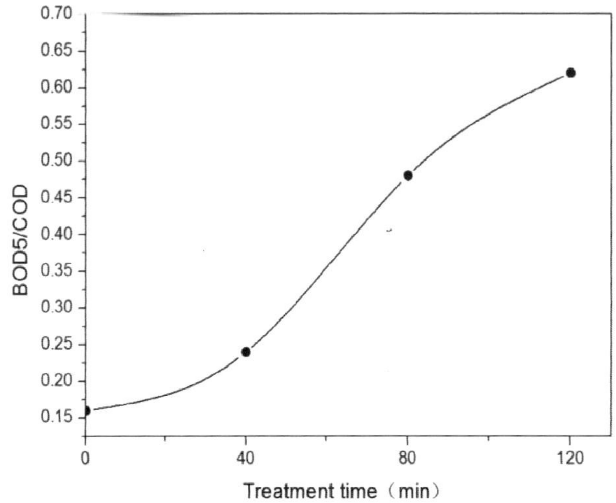

Figure 4. Evolution of BOD_5/COD During the Treatment Time(power, 120w; T, 298K; pH_0, 7.4).

Furthermore, the effection of sludge reduction was observed by the concentration change of suspended matters. As is shown in figure 5, two suspended matters parameters declined as the experiment going. The reason is that complex structure matters be gradually decomposed and soluble components be gradually increased under the discharging. High temperature nearby electrodes kept volatile suspended matters in sludge sustainably volatilizing.

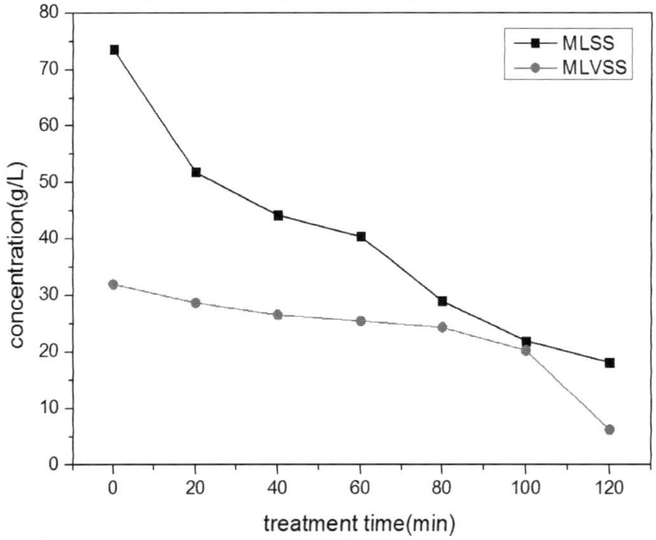

Figure 5. Concentration Variations of MLSS and MLVSS with Treatment Time(power, 120w; T, 298K; pH_0, 7.4).

3.2. pH and temperature variations during the discharging process

In order to research water and heat effect of RFDP in sludge treatment, pH and temperature changes during the discharging process were recorded. As shown in figure 6, sludge pH changed from initial 7.40 to 8.45 smoothly, one possible reason is that the death of microorganisms led to the conversion of metabolic method. This condition was beneficial to sludge digestion.

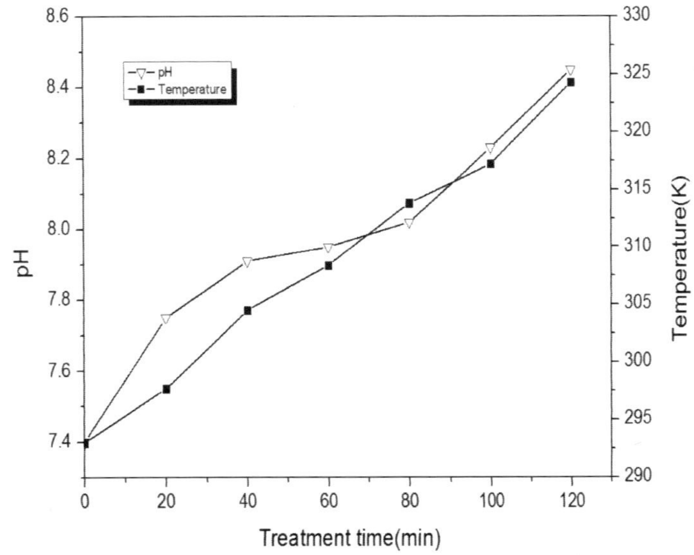

Figure 6. Temperature and pH Variation With Treatment Time(power,120w).

3.3. In supernatant, concentration of partial overproof metals during the discharging

Concentrations of heavy metals Cr, Ni, Zn in supernate after centrifugation at different discharging time were detected by ICP which are shown in figure 7 and figure 8, It was found that concentrations of these three representative heavy metals increased as the experiment going. The total amount of heavy metal in sludge is constant, therefore, only heavy metal in the solid phase leached to liquid phase, and its concentration would increase. Affected by RF discharge plasma, the following changes occurred to metals in sludge:

(1) Metal cation participated in free radical reactions, changed from high valence to low valence[9]:

$$\cdot O_2^- / \cdot HO_2 + M \rightarrow \text{Products} \quad (7)$$

$$\cdot H + M \rightarrow \text{Products} \quad (8)$$

(2) The destruction of the cell structure of sludge caused the intracellular material be released into the liquid phase, therefore, increasing the content of metal elements in the supernatant.

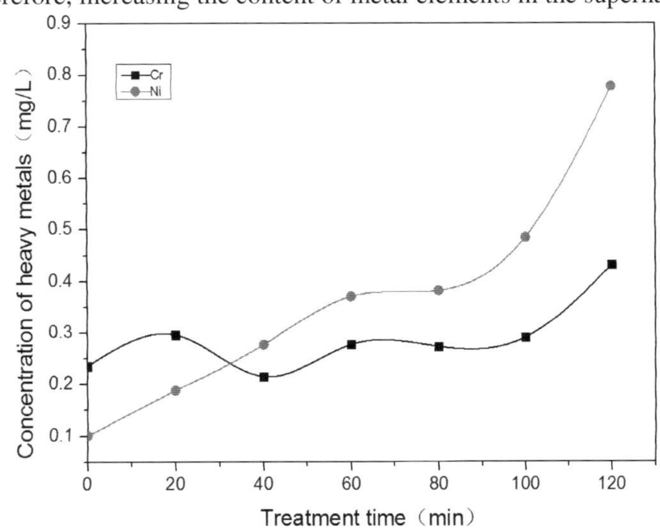

Figure 7. Evolution of Partial Heavy Metal Concentration During the Treatment Time(power, 120w; T, 298K; pH_0, 7.4).

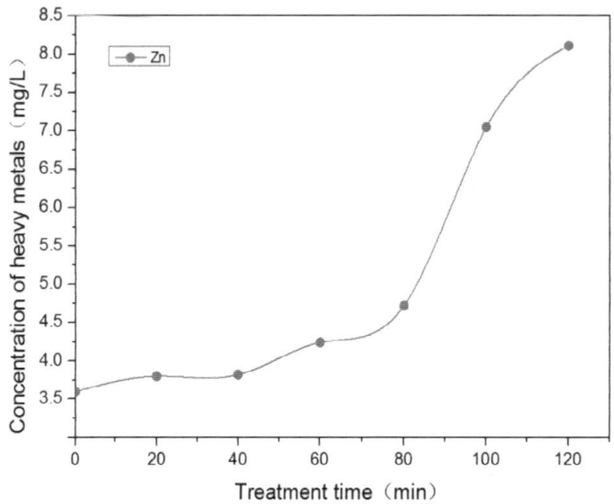

Figure 8. Evolution of Partial Heavy Metal Concentration During the Treatment Time(power, 120w; T, 298K; pH_0, 7.4).

4. Conclusion

The effect of RFDP treatment of sludge is obvious. The heat generated during discharge is well utilized. The oxidation and reduction of free radicals were used in the reaction. The effect of destroying the cell structure of sludge is remarkable, chemical active matters gradually degrade organic matters in sludge, supernatant COD concentration increased more than 6 times. Heavy metals precipitate from the sludge cells into the supernatant, among them, the zinc content increased 125%, the chromium content increased 600% , the nickel content increased 80%.

Acknowledgments

This work was supported by the National Natural Science Foundation of China (51008262), the Natural Science Foundation of Fujian province, China (2015J01651), and the scholarship for overseas study of Fujian Province, China.

References

[1] Hara K, Mino T 2008 *Environmental Assessment of Sewage Sludge Recycling Options and Treatment Processes in Tokyo J. Waste Management* **28(12)** 2645 -2652

[2] Zhipeng Z, Saeid B, M.T. Munir and Brent R 2017 *Variation in Metals During Wet Oxidation of Sewage Sludge J. Bioresource Technology* **245(12)** 234-241

[3] WanQian G, QingLian W, ShanShan Y, SiMai P and HaiChao L 2014 *The Promising Resource Utilization Methods of Excess Sludge: A Review J.Applied Mechanics and Materials* **2972** 507

[4] Qiong Z 2013 *Effect of Chironomid Larvae on Sludge Reduction and Activated Sludge Settling Performance. In Conf. on Materials for Renewable Energy and Environment (Beijing)* vol 1 pp 754-758

[5] Nina Van de Moortel, Rob Van den Broeck, Jan Degr Г Ë ve, Raf Dewil 2017 *Comparing glow discharge plasma and ultrasound treatment for improving aerobic respiration of activated sludge J. Water Research* **122** 109

[6] Igor B, Serhiy I, and Nikolay V 2017 *New Combined-Cycle Gas Turbine System for Plasma-Assisted Disposal of Sewage Sludge J. Plasma Sci.* **45(12)** 3100-3104

[7] Yu G, Yongdi D and Yekun M 2016 *Disruption of Microbial Cell Within Waste Activated Sludge by DC Corona Assisted Pulsed Electric Field J. Plasma Sci.* **44(11)** 2682-2691

[8] Shuangyan S, XingJ and Peng Z 2016 *Measurement of Emission Spectraof Air Dielectric Barrier Discharge and Particle Analysis of Discharge Process J. Spectroscopy and Spectral Analysis* **36(02)** 359-363

[9] HaoJie F 2007 *Study on the Microscopic Reaction Mechanism of Sulfur-containing Compounds with Oxidizing Radicals in the Aqueous Phase* PhD thesis. Fudan University

IOP Conf. Series: Earth and Environmental Science **146** (2018) 012070 doi:10.1088/1755-1315/146/1/012070

Statistical Research on Effect of Desulfurated Parameters on Desulfurization Efficiency

Shuli Xiao[1, a] and Yingxiang Jiang[1, b]

[1]Faculty of Environmental Science and Engineering, Kunming University of Science and Technology, Yunnan, China 650032.

[a]kmustxsl@163.com; [b]jiangyingxiang@kmust.edu.cn

Abstract. Aluminum electrolysis is an important environmental source of pollution, including carbon anode for aluminum electrolytic aluminum industry. The calcined petroleum coke in the process of burning not only produces the number of amazing soot, but also emits SO_2 emissions. At present, most aluminum carbon, petroleum coke and calcined flue gas require desulfurization treatment, but little of the effect of desulfurization has been reported in the literature. In order to promote the healthy development of aluminum electrolytic green and environmental protection career progress, Origin and SPSS software was used to analyze the calcined flue gas data of aluminum charcoal coke rotary kiln in one aluminum plant in Yunnan Province. We obtained the correlation between the monitoring parameters and SO_2 removal efficiency, and gave some advice on controlling the flue gas flow rate, temperature, humidity or oxygen content to increase flue gas desulfurization efficiency.

1. Introduction

At present, China has entered the period of promoting the overall modernization, accelerating the transformation of economic system reform, especially with the further accelerating the process of modernization. Environmental issues have become increasingly prominent, one of the most notable is the air pollution. More and more researchers focus on flue gas desulfurization, but the studies on influence of flue gas parameters on the desulfurization efficiency are not definite enough.

Some researchers found that increasing the flue gas flow rate can reduce the flue gas residence time in the absorption tower, resulting in SO_2 removal rate decreased. When the flue gas temperature is too high, the absorption liquid temperature also increased and increasing ammonia escape, then this can promote the dissolution of $(NH_4)_2SO_4$, while accelerating the decomposition of $(NH_4)_2SO_3$ and making desulfurization efficiency reduced [1]. However, Wang et al pointed out that increasing the gas velocity can increase the mass transfer rate and mass transfer coefficient in the absorption tower, so as to increase the ability of the absorption liquid to absorb SO_2. Therefore he thought increasing the gas velocity can improve the SO_2 removal rate [2].Jiang et al found that the addition of moist air can increase the amount of condensable water and improve the desulfurization effect [3]. In addition, Yang et al studied the effects of humidity and temperature of flue gas from lime-gypsum boiler desulfurization on the composition of flue gas pollutants in many articles [4-6]. Their research shows that humidity and temperature can indeed affect the dust and SO_2 removal rate, but they also did some quantitative studies on the extent of the impact. Some literatures showed that the addition of oxygen to the desulfurization solution can promote the oxidation reaction of SO_3^- and HSO_3^-, then the oxidation products would generate precipitation with Ca^{2+}, reduce the mass transfer resistance between gas and liquid phase, and promote the dissolution of SO_2 in water, indicating that the oxygen content of flue gas can also affect the removal of SO_2 [7-9].

He B. et al in the study of the effect of temperature on the efficiency of ammonia desulfurization found that SO_2 removal rate at low flue gas temperature is relatively high. With the increase of flue gas temperature, SO_2 removal rate showed a decreasing trend and reached the lowest at 70°C.Then, while the temperature continues to rise, SO_2 removal rate showed an upward trend [10]. However, some researches got the opposite result from He B. They indicated that the initial reaction rate of desulfurization increased with increasing temperature, after a period of time, increasing the temperature resulting in desulfurization reaction rate decreases [11-14]. It shows that flue gas temperature is a complicated parameter, and it may have interaction and restriction relationship with other flue gas parameters [15]. In other words, the study of desulfurization efficiency from a single perspective is often not comprehensive.

By analyzing previous studies, we found that flue gas parameters such as temperature, humidity, oxygen content or flow can indeed affect the composition of pollutants in flue gas. Therefore, it is feasible to control SO_2 emission by artificially controlling flue gas parameters.

2. Data Analysis

2.1. Data sources

This study selected the inlet and outlet flue gas data of ammonia desulfurization system from aluminum charcoal coke rotary kiln in one aluminum plant in Yunnan Province, monitoring gas parameters to get a set of sample data each hour, including flow, temperature, oxygen content and moisture content in outlet flue gas and SO_2 concentration in input and output flue gas during 0 o'clock on January 1st to 23 o'clock on January 5th, 2017.

The anode back calcined kiln specification is Φ2.2×45, designed capacity of 5.5t/h, it forms an annual output of 400,000 tons of electrolytic aluminum production scale and an annual output of 180,000 tons of prebaked anode production line.

2.2. Analytical method

We use Origin 8.0 to indicate the trend of changes in sample emissions data. The mathematical expectation and confidence interval of sample emission data were analyzed by SPSS 19.0, then we obtained the normal range of sample emission data for normal system operation. The significance of the correlation between flue gas parameters and SO_2 emissions is explained by the Pearson correlation coefficient. And SO_2 removal rate equals the ratio of reduction of SO_2 concentration and SO_2 concentration in inlet flue gas.

2.3. Analysis of theoretical basis

Confidence interval refers to the estimated interval of the overall parameters constructed by the sample statistics. In statistics, the confidence interval of a probabilistic sample is the interval estimate of some overall parameter of the sample. Confidence interval shows the true value of this parameter has a certain probability of falling around the measurement results. The confidence interval gives the confidence level of the measured value of the measured parameter, which is the "one probability" previously required.

For a given set of sample data with an average of μ and a standard deviation of σ, the 100 $(1-\alpha)\%$ confidence interval for the mean of the overall data is $(\mu-Z_{\alpha/2}\sigma,\ \mu + Z_{\alpha/2}\sigma)$, where α is the coverage area of the non-confidence level in the normal distribution, and $Z_{\alpha/2}$ is the corresponding standard score.

In statistics, the Pearson correlation coefficient, also known as the Pearson product-moment correlation coefficient (PPMCC or PCCs), is used to measure the difference between two variables, X and Y correlated (linearly related) with values between -1 and 1. The Pearson correlation coefficient can be expressed as follows:

$$\rho_{X,Y} = \frac{cov(X,Y)}{\sigma_X \sigma_Y} = \frac{E[(X-\mu_X)(Y-\mu_Y)]}{\sigma_X \sigma_Y} \tag{1}$$

Formula (1) defines the overall correlation coefficient, commonly used in lowercase Greek letter ρ as a representative symbol. Estimating the sample covariance and standard deviation, Pearson correlation coefficient can be obtained:

$$\mathrm{r} = \frac{\sum_{i=1}^{n}(X_i-\bar{X})(Y_i-\bar{Y})}{\left[\sum_{i=1}^{n}(X_i-\bar{X})^2\right]^{1/2}\left[\sum_{i=1}^{n}(Y_i-\bar{Y})^2\right]^{1/2}} \tag{2}$$

It can also be estimated from the standard point average of the sample point (X_i, Y_i) to obtain an equivalent expression to the formula (2):

$$\mathrm{r} = \frac{1}{n-1}\sum_{i=1}^{n}\left(\frac{X_i-\bar{X}}{\sigma_X}\right)\left(\frac{Y_i-\bar{Y}}{\sigma_Y}\right) \tag{3}$$

In the above formulas, $\overline{X}$ and $\overline{Y}$ are the average of the sample, and σ_X and σ_Y are the standard deviation of the sample. More generally, the correlation coefficient is positive if both X_i and Y_i tend to be larger or at the same time tend to be smaller than their respective mean. If X_i and Y_i tend to fall on the opposite side of their mean, the correlation coefficient is negative.

3. Results and discussion

3.1. Trends of samples

There are 120 sets of flue gas parameters and SO_2 samples respectively, which is relatively representative. By analyzing the variety trend of flue gas parameters and SO_2 emission, we can get a general understanding of the flue gas entering the desulfurization system.

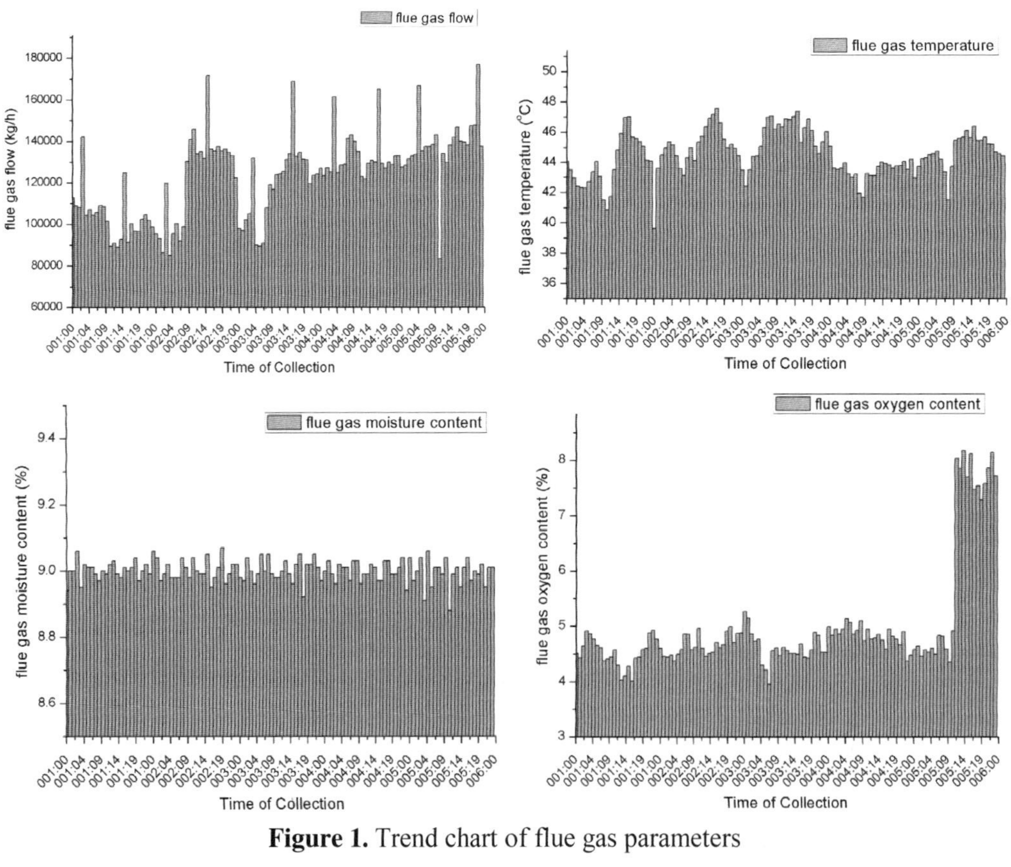

Figure 1. Trend chart of flue gas parameters

As shown in Figure 1, the flue gas parameters of the desulfurization system show less obvious correlation with the time change and did not show the expected change with time except flue gas oxygen content. Flue gas flow roughly fluctuates from 80000 m^3/h to 140000m^3/h, flue gas temperature is relatively stable distribution from 40°C to 50°C, in addition, flue gas moisture content is relatively stable from 8.9% to 9.1%. As for the flue gas oxygen content, excluding suddenly increased 3.11% at 12 o'clock on the 5th resulting in continued to reach 7.28%~8.18% till 23 o'clock on the 5th, the flue gas oxygen content is generally more stable distribution between 4%~6%.

We can see that there is also no obvious correlation between SO_2 concentration and time from Figure 2. SO_2 concentration in inlet flue gas is mostly concentrated in 2500 mg/m^3~5500 mg/m^3. However, the variation of SO_2 concentration in outlet flue is more obvious and varies greatly from 0 to 400 mg/m^3.

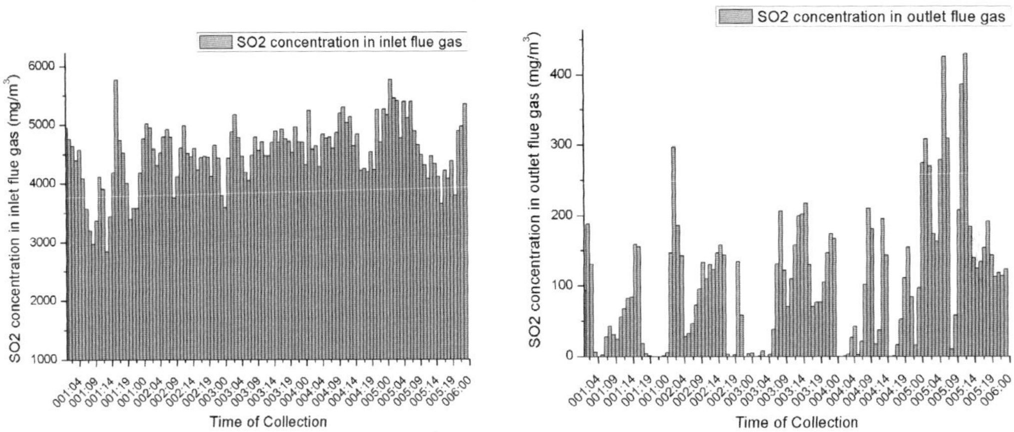

Figure 2. Trend chart of SO_2 concentration

3.2. Confidence interval analysis

Confidence interval refers to the estimated interval of the overall parameters constructed by the sample statistics. In statistics, the confidence interval for a probabilistic sample is the interval estimate for one overall parameter of that sample. In this article, the confidence level is 0.95.

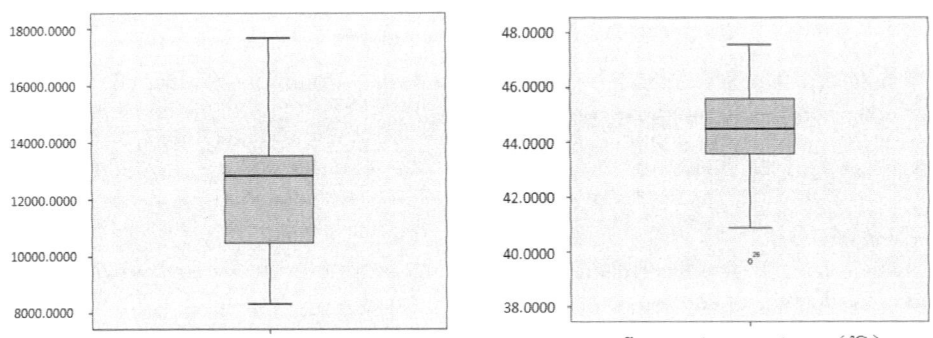

IOP Conf. Series: Earth and Environmental Science **146** (2018) 012070 doi:10.1088/1755-1315/146/1/012070

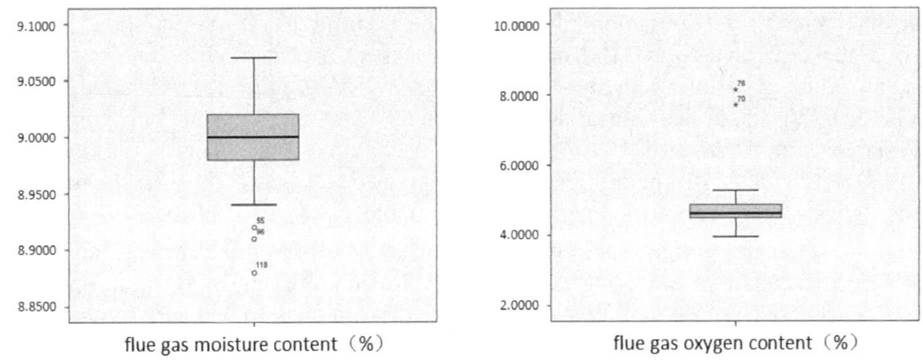

Figure 3. Confidence interval analysis of flue gas parameters

After removing a very small number of outliers, the parameters vary around their respective mean during respective confidence intervals when desulfurization system is working properly. As shown in Figure 3, the confidence interval of flue gas flow is 119414.6177 kg/h~126728.7723 kg/h and 5% trimmed mean is 122551.92204 kg/h. the confidence interval of flue gas temperature is from 44.2866°C to 44.8342°C, 44.6021°C is its 5% trimmed mean. In addition, the confidence interval of flue gas moisture content is from 8.9940% to 9.0058% and the confidence interval of flue gas oxygen content is from 4.7863% to 5.1401%, their 5% trimmed mean is 9.0009% and 4.6622% respectively.

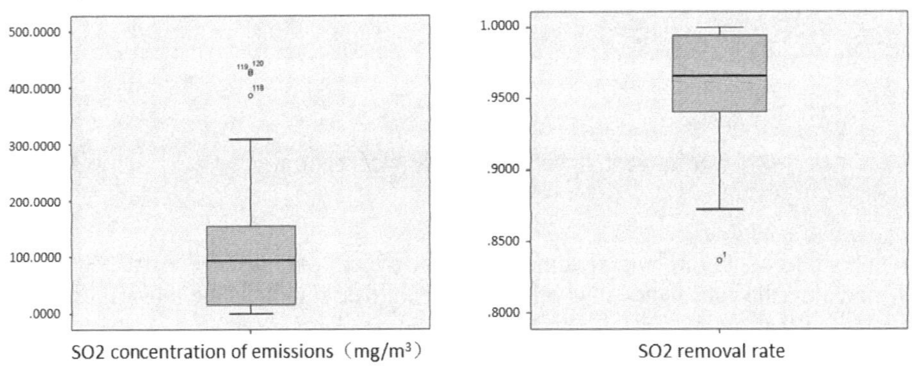

Figure 4. Confidence interval analysis of SO_2

Figure 4 indicates that SO_2 emission concentration varies around its 5% trimmed mean (92.54 mg/m^3) during its confidence interval (from 84.0930 mg/m^3 to 118.3852 mg/m^3)when desulfurization system is working properly. In addition, SO_2 removal rate ranges from 95.28% to 96.74%, and the average desulfurization rate is 96.42%.

3.3. Correlation analysis

In this paper, we used Pearson correlation analysis in SPSS software to analyze the dependence between flue gas parameters and SO_2 emissions. We all know that Pearson correlation coefficient is used to measure whether two data sets are in a line and the linear relationship between distance variables.

3.3.1. Correlation analysis of flow to SO_2 emissions. As shown in Table 1, there is a significant positive correlation between flue gas flow and SO_2 emission concentration, it indicates that increasing the flue gas flow can increase the SO_2 emission concentration and vice versa. The same method is used to the dependence of flue gas flow to SO_2 removal rate, this indicates there is a significant negative correlation between flue gas flow and SO_2 removal rate.

Table 1. Correlations of flue gas flow to SO_2

	SO_2 concentration	SO_2 removal rate
Pearson correlation	0.222*	-0.234*
Sig. (2-tailed)	0.015	0.010
N	120	120

*Correlation is significant at the 0.05 level (2-tailed).

We speculate that the increase of flue gas flow leads to the increase of oxygen supplied to the desulfurization system, more sulfur-containing components are oxidized to generate SO_2. At the same time, the larger the flue gas flow, the shorter the time of flue gas stay in the absorption tower. Therefore, increasing the flue gas flow will increase the SO_2 concentration of outlet flue gas and decrease the SO_2 removal rate.

3.3.2. Correlation analysis of temperature to SO_2 emissions. Table 2 suggests that flue gas temperature is significantly related to both SO_2 concentration and SO_2 removal rate. The different part is that flue gas temperature is positively correlated with SO_2 concentration but negatively correlated with SO_2 removal rate.

Table 2. Correlations of flue gas temperature to SO_2

	SO_2 concentration	SO_2 removal rate
Pearson correlation	0.253**	-0.259**
Sig. (2-tailed)	0.005	0.004
N	120	120

**Correlation is significant at the 0.01 level (2-tailed).

Increasing the flue gas temperature can provide more energy for the desulfurization reaction and oxidize sulfur component more thoroughly, then produce more SO_2. In addition, increasing flue gas temperature can accelerate decomposition of H_2SO_3. On the other hand, the increase of gas temperature leads to higher temperature of the desulfurization solution with the solubility of ammonia and SO_2 in the desulfurization solution decreasing, this releases more SO_2.

3.3.3. Correlation analysis of moisture content to SO_2 emissions. From Table 3, neither the correlation of moisture content to SO_2 concentration nor the correlation of moisture content to SO_2 removal rate is significant. It means that flue gas moisture content has no significant effect on SO_2 concentration and SO_2 removal rate.

Table 3. Correlations of flue gas moisture content to SO_2

	SO_2 concentration	SO_2 removal rate
Pearson correlation	-0.117	0.050
Sig. (2-tailed)	0.205	0.587
N	120	120

In theory, the higher moisture content of the flue gas results in a more moisturized absorbent surface, then more SO_2 will be absorbed. However, in practical applications, there are many other factors may restrict the desulfurization effect besides the moisture content, such as the temperature mentioned above. High temperature can promote the evaporation of desulfurization solution, thereby change the flue gas moisture content and may cover up the impact of moisture content. Therefore, the direct impact of moisture content is not significant.

3.3.4. Correlation analysis of oxygen content to SO_2 emissions. There is no significant correlation between flue gas oxygen content and SO_2 emissions (see Table 4).

Similar to the analysis of moisture content, the change of the gas flow caused by oxygen content has a stronger effect on SO_2 emission than oxygen content itself. Therefore, the effect of oxygen content on the desulfurization efficiency is not obvious.

Table 4. Correlations of flue gas oxygen content to SO_2

	SO_2 concentration	SO_2 removal rate
Pearson correlation	-0.053	0.086
Sig. (2-tailed)	0.586	0.370
N	120	120

4. Conclusions

From the above analysis, it can be seen that Desulfurization system flue gas parameters can indeed directly affect the desulfurization efficiency, in other words, it is feasible to improve the desulfurization efficiency by controlling the flue gas parameters of the desulfurization system.

(1) The SO_2 removal rate is between 95.28%~96.74% and its 5% trimming average value is 96.42%, this industry requires SO_2 removal rate not less than 95%, so it seems that the desulfurization efficiency of the aluminum plant is relatively satisfactory.

(2) The results of Pearson correlation analysis indicated that SO_2 removal efficiency can be improved by reducing the flue gas flow rate or reducing the flue gas temperature, while changing the flue gas oxygen content or flue gas moisture content has no obvious effect on the flue gas SO_2 removal efficiency.

(3) This paper only qualitatively analyzes the correlation of desulfurization system gas parameters on desulfurization, the quantitative results require further study.

Acknowledgments

Our thanks to YL Aluminum Plant for allowing us to use the exhaust gas data of online monitoring.

References

[1] Chen Z 2013 Efficiency prediction of ammonia flue gas desulfurization based on partial least squares regression *Journal of Chinese Society of Power Engineering* 33(3):205-209.

[2] Wang Q and Su B 2005 Analysis of influence on wet FGD system by applying high gas velocity Energy Conservation Technology 23(5): 450-453.

[3] Jiang Y, Wu H and Luo F 2016 Performance study on fine particle removal from wet desulfurized flue gas by adding humid air *Journal of Southeast University* 46(1):81-86.

[4] Bao J, Mao L and Li Z 2016 Temperature and humidity variation characteristics of desulfurization flue gas at outlet of wet flue gas desulfurization system *Transactions of the Chinese Society of Agricultural Engineering* 32(8):231-237.

[5] Pan D, Wu H and Huang R 2016 Removal properties of sulfuric acid mist during limestone-gypsum flue gas desulfurization process *Journal of Southeast University* 46(2):311-316.

[6] Bao J, Yang L and Yan J 2009 Performance of removal of fine particles by WFGD system Journal *of the Chemical Industry & Engineering Society of China* 60(5):1260-1267.

[7] Pang X, Xue Y and Yuan C 2008 Effect of controlled oxidation on wet FGD process Environmental *Science & Technology* 31(5):87-90.

[8] Pan G, Li H and Zhang M 2010 The influence of relative humidity on desulfurization and monitoring of flue gas *2010 Chinese Society of Environmental Sciences Academic Annual Proceedings* 4:10-12.

[9] Pan D, Wu H and Jiang Y 2016 Improvement in removal of fine particles and SO_3 acid mist from desulfurized flue gas with heterogeneous condensation *Journal of Fuel Chemistry and Technology* 44(1):113-119.

[10] He B, Zheng X and Chang D 2002 Experimental study of temperature impacts on the SO_2 reduction ratio by ammonia gas scrubbing *Acta Scientiae Circumstantiae* 22(3):412-416.

[11] Song Z, Zhang M and Ma C 2007 Effect of temperature on the sulfidation of limestone at reduction desulfurization *Journal of Combustion Science and Technology* 13(4):370-375.

[12] Li L and Su B 2004 Analyzing influence of temperature of absorbent liquid in lime wet flue gas desulfurization *Shanxi Architecture* 30(19):118-119.

[13] Wang H, Li S and Lei Yu 2012 Effect of operating temperature on circulating fluidized bed-flue gas desulfurization efficiency *Journal of Northeast Electric Power University* 32(2):21-25..

[14] Wu H 2013 Effect of operating temperature on circulating fluidized bed-flue gas desulfurization efficiency *Chemical Enterprise Management* (6):128-128.

[15] Huang Y. 2004. Research and optimization of operating parameters of lime wet flue gas desulfurization process system *Graduate School Taiyuan University of Technology*.

IOP Conf. Series: Earth and Environmental Science **146** (2018) 012071 doi:10.1088/1755-1315/146/1/012071

Research on the Ultrafiltration and Removal of Aniline via the Compound of Sophorolipid and Rhamnolipid

H Y Fu[1,2*], D Zhao [1], M Xu[3], Y P Li [1,4], J Liu [1,2], Z B Zhang[5], H Z Yan[1] and H D Zhu [1]

[1]College of Environmental Science and Engineering, Xiamen University of Technology, Xiamen 361024, China
[2]Key Laboratory of Environmental Biotechnology(XMUT), Fujian Province University
[3]Chinese Academy for Environmental Planning, Beijing, 100012, China
[4]Fujian Engineering Research Center of Rural Sewage Treatment and Water Safety
[5] XiaMen Lianchuangda Technology Co., Ltd, Xiamen 36100p, China

[*]Corresponding Author. Phone: 0592-6291151, e-mail:fuhy@xmut.edu.cn

Abstract: In this paper, a biological surfactant composed of sophorolipid and rhamnolipid is used as an enhanced ultrafiltration treatment for simulated aniline wastewater and the impact of the surfactant concentration, feed liquid aniline concentration, surfactant compound ratio, solution pH value and operating pressure on the aniline rejection rate and permeation flux is analyzed. The research results showed that the biological surfactant compound was able to remove aniline. The removal effect of aniline was considerable when the aniline concentration was 150 mg/L, the surfactant concentration was 0.8 mmol/L and the surfactant compound ratio was 0.1. The pH value had a significant impact on the aniline rejection rate. However, the operating pressure had little impact on the aniline rejection rate, although it exerted a considerable influence on the permeation flux.

1. Introduction

Chemical production processes related to pharmaceutical production, printing, dyeing, petroleum refining, etc. produce a considerable amount of aniline wastewater that can cause serious pollution hazards to humans and animals in the case of direct discharge [1-3]. Aniline wastewater has a considerable influence on wild animals and humans because of its obviously toxic, carcinogenic and mutagenic properties. In addition, aniline wastewater can further accumulate in the body because of its biological insolubility and high water solubility. Aniline wastewater is classified as a persistent organic pollutant [3-4]. Therefore, the treatment of aniline wastewater is of great practical significance. The treatment methods of aniline wastewater mainly include chemical oxidation, biochemical treatment, membrane separation, light degradation, etc. [5-10]. Among membrane technologies, micellar enhanced ultrafiltration (MEUF) technology is a high-efficiency hot water treatment technology. MEUF technology presents low energy consumption, is easy and simple to implement, and is capable of effectively removing refractory organics in wastewater [11-12]. MEUF technology has broad application prospects. Compared with chemical surfactants, biological surfactants are more suitable for application in MEUF because of their characteristics, including biological degradation, renewability, environmental protection, etc. [13]. In this study, an environmentally friendly biological surfactant compound of sophorolipid and rhamnolipid was used to remove the organic pollutant aniline and the impacts of the biological surfactant concentration (surfactant concentration for short), feed liquid aniline concentration, surfactant compound ratio (i.e., the proportion of sophorolipid in the total concentration of surfactants (α)), solution pH value and pressure on the aniline rejection rate and permeation flux were evaluated.

2 Experiments

2.1 Experimental reagents and devices

Reagents: Aniline was purchased from Shanghai Hushi Chemical Co., Ltd. and used as the analytical reagent; rhamnolipid was purchased from Huzhou Zijin Biotechnology Co., Ltd.; and sophorolipid was purchased from Xi'an Kono Chemical Co., Ltd.

Devices: The membrane device used in the experiment was the FlowMem-0015 ultrafiltration flat membrane small test equipment, which was provided by Xiamen Fumei Technology Co., Ltd. See [14] for the main performance parameters of the FlowMem-0015 ultrafiltration flat membrane small test equipment and ultrafiltration membrane.

2.2 Experimental process

At room temperature (25℃), a certain amount of rhamnolipid and sophorolipid were weighed and added to the simulated aniline wastewater, which was then evenly stirred and set aside for 20 minutes for ultrafiltration. Simulated wastewater entered the membrane module through a membrane pump. The concentrated solution reflowed to the feed chute for further ultrafiltration. The penetrating fluid was collected by a special container. Under a certain pressure, the permeation flux was measured after 20 minutes of operation, and then a moderate amount of penetrating fluid was collected to measure the concentration of aniline. See Figure 1. for the experimental process.

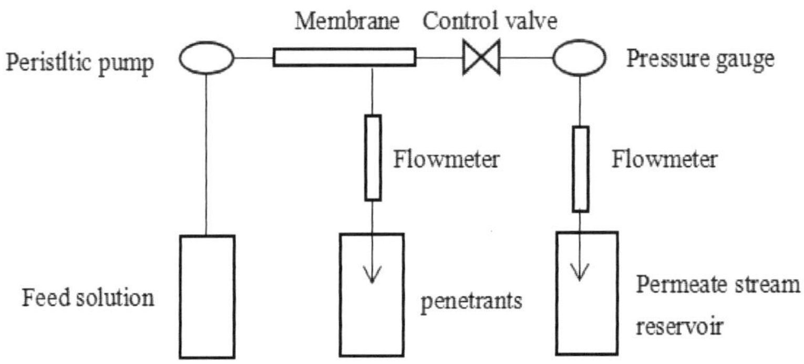

Figure 1. Schematic of micelle enhanced ultrafiltration process.

3 Experimental results and discussion

3.1 Impact of aniline concentration on MEUF

When the surfactant concentration in the feed liquid was 0.5 mmol/L, the α value was 0.1. The change in aniline rejection rate based on the initial concentration of aniline in the feed liquid is shown in Figure 2.

As the influent aniline concentration increased, the aniline rejection rate (R_{AN}) first increased and then decreased. When the initial concentration of aniline was less than 150 mg/L, the R_{AN} showed an increasing trend from 7.4% to 28.6% because a number of micelles were observed in the solution when the surfactant concentration was 0.5 mmol/L. When the initial concentration of aniline increased from 10 mg/L to 150 mg/L, the excessive micelles in the wastewater solubilized the aniline molecules and rejected more aniline. When the aniline concentration was greater than 150 mg/L, the interiors of the micelles in wastewater tended to be saturated. Therefore, the number of aniline molecules that could not be solubilized by micelles increased, which led to a decrease in the R_{AN}.

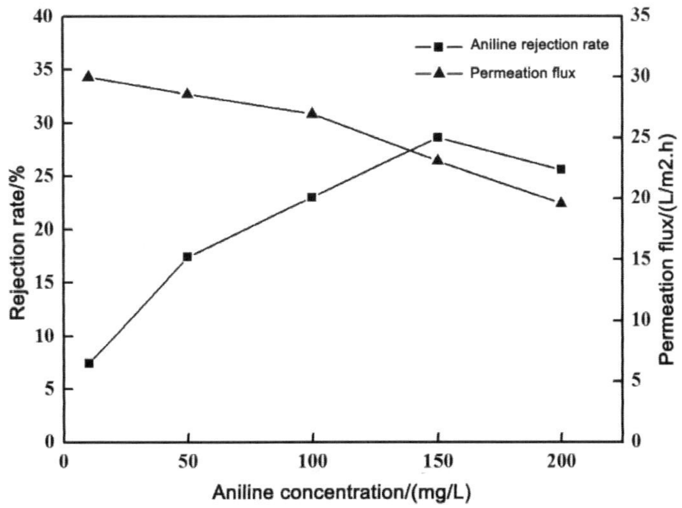

Figure 2. Impact of the feed liquid aniline concentration on the aniline rejection rate and permeation flux.

The permeation flux (J) gradually decreased as the initial concentration of aniline in solution increased. When the aniline concentration changed between 10 mg/L and 200 mg/L, the J decreased from 30 L/m^2·h to 19.6 L/m^2·h. Moreover, an increase in the initial concentration of aniline aggravated the membrane pollution.

3.2 Impact of surfactant concentration on MEUF

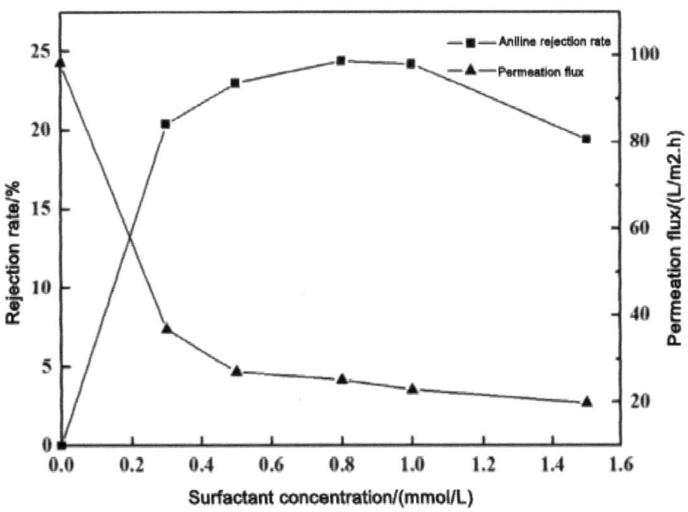

Figure 3. Impact of the surfactant concentration on the aniline rejection rate and permeation flux.

As displayed in Figure 3, the R_{AN} first increased rapidly and then decreased as the surfactant concentration increased. The molecular weight of aniline in wastewater was much smaller than the molecular weight cut off (MWCO) of the ultrafiltration membrane. Therefore, aniline molecules would outflow through the ultrafiltration membrane when surfactants were not added to wastewater.

As a result, the R_{AN} was close to 0, although it increased to 20.4% when the surfactant concentration increased from 0 to 0.3 mmol/L. The R_{AN} increasing trend was gentle with the further increases in the surfactant concentration. When the surfactant concentration was 0.8 mmol/L, the R_{AN} reached the maximum value of 24.4%. When the surfactant concentration varied from 1.0 mmol/L to 1.5 mmol/L, the R_{AN} started to show a downward trend, which might have been related to a change in the shape of the micelles in wastewater under high surfactant concentrations. The particle size reduced from the previous spherically shaped micelles to slender rod-shaped micelles. Therefore, the micelles solubilizing aniline molecules could not be rejected by the membrane and would outflow along with water.

The J was reduced by more than half when surfactants were initially added to the feed liquid. The declining rate of the J decreased when the surfactant concentration was greater than 0.5 mmol/L. Initially, the J presented a sharp drop, which might have been related to the absorption of the added surfactants by the ultrafiltration membrane or membrane pollution caused by the effect of concentration polarization, which then led to a decline in the membrane flux[15]. The particle size of the micelles gradually decreased with further increases in the surfactant concentration and might have led to the blockage of membrane pores by micelles with a similar size as the pores, which thus affected the J.

3.3 Impact of the surfactant compound ratio on MEUF

The impact of different α values on the R_{AN} when the surfactant concentration was fixed at 0.8 mmol/L is shown in Figure 4. The R_{AN} first increased and then decreased with an increase in the proportion of sophorolipid in the total concentration of surfactants. The R_{AN} rate rose from 17.4% to 24.4% when the proportion of sophorolipid in the total concentration of surfactants increased from 0 to 0.1. The R_{AN} gradually decreased when the proportion of sophorolipid in the total concentration of surfactants was greater than 0.1 because the dosage of added sophorolipid was too large, which produced a stereo-hindrance effect. Sophorolipid was rejected by the mixed system, which led to an increase in the CMC of the mixed system and a decline in the number of formed micelles; in addition, such conditions would reduce the diameter of the micelles in solution and cause part of the micelles to outflow through the membrane pores, thereby reducing the R_{AN}. The R_{AN} was the largest when the sophorolipid concentration accounted for 0.1% of the total concentration of surfactants, which might have been related to the good combination of rhamnolipid and sophorolipid generating a small CMC in the compound system. Therefore, additional micelles could be generated and additional aniline could be solubilized under the condition of the same surfactant concentration.

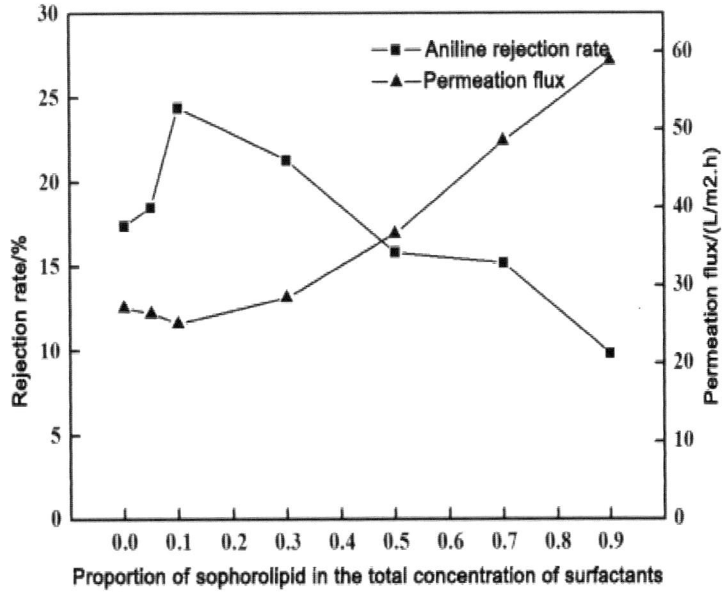

Figure 4. Impact of the surfactant compound ratio on the aniline rejection rate and permeation flux.

The J first decreased and then increased as the α value increased, and it subsequently increased from 25.2 L/m^2·h to 58.9 L/m^2·h because of an increase in the proportion of sophorolipid in the total concentration of surfactants, a decrease in the number of micelles in solution and a reduction in the membrane pollution.

3.4 Impact of pH value on MEUF

As shown in Figure 5, the R_{AN} first decreased and then slowly increased as the pH value of the feed liquid increased. The R_{AN} presented a downward trend when the pH value varied from 4 to 7. The MEUF process had the best rejection effect on aniline, and the R_{AN} was 41.1% when the pH value was 4 because aniline can hydrolyze a considerable amount of H+ in water under acidic conditions and thereby generate positively charged aniline ions. Therefore, micelles in wastewater could not only remove aniline molecules through a solubilization effect but also absorb aniline ions on the surface of micelles via electrostatic adsorption and reject more aniline. Under alkaline conditions, the R_{AN} increased, which might have been related to the good solubility of rhamnolipid under alkaline conditions and the subsequent increase in micelle formation and aniline solubilization.

The figure clearly shows that the J value continually increased as the pH value increased. When the pH value rose from 4 to 10, the J increased from 4.8 L/m^2·h to 33.7 L/m^2·h, which was mainly because the increase in pH value promoted the dissolution of rhamnolipid in solution and reduced the viscosity of the solution. Thus, the pH value had a significant impact on the J.

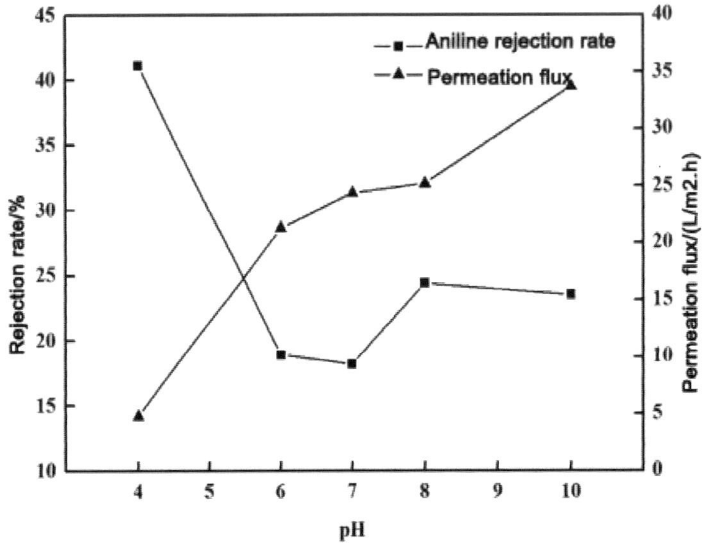

Figure 5. Impact of the pH value on the aniline rejection rate and permeation flux.

3.5 Impact of the operating pressure on MEUF

Changes in the R_{AN} and J value under changes in the operating pressure while the other operating conditions remained unchanged are shown in Figure 6 The R_{AN} showed no obvious changes when the pressure rose from 0.2 MPa to 0.4 MPa. Thus, the operating pressure had no significant impact on the R_{AN}. The J value also continued to increase as the operating pressure increased, and it presented a value of 25.2 L/m^2·h when the pressure was 0.2 MPa and almost doubled to 51.1 L/m^2·h when the pressure rose to 0.4 MPa.

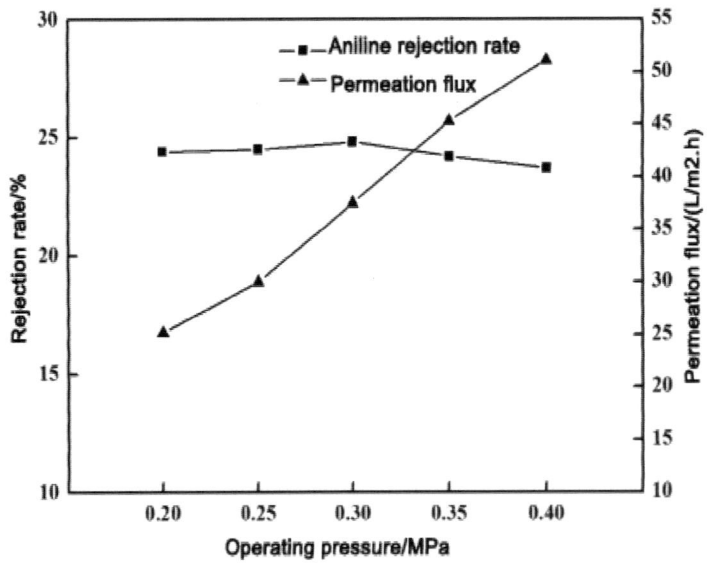

Figure 6. Impact of the operating pressure on the aniline rejection rate and permeation flux.

4 Conclusions

- The RAN increased as the initial concentration of aniline varied from 10 mg/L to 150 mg/L and presented a downward trend when the initial concentration of aniline in the feed liquid was greater than 150 mg/L. The J value continually decreased as the initial concentration of aniline in wastewater increased.
- The RAN continued to increase when the surfactant concentration varied from 0 mmol/L to 0.8 mmol/L. The RAN started to decrease when the surfactant concentration was greater than 0.8 mmol/L, and the J value rapidly decreased and then slowly reduced as the surfactant concentration increased.
- The RAN showed an increasing trend when α varied from 0 to 0.1 and began to decrease when the α value was greater than 0.1. The J initially decreased and then increased with increases of α.
- The RAN first decreased and then slowly increased as the pH value of the simulated water increased. Under acidic conditions, the removal effect of aniline was good because acidic conditions were beneficial to the hydrolysis of aniline and thus produced positively charged aniline ions, which were absorbed by micelles via electrostatic adsorption. Thus, additional aniline was removed. Moreover, the J value continued to increase with increases in the pH value.
- The RAN showed almost no changes as the operating pressure increased. Pressure had a significant impact on the J value, which linearly increased as the operating pressure increased.

Acknowledgements

This work was financially supported by Fujian Science and Technology Guiding Project (2018Y0079), National Natural Science Foundation of China(51109181).

5 References

[1] Zhang J, Guo J, Wu Y, Lan Y Q and Li Y Efficient activation of ozone by zero-valent copper for the degradation of aniline in aqueous solution 2017 *Journal of the Taiwan Institute of Chemical Engineers* **81** 335–342

[2] Beata S and Piotr S Photodegradation of aniline in water in the presence of chemically activated halloysite 2016 *Applied Clay Science* **124–125** 31–38

[3] Cui D Z, Shen D, Wu C C, Li C, Dongjin Leng and Zhao M Biodegradation of aniline by a novel bacterial mixed culture AC 2017 *International Biodeterioration & Biodegradation* **125** 86-96

[4] Jagtap N Ramaswamy VOxidation of aniline over titania pillared montmorillonite clays 2006 *Appl. Clay Sci.* **33** 289-98

[5] Pithan F Staudt-Bickel C and Lichtenthaler R N Synthesis of highly fluorinated copolyimide membranes for the removal of high boiling organics from process water and wastewater by pervaporation 2007 *Desalination* **148** 1-4

[6] Kumar A and Mathur N Photocatalytic degradation of aniline at the interface of TiO_2 suspensions containing carbonate ions 2006 *Colloid Interf. Sci.* **300** 1244-252

[7] Han Y, Quan X, Chen S and et al Electrochemically enhanced adsorption of aniline on activated

carbon fibers 2006 *Sep. Purif. Technol* **50** 3365-372

[8] Bahareh T, Mahdi P C, Nasser S, Mehrdad H, Timo L, Mari K, Mika S and Mika M Simultaneous removal of aniline and nickel from water by micellar-enhanced ultrafiltration with different molecular weight cut-off membranes 2014 *Separation and Purification Technology* **124** 26–35

[9] Wang L P, Zhang J Y, Wu Y X and Chang YZ Study on the dynamic adsorption of aniline in water by wheat straw 2011 *Journal of Environmental Engineering* **12** 52755-2760

[10] Navid O R, Mahdi P C, Hamed A N Removal of free active chlorine from synthetic wastewater by MEUF process using polyethersulfone/titania nanocomposite membrane 2017 *Separation and Purification Technology* **181** 213-222

[11] Rahmanian B, Pakizeh M, Mansoori S A A, Esfandyari M, Jafari D, Maddah H and Maskooki A Prediction of MEUF process performance using artificial neural networks and ANFIS approaches 2012 *Journal of the Taiwan Institute of Chemical Engineers* **43** 558-565

[12] Samal K, Das C and Mohanty K Application of saponin biosurfactant and its recovery in the MEUF process for removal of methyl violet from wastewater 2017 *Journal of Environmental Management* **203** 8-16

[13] Fu H Y, Zhang Z B, Chai T, Huang G H, Yu S J, Liu Z and Gao P F Study of the Removal of Aniline from Wastewater via MEUF Uisng Mixed Surfactants 2017 *Water* **9** 63-65

[14] Zaghbani N, Hafiane A, Dhahbi M. Separation of methylene blue fromaqueous solution by micellar enhanced ultrafiltration 2007 *J. Separation and Purification Technology* **55** 1117-124

[15] Huang J H *Mechanism and application of micellar enhanced ultrafiltration for treatment of cadmium containing wastewater* 2007 (Changsha:Hunan University)

Preparation of Polyvinylidene Fluoride Membrane with Functional Anthraquinones for Nitrogen Removal

Ye Qian, Wang Yuping, Xu Meilan, Liao Wenchao, Yan Bin*

1. School of Environmental Science and Engineering, Xiamen University of Technology, Xiamen 361024;
2. Key Laboratory of Environmental Biotechnology, Fujian Province University, Xiamen 361024;
3. Xiamen Key Laboratory of Membrane Research and Application, Xiamen 361024, China.

Corresponding author's : 2010111601@xmut.edu.cn

Abstract. This study relates to a method for preparation of polyvinylidene fluoride membrane with functional anthraquinones. The method is carried out according to the following steps: step 1: preparing 2- (1- hydroxy -3- butene) -1,4,5,8- tetramethoxyl naphthalene; step2: preparing polyvinylidene fluoride-aromatic ether copolymers: polyvinylidene fluoride was used as the initiator, 2-(1-hydroxy-3-butene) -1,4,5,8-tetramethoxynaphthalene was the monomer, N, N-dimethylformamide was solvent, cuprous chloride / Me6TREN was the catalytic, polyvinylidene fluoride-aromatic ether copolymer was synthesized by atomic transfer radical polymerization; step3: reducing the polyvinylidene fluoride-aromatic ether copolymer to quinone by demethoxy oxidation; step4: using the product of step3 and N, N-dimethylformamide as film-forming reagents, then scraping into a membrane. These functional membranes can effectively solve the problem of fixation of redox mediator and improve the efficiency of treatment of wastewater with high nitrogen concentration.

1. Introduction

Polyvinylidene fluoride(PVDF), because of its high mechanical strength, good chemical stability, good halogen resisting, excellent resistance to acid, alkali, oxidizing agents and anti-UV properties, has been widely used as raw materials for preparing the membrane material in the Environmental engineering. However, the disadvantages of high surface hydrophobicity and low surface energy of polyvinylidene fluoride would influence the service life of the membrane material. In order to further optimize the performance of polyvinylidene fluoride membrane, the researchers conducted a series of studies[1-4]. Such as blending and grafting, these method enable the functionalization of the polyvinylidene fluoride membrane, and the resulting membrane has good hydrophilicity and antifouling ability. Those researches have instructive significance to optimize the membrane performance and prolong the service life of the membrane, but their application principle is still based on physical separation, that is, the transfer and enrichment of pollutants, does not realize the degradation of pollutants, there is still possible harm to the environment. Therefore, it is of great significance to study and develop a membrane that can purify sewage and degrade pollutants [5-8].

The high concentration of nitrogen-containing domestic sewage, industrial wastewater and farmland surface water runoff into the lake, reservoir, river and bay waters, then cause some of the algae in the water to reproduce, and seriously deteriorate water quality , damage the ecological balance of water. Biological method is the most common method to solve the above-mentioned problem of water pollution, but it is limited by the electron transport rate in the process of biological denitrification, the effect of biological treatment is unstable and the treatment efficiency is low. It has been found that the redox mediator can accelerate the electron transport rate in the process of biological denitrification and improve the efficiency of biological treatment.Anthraquinone was one kind of redox mediators, there were many researches which have confirmed that anthraquinone compounds can effectively promote the degradation of nitrogenous waste water. There were many reports that anthraquinones were put into use directly, which would cause the loss of anthraquinones and bring about the secondary pollution. In order to solve the above problems, the researchers carried out a series of studies to aviod the loss of anthraquinones. ZhiyuanMA et al.[9]have found that 1,5-dichloroanthraquinone which was immobilized in calcium alginate could promote the decolorization of acid red B, but it was only bound by the physical force on the carrier, easy to fall off from the carrier. Jing LIAN et al.[10] discovered a method for the fixation of anthraquinone sulfonate by using cyclic voltammetry. The result showed that the immobilized anthraquinone sulfonate can accelerate the process of nitrite biodegradation. However, the method of cyclic voltammetry for fixxing anthraquinone sulfonate was very complex as it was controlled by the preparation of polypyrrole membrane, which was affected by a variety of factors. Therefore, if the redox mediator is fixed on the membrane, it can effectively solve the problem of fixation of redox mediator and improve the efficiency of treatment of wastewater with high nitrogen concentration.

2. Experimental

2.1 Materials

PVDF powder (intrinsic viscosity = 1.11 dL/g, Mn = 431,000 g/mol, density = 1.77 g/cm^3) was bought from Shanghai 3F New Materials Co., China. N, N-dimethylformamide(DMF, reagent grade) was used as the solvent from Shanghai SSS Regent Co., China. (1-hydroxy-3-butene)-1,4,5,8-tetramethoxynaphthalene, cuprous chloride, Me6TREN were from Sinopharm Chemical Regent Co., China. All materials were used without further purification.

2.2. Membrane Preparation

Step1: preparing 2- (1- hydroxy -3- butene) -1,4,5,8- tetramethoxyl naphthalene;

Step2: preparing polyvinylidene fluoride-aromatic ether copolymers: polyvinylidene fluoride was used as the initiator, 2-(1-hydroxy-3-butene) -1,4,5,8-tetramethoxynaphthalene was the monomer, N, N-dimethylformamide was solvent, cuprous chloride / Me6TREN was the catalytic, polyvinylidene fluoride-aromatic ether copolymer was synthesized by atomic transfer radical polymerization;

Step3: reducing the polyvinylidene fluoride-aromatic ether copolymer to quinone by demethoxy oxidation;

Step4: using the product of step3 and N, N-dimethylformamide as film-forming reagents, then scraping into a membrane.

3. Results and Discussion

3.1. Preparing 2- (1- hydroxy -3- butene) -1,4,5,8- tetramethoxyl naphthalene

3.1.1. Preparing 1,4,5,8-tetramethoxynaphthalene. Adding naphthazarin, tetramethylammonium bromide and tetrahydrofuran to a round bottom flask, stirring to dissolve, then adding sodium dithionite aqueous solution and dimethyl sulfate solution, stirring evenly; then moving the round bottom flask to ice water bath, reacting for 1 h, then slowly dropping NaOH aqueous solution into the flask. After the drop adding, removing the ice bath, continue to react at room temperature for 30 min, and stirring continuously for 18 h until the reaction was complete. Then extracting the reaction solution with ethyl acetate, washing with saturated brine, drying by anhydrous magnesium sulfate, filtering, and recovering of ethyl acetate under reduced pressure. Finally, separating the solids by column chromatography to obtain the 1,4,5,8-tetramethoxynaphthalene ;1 H NMR (400 MHz,DMSO): δ 6.44 (d,4 H),3.37 (s,12 H, CH3).

The mass ratio of naphthazarin, tetrahydrofuran, sodium dithionite, dimethyl sulfate and sodium hydroxide is 1.5:75:55:110:125;

3.1.2. Preparing 1,4,5,8-tetramethoxynaphthalene-2-carboxaldehyde. Adding N, N-2-methylacetamide to the 2 mouth flask, removing the flask in an ice-water bath, slowly dropping phosphorus oxychloride and 0.063 mol/L 1,4,5,8-tetramethoxynaphthalene in chloroform solution, After the drop adding, removing the ice bath, heating and refluxing reaction for 5 h; then adding ice water to stop the reaction, extracting by chloroform, saturated brine washing, anhydrous magnesium sulfate drying, then filtering, and negative pressure recovery of chloroform, separating by column chromatography ;1 H NMR (400 MHz,DMSO): δ 6.49-6.51 (m,3 H),3.40 (s,9 H, CH3),3.43 (s,3 H, CH3), 10.11(s,1H,CHO).

The volume ratio of N, N-2-methylacetamide, phosphorus oxychloride, 1,4,5,8-tetramethoxynaphthalene in chloroform is 2.5:3:15.

3.1.3. Preparing 2- (1-hydroxy-3-butene) -1,4,5,8-tetramethoxynaphthalene. The schematic diagram of preparing of 2-(1-hydroxy-3-butene) -1,4,5,8-tetramethoxynaphthalene is shown in Figure 1.

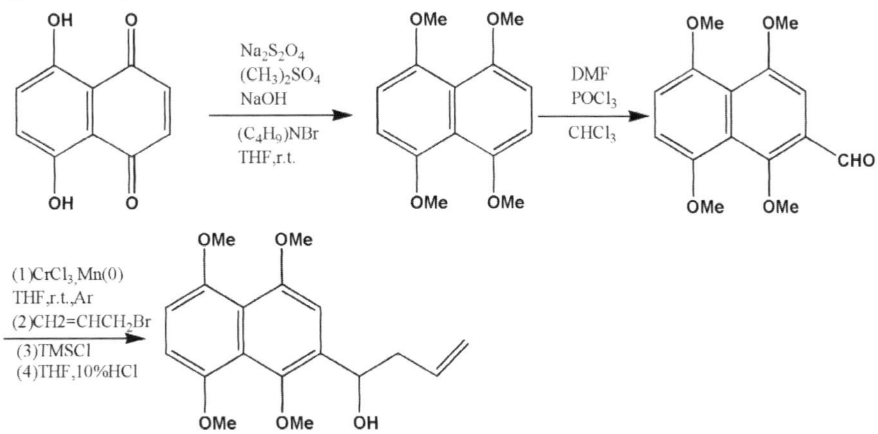

Figure 1. Schematic Diagram of Preparing of 2-(1-hydroxy-3-butene) -1,4,5,8-tetramethoxynaphthalene

Under the protection of argon, adding molecular sieve, anhydrous tetrahydrofuran, anhydrous chromium trichloride and manganese powder in turn in the dry 2 mouth flask, stirring until the color becomes black, then adding allyl bromide, adding 1,4,5, 8-tetramethoxynaphthalene-2-carbaldehyde and trimethylchlorosilane , reacting for 3 h, adding sodium bicarbonate to stop the reaction,washing with diatomite and ether, extracting by ether, saturated brine washing, anhydrous magnesium sulfate drying, and negative pressure recovery of the concentrated residue, dissolving in tetrahydrofuran, and hydrolyzing with 10% hydrochloric acid, stirring at room temperature for 10 min, extracting with ether, washing with saturated brine, anhydrous magnesium sulfate drying and concentrating under reduced pressure, then separating by column chromatography ;1 H NMR (400 MHz, DMSO): δ 6.47-6.51 (m, 3H),3.37 (s, 9H, CH_3),3.43 (s, 3H, CH_3), 8.45(s, 1H, OH), 4.83(t,1H,CH), 2.39(m,2H,CH_2), 4.92(d,2H,CH_2), 5.76(m,1H,CH).

The mass ratio of anhydrous tetrahydrofuran, anhydrous chromium trichloride, 1,4,5,8-tetramethoxynaphthalene-2-carbaldehyde, trimethylchlorosilane, allyl bromide, manganese powder is 20:20:50:60:50:700.

3.2. Preparing polyvinylidene fluoride-aromatic ether copolymers

Adding polyvinylidene fluoride and N, N-dimethylformamide to double glass reactor, stirring evenly, then adding Me6TREN and 2- (1-hydroxy-3-butene) -1,4,5,8-tetramethoxynaphthalene, injecting argon to remove oxygen for 30 min, adding cuprous chloride, removing oxygen for 1 hour, then sealing the reactor, removing the reactor to ice-water bath, reacting 1min under the ultraviolet irradiation in the magnetic stirring,filtering with the volume ratio of 1:1 of ethanol and water after the reaction, then extracting with chloroform several times, finally vacuum-drying to obtain the polyfluoroethylene-aromatic ether copolymer.

The mass ratio of 2- (1-hydroxy-3-butene) -1,4,5,8-tetramethoxynaphthalene, polyvinylidene fluoride, N, N-dimethylformamide, cuprous chloride / Me6TREN is.45:7:500:0.5.

The schematic diagram of preparing of polyvinylidene fluoride-aromatic ether copolymer is shown in Figure 2.

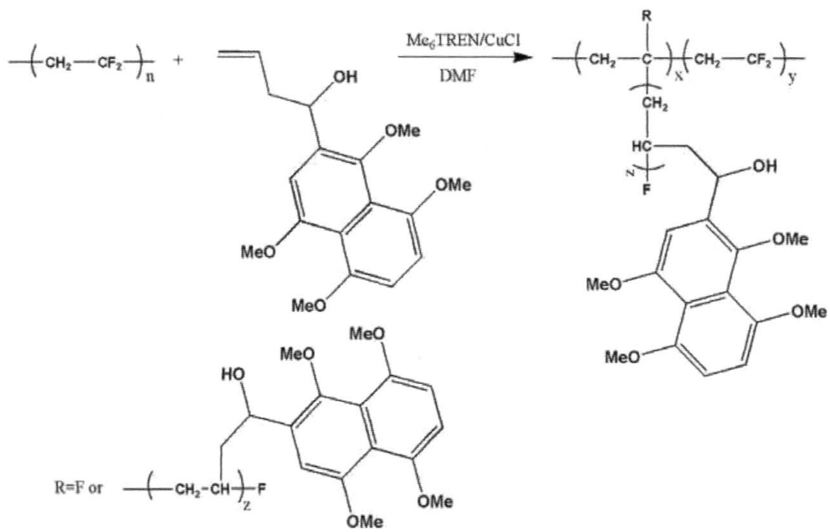

Figure 2. Schematic Diagram of Preparing of Polyvinylidene Fluoride-aromatic ether Copolymer

3.3. Reducing the polyvinylidene fluoride-aromatic ether copolymer to quinone by demethoxy oxidation

More specifically, adding the polyvinylidene fluoride-aromatic ether copolymer of acetonitrile solution in the 2 mouth flask, addding cerium ammonium nitrate aqueous solution at room temperature with stirring, reacting for 1 hour, negative pressure recovery of acetonitrile, extracting with chloroform, washing with water, and washing with saturated brine, anhydrous magnesium sulfate drying for 1.5 hour, negative pressure recovery of chloroform, separating the solids by column chromatography to obtain the mixture of 2-(1-hydroxy-3-butene) -5,8-dimethoxy-1,4-naphthoquinone and 6-(1-hydroxy-3-butene) -5,8-dimethoxy-1,4-naphthoquinone,finally vacuum drying.

The eluent of silica column chromatography is the mixture of the petroleum ether and acetone with a volume ratio of 3: 1.

3.4. Using the product of step3 and N, N-dimethylformamide as film-forming reagents, then scraping into a membrane.

The mass ratio of N, N-dimethylformamide and the product of step3 is 17:83. The sketch of double glass reactor is shown in Figure 3.

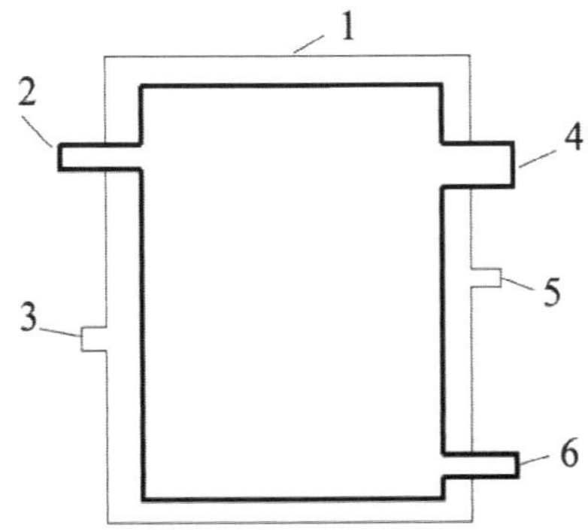

Figure 3. The Sketch of Double Glass Reactor: 1-the place of ultraviolet light, 2-vacuum orifice, 3-water inlet, 4-feed inlet, 5-water outlet, 6- inlert port.

3.5 Nitrogen Removal Effect of the PVDF membrane modified with anthraquinones

The application of polyvinylidene fluoride membrane with functional anthraquinones in the degradation of nitrogen - containing wastewater are shown in Table 1.

Table 1. Degradation of Nitrogen with membranes

Test items	Nitrogen content	Removal rate
The nitrogen-containing wastewater without treatment	200mg/L	0%
The nitrogen-containing wastewater treated by PVDF	56mg/L	72%
The nitrogen-containing wastewater treated by PVDF membrane with anthraquinones	16mg/L	92%

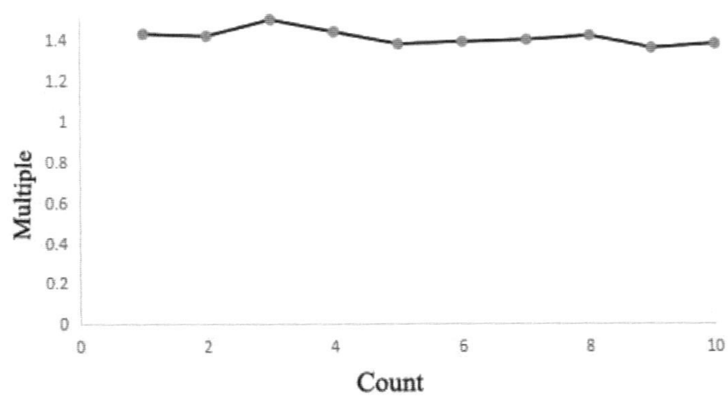

Figure 4. The Nitrogen Removal Effect of the Polyvinylidene Fluoride Membrane Modified with Anthraquinones

Figure 4 is the graph about the nitrogen removal effect of the polyvinylidene fluoride membrane modified with anthraquinones.The abscissa shows the times of circle application of the polyvinylidene fluoride membrane with functional anthraquinones, the ordinate shows the multiple of removal rate of nitrate. As can be seen from the graph, the ultrafiltration membrane prepared in this work has stable performance and can be recycled several times.

4. Conclusions

As pure anthraquinone compounds can not carry out the atomic radical polymerization, our work introduced the double bond of side chain into the naphthazarin by the NHK reaction, and contributed to facilitate the occurrence of the atomic transfer radical polymerization(ATRP) reaction between the polymer material and the anthraquinone. Further, the anthraquinone which fixed in the polyvinylidene fluoride membrane would not fall off.

The polyvinylidene fluoride membrane as a fixed carrier for the redox mediator can be adapted to various membrane processing apparatuses.

The membrane with functional anthraquinone can effectively promote the degradation of waste water with high concentration nitrogenous, and especially accelerate the degradation of printing and dyeing wastewater.

Acknowledgements

The authors acknowledge financial support for this work from the Fujian Provincial Science and Technology Key projects of China(2014N0031), and Fujian Provincial Natural Science Foundation of China (2016H0039).

References

[1] Maghsoud Z, Famili M H N, Madaeni S S, et al. Preparation of polyvinylchloride membranes from solvent mixture by immersion precipitation[J]. *Journal of Applied Polymer Science*, 2014, 131(8): 2-6.

[2] Ye Q, Cheng LH , Zhang L , Xing L, Chen HL,Preparation of Symmetric Network PVDF Membranes for Protein Adsorption via Vapor-Induced Phase Separation, *J. Macromolecular Sci.*, Part B: Physics, 50 (2011) 2004-2022.

[3] Ochoa N A, Masuelli M, Marchese J, et al. Effect of hydrophilicity on fouling of an emulsified oil wastewater with PVDF/PMMA membranes[J]. Journal of Membrane Science, 2003, 226(1): 203-211.

[4] Arezou A, Amir AY, Fatemeh R, Mahmood H, Maryam T, Aliakbar S. PVDF/PAN Blend Membrane: Preparation, Characterization and Fouling Analysis[J].*Journal of Polymers and the Environment.* 2017(4):1359-1360.

[5] Shen X, Yin X,Zhao Y, Chen L.Improved protein fouling resistance of PVDF membrane grafted with the polyampholyte layers[J].*Colloid and Polymer Science.* 2015(4):1205-1213.

[6] Hossein M, Farid S, Mehdi M.HBPE-g-PVDF/PVDF blend nanofiltration membrane: preparation and characterization[J]. *Journal of the Iranian Chemical Society.* 2016(9):1733-1740.

[7] Zhang W. Zhang Y, Fan R, Lewis R. A facile TiO2/PVDF composite membrane synthesis and their application in water purification[J]. *Journal of Nanoparticle Research.* 2016(18):31-33.

[8] Li M, Shi J, Chen C, Li N, Xu Z,et al.Optimized permeation and antifouling of PVDF hybrid ultrafiltration membranes: synergistic effect of dispersion and migration for fluorinated graphene oxide[J].*Journal of Nanoparticle Research.* 2013(19):114.

[9] Ma ZY, Xu ZF, Kang L, Zhang CX, Kang J, Guo JB. Enhanced bio-decolorrization of acid red B with immobilized redox mediator[J].*Heibei Journal of Industrial Science and Technology.* 2013(3): 133-137.

[10] Jing LIAN et al. Biological catalyzing denitrification of nitrite by immobilized redox mediator[J]. *Chinese Journal of Environmental Engineering.* 2012(6):1805-1809.

IOP Conf. Series: Earth and Environmental Science **146** (2018) 012073 doi:10.1088/1755-1315/146/1/012073

Optimization of H_2O_2 Modification Conditions of Bamboo-based Activated Carbon by Response Surface Methodology

Su Xu[1,2,*], WenChao Liao[2], Peng Zheng[2] and Yuan Fan[2]
1. Key Laboratory for Ecological Industry and Green Technology of Fujian Province, Wuyi University, Nanping 354300, China .
2. School of Environmental Sciences and Engineering, Xiamen University of Technology, Xiamen 361005, China

*xusu@xmut.edu.cn

Abstract. Using bamboo-based activated carbon as raw materials and H_2O_2 as modifying agent to enhance the Cr(VI) adsorption efficiency. Response surface methodology based on Box-Behnken design (BBD) was applied to investigate the effects and interactions of modified temperature, solid-liquid ratio, modified time and optimize the modification conditions. The optimum modification conditions for preparing bamboo-based activated carbon were modified temperature 38℃, solid-to-liquid ratio 1:10 and modified time 2 hours. The experimental and predicated results of the Cr(VI) adsorption value were 84.21% and 85%, respectively. This indicate the reliability of the model.

1. Introduction

In recent years, the damage to water resources from the rapid development of industrial technologies has become an important factor impacting economic and social development[1]. As defined by EPA, Chromium (Cr) is on the top priority list of toxic pollutants, and has been widely used as antifouling agent in cooling towers, chrome-plating, leather tanning, pigment, textile, and wood preservation[2]. In natural aquatic environments, Cr(VI) is one of the highly toxic metals and pose a serious threat to the environment and human health. To solve this problem, various methods have been developed to improve the removal efficiency of Cr(VI) from wastewater such as adsorption, chemical reduction and precipitation, coagulation, membrane filtration, and ion exchange. Among the various remediation technologies, absorption with activated carbons(ACs) was the most common method to removal Cr(VI) from effluents because of its high efficiency and low-cost process[3].

Cause the high specific surface area and well- modified surface, activated carbons are highly effective adsorbents for eliminating heavy metals from wastewater. A variety of carbonaceous materials such as biomass and their waster have been used to prepare ACs to removal of Cr(VI), for example Rai et al. prepared mango kernel with H_3PO_4 as activating agents[2] and Siboni et al. modified holly sawdust by using formaldehyde[4] ,Bamboo as a tropical plant is widely distributed in Southern Asia such as China, Thailand and Vietnam[4]. With the increase demand of bamboo-based products, great deals of bamboo wastes were left and became to an environmental issues. The conversion from bamboos waste to ACs has attracted tremendous economic and ecologic attention.

Several researchers have been focus on the development of modification process for bamboo-based ACs. Zhang et al. used bamboo as raw materials to prepare high surface area ACs by steam activation[5] and modified bamboo based ACs by using humic acid to absorbed Cr(VI) [6] or used in drink water[7]. H_2O_2 is a strong oxidant. Liu[8] et al. used H_2O_2 modified ACs to removal atrazine from water, Gao et al. modied ACs with H_2O_2 to adsorb sulfamethoxazole and ibuprofen[9]. H_2O_2 can change the pore structure to enhance the pore volume and specific surface area, modified ACs shows a high adsorb performance.

Response surface methodology(RSM) is useful for design of the experiments, for estimation the effects and interactions of process variables, and for predict the optimum operational conditions[10-12] . RSM can decrease the required experimental data to find the best operation conditions, and has been widely used in environmental science, biotechnology and chemical engineering fields.

No reports have been found by using RSM to optimize the modification conditions for bamboo-based ACs to removal Cr(VI) from waste water. Therefore, in the present study, RSM based on Box-Behnken design (BBD) has been used to optimize the modification process and determination the effects and interactions of the modification conditions (temperature, solid-fluid ratio and time) for the desirable response variable (Cr(VI) removal ratio). The main objectives were to optimize the modification process of bamboo-based ACs and investigate the influence of process variables on Cr(VI) removal ratio.

2. MATERIALS AND METHODS

Please follow these instructions as carefully as possible so all articles within a conference have the same style to the title page. This paragraph follows a section title so it should not be indented.

2.1. Materials

Commercial bamboo-based activated carbon from Guangshen(Xiamen, China) was used in this study. Hydrogen peroxide and potassium bichromate were provided from Runhe (Shanghai, China). Chromium(VI) standard was purchased from Sigma-Aldrich (St. Louis, MO,USA), all used chemical reagents in this study were of analytical grade. The stock solution of 1.0mg/L Cr(VI) was prepared by potassium dichromate and deionized water.

2.2. Modification of Bamboo-based ACs

Bamboo-based ACs was washed with deionized water and dried in an oven at 80℃ overnight. The washed ACs were treated by Hydrogen peroxide under modification variables including modification temperature, modification time and solid-liquid (AC/H_2O_2) ratio. Then the ACs were separated and washed with deionized water until the pH constant and dried in an oven at 80℃ for 24h..

2.3. Adsorption Experiments

The adsorption experiments were performed in 250ml Erlenmeyer flasks with 100ml of stock Cr(VI) solution, 0.1g modified ACs was added into the flasks. After shaking in an incubator at 298K to reach equilibrium, the adsorbent was isolated and the concentration of Cr(VI) solution was determined by measuring with 1.5-diphenylcarbohydrazide spectrophotometric method using a UV-vis spectrophotometer at 540nm. The calibration curve was prepared by measured of standard solution of potassium dichromate, and the adsorption efficiency of Cr(VI) on modified ACs was calculate by using the difference from Cr(VI) before/after adsorption and it was expressed by Eq. (1):

$$\text{Adsorption}(\%)) = \frac{C_0 - C_e}{C_0} \times 100\% \tag{1}$$

Where C_0(mgL-1) and C_e(mgL-1) are the initial and equilibrium concentration of Cr(VI),

2.4. Characterization of ACs

The surface area and pore volume were determined by using the Brunauer-Emmet-Teller (BET) . Boehm titration was performance to calculation the total acidic and basic groups.

2.5. Experimental Design

The modification process was optimization by using the response surface method (RSM). The experimental design and variances analysis were performed with Stat-Ease software (Design-Expert 8.0version). A three-levels three-factors Box-Behnken design (BBD) was employed and to optimized the modification conditions. Three variables are presented in Table 1.

Table 1. Factors and levels of the responding surface design

Factor	unit	symbo	Coded level		
			-1	0	+1
Modification temperature	℃	A	25	35	45
Solid-Fluid ratio	-	B	0.2	0.15	0.1
Modification time	h	C	2	3.5	5

Each variable was determined by preliminary experiments. An analysis of variance (ANOVA) table was generated to presented the effect and interaction of factors. The significances were judged by calculating the F-value at a probabilites (P-value) of 0.0001, 0.001 and 0.05.

3. RESULTS AND DISCUSSION

3.1. Characterization

It is generally verified that the porous structure of the ACs play an important role in their absorption process. Therefore, the specific surface area and pore volumes of the bamboo-based activated carbon before and after modification with H_2O_2 were characterized using N_2 adsorption-desorption analysis. The results are presented in Table 2.

Table 2. Specific surface areas of Bamboo-based activated carbon (before and after H_2O_2 modification)

ACs	Specific surface are (m2/g)	Pore volumes (cm3/g)
Before	867.2	0.513
After	1021.5	0.602

Based on the results, the surface area and pore volume of modified ACs are much higher than it before. Modification with H_2O_2 can efficiently increase the porous structure performance and enhance the adsorb ability of ACs.

Total acidic and basic groups were determined by using Boehm-titration and shows in Table 3. The content of total acidic groups was increased after modification process, H_2O_2 modification shows a effective ability to enhance the acidic groups for ACs and depress the basic groups.

Table 3. Boehm titration results of Bamboo-based activated carbon (before and after modification)

ACs	Total acidic groups (mmol/g)	Total basic groups (mmol/g)
Before	1.365	0.224
After	1.575	0.126

3.2. Analysis of Variances (ANOVA)Characterization

The experiments (17 runs) were designed by using BBD method. The Adsorption of Cr(VI) in % varies between 33% and 81%, respectively.

The analysis of variance (ANOVA) shows in Table 4, the model F-value of 33.72 and P-value is lower than 0.0001, which indicates that the model terms are highly significant, and there is only a 0.01%

chance that a model F-value this large could due to noise. The P-value of factors A and A^2 are lower than 0.0001 shows highly significant effect. Values of P great than 0.0001 and less than 0.005 indicate model terms are significant, in this case B and C^2 are significant model terms. Lack of fit is determining how well the model fits the data, in this study, the Lack of fit (P>0.05) is insignificant, the response were sufficiently represented.

Table 4 Variances analysis of the models

source	Sum of squares	df[a]	mean square	F-value	P
model	3663.50	9	407.06	33.72	< 0.0001
A	1984.50	1	1984.50	164.40	< 0.0001
B	98.00	1	98.00	8.12	0.0247
C	50.00	1	50.00	4.14	0.0831
AB	16.00	1	16.00	1.33	0.0287
AC	1.00	1	1.00	0.083	0.7818
BC	0.186	1	0.192	0.061	0.8765
A2	1216.84	1	1216.84	100.80	< 0.0001
B2	9.47	1	9.47	0.78	0.4051
C2	206.32	1	206.32	17.09	0.0044
residual	84.50	7	12.07		
Lack of fit	66.50	3	22.17	4.93	0.0788
Pure erro	18.00	4	4.50		
Cor total	3748.00	16			

[a] Degree of freedom

3.3. The Effect of Modification Conditions

In order to study the effect of modification conditions, the three dimensional (3D) surface responses plots were utilized to evaluate the relationship between the modification variables and responses. Figure 1(a-c) illustrate the fitted response surface plots of $A_{Cr(VI)}$ versus modification variables (modification temperature, solid-fluid ratio, modification time).

As show in Figure1a, the modification temperature demonstrates quadratic effects on the response. It is observed that the $A_{Cr(VI)}$ increase with temperature to obtain a maximum at 38℃ and thereafter decrease with a further increase in temperature for a given solid-fluid ratio. Modification temperature at about 38℃ is preferable for achieving a higher Cr(VI) adsorption efficiency. The increase in Cr(VI) adsorption efficiency as the modification temperature increase can be explained by the increase in specific surface area and pore volume. In contrast, the decrease in the Cr(VI) adsorption efficiency at higher temperature might attribute to the collapsing of pore structure.

Figure 1b presented the effects of modification temperature and modification time of $A_{Cr(VI)}$. As seen here, the Cr(VI) adsorption efficiency increase slightly with modification time, with modification time of about 3h, the $A_{Cr(VI)}$ reach maximum of above 81%. Followed the $A_{Cr(VI)}$ decreased with increase modification time.

The interaction between solid-fluid ratio and the modification time shows in Figure 1c. There is less interaction between solid-fluid ratio and modification time. Combine with the results of Boehm titration Table 3 and the analysis of variances Table 5, modification with H_2O_2 enhance only lightly the total acidic groups of ACs, the interaction is insignificant.

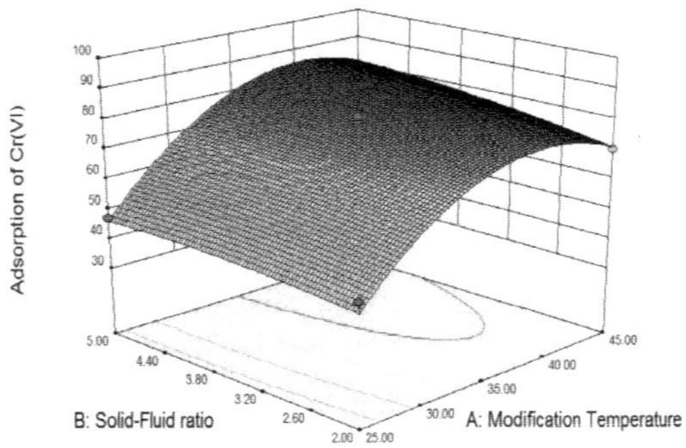

Figure 1-a. Response surface plots for A and B.

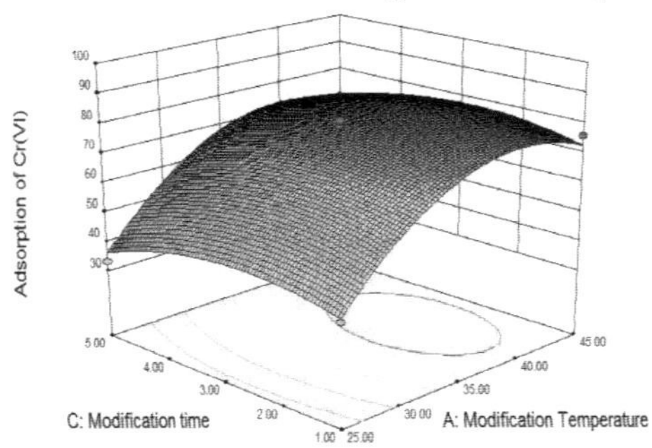

Figure 1-b. Response surface plots for A and C

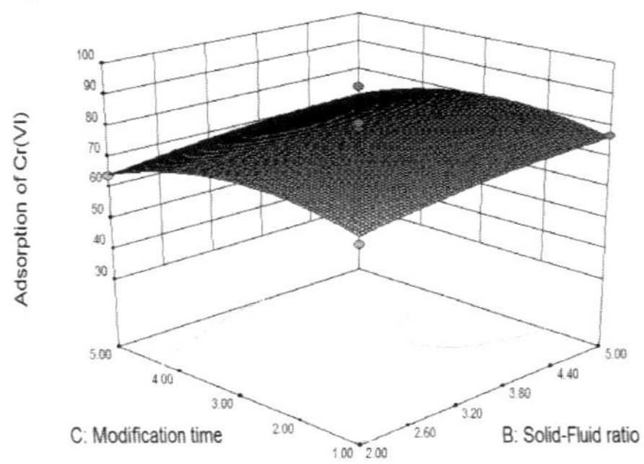

Figure 1-c. Response surface plots for B and C

3.4. Optimized Studies for Maximizing Cr(VI) Adsorption Efficiency

The Optimum modification conditions of the adsorption of Cr(VI) was carried out by using the Design-Expert software. According to the results above, the optimization equation shows in follow:

$$Y=79.00+15.75A+3.50B-2.50C+2.00AB+0.50AC+0.000BC-17.00A^2-1.50B^2-7.00C^2$$

Where A is the modification temperature ;B is the solid-fluid ratio and C is the modification time.

In addition, the modification temperature and time variables were selected as "minimize" to reduce the cost of the process and its energy demand. The optimal modification variables are presented in Table 5.

Table5. Optimization Results of the Modification variables

Modification variable	Optimum value
Modification temperature	38℃
Solid-Fluid ratio	1:10
Modification time	3

Three additional experiments were carried out under the optimal conditions, the average values of the triplicate experiments 84.61% are close to the predicted results 85.12, it indicate that the proposed model can well correlate the modification variables to the Cr(VI) adsorption efficiency.

4. Conclusion

In this work, modification conditions with H_2O_2 of bamboo-based activated carbon to improve the absorption of Cr(VI) were investigated and the modification variables have been optimized with response surface methodology. According to the BET and Boehm titration results, modification of ACs with H_2O_2 can increase the pore structure and total acidic groups. The influence of the modification variables including modification temperature, solid-fluid ratio and modification time on the adsorption efficiency of Cr(VI) was investigated using BBD combined with RSM. According to the obtained results, RSM model was an adequately application method for optimizing the variables for the bamboo-based activated carbon modification process. And the term for modification temperature shows the most significant effect to the absorption efficiency of Cr(VI). . With the optimized modification condition(modification temperature 38 ℃ , solid-fluid ratio 1:10 and modification time 3h) reach the absorption efficiency of Cr(VI) of 84.615, because of its high adsorption efficiency and low process cost, modification with H_2O_2 of bamboo-based activated carbon has a potential application in the treatment of Cr(VI) pollutant.

Acknowledgments

The authors wish to thank for the support of this study by research grants: Project of Key Laboratory for Ecological Industry and Green Technology of Fujian Province (WYKF2016-7) , Project of Education-Sciences Research of Fujian Province(JAT160352)and National Nature Science Foundation of China (51408518).

References

[1] XT Sun, LR Yang, TT Dong, et al. Removal of Cr(VI)from aqueous solution using amino-modified Fe3O4-SiO2-chitosan magnetic microspheres with high acid resistance and adsorption capacity. J.APPL POLYM.SCI.430(2016),1-11

[2] MK Rai, G Shahi, V Meena, R Meena, et al. Removal of hexavalent chromium Cr (VI) using activated carbon prepared from mango kernel activated with H3PO4. Resource Efficient Technologies 2(2016)863-870

[3] D Duranoglu, AW Trochimczuk, U Beker, Kinetics and thermodynamics of hexavalent chromium adsorption onto activated carbon derived from acrylonitrile-divinybenzene copolymer, Chem. Eng.J. 187(2012) 193-202

[4] MS Siboni, MR Samarghandi, S Azizian, et al. The removal of hexavalent chromium from aqueous solution using modified holly sawdust: Equilibrium and kinetics studies, Environ.Eng.Res 2011 16(2):55-60

[5] YJ Zhang, ZJ Xing, ZK Duan, et al. Effects of steam activation on the pore structure and surface chemistry of activated carbon derived from bamboo waste, Appl. Surface Sci. 315(2014):270-286

[6] YJ Zhang, JL Ou, ZK Duan, et al. Adsorption of Cr(VI)on bamboo bark-based activated carbon in the absence and presence of humic acid, Colloids and Surface A. 481(2015): 108-116
[7] HW Chen, YS Chuang, CF Hsu, et al. Adsorption characteristics of trace levels of bromate in drinking water by modified bamboo-based activated carbons, Journal of environmental science and health, 2017, 52:1055-1062
[8] L Liu, BY Shi, K Gai, et al. Adsorption removal of atrazine from water by chemically modified activated carbons, Chinese Journal of Environ. Eng. 6 (2012):2483-2488
[9] SS Gao, JB Zhao, JY Tian, et al. Influence of chemical modification on activated carbon for adsorption of sulfamethoxazole and ibuprofen, Chinese Journal of Environ. Eng. 9 (2015):4650-4654
[10] ZA Ghani, M Yusoff S, NQ Zaman, et al. Optimization of preparation conditions for activated carbon from banana pseudo-stem using response surface methodology on removal of color and COD from landfill leachate, Waste Management. 62(2017):177-187
[11] A Mohsen, S Shahrokh, G Ahad, Optimization of CO2 capture process from simulated flue gas by dry regenerable alkli metal carbonate based adsorbent using response surface methodology, Energy Fuels. 31(2017):5286-5296
[12] S Mohammad, B Salar, K Komeil, et al. Optimizing the preparation of meso- and microporous canola stalk derived hydrothermal carbon via response surface methodology for methylene blue removal, Energy Fuels 31(2017):12327-12338

Nonlinear Absorption Properties of Quantum Dots of CdTe / ZnS and CdTe / CdS

Chang Qing*, Guan Jing and Zhang Dongshuai

College of Electronics Engineering, Heilongjiang University, Harbin 150080, China

Email: 15545011280@163.com

Abstract. Herein, the nonlinear absorption properties of CdTe / ZnS and CdTe / CdS were investigated. The nonlinear absorption characteristics of these two kinds of quantum dots (QDs) were measured by Z-scan experimental apparatus under the wavelength of 532 nm, pulse widths of 4 ns and 21 ps, respectively, and a repetition frequency of 10 Hz. The results of experiment indicate the nonlinear absorption of CdTe / ZnS QDs and CdTe / CdS QDs show saturation absorption in nanosecond and picosecond laser pulses. By theoretical data fitting, we can draw a conclusion that the nonlinear absorption coefficients of CdTe / ZnS QDs and CdTe / CdS QDs are -2.57×10^{-10} m / W and -2.08×10^{-10} m / W,respectively;at a pulse width of 21 ps, the nonlinear absorption coefficients of CdTe / ZnS QDs and CdTe / CdS QDs are -1.38×10^{-11} m / W and -1.14×10^{-11} m / W,respectively. From the final data analysis, it can be seen that the nonlinear absorption properties of CdTe / ZnS QDs and CdTe / CdS QDs in the effect of nanosecond laser pulses are stronger than those under the condition of picosecond laser pulses. At the same time, it was found that under the different laser pulses of nanosecond and picosecond, CdTe / ZnS QDs showed stronger nonlinear absorption than that of CdTe / CdS QDs.

Keywords: Semiconductor core-shell QDs; Z-scan technique; Nonlinear absorption;

1. Introduction

Quantum dots(QDs) are important low dimensional semiconductor materials. Their dimensions are less than two times of the exciton Bohr radius of their corresponding semiconductor materials[1]. Quantum dots are generally spherical or spheroidal, with diameters often between 2 and 20 nanometers. Common quantum dots consist of IV, II-VI, IV-VI or III-V elements. Semiconductor quantum dots have a lot of superior fluorescence characteristics[2], such as: wide excitation spectrum with continuous distribution, narrow and symmetrical emission spectrum, adjustable color, high photochemical stability, as well as long fluorescence life. The unique nature of quantum dots stems from its own quantum effect. If the size of a particle reaches the nanometre, it will lead to the size effect[3], the quantum confinement effect, the macroscopic quantum tunneling effect and the surface effect, The different low-dimensional physical properties of the system and the microscopic system show many physicochemical properties different from that of the macroscopic materia[4]. Modern quantum dots technology dates back to the mid-1970s, initially to address the global energy crisis.

IOP Conf. Series: Earth and Environmental Science **146** (2018) 012074 doi:10.1088/1755-1315/146/1/012074

Through the study of photoelectrochemistry, the interface between the semiconductor and the liquid is generated, so that the energy is generated by utilizing the excellent surface area ratio of the nanocrystalline particles[5]. The initial semiconductor quantum dots study began with Louis Brus, Bell Labs, in the early 1980s and Alexander Efros and Victor.I. Klimov, the former Yoffe Institute of the Soviet Union[6]. Further research on subsequent quantum dots started in the early 1990s with CdSe quantum dots embedded in glass[7]. In 1993, the preparation of CdSe nanocrystals was a very successful case, efficient luminescence of CdSe quantum dots was first synthesized by Bawendi et al.Since its development in 1997, with the constantly enhancement of quantum dots manufacture technology, quantum dots have been gradually used in biomarkers and other aspects[8]. In 1998, Alivisatos and Nie respectively published the thesis about the quantum dots as biological probes respectively in Science. For the first time, they applied quantum dots to living cell systems for biomarker, which has caused the research craze of quantum dots[9-10].

In this paper, using the Z-scan technique, under the action of 532 nm laser wavelength, 4 ns pulse width and 21 ps pulse width, the nonlinear absorption properties of CdTe / ZnS quantum dots and CdTe / CdS quantum dots were investigated when the laser pulses widths were different. The results of test indicate the nonlinear absorption type of CdTe / ZnS and CdTe / CdS semiconductor quantum dots show saturation absorption under nanosecond and picosecond laser pulses. The nonlinear absorption properties of CdTe / ZnS and CdTe / CdS in the effect of nanosecond laser pulses are stronger than those in the effect of picosecond. The nonlinear absorption properties of CdTe / ZnS under nanosecond laser pulses and picosecond laser pulses are both stronger than that of CdTe / CdS.

2. Experiment

The samples used in this experiment were purchased.Under the nanosecond laser pulse width, the linear transmittances of CdTe / CdS QDs and CdTe / ZnS QDs were 59% and 69%, respectively. Under the picosecond laser pulse width, the linear transmittances of CdTe / CdS QDs and CdTe / ZnS QDs were 59% and 57%, respectively. The UV-visible absorption spectra of CdTe / CdS QDs and CdTe / ZnS QDs are shown in Figure 1.

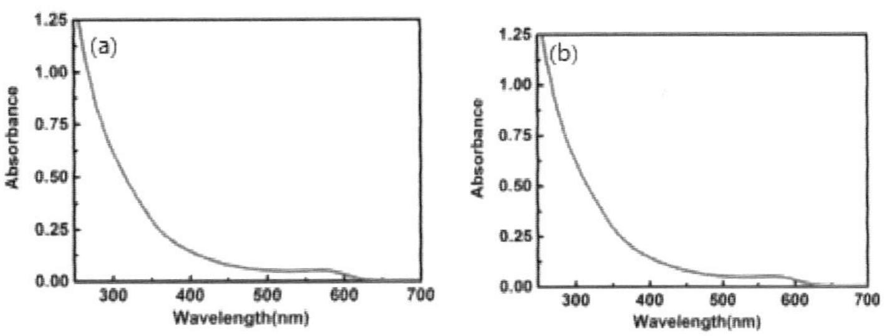

Figure 1. Absorption spectrums of **(a)** CdTe / CdS quantum dots and **(b)** CdTe / ZnS quantum dots

The laser source used in this experiment is Nd: YAG ns / ps laser. The output wavelength of the laser light source is 532 nm. The output light source has the pulses width of 4 ns and 21 ps, respectively. The repetition frequency is 10 Hz. Nanosecond output single pulse energy is 1uJ, picosecond output single pulse energy is 0.35uJ. In the experiment, we employed the Z-scan technique to measure the nonlinear properties of CdTe / CdS QDs and CdTe / ZnS QDs under nanosecond and picosecond laser pulses. The simple schematic diagram of the optical path of experimental equipment is shown in Figure 2, the attenuated laser pulses through the lens to focus on the sample, and change the position of the sample on the optical axis , The light passing through the sample is fall into two beams by the beam splitter, and the reflected light enters detector D_1 to measure the nonlinear absorption characteristics of the material, that is, an Open aperture Z-scan; the light penetrated through the beam splitter passes through an adjustable size of the hole, is received by the detector D_2 to detect the nonlinear refraction characteristics, namely Closed aperture Z-scan. Open aperture and Closed aperture Z-scan characteristic diagrams are shown in Figure 3-1, Figure 3-2, Figure 4-1 and Figure 4-2.

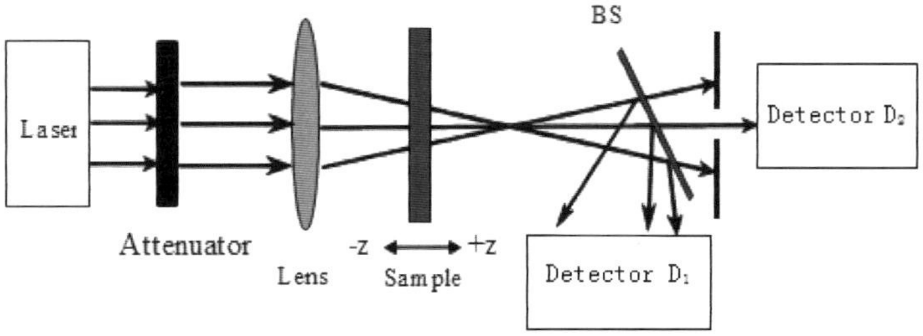

Figure 2. Structure of experimental device of Z-scan

3. Theoretical analysis

The open aperture Z-scan technique is one way could be applied on the study of the optical nonlinear absorption characteristic.In a general way, nonlinear materials absorption properties incorporating saturable absorption and reverse saturable absorption. The reverse saturable absorption is that nonlinear absorption coefficient increase with the beam intensity increases, and the curve shows as a valley, otherwise known as the saturable absorption. Saturated absorption as shown in Figure 3-a. Reverse saturation absorption as shown below in Figure 3-b.

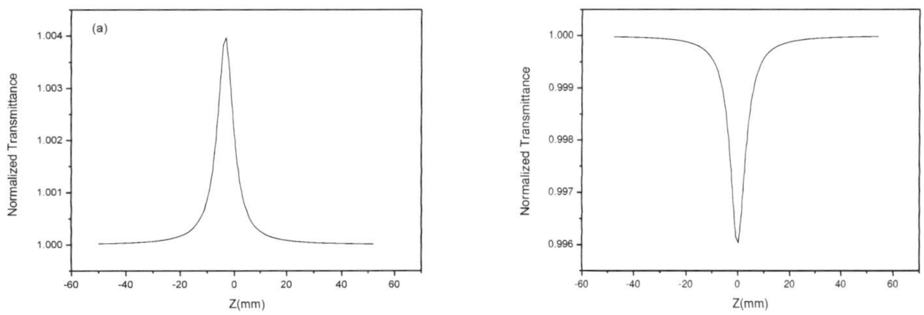

Figure 3 (a) Saturated absorption **(b)** Reverse saturation absorption

The closed aperture Z-scan technique,which is usually used to study the optical nonlinear refraction characteristic. Nonlinear refraction properties incorporating self-focusing and self-defocusing. If the consequence indicates just like valley-peak shape, the refraction properties are self-focusing with a positive nonlinear refractivity. If the picture indicates like peak-valley shape, the refraction properties is self-defocusing with a negative nonlinear refractivity. Self-focusing effect as shown in Figure 4-a. Self-defocusing effect as shown in Figure 4-b.

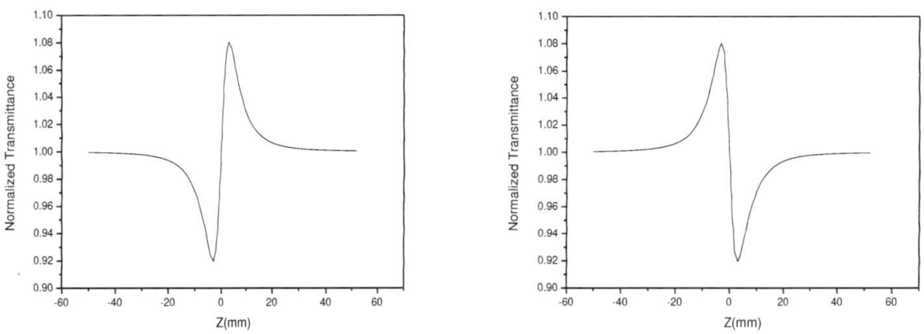

Figure 4 (a) Self-focusing effect **(b)** Self-defocusing effect

The theoretical analysis of Z-scan experimental results are presented by the following formula. The absorption of the material is equal to the linear absorption part plus the nonlinear absorption part, as follows:

$$\alpha(I)=\alpha_0+\beta I \tag{3-1}$$

Where α_0 is the linear absorption coefficient, I is the light intensity, and β is the nonlinear absorption coefficient.

The normalized energy transmission of open aperture Z-scan can be expressed as[11]:

$$T(z)=\frac{1}{\sqrt{\pi}q_0(z,0)}\int_{-\infty}^{\infty}\ln\left[1+q_0(z,0)e^{-r^2}\right]dr \tag{3-2}$$

$$q_0(z,0)=\beta I_0 L_{eff}/(1+z^2/z_0^2) \tag{3-3}$$

Where I_0 is the on-axis peak intensity at the focus (z=0), E is single pulse energy of the pulsed laser, ω_0 is beam radius size of the focus, τ is pulse width, β is the nonlinear absorption coefficient, Z_0 is the diffraction length of the beam, $L_{eff}=(1-T_0)\cdot l/(-\ln T_0)$ is the effective thickness of the sample, l is the thickness of sample cell, T_0 is linear transmittance of the solution.

When $|q_0|$ less than 1,the transmittance of medium can be expressed as:

$$T(z)=\sum_{m=0}^{\infty}\frac{[-q_0(z,0)]^m}{(m+1)^{3/2}} \tag{3-4}$$

When m is 0 and 1, Equations (3-4) is changed to:

$$T(Z)=1-\frac{q_0(z,0)}{2^{3/2}}=1-\frac{\beta I_0 L_{eff}}{\sqrt{8}(1+Z^2/Z_0^2)} \tag{3-5}$$

$$A=\beta I_0 L_{eff} \tag{3-6}$$

The normalized transmittance of closed aperture Z-scan curve is[12]:

$$T(z,\Delta\phi)=1+\frac{4x\Delta\phi(t)}{(x^2+1)(x^2+9)} \tag{3-7}$$

Where $\Delta\phi(t)=kI_0(t)n_2L_{eff}$ is the nonlinear phase shift at the focus, n_2 is the nonlinear refractive index coefficient, $x=Z/Z_0$.

When we consider only the third-order nonlinear of the material, the relationship between the real and the imaginary parts of the third-order nonlinear polarizability and the nonlinear refractive index and the nonlinear absorption coefficient is:

$$\mathrm{Re}\,\chi^{(3)}(esu)=\frac{cn_0^2}{720\pi^2}\cdot n_2 \tag{3-8}$$

$$\mathrm{Im}\,\chi^{(3)}(esu)=\frac{n_0^2\varepsilon_0 c\lambda}{2\pi}\cdot\beta\times\frac{9}{4\pi}\times10^8 \tag{3-9}$$

The third-order nonlinear polarizability expression is[13]:

$$\left|\chi^{(3)}\right|=[(\mathrm{Re}(\chi^{(3)}))^2+(\mathrm{Im}(\chi^{(3)}))^2]^{\frac{1}{2}} \tag{3-10}$$

4. Experimental results and analysis

Experiment results show that the summit wavelength of CdTe / CdS QDs is 565 nm and the summit wavelength of CdTe / ZnS QDs is 580 nm.The bandwidth of CdTe / CdS QDs is 2.19 eV,the bandwidth of CdTe / ZnS QDs is 2.14 eV, based on the UV-visible absorption spectra of these ones.

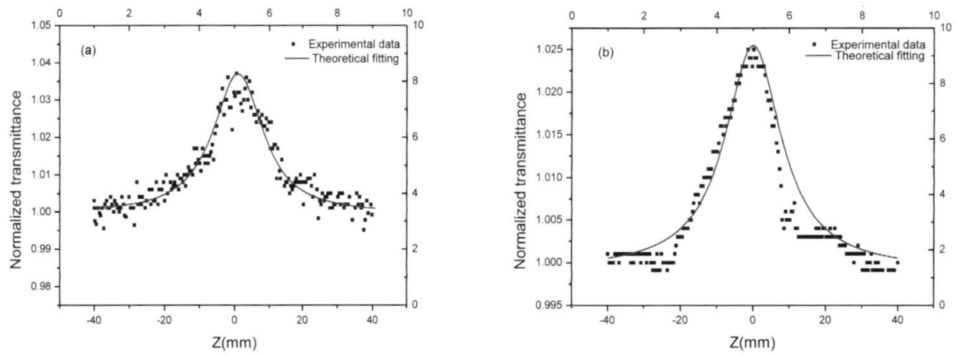

Figure 5. Nonlinear absorption normalized transmittance curves of **(a)** CdTe / CdS quantum dots and **(b)** CdTe / ZnS quantum dots for nanosecond laser pulses

It can be clearly seen from Figure 5 that the nonlinear absorption curves of CdTe / CdS QDs and CdTe / ZnS QDs all exhibit the peak shape under the action of a nanosecond laser pulses. These show that the nonlinear absorption showed saturated absorption, with the increase of light intensity, its transmittance will increase, the absorption will decrease.

The theoretical fitting of Z-scan data using the formulas 3-2 and 3-3 shows that under the condition of nanosecond laser pulse, the nonlinear absorption coefficient of the CdTe / CdS QDs is -2.08×10^{-10} m / W, the nonlinear absorption coefficient of CdTe / ZnS QDs is -2.57×10^{-10} m / W.

Obviously, the nonlinear absorption coefficient of CdTe / ZnS QDs is larger than CdTe / CdS QDs.

Similarly, Repeating the above experiment, we get the normalized curves of the open aperture Z-scan of CdTe / ZnS QDs and CdTe / CdS QDs under picosecond laser pulses, as shown in Figure 6.

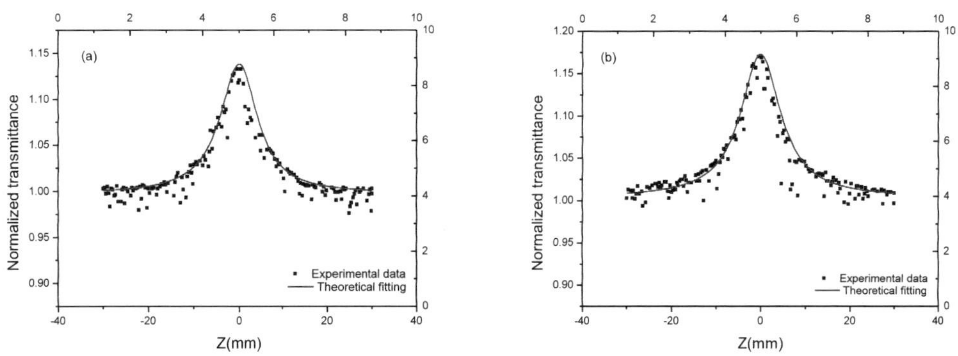

Figure 6. Nonlinear absorption normalized transmittance curves of **(a)** CdTe / CdS quantum dots and **(b)** CdTe / ZnS quantum dots for picosecond laser pulses

From the experimental data fitting, we can see that CdTe / ZnS QDs and CdTe / CdS QDs exhibit a peak shape,and both of samples show saturated absorption.The materials show the decrease of absorption when the increase of incident light intensity. By fitting the experimental data, we can conclude that the nonlinear absorption coefficient of CdTe / CdS QDs is 1.14×10^{-11}m / W and CdTe / ZnS QDs is -1.38×10^{-11}m / W. it can be seen that the nonlinear absorption of CdTe / ZnS QDs is stronger than that of CdTe / CdS QDs under a condition of picosecond laser pulses.

5.Conclusion

In this paper, CdTe / CdS and CdTe / ZnS semiconductor quantum dots as the experimental object,the nonlinear optical characteristics of Z-scan technology were compared and analyzed under nanosecond and picosecond laser pulses.The results of experimental indicate that the nonlinear absorption are saturated absorption under the action of 532 nm nanosecond and picosecond laser pulses. The nonlinear absorption coefficients of CdTe / CdS QDs and CdTe / ZnS QDs under the nanosecond laser pulse are -2.08×10^{-10} m / W and -2.57×10^{-10} m/W, respectively;the nonlinear absorption coefficients in the effect of picosecond laser pulse are -1.14×10^{-11} m/W and -1.38×10^{-11} m/W, respectively.The consequences indicate the nonlinear absorption characteristics of CdTe / ZnS QDs and CdTe / CdS QDs under the condition of nanosecond laser pulse are stronger than that in the effect of picosecond laser pulses. In the meantime, Under these two different laser pulses widths, the nonlinear absorption properties of CdTe / ZnS QDs are both stronger than that of CdTe / CdS QDs.

6.Reference

[1] Geng Rui, Chen Qingshan, and Lyu Yong 2017 *Journal of Applied Optics*. 38 732-739
[2] HU Lian and WU Hui-zhen 2015 *Chinese Journal of Luminescence*. 36 1106-1112
[3] Jin X, Shui M, Wang Y X, et al 2010 *Chin Phys B*.19 074203
[4] Chen O, Zhao J, Chauhan V P, et al 2013 *Nature Materials*
[5] Cheng Cheng and Cheng Xiaoyu 2017 *Beijing. Science Press*
[6] C.Zong, K.Ai, G.Zhang, H.Li and L.Lu 2011 *Anal.Chem.* 83 3126
[7] Murray C B, Norris D J and Bawendi M G 1999 *J. Am. Chem. Soc.* 115 8706-8715
[8] P. Reiss, M. Protie're and L. Li 2009 *Small*
[9] Y.H.Chan, Y.Jin, C.Wu and D.T.Chiu 2011 *Chem.Commun.* 47 2820
[10] A.P.Silva, H.Q.N.Gunaratne, T.Gunnlagsson, A.J.M.Huxley and C.P.McCoy 1997 *Chem.Rev.* 97 1515
[11] Y.T.Zhang 2013 *Huazhong University of Science and Technology*. 26-29
[12] T.Xia, D.J.Hagan, M.S.Bahae and E.W.VanStryland 1994 *Opt.Lett.* 19 317-319
[13] Sheik-Bahae M, Said A A, Wei T H, et al 2002 *IEEE Journal of Quantum Electronics*, 26 760-769

IOP Conf. Series: Earth and Environmental Science **146** (2018) 012075 doi:10.1088/1755-1315/146/1/012075

Discrimination of Ship's Bow in SAR Image

Yang Hu, Boli Xiong and Gangyao Kuang

National University of Defence Technology, College of Electronic Science, Changsha, Hunan, China

Email: 249219052@qq.com

Abstract. Discrimination of ship's bow is a great application for marine surveillance in synthetic aperture radar (SAR) image. The real ship contour is approximately elliptical and the bow's curvature is significantly larger than the stern's curvature. The increasing resolution of SAR image makes ship target appear to be more structured and shaped which makes discrimination of ship's bow possibly in SAR image. In this paper, we advance a method for discrimination of ship's bow which includes two important parts. The first part, we use the gradient vector flow (GVF) snake model with elliptical constraint (EC) to get a more realistic ship contour for curve fitting. The second part, we use the least squares quadratic polynomial curve fitting to fit the two ends of the contour along the ship's main shaft. The bigger absolute value of second order polynomial fitting coefficients, the more curved the rigid curve. The more curved end is identified as the ship's bow. Experiments performed on detected small-size ship slices which also called regions of interest (ROI).These ROI slices which come from RADARSAT-2 data and one ROI slice contains only one ship target. Experiment results illustrate the effective performance of proposed method on discrimination for the ship's bow in SAR image.

1. Introduction

SAR, independent with weather and illumination conditions, is widely used in maritime surveillance. Ship target supervisor using SAR data plays an important role in the field of maritime surveillance [1].In order to get the ship's moving trend, we need to discriminate which end is the ship's bow. Earlier works in the literature focus on ship's bow using wake signatures. However, it is found that ship backscatter is robust and independent of the sea state whereas wakes are often not visible at large incident angles and long wavelengths [2]. Because of these reasons along with the fact that stationary and slow moving ships do not create wakes, the focus of the later literature move on to the feature of the ship target itself. The increasing resolution of SAR images make the ship targets appear to be more structured and shaped.

The real ship target's contour is regular. So, we can use the least squares quadratic polynomial curve fitting to fit two ends of the contour along the ship's main shaft, compare two the second order polynomial fitting coefficients of the ends to get the bow.

Before fitting step, we need to get a good contour of the ship's edge. In high resolution SAR image, the ship target general close fitting for the elliptical area of high brightness, "ghosting" and "sidelobe" will only in the local change of ship target geometry, and less influence on the whole contour. So, we can use the GVF snake model with EC to extract the ship target precisely and the get its contour [3].

Section 2 of this paper demonstrates the GVF snake model with EC which can get the precise ship's contour. Section 3 details using the least squares quadratic polynomial curve fitting to fit the two ends of the contour along the ship's main shaft and the whole process of discrimination of ship's bow. Section 4, Experimental Results. Section 5, Conclusion.

2. GVF snake with EC

2.1. Snake model and GVF

Snake model is the classic parameter active contour model. Snake model gets the initial contour of a closed curve near the target's border, internal force and external force generated when we minimize the energy. The contour curve has some deformations under the joint action of its internal forces and image data's external forces. The final result of deformation is to make the contour curve be the most similar to the real target's contour. Snake contour usually be represented by a curve $M(s)=(x(s),y(s)),s\in[0,1]$.The snake contour deforms to minimize the following energy functions:

$$E(M)=E_{\text{int}}+E_{ext} \tag{1}$$

$$E_{\text{int}}=\frac{1}{2}\int_0^1\left[\alpha\left|\frac{\partial M(s)}{\partial s}\right|^2+\beta\left|\frac{\partial M(s)}{\partial^2 s}\right|^2\right]ds \tag{2}$$

Where E_{int} is the internal energy, in which α is one weighting parameter that control elasticity and β is the other one weighting parameter that control rigidity of the snake curve. With the work of E_{int}, the contour curve stretch and bend during deformation. E_{ext} is the external energy that guides the contour toward edges. *Xu and Prince* improved the term by introducing GVF field $V(x,y)=[u(x,y),v(x,y)]$ [4], which minimizes the energy function:

$$E_{ext}=\iint[\mu\left|\nabla V\right|^2+\left|\nabla f\right|^2(V-\nabla f)^2]dxdy \tag{3}$$

Where μ is a regularization parameter, and f is the edge map derived from the raw image. The final contour of The GVF snake model is closed, but the shape anomalies make the contour unlike ships.

2.2. Elliptical constraint

Since the GVF snake model just utilizes the image information, the snake contour always seems not like the real ship's contour. As mentioned earlier, the contour of the real ship has a near-elliptical outline. Therefore, the EC is incorporated into the energy function of the GVF snake to restrict the shape of the snake contour during deformation. Within $x-y$ domain, an ellipse can be expressed as the quadratic polynomial:

$$ax^2+bxy+cy^2+dx+ey+f=0,b^2-4ac<0 \tag{4}$$

Where a, b, c, d, e, f are ellipse parameters. The algebraic distance of a point (x_i,y_i) to the ellipse is:

$$D(x_i,y_i)=ax_i^2+bx_iy_i+cy_i^2+dx_i+ey_i+f \tag{5}$$

The next, we need to minimize the sum of squared algebra distances to fit the ellipse. *Fitzgibbon et al* had introduced the equality constraint $4ac-b^2=1$ to solve the problem with the inequation constraint $b^2-4ac<0$ which is difficult to solve in general [5].With the equality constraint, the unique solution of the parameters can be obtained according to the reference [5].

The ellipse can be fit by some points of the snake contour by *Fitzgibbon algorithm*. The distance from a point to the ellipse can be presented by $|D(M(s))|$, When the point is further away from the ellipse, the value of $|D(M(s))|$ is larger. On the contrary, the value is smaller. When the point is on the ellipse, $|D(M(s))|=0$. Therefore EC can be applied to the snake curve and $|D(M(s))|$ accords with the variation law of the snake energy function [6].Thus the energy function of the GVF snake under the EC can be presented as follow:

$$E(M(s)) = E_{\text{int}} + \gamma E_{ext} + \delta E_{EC}, E_{EC} = \int_0^1 |D(M(s))|^2 ds \quad (6)$$

Where E_{ext} is the energy term of GVF, E_{EC} is the energy term of EC, γ is weighting parameter that control the strength of GVF field and δ is weighting parameter that control the strength of EC field.

The whole deformation should get through some iterations. In each iteration, the GVF snake with EC is explored by two independent operations:(1)The snake curve is fixed and the ellipse parameters are estimated by ellipse fitting.(2)The ellipse is fixed and the snake curve deforms under the GVF with the EC. During deformations, each point of the snake moves under the influence of the external force field and stops when it undertakes balanced forces. When the change rate f the energy function is less than 1%, it is supposed to be converged and the new balance is created. By the way, the shape anomalies are filtered and the target contours are recovered.

3. Discrimination of ship's bow

In order to reduce the navigation resistance and increase the stability of navigation, the most of ship targets are designed as the bow pointed, stern rules shape characteristics of wide. Especially in warship target, the structure is more obvious. With the improvement of SAR image resolution, the target contour feature becomes more and more clear. The GVF snake model with EC can give us a good target contour which can get good fitting results with least square quadratic polynomial curve fitting.

3.1. The least squares quadratic polynomial curve fitting

Polynomial fitting method is the most commonly used method in curve fitting. The method can be used to represent the nonlinear relationships between the independent variable q and the dependent variable x.

$$q(x) = \sum_{i=0}^{N} a_i x^i = a_0 + a_1 x + a_2 x^2 + a_3 x^3 + \cdots a_N x^N,\ N \geq 2 \quad (7)$$

Where N is the order of polynomial fitting and a_0, a_1,$\cdots$ a_N are the parameters of polynomial fitting.

The least square method is a mathematical optimization technique [7]. It matches the best function of the data by minimizing the squared error. In the given function $y = f(x)$, the values are y_1, $y_2 \cdots$ y_N of N points x_1 , x_2 $\cdots$ x_N ,and each point is obtained by the formula $\sum_{i=1}^{N}(p(x_i) - y_i)^2 = \min$ to obtain the fitting curve.

Considering that it is necessary to compare the curvature of the two ends of the ship's main shaft, the least squares quadratic polynomial curve fitting can be obtained. When $N = 2$, the fitting curve can be expressed as this:

$$q(x) = \sum_{i=0}^{N} a_i x^i = a_0 + a_1 x + a_2 x^2,\ N = 2 \quad (8)$$

The absolute value of the quadratic coefficient of the bow's end is bigger than the stern's end. According to that, we can discriminate the ship's bow.

3.2. The bow discrimination process

The bow discrimination process is shown in figure 1.The first step, we need to input the ROI slice which come from RADARSAT-2 data usually contains only one target, the ROI slice has shown in figure 2(a). The second step, the canny operator is used to extract edge from the ROI slice which is a binary map, the edge map has shown in figure 2(b). One important step is the next step, let the edge to deform under the GVF with the EC and get a good ship's contour, the ship's contour under the GVF with the EC has shown in figure 2(c). From the contour, we could calculate some ship's geometric parameters, such as length, width, area. For easy to get the least squares quadratic polynomial curve fitting, we could rotate the whole ship's contour to make its main shaft be perpendicular to the horizontal axis, the rotated contour map has shown in figure 2(d). Next step, the top least squares quadratic polynomial curve fitting is made by top eighth contour points and the second order polynomial fitting coefficient is represented as p_top. Beside, the bottom least squares quadratic polynomial curve fitting is made by bottom eighth contour points and the second order polynomial fitting coefficient is represented as p_bottom. The top-and-bottom fitting curve map has shown in figure 2(e). The final step is to compare the absolute value of p_top and p_bottom. If the absolute value of p_top is bigger than the absolute value of p_bottom, the top part of contour is the bow of the ship. Otherwise, the bottom part of contour is the bow of the ship. The discriminated bow map has shown in figure 2(e).

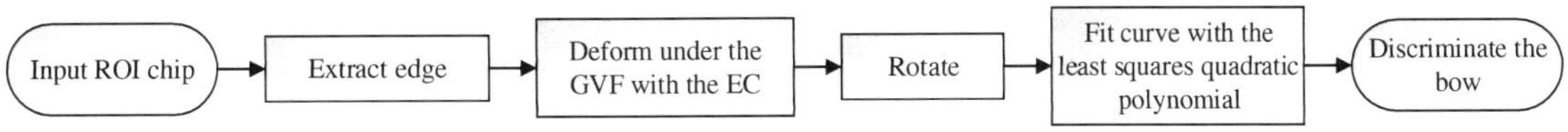

Figure 1. Flowchart of the discrimination of ship's bow

4. Experimental Results

According to the method described above, twenty sets of experiments had conducted, and each experiment result had shown that the bow was found correctly. And one of the results is shown in figure 2.

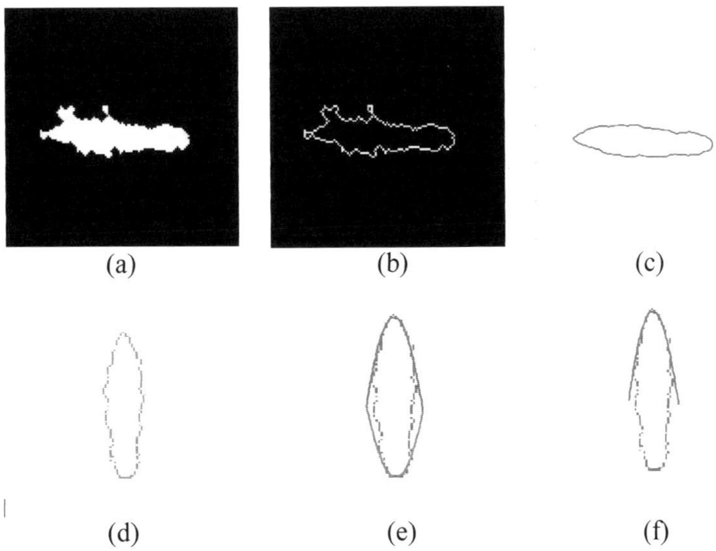

Figure 2. Each step's result map of the discrimination of ship's bow.(a) The ROI slice.(b)Edge map.(c)Ship's contour map under the GVF with the EC.(d)Rotated contour map.(e)Top-and-bottom fitting curve map.(f)The discriminated bow map.

Beside this set of map result, we also got fitting curve's coefficients. For second order polynomial fitting coefficients, p_top is 0.329 and p_bottom is -0.248. The absolute value of p_top is bigger than the absolute value of p_bottom, so the top end of the contour is the bow of the ship. The other nineteen sets of experiments also got the right discrimination results. The whole results of discrimination are shown in table 1.In order to make the compared result of the second order polynomial fitting coefficients more clear, we have highlighted p_top or p_bottom which has a bigger absolute value in table 1.Combining the discriminated bow and the true ship's bow, we get the discrimination result which is right or false. Due to the space impact of this paper, there are only shown four sets of experimental map results which corresponded to experiment_3, experiment_4, experiment_14 and experiment_18 of table 1.These map results have shown in figure 3.

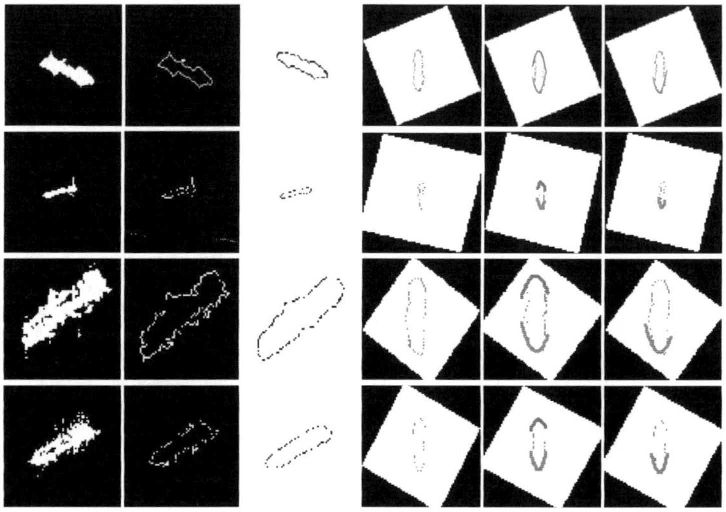

Figure 3. Four sets of experiment's map results which corresponded to experiment_3, experiment_4, experiment_14 and experiment_18 of table 1 row by row. And the five columns of experiment's map results correspond to the ROI slice map, edge map, ship's contour map under the GVF with the EC, rotated contour map, top-and-bottom fitting curve map and the discriminated bow map column by column.

Table 1. The whole results of discrimination.

Number of experiment	p_top	p_buttom	True ship's bow	Discrimination result
1	0.329	-0.248	top	right
2	0.252	-0.491	bottom	right
3	0.404	-0.405	bottom	right
4	0.388	-0.420	bottom	right
5	0.078	-0.146	bottom	right
6	0.135	-0.656	bottom	right
7	0.155	-0.048	top	right
8	0.255	-0.207	top	right
9	0.166	-0.045	top	right
10	0.291	-0.277	top	right
11	0.079	-0.030	top	right
12	0.045	-0.031	top	right
13	0.144	-0.255	bottom	right
14	0.097	-0.103	bottom	right
15	0.244	-0.047	top	right
16	0.142	-0.243	bottom	right
17	0.108	-0.159	bottom	right
18	0.087	-0.164	bottom	right
19	0.083	-0.150	bottom	right
20	0.450	-0.300	top	right

5. Conclusion

Thanks to the improvement of image resolution and the method of GVF snake model with EC, we can get a good contour which is similar to real ship's contour. And that makes it possible to use the least squares quadratic polynomial curve fitting to discriminate the ship's bow. If we get more information, we could use ship's bow to predict the future position of the ship. This method still has a serious weakness, which needs a complete and clear contour. But it is difficult to get a good contour when there are a lot of strong noises and interferences in the original SAR image. So, there are still some works need to be done for improving its robustness.

6. References

[1] Ouchi K. Recent Trend and Advance of Synthetic Aperture Radar with Selected Topics[J]. Remote Sensing, 2013, 5(2):716-807.
[2] Eldhuset K. An automatic ship and ship wake detection system for spaceborne SAR images in coastal regions[J]. Geoscience & Remote Sensing IEEE Transactions on, 1996, 34(4):1010-1019.
[3] Zhang X, Dong G, Xiong B, et al. Refined segmentation of ship target in SAR images based on GVF snake with elliptical constraint[J]. Remote Sensing Letters, 2017, 8(8):791-800.
[4] Xu C, Prince J L. Gradient Vector Flow: A New External Force for Snakes[C]// Conference on Computer Vision and Pattern Recognition. IEEE Computer Society, 1997:66.
[5] Fitzgibbon A, Pilu M, Fisher R B. Direct Least Square Fitting of Ellipses[J]. IEEE Transactions on Pattern Analysis & Machine Intelligence, 1998, 21(5):476 - 480.
[6] Delgado-Gonzalo R, Schmitter D, Uhlmann V, et al. Efficient Shape Priors for Spline-Based Snakes[J]. IEEE Transactions on Image Processing A Publication of the IEEE Signal Processing Society, 2015, 24(11):3915-26.
[7] Chen G, Ren Z L, Sun H Z. Curve Fitting in Least-Square Method and Its Realization with Matlab[J]. Ordnance Industry Automation, 2005.

IOP Conf. Series: Earth and Environmental Science **146** (2018) 012076 doi:10.1088/1755-1315/146/1/012076

Study on the Effect of Coagulant Ratio on Dehydration Performance of Acrylic Sludge

ZHANG Cheng, ZHOU Cai-ling, ZHOU You-chao, SHEN Yue, WANG Yi-wei, GE Shi-fu*

Department of Energy and Environment, Southeast University, Nanjing 210096 China

E-mail: ZHANG Cheng 15062203880@163.com;*GE Shi-fu ge1962@126.com

Abstract. In this research, the coagulants mass ratio affecting the moisture content of acrylic sludge cake was studied. The effects of APAM, CPAM and PAC mass ratio on the moisture content of pressure filter cake and the specific resistance of sludge were studied by orthogonal experiment. The results showed that the specific resistance of sludge presented a highly positive correlation with the moisture content of the filter cake after the treatment of acrylic sludge with CPAM, APAM and PAC. The multivariate regression model for moisture content of the filter cake with coagulants mass ratio was established by response surface methodology. And the model was accurate and reliable in the experimental range which could predict the relationship between the moisture content of the filter cake and the coagulants. In addition, the order of coagulants influencing the moisture content of the filter cake was: CPAM> PAC> APAM. The optimal mass ratio of coagulants to sludge content were 1.5 ‰ (APAM), 1 ‰ (CPAM) and 1.9 ‰ (PAC); the moisture content of the filter cake was 46.45%; and the specific resistance of sludge was $9.65 \times 10^{-7} s^2$ / cm. The study about the effects of coagulants addition order on dewatering performance indicated that the optimal order was PAC-CPAM-APAM. This work could provide a theoretical basis for the design and optimal control of acrylic sludge.

1. Introduction

Large amount of fine granular and high moisture content sludge was produced after the treatment of sewage, and the sludge had greater impact on the energy consumption in subsequent treatment [1]. The dehydration of sludge has become a vital part of the sludge treatment process [2].While the type and proportion of coagulants are the main factors affecting the sludge dehydration effect. Using two or more categories of coagulants to reduce sludge specific resistance and to improve the sludge dehydration performance is feasible.

One petrochemical plant in Anqing is now stocking nearly tons of acrylic sludge. The moisture content of the sludge is nearly 90%. And it is characterized as hazardous waste, which has great environmental risks and is in urgent need of safe disposal. In this paper, orthogonal experiment and response surface optimization test ware conducted to determine the optimal coagulants ratio in order to achieve the best performance of dehydration. And it ultimately provides a safe basis for reduction and disposal of acrylic fiber sludge.

2. Experimental materials and equipment

2.1. Experimental materials and methods

The acrylic sludge used in the experiment is from the petrochemical plant in Anqing. The initial moisture content and specific resistance of sludge are 90.62% and $8.49 \times 10^8 s^2/cm$. Figure.1 presents two kinds of sludge specific resistance measuring devices.

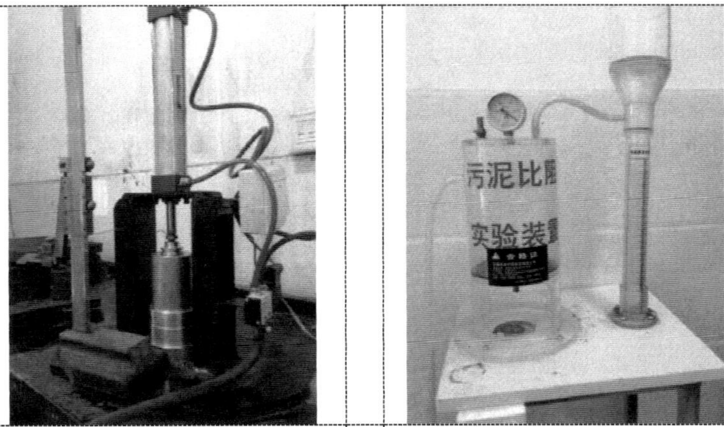

a. Cylinder pressure filtration device. **b.** Buchner funnel device.

Figure 1. Sludge specific resistance measuring devices.

2.1.1. Cylinder pressure filtration method. Mix the sludge and coagulants thoroughly, then put them into the equipment. Under 2Mpa air pressure, record the time interval t (s) and the piston decreasing displacement H (cm). According to the study[3], use the formula (1) to calculate the specific resistance.

$$\frac{t}{H} = \frac{\mu C_S}{2} \frac{R_C}{\Delta P} H + \frac{\mu R_m(\varepsilon)}{\Delta P} \quad (1)$$

2.1.2. Buchner funnel method . Set the vacuum pressure P, then record the time t (s)and the filtrate volume V (ml), According to the study [4], use the formula (2) to calculate the specific resistance.

$$r = \frac{2PA^2}{\mu} \cdot \frac{b}{C} \quad (2)$$

In the formula, μ-dynamic viscosity of the filtrate, $N \cdot s/m^2$; P-filter pressure, N/m^2, that's vacuum gauge numerical readings; A-filter area, m^2, the diameter of the circle area is Buchner funnel inner diameter; b-the slope of Specific resistance formula, calculated by the coordinate value; c-filter cake dry weight per unit volume of filtrate, kg/m^3, that's the filtered mud cake weight.

The Buchner funnel method and cylinder pressure filtration method are used to measure the specific resistance of sludge after conditioning; and measure pressure filter cake moisture content by the SFY-20AHalogen fast moisture analyzer.

2.2. Experimental materials and methods

Niu et al. [1]had studied the dosage of coagulants used in the remaining sludge from Beijing Beixiaohe Sewage Treatment Plant. The result shows: when the mass ratio of coagulants to dry sludge mass is 10%PAC + 0.5%PAM, the dehydration performance is optimal. Li et al. [5] had also studied the dosage of coagulants used in the sludge from Urban sewage treatment plant. The result shows: the optimal ratio of coagulant CPAM and dry sludge mass is $4.28 \sim 7.13 g \cdot kg^{-1}$.

This article has investigated three similar and commonly used coagulants in Nanjing. One chemical plant pretreated sewage into sludge by APAM, while used CPAM and PAC in improving dehydration performance. The mass ratio of APAM to sewage is 3‰; the mass ratio of CPAM to sludge is 1‰;.

and the PAC solution is mixed with the sludge at a volume of 1: 1. The moisture content of pressure-filtered sludge is about 80%.

Therefore, this paper studies three coagulants effect on the sludge dewatering performance. They are APAM solid, CPAM solid and PAC solution. For acrylic sludge, determine the mass ratio of three coagulants to sludge wet weight as 0.25×10^{-3},0.5×10^{-3},1×10^{-3},2×10^{-3},3×10^{-3}.

2.3. Response surface optimization experiment

In order to study the effect of three kinds of coagulants on the moisture content of filter press sludge cake and the specific resistance of the sludge, select filter cake moisture content as the response value. And use statistical analysis software Design-Expert 8.0.6.1 to establish three factors and three levels Box-Behnen model [6]. Determine optimal conditions of the filter press cake moisture content by response surface analysis. Experimental factors and levels are -1, 0, +1, the corresponding addition amounts are 1×10^{-3}, 2×10^{-3} and 3×10^{-3}.

2.4. Different coagulants addition sequence experiment

To study the effects of three coagulants addition sequences on sludge dehydration, the moisture content of filter press cake and sludge specific resistance were measured under different coagulants addition sequence. According to the results, get the optimal coagulant addition sequence make the moisture content and sludge specific resistance the lowest.

3. Results and Analysis

3.1. Orthogonal experiment analysis

Figure.2 presents the orthogonal experimental results of the moisture content of pressure filtration cake and the sludge specific resistance.

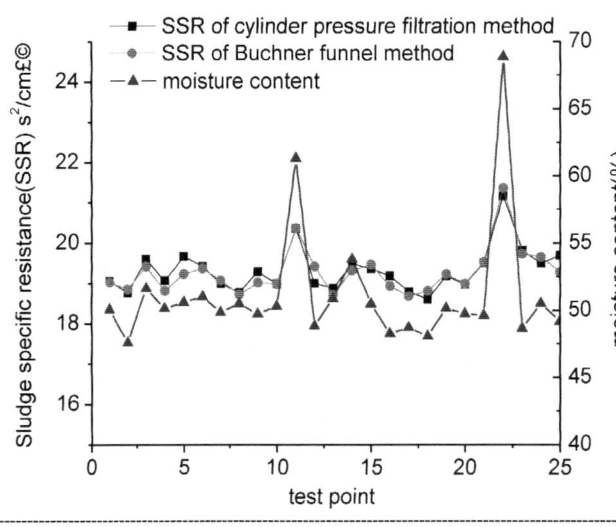

Figure 2 results of the moisture content and the sludge specific resistance

Figure.2 demonstrates that the specific resistance of sludge measured by pressure filtration and Buchner funnel method are basically the same and the average relative error is 13.4%. The sludge pressure filter cake moisture content and sludge specific resistance are positively correlated that the specific resistance increases with the increase of the moisture content of filter cake. Buchner or cylinder method to measure specific resistance can be both used in the acrylic sludge treatment project.

The mass ratio of three coagulants to sludge wet weight making press filter cake moisture content the lowest is 2‰, 1‰ and 2‰.

3.2. Response surface optimization experiment analysis

3.2.1. Model establishment and significance test. Response surface experimental design and test results are shown in Table 1. The software Design-Expert 8.0.6.1 was used for the second response surface regression analysis, and a multivariate quadratic regression model was obtained on the moisture content of filter press cake (Y):

$$Y=47.94+0.19A+2.42B-1.46C-1.55AB+0.28AC-2.17BC+1.81A2+1.77B2+1.67C2 \quad (3)$$

Table 1. A slightly more complex table with a narrow caption

Test No.	A	B	C	Y(MC)
1	0	0	0	47.95
2	-1	0	1	49.22
3	0	0	0	47.95
4	1	1	0	50.76
5	0	-1	1	48.13
6	1	0	-1	53.06
7	0	1	1	50.3
8	0	0	0	47.95
9	-1	0	-1	51.3
10	-1	1	0	55.44
11	-1	-1	0	49.18
12	0	1	-1	58.98
13	1	-1	0	50.7
14	0	-1	-1	48.13
15	1	0	1	52.1
16	0	0	0	49.5
17	0	0	0	46.35

The result of variance analysis about formula (3) was shown in Table 2. The p value on the Table 2 is the significance level of the model[7]. The model $p=0.0255<0.05$ shows that the multivariate model is significant. Misjudgment item value is 0.0867 and greater than 0.05,which shows that the experimental results fit well with the mathematical model and means the results can be conjectured by formula (3). The interrelated coefficient of the regression equation is $R^2=0.8655$, which means that 86.55% of the data can be explained by this equation and the experimental deviation is small. The coefficient of variation (CV) reflects the confidence of the model, the lower the CV value , the higher the model confidence [8]. The CV value of this model is 3.21%, representing higher confidence and changes in response values can be analyzed by this model. While the value p of the CPAM(B) is Less than 0.01, indicating that CAPAM(B) has extremely significant impact on the response value. The value p of the BC is between 0.01 and 0.05, meaning that BC have significant impact on the response value. And the value p of the A, C, AB, AC, A^2, B^2 and C^2 are greater than 0.05, indicating that their impact on the response value is not significant. According to the value F, the various influences of factors on the pressure filter cake moisture content are in this order: CPAM(B)>PAC(C)>APAM(A).

Table 2. Analysis of variance results.

Sources of	ANOVA	DOF	MS	F	p	Significance

variance						
model	136.27	9	15.14	4.78	0.0255	*
A	0.27	1	0.27	0.09	0.7772	
B	46.75	1	46.75	14.77	0.0063	**
C	17.17	1	17.17	5.43	0.0527	
AB	9.61	1	9.61	3.04	0.1249	
AC	0.31	1	0.31	0.10	0.7621	
BC	18.84	1	18.84	5.95	0.0448	*
A2	13.76	1	13.76	4.35	0.0755	
B2	13.23	1	13.23	4.18	0.0802	
C2	11.78	1	11.78	3.72	0.0950	
Residual	22.15	7	3.16			
Misjudgement	17.19	3	5.73	4.62	0.0867	
Pure error	4.96	4	1.24			
sum	158.43	16				

**, P<0.01, extremely significant;
*, 0.01<P<0.05, significant

3.2.2. Response surface interaction and optimization. Figure.3~5 show the response surface and contour line figures of the coagulants influences on the press filter cake moisture content. The effect of any two factors and their interactions can be analyzed and evaluated through these figures and the range of the optimal factor level can be also determined [7-9]. The more the contour line shape tends to the ellipse, the stronger the interaction is. The more rounded the contour line is, the weaker the interaction is [7]. According to Fig.3, the interaction of APAM(A) and CPAM(B) is not significant(p>0.05), since the contour line tends to the round; According to Fig.4, the interaction of APAM(A) and PAC(C) is not significant(p>0.05), since the contour line tends to the round; According to Fig.5, the interaction of CPAM(B) and PAC(C) is relatively significant(p<0.05), since the contour line tends to the ellipse.

Figure.3~5 Show that the optimal point of interaction between different factors falls within the experimental rang. And the optimal mass ratio of APAM,CPAM,PAC to the sludge to get the minimum moisture content of sludge is: 1.5‰, 1‰ and 1.9‰. The predictive value of moisture content is 46.83%. Under this optimized conditions, three parallel verification tests are carried out, and the moisture content of sludge is 46.455%. The verification results are in line with the theoretical predictions and the ratio is more accurate than orthogonal experiments.

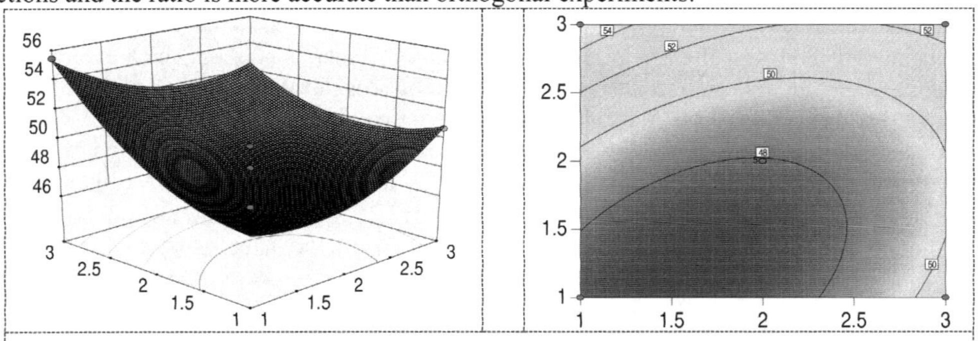

Figure 3 the effect of CPAM, APAM on moisture content(MC).

IOP Conf. Series: Earth and Environmental Science **146** (2018) 012076 doi:10.1088/1755-1315/146/1/012076

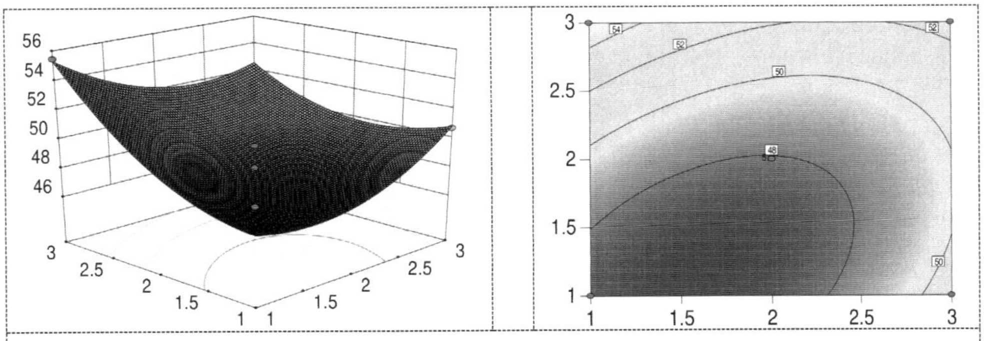

Figure 4 the effect of PAC, APAM on moisture content(MC).

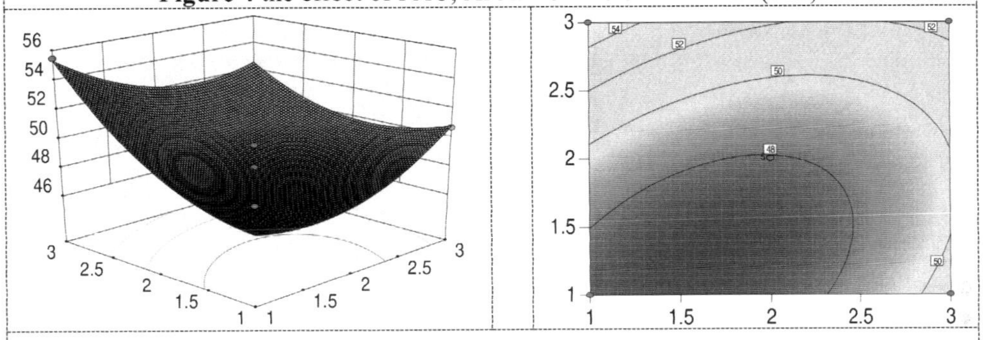

Figure 5 the effect of PAC, CPAM on moisture content(MC).

3.3. Effect of coagulants addition sequence on dehydration performance

According to the results of response surface and orthogonal experiment, the optimal mass ratio of three kinds of coagulants is applied. The mass ratio of APAM, CPAM and PAC to sludge mass is 1.5‰, 1‰ and 1.9‰. And under this ratio measure the moisture content of sludge cake in different coagulants addition sequence and the results are shown in Table 3.

Table 3. Effect of coagulants addition sequence on the moisture content and the specific resistance of sludge.

Order	addition sequence	moisture content %	specific resistance (×107) s^2/cm
1	APAM-CPAM-PAC	46.77	9.61
2	APAM-PAC-CPAM	46.82	9.76
3	CPAM-APAM-PAC	46.13	8.55
4	CPAM-PAC-APAM	46.28	8.63
5	PAC-CPAM-APAM	45.11	7.77
6	PAC-APAM-CPAM	45.79	7.84

The moisture content and specific resistance of sludge cake are the smallest when the addition order is PAC-CPAM-APAM. While the moisture content and specific resistance are 5.11% and $7.73\times10^7 s^2$/cm. The values are smaller when adding PAC first than CPAM, while adding APAM first is the biggest. Inorganic coagulant PAC forms various forms of water and complex in water and can adsorb negatively charged colloidal particles, destabilizing the colloid by compressing double layer, electrical neutralization and bridging between hydroxyl groups. Then add polymer coagulant PAM, which has better adsorption bridging function. The long-chain molecular structure can adsorb, net catch and sweeping the sludge particles, to promote the formation of larger flocculation and its separation from the water, eventually achieving solid-liquid separation [10]. However, adding PAM firstly, the macromolecular structure of it makes sludge flocculation structure loose. Then adding PAC after that just neutralizes part of the electric charge, and weakens the function of compressed double layer, which results in worsened effect of sludge dehydration [11].

4. Conclusion

(1)After coagulants conditioning , the pressure filter cake moisture content of acrylic sludge is positive correlated with sludge specific resistance. The specific resistance increases with the increase of the moisture content of filter cake. The significant order of the effect on the pressure filter cake moisture content is: CPAM>PAC>APAM.

(2)The optimal mass ratio of three coagulants to sludge wet weight making press filter cake moisture content the lowest is: (APAM) 1.5‰, (CPAM)1‰ and (PAC) 1.9‰. Under this condition, the press filter cake moisture content is 46.45% and the sludge specific resistance is 9.65×10^{-7} s^2/cm.

(3)The optimal coagulants addition sequence is PAC-CPAM-APAM.

5. References

[1] Niu M Q, Zhang W J, Wang D S. Study on effect of chemical conditioning using different coagulants on sludge dewatering performance[J]. Acta Scientiae Circumstantiae, 2012, 32(9): 2126-2133.

[2] Zheng L Q, Cui H S, Huang R J, Wang X F . Synthesis of novel modified polyacrylamide and its effect on dewatering performance of sludge [J] . Chinese Journal of Environmental Engineering,2011,5 (5) : 1166-1170.

[3] Zhao P T, Ge S F, Huang Y, Chen Z Q; Li X. Measurement of specific resistance to filtration of sludge by pressure filtration[J].Journal of Southeast University (Natural Science Edition),2011,41(1): 155-159.

[4] Dong Y, Hang X G, Niu Y, Jiang X C, Zhuang Z K, Zhu S K, Lu Y S. Experiment on improving sludge dewatering ability using CPAM and CPAM coupled with inorganic coagulant [J]. EnvironmentalProtectionandTechnology,2015,(05): 24-28.

[5] Li T, Wang Y L, Feng J, Xu Meng. Relationship Between Physical and Chemical Properties of Activated Sludge and Flocculation Conditioning Dosing [J]. Acta Scientiae Circumstantiae,2012,(03): 889-895.

[6] Guo R, Li J J, Guo L, Cheng G X. Optimization of the ultrasonic extraction process of glycyrrhizic acid by using response surface methodology [J]. Jour. of Northwest Sci2Tech Univ. of Agri. and For. (N at. Sci. Ed.),2006,34(9): 187-192.

[7] GHORBEL B O,HAJJI S,YOUNES I,et al. Optimization of chitin extraction from shrimp waste with Bacillus pumilus A1 using response surface methodology[J]. International Journal of Biological Macromolecules,2013,61(10): 243-250.

[8] WANG X,WU Q,WU Y,et al. Response surface optimized ultrasonic-assisted extraction of flavonoids from Sparganii rhizoma and evaluation of their in vitro antioxidant activities[J]. Molecules,2012,17(6): 6769-6783.

[9] Zhong K, Wang Q. Optimization of hot water extraction of polysaccharide from dried longan pulp using Response Surface Methodology [J]. Chemical Industry and Engineering Process,2010,29(4): 739-744.

[10] Tu Y. Influence of coagulant on sludge dewatering performance in sludge conditioning [D]. Nanchang University,2008.

[11] Yang X Y, Zhou L. Application and research progress of coagulation process for treatment of dyeing wastewater [J]. Environmental Protection of Chemical Industry,2016,36(01): 1-4. [2017-10-09].

IOP Conf. Series: Earth and Environmental Science **146** (2018) 012077 doi:10.1088/1755-1315/146/1/012077

Study of Microwave Synergistic Enzyme Method for Extraction from Laminaria Japonica by Response Surface Methodology

Su Xu[1], WenChao Liao[1], Wei Chen, Bo Kang, Jingpin Chen, Yitao Lin[1]

[1]School of Environmental Sciences and Engineering, Xiamen University of Technology, Xiamen 361005, China

Corresponding author e-mail: schnell19@qq.com

Abstract. Fucoxanthin is a carotenoid which occurs in marine brown algae and exerts multiple beneficial effects on human health, including the anti-diabetic, anti-obesity, anti-cancer, and antioxidant. The aim of this study was to maximise the extraction yield of fucoxanthin from laminaria japonica. By using microwave synergistic enzyme method, the extraction condition, namely the enzyme amount, complex enzyme ratio and microwave temperature were investigated and optimized with Response surface methodology (RSM) combined with Box-Behnken design (BBD). Microwave temperature was found to have the most significant effect on fucoxanthin extraction yield. With the optimized process parameter of extraction in this work, extraction yield of 0.443mg $_{fucoxanthin}$/g $_{dry\ algae}$ was obtained by 0.14g enzyme, complex enzyme ratio (pectinase : cellulase) of 1.12 and microwave temperature at 42.6°C. Therefore, with higher efficiency and low energy consumption system this method can be realized simpler.

1. Introduction

Fucoxanthin(FX) known as a carotenoid derivative found in brown seaweeds. Recently, several researches have demonstrated that FX shows anti-obesity, anti-cancer effects and can be used as free radical scavenger [1]-[3]. Fucoxanthin is estimated to account for more than 10% of the total production of carotenoids in nature, and is responsible for the brown to yellow colour of brown algae [4].

Laminaria japonica, is one species of brown seaweeds and was widely cultured in the northwest Pacific. In many countries, especial in Asian such as China, Korea and Japan, cause the high contents of minerals, vitamins and non-caloric dietary fiber, Laminaria japonica has been consumed as a popular marine vegetable and used as a health food [5].

Suitable operation condition during extraction stage was required due to the thermal sensitivity of fucoxanthin [6]. Extraction of fucoxanthin from laminaria japonica using conventional solvents (methanol, acetone, ethanol) [7] and supercritical CO_2 [8], from Eisenia bicyclis using pressurized liquid extraction [9] and from Hincksia mitchellae using supercritical anti-solvent crystallization[10] have been reported. However, the high operation cost and energy consumption inhibit the industrial production efficiency and popularize.

Microwave can effective break cell wall and has been widely used in environmental, food or pharmacy fields. Enzyme process has been reported to improve the extractions efficient of fucoxanthin from laminaria japonica. Response surface methodology(RSM) is useful for design of the experiments, for estimation the effects and interactions of process variables, and for predict the optimum operational conditions. No reports have been found to extraction fucoxanthin from laminaria japonica by using microwave synergistic enzyme method and optimize the extraction condition with RSM. Therefore, in the present study, RSM based on Box Behnken (BBD) has been used to optimize the microwave synergistic enzyme extractions condition, and determination the effects and interaction of the extraction variables (enzyme amount, complex enzyme ratio, microwave temperature) for extraction yield of fucoxanthin from dry laminaria japonica powder.

2. Materials and methods

2.1. Sample preparation and materials

Fresh Laminaria japonica was used as raw materials in this study and provided from LINHAI Seaweed, Co. Ltd (Xiamen China). Sample was cleaned with deionized water to removed sand, organic debris and epiphytes, the fresh algae was freeze-dried and ground to pieces of about 1mm then stored at -4°C in darkness until use in order to prevent degradation. Ethanol, pectinase (1000U/g) and cellulase (30000U/g) were purchased from Aladdin biochemical Co. Ltd (Shanghai China). Fucoxanthin standard was from Sigma-Aldrich (Rep. of Ireland).

2.2 Microwave synergistic enzymatic Extraction

Freeze-dried algae (1g) was mixed with pectinase and cellulase, then extracted by using ethanol 95%. A Microwave extractor (APEX YIYAO Shanghai, China) was assisted with enzyme extraction of fucoxanthin. After extraction process, the solution was filtered through a filter (0.45μm) and the solvent was removed by using a vacuum evaporator. Microwave synergistic enzyme extraction was performed in darkness due to the photo sensitivity of fucoxanthin. The residues were analysis with a UV spectrometer under 449nm. All experiments were conduced in triplicate.

2.3 Extraction yield

The extraction yield was expressed by Eq. (1).

$$X = AV/(A_{1\mathrm{cm}}^{1\%} \times 100) \tag{1}$$

where A is the absorption value of sample by 449nm, Abs; $A_{1\mathrm{cm}}^{1\%}$ is the theoretical absorption value of sample with concentration of 1g/L in 1cm optical length, here is 1600, Abs; V is the sample volume, ml; X is the extraction yield of fucoxanthin, mg fucoxanthin/g dry laminaria japonica.

2.4 Experimental design

To optimized the extraction process, a response surface method (RSM) was used and with Stat-Ease software (Design-Expert 8.0version) to performed the experimental design and the variances analysis. A three-levels three-factors Box-Behnken design (BBD) was employed and to optimize the extraction conditions. Three variables are presented in Table 1.

Table 1. Factors and levels of the responding surface design

Factor	unit	symbol	Coded level	
			-1	+1
Enzyme amount	g	A	0.07	0.15
Complex enzyme ratio (pectinase : cellulase)	-	B	0.5	2
Microwave temperature	°C	C	40	50

Each variable and rang was determined by preliminary experiments. An analysis of variance (ANOVA) table was generated to presented the effect and interaction of factors. The significances were judged by calculating the F-value at a probabilites (P-value) of 0.0001, 0.001 and 0.05.

3. Results and Discussion

The UV-spectrometer results from fucoxanthin standard and extracted powder indicated that the extracted powder was fucoxanthin.

3.1 Analysis of Variance (ANOVA)

The designed experiments (17 runs) by using BBD method are shown in Table 2. The extraction yield varies between 0.27 mg/g and 0.45 mg/g, respectively.

Table 2. The experimental results of Box-Benhnken Design

STD	A	B	C	Extraction yield /(mg/g)
1	0.11	1.25	45	0.43
2	0.15	1.25	50	0.31
3	0.15	2.00	45	0.35
4	0.11	1.25	45	0.41
5	0.15	1.25	40	0.43
6	0.11	0.50	50	0.28
7	0.07	0.50	45	0.32
8	0.11	1.25	45	0.43
9	0.15	0.50	45	0.38
10	0.11	1.25	40	0.37
11	0.11	2.00	45	0.42
12	0.11	2.00	50	0.29
13	0.07	1.25	45	0.31
14	0.11	1.25	50	0.45
15	0.07	1.25	50	0.31
16	0.11	2.00	40	0.32
17	0.07	1.25	40	0.33

The analysis of variance (ANOVA) shows in Table 3, the model F-value of 43.03 implies the model is significant. There is only a 0.01% chance that a "Model F-Value" this large could due to noise. P-value is lower than 0.0001, which indicates that the model terms are highly significant. The P-value of factors C, B^2 and C^2 are lower than 0.0001 shows highly significant effect. Values of P great than 0.0001 and less than 0.005 indicate model terms are significant, in this case A, AC and A^2 are significant model terms. Lack of fit is determining how well the model fits the data, in this study, the Lack of fit (P>0.05) is insignificant, the response was sufficiently represented.

Table 3. Anova of regression model

source	Sum of squares	d_f^a	mean square	F	P
Model	0.051	9	5.716E-003	43.03	< 0.0001
A	5.000E-003	1	5.000E-003	37.63	0.0005
B	8.000E-004	1	8.000E-004	6.02	0.0439
C	8.450E-003	1	8.450E-003	63.60	<0.0001
AB	1.000E-004	1	1.000E-004	0.75	0.4144
AC	2.500E-003		2.500E-003	18.82	0.0034
BC	9.000E-004	1	9.000E-004	6.77	0.0353
A^2	3.541E-003	1	3.541E-003	26.65	0.0013
B^2	0.015	1	0.015	110.32	<0.0001
C^2	0.012	1	0.012	92.41	<0.0001
Residual	9.300E-004	7	1.329E-004		
Lack of fit	5.000E-005	3	1.667E-005	0.076	0.9698
Pure erro	8.800E-004	4	2.200E-004		
Cor total	0.052	16			

The value of R^2 (0.9822) was close to that of R_a^2 (0.9894). This suggests that the predicated values (response) match well with the observed value, supporting that the regression model is adequate to explain most of the variability for extraction yield of fucoxanthin from Laminaria japonica in the given experimental domain.

3.2 Effect of Extraction Conditions

The three dimensional (3D) surface responses plots were utilized to study the effect and relationship of microwave synergistic enzymatic extraction condition for fucoxanthin from Laminaria japonica. Figure 1(a-c) shows the fitted response surface plots of extraction yield versus extraction variables (enzyme amount, complex enzyme ratio and microwave temperature)

a)

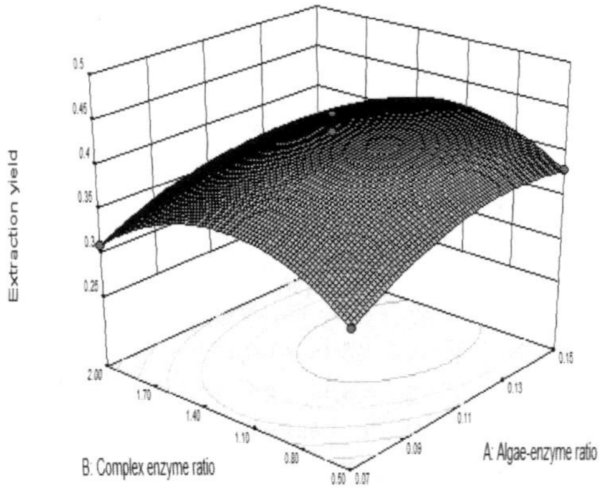

b)

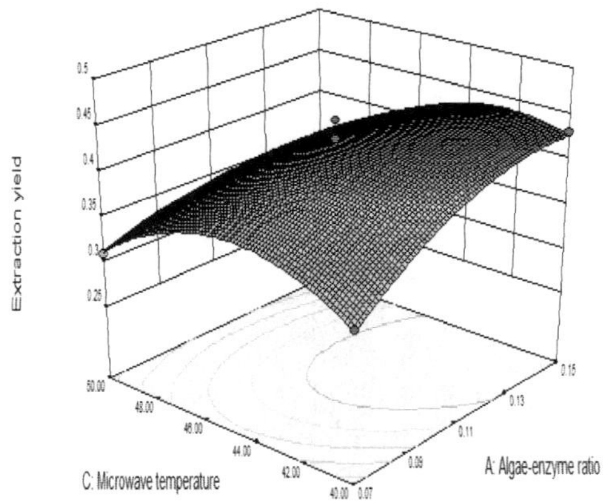

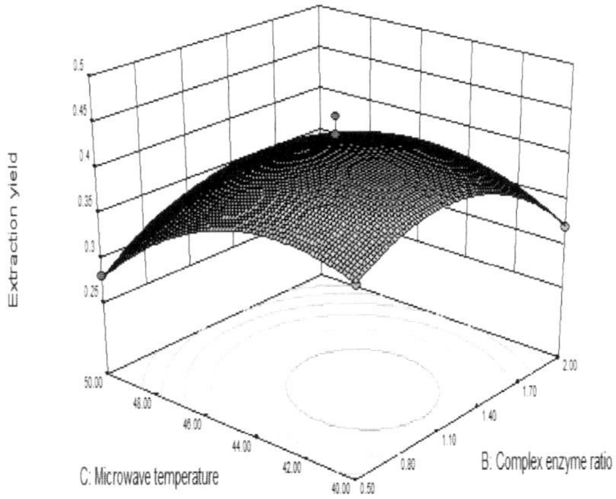

c)

Figure.1 Response surface plots

It is observed that the extraction yield is directly correlated with these variables, As shown in Figure 1a and 1b, the plot is not perfectly convex. This indicates that there is less interaction between the two variables. Moreover, the extraction yield increased slightly with the enzyme amount for a given complex enzyme ratio. By enzyme amount at 0.15 g the extraction yield reached a highest value. The increase in extraction yield can be explained by the increase in the amount of enzyme and the effective broken of algae cell wall, thus bringing about an increase in the extraction yield. It is also observed that the extraction yield increase with complex enzyme ratio to obtain a maximum at 1.2 and thereafter decrease with a further increase in complex enzyme ratio for a given enzyme amount. Therefore, a suitable complex enzyme ratio (pectinase: cellulase) about 1.4 is preferable for achieving a higher extraction yield. This trend is consistent to explained as follows: fucoxanthin in laminaria japonica surround mainly by cellulose, semi-cellulose and pectin substance, in this study used enzyme were pectinase (1000U/g) and cellulase (30000U/g), cellulase has higher activity unit than pectinase, with suitable ratio of pectinase and cellulase achieved extraction yield the maximum. Figure 1c presented the effects of complex enzyme ratio and microwave temperature on the extraction yield. As indicated, the extraction yield achieved maximum by microwave temperature at 45°C for a given complex enzyme ratio, thereafter decrease with increase temperature. This can be attributed to the fact that the fucoxanthin and enzyme were thermal sensitivity. Microwave temperature had an antagonistic effect of extraction yield due to the thermal sensitivity. By microwave temperature at 45°C the enzymes had a suitable environment, in contrast, the enzymes have less activity by higher temperature.

3.3 Optimized Studies for Maximizing Extraction Yield

The Optimum modification conditions of the extraction yield was carried out by using the Design-Expert software. According to the results above, the optimization equation shows in follow:

$$Y=0.43+0.025A-0.01B-0.033C-5E\text{-}003AB-0.025AC+0.015BC-0.029A^2-0.059B^2-0.054C^2$$

where A is the enzyme amount; B is the complex enzyme ratio and C is the microwave temperature.

Table4. Optimization Results of the Modification variables

Modification variable	Optimum value
Enzyme amount	0.14
Complex enzyme ratio	1.12
Microwave temperature	42.6°C

The aim of this study was to determine the interaction of extraction variables and maximize the extraction yield with the extraction condition from the model. Enzyme amount, complex enzyme ratio and microwave temperature were selected by applying the regression model. To confirm the model adequacy three additional experiments were carried out under the optimal conditions, the average values of the triplicate experiments 0.443 mg/g (fucoxanthin/dry algae) are close to the predicated results 0.452 mg/g, it indicate that the proposed model can well correlate the microwave assisted enzymatic extraction variables to the extraction yield.

4. Conclusion

In this work, the extraction condition for fucoxanthin from laminaria japonica with microwave synergistic enzyme method were investigated and optimized to improve the extraction yield by using response surface methodology. According to the BBD analysis results, the effect and interaction of the extraction variables including enzyme amount, complex enzyme ratio and microwave temperature on the extraction yield of fucoxanthin was investigated. The term for microwave temperature shows the most significant effect to the extraction yield of fucoxanthin. With the optimized condition (enzyme amount 0.14 g, complex enzyme ratio 1.12 and microwave temperature 42.6°C) achieve the extraction yield of fucoxanthin to 0.443, and shows a higher extraction yield and lower process cost compare to other methods. Therefore, this method can be used in extraction of thermal sensitive materials.

5. Acknowledgement

The authors wish to thank for the support of this study by research grants: Project of Education-Sciences Research of Fujian Province（JAT160352）and National Nature Science Foundation of China (51408518).

6. References

[1] MA Gammone, N D'Orazio, Anti-obesity activity of the marine carotenoid fucoxanthin, Mar. Drugs 13 (2015) 2196–2214.

[2] SZ Moghadamtousi, H Karimian, R Khanabdali, et al., Anticancer and antitumor potential of fucoidan and fucoxanthin, two main metabolites isolated from brown algae, Sci. World J. (2014).

[3] A Fung, N Hamid, J Lu, Fucoxanthin content and antioxidant properties of Undaria pinnatifida. Food Chem136 (2013):1055–1062

[4] CL Hurd, PJ Harrison, K Bischof, et al. Seaweed ecology and physiology. Cambridge University Press, Cambridge (2014)

[5] PS Saravana, AT Getachew, YJ Cho, et al. Influence of co-solvents on fucoxanthin and phlorotannin recovery from brown seaweed using supercritical CO2, J. of Supercritical Fluids 120 (2017) 295-303

[6] A Piovan, R Seraglia, B Bresin, et al. Fucoxanthin from Undaria pinnatifida: photostability and coextractive effects. Molecules 18(2013):6298–6310

[7] K Kanazawa, Y Ozaki, T Hashimoto, et al. Commercialscale preparation of biofunctional fucoxanthin from waste parts of brown sea algae Laminaria japonica. Food Sci Technol Res 14(2008): 573–582

[8] SS Periaswamy, TG Adane, JC Yeon, et al. Influence of co-solvents on fucoxanthin and phlorotannin recovery from brown seaweed using supercritical CO_2. J. of Supercritical Fluids 120 (2017): 295-303

[9] YF Shang, SM Kim, WJ Lee, et al. Pressurized liquid method for fucoxanthin extraction from Eisenia bicyclis (Kjellman) Setchell. J Biosci Bioeng 111(2011):237–241

[10] CR Chen, DM Lin, CM Chang, et al. Supercritical carbon dioxide anti-solvent crystallization of fucoxanthin chromatographically purified from Hincksia mitchellae P.C.Silva. J. of Supercritical Fluids 119 (2017):1-8

IOP Conf. Series: Earth and Environmental Science **146** (2018) 012078 doi:10.1088/1755-1315/146/1/012078

The Integrative Design in Assemble Habitation and Its Circumstance in Mountainous Region

Xiaowei HUANG [1, 3] and Zhaohui TANG[2]

[1] School of Architecture, South China University of Technology, Guangzhou, China, 510006

[2] Architectural Design & Research Institute of SCUT, Guangzhou, China, 510006

[3] School of Architecture, Changsha University of Science & Technology, Changsha, China, 410004

Email: june_hxw@163.com

Abstract. This paper researches the integrative design in assemble habitation and its circumstance in mountainous region, taking example for Xinhua mountainous region of Hunan province. First, the paper analyses the actuality of the mountainous assemble habitation according as the state policy and criterion which can ensure the boongs building to adapting the change along with development in mountainous region. Furthermore, the research expatiates on three basic theories -- the geomantic theory of China, the Paolo Soleri's theory—arcology, and the mountainous bioclimatology theory. The geomantic theory of China is a simple environment concept in Chinese antiquity. It guides people who can carefully seeing about the nature surroundings to create the all right environment of mountainous resident. Arcology is consistent with the geomantic theory of china about the ecological environment of residential buildings. In nature, as an organism evolves it increases in complexity and it also becomes a more compact or miniaturized system. The mountainous bioclimatology architecture should inosculate area, visual sensation and vitality of scenery together to cause a sensibility resonance. Then, the paper brings forward for integrative design in four patterns of assemble habitation with its circumstance in mountainous region. Such as the pattern of protection, the pattern of renovation, the pattern of derivation and the pattern of restoration. This part is the most important one which will set up a suit of the pattern truss for mountainous boongs. Through these four patterns reseaches, we can see a new phenomenon of settlements in mountainous region. The settlements in mountainous region will become suitable, geomantic, arcological and bioclimatological. In the end, the research will approve the new integrative design pattern which is feasible and continuable in development foreground. So, we should utilize the intelligent system to construct a integrative and an ecological residence follow a new mountainous architecture culture.

1. The Background

Through these years rapid development, the construction in our country has also progressed everywhere . The mountainous region is one important part of our country. So the research on the change of assemble habitation in mountainous region is also very significant. Currently, in the mountainous region, the glebe is changed to belongs with industry, commerce, inhabitation and some other function. It brings on immingle inhabitancy not only in boongs, but also in immigrations[1]. The mountainous assemble is an intergradation between city and village, and it is the village which has to be provided with some city's characters. In this area, the assemble habitation forms strong contrast and coexistence all round, such as different occupation, different life style, different belief, different conception, different demand and different civilization diathesis. Therefore, the paper will according as the state policy and criterion, based on normal theories, analyse the actuality of the mountainous assemble habitation, adapt the mountainous circumstance. Accordingly we researches the plan and technique which is the integrative design to improve the assemble habitation and its circumstance of the boongs in mountainous region.

2. The Actuality of The Mountainous Assemble Habitation

On account of the unique mountainous terrain, the mountainous assemble habitation is belongs to close, laggard and original. But, during the development with the city construction, the mountainous area has to be empoldered out of order. The village government can not manage availably. The mountain soil is exploited cosmically ,not according as the state policy and criterion. In the mountainous region today, so much soil forms dispersedly. More and more soil is hired to the developer. But the developer can not open out the hired soil in time, so it forms a popular phenomenon that the mountainous area is left unused[2] . Whereas the boongs lose the soil, their habitation circumstance occurs strictness problem. So the circumstance in the mountainous assemble is destroyed by city's expanding.

2.1. The Traditional Mountain Circumstance Havoced

The title is set 17 point Times Bold, flush left, unjustified. The first letter of the title should be capitalized with the rest in lower case. It should not be indented. Leave 28 mm of space above the title and 10 mm after the title.

Through almost 30 years development, the mountainous region has been faced with a breakthrough period. Much more soil resource has been impropriated. And most of them has been wasted. For example, in Xinhua area of Hunan province the actuality of the assemble circumstance presents confusedly in many aspects, such as:

- The soil resource: the plowland has degenerated and has been wasted badly.
- The habitation circumstance: the chaotic sky, the disafforestation, the mountainous region has been pushed planely, and the water has been polluted strictly.
- The mountainous assemble crowds, the path lacked, the building out of order.
- The boong's estate : lowness efficiency, lowness return, highness energy sources consume.
- The management of government: segmentation, short-term and not the holistic plan.

The figure 1 is a true portrayal of the circumstance.

Figure 1. The Xinhua Zhenglong traditional circumstance confused.

2.2. The Pattern of Mountainous Boong Has Been Broken

As the life style changed, the pattern of mountainous boong has been broken. In the course of city's expanding, the traditional assemble habitation is reducing step by step. More and more mountainous boongs have lost the soil for national purchase. Then they have to hire self-house to make a living. They build more house in unit of doors on their remained soil. These house layout dispersedly, construct random, establishment behindhand. For example, the boongs in Xinhua mountainous area still use traditional brick skill to build their house more than 70%: clay brick for wall, handwork for building, few prefabrication, modulus lowly, modernize poor. And it must be changed that the assemble intensivism has to develop the living level of boongs.

It is the other reason that the family configuration has been transformated. For example in hunan's mountainous area: Ten years ago, there is 54% three era in one family and 12% more era in one family. Untill 2008, these proportion descended to 27% and 8%. It is indicated that the traditional family configuration has not adapt in today like a fish out of water. Some new configurations appear, such as young couple family, young bachelordom family, aged couple family, old bachelordom family ets. Now we can see the principle of openning an account in police department which also takes the change place gradually. The singleton all serves as one family without any era. If there are more than tow children with agro-residence and the children reach marriage age, the family can dispart. Currently, many hunan's mountainous tenement has three or four fellows. Also it is the norm tenement that their residence building can not occupy over 80 centiare. Then the abruption from old big family has induced the traditional pattern of mountainous assemble residence broken, as the figure 2 displayed.

Figure 2. The Xinhua Shangmei traditional assemble habitation is reducing step by step.

3. Integrative Design in Assemble Habitation And Its Circumstance in Mountainous Region

Human being's activity have conflicted with entironment. It leads to the deterioration between entironment and residence. Also the assemble habitation and its circumstance deteriorate day by day through the expanding from city to mountain. So we must scan renewedly the connection with hunan and nature, buildings and entironment. In our country academician Wuliangyong bring forward "the sciences of human settlements in china". Mr wu said the human settlements include all the assenble form as villages, towns,and cities. It is emphasized that the research of human settlements is intergrative, roundly, systemic and compositive. And it aims to control the impersonality rule of development in human settlements[3]. Then we can design appropriate settlement for human. This paper based on Wu's theory. On emphases, we combine other theories ---- Geomantic Theory of China, Arcology, Intergrated Bioclimatic Architecture theory. We research the integrative design in the assemble habitation and its circumstance in mountainous region. The figure 3 is an example of the integrative design.

Figure 3. The Xinhua new assemble settlements with integrative design.

3.1. The Basic Theories

This paper provides an all-round new ecological concept of residential buildings which is the integrative design in the assemble habitation and its circumstance in mountainous region. According as these three theories, which are the most popular in the field of modern architecture, the research on residential greening may hold the accurate direction to develop. In addition, the paper researches the structural support to ensure the integrative design in the assemble habitation and its circumstance in mountainous region achieved. By way of testing, the quality of air, the temperature and the humidity of the residential buildings are improved evidently. The integrative design will truly become an important part of the mountainous residence.

- Geomantic Theory of China

The geomantic theory of China is a simple environment concept in Chinese antiquity. It guides people who can carefully seeing about the nature surroundings to create the all right environment of resident. From the theory of modern science, the geomantic theory is an all-around science, involving the Geophysics, the Hydrology, the Euthenics, the Arcology, the Cosmoplastism and the Somatology. The hard core of the geomantic theory of china has close connection with residential environment probably in three aspects which are the physiognomic theory, the landscape theory and the storing wind and assembling climate theory. This is the originally ecological theory on architecture in archaic china, and is the reference to choose appropriate method for mountainous design[4].

- Arcology

Arcology, the concept of ecological human habitats was introduced by Paolo Soleri in late 1960's, exploring an alternative urban development in the age of environmental crisis .In nature, as an organism evolves it increases in complexity and it also becomes a more compact or miniaturized system. Arcology, architecture and ecology as one integral process, is capable of demonstrating positive response to the many problems of urban civilization, population, pollution, energy and natural resource depletion, food scarcity and quality of life. Overall, arcology seeks to embody a "Lean Alternative" to hyper consumption and wastefulness through more frugal, efficient and intelligent design, especially in terms of heating, lighting and cooling[1]. Therefore, arcology is consistent with the geomantic theory of china about the ecological environment of mountainous residential buildings[5].

- Intergrated Bioclimatic Architecture theory

Because of the different region climate, the bioclimatology theory is used differently and widely in modern architecture. The famous Spaish architect--J.Barebar founded the theory of Intergrated Bioclimatic Architecture, which also is called as continuable architecture. J.Barebar considers that the

Intergrated Bioclimatic Architecture profits from the landscape. "The bioclimatic architecture" incarnates that the buildings have ability of applicability, adjustment and perfection by itself. Accordingly, the mountainous design will combine with nature, obey the bionics principle and hold the continuable principia[2]. In addition, the bioclimatic architecture also makes deep attention with healthiness of residential condition. Anyway, the Intergrated Bioclimatic Architecture should inosculate area, visual sensation and vitality of scenery together to cause a sensibility resonance[6].

3.2. Integrative Design Tactic in Four Patterns of The Assemble Mountainous Habitation

How will we bring planning of the mountainous habitation into city planning. So we can have criterion to manage as a whole, design as integer. This is the core of the research on the integrative design in the assemble habitation and its circumstance in mountainous region[7]. Concretely, we research four new patterns of the assemble mountainous habitation as follows.

- The Pattern of Protection

Based on the intergrative design, we will protect the original patterns of the assemble mountainous habitation which have national characteristic or belong to history style and features, as the figure 5 showed. Also we must revert the history appearance to retain local color, as the figure 4 showed. Therefore, we don't move the basic assemble room. We only maintain, amend and revert the characteristic room. On this way, the intrinsic mountainous habitation will have renew life.

Figure 4. The protection patterns retain local color.

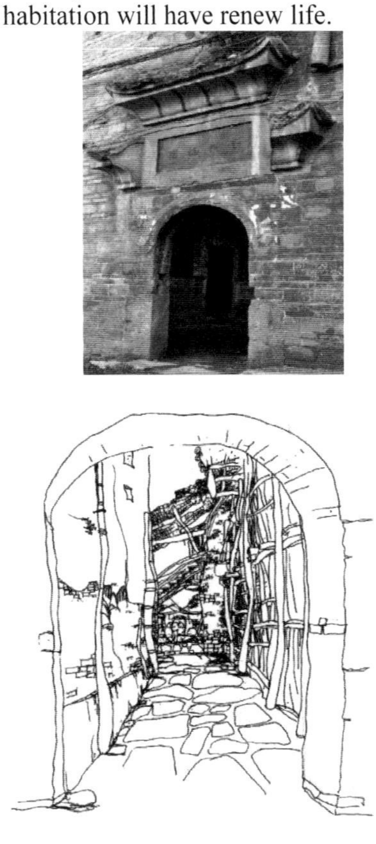

Figure 5. The protection patterns belong to history style.

- The Pattern of Renovation

Along with the change in boogns family, we bring forward the pattern of renovation especially for old residenter who will not adapt more movement for their life. Also it can debase efficiently the society cost to reconstruct the assemble residence. Furthermore, based on integrative design we can renovate

IOP Conf. Series: Earth and Environmental Science **146** (2018) 012078 doi:10.1088/1755-1315/146/1/012078

the quality of original residence. And we can combine the original residence with the mountainous circumstance perfectly. Figure 6 showed how the renovation patterns adapt mountainous circumstance

Figure 6. The renovation patterns adapt mountainous circumstance.

- The Pattern of Derivation

In the course of the assemble development, the boogns improve their independence consciousness more and more. The structure of original family has been changed within. Then a new pattern of residence derivates inside the original assemble room. This pattern of derivation will be based on village group, as the figure 7 showed. Besides it has associated and coherence in the extreme. On the principle of integrative design, we will build the residence with conjunction each other. Also the pattern of derivation will be more appetent in big old family. The boogn's relation will be more harmonious than before.

Figure 7. The derivation patterns base on village group.

- The Pattern of Restoration

During the mountainous householder juvenscence, the youngling of the boogn pursues new pattern of residence. So the pattern of restoration will satisfy with different aims of different person. And this pattern can keep stabilization of assemble together with development benign. When we design the pattern of restoration, we must follow the mountainous characteristic, as the figure 8 displayed. We also must adjust measures to local conditions, as the figure 9 displayed. We will obey the nature rule to rebuild new assemble residence. Of course, the new pattern will be seasoned with standardization development.

Figure 8. The restoration patterns follow the mountainous characteristic.

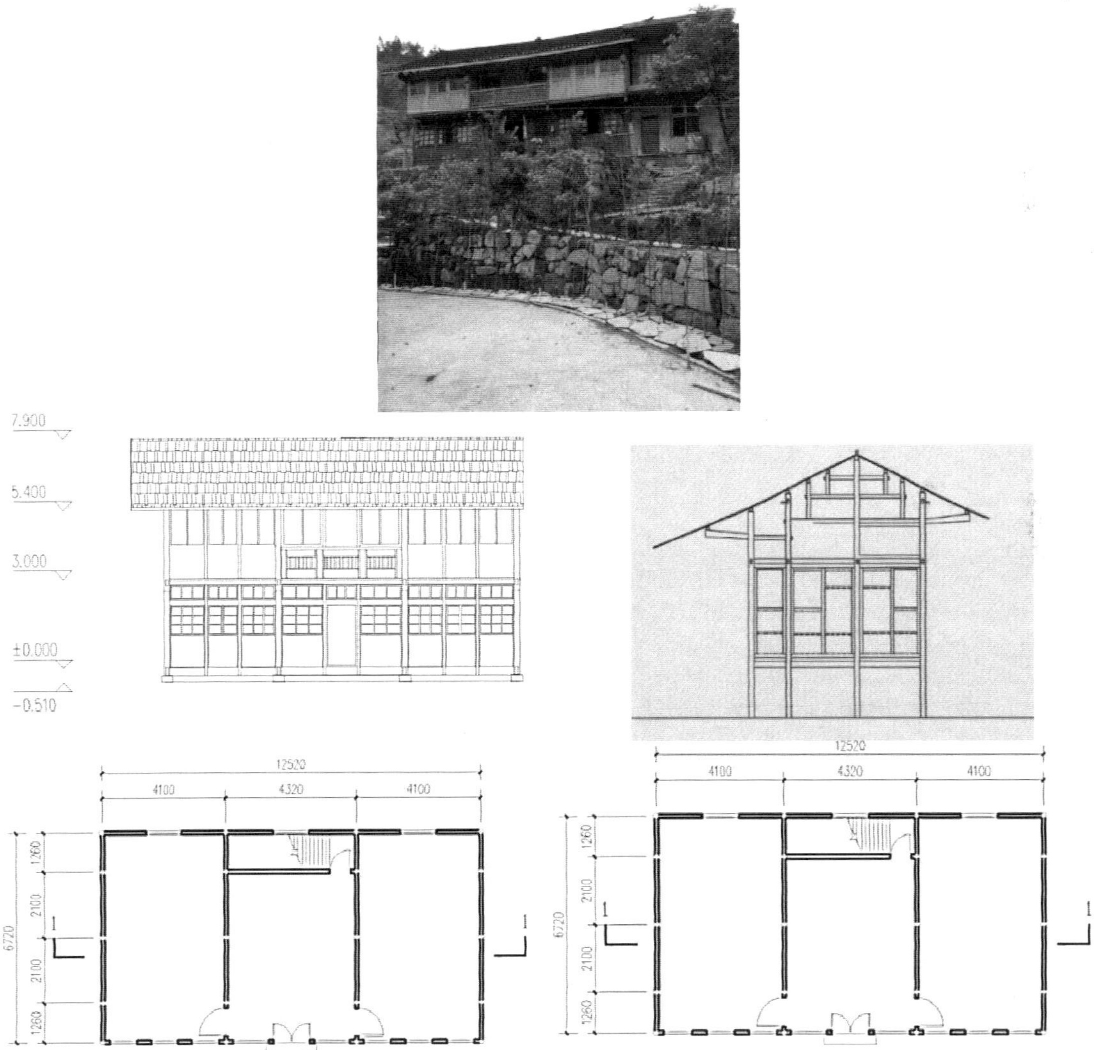

Figure 9. The restoration patterns adjust to local conditions.

4. Conclusion

Today, all kinds of residence have developed rapidly. The mountainous assemble is also one important kind of human settlement. The mountainous boogn has changed their life style unprecedentedly. We research the integrative design in the assemble habitation and its circumstance of the boongs in mountainous region to solve some problem of this especial assemble habitation. This paper provides an all-round new integrative design concept of residential buildings. According as the theories, which are the most popular in the field of modern architecture, the research on mountainous residence may hold the accurate direction to develop. In addition, the paper researches the structural support to ensure the integrative design achieved. By way of testing, the quality of air, the temperature and the humidity of the residential buildings are improved evidently.

In future, if just as been hoped for the framework and maintenance system can regulate itself according to the weather and comfort, the buildings get intelligence[8]. Furthermore, due to the rapid developing of the technology of computer, communications, and automatic control, we can intelligentize the futural assemble habitation and its circumstance of the boongs in mountainous region to meet the integrative idea. In the future, mountainous residential subzone would be an architecture complex with the self-adjusting function commanded by high-tech, and the mountainous settlement would be a intelligent and a tri-dimensional one. The intelligent assemble should integrate the inside environment and outside by an intelligentized design, which may introduce the using of self-control device, intelligent management system ,CATV, Intranet, and the intelligent household, etc, to realize the dust prevention, germ destroyed, noise control, climate conditioning and environment prettified[9].

So, it is concluded that to build genuine mountainous architecture, we should utilize the intelligent system to construct an integrative and an ecological residence follow a new mountainous architecture culture[10]

5. References

[1] Lingliang ZHOU 2005 *The Theory and Practice of the Town Development* (.Beijing: China Economy Press)
[2] Jingxing WANG 2005 *The Farmland System and Its Pattern in Modernization course* (Beijing: China. Economy Press)
[3] Liangyong WU 2001 *The Sciences of Human Settlements in China* (Beijing: China Architecture & Building Press)
[4] Paolo Soleri 1969 *Arcology: the city in the image of man* (Cambridge: MIT Press)
[5] Barebar J 1993 *Intergrated Bioclimatic Architecture* (Barcelona: Goregy wright forum)
[6] Qi-bin LIN Yan LIN Zhi-ya ZHAO 1999 *The residence design on human habit* (Guangzhou: Huanan University of Technology Publishing House)
[7] Xiaowei HUANG 2007 The Tri-dimensional Greening Research on Ecological Environment of Residential Buildings *Journal of Harbin Institute of Technology* L1
[8] Chang-qing WANG and Peng-fei DU 1999 *American Greening Building Technician* (Beijing: Chinese Industry Building Press)
[9] Jing-qi HUANG 2000 *The Design and Construction on Roof Garden* (Beijing: China Forestry Publishing House)
[10] The Technology Committee of Ministry of Construction 2001 *Intelligent Building Technician and Utilization* (Beijing: China Architecture & Building Press)

Examining the Phenomenology of Human Experience in Design Process and Characteristics of Architectural Approaches

Stephen T. F. Poon

School of Media, Arts & Design, Asia Pacific University of Technology & Innovation, Malaysia

stephentfpoon@aol.com

Abstract. Research into phenomenological contexts of architecture has been shaped by shifting dynamics, the influences of changes within societies and their relationship to human perception. The sensory construct of Steven Holl's architectural design heritage can be ascribed to 'natural phenomena' theories that seek philosophical revelation in the process of understanding architectural design. Phenomenological theories are characterised by the relational premise between architecture's sensitivity to the focal design inspiration and the innate human sensory perceptions of atmosphere and material culture, resulting in an aesthetically balanced, fluid and organic framework, enriching building character and positioning humanity rather than spatial rationality. To evaluate the contribution of phenomenological architecture, an outstanding project work of American architect Steven Holl in planning and creating the building design for St Ignatius Chapel, an institution of religious worship in United States was analysed to understand how the use of forms, textures, light and shadow affect the spatial perception of architectural space. The use of visual perspective to reflect on the phenomenology of spatial and material suggests the importance of intrinsic sensory experiences intertwined the architectural design process, as they weave in 'human' content, our emotions, intuitions and organic dimensions of relations with others, providing buildings and places depth in aesthetic meaning.

Keywords: *phenomenology, multi-sensory, perception, spatial, material*

1. Introduction

Phenomenology of architecture is theoretically constructed on the paradigm of observed and interpreted human experiences and behaviours towards physical sites through the analysis of sensory influences. As a scholarly discipline, this is significantly different from research based on studying design functionality, efficiency and performance of buildings, as principled on rationalist and socioeconomic contexts. Phenomenology's approach acknowledges the complexity of double-fold levels of knowledge: *ontological* or objective, foundational characteristics of the life-world, and *epistemological*, where lived experiences become a primary source of self-knowing, developing richer, authentic perspectives through our emotions, sentiments and memories [1].

Alberto Perez-Gomez relates the role of modern architecture itself as a lived experience at the epistemological level; affirming the belief by architectural academics like Joseph Rykwert that built structures are theatres of our memories and meaning-making, since geometrical arrangements are similar to human experiences with spatiality and temporality [2].

This paper gives insights on how phenomenology influences the philosophic intent behind architecture, through a critical discourse on Steven Holl's phenomenological approach as the intertwining of the sensorial and experiential, and argues for his discernment of the spirit of modernist structural design that leads to existential revelation of the resulting work.

2. Literature Review

Merleau-Ponty [3] pioneered applied phenomenology, founded on the nature of perception expounded by Edmund Husserl and Martin Heidegger, who theorised that conscious experience from the first person's point of view, derives from perceptual dimensions.

Perceptions, the instrument connecting our physical bodies through sensory organs (i.e. sight, sound, taste, touch and smell), enables us to experience the world intelligently and intimately, form the foundation of our creative receptivity and the subjective intertwining of each person's *self* with others [4]. Seamon [5] elaborates the development of consciousness through perceptual awareness as the lived aspect of innate material experiences, and researchers' need to be constantly involved in some aspect of the world as their object. Finnish architect Juhani Pallasmaa [6] focuses architectural phenomenology discourse as an intellectual form of 'inner language' of spaces, by philosophically examining the experiences preceding from authentic "emotional forces" through introspection. This field of research, he states, is fraught with the challenge of *ocularcentrism*: looking at the surface, rather than seeing to understand the metaphysical.

Hsu, Chang and Lin [7], seeking a comparative design syntax by studying various contexts in the development of modern architectural design philosophy, present literature arguing that phenomenological perception of spaces instils sensory experiences through focusing on the emotional, poetical realms of geometric aesthetics being the metaphysical foundation of the designing process which enables inhabitants to relate to the *genius loci*, or unique character, of a place in their own meaningful ways. Current literature pushes forward these debates, describing and investigating the aesthetics of spatial and material beauty [1], and what has emerged are flawed understanding about our place within landscapes and sites, as the quality of what we perceive is affected by "shadows, voids, opacities, transparencies and translucencies" [8]. In questioning the perceptual value of design to contemporary architecture, Pallasmaa [9] point out: "Every touching experience of architecture is multi-sensory; qualities of matter, space and scale are measured equally by the eye, ear, nose and skin, tongue and skeleton and muscle". Haptic architectural forms thus include sensorial depths, embracing visual, tactile, aural qualities; even a sense of odour can be conceived to be presented as a physicality of the building form [8].

Achieving this form of *hapticity*, Pallasmaa [10] believes, requires that architects think with their bodies in the designing process. By recording imageries and embodying the sensory wisdom of what they touch and feel in their muscular memories, they could lucidly present these images to communicate ideologies of designing *with* and *for* a sensual body. Many philosophical treatises have been produced to examine contradictory outcomes with regards lighting and materials as the basis of perceptual design dimensions; these projects have also been subject to critical phenomenology research. For instance, Peter Zumthor arranges the physical framework of spaces to seduce, kindle and extend human perceptiveness. Experimentation with memories of his childhood, vividly sought these stories as the "reservoirs of architectural atmospheres and images" which directly inspired his projects [11]. In *Atmospheres*, he states his idea with materiality with light experimentation: "So the first of my favourite ideas is this to plan the building as a pure mass of shadow then afterwards to put in light as if you were hollowing out the darkness". Zumthor [11], [12] conceives the tectonic of light in dark spaces, such as the *Therme Vals* (Figure 1), describing it as "(an initial) feeling ... for darkness and light, for the reflection of light upon water, for the diffusion of light though steam-filled air … for the ritual of bathing".

Figure 1. Therme Vals, Switzerland by Peter Zumthor [19]

In summary, multi-sensory experiences are considered "ambitious" theorising in principle, as the scale of interpretations for framing and referencing perceptions seem scholarly impossible to pinpoint, review or determine in terms of value and effects since they are formed out of physical limitations, obstacles, change and biases. The notion that human experience is at the core of architectural phenomenology movement has produced vocal conflicts of opinions among scholars of contemporary architecture. Contemporary researchers, as Bognar [13] and later, Seamon [5] point out, have reached a point of concurring that integrating human experience phenomenologically enjoins theories with application in architectural design process, reminding us that people and surroundings comprise "an indivisible whole".

3. Methodology

In enlarging the scope of discussions on the objectives, applications, issues, methodologies and implications of qualitative phenomenological research, the body of critical phenomenological research establish their position to provide relational interpretations between studied variables and elements [14].

Figure 2. Interior view of St Ignatius Chapel [20]

To consider how phenomenology influences the philosophic intent behind architecture, the Chapel of St. Ignatius (Figure 2) located in the University of Seattle, designed and constructed between 1994 and 1997, will be presented for case study analysis in this study. To map the ideological derivation for the design of this Jesuit place of worship considers how watercolour sketches were translated into built form. A critical discussion of Holl's *eidos* (intellectual form), follows the case study.

4. Findings and Analysis

Holl introduced the concept of 'Seven Bottles of Light in a Stone Box' (Figure 3) as a metaphorical visual of spiritual life-world awareness to acknowledge the seven sacraments symbolising God's divine covenant with His believers. Thematically, the seven lights are oriented towards the chapel's roof tectonics while the stone box represents the brutal expression of the building as in Figure 4 and 5 [15]. Conceptually, each of the bottles represents the seven Jesuit orders closely identifying each religious exercise, with the spatial environment organised accordingly. To capture pure meditative serenity, Holl uses ethereal lighting reflected from coloured windows, baffles and clear glass panes as the 'structural material' of architecture; natural daylight sheds a sacred atmosphere, both intimate and warm, to invite the congregation as well as individuals to present their prayer offerings.

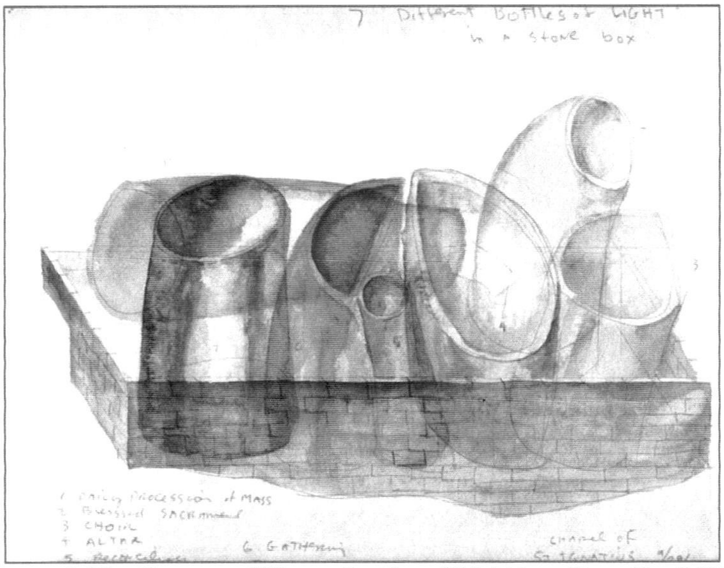

Figure 3. The Concept of "Seven Bottles of Light in a Stone Box" [21]

4.1. Light as structure

Light becomes fundamental in phenomenological architecture. As Holl [16] explains, "Space is oblivion without light. A building speaks through the silence of perception orchestrated by light". St. Ignatius Chapel shows the uses of light as a phenomenon of personalised experience, as the interplay of darkness and shadow imputes one's inner spiritual thoughts, referred to as *consolations* and *desolations* [17]. The emphasis shifts from tectonic logic to light and shadow to illuminate the intertwining of images and emotions in the lives of worshippers partaking in religious devotional rites. The idea of "intertwining", as pointed out by Yorgancioğlu [15] in his thesis, was initially borrowed from Merleau-Ponty:

> *"In a three-dimensional triad, the reciprocal insertion of the body – oneself – in the interwoven landscapes of architecture yields identity and difference. This insertion of oneself is an intertwining of one with the "others". Without purpose as an object, without recourse as a style, architecture depends on this reciprocal insertion of the other, oneself".*

IOP Conf. Series: Earth and Environmental Science **146** (2018) 012079 doi:10.1088/1755-1315/146/1/012079

Theoretically, Holl's phenomenological constructs corresponds with Pallasma's perception of multi–sensory experiences of physical space, colours, light and textures. Perception of spatial depth, as argued by Pallasma [6] cumulates in the use of haptic memory through contact with light as the material of structure and distance to intensify the *act of believing* from what is understood visually.

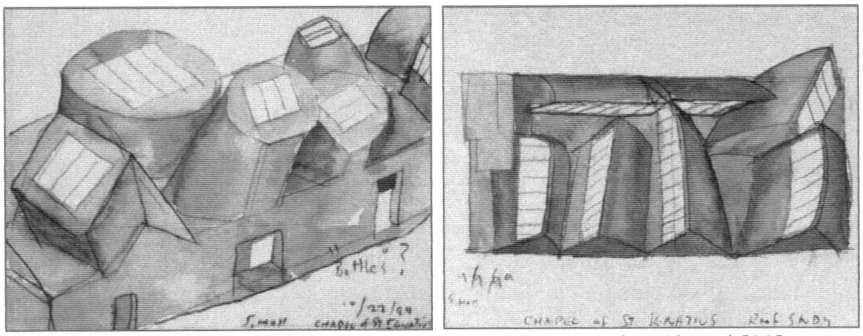

Figure 4 and 5. Watercolour sketches of St. Ignatius Chapel [22]

Holl further illustrates phenomenological hapticity in the chapel's exterior. Holl [17] explains that tactility, represented through interlocking concrete slabs for the chapel's windows and floor, invite worshippers to engage deeply with their faith. While some critiques consider the exterior's drab "stoniness" as uncomplimentary of the breath-taking interior, the construction of the building in "tilt-up" style is Holl's signature parallax concept, an attempt to fuse the physical tactility and intelligence of material, space and the church (building owner)'s outreach mission into the result.

5. Discussion and Interpretation

To critically acknowledge perceptual living experiences as central in the development of depth (rather than breadth) in spatial architecture, an evaluation of its unique approach is necessitated. Holl uses the technique he calls 'parallax', a series of perspectives aimed at integrating sensorial depth by imagining how 'the spaces of eternity' feels like through the art of watercolour to express phenomenological concepts, drawn from the writings of Merleau-Ponty and other philosophers [18]. In Holl's watercolour perspective sketches, the experiential potential of concept is sought. By beginning the reflective process of memorialising his perceptions, the transference of intangible experiences from sketches to real buildings clarifies the exactitude of experiences he wanted to capture and immortalise spatially [12].

Imagining thus begets the design process for architecture for Holl, reflecting Pallasmaa's [10] theory of multi-sensory perception as an essential exploration, reinvigorating our process of understanding, helping us gain subjective yet substantial insights of architecture as meaningful representation of valuable human emotions. As a water-colourist, Holl sketched to visually articulate an indeterminate idea into a more definite concept. From the review of literature and discussion, it could be critically argued that phenomenology architecture emphasise on the role of light to create perceptual awareness, humanising places and buildings through conscious design. From this perspective, there is a definite resemblance to Pallasma's [9] theory, emphasised in the latter through reinforcing multi-sensory perception to create a rich embodiment of material and spiritual construct. At the same time, Holl refined the collaborative phenomenological construct using the tectonic of light to uniquely portray his ideology of human perception of a chapel as a site for deepening one's relationship with God. Holl illustrates a set of intentional values to reveal a heightened state of awareness. Employing poetic spaces, and lighting nuances in building design "diminishes" our human physicality, metaphorically "dislocates" our traditional rationality, expresses our imagination and augments the atmosphere of the place [12]. Essentially, Holl's precise emphasis on the existential potential of architecture conceptualises his belief that our lived experiences contribute metaphorically towards thoughtful architecture. The chapel's interior perspectives manifest eternities through the artful effects of light and shadow to abstract perceptions into spiritual emotions.

6. Conclusion

The essence of change itself encourages us to keep searching for humanity's "wonderful possibilities" [1], however, architects' intellectualised removal of personalities from the process of designing, much favoured in the modernist movement of the 20th-century until the recent decade, has resulted in an emotive void, hampering aesthetics from becoming an integral part of the intelligent, intimate cognition of places, landscapes and topography.

Phenomenological architecture visited in this research shows the importance of metaphysical interpretations of "pure looking" [10], where the aim is to appreciate the simplicity of structural design, light, shadow and space as the context of perceptual experiences that move, change and even fundamentally orchestrate landscapes, revealing fully the human spirit [8]. Phenomenological association between human perceptions, space and form can be studied in various architectural concepts. Diagrammatical sketches and the application of haptic/tactile elements such as light and textures nevertheless continue as problematic scholarly treatise in the broader context of contemporary architecture practice. Acknowledging phenomenology's application beyond the domain of theoretical scholarship is crucial; human perceptions must not be neglected but instead integrated into spatial architectural designs bearing our sensory experiences. Pallasmaa [9] believes that for buildings to embellish meaning and embody memories, they should do more than reflect practical environmental realities of planning geographical and physical spaces to fit surroundings. By abstracting our emotional connections and perceptual awareness, phenomenological architecture expresses the sanctity of human contexts, capturing empathy, joys, sorrows and the shared experiences of people's lives, their feelings, memories and consciousness.

7. References

[1] Reisner-Cook Y 2009 *The Troubled Relationship between Architecture and Aesthetic: Exploring the Self and Emotional Beauty in Design* (Dissertation) (Australia: Royal Melbourne Institute of Technology)
[2] Perez-Gomez A, Pallasmaa J and Holl S 2006 *Questions of Perception: Phenomenology of Architecture* 2nd ed (San Francisco: William K Stout Publishing)
[3] Internet Encyclopedia of Philosophy n.d. *Maurice Merleau-Ponty*. Available from: http://www.iep.utm.edu/merleau/#SH3a(Accessed 05 July 2017)
[4] Frampton K 2002 'Towards a Critical Regionalism: Six Points for an Architecture of Resistance' In Frampton K ed *Labour, Work and Architecture: Collected Essays on Architectural Design* (London: Phaidon Press)
[5] Seamon D and Mugerauer R (eds) 2000 *Dwelling, Place and Environment* (London: Croom Helm Publishers)
[6] Pallasmaa J 2012 *The Eyes of The Skin: Architecture and The Senses* (West Sussex: John Wiley & Sons, Ltd)
[7] Hsu H L, Chang Y L and Lin H H 2015 'Emotional Architecture: A Study of Tadao Ando's Genius Loci Design Philosophy and Design Syntax'. In *International Journal of Chemical, Environmental & Biological Sciences*, vol 3 no 6 pp 456-463
[8] Holl S, Pallasma J and Gomez A 2006 'Questions of Perception: Phenomenology of Architecture' in *Questions of Perception* (Tokyo: A = U)
[9] Pallasmaa J 1996 'The Geometry of Feeling: A Look at the Phenomenology of Architecture'. In Nesbitt K (ed.) *Theorising a New Agenda for Architecture: An Anthology of Architectural Theory 1965-1995* (New York: Princeton Architectural Press)
[10] Pallasmaa J 2009 *The Thinking Hand: Existential and Embodies Wisdom in Architecture* (West Sussex: John Wiley & Sons, Ltd)
[11] Zumthor P 2006 *Atmospheres* (Birkhauser: Switzerland)
[12] Hawkes D 2008 *The Environmental Imagination: Technics and Poetics of the Architectural Environment* (Abingdon, Oxon: Routledge)

[13] Bognar B 1985 'A Phenomenological Approach to Architecture and Its Teaching in the Design Studio' in Seamon D and Mugerauer R (eds.) 2000 *Dwelling, Place and Environment* (London: Croom Helm Publishers) pp 183-200
[14] Lester S 1999 An Introduction to Phenomenological Research. *Stan Lester Developments*, Taunton. Available from http://www.rgs.org/nr/rdonlyres/f50603e0-41af-4b15-9c84-ba7e4de8cb4f/0/seaweedphenomenologyresearch.pdf (Accessed 23 June 2017)
[15] Yorgancioğlu D 2004 *Steven Holl: A Translation of Phenomenological Philosophy into the Realm of Architecture* Graduate School of Natural and Applied Science (Ankara, Turkey: Middle East Technical University)
[16] Holl S 2006 *Luminosity/Porosity* (Tokyo: Toto Shuppan Publishing)
[17] Holl S 1999 *The Chapel of St Ignatius* (New York: Princeton Architectural Press)
[18] Schoof J 2016 *International VELUX Award 2018*. Available from: http://iva.velux.com/competitions/international-velux-award/posts/designed-for-all-senses-the-architecture-of-steven-holl (Accessed 12 May 2017)
[19] Figure 1: Therme Vals, Switzerland by Peter Zumthor. Available from http://www.veluxstiftung.ch/en/index.php?page_id=58 (Accessed 01 June 2017)
[20] Figure 2: Interior view of St Ignatius Chapel. Available from http://www.scofield.com/showcase_public2c.html (Accessed 30 July 2017)
[21] Figure 3: Seven Bottle of Light in a Stone Box. Retrieved from http://www.stevenholl.com/project-detail.php?id=40(Accessed 29 May 2017)
[22] Figure 4 and 5: Watercolor sketches of St. Ignatius Chapel. Available from http://redchalksketch.wordpress.com/2011/03/03/chapel-of-st-ignatius-steven-holl-architects/ (Accessed 29 June 2017)

IOP Conf. Series: Earth and Environmental Science **146** (2018) 012080 doi:10.1088/1755-1315/146/1/012080

Value of inspection in steel structural integrity management

Guang Zou[1, 2], **Kian Banisoleiman**[1], **Arturo González** [2]

[1]Lloyd's Register GTC, Lloyd's Register Group Limited, Southampton, UK
[2]Departement of Civil Engineering, University College Dublin, Dublin, Ireland

guang.zou@lr.org

Abstract. Fatigue cracking is a common problem that needs to be managed in the life cycles of steel structures. Operational inspections and repairs are important means of fatigue crack management. Driven by high relevance in safety control and budget saving, inspection and maintenance planning has been widely studied. However, the value of inspection and repairs has typically not been fully appreciated and quantified rationally before they are implemented. The basic idea of this paper is to address the planning problem with focus on repair other than on inspection. A maintenance strategy without inspection is studied and serves as comparison of a maintenance strategy with inspection. Then the value of repair and the value of inspection relative to repair can be evaluated respectively. An illustrative example is performed on a typical fatigue-prone detail in steel structures.

1. Introduction

Fatigue cracking is one of the most common failure mechanisms for steel structures subjected to cyclic loading. It cannot be avoided absolutely by design check due to inherent uncertainties associated with the fatigue process, and inspections and repairs are necessary. Inspection and maintenance planning has gained wide attention due to heavy costs involved and various uncertainties associated with material and loading characteristics, deterioration model, inspection quality and repair effect. Generic inspection planning is extensively studied based on Bayesian decision theory [1-3]. The problem is also formulated as a multi-objective optimisation problem based on event tree analysis and solved with optimisation techniques [4-7]. The primary focus of previous studies is on how inspection should be planned, and repair decision is dependent on a repair criterion and inspection result.

This paper addresses the problem from a different perspective with focus primarily on repair. The reason is that it is the repair activity that ultimately changes the reliability of structure. From this point, a maintenance strategy without any inspection is studied and serves as comparison of a maintenance strategy with inspection. On the basis of a case study on a typical fatigue-prone structural detail, the value of repair and the value of inspection relative to repair is discussed and evaluated.

2. Fatigue deterioration modelling

Based on fracture mechanics, fatigue process can be divided into three stages: crack initiation, crack propagation and final fracture. The crack size in the crack initiation stage is not critical as it is too small and is hardly detectable by typical non-destructive testing (NDT) methods. In practical, the crack initiation stage is often negligible comparable with the crack propagation stage because of the presence of initial flaws/cracks. Also, the final fracture usually occurs very quickly, and the crack propagation stage is thus the focus of crack control.

Paris' equation relates the crack propagation rate to the range of stress intensity factor, and is given by

$$\frac{da}{dN} = C\Delta K^m \tag{1}$$

where da is increment in crack propagation for dN stress cycles, C and m are material parameters, ΔK is stress intensity factor range, given by

$$\Delta K = \Delta\sigma Y(a)\sqrt{\pi a} \tag{2}$$

where $Y(a)$ is geometry function and $\Delta\sigma$ is stress range. The stress range is normally determined by design S-N curve, which is given by

$$\begin{cases} N_f \Delta\sigma^{m_1} = \overline{a_1} & N_f \leq 10^7 \\ N_f \Delta\sigma^{m_2} = \overline{a_2} & N_f \geq 10^7 \end{cases} \tag{3}$$

where m_1 and m_2 are the fatigue strength exponents, and $\overline{a_1}$ and $\overline{a_2}$ are the fatigue strength coefficients.

By integration of equation (1), the number of cycles for the crack to develop from the initial crack size a_0 to the critical size a_c can be expressed as

$$N = \frac{1}{\pi^{m/2} C \Delta\sigma^m} \int_{a_0}^{a_c} \frac{da}{a^{m/2} Y(a)^m} \tag{4}$$

If the geometry function $Y(a)$ is known, it is also possible to calculate the crack size $a(t)$ at time t when the structural detail has exposed to $N(t)$ cycles of fatigue loading.

3. Maintenance strategy

Three maintenance strategies are tested. The first one is that no human intervention is planned, and the structure has a failure probability which is determined by the design and construction quality. The second one is that one repair is planned during the life cycle. The timing for the repair t_r is optimized and shown in the following section. If the structure is survived at the time of the planned repair, one repair will be implemented. The third one is that one inspection is planned during the life cycle. The timing for the inspection t_i is optimised in the following section. If the structure is survived at the time of the planned inspection, one inspection will be implemented. If a crack is detected, it will be repaired. The maintenance strategies are summarized in Table 1.

Table 1. Three strategies for operational maintenance.

Case	Maintenance activity
Case1	No inspection & no repair
Case2	Repair is planned without inspection
Case3	Repair would be done following plan inspection

4. Failure probability and probability of repair

4.1. Failure probability without repair

Failure probability is calculated based on formulation of a limit state, in which the structural capacity is defined by a failure criterion. The limit state formulation in this paper is based on serviceability analysis. The basic idea is that a structural detail is not serviceable if a through thickness crack exists, so a critical crack size equals to plate thickness is used as a failure criterion. Limit-state function can be formulated as

$$M(t) = a_c - a(t) \tag{5}$$

where a_c is the critical crack size, $a(t)$ is the crack size at time t under fatigue loading.

The probability of failure is given by

$$P_f(t) = P(M(t) < 0) \tag{6}$$

4.2. Probability of repair

The probability of inspection is actually implemented at time t is the probability that the structure is survived at that time, and is given by

$$P_i(t) = 1 - P_f(t) \tag{7}$$

In Case 2, the probability of repair is equal to the probability of inspection. In Case 3, the repair strategy that all detected cracks are repaired is used. This means that the probability of repair is equal to the probability of detection. If the smallest detectable crack size of an NDT method is a_d, then the limit state function for detection or no detection is defined as

$$D(t) = a_d - a(t) \tag{8}$$

The function is negative when a crack is detected and it is positive when no crack is detected. The probability of repair is given by

$$P_r(t) = \left(1 - P_f(t)\right) \cdot P(D(t) < 0) \tag{9}$$

4.3. Failure probability with planned repair
The failure probability before planned repair time t_r is equal to the initial failure probability without repair. The failure probability after planned repair time t_r should take into account the influence of planned repair.

For Case 2, the failure probability after planned repair is given by

$$P_f^1(t) = P_f(t_r) + P(M(t_r) > 0 \cap M(t - t_r) < 0) \tag{10}$$

For Case 3, the failure probability after planned repair is given by

$$P_f^1(t) = P_f(t_r) + P(M(t_r) > 0 \cap D(t_r) > 0 \cap M(t) < 0) + P(M(t_r) > 0 \cap D(t_r) < 0 \cap M(t - t_r) < 0) \tag{11}$$

5. Illustrative example
An illustrative example is performed on fatigue-prone T joints subjected to cyclic fatigue loading, which are common connections in steel structures. Typical structure is a plate with stiffeners. The stability of the plate is improved by stiffeners, but cracks are highly likely to initiate and propagate along the weld toes of the joints. Fatigue reliability of such joints is a problem that needs to be addressed during the life cycle of the structure detail. The geometry and critical location of a T joint is shown in Figure 1.

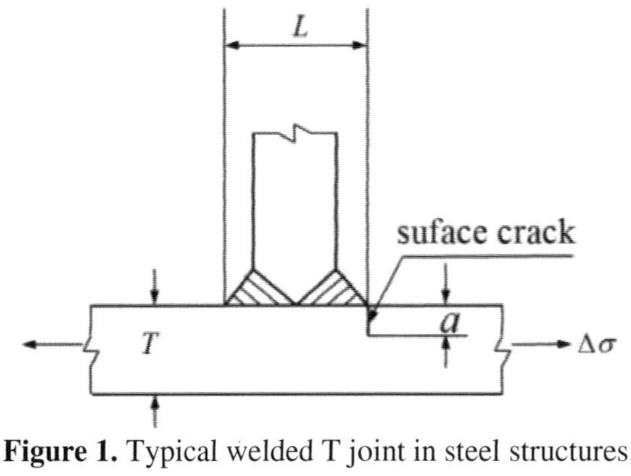

Figure 1. Typical welded T joint in steel structures.

5.1. Probabilistic model
Fatigue performance of such structural detail is given by a bi-linear S-N curve [8]. The frequency of fatigue loading is approximately 0.16Hz, which corresponds to $N_0 = 5 \times 10^6$ cycles per year. The uncertainties associated with loads and stress calculations are modelled with a normally distributed variable B. The plate thickness is T =25mm. The required service life is T_{SL} =20 years. The parameters and variables used in the model are summarised in tables 2 and 3.

Table 2. Parameters used in the model.

Parameter	Unit	Value
T_{SL}	year	20
N_0	cycle	5×10^6
$\log_{10} \overline{a_1}$	$N^4 \cdot mm^{-6}$	11.855
$\log_{10} \overline{a_2}$	$N^4 \cdot mm^{-6}$	15.091
T	mm	25
a_d	mm	0.89
m_1	-	3
m_2	-	5

Table 3. Variables used in the model.

Parameter	Distribution	Unit	Mean	Standard Deviation
a_0	Exponential	mm	0.043	0.043
$\log_{10} C$	Normal	$N^{-4} \cdot mm^{5.5}$	-12.74	0.11
B	Normal	-	1	0.15

5.2. Results and discussions

In this study, Case 1 represents initial structural reliability, which is determined by design check and manufacture quality control, without any operational maintenance. Case 2 reflects the influence of repair on the life cycle reliability (the reliability at the end of service life), while Case 3 reflects the influence of both inspection and repair. The timing for the inspection t_i and for the repair t_r in Case 2 and 3 are optimised with the objective to maximise the life cycle reliability. The optimum maintenance plans in Case 2 and 3 are analysed with respect to life cycle reliability.

Table 4 summarizes calculation results for Case 1, 2 and 3, which include the reliability index β, optimum time for inspection and repair t_i and t_r, the probability for inspection and repair and P_i and P_r. Based on the results, the following discussions are made:

Table 4. Optimised maintenance plan.

	Case1	Case2	Case3
β	1.12	2.40	2.62
t_i (year)	n/a	n/a	9
t_r (year)	n/a	10	9
P_i	n/a	n/a	0.998
P_r	n/a	0.996	0.311

1) In Case 2, the structure is intervened by repair in the middle of its service life when the repair is optimum from the point of increasing life cycle reliability. Compared with Case 1, the life cycle reliability increases from 1.12 to 2.40 owe to repair, by which the structure is physically changed.
2) Comparing Case3 with Case1, it can be seen that with the adoption of inspection and possible repair (if detected by inspection), life cycle reliability increases significantly from 1.12 to 2.62. The increase in reliability is benefited from both repair and inspection.

3) An interesting finding is found by comparing Case 3 with Case2 that repairs do not always increase structural reliability. The probability of repair in Case 2 is much higher than that in Case 3, but the structural reliability in Case 2 is lower than Case 3.
4) The value of inspection in Case 3 lies in two aspects. On one hand, the cracks that grow fast are detected and their failure probability is lowered by physically changing the structure. On the other hand, the cracks that grow slow are identified and their failure probability is lowered according to the new information gained by inspection (no detection).

Acknowledgments
The authors would like to express their gratitude to the European Union's Horizon 2020 research and innovation programme for their funding toward this project under the Marie Sklodowska-Curie grant agreement No. 642453 (http://trussitn.eu).

References

[1] Faber MH. Risk-based inspection: The framework. Structural engineering international. 2002;12(3):186-195.
[2] Straub D, Faber MH. Risk based inspection planning for structural systems. Structural safety. 2005;27(4):335-355.
[3] Kim S, Soliman M, Frangopol DM. Generalized Probabilistic Framework for Optimum Inspection and Maintenance Planning. Journal of Structural Engineering. 2013;139(3):435-447.
[4] Soliman M, Frangopol DM, Mondoro A. A probabilistic approach for optimizing inspection, monitoring, and maintenance actions against fatigue of critical ship details. Structural Safety. 2016;60:91-101.
[5] Kim S, Frangopol DM. Efficient multi-objective optimisation of probabilistic service life management. Structure and Infrastructure Engineering. 2017;13(1):147-159.
[6] Dong Y, Frangopol DM. Incorporation of risk and updating in inspection of fatigue-sensitive details of ship structures. International Journal of Fatigue. 2016;82:676-688.
[7] Lotsberg I, Sigurdsson G, Fjeldstad A, Moan T. Probabilistic methods for planning of inspection for fatigue cracks in offshore structures. Marine Structures. 2016;46:167-192.
[8] DNV G. DNVGL-RP-0005: Fatigue design of offshore steel structures. Det Norske Veritas AS, Oslo, Norway; 2014.

Determinant of FDI inflow from Asian Economies to Lao PDR

Visansack Khamphengvong [1] **, Enjun Xia** [2] **and Khaysy Srithilat**[2]

[1] School of Management and Economics, Beijing Institute of Technology.
[2] University of Finance and Economics, Dalian, China

Visansack_kp@yahoo.com

Abstract. FDI inflow plays an essential role in economic growth in any countries. Thus FDI inflow and its impact on economic growth deserve attention in the economics literatures. This article intends to analyses the main determinant of FDI inflow from some Asian countries that contributed a high value of investment in Lao PDR. The static and dynamic gravity models were applied to panel data over 1995-2015 time periods. The findings conclude that the natural resource (hydropower, mining), FDI in the previous year, market size (home countries), inflation rate, infrastructure, and WTO have the positive impact on FDI inflow. While real exchange rate, trade openness, ASEAN, distance, and crisis are not the critical determinant to attract FDI from Asian countries.

Keywords: FDI inflow; Asian Economies; Dynamic gravity model; Lao PDR

1. Introduction

Foreign direct investment (FDI) is one vital source of economic development in developing countries, which lack of financial resources and under development of financial market. Insufficient domestic saving might cause the gap between the investment and saving in the nation. Many developing countries tried to attract FDI by promoting state its self to global because the volume of FDI would bring a massive capital to the nation, increase job opportunity, technology spillover, and the increase of FDI can effect to infrastructure development to the state [1, 2]. FDI not only performance as a significant source of funding for bank loans, but also expand capital accumulation and more intensifies the quality of capital stock [3]. Moreover, inwards of FDI in the host country can generate the potential for export volume and further opportunities accessing the global markets [4].

Numerous of empirical studies have been conducted on foreign direct investment. Vernon [5] developed econometric framework as production cycle and suggests that production function divided into four elements such as innovation, production growth, reach to maturity, and decline. The research of Dunning [6] represented the Eclectic framework which discusses on activities of the investment abroad; the context describes FDI behaviour. The eclectic theory provides an extensive framework and the concept reviewed to comprehend the complexities of multinational enterprises (MNEs) activities. It commonly knows that the framework regularly recognised as the OLI paradigm as it brings together advantage of ownership, location and internalisation (OLI) [7]. Hymer [8] conducted the theory of the production of foreign firms. The MNEs activities abroad would have the advantage while investments overseas, which have specific ownership advantage. When invests abroad, the MNEs faced with high competition, thus the property of benefits could be sufficient than a disadvantage. The international production framework requires to explicitly considering the competitive advantages that arise from the way corporations organise their business transactions. The increased of interdependence on the mid-market, expansion of regional and country asset portfolios to create an external economy of dependence [9]. Cave [10] established the Multinational Enterprises (MNEs) and tried to clarify that when multinational enterprises access to the host country, they have an excellent opportunity changing monopoly power from the technology spillover, knowledge dissemination, production technique, cost reduction, and other component associated. The evidence form Buckley and Peter Casson [11] constructed internationalisation framework to understand deeply on the production and impact of multinational corporations. Later, Helpman and Makusen [12, 13] also developed the same theory and counselled that MNEs have to understand clearly before invest abroad such as market-searching of the horizontal FDI. It is necessary for the investors to monitor the economic environment, domestic market, as much as to focus on the local manufacturing in the recipient countries. Another indispensable called horizontal FDI which represented research-seeking activities of the MNEs on how to access to infrastructure obtains to low labour cost, natural resources, and another component like exchange rate.

2. Data and econometrics methodology

The dataset of this work was obtained from various sources from Lao PDR such as Ministry of planning and investment, National Statistic Centre, the Bank of Lao of PDR, and the World development indicators (WDI 2017). The section applies the panel data from 1995 to 2015, by involving the gravity model to investigate the primary determinant of Asian FDI flow into Lao PDR. This article takes seven Asian countries which have a tremendous amount of investment in the recipient country including China, Vietnam, Thailand, Singapore, Malaysia, Korea and Japan

2.1. Econometric methodology

The present research work employs the gravity model proposed by [14–16] to capture the determinant of foreign direct investment in selected Asian countries. The gravity model widely used along the line of economic to solve the problem of international trade and foreign direct investment in their countries, the present work performed both dynamic and static model simultaneously by applying a fixed effect, random effect and Generalized Method of Movement (GMM) suggested by [17, 18]. The estimation acquired from fixed effect approach produces consistent estimate across the selected group of countries irrespective the correlation between individual effect and explanatory variable. Fixed effect model performed under Ordinary Least Square (OLS) which expressed follow

Static gravity equation:

$$lnFDI_{ijt} = \alpha + \beta_1 Lnhydro_{jt} + \beta_2 Lnmin_{jt} + \beta_3 lnMarket_{it} + \beta_4 lnMarket_{jt} + \beta_5 lnGDPpt_{it} + \beta_6 lnGDPpt_{jt} + \beta_7 Lnexch_{ijt} + \beta_8 openess_{jt} + \beta_9 INFL_{jt} + \beta_{10} INFRAS_{jt} + \beta_{11} LnDIST_{ij} + +\beta_{12} ASEAN_{jt} + \beta_{13} Crises_{jt} + \beta_{14} WTO_{jt} + \varepsilon_{ijt} \quad (1)$$

Where FDI_{ijt} represents the foreign direct investment inflow from the country i (source country) to the country j (host country) at time period t, $hydro_{it}$, min_{it}, are the total investment in hydropower and mining represented as natural resource density in the host country j, $Market_{it}$ and $Market_{jt}$stand for the market size between home country i and host country j represented by GDP between both countries, $GDPpt_{it}$ and $GDPpt_{jt}$ are presented GDP per capita which is a proxy of labour cost of home country i and host country j [19], $exch_{ijt}$ stand for the real exchange rate between host country j and home country i, $Openess_{jt}$ is a trade openness of the host country j estimated of total import and export divide as percentage of GDP, $infla_{jt}$ is an inflation rate of the

host country j, $infras_{jt}$ is air transport carrier departure worldwide in the host country j, $dist_{ij}$ is denote as the geographic distance among the capital city of the home country i and the host country j, ASEAN is the dummy of the years that Lao PDR became a member of ASEAN (1997), Crisis is dummy of years that host country j faced to the global financial crisis (2008-2009), defined as 1 for 2008 and 2009, and zero for other year. WTO represented the dummy of year accession to the World Trade Organization of the host country j in 2013 (WTO), defined as 0 prior accession to WTO and 1 after accession to WTO. To resolve the issue [18] proposed a system GMM estimator that combines a system with the first difference and level regression. However, the vast majority of empirical research included in the necessary explanatory variables related to distance and geographic information between home and host countries. Therefore, the observed model of the Lao PDR determinant in selected Asian countries can be expressed as follows:
Dynamic gravity equation

$$lnFDI_{ijt} = \alpha + \beta_1 LnFDI_{it,t-1} + \beta_2 Lnhydro_{jt} + \beta_3 Lnmin_{jt} + \beta_4 lnMarket_{it} + \beta_5 lnMarket_{jt} + \beta_6 lnGDPpt_{it} + \beta_7 lnGDPpt_{jt} + \beta_8 Lnexch_{ijt} + \beta_9 openess_{jt} + \beta_{10} INFL_{jt} + \beta_{11} INFRAS_{jt} + \beta_{12} LnDIST_{ij} + +\beta_{13} ASEAN_{jt} + \beta_{14} Crises_{jt} + \beta_{15} WTO_{jt} + \varepsilon_{ijt} \quad (2)$$

3. Results of the static models

3.1. Results of the static models

In the static models suggests that the coefficient of natural resource (hydropower, mining) is positive and statistically at 1% significantly (table1), which indicate that the country where have wealth natural resource is critical determinant FDI inflow into the country [20, 21]. The GDP as a market size between home and host state explains the growth of economics. The result of static models found that the size of the origin country is the crucial consideration for them to access to the host country [22]. The negative coefficient in the host country is the labour-intensive, and the result suggests that low cost of wage in the host country is an opportunities to support FDI [23]. However, the estimated of real exchange rate on this models is not confirmed the relationship between FDI and exchange rate. Similar to the coefficient result of trade openness in the host country that might not the essential central support the inflow of FDI. The estimated coefficient of inflation rate suggests that high inflation rate in the host country might decrease the number FDI into the recipient country [24]. The result of infrastructure in the host country indicated that the consequence of the coefficient is positive and statistically significant to support FDI in the host country [25]. Nevertheless, the result of the distance indicated that the geographic distance between the origin and host country is not positive and insignificant to support FDI. In the same order, the outcome of dummy variables including ASEAN, crisis, and WTO are insignificant, which meaning that ASEAN, crisis, and WTO is not the impact essential to support FDI in the host country.

3.2. Results of GMM dynamic models

In the GMM estimator, the authors applied the lagged FDI to present the investment condition in the previous year of the host country. The estimation of lagged variable expected the result to be confident and significant of statistically, the gravity model on FDI result should be dynamical. According to the coefficient, the result has confirmed the view which indicates the effect of FDI is positive and significant at 1 %. The result suggests that the environment of investment in the previous year is critical for investor's consideration before access to the host country. The finding of prior studies also recommended that lagged FDI variable has played an essential role in FDI stocks, and even conclude that FDI during the last year of the host country has vital for investors' decision invest aboard [26, 27]

IOP Conf. Series: Earth and Environmental Science **146** (2018) 012081 doi:10.1088/1755-1315/146/1/012081

Table 1. The estimation result of static and dynamic gravity models

Variables	Dependent: FDI			
	Static estimations		Dynamic estimations	
	FE	RE	Diff. GMM	Sys.GMM
FDI_{ijt}	-	-	0.9999989***	0.9997553***
$hydro_{it}$,	97.18341***	0.2358195***	0.0120843***	-0.0395367***
min_{it}	0.5803433***	0.5803797**	-0.0029924***	-1.594662***
$Market_{it}$	-159.6494*	-176.516*	-	1.373132***
$Market_{jt}$	1.878592	-0.8658451	0.00000192	0.0010355
$GDPpt_{it}$	-2.33072	-2.702684	-0.00000293	-0.00167
$GDPpt_{jt}$	202.1326*	2179357*	-	-
$exch_{ijt}$	0.0343437	-0.0515568	0.000000029	0.0000216
$Openess_{jt}$	-0.0872791	0.09002735	-0.0089946***	-0.0539013***
$infla_{jt}$	0.0522729**	0.0528511***	0.000359***	-0.0751527***
$infras_{jt}$	0.0005923**	0.0005957**	0.0000269***	0.0002825***
$dist_{ij}$	-	-4.277381	-	-0.0009069
$ASEAN$	-0.8893869	0.6990475	-	-7.239249***
$Crisis$	1.10539	0.2725493	-0.1400701***	1.714349***
WTO	-2.325262	-0.0131999	0.39333784***	3.631128***
Constant	-788.1586***	-341.9257	-	-
R	0.085	0.6145	-	-
N/Group No:	63/21	63/22	60/20	63/21
F-Statistique	0.000	0.000	-	-
Normality Test	0.000	0.000		
AR(2)	-	-	0.158	0.001
Hansen test	-	-	19	21

Note:***,**,* indicate the significant level at 1%, 5% 10% respectively. The Different GMM and System applied the robust standard errors.

The coefficients result also confirmed that the abundance of natural resource in the recipient country can attract more FDI inflow [28]. In the same context, The GDP coefficient indicates the positive and significant of the origin countries to encourage the FDI growth in the host country to promote the FDI growth, while the GDP per capita proxy as the labour cost does not become essential to support FDI in both homes, and the host country. Similarly, the exchange rate and trade openness are not the primary critical attract FDI into the host country. However, FDI and trade openness between home and host country are substitutable [29, 30]. GMM models suggest that high inflation rate in the recipient country may discourage the inflow to FDI [26]. However, the result of infrastructure is positive and indicates that infrastructure in the host country could bring numerous of FDI into the state [31]. Interestingly, the result of the global economic crisis was not the serious affected to FDI in the host country. Lastly, The GMM estimators recommend that the more opens economic as accession to WTO in the country is more attract foreign direct investment into the country [32, 33].

4. Conclusion

The result of gravity model illustrates that the natural resource (hydropower, mining) in the recipient country is the primary source to attract FDI inflow from Asian (selected) countries to Lao PDR. In addition, the previous environment of FDI plays a significant role in the host country to support the consideration of the investment partners. In the same context, the statistic of GMM model suggests shown positive of the market size in the origin countries at 1% significantly. This result explains that the economic growth in the Asian nations much causes the increasing flow of FDI into the host country. Moreover, the infrastructure in the host country is crucial that can attract investment from Asian countries. Other critical including inflation rate and WTO also contributes the impact of FDI inflow into Lao PDR. However, the real exchange rate, trade openness, ASEAN, distance, and crisis are not the primary effect to inward FDI in the host country.

The principal impact the FDI growth in the host country depends on the various elements which from both origin and host country. However, the primary influences FDI inflow could be the factor of the recipient country. It is necessary for the host state to improve and promote the programs to support the FDI. According to the result of this article which gives some recommends to the policymaker to enhance the host country attracting more increase the investment from abroad. Similarly, the government should strengthen the human capacity and improve the ability and skill of labour. Lao PDR should develop the infrastructure, the network connecting to whole areas of the investment including road, air transportation, and railways. Importantly, the government should control economic structure and secure with stability in the long-term because the economic stability and growth can contribute the inflow of FDI. According to the empirical results of this article confirm that the leading inwards of FDI driven from the natural resource. Consequently, the host country should increase more projects and add more variety of the resource to enhance the flow of FDI into the country.

5. Acknowledgments

The authors are thankful to Mr Danish and Mr Muhammad Awais Baloch for revising and providing valuable comments to improve the manuscript.

Reference

[1] Alfaro L, Chanda A, Kalemli-Ozcan S, et al. Does foreign direct investment promote growth? Exploring the role of financial markets on linkages. Journal of Economic development 2010; 91: 242–256.
[2] Gulcin Tapsin. The Relationship Between Foreign Direct Investment and Economic Growth in Togo [1991-2009]. J Bus Manag Econ 2009; 1269–1273.
[3] Ajayi S. FDI and Economic Development in Africa. In: International Conference on Accelerating Africa's Development Five years into Twenty-First Century, Tunis. 2006.
[4] Adefabi RA. Effects of FDI and Human Capital on Economic Growth in Sub-saharan Africa.pdf. Pakistan J Soc Sci.
[5] Vernon R. International Investment and International Trade in the Product Cycle. J Econhttp://bev.berkeley.edu/ipe/readings/international investment and international trade in the product cycle.pdf (1966).
[6] Dunning, J.D. The eclectic paradigm of international production: a restatement and some possible extensions. J Int Bus Stud 1988; 279–288.
[7] Dunning JH. The Eclectic (OLI) Paradigm of International Production: Past, Present and Future. Int J Econ Bus 2001; 8: 173–190.
[8] Hymer SH. The international operations of national firms: A study of direct foreign investment. MIT Press Cambridge, MA; 14.

[9] Dunning JH. Reappraising the eclectic paradigm in an age of alliance capitalism. J Int Bus Stud 1995; 111–142.
[10] Caves, Richard E. International corporation: The industrial economics of foreign investment,. Economica 1971; 1–27.
[11] BUCKLEY, Peter E CASSON M. The future of the Multinational Enterprise. Mac Millan.
[12] Helpman E. A Simple Theory of International Trade with Multinational Corporations Elhanan Helpman. J Polit Econ 1984; 92: 451–471.
[13] Makusen J. Multinationals, Multi-Plant Economies and the Gains from Trade. J Int Econ.
[14] Desbordes R, Vicard V. Foreign direct investment and bilateral investment treaties : An international political perspective. J Comp Econ 2009; 37: 372–386.
[15] Buch CM, Kokta RM, Piazolo D. Foreign direct investment in Europe : Is there redirection from the South to the East ? J Comp Econ 2003; 31: 94–109.
[16] Kahouli B, Maktouf S. Regional Trade Agreements in Mediterranean Area : Econometric Analysis by Static Gravity Model Regional Trade Agreements in Mediterranean Area : Econometric Analysis by Static. J East-West Bus 2013; 37–41.
[17] Arellano M, Bond S. Some tests of specification for panel data: Monte Carlo evidence and an application to employment equations. Rev Econ Stud 1991; 58: 277–297.
[18] Blundell R, Bond S. Initial conditions and moment restrictions in dynamic panel data models. J Econom 1998; 87: 115–143.
[19] Evenett SJ, Keller W. On Theories Explaining the Success of the Gravity Equation. J Polit Econ; 110.
[20] Asiedu E. Foreign Direct Investment in Africa: The Role of Natural Resources, Market Size, Government Policy, Institutions and Political Instability. World Econ 2006; 29: 63–77.
[21] Kolstad I, Wiig A. What determines Chinese outward FDI ? J World Bus 2012; 47: 26–34.
[22] Medvedev D. Beyond Trade: The Impact of Preferential Trade Agreements on FDI Inflows. World Dev 2012; 40: 49–61.
[23] Hailu ZA. Demand Side Factors Affecting the Inflow of Foreign Direct Investment to African Countries: Does Capital Market Matter? Int J Bus Manag 2010; 5: 104–116.
[24] Kleinert, Jorn TF. Gravity of FDI. Rev Int Econ 2010; 18: 1–13.
[25] Union S. ECONOMIC Survey of Europe. 2001; 185–225.
[26] Mina WM. The Institutional Reforms Debate and FDI Flows to the MENA Region: The ' Best' Ensemble. World Dev 2012; 40: 1798–1809.
[27] Hoang HH, Goujon M. Determinants of foreign direct investment in Vietnamese provinces: a spatial econometric analysis. Post-Communist Econ 2014; 26: 103–121.
[28] Aleksynska M, Havrylchyk O. European Journal of Political Economy FDI from the south : The role of institutional distance and natural resources. Eur J Polit Econ 2013; 29: 38–53.
[29] Kalaivani M. Determinants of Foreign Institutional Investment in India: An Empirical Analysis. Glob Bus Rev 2013; 16: 364–376.
[30] Kahouli B, Maktouf S. The determinants of FDI and the impact of the economic crisis on the implementation of RTAs: A static and dynamic gravity model. Int Bus Rev 2015; 24: 518–529.
[31] Rjoub H, Aga M, Alrub AA, et al. Financial Reforms and Determinants of FDI : Evidence from Landlocked Countries in Sub-Saharan Africa. MDPI 2017; 1–12.
[32] Farshid P, Ali S, Gholamhosein S. The impact of foreign direct investment and trade on economic growth — Taking China, Korea, Malaysia, Philippines & Thailand for example. China-USA Bus Rev 2009; 8: 37–43.
[33] Aw T, Tang T-C. The Determinants of Inward Foreign Direct Investment: The Case of Malaysia. Int J Bus Soc 2010; 11: 59–76.